Problem Books in Mathematics

Edited by P.R. Halmos

Bernard R. Gelbaum

Problems in Real and Complex Analysis

With 18 Figures

Springer

Bernard R. Gelbaum
Department of Mathematics
State University of New York (SUNY)
Buffalo, Ny 14214-3093
USA

Series Editor
Paul R. Halmos
Department of Mathematics
University of Santa Clara
Santa Clara, CA 95053
USA

Mathematics Subject Classification (1991): 26-01, 28-01, 30-01, 31-01

Library of Congress Cataloging-in-Publication Data
Gelbaum, Bernard R.
Problems in real and complex analysis/Bernard R. Gelbaum.
p. cm.—(Problem books in mathematics)
Includes bibliographical references and index.
ISBN 0-387-97766-X
1.Mathematical analysis—Problems, exercises, etc. I. Title.
II. Series.
QA301.G46 1992
515 – dc20 91-43908

Gelbaum: Problems in Real and Complex Analysis

First Indian Reprint 2010
Second Indian Reprint 2013

ISBN 978-81-8489-627-5

This edition is published by Springer (India) Private Limited, (a part of Springer Science+Business Media), Registered Office: 7th Floor, Vijaya Building, 17 Barakhamba Road, New Delhi – 110 001, India

Printed in India by Rakmo Press (P) Ltd., New Delhi.

Preface

In the pages that follow there are:

A. A revised and enlarged version of *Problems in analysis* (PIA). (All typographical, stylistic, and mathematical errors in PIA and known to the writer have been corrected.)
B. A new section COMPLEX ANALYSIS containing problems distributed among many of the principal topics in the theory of functions of a complex variable.
C. A total of 878 problems and their solutions.
D. An enlarged Index/Glossary and an enlarged Symbol List.

Notational and terminological conventions are to be found for the most part under **Conventions** at the beginnings of the chapters. Special items not included in **Conventions** are completely explained in the Index/Glossary.

The audience to which the current book is addressed differs little from the audience for PIA. The background of the reader is assumed to include a knowledge of the basic principles and theorems in real and complex analysis as those subjects are currently viewed. The aim of the problems is to sharpen and deepen the understanding of the mechanisms that underlie modern analysis.

I thank Springer-Verlag for its interest in and support of this project.

State University of New York at Buffalo B. R. G.

Contents

The symbol a/b under Pages below indicates that the Problems for the section begin on page a and the corresponding Solutions begin on page b. Thus 3/139 on the line for Set Algebra indicates that the Problems in Set Algebra begin on page 3 and the corresponding Solutions begin on page 139.

REAL ANALYSIS:

PROBLEMS

1
Set Algebra and Function Lattices

1.1. Set Algebra

Conventions

When n is a cardinal number, the phrase "n objects" signifies n pairwise different objects.

The following are the notations for the standard sets of numbers: $\mathbb{N}$ (the set of natural numbers), $\mathbb{Z}$ (the set of integers), $\mathbb{Q}$ (the set of real rational numbers), $\mathbb{R}$ (the set of real numbers), $\mathbb{C}$ (the set of complex numbers), $\mathbb{T}$ (the set of complex numbers of absolute value 1), and $\mathbb{H}$ (the set of quaternions). Specialized notations are:

$\mathbb{I}_{\mathbb{R}}$	(the set of real irrational numbers);
$\mathbb{I}$	(the set of irrational numbers);
$\mathbb{A}_{\mathbb{R}}$	(the set of real algebraic numbers);
$\mathbb{A}$	(the set of algebraic numbers).

When S is a set of real numbers, $S^+ \stackrel{\text{def}}{=} S \cap [0, \infty)$, e.g.,

$$\mathbb{Z}^+ = \{0, 1, 2, \ldots\} = \mathbb{N} \uplus \{0\}.$$

When $-\infty \leq a \leq b \leq \infty$, there are the following one-dimensional intervals:

$$(a, b) \stackrel{\text{def}}{=} \{ x \ : \ a < x < b \}, \text{ an open interval;}$$

$$[a, b) \stackrel{\text{def}}{=} \{ x \ : \ a \leq x < b \}, \text{ a half-open interval;}$$

$$(a, b] \stackrel{\text{def}}{=} \{ x \ : \ a < x \leq b \}, \text{ a half-open interval;}$$

$$[a, b] \stackrel{\text{def}}{=} \{ x \ : \ a \leq x \leq b \}, \text{ a closed interval.}$$

More generally, an interval I in $\mathbb{R}^n$ is either $\emptyset$ or the Cartesian product of n one-dimensional intervals each of which has a nonempty interior. In $\mathbb{R}^n$ a half-open n-dimensional interval is a set of the form $\times_{k=1}^{n} [a_k, b_k)$. If $b_k - a_k$ is k-free, the half-open n-dimensional interval is a half-open n-dimensional cube. When $n > 1$, elements of $\mathbb{R}^n$ or $\mathbb{C}^n$ are regarded as vectors and are denoted by **boldface** letters: $\mathbf{x}$, $\mathbf{O}$,... .

When X is a set and E is a subset of 2^X, $\mathsf{R}(\mathsf{E})$, resp. $\sigma\mathsf{R}(\mathsf{E})$, resp. $\mathsf{A}(\mathsf{E})$, resp. $\sigma\mathsf{A}(\mathsf{E})$ is the intersection of the (nonempty) set of rings, resp. σ-rings, resp. algebras, resp. σ-algebras containing E.

The cardinality of X is $\#(X)$, e.g., $\#(\mathbb{N}) \stackrel{\text{def}}{=} \aleph_0$, $\#(\mathbb{R}) \stackrel{\text{def}}{=} \mathfrak{c}$. The ordinal number of the well-ordered set of equivalence classes of well-ordered countable sets is Ω.

For a topological space X (cf. **Chapter 2**), $\mathsf{O}(X)$ resp. $\mathsf{F}(X)$ resp. $\mathsf{K}(X)$ is the set of open resp. closed resp. compact subsets of X. [When misinterpretation is unlikely, the notations O resp. F resp. K resp. S_β, resp. S_λ serve for $\mathsf{O}(X)$ resp. $\mathsf{F}(X)$ resp. $\mathsf{K}(X)$ resp. $\mathsf{S}_\beta(X)$ resp. $\mathsf{S}_\lambda(\mathbb{R}^n)$.]

When E is a subset of a set X, the characteristic function of E is $\chi_E(x) \stackrel{\text{def}}{=} \begin{cases} 1 & \text{if } x \in E \\ 0 & \text{otherwise} \end{cases}$.

For a sequence $\mathcal{E} \stackrel{\text{def}}{=} \{E_n\}_{n\in\mathbb{N}}$ of sets there are:

$$\overline{\mathcal{E}} \stackrel{\text{def}}{=} \overline{\lim}_{n\to\infty} E_n \stackrel{\text{def}}{=} \bigcap_{n=1}^{\infty} \bigcup_{m=n}^{\infty} E_m;$$

$$\underline{\mathcal{E}} \stackrel{\text{def}}{=} \underline{\lim}_{n\to\infty} E_n \stackrel{\text{def}}{=} \bigcup_{n=1}^{\infty} \bigcap_{m=n}^{\infty} E_m.$$

When $\overline{\mathcal{E}} = \underline{\mathcal{E}}$, $\lim_{n\to\infty} E_n \stackrel{\text{def}}{=} \overline{\mathcal{E}}$.

A subset M of 2^X is monotone if it is closed with respect to the formation of limits of monotone sequences of sets in M. For a subset E of 2^X, $\mathsf{M}(\mathsf{E})$ is the monotone class generated by E, i.e., $\mathsf{M}(\mathsf{E})$ is the intersection of all monotone classes containing E.

For any map f in Y^X, $\operatorname{im}(f) \stackrel{\text{def}}{=} f(X)$.

When X and Y are topological spaces, $C(X,Y)$ is the set of continuous maps from X to Y: $C(X,Y) \stackrel{\text{def}}{=} \{f : f \in Y^X,\ f \text{ continuous}\}$. When X is locally compact $C_0(X,\mathbb{C})$ is the set of continuous functions vanishing at infinity; $C_{00}(X,\mathbb{C})$ is the set of continuous functions having compact support: $C_{00}(X,\mathbb{C}) \stackrel{\text{def}}{=} \left\{f : f \in \mathbb{C}^X,\ \operatorname{supp}(f) \in \mathsf{K}(X)\right\}$.

The symbol (X,d) denotes a metric space X endowed with the metric d.

The measure situation (X,S,μ) consists of a set X, a σ-ring S in 2^X, and a countably additive set function $\mu : \mathsf{S} \ni E \mapsto \mu(E) \in \mathbb{C}$. When the range of μ is a subset of $[0,\infty]$, μ is a positive measure and (X,S,μ) is positive; when the range of μ is a subset of $(-\infty,\infty]$ or of $[-\infty,\infty)$, μ is a signed measure and (X,S,μ) is signed; when the range of μ is a subset of $\mathbb{C}$, μ is a complex measure and (X,S,μ) is complex. [The abuse of language just exemplified is rather general: an adjective applied to μ is applied as well to (X,S,μ).] Unless otherwise qualified, μ in (X,S,μ) is to be taken as positive.

When $\mu(E) < \infty$ for every E in S, (X, S, μ) is finite; when $X \in \mathsf{S}$ and $\mu(X) < \infty$, (X, S, μ) is totally finite.

For (X, S, μ), a set that is the countable union of sets of finite measure is σ-finite. When every E in S is σ-finite, (X, S, μ) is σ-finite; when X is σ-finite (whence $X \in \mathsf{S}$), (X, S, μ) is totally σ-finite.

When X is a topological space, μ in $(X, \mathsf{S}_\beta, \mu)$ is a Borel measure; in particular, $\mu(K) < \infty$ if $K \in \mathsf{K}(X)$.

To minimize the use of imperatives, e.g., "show, prove," etc., most **Problems** are given as assertions to be demonstrated.

1.1. If M is a monotone subset of 2^X and if M is closed with respect to the formation of finite unions and intersections, then M is closed with respect to the formation of countable unions and intersections.

1.2. If M is monotone, R is a ring of sets, and $\mathsf{M} \supset \mathsf{R}$ then $\mathsf{M} \supset \sigma\mathsf{R}(\mathsf{R})$.

1.3. For (X, d), if $2^X \supset \mathsf{M} \supset \mathsf{O}(X)$ and M is monotone then M contains $\mathsf{F}(X)$.

1.4. If $2^{\mathbb{R}} \supset \mathsf{M} \supset \mathsf{O}(\mathbb{R})$ and M is monotone then

$$\mathsf{M} \supset \sigma\mathsf{R}(\mathsf{O}) \text{ and } \sigma\mathsf{R}(\mathsf{O}) = \sigma\mathsf{R}(\mathsf{F}) = \sigma\mathsf{R}(\mathsf{K}).$$

1.5. If S is a σ-ring then $\#(\mathsf{S}) \neq \aleph_0$.

1.6. a) If $\mathsf{E} \subset 2^X$ and $\#(\mathsf{E}) \geq 2$ then $\#(\sigma\mathsf{R}(\mathsf{E})) \leq (\#(\mathsf{E}))^{\aleph_0}$. b) What is $\#(\mathsf{S}_\beta(\mathbb{R}^n))$? c) For M, the set of all σ-finite $(\mathbb{R}^n, \mathsf{S}_\beta, \mu)$ such that for each $\mathbf{x}$ in $\mathbb{R}^n$, $\mu(\{\mathbf{x}\}) \equiv 0$, what is $\#(\mathsf{M})$?

1.7. If $A \in \sigma\mathsf{R}(\mathsf{E})$ then for some finite or countable subset E_0 in E, $A \in \sigma\mathsf{R}(\mathsf{E}_0)$.

1.8. If $\mathsf{E} \subset 2^X$ and for each sequence $\{p, q, r, \ldots\}$ in $\mathbb{N}$, 2^X contains a sequence $S \stackrel{\text{def}}{=} \{A_p, A_{pq}, A_{pqr}, \ldots\}$ such that:

i. $A_p = \bigcap_q A_{pq}$, $A_{pq} = \bigcup_r A_{pqr}$, $\ldots$,
ii. for each S there is in $\mathbb{N}$ an $m(S)$ such that each member of S with more than $m(S)$ indices is in E,

then the set A of countable unions of sets A_p is closed with respect to the formation of countable unions and intersections of its members. If, furthermore,

iii. $\{E \in \mathsf{E}\} \Rightarrow \{X \setminus E \in \mathsf{E}\}$ (E is closed with respect to complementation),

then $\mathsf{A} = \sigma\mathsf{R}(\mathsf{E})$.

1.9. For a sequence $\mathcal{E} \stackrel{\text{def}}{=} \{E_n\}_{n\in\mathbb{N}}$: a)

$$\begin{aligned}
\overline{\mathcal{E}} &= \{\, x \;:\; x \text{ is in infinitely many } E_n \,\}, \\
\underline{\mathcal{E}} &= \{\, x \;:\; x \text{ is in all but finitely many } E_n \,\}, \\
\underline{\mathcal{E}} &\subset \overline{\mathcal{E}};
\end{aligned}$$

b) if $E_n \subset E_{n+1}$ resp. $E_n \supset E_{n+1}$, $n \in \mathbb{N}$, then $\lim_{n\to\infty} E_n$ is $\bigcup_{n\in\mathbb{N}} E_n$ resp. $\bigcap_{n\in\mathbb{N}} E_n$; c) $\chi_{\overline{\mathcal{E}}} = \overline{\lim}_{n\to\infty}\chi_{E_n}$, $\chi_{\underline{\mathcal{E}}} = \underline{\lim}_{n\to\infty}\chi_{E_n}$.

1.10. If $\mathcal{U}$ is a subalgebra of $\mathbb{C}^X$, $\mathbf{1} : \mathbf{X} \ni \mathbf{x} \mapsto \mathbf{1}$ is in $\mathcal{U}$, and $\mathcal{U}$ is closed with respect to the formation of pointwise sequential limits then: a) $\mathcal{F} \stackrel{\text{def}}{=} \{\, E \;:\; \chi_E \in \mathcal{U}\}$ is a σ-algebra; b) if (X, d) is a locally compact space and $\mathcal{U} \supset C_{00}(X, \mathbb{C})$ then $\sigma\mathsf{R}(\mathsf{K}(X)) \subset \mathcal{F}$.

For a set E of sets and $\mathcal{F}$, the set of all maps

$$f : \mathbb{N}^{\mathbb{N}} \ni \nu \stackrel{\text{def}}{=} \{n_1, n_2, \ldots\} \mapsto \{f(\nu)_k\}_{k\in\mathbb{N}} \in \mathsf{E}^{\mathbb{N}},$$

the operation $\mathcal{A}$ yields, for each f in $\mathcal{F}$, the set

$$M_f \stackrel{\text{def}}{=} \bigcup_{\nu\in\mathbb{N}^{\mathbb{N}}} \bigcap_{k\in\mathbb{N}} f(\nu)_k,$$

and for E, the set $\mathcal{A}(\mathsf{E}) \stackrel{\text{def}}{=} \{M_f \;:\; f \in \mathcal{F}\}$. It is convenient to denote $f(\nu)_k$ by $M_{n_1,n_2,\ldots,n_k}$ and thereby to indicate not merely the kth element of $f(\nu)$ but also the first k terms of the sequence from which the kth element is derived. The map f is regular iff for all ν and all k, $f(\nu)_{k+1} \subset f(\nu)_k$.

1.11. By abuse of language, $\mathcal{A}$ is idempotent; by abuse of notation, $\mathcal{A}^2 = \mathcal{A}$, i.e., $\mathcal{A}(\mathcal{A}(\mathsf{E})) = \mathcal{A}(\mathsf{E})$.

1.12. $\mathcal{A}(\mathsf{E})$ is closed with respect to the formation of a) countable intersections and b) countable unions of elements of $\mathcal{A}(\mathsf{E})$.

1.13. If E is closed with respect to the formation of finite intersections, then for every f in $\mathcal{F}$ and every M in E there is a regular g such that $M_g = M_f$.

1.14. If f is regular then: a)

$$\begin{aligned}
&\bigcup_{m\in\mathbb{N}} \bigcup_{\nu\in\mathbb{N}^{\mathbb{N}}} \bigcap_{k\in\mathbb{N}} M_{n_1,\ldots,n_i,m,n_{i+1},\ldots,n_{i+k}} \\
&\qquad = \bigcup_{\nu\in\mathbb{N}^{\mathbb{N}}} \bigcap_{k\in\mathbb{N}} M_{n_1,\ldots,n_i,n_{i+1},\ldots,n_{i+k}};
\end{aligned} \tag{1.1}$$

b) $\bigcup_{\nu\in\mathsf{N}^{\mathsf{N}}}\bigcup_{k\in\mathsf{N}} M_{n_1,n_2,\ldots,n_k}$ is the countable union of sets in E; c) M denoting $M_{n_1,n_2,\ldots,n_k}$ when $k=0$,

$$M\setminus M_f \subset \bigcup_{\nu\in\mathsf{N}^{\mathsf{N}}}\bigcup_{k=0}^{\infty}\left(M_{n_1,n_2,\ldots,n_k}\setminus\bigcup_{m=1}^{\infty}M_{n_1,n_2,\ldots,n_k,m}\right). \tag{1.2}$$

1.15. If $\mathsf{E}=\mathsf{F}([0,1])$ then $\mathsf{S}_\beta([0,1])\subset\mathcal{A}(\mathsf{E})$, i.e., every Borel set in $[0,1]$ is an analytic (Suslin) set.

1.16. If $\mathsf{E}=\mathsf{F}([0,1])$, $h\in C([0,1],\mathbb{R})$, and $E\in\mathsf{E}$ then $h(E)\in\mathcal{A}(\mathsf{F}(\mathbb{R}))$.

1.17. The set of Lebesgue measurable sets is invariant under $\mathcal{A}$, i.e., $\mathcal{A}(\mathsf{S}_\lambda([0,1]))=\mathsf{S}_\lambda([0,1])$.

1.18. If $\mathfrak{a}$ is an infinite cardinal number, there is an E such that for each M in E, $\#(\mathcal{A}(M))=\mathfrak{a}$.

1.19. If $E\in\mathsf{S}_\beta([0,1])$ and $h\in C([0,1],\mathbb{R})$ then $h(E)$ is Lebesgue measurable.

1.2. Function Lattices

Conventions

The next discussion provides the basis for the Daniell-Stone extension of a nonnegative linear functional I defined on a function lattice L. Limiting operations entail the use of extended $\mathbb{R}$-valued functions, i.e., functions that may assume the "values" $\pm\infty$. Consequently the extended real number system $\overline{\mathbb{R}}\stackrel{\text{def}}{=}\mathbb{R}\cup\{-\infty\}\cup\{\infty\}$ is introduced. In $\overline{\mathbb{R}}$, $0\cdot\infty=0$.

For a set X let V be a function lattice in $\overline{\mathbb{R}}^X$, i.e., a)

$$\{f,g\in V\}\Rightarrow\{f\wedge g,f\vee g\in V\}$$

(V is a lattice); b) when $f,g\in V$ and $a\in\mathbb{R}$ then $af\in V$ and, when $f+g$ is meaningful, i.e., when for each x, $f(x)+g(x)$ is not of the form $\infty+(-\infty)$ or $-\infty+\infty$, then $f+g$ is in V.

In what follows, the formulæ $f\pm g$ signify that $f(x)\pm g(x)$ are meaningful for all x.

Thus V fails to be a vector space only insofar as addition of functions in V is not always defined.

For any triple a,b,c of real numbers, the middle number is

$$\operatorname{mid}(a,b,c)\stackrel{\text{def}}{=}(a\wedge b)\vee(a\wedge c)\vee(b\wedge c).$$

For functions f,g,h (in $\mathbb{R}^X$), the middle function is

$$\operatorname{mid}(f,g,h)(x)\stackrel{\text{def}}{=}\operatorname{mid}(f(x),g(x),h(x)).$$

1.20. a) $\mathrm{mid}(a,b,c) = (a\vee b)\wedge(a\vee c)\wedge(b\vee c)$; b) $\mathrm{mid}\,(a,b,c) \in \{a,b,c\}$; c) for some functions f,g,h, $\mathrm{mid}\,(f,g,h) \notin \{f,g,h\}$; d) $f\vee g + f\wedge g = f+g$.

1.21. If V is a function lattice then V is mid-closed, i.e.,

$$\{f,g,h \in V\} \Rightarrow \{\mathrm{mid}\,(f,g,h) \in V\}.$$

1.22. If $a \in \mathbb{R}$ and f,g,p,f_n are in V then:

$$\{a \neq 0\} \Rightarrow \left\{\mathrm{mid}(af,g,h) = a\,\mathrm{mid}\left(f,\frac{g}{a},\frac{f}{a}\right)\right\}; \tag{1.3}$$

$$p+\mathrm{mid}(f,g,h) = \mathrm{mid}(p+f,p+g,p+h); \tag{1.4}$$

$$\{f_1 \leq f_2\} \Rightarrow \{\mathrm{mid}\,(f_1,g,h) \leq \mathrm{mid}\,(f_2,g,h)\}; \tag{1.5}$$

$$\mathrm{mid}\left(f_1 \overset{\vee}{\wedge} f_2, g, h\right) = \mathrm{mid}\,(f_1,g,h) \overset{\vee}{\wedge} \mathrm{mid}\,(f_2,g,h); \tag{1.6}$$

$$\mathrm{mid}\left(\lim_{n\to\infty} f_n, g, h\right) = \lim_{n\to\infty} \mathrm{mid}\,(f_n,g,h). \tag{1.7}$$

If $g(x) \leq h(x)$ then

$$h\wedge f(x) - g\wedge f(x) = \begin{cases} h(x)-g(x) & \text{if } \mathrm{mid}(f,g,h)(x) = h(x) \\ f(x)-g(x) & \text{if } \mathrm{mid}(f,g,h)(x) = f(x) \\ 0 & \text{otherwise.} \end{cases} \tag{1.8}$$

The **Problems** that follow are posed in the context of a set X, a function lattice V contained in $\overline{\mathbb{R}}^X$, and a nonnegative linear functional

$$J : V \ni f \mapsto J(f) \in \mathbb{R}.$$

Thus: a) $\{f \geq 0\} \Rightarrow \{J(f) \geq 0\}$; b) when $a,b \in \mathbb{R}$ and $f,g \in V$, then

$$J(af+bg) = aJ(f) + bJ(g)$$

so long as the left member of the equation is meaningful.

A function f in $\overline{\mathbb{R}}^X$ is Daniell measurable, i.e., $f \in \mathcal{M}$, iff

$$\{V \ni g \leq h \in V\} \Rightarrow \{\mathrm{mid}(f,g,h) \in V\}.$$

1.23. If

$$\{\{f_n \uparrow f\} \wedge \{J\,(f_n) \leq M < \infty,\ n \in \mathbb{N}\}\} \Rightarrow \{\{f \in V\} \wedge \{J\,(f_n) \uparrow J\,(f)\}\}$$

then: a) $\{f \in V\} \Rightarrow \{|f| \in V\}$; b) $\mathcal{M}$ is a function lattice; c) $\mathcal{M}$ is a monotone class of functions; d) $\mathcal{M}$ is closed with respect to $\overline{\lim}$, $\underline{\lim}$, and lim applied to sequences in $\mathcal{M}$; e) if $f \in \mathcal{M}$ and $|f| \in V$ then $f \in V$; f) $\{f \in \mathcal{M}\} \Leftrightarrow \{\{0 \leq g \in V\} \Rightarrow \{\text{mid}(f, -g, g) \in V\}\}$.

A set E is a Daniell measurable subset of X iff $\chi_E \in \mathcal{M}$. The set of Daniell measurable subsets of X is denoted A.

1.24. The set A is a σ-ring and if $\mathbf{1} \in \mathcal{M}$ then A is a σ-algebra.

When $f \in \mathbb{R}^X$, $\alpha \in \mathbb{R}$, and $\circ$ is one of $<, \leq, >, \geq, =, \neq$,

$$E_\circ(f, \alpha) \stackrel{\text{def}}{=} \{x \,:\, x \in X, f(x) \circ \alpha\}.$$

1.25. a) $f^{-1}(\mathsf{S}_\beta(\mathbb{R})) \subset \mathsf{A}$ iff one of $E_<(f,\alpha)$, $E_\leq(f,\alpha)$, $E_>(f,\alpha)$, and $E_\geq(f,\alpha)$ is in A for all real α; b) for some (X, S, μ) there is an f such that for all real α, $E_=(f,\alpha) \in \mathsf{A}$ and $f^{-1}(\mathsf{S}_\beta(\mathbb{R})) \not\subset \mathsf{A}$.

1.26. If $\mathbf{1} \in \mathcal{M}$ then $f \in \mathcal{M}$ iff for all real α, $E_<(f,\alpha) \in \mathsf{A}$, i.e., f is Daniell measurable iff f is A-measurable.

Let L be a function lattice in $\mathbb{R}^X$ and let $I : L \ni f \mapsto I(f) \in \mathbb{R}$ be a Daniell functional, i.e., a nonnegative linear functional satisfying

$$\{L \ni f_n \downarrow 0\} \Rightarrow \{I(f_n) \downarrow 0\}.$$

The **Problems 1.27 – 1.36** lead to the construction in $\overline{\mathbb{R}}^X$ of a function lattice V containing L and to which I can be extended as a nonnegative linear functional, denoted J, $J : V \mapsto \mathbb{R}$. There emerges a context for **1.20–1.26**, a measure situation (X, S, μ) for which $L^1(X, \mu) = V$, and a formula $J(f) = \int_X f(x)\, d\mu(x)$.

For a monotone sequence $\{a_n\}_{n \in \mathbb{N}}$ in $\mathbb{R}$, $\lim_{n\to\infty} a_n$ exists as an element of $\overline{\mathbb{R}}$. Let L_u be the set of $\overline{\mathbb{R}}$-valued functions that are limits of monotonely increasing sequences of functions in L.

1.27. If $\{f_n\}_{n\in\mathbb{N}}, \{g_m\}_{m\in\mathbb{N}} \subset L$, and $f_n \uparrow f$, $g_m \uparrow f$ then

$$\lim_{n\to\infty} I(f_n) = \lim_{m\to\infty} I(g_m).$$

[**Note 1.1:** Thus $\widetilde{I} : L_u \in f \mapsto \lim_{n\to\infty} I(f_n)$ is independent of the sequence $\{f_n\}_{n\in\mathbb{N}}$. For simplicity, the notation $\widetilde{I}$ is dropped in favor of I. For f in L_u, $I(f)$ is $\overline{\mathbb{R}}$-valued.]

1.28. If $f, g \in L_u$ and $\alpha, \beta \geq 0$ then

$$\alpha f + \beta g \in L_u \text{ and } I(\alpha f + \beta g) = \alpha I(f) + \beta I(g).$$

1.29. If $L_u \ni f_n \uparrow f$ then $f \in L_u$ and $I(f_n) \uparrow I(f)$.

Let L_{ul} be the set of all limits of monotonely decreasing sequences $\{g_n\}_{n\in\mathbb{N}}$ such that each $g_n \in L_u$, $-\infty < \lim_{n\to\infty} I(g_n)$, and $I(g_1) < \infty$.

1.30. If $f \in L_u$ then for some p in L_u and some h in L, $p \geq 0$ and $f = p + h$.

For $f \in \overline{\mathbb{R}}^X$:

$$\overline{I}(f) \stackrel{\text{def}}{=} \begin{cases} \infty & \text{if } \{g \ : \ g \in L_u, g \geq f\} = \emptyset \\ \inf\{I(g) \ : \ g \in L_u, g \geq f\} & \text{otherwise} \end{cases};$$

$$\underline{I}(f) \stackrel{\text{def}}{=} -\overline{I}(-f).$$

1.31. The functional $\overline{I}$ is subadditive and

a) $\{f \leq g\} \Rightarrow \{\{\overline{I}(f) \leq \overline{I}(g)\} \wedge \{\underline{I}(f) \leq \underline{I}(g)\}\}$;
b) $\forall f \ \{\underline{I}(f) \leq \overline{I}(f)\}$;
c) $\{f \in L_u\} \Rightarrow \{\overline{I}(f) = \underline{I}(f) = I(f)\}$;
d) $\left\{\{0 \leq f_n\} \wedge \left\{f \stackrel{\text{def}}{=} \sum_{n=1}^{\infty} f_n\right\}\right\} \Rightarrow \left\{\overline{I}(f) \leq \sum_{n=1}^{\infty} \overline{I}(f_n)\right\}$;
e) $\{0 \leq f_n \uparrow f\} \Rightarrow \{\overline{I}(f_n) \uparrow \overline{I}(f)\}$.

[**Note 1.2:** The last inequality is a precursor of the monotone convergence theorem of Lebesgue.]

Let L^1 be $\left\{f \ : \ f \in \overline{\mathbb{R}}^X, \ -\infty < \underline{I}(f) = \overline{I}(f) < \infty\right\}$ and for f in L^1 let $J(f)$ be the common value of $\underline{I}(f)$ and $\overline{I}(f)$. In what follows, L^1 plays the rôle of V in **1.23–1.31**. In particular, for L^1 there are the sets $\mathcal{M}$, A, etc.

A measure $\mu : \mathsf{A} \ni E \mapsto \mu(E) \in [0, \infty]$ is defined according to:

$$\mu(E) = \begin{cases} \infty & \text{if } \chi_E \notin L^1 \\ J(\chi_E) & \text{otherwise} \end{cases}.$$

1.32. a) $L \subset \{f \ : \ f \in L_u, \ I(f) < \infty\} \cup L_{ul} \subset L^1$; b) $f \in L^1$ iff for every positive ϵ, L_u contains a g and a $-h$ such that $h \leq f \leq g$ and $I(g + (-h)) < \epsilon$; c) L^1 is a function lattice; d) J is a nonnegative linear functional on L^1 and $J\big|_L = I$; e) if $L^1 \ni f_n \uparrow f$ then $f \in L^1$ iff for some M, $J(f_n) \leq M < \infty$ [and then $J(f) = \lim_{n\to\infty} J(f_n)$]; f) μ is a measure on A, i.e., (X, A, μ) is a measure situation; g) $f \in L^1$ iff $f \in \mathcal{M}$ and $|f| \in L^1$; h) $\{f \in L^1\} \Rightarrow \{J(f) = \int_X f(x)\, d\mu(x)\}$.

1.33. An f is in L^1 iff for some g in L_{ul} and some nonnegative function p in L^1, $J(p) = 0$ and $f = g - p$.

1.34. A nonnegative f is in $\mathcal{M}$ iff: a) for all g in L^1, $f \wedge g \in L^1$; b) iff for all h in L, $f \wedge h \in L^1$.

Customarily (X, S, μ) is enlarged, via the set

$$\mathsf{N} \stackrel{\text{def}}{=} \{N : N \in \mathsf{S}, \mu(N) = 0\}$$

of null sets, to include all sets of the form $(E \setminus N_1) \cup N_2$, $E \in \mathsf{S}$, $N_1, N_2 \in \mathsf{N}$. The resulting completion $\widetilde{\mathsf{S}}$ is again a σ-ring (σ-algebra) according as S is a σ-ring (σ-algebra). For E in S and $\mu((E \setminus N_1) \cup N_2) \stackrel{\text{def}}{=} \mu(E)$. The measure situation (X, S, μ) is complete iff $\widetilde{\mathsf{S}} = \mathsf{S}$.

When X is a locally compact topological space and L is $C_{00}(X, \mathbb{R})$ then for each nonnegative linear functional $I : L \mapsto \mathbb{R}$ the Daniell approach can be applied to L to derive a measure situation (X, A, μ):

$$\mu : \mathsf{A} \ni E \mapsto \mu(E) \in [0, \infty].$$

For example, if $X \stackrel{\text{def}}{=} \mathbb{R}^n$ and $I(f)$ is the Riemann integral of f then $\mathsf{A} = \mathsf{S}_\lambda(\mathbb{R}^n)$, the σ-algebra of Lebesgue measurable sets in $\mathbb{R}^n$ and μ is n-dimensional Lebesgue measure: $\mu = \lambda_n$.

Let two functions in L^1 be regarded as equal iff they differ only on a null set. Then

$$d : L^1 \times L^1 \ni \{f, g\} \mapsto d(f, g) \stackrel{\text{def}}{=} \int_X |f(x) - g(x)| \, d\mu(x)$$

is a metric for L^1.

1.35. In the context of the Daniell construction, L is dense in L^1 metrized by d as described above.

Define an outer measure $\mu^* : 2^X \ni E \mapsto \mu^*(E) \stackrel{\text{def}}{=} \overline{I}(\chi_E) \in [0, \infty]$. Call a set E Caratheodory measurable iff for all A in 2^X

$$\mu^*(A) = \mu^*(A \cap E) + \mu^*(A \setminus E)$$

(E μ^*-additively splits every set A.)

1.36. If $\mathbf{1} \in \mathcal{M}$ then E in 2^X is Daniell measurable iff E is Caratheodory measurable.

2

Topology, Limits, and Continuity

2.1. Topology

Conventions

A topological space is a pair $(X, \mathcal{T})$ (or simply X) consisting of a set X and a subset $\mathcal{T}$ of 2^X: $\mathcal{T}$ is the topology for X. The elements of $\mathcal{T}$ are the open sets of X, $\emptyset, X \in \mathcal{T}$, and $\mathcal{T}$ is closed with respect to the formation of arbitrary unions and finite intersections.

When $\mathcal{T}$ and $\mathcal{T}'$ are topologies for X and $\mathcal{T} \subset \mathcal{T}'$ then $\mathcal{T}'$ is stronger than $\mathcal{T}$ while $\mathcal{T}$ is weaker than $\mathcal{T}'$. The weakest or trivial topology is $\{\emptyset, X\}$ and the strongest or discrete topology is 2^X.

For a topological space X, a subset $\mathcal{B} \stackrel{\text{def}}{=} \{U_\lambda\}_{\lambda \in \Lambda}$ of $\mathcal{T}$ is a base for $\mathcal{T}$ iff every element (open set) in $\mathcal{T}$ is the union of (some) elements of $\mathcal{B}$.

If $\mathcal{B}$ is an arbitrary subset of 2^X then $\mathcal{B}$ is contained in the discrete topology 2^X, whence the set of all topologies containing $\mathcal{B}$ is nonempty. The intersection of all topologies containing $\mathcal{B}$ is the topology for which $\mathcal{B}$ is a base.

When $(X_1, \mathcal{T}_1)$ and $(X_2, \mathcal{T}_2)$ are topological spaces and $f \in X_2^{X_1}$ then f is:

- ▷ continuous iff $f^{-1}(\mathcal{T}_2) \subset \mathcal{T}_1$;
- ▷ open iff $f(\mathcal{T}_1) \subset \mathcal{T}_2$.

The set of continuous maps in $X_2^{X_1}$ is $C(X_1, X_2)$. A γ in $C([0,1], X)$ is a curve and $\gamma^* \stackrel{\text{def}}{=} \gamma([0,1])$ is the corresponding curve-image. When $\gamma(0) = \gamma(1)$, the curve is closed. If, to boot, γ is injective on $(0,1)$, the curve is a simple closed (Jordan) curve.

When $A \subset X$ and

$$\mathcal{A} \stackrel{\text{def}}{=} \{U \ : \ U \text{ open and } U \subset A\}$$

then A° is $\bigcup_{U \in \mathcal{A}} U$, the (possibly empty) interior of A. For a nonempty subset A of X, a neighborhood $N(A)$ is a set such that $A \subset N(A)^\circ$. For the set $\mathcal{F} \stackrel{\text{def}}{=} \mathcal{N}(A)$ of neighborhoods of a nonempty set A, the following obtain:

i. $\mathcal{F} \neq \emptyset$, $\emptyset \notin \mathcal{F}$;
ii. $\{F, F' \in \mathcal{F}\} \Rightarrow \{F \cap F' \in \mathcal{F}\}$;
iii. $\{\{F \in \mathcal{F}\} \wedge \{F \subset G\}\} \Rightarrow \{G \in \mathcal{F}\}$.

A base of neighborhoods at a point x in a topological space X is a subset $\mathcal{B}$ of the set $\mathcal{N}(\{x\})$ [for simplicity, $\mathcal{N}(x)$] and is such that every $N(\{x\})$ [for simplicity, $N(x)$] is a union of elements of $\mathcal{B}$.

For a set $X_\gamma, \gamma \in \Gamma$, $\times_{\gamma\in\Gamma} X_\gamma$ is the Cartesian product of the sets X_γ, i.e., $\times_{\gamma\in\Gamma} X_\gamma \stackrel{\text{def}}{=} \{f : f : \Gamma \ni \gamma \mapsto f(\gamma) \in X_\gamma\}$. When, for some X, $X_\gamma \equiv X$, then $\times_{\gamma\in\Gamma} X_\gamma = X^\Gamma$ and the choice of notation is a matter of taste.

A map $f : \times_{\gamma\in\Gamma} X_\gamma \mapsto \times_{\gamma\in\Gamma} Y_\gamma$ between Cartesian products is finitely (countably) determined iff for some finite (countable) subset $\{\gamma_k\}_{k\in K}$ in Γ, each $\mathbf{x} \stackrel{\text{def}}{=} \{x_\gamma\}_{\gamma\in\Gamma}$ and $\mathbf{x}' \stackrel{\text{def}}{=} \{x'_\gamma\}_{\gamma\in\Gamma}$,

$$\left\{f(\mathbf{x})_{\gamma_k} = f(\mathbf{x}')_{\gamma_k}, \; k \in K\right\} \Rightarrow \{f(\mathbf{x}) = f(\mathbf{x}')\}.$$

When Δ is a subset of Γ and for each δ in Δ, A_δ is a subset of X_δ then the pair $\left(\Delta, \{A_\delta\}_{\delta\in\Delta}\right)$ determines the cylinder

$$\mathfrak{Z}\left(\{A_\delta\}_{\delta\in\Delta}\right) \stackrel{\text{def}}{=} \times_{\delta\in\Delta} A_\delta \times \times_{\gamma\in\Gamma\setminus\Delta} X_\gamma.$$

When each X_γ is a topological space, a basic neighborhood B for the product topology of $\times_{\gamma\in\Gamma} X_\gamma$ is a cylinder determined by:

a finite subset $\Delta \stackrel{\text{def}}{=} \{\gamma_i\}_{1\le i\le n}$ of Γ;
points x_{γ_i} in $X_{\gamma_i}, 1 \le i \le n$;
neighborhoods $N_{\gamma_i}(x_{\gamma_i}), 1 \le i \le n$;

whereupon

$$B \stackrel{\text{def}}{=} \mathfrak{Z}\left(\{N_{\gamma_i}(x_{\gamma_i})\}_{\gamma_i\in\Delta}\right).$$

The set of all finite unions of such cylinders is closed with respect to the formation of finite intersections. It follows that the set $\mathfrak{B}$ of all basic neighborhoods is a base for the product topology of $\times_{\gamma\in\Gamma} X_\gamma$.

More generally, a base for a topology of $\times_{\gamma\in\Gamma} X_\gamma$ is a set Σ of cylinders such that the set of all unions of elements of Σ is closed with respect to the formation of finite intersections.

A partially ordered set (poset) is a pair $\left(\Gamma \stackrel{\text{def}}{=} \{\gamma\}, \prec\right)$ (or simply Γ) in which the order $\prec$ is transitive and the relation $\gamma \prec \gamma'$ (also written $\gamma' \succ \gamma$) obtains for a (possibly empty) set of pairs (γ, γ'). The partially ordered set Γ is directed iff for each pair (γ, γ') for some g'' in Γ, $\gamma'' \succ \gamma$ and $\gamma'' \succ \gamma'$. A directed set Γ is a diset and a subdiset of Γ is a subset that is a diset with respect to the partial order in Γ.

A map $n : \Lambda \ni \lambda \mapsto n(\lambda) \in X$ of a diset $\Lambda \stackrel{\text{def}}{=} \{\lambda, \prec\}$ is a net. The net n converges to a in X iff for each $N(a)$ and some $\lambda_0(N(a))$ in Λ, $\{\lambda_0(N(a)) \prec \lambda\} \Rightarrow \{n(\lambda) \in N(a)\}$.

A point a is a limit point of a set A ($a \in A^\bullet$) iff every $N(a)$ meets $A \setminus \{a\}$; the closure $\overline{A}$ of A is $A \cup A^\bullet$; A is closed iff $X \setminus A$ is open iff $A = \overline{A}$; A is dense in itself or self-dense iff $A \subset A^\bullet$; the union of all the self-dense subsets of A is the self-dense kernel of A; A is perfect iff A is closed and dense in itself.

A point a in A is isolated iff for some $N(a)$, $N(a) \cap A = \{a\}$.

Subsets A and B of X are separated iff $(\overline{A} \cap B) \cup (A \cap \overline{B}) = \emptyset$. The space X is connected iff X is not the union of a pair of separated and nonempty sets, i.e., iff X contains no nonempty proper subset that is both open and closed, i.e., iff X is not the union of two nonempty and disjoint open subsets.

A point adheres to or is adherent to or is in the adherence of a net n iff x is in the closure of the range of n: $x \in \overline{n(\Lambda)}$.

A net $n : \Lambda \mapsto X$ is eventually in A iff for some λ_0,

$$\{\lambda \succ \lambda_0\} \Rightarrow \{n(\lambda) \in A\};$$

n is frequently in A iff for every λ and some a λ', $\lambda' \succ \lambda$ and $n(\lambda') \in A$. Two nets $n : \Lambda \ni \lambda \mapsto X$ and $n' : \Lambda' \ni \lambda' \mapsto X$ are essentially equal ($n \doteq n'$) iff for some λ_0 resp. λ'_0 in Λ resp. Λ',

$$\{\{\lambda \succ \lambda_0\} \wedge \{\lambda' \succ \lambda'_0\}\} \Rightarrow \{n(\lambda) = n(\lambda')\};$$

$\doteq$ is an equivalence relation.

When Λ is a diset and $\Lambda' \subset \Lambda$ then Λ' is cofinal with Λ iff for each λ in Λ and some $\lambda'(\lambda)$ in Λ', $\lambda'(\lambda) \succ \lambda$. If Λ' is cofinal with Λ and $n : \Lambda \to X$ is a net, the net $n\big|_{\Lambda'}$, the restriction of n to Λ', is cofinal with n.

In general, for a set X, a filter in 2^X is a subset $\mathcal{F} \stackrel{\text{def}}{=} \{F\}$ conforming to i – iii, page 12, (for the set of neighborhoods of a point). A filter $\mathcal{F}'$ refines or is a refinement of or is finer than a filter $\mathcal{F}$ iff $\mathcal{F}' \supset \mathcal{F}$, in which case one writes $\mathcal{F}' \succ \mathcal{F}$. When $\mathcal{F}'$ is finer than $\mathcal{F}$, $\mathcal{F}$ is coarser than $\mathcal{F}'$.

The set of all filters in 2^X is partially ordered by $\succ$ and for a given filter $\mathcal{F}$ there is, by virtue of any one of the equivalents (Hausdorff maximality principle, Axiom of Choice, Well-ordering Axiom, etc.) of Zorn's lemma, a maximal filter $\mathcal{U}$ such that $\mathcal{U} \succ \mathcal{F}$. A maximal filter $\mathcal{U}$ is an ultrafilter.

The filter consisting of the single set X is coarser than any filter. For x in X, the filter consisting of all the supersets of $\{x\}$ is an ultrafilter.

When $\mathcal{B}$ is a subset of 2^X, the set of all supersets of elements of $\mathcal{B}$ is a filter iff the intersection of every pair of elements of $\mathcal{B}$ is nonempty, in which case $\mathcal{B}$ is a filter base. The intersection of all filters containing a filter base is the filter generated by $\mathcal{B}$. Every filter $\mathcal{F}$ is a filter base and $\mathcal{F}$ generates itself. For a map $f : X \mapsto Y$ and a filter base $\mathcal{B}$ in 2^X, $\{f(B)\}_{B\in\mathcal{B}} \stackrel{\text{def}}{=} f(\mathcal{B})$ is a filter base in 2^Y.

When X is a topological space, a filter $\mathcal{F}$ converges to x in X iff every $N(x)$ contains an element of $\mathcal{F}$: $\mathcal{F}$ invades every $N(x)$. Because $\mathcal{N}(x)$ is itself a filter, $\mathcal{F}$ converges to x iff $\mathcal{F}$ refines $\mathcal{N}(x)$).

A point x in X adheres to or is adherent to or is in the adherence of the base $\mathcal{B}$ of a filter iff each $N(x)$ meets every element of $\mathcal{B}$, i.e., iff x is in the closure of each element of $\mathcal{B}$.

A filter $\mathcal{F} \stackrel{\text{def}}{=} \{F\}$ is a diset with respect to the partial order induced by reversed inclusion among the elements of $\mathcal{F}$: $F' \succ F$ iff $F' \subset F$. A map $n : \mathcal{F} \ni F \mapsto n(F) \in F$ is thus a net corresponding to the filter $\mathcal{F}$. Conversely, when Λ is a diset and $n : \Lambda \ni \lambda \mapsto n(\lambda) \in X$ is a net, let B_λ be $\{ n(\mu) \ : \ \mu \succ \lambda \}$. Then $\{B_\lambda\}_{\lambda\in\Lambda}$ is a filter base that generates the filter $\mathcal{F}_n$ corresponding to the net n. The correspondence net $\to$ filter is injective. The correspondence filter $\to$ net is not a map because there can be more than one net corresponding to a given filter.

[**Note 2.1:** Owing to the correspondences between nets and filters, any notion formulable in terms of nets is equally formulable in terms of filters.]

The boundary ∂S of a set S in a topological space X is $\overline{S} \cap \overline{X \setminus S}$.

A topological group G is a Hausdorff space and a group for which the map $G^2 \ni \{x, y\} \mapsto xy^{-1} \in G$ is continuous.

A uniformity for a set X is defined by a nonempty subset U of $2^{X\times X}$. The assumptions about U are:

i. $\{U, V \in \mathsf{U}\} \Rightarrow \{\exists W \{\{W \in \mathsf{U}\} \wedge \{W \subset U \cap V\}\}\}$;

ii. $\{U \in \mathsf{U}\} \Rightarrow \left\{U \supset \{ (x,x) \ : \ x \in X \} \stackrel{\text{def}}{=} \Delta\right\}$;

iii. $\{U \in \mathsf{U}\} \Rightarrow \left\{U^{-1} \stackrel{\text{def}}{=} \{ (y,x) \ : \ (x,y) \in U \} \in \mathsf{U}\right\}$;

iv. $W \circ W$ denoting $\{ (x,z) \ : \ \exists y \{(x,y),(y,z) \in W\} \}$, $\{U \in \mathsf{U}\} \Rightarrow \exists W \{\{W \in \mathsf{U}\} \wedge \{W \circ W \subset U\}\}$.

(Occasionally it is assumed as well that:

v. $\Delta = \bigcap_{U\in\mathsf{U}} U$.)

Note that a uniformity is a filter base in $2^{X\times X}$ and generates a filter with some added properties, *iii*, *iv*, and occasionally *v*.

A subset $\mathcal{B}$ of $2^{X\times X}$ is a base for a uniformity iff the set $\mathsf{U}(\mathcal{B})$ of all supersets of elements of $\mathcal{B}$ is a uniformity.

A uniformity situation (X, U) is a set X and a uniformity U for X.

Two important examples of uniformity situations are: a) a metric space (X, d) for which $\mathsf{U} = \{ \{ (a,b) \ : \ d(a,b) < \epsilon \} \ : \ \epsilon > 0 \}$; b) a topological group G in which $\mathsf{U} = \left\{ \left\{ (a,b) \ : \ ab^{-1} \in V \right\} \ : \ V \in \mathcal{N}(e) \right\}$.

If (X, d) is a metric space, $i - v$ obtain for the corresponding uniformity U, which has a countable base. Conversely, if (X, U) is a uniformity situation, U satisfies $i - v$, and U has a countable base then for some metric $d : X^2 \ni \{x, y\} \mapsto d(x, y) \in [0, \infty)$, U is the uniformity corresponding to d [**GiJe, RoDi, We1**].

When (X, d) is a metric space, $a \in X$, and $r \geq 0$, the closed ball $B(a, r)$ centered at a and of radius r is $\{ x : x \in X, d(x, a) \leq r \}$; the corresponding open ball is $B(a, r)^\circ \stackrel{\text{def}}{=} \{ x : x \in X, d(x, a) < r \}$ (which is empty if $r = 0$).

For a uniformity situation (X, U), a net $n : \Lambda \ni \lambda \mapsto n(\lambda) \in X$ is a Cauchy net iff for each U in U and some $\lambda(U)$,

$$\{\lambda, \lambda' \succ \lambda(U)\} \Rightarrow \{(n(\lambda), n(\lambda')) \in U\}.$$

The set X in (X, U) is Cauchy complete iff every Cauchy net converges. If X is not Cauchy complete, an analog of the Cantor completion of a metric space leads to the Cauchy completion $\widetilde{X}$ of X.

For f in $\mathbb{R}^X$, a in X and $\mathcal{N}(a)$,

$$\overline{\lim}_{x=a} f(x) \stackrel{\text{def}}{=} \inf_{N \in \mathcal{N}(a)} \sup_{x \in N} f(x), \ \underline{\lim}_{x=a} f(x) \stackrel{\text{def}}{=} \sup_{N \in \mathcal{N}(a)} \inf_{x \in N} f(x).$$

According as $\overline{\lim}_{x=a} f(x) = f(a)$ resp. $\underline{\lim}_{x=a} f(x) = f(a)$ for all a f is upper semicontinuous (usc) resp. lower semicontinuous (lsc).

When A, B are subsets of a group G,

$$AB = \{ ab : a \in A, \ b \in B \}, \ A^{-1} = \{ a^{-1} : a \in A \}.$$

The set

$$\left\{ a : a = \sum_{n=1}^{\infty} \alpha_n 3^{-n}, \ \alpha_n = 0 \text{ or } 2, n \in \mathbb{N} \right\} \tag{2.1}$$

is **the** Cantor set C_0, i.e., the complement in $[0, 1]$ of the union of groups of "middle third" open intervals

$$\left\{ \left(\frac{1}{3}, \frac{2}{3} \right) \right\}, \left\{ \left(\frac{1}{9}, \frac{2}{9} \right), \left(\frac{7}{9}, \frac{8}{9} \right) \right\}, \ldots$$

(the nth group consists of 2^{n-1} intervals).

If $0 \leq \alpha < 1$ and the intervals described above are replaced by similarly situated intervals of lengths contracted by the factor $1 - \alpha$, the complement of their union is **the** Cantor-like set C_α. The Lebesgue measure of C_α is α. If each interval in the nth group is replaced by a similarly situated interval of length a_n and $\sum_{n=1}^{\infty} 2^{n-1} a_n = 1 - \alpha$, the complement of the union of the intervals is homeomorphic to C_α. It, like each C_α, is a nowhere dense perfect subset of $[0, 1]$ and its Lebesgue measure is α.

When K is a compact subset of an open set V in a topological space X, the notation $K \prec f \prec V$ signifies that f is a continuous function, for all x, $0 \le f(x) \le 1$, $f(K) = \{1\}$, and $f(X \setminus V) = \{0\}$.

2.1. a) $a \in \overline{A}$ iff some net n with range in A converges to a; b) $a \in A^\bullet$ iff some net n with range in $A \setminus \{a\}$ converges to a; c) A is closed iff $A = \overline{A}$ iff $A^\bullet \subset A$; d) a is an isolated point of A iff $a \in A \setminus A^\bullet$; e) A is self-dense iff $A \subset A^\bullet$; f) A is perfect iff $A = A^\bullet$ $(= \overline{A})$.

2.2. If, for some x, the net $n : \Lambda \mapsto X$, is frequently in each $N(x)$ then Λ contains a subdiset Γ such that some net $m : \Gamma \mapsto X$ converges to x.

2.3. A filter $\mathcal{F}$ converges to x iff every net n corresponding to $\mathcal{F}$ converges to x.

2.4. If the filter $\mathcal{F}$ corresponds to the net n then n converges to x iff $\mathcal{F}$ converges to x.

2.5. a) f is usc iff $-f$ is lsc iff for all real α, $E_<(f,\alpha)$ is open; b) f is lsc iff $-f$ is usc iff for all real α, $E_>(f,\alpha)$ is open.

2.6. If $f \in \mathbb{R}^X$, $a \in X$, and $\overline{\lim}_{x=a} f(x) \stackrel{\text{def}}{=} M \in \mathbb{R}$ or $\underline{\lim}_{x=a} f(x)$ then for some net n converging to a, $f \circ n$ converges to M.

2.7. When the set of all sequences, regarded as nets with the common range $\mathcal{D} \stackrel{\text{def}}{=} \{0,1\}$ is divided into $\doteq$-equivalence classes, the cardinality of the resulting set S of equivalence classes is $\mathfrak{c}$.

2.8. If X is a topological space then x is adherent to a filter $\mathcal{F}$ in 2^X iff some filter $\mathcal{F}'$ finer than $\mathcal{F}$ converges to x.

2.9. A topological space X is compact iff every ultrafilter in 2^X converges. Similarly, X is compact iff for every net $n : \Lambda \mapsto X$, there is a convergent cofinal net $n\big|_{\Lambda'}$.

2.10. True or false: $\mathbb{R}$ is the continuous image of $[0,1)$?

2.11. For some uncountable subset E in $[0,1]$, $(E-E)^\circ = \emptyset$. [Note that if $\lambda(E) > 0$ then $(E-E)^\circ \ne \emptyset$, cf. **4.34**.]

2.12. With respect to the metric $d : [0,1)^2 \ni \{x,y\} \mapsto |x-y|$, $[0,1)$ is not complete. Is there a second metric D for $[0,1)$ so that $([0,1),d)$ and $([0,1),D)$ are homeomorphic and $([0,1),D)$ is complete?

2.13. If $f : [0,1] \stackrel{\text{sur}}{\mapsto} A \times B$ is a homeomorphism then one of A and B consists of a single point.

2.14. a) For each a in C_0 the ternary representation specified in (2.1) is unique; b) $\mathcal{D}^{\mathbb{N}}$ and C_0 are homeomorphic; c) the maps

$$f_n : C_0 \ni a \mapsto (-1)^{\frac{a_n}{2}},\ n \in \mathbb{N},$$

are continuous; d) $\phi_0 : C_0 \ni a \mapsto \sum_{n=1}^{\infty} \frac{a_n}{2} 2^{-n} \in [0,1]$, **the** Cantor function, is a continuous surjection; e) $C_0 + C_0 = [0,2]$, $C_0 - C_0 = [-1,1]$; f) ϕ_0 is not injective.

For a topological space X the weight W of X is the least of the cardinalities of the bases of X. A base $\mathcal{B}$ of X is minimal iff $\#(\mathcal{B}) = \mathsf{W}$. Let $\mathcal{D}$ denote the set $\{0,1\}$ endowed with the discrete topology. For a set M, $\mathcal{D}^M$ with the product topology is a dyadic space.

2.15. a) If X is a compact Hausdorff space of weight W and $\#(M) = \mathsf{W}$ then X is the continuous image of a closed subset of $\mathcal{D}^M$. b) If B is a closed subset of **the** Cantor set C_0 there is a continuous surjection $h : C_0 \overset{\text{sur}}{\mapsto} B$. c) If $\mathsf{W} = \aleph_0$ then X is the continuous image of $\mathcal{D}^{\mathbb{N}}$.

2.16. When $\#(X) > \aleph_0$, $x_0 \in X$, and

$$\mathcal{T} \overset{\text{def}}{=} \{ A \ : \ x_0 \notin A \text{ or } \#(X \setminus A) \in \mathbb{Z}^+ \}$$

then: a) $(X, \mathcal{T})$ is a compact Hausdorff space; b) if $x \neq x_0$ then $\{x\} \in \mathcal{T}$; c) the weight of X is $\#(X)$.

2.17. If $\mathcal{O} \overset{\text{def}}{=} \{O_\gamma\}_{\gamma \in \Gamma}$ is a set of pairwise disjoint open sets in the dyadic space $X \overset{\text{def}}{=} \mathcal{D}^M$ then Γ is empty, finite, or countable.

2.18. If $\{X_\lambda\}_{\lambda \in \Lambda}$ is a set of separable spaces then a set $\mathcal{O}$ of pairwise disjoint open sets in $X \overset{\text{def}}{=} \times_{\lambda \in \Lambda} X_\lambda$ is empty, finite, or countable.

2.19. For X in **2.16**, if $f : \mathcal{D}^M \mapsto X$ is continuous then $f\left(\mathcal{D}^M\right) \neq X$.

2.20. If a closed subset F of $\mathbb{R}$ is the union of half-open intervals:

$$\mathsf{F}(\mathbb{R}) \ni F = \bigcup_{\lambda \in \Lambda} (a_\lambda, b_\lambda],$$

then it is the union of countably many of those half-open intervals.

2.21. True or false: $\mathbb{R}$ metrized according to

$$d(x,y) \overset{\text{def}}{=} |\arctan x - \arctan y|$$

is complete?

2.22. If $S \subset (0,\infty)$, $\sup(S) \overset{\text{def}}{=} u < 1$, and

$$\{\{x \in S\} \wedge \{y \in S\}\} \Rightarrow \{xy^{-1} \in S\}$$

then $u \in S$.

2.23. True or false: $\mathbb{R} \setminus \mathbb{Q}$ and $(\mathbb{R} \setminus \mathbb{Q}) \cap (0,1)$ are homeomorphic?

2.24. If X is a complete metric space without isolated points and U is a nonempty open subset of X then $\#(U) \geq \mathfrak{c}$.

2.25. A nonempty, countable, and compact subset K of a complete metric space X contains an isolated point.

2.26. If $\{F_k\}_{k\in\mathbb{N}}$ is a sequence of closed subsets of $\mathbb{R}^n$ and $\mathbb{R}^n = \bigcup_{k\in\mathbb{N}} F_k$ then $\bigcup_{k\in\mathbb{N}} F_k^\circ$ is dense in $\mathbb{R}^n$.

2.27. If A is an uncountable subset of $\mathbb{R}^n$ then: a) for some a in A and every $N(a)$, $\#(N(a) \cap A) > \aleph_0$; b) $S \stackrel{\text{def}}{=} \{b : b \in A,\ \#(N(b) \cap A) > \aleph_0\}$ is closed; c) $\#(S) > \aleph_0$; d) when $n = 1$, for some b resp. c in A,

$$\{\epsilon > 0\} \Rightarrow \{\#(A \cap (b, b+\epsilon)) > \aleph_0\}$$

resp.

$$\{\epsilon > 0\} \Rightarrow \{\#(A \cap (c-\epsilon, c)) > \aleph_0\}.$$

2.28. (Cantor-Bendixson) If X contains a countable dense subset D then for some perfect subset P and some countable subset C for which the self-dense kernel is empty, $X = C \cup P$.

2.29. If $S \in \mathcal{A}(\mathsf{F}([0,1]))$ then $\#(S) \leq \aleph_0$ or $\#(S) = \mathfrak{c}$.

When $f : X \mapsto Y$ is a map between metric spaces (X, d) and (Y, δ), f is expansive, an expansion resp. contractive, a contraction iff for all $\{x, x'\}$ in X^2, $\delta(f(x), f(x')) \geq d(x, x')$ resp. $\delta(f(x), f(x')) \leq d(x, x')$.

2.30. For a compact metric space (X, d), if f in X^X is expansive then: a) f is an isometry, i.e., for all x, y in X, $d(f(x), f(y)) = d(x, y)$; b) f is an auteomorphism of X; c) if f is contractive and $f(X) = X$ then f is an isometry.

2.31. For some compact metrizable space X there is an auteomorphism $T : X \overset{\text{auteo}}{\mapsto} X$ that is an isometry for no metric d compatible with the topology of X.

2.32. For a compact metric space (X, d) and an equicontinuous subset E of $C(X, X)$ metrized according to $D(f, g) = \sup\{d(f(x), g(x)) : x \in X\}$, $\overline{E}$ is D-compact.

2.33. If (X, d) and (Y, δ) are compact metric spaces then f in Y^X is continuous iff the graph $\mathcal{G}(f) \stackrel{\text{def}}{=} \{(x, f(x)) : x \in X\}$ of f is closed in $X \times Y$.

2.34. If X is a separable space then $\#(\mathsf{O}) = \#(\mathsf{F}) \leq \mathfrak{c}$.

2.35. True or false: for $A \stackrel{\text{def}}{=} C([0,1],[0,1])$ metrized according to

$$d(f,g) \stackrel{\text{def}}{=} \sup_{x\in[0,1]} |f(x)-g(x)| \stackrel{\text{def}}{=} \|f-g\|_\infty,$$

A_i the set of injective elements of A, A_s the set of surjective elements of A, and A_b the set of bijective elements of A ($A_b \stackrel{\text{def}}{=} A_i \cap A_s$), a) A_i is closed? b) A_s is closed? c) A_b is closed? d) A is connected? e) A is compact?

2.36. For compact Hausdorff spaces X and Y and f in $C(X,Y)$, if $f^{-1}(y)$ is connected for every y in Y then $B \stackrel{\text{def}}{=} f^{-1}(A)$ is connected for every connected subset A of Y. If Y is not a Hausdorff space and f is as described, there can be a connected set A such that $f^{-1}(A)$ is not connected.

2.37. For a set $\{X_\gamma\}_{\gamma\in\Gamma}$ of compact Hausdorff spaces and

$$X \stackrel{\text{def}}{=} \times_{\gamma\in\Gamma} X_\gamma,$$

if $f \in C(X,\mathbb{R})$ and $\epsilon > 0$, then for some finitely determined g in $C(X,\mathbb{R})$, $\sup_{\mathbf{x}\in X} |f(\mathbf{x}) - g(\mathbf{x})| < \epsilon$. Furthermore f is countably determined.

2.38. If X is compact, $f \in \mathbb{R}^X$, and for each t in $\mathbb{R}$, $f^{-1}([t,\infty))$ is closed then for some x_0 in X, $f(x_0) = \sup_{x\in X} f(x) < \infty$.

2.39. For K compact and U_1, U_2 open subsets of a Hausdorff space, if $K \subset U_1 \cup U_2$ then U_i contains a compact subset K_i, $i = 1,2$, such that $K = K_1 \cup K_2$.

2.40. If $\#(\Gamma) \le \aleph_0$ then $\Xi \stackrel{\text{def}}{=} [0,1]^\Gamma$ is separable.

2.41. Endowed with the product topology, $X \stackrel{\text{def}}{=} [0,1]^{[0,1]}$ is not metrizable but as the function space $\{f \ : \ f : [0,1] \mapsto [0,1]\}$, X is metrizable.

2.42. If Y^n and Y are homeomorphic for all n in $\mathbb{N}$, need $Y^{\mathbb{N}}$ and Y be homeomorphic?

2.43. Show $[0,1]$ is not the union of countably many pairwise disjoint and nonempty closed sets.

2.44. If $n \ge 2$, F is a closed proper subset of $\mathbb{R}^n$, and the boundary ∂F contains no nonempty perfect subset then F is empty, finite, or countable.

2.45. If $n > 1$ and $\gamma : [0,1] \mapsto \mathbb{R}^n$ is rectifiable then $\gamma([0,1])$ is a null set (λ_n) [whence $(\gamma([0,1]))^\circ = \emptyset$].

2.46. For $\mathbb{R}$, the set of all half-open intervals $(a,b]$ is a base for a topology $\mathcal{T}$ stronger than the Euclidean topology $\mathcal{E}$ for $\mathbb{R}$. Is $\mathcal{T}$ a separable topology for $\mathbb{R}$?

2.47. True or false: $\mathbb{Q}$ is a G_δ?

2.48. a) For (X, U) and U in U, $x \in V_x(U) \stackrel{\text{def}}{=} \{ y : (x, y) \in U \}$ and $\mathcal{V}(x) \stackrel{\text{def}}{=} \{ V_x(U) : U \in \mathsf{U} \}$ is a filter base. b) If $\mathcal{T}$ is the topology for which $\{ V(x) : V(x) \in \mathcal{V}(x), x \in X \}$ is a base then $\mathcal{T}$ is a Hausdorff topology for X iff $\Delta = \bigcap_{U \in \mathsf{U}} U$. c) If X is compact in the topology $\mathcal{T}$ then X is Cauchy complete.

If X resp. Y is a uniform space with a topology induced by a uniformity U resp. V a map $f : X \mapsto Y$ is uniformly continuous iff for every vicinity V in V and some vicinity $U(V)$ in U, $(f(x), f(y)) \in V$ whenever $(x, y) \in U(V)$. (A uniformly continuous map is continuous.)

[**Note 2.2:** Every uniform space X is completely regular. Furthermore, if $(X, \mathcal{T})$ is a compact Hausdorff space then there is a unique uniformity U that induces $\mathcal{T}$ [**Bou, Wel**].]

For each U of a uniformity U, let $V_x(U)$ be as in **2.48**. Then for the corresponding topology $\mathcal{T}$, $\mathcal{O}_U \stackrel{\text{def}}{=} \{ V_x(U)^\circ : x \in X \}$ is an open cover of X.

On the other hand, for a given set X, a given topology $\mathcal{T}$ for X gives rise to a set of open covers of X. The open covers of X are partially ordered by refinement, i.e., $\mathcal{O}' \succ \mathcal{O}$ iff each element of $\mathcal{O}'$ is contained in an element of $\mathcal{O}$.

2.49. If $\mathcal{T}$ is a topology induced by a uniformity U for X then with respect to $\succ$ as defined above, the set $\{ \mathcal{O}_U : U \in \mathsf{U} \}$ of open covers is a diset.

[**Note 2.3:** The contents of **2.49** may be viewed as a motivation for the interesting, alternative, and equivalent approach to the notion of a uniformity discussed in [**Tu, Wel**].]

2.50. If X is a topological space then every open cover of X contains a subcover of cardinality not exceeding $\#(X)$.

2.51. If X is a Hausdorff space for which $\mathcal{N}$ is a base and for each N in $\mathcal{N}$, V_N is an open set containing ∂N then $D \stackrel{\text{def}}{=} X \setminus \bigcup_{N \in \mathcal{N}} V_N$ is empty, a single point, or totally disconnected.

2.52. If G is an unbounded open set in $[0, \infty)$ then

$$D \stackrel{\text{def}}{=} \{ x : x \in (0, \infty),\ nx \in G \text{ infinitely often} \}$$

is G_δ dense in $[0, \infty)$.

2.2. Limits

Conventions

The series $\sum_{n=1}^{\infty} a_n$ may or may not converge; when it does, its sum is $\sum_{n=1}^{\infty} a_n$. The degree of a polynomial p is $\deg(p)$.

For $z \in \mathbb{C}$, the function $\operatorname{sgn}(z)$ is defined so that

$$z\operatorname{sgn}(z) \equiv |z|,\ \overline{\operatorname{sgn}(z)}|z| = z,\ \operatorname{sgn}(z)\overline{\operatorname{sgn}(z)} = \begin{cases} 1 & \text{if } z \neq 0 \\ 0 & \text{otherwise.} \end{cases}$$

2.53. If $a_n \geq a_{n+1} \geq 0, n \in \mathbb{N}$, then $\sum_{n\in\mathbb{N}} a_n$ converges iff $\sum_{n\in\mathbb{N}} 2^n a_{2^n}$ converges.

2.54. (Abel) If, for all n in $\mathbb{N}$, $0 < A_n \stackrel{\text{def}}{=} \sum_{k=1}^n a_k$ and $\sum_{k=1}^\infty a_k = \infty$ then $\sum_{k=1}^\infty \frac{a_k}{A_k} = \infty$ and if $r > 1$ then $\sum_{k=1}^\infty \frac{a_k}{A_k^r} < \infty$.

2.55. If $b_n \downarrow 0$ and $\sum_{n=1}^\infty b_n = \infty$ then for some sequence $\{a_n\}_{n\in\mathbb{N}}$ in $\mathbb{R}^+$, $\lim_{n\to\infty} \frac{a_n}{b_n} = 1$ and $\sum_{n=1}^\infty (-1)^n a_n$ diverges.

2.56. If $a_n, b_n \in \mathbb{R}$, $(a_n + b_n)\, b_n \neq 0$, $n \in \mathbb{N}$, and both $\sum_{n=1}^\infty \frac{a_n}{b_n}$ and $\sum_{n=1}^\infty \left(\frac{a_n}{b_n}\right)^2$ converge then $\sum_{n=1}^\infty \frac{a_n}{(a_n + b_n)}$ converges.

Let E be the dyadic space $\mathcal{D}^{\mathbb{N}}$. An element E of E is a sequence $\{\epsilon_n\}_{n\in\mathbb{N}}$ such that $\epsilon_n = 0$ or $\epsilon_n = 1$, $n \in \mathbb{N}$. For $\sum_{n=1}^\infty a_n$ and E in E, $\sum_{n=1}^\infty \epsilon_n a_n$ is a subsum of $\sum_{n=1}^\infty a_n$.

2.57.

i. If $a_n > 0$, $n \in \mathbb{N}$, $\sum_{n=1}^\infty a_n = 1$ then the set S of all subsums is $[0, 1]$ iff for some autojection $\pi : \mathbb{N} \ni n \stackrel{\text{aut}}{\mapsto} \pi(n) \in \mathbb{N}$,

$$\{n \in \mathbb{N}\} \Rightarrow \left\{ a_{\pi(n)} \leq \sum_{m=n+1}^\infty a_{\pi(m)} \right\}.$$

ia. If the hypothesis $\sum_{n=1}^\infty a_n = 1$ in *i* is dropped and the hypothesis $\lim_{n\to\infty} a_n = 0$ is added then $S = [0, \sum_{n=1}^\infty a_n]$ iff there is an autojection π as described.

ii. In the context of *i*, if $a_n > \sum_{k=n+1}^\infty a_k$, $n \in \mathbb{N}$ then S is homeomorphic to the Cantor set C_0.

iia. If $r \in (0, 1)$, $a = 1 - r$, and $a_n \stackrel{\text{def}}{=} ar^{n-1}$, $n \in \mathbb{N}$, then S is a Cantor-like set iff $0 < r < \frac{1}{2}$, whereas $S = [0, 1]$ iff $\frac{1}{2} \leq r < 1$.

iii. In the context of *iia* what is $\lambda(S) \stackrel{\text{def}}{=} \alpha$ as a function of r?

iv. If $\{a_n\}_{n\in\mathbb{N}} \subset (0, \infty)$ and $\lim_{n\to\infty} a_n = 0$ then $\{a_n\}_{n\in\mathbb{N}}$ contains a finite or infinite sequence $\left\{\{a_{pq}\}_{q\in\mathbb{N}}\right\}_{1\leq p<P}$ of subsequences such that $\{a_n\}_{n\in\mathbb{N}} = \bigcup_{1\leq p<P} \{a_{pq}\}_{q\in\mathbb{N}}$ and for all p, $a_{pq} > \sum_{k=1}^\infty a_{p,q+k}$.

v. In the context of *iia* and *iv*, if $0 < r < \frac{1}{2}$ there is no analogous decomposition for which $a_{pq} \le \sum_{k=1}^{\infty} a_{p,q+k}$, $1 \le p < P$.

2.58. If $\{q_n\}_{n\in\mathsf{N}} \subset \mathbb{R}$ how are the convergence/divergence of $\sum_{n=1}^{\infty} \frac{1}{n^{q_n}}$ related to $\overline{\lim}_{n\to\infty} q_n$, $\underline{\lim}_{n\to\infty} q_n$, and, if it exists, $\lim_{n\to\infty} q_n$?

2.59. There are constants C, D such that if $n \ge 2$ then

$$C \ln n \le \sum_{k=1}^{\infty} \left(1 - \left(1 - 2^{-k}\right)^n\right) \le D \ln n.$$

2.60. For $\{p_n\}_{n\in\mathsf{N}}$ a sequence of polynomials for which the degrees are bounded: $\deg(p_n) \le M < \infty$, $n \in \mathbb{N}$, if the sequence converges to a function f at $M+1$ distinct points then f is a polynomial. If $M = \infty$ and convergence is uniform on $[0,1]$, need f be a polynomial?

2.61. a) $\chi_{\mathbb{Q}}(x) = \lim_{m\to\infty} [\lim_{n\to\infty} \{\cos(m!x\pi)\}^n] \stackrel{\text{def}}{=} \lim_{m\to\infty} s_m$; b) $\operatorname{sgn}(x) = \lim_{n\to\infty} \frac{2}{\pi} \arctan(nx)$; c) $\lim_{n\to\infty} n \sin(2\pi e n!) = 2\pi$ [whence $e \notin \mathbb{Q}$]; d) $1 - \chi_{\mathbb{Q}}(x) = \lim_{m\to\infty} \operatorname{sgn}\left(\sin^2(m!\pi x)\right)$.

2.62. $F(x) \stackrel{\text{def}}{=} e^{\frac{x^2}{2}} \int_x^{\infty} e^{-\frac{t^2}{2}}\,dt \downarrow 0$ as $x \uparrow \infty$.

2.63. If $\{r_n\}_{n\in\mathsf{N}} \subset \mathbb{R}$ then $\lim_{n\to\infty} \int_0^{\infty} e^{-x} \left[\sin\left(x + \frac{r_n\pi}{n}\right)\right]^n dx = 0$.

2.64. $\lim_{\epsilon\to 0} \int_0^{\infty} \left(1 - e^{(\epsilon x)^2}\right) e^{-x^3} \sin^4 x\,dx = 0$.

2.65. $\lim_{n\to\infty} \int_{-\infty}^{\infty} \left(1 - e^{-\frac{t^2}{n}}\right) e^{-|t|} \sin^3 t\,dt = ?$

2.66. If

$$f(x) = \begin{cases} \dfrac{(x \ln x)}{x-1} & \text{if } x \in (0,1) \\ 0 & \text{if } x = 0 \\ 1 & \text{if } x = 1 \end{cases}$$

then $\displaystyle\int_0^1 f(x)\,dx = 1 - \sum_{n=2}^{\infty} \frac{1}{n^2(n-1)}$.

2.67. If

$$\lim_{m\to\infty} t_{mn} = 0,\ n \in \mathbb{N},\ \sum_{n=1}^{\infty} |t_{mn}| \le M < \infty,\ m \in \mathbb{N},\ \lim_{m\to\infty} \sum_{n=1}^{\infty} t_{mn} = 1,$$

[$T \stackrel{\text{def}}{=} (t_{mn})_{m,n=1}^{\infty}$ is a Toeplitz matrix], and $\lim_{m\to\infty} \sum_{n=1}^{\infty} t_{mn}^2 = 0$ then $e^z = \lim_{m\to\infty} \prod_{n\in\mathsf{N}} (1 + t_{mn} z)$, e.g., $e^z = \lim_{m\to\infty} \left(1 + \frac{z}{m}\right)^m$.

2.68. For the Toeplitz matrix $T \stackrel{\text{def}}{=} (t_{mn})_{m,n=1}^{\infty}$, such that $t_{mn} \geq 0$ and $\sum_{n=1}^{\infty} t_{mn} \equiv 1$, if $s_n \uparrow s$ then $\lim_{m\to\infty} \sum_{n=1}^{\infty} t_{mn}s_n \stackrel{\text{def}}{=} \sigma_m \to s$.

2.69. (Kronecker) If $\{a_n\}_{n\in\mathbb{N}} \subset \mathbb{C}$ and the series $\sum_{n=1}^{\infty} \frac{a_n}{n}$ converges then $\lim_{N\to\infty} \frac{\sum_{n=1}^{N} a_n}{N} \stackrel{\text{def}}{=} \lim_{N\to\infty} A_N = 0$.

2.70. If $a \in \mathbb{R}$, $u_0 = a$, $u_{n+1} = \sin u_n$, $n \in \mathbb{Z}^+$, then $\lim_{n\to\infty} u_n = 0$.

2.71. If, for all x in $(0,\infty)$, $f_1(x) = x$, $f_{n+1}(x) = x^{f_n(x)}$, $n \geq 2$, and $\lim_{n\to\infty} f_n(x) \stackrel{\text{def}}{=} \alpha$ exists and is finite then $0 < \alpha^{\frac{1}{\alpha}} = x \leq e^{\frac{1}{e}}$.

2.72. If $\alpha \in \mathbb{R}$ and

$$\binom{\alpha}{n} \stackrel{\text{def}}{=} \begin{cases} 1 & \text{if } n = 0 \\ \dfrac{\alpha(\alpha-1)\cdots(\alpha-n+1)}{n!} & \text{if } n \in \mathbb{N} \end{cases}$$

then on $(-1,1)$, $\sum_{n=0}^{\infty} \binom{\alpha}{n} x^n = (1+x)^\alpha$ and if $\alpha > 0$ the series converges absolutely and uniformly on $[-1,1]$.

2.3. Continuity

Conventions

For $f : X \mapsto X$ the compositions $\underbrace{f \circ f \circ \cdots \circ f}_{n \text{ times}}$, $n \in \mathbb{N}$, are well-defined and are denoted $f^{\{n\}}$.

An ideal I in a ring R is a principal ideal iff for some x, I is the intersection of all R-ideals containing x.

For (X, S, μ) and p in $(0,\infty]$, $L^p(X, d\mu)$ is the set of (equivalence classes of) measurable functions f for which $\|f\|_p^p \stackrel{\text{def}}{=} \int_X |f(x)|^p \, d\mu < \infty$. When f is an essentially bounded function, $\|f\|_\infty$ is the essential supremum of f: $\|f\|_\infty \stackrel{\text{def}}{=} \inf\{ M : |f(x)| \leq M \text{ a.e.}\}$ $(= \lim_{p\to\infty} \|f\|_p)$.

The symbol for Lebesgue measure in $\mathbb{R}^n$ is λ_n or, absent ambiguity, simply λ. For simplicity, to represent an integral over a (measurable) subset E resp. $\mathbf{E}$ of $\mathbb{R}$ resp. $\mathbb{R}^n$, $\int_E f(x)\,dx$ resp. $\int_{\mathbf{E}} f(\mathbf{x})\,d\mathbf{x}$ is used rather than $\int_E f(x)\,d\lambda(x)$ resp. $\int_{\mathbf{E}} f(\mathbf{x})\,d\lambda_n(\mathbf{x})$.

When E is a topological vector space over a topological field $\mathbb{K}$ (usually $\mathbb{R}$ or $\mathbb{C}$), $[E,F]$ is the set of morphisms in the category *TVS* of topological vector spaces and continuous linear maps among them; $[E,\mathbb{K}]$ is the dual space and is denoted E^*.

In a normed vector space $(E, \|\ \|)$ the norm of $\mathbf{x}$ is $\|\mathbf{x}\|$. The norm induces a metric d in E according to the formula: $d(\mathbf{x},\mathbf{y}) \stackrel{\text{def}}{=} \|\mathbf{x}-\mathbf{y}\|$. When E, F are normed vector spaces, and $T \in [E,F]$,

$$\|T\| \stackrel{\text{def}}{=} \sup\{ \|T\mathbf{x}\| : \mathbf{x} \in B(\mathbf{O},1)\}$$

is the operator norm of T. A compact T in $[E, F]$ maps each bounded set B into a precompact set, i.e., $\overline{T(B)}$ is compact.

An extreme point $\mathbf{x}$ in a convex subset K of a vector space E is a vector not in the convex hull of two other vectors in K, i.e., $\mathbf{x}$ is not of the form $t\mathbf{y} + (1-t)\mathbf{z},\ t \in (0,1),\ \mathbf{y},\ \mathbf{z} \in K,\ \mathbf{x} \notin \{\mathbf{y}, \mathbf{z}\}$.

2.73. If A is an F_σ in $\mathbb{R}$ and $f \in C(\mathbb{R}, \mathbb{R})$ then $f(A)$ is an F_σ.

2.74. If $f \in C(\mathbb{R}, \mathbb{R})$ and V is open in $\mathbb{R}$ then $f(V)$ is a Borel set.

2.75. If $f \in C(\mathbb{T}, \mathbb{R})$ for some z in $\mathbb{T}$, $f(z) = f(-z)$, i.e., for some x in $\mathbb{R}$, $f^{-1}(x)$ contains two antipodal points.

2.76. If $f, g \in C([0,1], \mathbb{R})$ and $\{f(a) = f(b)\} \Rightarrow \{g(a) = g(b)\}$ then there is a sequence $\{p_n\}_{n\in\mathbb{N}}$ of polynomials such that $p_n \circ f \xrightarrow{\text{u}} g$ on $[0,1]$.

2.77. For

$$f(x) = \begin{cases} \dfrac{3x+1}{2} & \text{if } -1 \le x \le -\dfrac{1}{3} \\ 0 & \text{if } -\dfrac{1}{3} < x \le \dfrac{1}{3} \\ \dfrac{3x-1}{2} & \text{if } \dfrac{1}{3} < x \le 1 \end{cases},$$

i. if $L \in (C([0,1], \mathbb{C}))^*$, $L(f^n) = 0$, $n \in \mathbb{N}$, $g \in C([-1,1], \mathbb{C})$, and $g\left(\left[-\frac{1}{3}, \frac{1}{3}\right]\right) = 0$ then $L(g) = 0$;

ii. if $L \in C([0,1], \mathbb{C})^*$, $L\left(f^{\{n\}}\right) = 0$, $n \in \mathbb{N}$, $h \in C([-1,1], \mathbb{C})$, h is odd $(h(x) = -h(-x))$, and $h\left(\left[-\frac{1}{3}, \frac{1}{3}\right]\right) = \{0\}$ then $L(h) = 0$.

2.78. For $f_n : [0,1] \ni x \mapsto e^{nx}$, $n \in \mathbb{N}$, and $A_N \stackrel{\text{def}}{=} \{f_n : n \ge N\}$, the linear span S_N of A_N is $\|\ \|_\infty$-dense in $C([0,1], \mathbb{C})$ for all N in $\mathbb{N}$.

2.79. If $f \in C([0,1], \mathbb{R})$ and

$$\int_0^1 x^n f(x)\,dx = 0 \text{ or } \int_0^1 e^{\pm 2\pi i n x} f(x)\,dx = 0,\ n \in \mathbb{Z}^+,$$

then $f(x) \equiv 0$.

2.80. For some $\{a_n\}_{n\in\mathbb{N}}$ contained in $\mathbb{C}$, if $f \in C([0,1], \mathbb{C})$ and $f(0) = 0$ then there is a sequence $\{m_k(f)\}_{k\in\mathbb{N}}$ for which

$$\lim_{k\to\infty} \left\| \sum_{n=1}^{m_k(f)} a_n x^n - f(x) \right\|_\infty = 0.$$

2.81. If $f \in C((0,\infty),\mathbb{R})$, $0 < a < b < \infty$, and, for all h in (a,b), $f(nh) \to 0$ as $n \to \infty$ then $f(x) \to 0$ as $x \to \infty$.

2.82. If $f_0 \in C([0,1],\mathbb{R})$, $f_n(x) \stackrel{\text{def}}{=} \int_0^x f_{n-1}(t)\,dt$, $n \in \mathbb{N}$, and for each x in $[0,1]$ and some n_x in $\mathbb{N}$, $f_{n_x}(x) = 0$ then: a) $[0,1]$ contains a nonempty open set on which f_0 is 0; b) for every n and every b in $(0,1]$, f_n has infinitely many zeros in $(0,b)$.

2.83. If $f \in C^\infty((0,1),\mathbb{C})$ and for each x in $(0,1)$ there is an n_x such that $f^{(n_x)}(x) = 0$ then $(0,1)$ contains a nonempty open interval (a,b) and for some polynomial p, $f\big|_{(a,b)} = p$.

2.84. If $g \in C([0,1],[0,1])$, $g(0) = 1 - g(1) = 0$, and for some m in $\mathbb{N}$, $g^{\{m\}}(x) \equiv x$, then $g(x) \equiv x$.

2.85. True or false: a) if for all x in $\mathbb{R}$, $\lim_{y\uparrow x} f(y) = f(x)$, i.e., if f is left-continuous, then f is bounded on every finite interval? b) if for all x in $[0,1]$, $\overline{\lim}_{y=x} f(y) = f(x)$, i.e., if f is upper semicontinuous, then f is bounded above?

2.86. If $f \in C([0,1],\mathbb{R})$ and $f(0) = 0$ then the sequence $S \stackrel{\text{def}}{=} \{f^n\}_{n\in\mathbb{N}}$ is equicontinuous iff $\|f\|_\infty < 1$.

2.87. If $f \in C([0,1],\mathbb{C})$ then $\lim_{n\to\infty} \dfrac{\int_0^1 x^n f(x)\,dx}{\int_0^1 x^n\,dx} = f(1)$.

2.88. If $A \stackrel{\text{def}}{=} \left\{ f \ : \ f \in C^3([0,1],\mathbb{R}),\ \|f\|_\infty,\ \left\|f'''\right\|_\infty \le 1 \right\}$ then

$$M \stackrel{\text{def}}{=} \sup_{f\in A} \left(\|f'\|_\infty + \left\|f''\right\|_\infty \right) < \infty.$$

2.89. Let P be the statement: A sequence $\{p_n\}_{n\in\mathbb{N}}$ of polynomials converges uniformly on X to f. Prove or disprove P if $f \in C_0(X,\mathbb{C})$ and a) $X = [0,1]$, b) $X = [0,\infty)$, or c) $X = D(0,1) \stackrel{\text{def}}{=} \{ z \ : \ z \in \mathbb{C},\ |z| \le 1 \}$.

2.90. What is a useful necessary and sufficient condition that a set F closed in $[0,\infty)$ be such that every f in $C([0,\infty),\mathbb{R})$ is uniformly approximable on F by polynomial functions of $x^n, n \in \mathbb{N}$?

2.91. True or false: if f is uniformly continuous on a bounded interval (a,b) then f is bounded on (a,b)?

For f in $C(\mathbb{T},\mathbb{C})$ and a in $\mathbb{T}$, the translate $f_{[a]}$ of f is the map

$$\mathbb{T} \ni z \mapsto f(az).$$

The translation-invariant Haar measure τ is defined on the σ-algebra of Borel subsets of $\mathcal{T}$: τ is normalized so that $\tau(\mathbb{T}) = 1$, i.e., if $0 \le t_1 < t_2 < 2\pi$ then $\tau\left(\{ e^{it} \ : \ t_1 \le t < t_2 \}\right) \stackrel{\text{def}}{=} \frac{t_2 - t_1}{2\pi}$.

2.92. If $f, g \in C(\mathbb{T}, \mathbb{C})$ then

$$\lim_{n\to\infty} \int_{\mathbb{T}} f(z) g(z^n)\, d\tau(z) = \int_{\mathbb{T}} f(z)\, d\tau(z) \int_{\mathbb{T}} g(z)\, d\tau(z).$$

2.93. If $f \in C(\mathbb{T}, \mathbb{C})$ and $A \stackrel{\text{def}}{=} \{ f_{[a]} \ : \ a \in \mathbb{T} \}$ then the $\| \ \|_\infty$-closure $\overline{\text{Conv}(A)}$ of the convex hull of A contains precisely one constant function $C(f)$. What is $C(f)$?

2.94. Let X be a complete metric space and in $C(X, \mathbb{C})$ let $\mathcal{F}$, a subset of $C(X, \mathbb{C})$, be such that for each x in X, $\sup_{f\in\mathcal{F}} |f(x)| < \infty$. There is a positive M and in X a nonempty open subset V such that for all x in V, $\sup_{f\in\mathcal{F}} |f(x)| \le M$.

2.95. For X a compact Hausdorff space, if $\{f_n\}_{n\in\mathbb{N}}$ is bounded in $C(X, \mathbb{C})$ and $\lim_{n\to\infty} f_n(x)$ exists for all x in X then

$$\{L \in C(X, \mathbb{C})^*\} \Rightarrow \left\{\lim_{n\to\infty} L(f_n) \text{ exists}\right\}.$$

2.96. If X, Y are compact Hausdorff spaces and

$$A \stackrel{\text{def}}{=} \left\{ \sum_{i=1}^{n} f_i g_i \ : \ \{f_i, g_i\} \in C(X, \mathbb{C}) \times C(Y, \mathbb{C}),\ n \in \mathbb{N} \right\}$$

then A is $\| \ \|_\infty$-dense in $C(X \times Y, \mathbb{C})$.

2.97. For $(X, \mathcal{T})$, (Y, d), and $f \in C(X, Y)$: a) $\text{Cont}(f)$ is a G_δ; b) if $\{f_n\}_{n\in\mathbb{N}} \subset C(X, Y)$, and $\lim_{n\to\infty} f_n \stackrel{\text{def}}{=} f$ then

$$\text{Cont}(f) \supset \bigcap_{k\in\mathbb{N}} \bigcup_{m\in\mathbb{N}} \left[\bigcap_{n=m}^{\infty} \left\{ x \ : \ x \in X, d(f_n(x), f_m(x)) \le \frac{1}{k} \right\} \right]^\circ;$$

c) if X is a complete metric space then $\text{Cont}(f)$ is *-dense in X.

2.98. What is the set of extreme points of $B(\mathbf{O}, 1) \stackrel{\text{def}}{=} B_1(E)$ for: a) $E \stackrel{\text{def}}{=} C([0,1], \mathbb{C})$? b) $E \stackrel{\text{def}}{=} C_0(X, \mathbb{R})$ when X is a connected locally compact Hausdorff space?

2.99. If A is a closed subset of $[0,1]$ and for every compact metric space X there is a continuous surjection $f_X : A \stackrel{\text{sur}}{\mapsto} X$ then: a) the cardinality of the set C of components of A is $\mathfrak{c}$; b) there is a set P homeomorphic to the Cantor set C_0, a set D such that $\#(D) \le \aleph_0$, and an open set U such that $A = P \cup D \cup U$; c) for each compact metric space X there is a continuous surjection $g_X : A \setminus A^\circ \mapsto X$.

2.100. True or false: if $\{f_n\}_{n=0}^\infty \subset C([0,1],\mathbb{C})$ and $L(f_n) \to L(f_0)$ for all L in $C([0,1],\mathbb{C})^*$, then $f_n(x_n) \to f_0(x_0)$ whenever $x_n \to x_0$?

2.101. True or false: if $L \in C([0,1],\mathbb{C})^*$ and $\|L\| = 1$ then for some f in $C([0,1],\mathbb{C})$, $\|f\|_\infty \le 1$ and $L(f) = 1$?

2.102. a) $A \stackrel{\text{def}}{=} C([0,1],\mathbb{C}) \cap \{f : f(\mathbb{Q}) \subset \mathbb{Q}\}$ is a $\| \ \|_\infty$-dense Borel subset of $C([0,1],\mathbb{R})$. b) What is A°?

2.103. If A is a closed subspace of $C([0,1],\mathbb{C})$, $g \in \mathbb{R}^{\mathbb{R}}$, and $g \cdot A \subset A$ then $M_g : A \ni f \mapsto gf \in A$ is linear and continuous.

2.104. True or false: for $A \stackrel{\text{def}}{=} \{f : f \in C([0,1],\mathbb{C}),\ f\left(\frac{1}{2}\right) = 0\}$: a) A is a principal ideal? b) for some g_0 in A, $\overline{g_0 \cdot C([0,1],\mathbb{C})} = A$?

2.105. If K is a compact subset of $[0,1]$,

$$A_K \stackrel{\text{def}}{=} \{f : f \in C([0,1],\mathbb{C}),\ f(K) = \{0\}\}, \text{ and}$$
$$B_K \stackrel{\text{def}}{=} \bigcup_{K \subset V \in \mathcal{O}([0,1])} \{f : f \in C([0,1],\mathbb{C}),\ f(V) = \{0\}\}$$

then: a) $\overline{B_K} = A_K$; b) A_K is an ideal but A_K is not a principal ideal.

2.106. a) If $S \subset C(X,\mathbb{R})$ then $h \stackrel{\text{def}}{=} \bigvee\{f : f \in S\}$ is lsc while $\bigwedge\{f : f \in S\}$ is usc. b) When X is locally compact a nonnegative resp. nonpositive h in $\mathbb{R}^X$ is lsc resp. usc iff for some subset S of $C_{00}(X,\mathbb{R})$,

$$h = \bigvee\{f : f \in S\} \text{ resp. } h = \bigwedge\{f : f \in S\}.$$

2.107. For P, Q, R one of the predicates lsc, usc, semicontinuous (sc), what combinations (if any) of the predicates make the following true:

a) $\{P(f) \wedge Q(g)\} \Rightarrow \{R(f \pm g)\}$;
b) $\{P(f) \wedge Q(g)\} \Rightarrow \{R(f \cdot g)\}$;
c) $\{P(f) \wedge Q(g)\} \Rightarrow \{R(f \circ g)\}$?

2.108. If X and Y are topological spaces and $f \in Y^X$ then f is continuous at x iff: a) for every filter $\mathcal{F}$ converging to x, $f(\mathcal{F})$ is the base of a filter converging to $f(x)$; b) for every net n converging to x the net $f(n)$ converges to $f(x)$.

2.109. If (X, U) and (Y, W) are uniform spaces and X is compact then every map f in $C(X, Y)$ is uniformly continuous.

A property P of objects in a category $\mathcal{C}$ with a zero object is a Quotient Lifting (QL) property iff for objects A, B in $\mathcal{C}$ and a quotient object A/B, $\{P(B) \wedge P(A/B)\} \Rightarrow \{P(A)\}$.

2.110. For each category listed below prove or disprove that the associated property P is a QL property:

i. $\mathcal{G}$, the category of groups and coset spaces, P is *countability;*
ii. $\mathcal{G}$, P is *abelianity*;
iii. $\mathcal{LCG}$, the category of locally compact groups and coset spaces of closed subgroups, P is *compactness*;
iv. $\mathcal{MTG}$, the category of metric topological groups and coset spaces of closed subgroups, P is *separability*;
v. $\mathcal{TG}$, the category of topological groups and coset spaces of closed subgroups, P is *separability*.

[**Note 2.4:** In *iv* above, separability is equivalent to the existence of a countable dense subset. In *v* separability means simply the existence of a countable base for the topology.]

2.111. The set Λ of all nonempty open intervals U containing the real number a is a diset when it is partially ordered by reversed inclusion, i.e., $U' \succ U$ iff $U' \subset U$. For f in $\mathbb{R}^{\mathbb{R}}$, U in $\mathcal{N}(x)$, and a y in U, let $n(U)$ be $f(y)$. Let U in $\mathbb{R}^2$ be the uniform structure for which a typical vicinity is, for some positive ϵ, $V(\epsilon) \stackrel{\text{def}}{=} \{\{u, v\} : |u - v| < \epsilon\}$. a) $L \stackrel{\text{def}}{=} \lim_{x \to a} f(x)$ exists iff n is a Cauchy net; b) f is continuous at a iff the net n converges to $f(a)$.

2.112. Let Λ be the set of all complements C in $\mathbb{N}$ of finite subsets of $\mathbb{N}$. Partially ordered by reversed inclusion, i.e., $C' \succ C$ iff $C' \subset C$, Λ is a diset. For f in $\mathbb{R}^{\mathbb{N}}$ and C in Λ, let $n(C)$ be $f(\inf(C))$. Show $\{f(k)\}_{k \in \mathbb{N}}$ is a Cauchy sequence iff n is a Cauchy net.

2.113. For a net $n : \Lambda \ni \lambda \mapsto n(\lambda) \in \mathbb{R}$,

$$\overline{\lim}(n) \stackrel{\text{def}}{=} \inf_{\mu \in \Lambda} \sup_{\lambda \succ \mu} n(\lambda), \text{ and } \underline{\lim}(n) \stackrel{\text{def}}{=} \sup_{\mu \in \Lambda} \inf_{\lambda \succ \mu} n(\lambda),$$

show: a) n is a Cauchy net iff $\overline{\lim}(n) = \underline{\lim}(n)$; b) for f and Λ as in **2.111**, $\overline{\lim}_{k \to \infty} f(k) = \overline{\lim}(n)$, and $\underline{\lim}_{k \to \infty} f(k) = \underline{\lim}(n)$.

2.114. For a complete metric space (X, d) without isolated points, if each f in $C(X, \mathbb{R})$ is uniformly continuous then X is compact.

3
Real- and Complex-valued Functions

3.1. Real-valued Functions

Conventions

A map $\phi : K \mapsto \mathbb{R}$ of a convex open subset K of a vector space V is convex iff for all t in $[0,1]$ and any $\mathbf{x}$ and $\mathbf{y}$ in K,

$$\phi(t\mathbf{x} + (1-t)\mathbf{y}) \le t\phi(\mathbf{x}) + (1-t)\phi(\mathbf{y}).$$

When K is an open interval in $\mathbb{R}$ and $a \in K$, a line l through $(a, \phi(a))$ is a supporting line iff l lies below the graph of ϕ: $\{(p,q) \in l\} \Rightarrow \{q \le \phi(p)\}$.

A topological vector space V is an abelian topological group that is also a module over a topological field $\mathbb{K}$ (usually $\mathbb{C}$ or $\mathbb{R}$). In particular the maps $V \times V \ni \{\mathbf{x},\mathbf{y}\} \mapsto \mathbf{x}+\mathbf{y} \in V$, $\mathbb{K} \times V \ni \{t, \mathbf{x}\} \mapsto t\mathbf{x} \in V$ are continuous.

For topological vector spaces V, W, $[V,W]$ resp. $[V,W]_e$ is the set of continuous homomorphisms resp. continuous epimorphisms from V to W; $[V]$ resp. $[V]_a$ is the set of continuous endomorphisms resp. continuous automorphisms of V; the identity in $[V]$ and $[V]_a$ is id.

When $\mathbf{x} \stackrel{\text{def}}{=} (x_1, \ldots, x_n) \in \mathbb{R}^n$, $\|\mathbf{x}\|$ is the Euclidean length of $\mathbf{x}$:

$$\|\mathbf{x}\| = \sqrt{\sum_{i=1}^{n} x_i^2}.$$

For $\mathbf{f}$ in $(\mathbb{R}^m)^{\mathbb{R}^n}$ and $\mathbf{x}_0$ in $\mathbb{R}^n$, when there is in $[\mathbb{R}^n, \mathbb{R}^m]$ a T such that

$$\lim_{\substack{\mathbf{h}\neq\mathbf{0}\\ \|\mathbf{h}\|\to 0}} \frac{\|\mathbf{f}(\mathbf{x}_0+\mathbf{h}) - \mathbf{f}(\mathbf{x}_0) - T(\mathbf{h})\|}{\|\mathbf{h}\|} = 0,$$

T is the differential or derivative $\mathbf{df}(\mathbf{x}_0)$ of $\mathbf{f}$ at $\mathbf{x}_0$. By induction there are defined higher differentials $\mathbf{d}^k\mathbf{f}$, $k = 2, 3, \ldots$. Note: $\mathbf{df}(\mathbf{x}_0) \in [\mathbb{R}^n, \mathbb{R}^m]$ and $\mathbf{d}^2\mathbf{f}(\mathbf{x}_0) \in [\mathbb{R}^n, [\mathbb{R}^n, \mathbb{R}^m]]$, etc., while $\mathbf{d}^0\mathbf{f}(\mathbf{x}_0) = \mathbf{f}(\mathbf{x}_0)$. Correspondingly, for k in $\mathbb{N}$, $C^k(\mathbb{R}^n, \mathbb{R}^m)$ is the set of maps $f : \mathbb{R}^n \mapsto \mathbb{R}^m$ such that $\mathbf{d}^k\mathbf{f}(\mathbf{x})$ exists and is continuous for all $\mathbf{x}$; $C^\infty(\mathbb{R}^n, \mathbb{R}^m) \stackrel{\text{def}}{=} \bigcap_{k=0}^{\infty} C^k(\mathbb{R}^n, \mathbb{R}^m)$. Similar definitions apply for $C^k(\mathbb{R}^n, \mathbb{C}^m)$ and $C^\infty(\mathbb{R}^n, \mathbb{C}^m)$ and even more generally when $\mathbb{R}^n$ is replaced by an open subset U of $\mathbb{R}^n$.

An $\mathbf{f}$ in $(\mathbb{R}^m)^{\mathbb{R}^n}$ is in Lip(α) iff for some constant K and all $\mathbf{x}$, $\mathbf{y}$,

$$\|\mathbf{f}(\mathbf{x}) - \mathbf{f}(\mathbf{y})\| \leq K\|\mathbf{x} - \mathbf{y}\|^\alpha.$$

An f in $\mathbb{R}^{\mathbb{R}}$ enjoys the intermediate value property iff whenever y is between $f(a)$ and $f(b)$ there is, between a and b, an x such that $f(x) = y$.

For f, g in $\mathbb{R}^{\mathbb{R}}$,

$$f^+ \stackrel{\text{def}}{=} f \vee 0 = \frac{f + |f|}{2}, \; f^- \stackrel{\text{def}}{=} f \wedge 0 = \frac{f - |f|}{2}$$
$$f = f^+ + f^-, \; |f| = f^+ - f^-$$
$$\max\{f, g\} \stackrel{\text{def}}{=} f \vee g = g + (f - g)^+, \; \min\{f, g\} \stackrel{\text{def}}{=} f \wedge g = g + (f - g)^-.$$

The determinant of a square matrix M is $\det(M)$.

When G is a locally compact group, $a \in G$, and $f \in Y^G$ then

$$f_{[a]}(x) \stackrel{\text{def}}{=} f(ax), \; f^{[a]}(x) \stackrel{\text{def}}{=} f(xa).$$

When $(X_i, d_i), i = 1, 2$, are metric spaces, $\epsilon > 0$, and $f \in C(X_1, X_2)$, the modulus of continuity of f at x is

$$\omega(f, x; \epsilon) \stackrel{\text{def}}{=} \sup\{\delta \; : \; \{d_1(x, y) < \delta\} \Rightarrow \{d_2(f(x) - f(y)) < \epsilon\}\}.$$

A function f in $\mathbb{C}^{\mathbb{R}}$ is real analytic iff for each a in $\mathbb{R}$ there is a positive $r(a)$ and in $\mathbb{C}$ a sequence $\{c_n(a)\}$ such that

$$f(x) = \sum_{n=0}^{\infty} c_n(a)(x - a)^n, \; |x - a| < r(a).$$

When S is a subset of a ring, $S[x]$ is the set of polynomials

$$\left\{\sum_{n=1}^{N} s_n x^n, s_n \in S, N \in \mathbb{Z}^+\right\}.$$

When S is a ring then $S[x]$ is also a ring.

For a finite subset $\{a_n\}_{n=1}^N$ in $\mathbb{R}$ and

$$s_0 \stackrel{\text{def}}{=} 0, \; s_m \stackrel{\text{def}}{=} \sum_{n=1}^{m} a_n, \; m = 1, \ldots, N,$$

a chain of indices is a sequence $n, n+1, \ldots, n+k$. An index n is distinguished iff for some n' greater than n, $s_{n'} > s_{n-1}$; a maximal chain of distinguished indices is a block.

3.1. (F. Riesz) If $\{a_n\}_{n=1}^N \subset \mathbb{R}$, $D \stackrel{\text{def}}{=} \{n_k\}_{k=1}^K$, $n_1 < n_2 < \cdots < n_K$, is the (nonempty) set of distinguished indices, n is in a block, $n^\#$ is the greatest index in the block, and $n^\# > n$ then $s_{n^\#} > s_{n-1}$ (whence $\sum_{n\in D} a_n > 0$).

For f in $\mathbb{R}^{(0,1)}$ and bounded, an x in $(0,1)$ is distinguished if for some x' in $(x,1)$, $\overline{\lim}_{y=x} f(y) < f(x')$.

3.2. (F. Riesz) The set D of distinguished elements in $(0,1)$ is open and if $\emptyset \neq D \stackrel{\text{def}}{=} \dot{\bigcup}_{n=1}^M (a_n, b_n)$, $1 \le n < M \le \infty$, and $x \in (a_n, b_n)$ then $f(x) \le \overline{\lim}_{y=b_n} f(y)$.

3.3. If $\{a_r\}_{r=1}^R$ and $\{b_s\}_{s=1}^S$, $S \le R$, are sequences in $\mathbb{C}$ then for some p in $\mathbb{C}[x]$,

$$\deg(p) \le R+S-1,\ p^{(r-1)}(1) = a_r,\ 1 \le r \le R,\ p^{(s-1)}(2) = b_s,\ 1 \le s \le S.$$

3.4. If $a_n > 0, n \in \mathbb{N}$, $\{b_n\}_{n\in\mathbb{N}} \subset \mathbb{R}$, and $\sum_{n=1}^\infty a_n < \infty$ then for some monotonely increasing function f, $f(b_n+0) - f(b_n-0) = a_n$, $n \in \mathbb{N}$, and $\operatorname{Cont}(f) \stackrel{\text{def}}{=} \mathbb{R} \setminus \{b_n\}_{n\in\mathbb{N}} \stackrel{\text{def}}{=} S$.

3.5. If $-\infty < x_1 < x_2 < \cdots < x_n < \infty$, $s(\theta) \stackrel{\text{def}}{=} \sum_{i=1}^n |x_i - \theta|$, and

$$m = \begin{cases} x_{\frac{n+1}{2}} & \text{if } n \text{ is odd} \\ \widetilde{m} & \text{if } n \text{ is even and } x_{\frac{n}{2}} \le \widetilde{m} \le x_{\frac{n}{2}+1} \end{cases} \quad \text{then } \min_{\theta\in\mathbb{R}} s(\theta) = s(m).$$

3.6. If $f \in \mathbb{R}^{\mathbb{R}}$ and for each x and some positive $\delta(x)$,

$$\{x - \delta(x) < a < x < b < x + \delta(x)\} \Rightarrow \{f(a) \le f(x) \le f(b)\}$$

then $\{p < q\} \Rightarrow \{f(p) \le f(q)\}$: if f is locally monotone then f is monotone.

3.7. If $-\infty \le u < v \le \infty$, then ϕ in $\mathbb{R}^{(u,v)}$ is convex iff

$$\{u < x \le x' < y' \le y < v\} \Rightarrow \left\{ \frac{(\phi(y) - \phi(x))}{y - x} \le \frac{(\phi(y') - \phi(x'))}{y' - x'} \right\}.$$

What is the geometric meaning of the criterion?

3.8. If ϕ in $\mathbb{R}^{\mathbb{R}}$ is convex then in $[-\infty, \infty]$ there are p and q such that $p \le q$ and such that on:

$$\begin{cases} (-\infty, p) & \phi \text{ is monotonely decreasing;} \\ [p, q] & \phi \text{ is constant;} \\ (q, \infty) & \phi \text{ is monotonely increasing.} \end{cases}$$

Give examples of convex functions for which

a) $-\infty = p = q$, b) $-\infty = p < q < \infty$, c) $-\infty = p < q = \infty$,
d) $p = q = \infty$, e) $-\infty < p < q = \infty$, f) $-\infty < p < q < \infty$.

3.9. If ϕ in $\mathbb{R}^{(u,v)}$ is convex then $\phi \in \text{Lip}(1)$, f has left- and right-hand derivatives $\phi'_l(x)$ and $\phi'_r(x)$ for all x in (u,v), and f is differentiable almost everywhere on (u,v).

3.10. True or false: if $\phi, \psi \in \mathbb{R}^{\mathbb{R}}$ and both are convex then so is $\phi \circ \psi$ convex?

3.11. If ϕ'' exists on (u,v) then: a) ϕ is convex if $\phi'' > 0$ on (u,v); b) if ϕ is convex then $\phi'' \geq 0$ on (u,v).

3.12. a) If $\phi \in \mathbb{R}^{\mathbb{R}}$, $\phi > 0$, and $\ln(\phi)$ is convex then ϕ is convex; b) the converse assertion is false.

3.13. True or false:

i. if $g \in \mathbb{R}^{(0,1)}$, $g \geq 0$, and $g(x) \to \infty$ as $x \to 0$ then for some convex function ϕ in $\mathbb{R}^{(0,1)}$, $\phi \leq g$ and $\phi(x) \to \infty$ as $x \to 0$?
ii. if $g \in \mathbb{R}^{(0,\infty)}$, $g \geq 0$, and $g(x) \to \infty$ as $x \to \infty$ then for some convex function ϕ in $\mathbb{R}^{(0,\infty)}$, $\phi \leq g$ and $\phi(x) \to \infty$ as $x \to \infty$?

3.14. If a, b, c, y are nonnegative functions in $C([0,\infty), \mathbb{R})$ and for all t in $[0,\infty)$, $y(t) \leq \int_0^t [a(s)y(s) + b(s)]\, ds + c(t)$ then

$$y(t) \leq \left[\int_0^t b(s)\, ds + \max_{0\leq s\leq t} c(s)\right] \exp\left(\int_0^t a(s)\, ds\right).$$

3.15. If $q \in L^1([0,\infty), \lambda)$ and $y'' + y = -qy,\ y(0) = 0,\ y'(0) = 1$ then for some finite M, $\|y\|_\infty \leq M$.

3.16. If $f \in C^2([0,1], \mathbb{R})$ and $\lambda\left(f^{-1}(0)\right) = 0$ then: a) $|f|''$ exists a.e.; b) if

$$h(x) \stackrel{\text{def}}{=} \begin{cases} |f|''(x) & \text{if } |f|''(x) \text{ exists} \\ 0 & \text{otherwise} \end{cases}$$

then $h \in L^\infty([0,1], \lambda)$; c) if $0 \leq g \in C^2([0,1], \mathbb{R})$ then

$$\int_0^1 g(x)h(x)\, dx \leq \int_0^1 |f(x)|g''(x)\, dx.$$

3.17. True or false: for f in $(0,1)^{(0,1)}$,

i. if f is continuous and $\{a_n\}_{n\in\mathbb{N}}$ is a Cauchy sequence then $\{f(a_n)\}_{n\in\mathbb{N}}$ is a Cauchy sequence?
ii. if f maps every Cauchy sequence into a Cauchy sequence then f is continuous?

3.18. If $f, g \in \mathbb{R}^{(0,\infty)}$, $\lim_{x\to 0} g(x)$ exists, and for all a, b in $(0,\infty)$, $|f(b) - f(a)| \leq |g(b) - g(a)|$ then $\lim_{x\to 0} f(x)$ exists.

3.19. For some monotonely decreasing g in $\mathbb{R}^{[0,4\pi]}$ and all real r,

$$\lambda(\{x : x \in [0,4\pi],\ \sin x > r\}) = \lambda(\{x : x \in [0,4\pi],\ g(x) > r\}).$$

3.20. a) For some sequence $\{f_k\}_{k\in\mathbb{N}}$ of monotonely increasing functions in $\mathbb{R}^{[0,1]}$, $\lim_{k\to\infty} f_k(x) \equiv 1$ and for some x, $\lim_{k\to\infty} f_k'(x) = \infty$. b) If $\{f_k\}_{k\in\mathbb{N}} \subset \mathbb{R}^{[0,1]}$ and each f_k is monotonely increasing then

$$\left\{\lim_{k\to\infty} f_k(x) \equiv 1\right\} \Rightarrow \{\underline{\lim}_{k\to\infty} f_k'(x) \doteq 0\}.$$

3.21. Is there in $\mathbb{R}^{\mathbb{R}}$ an f such that $\mathrm{Cont}(f) = \mathbb{A}_{\mathbb{R}}$? $= \mathbb{R}\setminus\mathbb{A}_{\mathbb{R}}$?

3.22. True or false: if $\{f_n\}_{n\in\mathbb{N}} \subset C^2(\mathbb{R},\mathbb{R})$, $\|f_n''\|_\infty \le M'' < \infty$, and $f_n \stackrel{\mathrm{u}}{\to} 0$ then for some finite M' and all n in $\mathbb{N}$, $\|f_n'\|_\infty \le M'$?

3.23. If $f \in \mathbb{R}^{\mathbb{R}}$ and f is continuous a.e. then f is Lebesgue measurable.

3.24. There is in $\mathbb{R}^{\mathbb{R}}$ no function f for which: a) f' exists on an open set U containing $\{0\}$, b) $f(0) = 0$, and c) $f'(x) = \chi_{(-\infty,0]} \circ f(x)$ on U.

3.25. If $f \in C(\mathbb{R},\mathbb{R})$ and $\overline{\lim}_{h\downarrow 0}\dfrac{f(x+h)-f(x)}{h}$ is never negative then f is monotonely increasing.

3.26. For some f in $C(\mathbb{R},\mathbb{R})$, $\overline{\lim}_{h\downarrow 0}\dfrac{f(x+h)-f(x)}{h} \ge 0$ a.e. and yet f is monotonely increasing on no nonempty open interval.

3.27. For $f \in C((0,1),\mathbb{R})$, there is a metric d for $(0,1)$ so that f is d-uniformly continuous while the topology induced by d is the standard topology of $(0,1)$.

3.28. If $f \in \mathbb{R}^{\mathbb{R}^n}$ and for all $\mathbf{x}$ in some nonempty open ball $B(\mathbf{O},r)^\circ$ in $\mathbb{R}^n$, $f(\mathbf{x}) \stackrel{\mathrm{def}}{=} f(x_1,\ldots,x_n) = \sum_{k_1+\cdots+k_n\ge 0} a_{k_1,\ldots,k_n} x_1^{k_1}\cdots x_n^{k_n}$ then either $f = 0$ in $B(\mathbf{O},r)^\circ$ or $\lambda_n\left(f^{-1}(0)\cap B(\mathbf{O},r)^\circ\right) = 0$.

3.29. a) In $C(\mathbb{R}^n,\mathbb{R})$, the set V of all f such that for some M_f, k_f, and all $\mathbf{x}$ in $\mathbb{R}^n$, $|f(\mathbf{x})| \le M_f(1+\|\mathbf{x}\|)^{k_f}$ is a function lattice. b) If $P : V \mapsto \mathbb{R}$ is a nonnegative linear functional and μ is a Borel measure such that for all f in $C_0(\mathbb{R}^n,\mathbb{R})$,

$$P(f) = \int_{\mathbb{R}^n} f(x)\,d\mu(x), \tag{3.1}$$

then (3.1) obtains for all f in V.

3.30. For $\Sigma \stackrel{\text{def}}{=} \{\mathbf{x} : \mathbf{x} \in \mathbb{R}^n, \|\mathbf{x}\| = 1\}$ (the surface of the unit sphere in $\mathbb{R}^n$), if $f, \dfrac{\partial f}{\partial t} \in C(\Sigma \times \mathbb{R}, \mathbb{R})$ and $f^2 + \left(\dfrac{\partial f}{\partial t}\right)^2 > 0$ then for each t_0 in $\mathbb{R}$, there is an open set $U(t_0)$ containing t_0 and such that for all $\mathbf{x}$ in Σ, there is in $U(t_0)$ at most one t such that $f(\mathbf{x}, t) = 0$.

3.31. If $\{f, g\} \subset C(\mathbb{R}^n, \mathbb{R})$, for some m, $f(t\mathbf{x}) \equiv t^m f(\mathbf{x})$, $g(t\mathbf{x}) \equiv t^m g(\mathbf{x})$, $f \geq 0$, $g(\mathbf{x}) > 0$ whenever $\mathbf{x} \neq \mathbf{O}$, and $f(\mathbf{x}) = 0$ then for some constants C, D and all $\mathbf{x}$, $Cf(\mathbf{x}) + Dg(\mathbf{x}) \geq \|\mathbf{x}\|^m$.

3.32. If $\mathbf{f} \in C^1(B_1(\mathbb{R}^n), \mathbb{R}^n)$ then for some positive δ, if

$$\sup_{\mathbf{x}} \|\mathbf{df}(\mathbf{x}) - \mathrm{id}\| < \delta$$

then $\mathbf{f}$ is injective on $B_1(\mathbb{R}^n)$.

3.33. For f in $C(\mathbb{R}^n, \mathbb{R})$ if $\mathbf{d}f(\mathbf{x}_0) = \mathbf{O}$ and $\left(\mathbf{d}^2 f(\mathbf{x}_0)\right)^{-1}$ exists then for some $N(x_0)$ and all $\mathbf{y}$ in $N(x_0) \setminus \{\mathbf{x}_0\}$, $\mathbf{d}f(\mathbf{y}) \neq \mathbf{O}$.

3.34. If $f \in C([0,1], \mathbb{C}) \cap BV([0,1], \mathbb{C})$ and $\epsilon > 0$, there is a positive a such that for a partition P, $\{|P| < a\} \Rightarrow \left\{\mathrm{var}_{[0,1],P}(f) > \mathrm{var}_{[0,1]}(f) - \epsilon\right\}$. Is the conclusion valid if f is not continuous?

3.35. If $f \in BV([0,1], \mathbb{C})$ then $f' \in L^1([0,1], \lambda)$.

3.36. For f in $\mathbb{R}^{[0,1]}$ and a in $[0,1]$, $V : [0,1] \ni x \mapsto \mathrm{var}_{[0,x]} f$ is left-continuous resp. right-continuous resp. continuous at a iff f is of bounded variation on $[0,a]$ resp. an interval $[0,b)$ containing $[0,a]$ resp. an interval $[0,b)$ containing $[0,a]$ and f is left-continuous resp. right-continuous resp. continuous at a.

3.37. If f is of bounded variation on an interval and enjoys the intermediate value property there then f is continuous. [For a nowhere continuous function enjoying the intermediate property on every interval (cf. **[GeO]**).]

3.38. If $g \in BV([-1,1], \mathbb{C})$, g is continuous at $\pm a$, and for every even continuous f, $\int_{-1}^{1} f(x)g(x)\,dx = 0$, then $g(a) + g(-a) = 0$.

3.39. Some f is in $BV([0,1], \mathbb{C}) \cap AC([0,a), \mathbb{C})$ for all a in $(0,1)$ and yet $\lim_{x \to 1} f(x)$ exists and is not $f(1)$.

3.40. For what values of a is

$$f_a : [0,1] \ni x \mapsto \begin{cases} 0 & \text{if } x = 0 \\ x^a \cos\left(\dfrac{1}{x}\right) & \text{if } 0 < x \leq 1 \end{cases}$$

of bounded variation? absolutely continuous?

3.41. Some f in $C([0,1],\mathbb{C}) \setminus AC([0,1],\mathbb{C})$ is in $AC([0,a),\mathbb{C})$ for all a in $(0,1)$.

3.42. In $AC([0,1],\mathbb{R})$ there is a strictly monotonely increasing function f such that $f'(x) = 0$ on a set of positive (Lebesgue) measure.

3.43. For r in $\left(0, \frac{1}{2}\right)$ and S the corresponding Cantor-like set discussed in **2.57***ii* – *iii*, ϕ_S is the analog of **the** Cantor function, what is: a) $\int_{[0,1]} \phi_S(x)\, dx$; b) the length of the graph $\mathcal{G}_S$ of $y = \phi_S(x)$? For α in $(0,1)$, what are the solutions to a) and b) for **the** Cantor-like set C_α and its corresponding function ϕ_α?

3.44. For R symbolizing any of $<, >, =$ and p a polynomial of degree n, if $E_{R,j}(p) \stackrel{\text{def}}{=} \{ x : p^{(j)}(x) R 0 \}$ then $E_R(p) \stackrel{\text{def}}{=} \bigcap_{j=0}^{n} E_{R,j}(p)$ is $\emptyset$ or an interval.

3.45. An f in $\mathbb{R}^{[0,1]}$ is Riemann integrable iff $\lambda(\text{Discont}(f)) = 0$ and f is bounded.

3.46. For some uniformly bounded sequence $\{f_n\}_{n\in\mathbb{N}}$ of functions Riemann integrable on $[0,1]$, $\lim_{n\to\infty} f_n$ is not Riemann integrable on $[0,1]$.

3.47. There is a Riemann integrable function f and a continuous function g such that for no Riemann integrable function h, $h \doteq f \circ g$.

3.48. If f' exists everywhere on $\mathbb{R}$ then f is strictly increasing iff $f' \geq 0$ everywhere and $D \stackrel{\text{def}}{=} \{ x : f'(x) = 0 \}$ is totally disconnected.

3.49. If $f \in \mathbb{R}^{\mathbb{R}}$ then the set S of sites of strict local maxima of f is empty, finite, or countable.

3.50. For f in $C(\mathbb{R},\mathbb{R})$, n in $\mathbb{N}$, and

$$\Delta_n : \mathbb{R} \ni x \mapsto 2^n \left(f\left(x + 2^{-n}\right) - f(x) \right),$$

if $\|\Delta_n\|_\infty \leq M < \infty$, $n \in \mathbb{N}$, and for all x, $\lim_{n\to\infty} \Delta_n(x) = 0$ then f is constant.

3.51. If $\{f,g\} \subset C([0,1][0,1])$ and $f \circ g = g \circ f$ then for some a in $[0,1]$, $f(a) = g(a)$. The conclusion is false if $\{f,g\} \subset C([0,1][0,1])$ is replaced by $\{f,g\} \subset C((0,\infty),(0,\infty))$.

3.52. If $f \in C([0,1],\mathbb{R})$ and for some c in $(0,1)$,

$$\lim_{h\in\mathbb{Q}\setminus\{0\},h\to 0} \frac{f(c+h) - f(c)}{h} \stackrel{\text{def}}{=} L$$

exists, then $f'(c)$ exists. For some f not in $C([0,1],\mathbb{R})$ the conclusion is false.

3.53. If $\phi, g \subset C(\mathbb{R},\mathbb{R})$ and for all compactly supported h in $C^\infty(\mathbb{R},\mathbb{R})$, $\int_{\mathbb{R}} f(x)h(x)\,dx = -\int_{\mathbb{R}} g(x)h'(x)\,dx$ then g is differentiable and $g' = f$.

3.54. If $a > 0$, $\mathbf{f} \in C(\mathbb{R}^n,\mathbb{R}^n)$, and, for all $\mathbf{x}$ and $\mathbf{y}$ in $\mathbb{R}^n$,

$$\|\mathbf{f}(\mathbf{x}) - \mathbf{f}(\mathbf{y})\| \geq a\|\mathbf{x} - \mathbf{y}\| \tag{3.2}$$

then $\mathbf{f}$ is an auteomorphism.

3.55. For f in $C(B(\mathbf{O},1),\mathbb{R})$ if

$$f(\mathbf{x}) = \frac{1}{2\pi r}\oint_{\|\mathbf{y}-\mathbf{x}\|=r} f(\mathbf{y})\,d\mathbf{y}$$

for all $\mathbf{x}$ in $B(\mathbf{O},1)^\circ$ and all r in $(0, 1 - \|\mathbf{x}\|)$, then f is constant iff

$$|f(\mathbf{O})| = \sup_{\mathbf{x}\in B(\mathbf{O},1)} |f(\mathbf{x})|.$$

3.56. If $f \in C(\mathbb{R},\mathbb{R})$, f is bounded, $\{t_n\}_{n\in\mathbb{N}} \subset \mathbb{R}$, and K is a compact subset of $\mathbb{R}$, then $\{f_{[t_n]}\}_{n\in\mathbb{N}}$ contains a subsequence converging uniformly on K.

3.57. If $f \in C([0,1],\mathbb{R})$ and $\{x \in [0,1]\} \Rightarrow \left\{|f(x)| \leq \int_0^x f(t)\,dt\right\}$ then $f(x) \equiv 0$.

3.58. If f' (exists and) is real analytic then so is f.

3.59. If $f \in C^\infty(\mathbb{R},\mathbb{R})$ and $f^{(n)}(x) \geq 0$, $n \in \mathbb{N}$, $x \in \mathbb{R}$ then f is real analytic.

3.60. a) If $a_n \downarrow 0$ then $\left|\sin\frac{x}{2}\right| \left|\sum_{n=m}^M a_n \sin nx\right| \leq a_m$ and if, furthermore, for some constant K, $na_n \leq K$ then for all x in $[-\pi,\pi]$, hence for all x in $[0,2\pi]$, $|S_N(x)| \stackrel{\text{def}}{=} \left|\sum_{n=1}^N a_n \sin nx\right| \leq K(1+\pi)$. b) If $\sum_{n=1}^\infty |a_n - a_{n+1}|$ converges and $\lim_{n\to\infty} a_n = 0$ then $\sum_{n=1}^\infty a_n \sin n\theta$ converges on $[0,2\pi]$ and uniformly on every closed subinterval of $(0,2\pi)$. c) If $a_n \downarrow 0$ then $\sum_{n=1}^\infty a_n \sin n\theta$ converges uniformly on $[0,2\pi]$ iff $na_n \to 0$ as $n \to \infty$.

3.2. Complex-valued Functions

Conventions

If U is an open subset of $\mathbb{R}$, $C^k(U,\mathbb{C})$ is the set of functions having continuous kth derivatives on U; $C^\infty(U,\mathbb{C}) \stackrel{\text{def}}{=} \bigcap_{k\in\mathbb{N}} C^k(U,\mathbb{C})$. For f in $C^k(U,\mathbb{C})$, $\|f\|^{(k)} \stackrel{\text{def}}{=} \sum_{j=0}^k \|f^{(j)}\|_\infty$ $(\leq \infty)$.

3.61. For f in $C([0,1],\mathbb{C})$, n in $\mathbb{N}$, and j in $\{0,1,\ldots,n-1\}$,

$$x_n(t) \stackrel{\text{def}}{=} \begin{cases} 0 & \text{if } t=0 \\ x_n\left(\frac{j}{n}\right) + \left(t-\frac{j}{n}\right) f\left(x_n\left(\frac{j}{n}\right)\right) & \text{if } \frac{j}{n} \le t \le \frac{j+1}{n}. \end{cases}$$

a) For some continuous function x, $x_n \stackrel{\text{u}}{\to} x$; b)

$$x(t) = \int_0^t f \circ x(s))\, ds, \; 0 \le t \le 1.$$

3.62. If n in $\mathbb{N}$, and N_n in $C([0,1],\mathbb{C})$ is the set of f such that for some x in $[0,1]$ and all h in $\mathbb{R}\setminus\{0\}$, $\left|\frac{f(x+h)-f(x)}{h}\right| \le n$ then N_n is closed and nowhere dense in the $\|\ \|_\infty$-induced topology of $C([0,1],\mathbb{C})$.

3.63. If $f_n \in C^1([0,1],\mathbb{C}), \|f_n'\|_\infty \le 1$, $n\in\mathbb{N}$, and for all g in $C([0,1],\mathbb{C})$, $\lim_{n\to\infty}\int_0^1 f_n(x)g(x)\,dx = 0$ then $\lim_{n\to\infty}\|f_n\|_\infty = 0$.

3.64. (Wirtinger) If $f\in C^1([0,\pi],\mathbb{C})$ and $f(0)=f(\pi)=0$ then for some K independent of f, $\int_0^\pi |f(x)|^2\,dx \le K\int_0^\pi |f'(x)|^2\,dx$.

3.65. The inclusion map $T: C^1([0,1],\mathbb{C}) \hookrightarrow C([0,1],\mathbb{C})$ is compact.

3.66. If X is a subspace of $C^1([0,1],\mathbb{C})$ and also a closed subspace of $C([0,1],\mathbb{C})$ then: a) X is closed in $C^1([0,1],\mathbb{C})$; b) there are positive constants k, K such that for all f in X, $k\|f\|^{(1)} \le \|f\|_\infty \le K\|f\|^{(1)}$; c) X is finite-dimensional.

3.67. If $f \in C^1(\mathbb{T},\mathbb{C})$ and $\int_{\mathbb{T}} f(z)\,d\tau(z) = 0$ then $\|f\|_2 \le \|f'\|_2$ and equality obtains iff for some constants a, b, $f(z) = az + b\overline{z}$.

3.68. If

$$\|\ \|' : C^1([0,1],\mathbb{C}) \ni f \mapsto \left(\int_0^1 |f(x)|^2\,dx\right)^{\frac{1}{2}}$$

$$\|\ \|'' : C^1([0,1],\mathbb{C}) \ni f \mapsto \left(\int_0^1 \left(|f(x)|^2 + |f'(x)|^2\right)\,dx\right)^{\frac{1}{2}}$$

are two norms and E_1 resp. E_2 is the completion of $C^1([0,1],\mathbb{C})$ with respect to $\|\ \|'$ resp. $\|\ \|''$ then the derivative operator D has a continuous extension $\widetilde{D}: E_2 \mapsto E_1$ and $\widetilde{D}^{-1}(\mathbf{O}) = \{\, f \;:\; f \text{ is constant}\,\}$.

3.69. If $\{z_n\}_{n=1}^N \subset \mathbb{T}$ then:

$$\text{a)}\ \left\{t_n \ge 0,\ \sum_{n=1}^N t_n = 1\right\} \Rightarrow \left\{\left|\sum_{n=1}^N t_n z_n\right| \le 1\right\};$$

$$\text{b)}\ \left\{\left|\sum_{n=1}^N t_n z_n\right| = 1\right\} \Rightarrow \left\{\prod_{n=1}^N (1-t_n) = 0\right\}.$$

3.70. If $a < b$ then $\left\{x \mapsto e^{inx}\right\}_{n\in\mathbb{Z}}$ is a linearly independent set on $[a, b]$.

3.71. (Hadamard) For the matrix $A \stackrel{\text{def}}{=} (a_{ij})_{i,j=1}^{n,n}$ with real entries,

$$|\det(A)| \le \prod_{i=1}^{n} \sqrt{\sum_{j=1}^{n} a_{ij}^2}\ .$$

The following **Problems** conclude with **3.75**, Brouwer's fixed-point theorem for $\mathbb{R}^2$

3.72. If $f \in C(D(0,1), \mathbb{T})$ then for some ϕ in $C(D(0,1), \mathbb{R})$, $f(z) = e^{i\phi(z)}$.

3.73. If $g \in C(\mathbb{T}, \mathbb{R})$ then there two antipodal points $p \stackrel{\text{def}}{=} e^{i\theta}$ and $\widetilde{p} \stackrel{\text{def}}{=} e^{i(\theta+\pi)}$ such that $g(p) = g(\widetilde{p})$.

3.74. If $h \in C(D(0,1), \mathbb{T})$ then for some $e^{i\theta}$, $h\left(e^{i\theta}\right) \neq e^{i\theta}$.

3.75. (Brouwer) If $f \in C(D(0,1), D(0,1))$ then for some p in $D(0,1)$, $f(p) = p$.

4
Measure and Topology

4.1. Borel Measures

Conventions

For $(X, \mathsf{S}_\beta, \mu)$, a Borel set E is inner resp. outer regular iff

$$\mu(E) = \sup\{\,\mu(K) \ : \ K \text{ compact}, \ K \subset E\,\}$$

resp.

$$\mu(E) = \inf\{\,\mu(U) \ : \ U \text{ open}, \ U \supset E\,\};$$

E is regular iff E is both inner and outer regular; μ is inner (outer) regular iff every Borel set is inner (outer) regular. (Similar conventions for *regular* apply to arbitrary nonnegative set functions.)

For (X, S, μ) there are defined two set functions:

$$\mu^* : 2^X \ni E \mapsto \mu^*(E) \stackrel{\text{def}}{=} \inf\{\,\mu(B) \ : \ B \in \mathsf{S}, \ E \subset B\,\}$$
$$\mu_* : 2^X \ni E \mapsto \mu_*(E) \stackrel{\text{def}}{=} \sup\{\,\mu(B) \ : \ B \in \mathsf{S}, \ B \subset E\,\},$$

the outer and inner measures defined by μ. An outer measure is a map $\mu^* : 2^X \ni E \mapsto \mu^*(E) \in [0, \infty]$ that is countably subadditive:

$$\mu^*\left(\bigcup_{n \in \mathbb{N}} E_n\right) \le \sum_{n=1}^{\infty} \mu^*(E_n).$$

An E in S is an atom iff $\mu(E) > 0$ and for all A in S, $\mu(A \cap E) = \mu(E)$ or $\mu(A \cap E) = 0$; (X, S, μ) is nonatomic iff there are no atoms in S. A discrete measure μ is one for which there is a map $f : X \ni x \mapsto f(x) \in [0, \infty)$ such that for any subset E of X, $\mu(E) = \sum_{x \in E} f(x)$; when $f(x) \equiv 1$, the discrete measure μ is counting measure and is denoted ζ.

When μ is complex and $E \in \mathsf{S}$ then the absolute value of μ is

$$|\mu| : \mathsf{S} \ni E \mapsto \sup\left\{\sum_{n=1}^{\infty} |\mu(E_n)| \ : \ E = \left(\dot{\bigcup}_{n \in \mathbb{N}} E_n\right), \ E_n \in \mathsf{S}\right\}.$$

A measurable partition of an E in S is a finite or countably infinite sequence $\{E_n\}_{1 \le n < N \le \infty}$ contained in S and such that $E = \dot{\bigcup}_{n=1}^{N} E_n$. Thus $|\mu|(E)$

is the supremum, taken over the set of all measurable partitions of E, of $\sum_{n=1}^{N} |\mu(E_n)|$.

When μ is complex, μ is inner (outer) regular iff $|\mu|$ is inner (outer) regular.

For (X, S, μ), $\mathcal{M}$ denotes the set of S-measurable $\mathbb{R}$-valued functions.

For x in $\mathbb{R}$, $\{x\}$ is the fractional part of x: $0 \leq \{x\} < 1$ and for a unique $n \stackrel{\text{def}}{=} [x]$ in $\mathbb{Z}$, $x = n + \{x\}$: $[x]$ is the greatest integer in x.

A function ψ in $\mathbb{R}^{\mathbb{R}}$ is concave iff $-\psi$ is convex.

4.1. a) (Jensen) For a convex function ϕ in $\mathbb{R}^{(u,v)}$, a set X, a vector space V contained in $\mathbb{R}^X$, and a nonnegative linear functional $L : V \mapsto \mathbb{R}$, if $\mathbf{1} \in V$, $L(\mathbf{1}) = 1$, and $f(X) \subset (u,v)$ then $\phi(L(f)) \leq L(\phi \circ f)$. [In particular, if (X, S, μ), $\mu(X) = 1$, $f \in L^1(X, \mu)$, and $f(X) \subset (u,v)$ then $\phi\left(\int_X f(x)\,d\mu(x)\right) \leq \int_X \phi \circ f(x)\,d\mu(x)$.]

b) A function ϕ in $\mathbb{R}^{(u,v)}$ is convex iff for every simple (Lebesgue measurable) function f in $\mathbb{R}^{(0,1)}$, $\phi\left(\int_{[0,1]} f(x)\,dx\right) \leq \int_{[0,1]} \phi \circ f(x)\,dx$ (*).

4.2. For a net $n : \Lambda \ni \lambda \mapsto n_\lambda \in \mathbb{R}^{\mathbb{R}}$, the map $\psi : \mathbb{R} \ni x \mapsto \underline{\lim}_{\lambda\in\Lambda} n_\lambda(x)$ is concave while $\phi : \mathbb{R} \ni x \mapsto \overline{\lim}_{\lambda\in\Lambda} n_\lambda(x)$ is convex.

4.3. For (X, S, μ) if $\mathcal{M} \ni f_n \geq 0, n \in \mathbb{N}$ and $\overline{\lim}_{n\to\infty} f_n \in L^1(X, \mu)$, the inequalities

$$\overline{\lim}_{n\to\infty} \int_X f_n(x)\,d\mu(x) \leq \int_X \overline{\lim}_{n\to\infty} f_n(x)\,d\mu(x)$$
$$\underline{\lim}_{n\to\infty} \int_X f_n(x)\,d\mu(x) \geq \int_X \underline{\lim}_{n\to\infty} f_n(x)\,d\mu(x)$$

(the latter is Fatou's lemma) flow from Jensen's inequality.

For a metric space (X, d), p, ϵ positive, and A a subset of X, let $\rho_\epsilon^p(A)$ be

$$\inf\left\{\sum_{k=1}^{\infty} (\operatorname{diam}(U_k))^p \ : \ U_k \in \mathcal{T}, \ \bigcup_{k\in\mathbb{N}} U_k \supset A, \ \operatorname{diam}(U_k) < \epsilon, \ k \in \mathbb{N}\right\}.$$

The p-dimensional Hausdorff measure of A is $\rho^p(A) \stackrel{\text{def}}{=} \sup_{\epsilon>0} \rho_\epsilon^p(A)$.

4.4. Show: a) $\rho^p(A) = \lim_{\epsilon\to 0} \rho_\epsilon^p(A)$; b) ρ^p is an outer measure on 2^X; c) if $\gamma : [0,1] \ni t \mapsto \gamma(t) \in X$ is a simple rectifiable curve then the length $\ell(\gamma)$ of γ is $\rho^1(\gamma([0,1])) \stackrel{\text{def}}{=} \rho^1(\gamma^*)$.

4.5. If $\rho^p(A) < \infty$ and $q > p$ then $\rho^q(A) = 0$.

4.6. If $p \in \mathbb{N}$, and $A \subset \mathbb{R}^p$, there is an A-free constant c_p, such that, λ_p^* denoting p-dimensional Lebesgue outer measure,

$$c_p \rho^p(A) \le \lambda_p^*(A) \le \rho^p(A).$$

4.7. If A and B are subsets of X and

$$\inf\{\, d(a,b) \;:\; a \in A,\ b \in B \,\} \stackrel{\text{def}}{=} \delta(A,B) > 0$$

(A and B are a positive distance apart) then $\rho^p(A \cup B) = \rho^p(A) + \rho^p(B)$.

4.8. For every closed subset F of X and any subset S of X,

$$\rho^p(S) = \rho^p(S \cap F) + \rho^p(S \setminus F),$$

i.e., every closed set F is Caratheodory measurable with respect to the outer measure ρ^p.

4.9. There is a constant K_p such that $\lambda_p^* = K_p \rho^p$.

4.10. If (X,d) is a metric space, $(X, \sigma\mathsf{R}(\mathsf{K}(X), \mu)$ is finite, and each x in X is the center of a σ-compact open ball then $(X, \sigma\mathsf{R}(\mathsf{K}(X), \mu)$ is regular.

4.11. For $(X, \sigma\mathsf{R}(\mathsf{O}(X)), \mu)$, if X is a separable, complete, metric space and μ is finite then μ is regular.

4.12. If μ is a Borel measure on $\mathbb{R}^n$ and $\mu(B(\mathbf{O}, r)) < \infty$ for every positive r then μ is regular.

4.13. If (X,d) is a compact metric space, $(X, \mathsf{S}_\beta, \mu)$ is finite, for each x in X, $\mu(\{x\}) = 0$, and $\epsilon > 0$ then there is a positive $\delta(\epsilon)$ such that for each Borel set E, $\{\mathrm{diam}(E) < \delta(\epsilon)\} \Rightarrow \{\mu(E) < \epsilon\}$.

4.14. If μ is a regular totally finite Borel measure on a compact Hausdorff space X there is a minimal closed subset F such that $\mu(X \setminus F) = 0$, i.e., if F_1 is closed and $\mu(X \setminus F_1) = 0$ then $F_1 \not\subset F$. If $f \in C(X, \mathbb{C})$ then $f(x) \doteq 0$ iff $f^{-1}(0) \supset F$.

4.15. If X is a compact Hausdorff space, ν is a finitely additive nonnegative set function defined on $\mathsf{S}_\beta(X)$, and for each E in $\mathsf{S}_\beta(X)$,

$$\begin{aligned} \nu(E) &= \inf\{\, \nu(U) \;:\; U \text{ open},\ U \supset E \,\} \\ &= \sup\{\, \nu(K) \;:\; K \text{ compact},\ K \subset E \,\} \end{aligned}$$

(ν is a regular nonnegative set function) then ν is countably additive.

4.16. For μ a finite Borel measure on a compact Hausdorff space X, there is a diset Γ and a net $\nu : \Gamma \ni \gamma \mapsto \nu_\gamma$ so that each ν_γ is a discrete measure and for each f in $C(X,\mathbb{C})$, the net $\Phi : \Gamma \ni \gamma \mapsto \int_X f(x)\,d\nu_\gamma(x)$ converges to $\int_X f(x)\,d\mu(x)$.

4.17. The difference of two regular complex measures is a regular complex measure.

4.18. If μ is a finite nonatomic Borel measure on a compact metric space (X,d) then for the topology of X, there is a countable base $\{U_n\}_{n\in\mathbb{N}}$ such that $\mu(\partial U_n) = 0,\ n \in \mathbb{N}$.

4.19. If $\{(X, \sigma\mathsf{R}(\mathsf{O}(X)), \mu_n)\}_{n\in\mathbb{N}}$ is a sequence of totally finite measure situations for a compact metric space (X,d),

$$\lim_{n\to\infty} \int_X f(x)\,d\mu_n(x) \equiv \int_X f(x)\,d\mu_1(x)$$

on $C(X,\mathbb{C})$, U is open, and $\mu_1(\partial U) = 0$ then $\lim_{n\to\infty} \mu_n(U) = \mu_1(U)$.

4.20. In the context of **4.19**, (conversely) if, for every open set U such that $\mu_1(\partial U) = 0$, $\lim_{n\to\infty}\mu_n(U) = \mu_1(U)$, then for every f in $C(X,\mathbb{C})$, $\lim_{n\to\infty} \int_X f(x)\,d\mu_n(x) = \int_X f(x)\,d\mu_1(x)$.

4.21. In the context of **4.20**, the assumption that each μ_n is totally finite is superfluous.

4.22. If $\{I_j\}_{j\in J}$ is a set of intervals in $\mathbb{R}$ and if $\lambda(I_j) > 0$ for all j then $E \stackrel{\text{def}}{=} \bigcup_{j\in J} I_j \in \mathsf{S}_\beta(\mathbb{R})$.

4.23. If $x \in \mathbb{I}_{\mathbb{R}}$ then: a) $\overline{\Delta(x)} \stackrel{\text{def}}{=} \overline{\{\{nx\} \;:\; n \in \mathbb{N}\}} = [0,1]$; b)

$$\{(a,b) \subset [0,1]\} \Rightarrow \left\{ \lim_{k\to\infty} \frac{\#(\{\{x\},\ldots,\{kx\}\} \cap [a,b])}{k} = b - a \right\},$$

i.e., $\{\{kx\}\}_{k\in\mathbb{N}}$ is equidistributed.

[**Note 4.1:** The result b) is a consequence of some classical work of Hermann Weyl [**Wey**].]

4.24. If (X, S, μ) is nonatomic then $\mu(\mathsf{S})$ is an interval $[0, M]$, $0 \le M \le \infty$.

For the remaining **Problems** in this **Section**, X is a locally compact space, $L = C_{00}(X,\mathbb{C})$, I is a Daniell functional, and $L_u, L_{ul}, \mathcal{M}, L^1, \mathsf{A}$, etc. have the meanings given them in **Section 1.2**.

If K is compact, U is open, and $K \subset U$, Urysohn's lemma implies that for some f in L, $K \prec f \prec U$.

4.25. If $S \subset L$ and $\{f, g \in S\} \Rightarrow \{f \wedge g \in S\}$, i.e., if S is $\wedge$-closed or directed downward, and $I : L \mapsto \mathbb{R}$ is a nonnegative linear functional then $\{\bigwedge\{f \;:\; f \in S\} = 0\} \Rightarrow \{\bigwedge\{I(f) \;:\; f \in S\} = 0\}$.

4.26. If $S \subset L_u$, and $\{f, g \in S\} \Rightarrow \{f \vee g \in S\}$ i.e., if S is $\vee$-closed or directed upward, then

$$F \stackrel{\text{def}}{=} \bigvee \{ f \; : \; f \in S \} \in L_u, \text{ and } A \stackrel{\text{def}}{=} \bigvee \{ I(f) \; : \; f \in S \} = I(F).$$

See **1.27**, of which **4.26** is, in the current context, a generalization.

4.27. If $U \in \mathsf{O}$, then

$$\begin{aligned} \mu^*(U) &= \sup \{ \mu^*(K) \; : \; \mathsf{K} \ni K \subset U \} \\ &= \sup \{ \mu^*(V) \; : \; V \in \mathsf{O}, \; U \supset \overline{V} \in \mathsf{K} \}. \end{aligned}$$

4.28. If E in 2^X then $\mu^*(E) = \inf \{ \mu^*(U) \; : \; U \in \mathsf{O}, \; E \subset U \}$.

4.29. Show: a) A is a σ-algebra and $\mathsf{O} \cup \mathsf{F} \subset \mathsf{A}$; b) for E in A,

$$\mu(E) = \inf \{ \mu(U) \; : \; E \subset U \in \mathsf{O} \};$$

c) if $E \in \mathsf{O}$ or $E \in \mathsf{A}$ and $\mu(E) < \infty$ then

$$\mu(E) = \sup \{ \mu(K) \; : \; \mathsf{K}(X) \ni K \subset E \}.$$

4.30. If, in the context of **4.29**, either X is σ-compact or (X, A, μ) is σ-finite then for all E in A, $\mu(E) = \sup \{ \mu(K) \; : \; \mathsf{K}(X) \ni K \subset E \}$.

4.31.

i. A subset E of $\mathbb{R}$ is a null set (λ) iff there is a sequence $\{I_n\}_{n \in \mathsf{N}}$ of intervals such that $\sum_{n=1}^{\infty} \lambda(I_n) < \infty$ and each point of E belongs to infinitely many I_n.

ii. For some null set E in $[0, 1]$, if f is Riemann integrable on $[0, 1]$ then $\text{Cont}(f) \cap E \neq \emptyset$.

iii. For a given null set E there is a monotonely increasing function f such that f' exists at no point of E.

4.32. If $\{f_n\}_{n \in \mathsf{N}}$ is a sequence of monotonely increasing functions in $\mathbb{R}^{[0,1]}$ and $f_n \stackrel{\text{meas}}{\rightarrow} f$ then on $\text{Cont}(f)$, $\lim_{n \to \infty} f_n(x) = f(x)$.

4.2. Haar Measure

Conventions

When G is a locally compact group, e is the identity of G, (G, S, μ) is the measure situation such that $\mathsf{S} = \sigma\mathsf{R}(\mathsf{K}(G))$ and μ is left-invariant (Haar) measure: $\{\{E \in \mathsf{S}\} \wedge \{x \in G\}\} \Rightarrow \{\{xE \in \mathsf{S}\} \wedge \{\mu(xE) = \mu(E)\}\}$. The set of Haar measurable sets or simply Haar sets is S and $\widetilde{\mathsf{S}}$ is the completion of S.

Because Haar measure is unique up to a multiplicative constant, when G is discrete, e.g., when $G = \mathbb{Z}$, since $0 < \mu(\{x\}) < \infty$, it is assumed that $\mu(\{x\}) \stackrel{\text{def}}{=} 1$ for (every) x in G; when G is compact, e.g., when $G = \mathbb{T}$, since $G \in \mathsf{S}$ and $0 < \mu(G) < \infty$, it is assumed that $\mu(G) \stackrel{\text{def}}{=} 1$.

4.33. For (G, S, μ), if $1 \le p \le \infty$, $f \in L^1(G, \mu)$, and $g \in L^p(G, \mu)$ then $f * g$ is well-defined, $f * g \in L^p(G, \mu)$, and $\|f * g\|_p \le \|f\|_1 \|g\|_p$.

4.34. If $E \in \mathsf{S}$ and $\mu(E) > 0$ then EE^{-1} contains some $N(e)$.

4.35. For some (G, S, μ) and a null set E, EE^{-1} contains some $N(e)$.

4.36. If $G \subset \mathbb{T}$ then $\#(G) \in \mathbb{N}$ or G is dense in $\mathbb{T}$.

4.37. If $x \in \mathbb{R}$ then $G \stackrel{\text{def}}{=} \{\exp(2\pi i n x)\}_{n \in \mathbb{Z}}$ is a subgroup of $\mathbb{T}$ and G is finite resp. dense in $\mathbb{T}$ if $x \in \mathbb{Q}$ resp. if $x \in \mathbb{I}_{\mathbb{R}}$.

4.38. If $\alpha \in \mathbb{I}_{\mathbb{R}}$ and $\xi \stackrel{\text{def}}{=} e^{i2\pi\alpha}$ then $G \stackrel{\text{def}}{=} \{\xi^n\}_{n \in \mathbb{Z}}$ and $H \stackrel{\text{def}}{=} \{\xi^{2n}\}_{n \in \mathbb{Z}}$ are dense subgroups of $\mathbb{T}$ and $G : H = 2$.

4.39. In the context of **4.38** let R be a complete set of $\mathbb{T}$-coset representatives for the subgroup G and let S be $HR \stackrel{\text{def}}{=} \{\, hr \ : \ h \in H, \ r \in R \}$.

i. For $(\mathbb{T}, \mathsf{S}, \tau)$, $S \notin \mathsf{S}$ and $\tau_*(S) = \tau_*(\mathbb{T} \setminus S) = 0$.
ii. For $w : \mathbb{R} \ni x \mapsto e^{ix} \in \mathbb{T}$, $\widetilde{S} \stackrel{\text{def}}{=} w^{-1}(S)$, and E in S_λ, $\lambda_*(E \cap \widetilde{S}) = 0$, $\lambda^*(E \cap \widetilde{S}) = \lambda(E)$.
iii. If P is a subset of $\mathbb{R}$, $P^\bullet \ne \emptyset$, $E \in \mathsf{S}_\beta(\mathbb{R})$, and for all p in P, $p + E = E$ then $\lambda(E) = 0$ or $\lambda(E) = \infty$.

4.40. Every connected subset of $\mathbb{R}$ is some kind of interval and hence is Borel measurable. Is every connected subset of $\mathbb{R}^2$ Lebesgue measurable?

4.41. If $H \stackrel{\text{def}}{=} \{h_n\}_{n \in \mathbb{N}}$ is a countable, infinite, dense, proper, subgroup of a locally compact group G and R is a complete set of coset representatives for H in G then $R \notin \widetilde{\mathsf{S}}$.

4.42. If H is a subgroup of a locally compact group G, G is a) compact or b) connected, and the quotient space G/H is countable and infinite then $H \notin \widetilde{\mathsf{S}}$.

4.43. If (X, S, μ) is totally finite, $\mu(X) > 0$, X is a group, μ is translation-invariant, and X contains subgroup H such that $\#(G/H) = \#(\mathbb{N})$ then $\widetilde{\mathsf{S}} \ne 2^X$.

4.44. If m, n in $\mathbb{Z}^+$ and $m + n > 0$ then the group $G \stackrel{\text{def}}{=} \mathbb{R}^m \times \mathbb{T}^n$ contains a subgroup that is not in $\mathsf{S}_{\lambda_m \times \tau_n}$.

4.45. If G is a compact group or a locally compact connected group for which there is a continuous open epimorphism $\phi : G \overset{\text{epi}}{\mapsto} \mathbb{T}$ then G contains a subgroup that is not in $\widetilde{\mathsf{S}}$.

4.46. a) If G is a locally compact group then every set E in S is the union of a σ-compact set and a null set. b) Some locally compact group G is not σ-compact [whence $G \notin \sigma\mathsf{R}(\mathsf{K}(G))$].

4.47. If a topological group G is connected then G is generated by each $N(e)$.

4.48. Let G be a locally compact group containing a closed proper subgroup H. a) If G is connected and $H \in \mathsf{S}$ then $\mu(H) = 0$. b) If G is σ-compact and connected then $\mu(H) = 0$. c) If H is normal, $H \in \mathsf{S}$, and $\mu(H) > 0$ then G/H is discrete in its quotient topology.

4.49. For locally compact groups G and H, Haar measure situations (G, S, μ) and (H, T, ν), and a continuous open epimorphism $\phi : G \overset{\text{epi}}{\mapsto} H$, show: a) $\mathsf{T} \subset \phi(\mathsf{S})$; b) if G is σ-compact then $\mathsf{T} = \phi(\mathsf{S})$; c) $\mathsf{T} = \phi(\mathsf{S})$ can obtain for some G that is not σ-compact; d)

$$\{\{E \in \mathsf{S}\} \wedge \{\mu(E) = 0\}\} \not\Rightarrow \{\nu(\phi(E)) = 0\};$$

e) if E is a subset of a null set in G, $\phi(E)$ need not be in T.

4.50. For a σ-compact locally compact group G, a closed normal subgroup H, and $K \stackrel{\text{def}}{=} G/H$, if K contains a subgroup E that is not a Haar set then G contains a subgroup that is not Haar set.

4.51. For the locally compact group G, (G, S, μ), E in S, and x in G, show $\nu : \mathsf{S} \ni E \mapsto \mu(Ex)$ is also a Haar measure on G. Hence $\mu(Ex) = \Delta(x)\mu(E)$. Show Δ, the modular function for G, is in $C(G, (0, \infty))$.

When $\Delta(G) = \{1\}$ the group G is unimodular.

4.52. If G is compact then G is unimodular.

4.53. Let H be a closed normal subgroup of the locally compact group G. View $G/H \stackrel{\text{def}}{=} K$ as the set $\{xH : x \in G\}$ of cosets of H. Let the Haar measures on G, H, and K be μ, ρ, and τ and let δ be the modular function for H. a) For f in $C_{00}(G, \mathbb{C})$, $F : G \ni x \mapsto \int_H f(xy)\, d\rho(y)$ is invariant on each coset of H and, as a function on K, $F \in C_{00}(K, \mathbb{C})$. b) For some constant k, and all f in $C_{00}(G, \mathbb{C})$,

$$\int_K \left(\int_H f(xy)\, d\rho(y) \right) d\tau(z) = k \int_G f(x)\, d\mu(x).$$

c) $\Delta\big|_H = \delta$.

4.54. If G is a locally compact group then $U \stackrel{\text{def}}{=} \Delta^{-1}(1)$ is a closed normal unimodular subgroup of G and U contains every normal unimodular subgroup of G.

4.55. If G is a compact infinite group then $\#(G) \geq \mathfrak{c}$.

A topological semigroup S is a Hausdorff space and a semigroup for which the map $S^2 \ni \{x, y\} \mapsto xy \in S$ is continuous. If

$$\{\{xy = xz\} \vee \{yx = zx\}\} \Rightarrow \{y = z\}$$

then S is a semigroup with a cancellation law.

4.56. If S is a semigroup with a cancellation law then S is a group if: a) for some (S, S, μ), $S \in \mathsf{S}$, $\mu(S) = M < \infty$, and both S and μ are left- and right-invariant; b) $\theta : S^2 \ni \{x, y\} \mapsto \{x, xy\}$ preserves (product) measurability **[GeK]**.

4.57. A compact topological semigroup S with a cancellation law is a compact topological group **[GKO]**.

5
Measure Theory

5.1. Measure and Integration

Conventions

When (X, S, μ) is signed μ is the difference of two positive measures $\mu^{\pm}$: $\mu = \mu^{+} - \mu^{-}$ such that for every E in S, at least one of $\mu^{\pm}(E)$ is in $\mathbb{R}$. There are in S two sets $P^{\pm}$ such that for E in S, $\mu^{\pm}(E) = \mu(P^{\pm} \cap E)$ and $|\mu| \stackrel{\text{def}}{=} \mu^{+} + \mu^{-}$.

For a set X and a σ-ring S, M_c is the vector space of complex measures μ defined on S and $\| \ \| : M_c \ni \mu \mapsto |\mu|(X)$ is a norm with respect to which M_c is a Banach space over $\mathbb{C}$.

For measure situations (X, S, μ_i), $i = 1, 2$, μ_1 is absolutely continuous with respect to μ_2 ($\mu_1 \ll \mu_2$) iff $\{\mu_2(E) = 0\} \Rightarrow \{\mu_1(E) = 0\}$; μ_1 and μ_2 are mutually singular ($\mu_1 \perp \mu_2$) iff there are in S disjoint sets A_1, A_2 such that for all E in S, $\mu_i(E) = \mu_i(E \cap A_i)$, $i = 1, 2$. The Lebesgue-Radon-Nikodým theorem **[HeS, Rud]** gives general conditions under there is, for μ_1, a Lebesgue decomposition with respect to μ_2 into a sum $\mu_{1a} + \mu_{1s}$ such that $\mu_{1a} \ll \mu_2, \mu_{1s} \perp \mu_2, \mu_{1s} \perp \mu_{1a}$, and for some integrable $h \stackrel{\text{def}}{=} \dfrac{d\mu_{1a}}{d\mu_2}$ (the Radon-Nikodým derivative), $\mu_{1a}(E) \equiv \int_E h(x)\, d\mu_2(x)$.

When X is a topological space, associated with each measure situation $(X, \mathsf{S}_\beta, \mu)$ is $\text{supp}(\mu)$, the support of the measure μ:

$$\text{supp}(\mu) \stackrel{\text{def}}{=} X \setminus \bigcup \{U \ : \ U \in \mathsf{O}(X),\ \mu(U) = 0\}.$$

Thus $\text{supp}(\mu)$ is the least closed set F such that if U is open and $F \cap U \neq \emptyset$ then $\mu(F \cap U) > 0$.

The symmetric difference of two sets is $A \triangle B \stackrel{\text{def}}{=} (A \setminus B) \cup (B \setminus A)$. For (X, S, μ), if $A, B \in \mathsf{S}$ then $A \doteq B$ iff $\mu(A \triangle B) = 0$.

For $\{(X_\gamma, \mathsf{S}_\gamma, \mu_\gamma)\}_{\gamma \in \Gamma}$, a product measure situation

$$(X, \mathsf{S}, \mu) \stackrel{\text{def}}{=} \bigtimes_{\gamma \in \Gamma} (X_\gamma, \mathsf{S}_\gamma, \mu_\gamma)$$

with product measure μ is defined as follows:

i. $X \stackrel{\text{def}}{=} \bigtimes_{\gamma \in \Gamma} X_\gamma$;

ii. S is the σ-ring generated by the set of all finite rectangles, i.e., sets of the form $E_{\gamma_1} \times \cdots \times E_{\gamma_n} \times \bigtimes_{\gamma \neq \gamma_1,\ldots,\gamma_n} X_\gamma$, $E_{\gamma_i} \in \mathsf{S}_{\gamma_i}$;
iii. when Γ is an infinite set, it is assumed that $\mu_\gamma(X_\gamma) \equiv 1$;
iv.

$$\mu\left(E_{\gamma_1} \times \cdots \times E_{\gamma_n} \times \bigtimes_{\gamma\neq\gamma_1,\ldots,\gamma_n} X_\gamma\right) \stackrel{\text{def}}{=} \bigtimes_{i=1}^{n} \mu_{\gamma_i}(E_{\gamma_i}) \times \bigtimes_{\gamma\neq\gamma_1,\ldots,\gamma_n} \mu_\gamma(X_\gamma).$$

By abuse of language and notation,

$$(X, \mathsf{S}, \mu) \stackrel{\text{def}}{=} \left(\bigtimes_{\gamma\in\Gamma} X_\gamma, \bigtimes_{\gamma\in\Gamma} \mathsf{S}_\gamma, \bigtimes_{\gamma\in\Gamma} \mu_\gamma\right),$$

while an element of X is a vector $\mathbf{x} \stackrel{\text{def}}{=} \{x_\gamma\}_{\gamma\in\Gamma}$. When $\Gamma' \subset \Gamma$, there is the projection

$$P_{\Gamma'} : X \ni \mathbf{x} \stackrel{\text{def}}{=} \{x_\gamma\}_{\gamma\in\Gamma} \stackrel{\text{sur}}{\mapsto} \mathbf{x}' \stackrel{\text{def}}{=} \{x_{\gamma'}\}_{\gamma'\in\Gamma'} \in X' \stackrel{\text{def}}{=} \bigtimes_{\gamma\in\Gamma'} X_\gamma.$$

When $\mathbf{x}' \in X'$, $\mathbf{x}'' \in P_{\Gamma\setminus\Gamma'}(X) \stackrel{\text{def}}{=} X''$, and $E \subset X$, the $\mathbf{x}'$-section of E is

$$E_{\mathbf{x}'} \stackrel{\text{def}}{=} P_{\Gamma\setminus\Gamma'}\left(P_{\Gamma'}^{-1}(\mathbf{x}') \cap E\right).$$

Furthermore, for some unique $\mathbf{x}$ in X,

$$P_{\Gamma'}(\mathbf{x}) = \mathbf{x}' \text{ and } P_{\Gamma\setminus\Gamma'}(\mathbf{x}) = \mathbf{x}''.$$

For this $\mathbf{x}$, and f in $\mathbb{R}^X$, $f_{\mathbf{x}'}(\mathbf{x}'') \stackrel{\text{def}}{=} f(\mathbf{x})$, $f_{\mathbf{x}''}(\mathbf{x}') \stackrel{\text{def}}{=} f(\mathbf{x})$. When μ', μ'' denote the product measures for X', X'', the Fubini-Tonelli theorems give conditions for the validity of

$$\int_X f(\mathbf{x})\, d\mu(\mathbf{x}) = \int_{X'} \left(\int_{X''} f_{\mathbf{x}'}(\mathbf{x}'')\, d\mu''(\mathbf{x}'')\right) d\mu'(\mathbf{x}').$$

A set $\mathcal{F}$ contained in $L^1(X,\mu)$ is uniformly integrable iff for each positive ϵ and some positive $\delta(\epsilon)$, if $\mu(E) < \delta(\epsilon)$ then for all f in $\mathcal{F}$, $\int_E |f(x)|\, d\mu(x) < \epsilon$.

When $t \in [0,1]$ and $1 < k \in \mathbb{N}$, t has one or two k-ary representations:

$$t \stackrel{\text{def}}{=} \sum_{n=1}^{\infty} \epsilon_n^{(i)} k^{-n}, \quad \epsilon_n^{(i)} \in \{0, 1, \ldots, k-1\}, \quad i = 1 \text{ or } i = 1, 2.$$

When there are two, the one for which $\sum_{n=1}^{\infty} \epsilon_n^{(i)} = \infty$ is **the** k-ary representation of t and then $\epsilon_n^{(i)}$ is **the** nth k-ary marker for t. More generally, the number $\epsilon_n^{(i)}$ is *an* nth k-*ary marker for* t. When $k = 2, 3$, $\epsilon_n^{(i)}$ is an nth binary, ternary marker for t; when $k = 10$, $\epsilon_n^{(i)}$ is an nth decimal marker.

5.1. A finite measure, e.g., a complex measure or its absolute value, is bounded.

5.2. If $\{a_n\}_{n\in\mathbb{N}} \subset \mathbb{C}^{\mathbb{N}}$ and, for every permutation π of $\mathbb{N}$, $\sum_{n=1}^{\infty} a_{\pi(n)}$ converges then: a) the map

$$\mu : 2^{\mathbb{N}} \ni E \mapsto \sum_{n\in E} a_n \tag{5.1}$$

is a complex measure and $|\mu|(E) = \sum_{n\in E} |a_n|$; b) if $\sum_{n=1}^{\infty} a_n$ is conditionally convergent then (5.1) defines a finitely additive measure on the ring of finite subsets of $\mathbb{N}$ but this measure is not extensible to a countably additive measure on $2^{\mathbb{N}}$.

5.3. a) If R is a ring of subsets of a set X, and for $(X, \sigma R(R), \mu_i)$, $i = 1, 2$,

$$\{E \in R\} \Rightarrow \{0 \le \mu_1(E) = \mu_2(E) < \infty\}$$

then $\mu_1 = \mu_2$ on $\sigma R(R)$.

b) If "$< \infty$" is replaced by "$\le \infty$," the conclusion in a) need not obtain.

5.4. If (X, S, μ) is a complex measure situation then $\nu : S \ni E \mapsto |\mu(E)|$ can fail to serve as a measure.

5.5. If (X, S, μ) is complex, $E \in S$, and

$$\nu(E) \stackrel{\text{def}}{=} \sup \{ |\mu(A)| \ : \ A \in S, \ A \subset E \}$$

then $\nu(E) \le |\mu|(E) \le 4\nu(E)$.

5.6. (Borel-Cantelli) For (X, S, μ), if $\{E_n\}_{n\in\mathbb{N}} \subset S$ and $\sum_{n=1}^{\infty} \mu(E_n) < \infty$ then $\mu\left(\overline{\lim}_{n\to\infty} E_n\right) = 0$ (cf. **5.155**).

5.7. For (X, S, μ) totally finite, if $f \in X^X$, $\{E \in S\} \Rightarrow \{f^{-1}(E) \in S\}$, and $\{\mu(N) = 0\} \Rightarrow \{\mu(f^{-1}(N)) = 0\}$ then for some h in $L^1(X, \mu)$,

$$\{g \in L^\infty(X, \mu)\} \Rightarrow \left\{ \int_X g \circ f(x)\, d\mu(x) = \int_X g(x)h(x)\, d\mu(x) \right\}.$$

5.8. For (X, S, μ), if $\{f_n\}_{n\in\mathbb{N}}$ is a sequence of nonnegative measurable functions, $f_n \stackrel{\text{a.e.}}{\to} f$, and $\lim_{n\to\infty} \int_X f_n(x)\, d\mu(x) = 0$ then $f \doteq 0$.

5.9. a) If

$$\{f_n\}_{n=0}^{\infty} \subset L^1([0,1],\lambda),\ f_n \overset{\text{meas}}{\to} f_0, f_n \geq 0,\ n \in \mathbb{N},$$
$$\text{and } \lim_{n\to\infty} \int_{[0,1]} f_n(x)\,dx = \int_{[0,1]} f_0(x)\,dx$$

then $\{E \in \mathsf{S}_\lambda\} \Rightarrow \left\{\lim_{n\to\infty} \int_E f_n(x)\,dx = \int_E f_0(x)\,dx\right\}$. b) For (X, S, μ), if

$$\{f_n\}_{n=0}^{\infty} \subset L^1([0,1],\lambda),\ f_n \overset{\text{a.e.}}{\to} f_0$$
$$f_n \geq 0,\ n \in \mathbb{N},\ \lim_{n\to\infty} \int_{[0,1]} f_n(x)\,dx = \int_{[0,1]} f_0(x)\,dx$$

then $\{E \in \mathsf{S}_\lambda\} \Rightarrow \left\{\lim_{n\to\infty} \int_E f_n(x)\,dx = \int_E f_0(x)\,dx\right\}$.

5.10. For (X, S, μ) totally finite and $\{f_n\}_{n\in\mathbb{N}}$ in $\mathcal{M}$, $f_n \overset{\text{meas}}{\to} 0$ iff

$$\lim_{n\to\infty} \int_X \left(\frac{|f_n(x)|}{1+|f_n(x)|}\right) d\mu(x) = 0.$$

5.11. If $\{\phi_n\}_{n\in\mathbb{N}}$ is orthonormal in $L^2(X,\mu)$,

$$g \in L^2(X,\mu),\ |\phi_n(x)| \leq |g(x)| \text{ a.e., } n \in \mathbb{N},$$

and $\sum_{n=1}^{\infty} a_n \phi_n(x)$ converges a.e. then $\lim_{n\to\infty} a_n = 0$. (This result generalizes the theorem of Cantor-Lebesgue.)

5.12. If $\{\phi_n\}_{n\in\mathbb{N}}$ is orthonormal in $L^2(X,\mu)$, and

$$E \overset{\text{def}}{=} \left\{ x \,:\, \exists \lim_{n\to\infty} \phi_n(x) \right\}$$

then a) $\phi(x) \overset{\text{def}}{=} \lim_{n\to\infty} \chi_E \cdot \phi_n(x) \doteq 0$; b) if $X = [0,1], \mathsf{S} = \mathsf{S}_\lambda$, and $\mu = \lambda$, then $\overline{\lim}_{n\to\infty} \text{var}_{[0,1]}(\phi_n) = \infty$.

5.13. If (X, S, μ) is totally finite, $\{f_n\}_{n\in\mathbb{N}} \subset L^\infty(X,\mu)$, $|f_n| \leq 1$, and $\lim_{n\to\infty} \int_X f_n(x)\,d\mu(x) = 0$ is there a subsequence $\{f_{n_k}\}_{k\in\mathbb{N}}$ such that $\lim_{k\to\infty} f_{n_k}(x) \doteq 0$?

5.14. For some (X, S, μ), and some sequence $\{f_n\}_{n\in\mathbb{N}}$ contained in $L^1(X,\mu)$: a) $f_n \overset{\text{a.e.}}{\to} 1$; b) $\int_X \left(f_n(x) - (f_n(x))^2\right) d\mu(x) \to \infty$ as $n \to \infty$.

5.15. If $f \in L^1(X,\mu)$, $b_n \uparrow \infty$, and $E_n \overset{\text{def}}{=} \{ x \,:\, |f(x)| \geq b_n \}$, $n \in \mathbb{N}$, then $\lim_{n\to\infty} b_n \mu(E_n) = 0$.

5.16. For (X, S, μ), ϵ, p in $(0,\infty)$, if $f \in \mathcal{M}$ then

$$\mu(\{ x \,:\, |f(x)| \geq \epsilon \}) \leq \frac{\int_X |f(x)|^p\,d\mu(x)}{\epsilon^p}.$$

5.17. If (X, S, μ_i) are totally finite, $i = 1,2,3$, then for the μ_3-Lebesgue decompositions $\mu_3 = \mu_{ja} + \mu_{js}$, $j = 1,2$, $\mu_{1s} \perp \mu_{2a}$.

5.18. If (X, S, μ_i) are totally finite, $i = 1,2$, then: a) for some E in S, if $\mu_{iE}(A)$ denotes $\mu_i(E \cap A)$, $i = 1,2$, for A in S, then $\mu_{iE} \ll \mu_{jE}$ and $\mu_{i\,X\setminus E} \perp \mu_{j\,X\setminus E}$, $i \neq j$; b) if, for all F in S, $\mu_{iF} \ll \mu_{jF}$ and $\mu_{i\,X\setminus F} \perp \mu_{j\,X\setminus F}$, $i \neq j$, then for all F in S and E as in a), $\mu_1(E \Delta F) + \mu_2(E \Delta F) = 0$.

5.19. If (X, S, μ_i), $i = 1,2$, are complex measure situations then $\mu_1 \perp \mu_2$ iff for all a_1, a_2 in $\mathbb{C}$, $|a_1|\,|\mu_1| + |a_2|\,|\mu_2| = |a_1\mu_1 + a_2\mu_2|$.

5.20. If (X, S, μ) is totally finite and $\mathcal{F} \stackrel{\text{def}}{=} \{f_\gamma\}_{\gamma \in \Gamma}$ is a set of S-measurable functions such that $\{f_{\gamma_1}, f_{\gamma_2} \in \mathcal{F}\} \Rightarrow \{f_{\gamma_1} \vee f_{\gamma_2} \in \mathcal{F}\}$ then: a) there is a S-measurable g such that for all γ, $f_\gamma \leq g$; b) if h is S-measurable and, for all γ, $f_\gamma \leq h$ a.e., then $g \leq h$ a.e. (g is a minimal measurable cover for $\mathcal{F}$).

5.21. If (X, S, μ) is totally σ-finite and $0 \leq f \in L^1(X, \mu)$ then

$$\int_X f(x)\,d\mu(x) = (\mu \times \lambda)(\{(x,y) \ : \ 0 \leq y \leq f(x)\})$$

("the integral is the area under the graph").

5.22. If (X, S, μ) is totally finite, $\mu(X) = 1$, and for f in $L^2(X, \mu)$,

$$\operatorname{Var}(f) \stackrel{\text{def}}{=} \int_X \left(f(x) - \int_X f(y)\,d\mu(y)\right)^2 d\mu(x),$$

then

$$\operatorname{Var}(f) = \int_X (f(x))^2\,d\mu(x) - \left(\int_X f(x)\,d\mu(x)\right)^2 \stackrel{\text{def}}{=} E\left(f^2\right) - (E(f))^2.$$

5.23. If f is as in **5.22**, $F_n : X^n \ni (x_1, \ldots, x_n) \mapsto \dfrac{\sum_{k=1}^n f(x_k)}{n}$, and $(X^n, \mathsf{S}^n, \mu^n)$, then $\operatorname{Var}(F_n) = \dfrac{\operatorname{Var}(f)}{n}$.

5.24. (Corollary to **5.23**.) If $0 \leq x \leq 1$ then

$$\sum_{k=0}^{n} \left(\frac{k}{n} - x\right)^2 \binom{n}{k} x^k (1-x)^{n-k} \leq \frac{1}{4n}.$$

5.25. True or false: for (X, S, μ) if f is measurable and positive a.e. then

$$\left\{\left\{\{E_n\}_{n\in\mathbb{N}} \subset \mathsf{S}\right\} \wedge \left\{\lim_{n\to\infty} \int_{E_n} f(x)\,d\mu(x) = 0\right\}\right\} \Rightarrow \left\{\lim_{n\to\infty} \mu(E_n) = 0\right\}?$$

5.26. If (X, S, μ) is totally finite, $f \in \mathbb{C}^{X\times[0,1]}$, for each x in X, f is a continuous function of t, and for each t in $[0,1]$, f is in $L^1(X,\mu)$ then the dominated convergence theorem obtains: if $g \in L^1(X,\mu)$ and for all t, $|f(x,t)| \le g(x)$ then $\int_X \lim_{t\to 0} f(x,t)\,d\mu(x) = \lim_{t\to 0}\int_X f(x,t)\,d\mu(x)$.

5.27. If (X, S, μ) is totally finite, f in $\mathbb{C}^{X\times[0,1]}$, for each x in X, f is a continuous function of t, and for each t in $[0,1]$, f is in $L^1(X,\mu)$, then Egorov's theorem obtains: if $\epsilon > 0$ for some E in S, $\mu(X \setminus E) < \epsilon$ and on E, $f(x,t) \stackrel{\mathrm{u}}{\to} g(x)$.

5.28. There is in $\mathbb{R}^{[0,1]}$ a sequence $\{f_n\}_{n\in\mathbb{N}}$ of Lebesgue measurable functions converging everywhere on $[0,1]$ and yet if $\lambda(E) = 1$, the convergence is not uniform on E.

The map $\mathbb{R} \ni x \stackrel{\text{epi}}{\mapsto} e^{i2\pi x} \in \mathbb{T}$ serves to identify $\mathbb{T}$ with $\mathbb{R}/\mathbb{Z} \stackrel{\text{def}}{=} \mathcal{P}$, which may be identified with $[0,1)$ according to the following rules:

i. Haar measure τ on $\mathbb{T}$ is identified with Lebesgue measure λ on $[0,1)$;
ii. multiplication in $\mathbb{T}$ is identified with addition (modulo 1) in $[0,1)$;
iii. the $\mathbb{R}^2$-induced topology of $\mathbb{T}$ is identified with the topology $\mathcal{T}$ for which a base consists of all open subintervals of $(0,1)$ together with the set of all intervals of the form $[0,a)$, $0 < a < 1$. [With respect to $\mathcal{T}$, $[0,1)$ is homeomorphic to $\mathbb{T}$ and, in particular, is compact.]

5.29. In the context of $\mathcal{P}$, and S of **4.37** – **4.39**, let $\{g_n\}_{n\in\mathbb{N}}$ be an enumeration of G. For $t \in (0,1)$, let k_t in $\mathbb{N}$ be such that $2^{-(k_t+1)} \le t < 2^{-k_t}$. If $f : \mathcal{P} \times (0,1) \ni (x,t) \mapsto \begin{cases} 1 & \text{if } x \in g_{k_t} + S \text{ and } x = 2^{k_t+1} - 1 \\ 0 & \text{otherwise} \end{cases}$ then for all x, $\lim_{t\to 0} f(x,t) = 0$ and for some positive p, $\lambda^*\left(\left\{x : f(x,t) > \frac{1}{2}\right\}\right) > p$.

[**Note 5.1: Problems 5.26 – 5.29** illustrate a limited degree of extensibility of classical results in measure theory.]

5.30. a) If $\mathcal{F} \subset L^1(X,\mu)$ and for each positive ϵ there is a positive $K(\epsilon)$ such that

$$\{f \in \mathcal{F}\} \wedge \{k > K(\epsilon)\} \Rightarrow \left\{\int_{E_{\ge}(|f|,k)} |f(x)|\,d\mu(x) < \epsilon\right\} \tag{5.2}$$

then $\mathcal{F}$ is uniformly integrable.

b) If (X, S, μ) is nonatomic and $\mathcal{F}$ is uniformly integrable then for each positive ϵ there is a $K(\epsilon)$ such that (5.2) holds.

c) Absent nonatomicity, $\mathcal{F}$ can be uniformly integrable while for some positive ϵ there is no $K(\epsilon)$ for which (5.2) holds.

5.31. a) If $f \in C(\mathbb{R}, \mathbb{R})$ and f is Riemann integrable on every bounded interval, so is $|f|$. b) If, to boot, $|f|$ is improperly Riemann integrable on $\mathbb{R}$ then f is improperly Riemann integrable, f is Lebesgue integrable, and $\int_{\mathbb{R}} f(x)\,d\lambda(x) = \lim_{\substack{t\to\infty\\ s\to-\infty}} \int_s^t f(x)\,dx$. c) There is a function g such that $|g|$ is Riemann integrable on $[0,1]$ although g is neither Riemann nor Lebesgue integrable on $[0,1]$.

5.32. True or false: for f in $C([0,1], \mathbb{R})$, if $\epsilon > 0$, there is a finite set of (Euclidean) rectangles that cover the graph $\mathcal{G} \stackrel{\text{def}}{=} \{(x, f(x)) : x \in [0,1]\}$ of f and are of total area less than ϵ.

When $f \in L^1(\mathbb{T}, \tau)$, the Fourier coefficients of f are

$$\hat{f}_n \stackrel{\text{def}}{=} \int_{\mathbb{T}} f(x)t^{-n}\,d\tau(t),\ n \in \mathbb{Z},$$

and the Fourier series for f is $\sum_{n\in\mathbb{Z}} \hat{f}_n t^n$. Furthermore,

$$S_N(f,x) \stackrel{\text{def}}{=} \sum_{n=-N}^{N} \hat{f}_n t^n \stackrel{\text{def}}{=} \int_{\mathbb{T}} D_N(y^{-1}x) f(y)\,d\tau(y)$$

$$\sigma_N(f,x) \stackrel{\text{def}}{=} \frac{1}{N+1}\sum_{n=0}^{N} S_n(f,x) \stackrel{\text{def}}{=} \int_{\mathbb{T}} F_N(y^{-1}x) f(y)\,d\tau(y).$$

The function D_N resp. F_N is Dirichlet's resp. Fejér's kernel.

[**Note 5.2:** In many treatments of classical analysis the notations $\hat{f}_n \stackrel{\text{def}}{=} \frac{1}{\sqrt{2\pi}} \int_0^{2\pi} f(x)\frac{e^{-inx}}{\sqrt{2\pi}}\,dx$ and $\sum_{n\in\mathbb{Z}} \hat{f}_n \frac{e^{inx}}{\sqrt{2\pi}}$, etc. are used instead of those introduced in the preceding paragraph. The transition is accomplished via the identification of $\mathbb{T}$ with $[0, 2\pi)$: $\theta : [0, 2\pi) \ni x \mapsto e^{ix} \stackrel{\text{def}}{=} \theta(x)$ and $dx \mapsto \frac{1}{2\pi}dx \stackrel{\text{def}}{=} d\tau(\theta(x))$. In this identification, the topology of $[0, 2\pi)$ is, as for $\mathcal{P}$ earlier, not the topology it inherits from $\mathbb{R}$.

The classical notation is not readily extended to cover the general discussion of harmonic analysis on locally compact abelian groups [**HeR, Loo, We2**], of which Fourier series, Fourier transforms, etc. are historically important parts.]

5.33. For $\mathbb{T}$ viewed as $[0, 2\pi)$ and τ as $\frac{\lambda}{2\pi}$, what are simple closed formulæ for Dirichlet's kernel D_N and Fejér's kernel F_N?

5.34. If A in $C(\mathbb{T},\mathbb{C})$ is the subset consisting of those functions f for which the Fourier coefficients $\hat{f}_n$ satisfy: $\sum_{n\in\mathbb{Z}} |\hat{f}_n|(1+|n|) \le 1$ then $\overline{A}$ is compact in the $\| \ \|_\infty$-induced topology of $C(\mathbb{T},\mathbb{C})$.

5.35. If $\{a_n\}_{n\in\mathbb{Z}} \subset \mathbb{C}$ and for some K in $[0,\infty)$, all n in $\mathbb{N}$, all t in $\mathbb{T}$, and all sequences $\{c_n\}_{n\in\mathbb{Z}}$ in $\mathbb{C}$, $\left|\sum_{n=-N}^{N} a_n c_n\right| \le K \left|\sum_{n=-N}^{N} c_n t^n\right|$ then there is a complex Borel measure μ such that $|\mu| \le K$ and for all n in $\mathbb{Z}$, $a_n = \int_{\mathbb{T}} t^n, d\mu(t)$.

Among the varieties of trigonometric series $\sum_{n=-\infty}^{\infty} c_n \dfrac{e^{int}}{\sqrt{2\pi}}$ are Fourier series, i.e., for some f in $L^1(\mathbb{T},\tau)$,

$$c_n = \int_{\mathbb{T}} f(t)t^{-n}\,d\tau(t) = \frac{1}{\sqrt{2\pi}}\int_0^{2\pi} f(x)\frac{e^{-inx}}{\sqrt{2\pi}}\,dx, n\in\mathbb{Z},$$

and Fourier-Stieltjes series, i.e., for some $(\mathbb{T}, \mathsf{S}_\beta, \mu)$ resp. $([0,2\pi), \mathsf{S}_\beta, \nu)$,

$$c_n = \int_{\mathbb{T}} t^{-n}\,d\mu(t) \text{ resp. } \int_0^{2\pi} e^{-inx}\,d\nu(x), n\in\mathbb{Z}.$$

The characterization in **5.36**d) of Fourier-Stieltjes series shows there are trigonometric series that are neither Fourier series nor Fourier-Stieltjes series.

5.36. a) The series

$$\sum_{n\in\mathbb{Z}} t^n \tag{5.3}$$

is a Fourier-Stieltjes series. b)

$$\left\{f,g\in L^1(\mathbb{T},\tau)\right\} \Rightarrow \left\{(f*g)\widehat{\ }_n \equiv \hat{f}_n\hat{g}_n\right\}. \tag{5.4}$$

c) If $f\in L^1(\mathbb{T},\tau)$ and

$$L_n(t) \stackrel{\text{def}}{=} \sum_{k=-n}^{n} \lambda_k t^k, n=0,1,\ldots,N,\ \Lambda_N(t) \stackrel{\text{def}}{=} \frac{1}{N+1}\sum_{n=1}^{N} L_n(t)$$

$$R_N(t) \stackrel{\text{def}}{=} \sum_{n=-N}^{N} \lambda_n \hat{f}_n t^n,\ \sigma_N^*(t) = \frac{1}{N+1}\sum_{n=0}^{N} R_n(t)$$

then $\sigma_N^*(t) = \int_{\mathbb{T}} f\left(s^{-1}t\right)\Lambda_N(s)\,d\tau(s)$. d) $\sum_{n=-\infty}^{\infty} c_n t^n$ is a Fourier-Stieltjes series iff for some M in $[0,\infty)$, and all N in $\mathbb{Z}$,

$$\left\|\sum_{n=-N}^{N} c_n\left(1-\frac{|n|}{N+1}\right)t^n\right\|_1 \le M. \tag{5.5}$$

5.37. For some measure situation $([0,1], \mathsf{S}_\beta, \mu)$, μ is nonatomic and $\text{supp}(\mu) = \{t : t = \sum_{n=1}^{\infty} \epsilon_n 10^{-n},\ \epsilon_n = 0 \text{ or } 7\}$.

5.38. If $F \in \mathsf{F}(\mathbb{R})$ then for some $(\mathbb{R}, \mathsf{S}_\beta, \mu)$, $\text{supp}(\mu) = F$.

5.39. For each compact perfect subset K of $\mathbb{R}^n$ and some nonatomic Borel measure μ, $\text{supp}(\mu) = K$.

5.40. Is there a $([0,1], \mathsf{S}_\beta, \mu)$ such that $\mu(\mathsf{S}_\beta)$ is **the** Cantor set C_0?

5.41. If $\mu([0,1]) = 1$, and f in $L^1_{\mathbb{R}}([0,1], \mu)$ then

$$\left\{\exp\left(\int_0^1 f(x)\, d\mu(x)\right) = \int_0^1 \exp(f(x))\, d\mu(x)\right\}$$
$$\Leftrightarrow \{f(x) \text{ is constant a.e. } (\mu)\}.$$

5.42. For $([0,1], \mathsf{S}_\beta, \mu)$, if

$$\{f \in C([0,1], [0,\infty))\} \Rightarrow \left\{f(0) \geq \int_0^1 f(x)\, d\mu(x)\right\}$$

then for some c in $[0,1]$, $\{f \in C([0,1], \mathbb{C})\} \Rightarrow \left\{\int_0^1 f(x)\, d\mu(x) = cf(0)\right\}$.

5.43. For $([0,1], \mathsf{S}_\beta, \mu)$ signed, if $\int_0^1 \sin^k \pi x\, d\mu(x) = 0$, $k \in \mathbb{N}$, then $\{E \in \mathsf{S}_\beta([0, \frac{1}{2}])\} \Rightarrow \{\mu(E) = -\mu(1-E)\}$.

5.44. In **5.43** with sin replaced by cos,

$$\left\{E \in \mathsf{S}_\beta\left([0,1] \setminus \left\{\frac{1}{2}\right\}\right)\right\} \Rightarrow \{\mu(E) = 0\}.$$

5.45. If $\{a_n\}_{n\in\mathbb{N}} \subset \mathbb{R}$, there is a signed $([0,1], \mathsf{S}_\beta, \mu)$ such that

$$\int_0^1 \cos^n \pi x\, d\mu(x) \equiv a_n.$$

5.46. For $([0,1], \mathsf{S}_\beta, \mu)$, if $\mathcal{F}$ is uniformly integrable then each sequence $\{f_n\}_{n\in\mathbb{N}}$ in $\mathcal{F}$ contains a subsequence $\{f_{n_k}\}_{k\in\mathbb{N}}$ such that for some g in $C([0,1], \mathbb{C})$, $\int_0^y f_{n_k}(x)\, d\mu(x) \overset{u}{\to} g(y)$.

5.47. Let P be the set of all $([0,1], \mathsf{S}_\beta, \mu)$ for which $\mu([0,1]) = 1$. What are the extreme points of P regarded as a subset of M, the Banach space of all signed measures defined on S_β?

5.48. If $\{a_n\}_{n=0}^{\infty}, \{t_n\}_{n\in\mathbb{N}} \subset \mathbb{R}$ and

$$\left\{\left\{\sum_{n=0}^{N} a_n e^{nx} \geq 0\right\} \wedge \{x \in [0,1]\}\right\} \Rightarrow \left\{\sum_{n=0}^{N} a_n t_n \geq 0\right\},$$

there is a unique $([0,1], \mathsf{S}_\beta, \mu)$ such that for all n in $\mathbb{N}$, $t_n = \int_0^1 e^{nx}\, d\mu(x)$.

5.49. For $X \stackrel{\text{def}}{=} \{(x,y) : 0 \le x \le y \le 1\}$ and $\mathsf{S}_\beta(X)$, are there measures μ_1, μ_2 such that whenever $0 \le a \le b \le 1$,

$$\mu_1 (([0,a] \times [0,b]) \cap X) = ab + a^2b^2, \ \ \mu_2 (([0,a] \times [0,b]) \cap X) = ab - a^3b^3?$$

5.50. If $(\mathbb{R}, \mathsf{S}_\beta, \mu)$ is complex, $f \in L^1(\mathbb{R}, \mu)$, and for all h in $C_{00}(\mathbb{R}, \mathbb{C})$, $-\int_{\mathbb{R}} f(x)h(x)\,dx = \int_{\mathbb{R}} h'(x)\,d\mu(x)$ then $\mu \ll \lambda$ and $\dfrac{d\mu}{d\lambda}(x) = \displaystyle\int_{-\infty}^{x} f(t)\,dt$.

5.51. If $E \in \mathsf{S}_\lambda$ and for all x in $\mathbb{R}$, $\lambda(E \Delta (x+E)) = 0$ then: a) for all x in $\mathbb{R}$, $\lambda((\mathbb{R} \setminus E) \Delta (x + (\mathbb{R} \setminus E))) = 0$; b) $\lambda(E) \cdot \lambda(\mathbb{R} \setminus E) = 0$.

5.52. For $(\mathbb{R}, \mathsf{S}_\beta, \mu)$ complex and $\hat{\mu}(x) \stackrel{\text{def}}{=} \int_{\mathbb{R}} e^{-itx}\, d\mu(t)$: a) if $\hat{\mu}(\mathbb{Z}) = 0$ then for all E in S_β, $\mu\left(\bigcup_{m \in \mathbb{Z}} (E + 2\pi m)\right) = 0$; b) for some μ and some E, $\hat{\mu}(\mathbb{Z}) = 0$ and $\sum_{m \in \mathbb{Z}} \mu(E + 2\pi m) \ne 0$.

5.53. For $([0,1], \mathsf{S}_\beta, \mu)$, if $\mu \perp \lambda$, $\mu([0,1]) < \infty$, and $f \in L^1([0,1], \lambda)$ then: a) $\int_0^1 f(x-y)\,d\mu(y) < \infty$ exists and is finite a.e. (λ); b) the hypothesis $\mu \perp \lambda$ is superfluous.

5.54. a) If $\{(\mathbb{R}, \mathsf{S}_\beta, \mu_n)\}_{n \in \mathbb{N}}$ is a sequence of complex (nonzero) measure situations then for some Borel measure ν and all n in $\mathbb{N}$, $\mu_n \ll \nu$. b) If $\{(\mathbb{R}, \mathsf{S}_\beta, \mu_n)\}_{n \in \mathbb{N}}$ is a sequence of positive measure situations, need there be a measure ν such that $0 \ne \nu \ll \mu_n$, $n \in \mathbb{N}$?

5.55. For $(\mathbb{R}, \mathsf{S}_\beta, \mu)$ totally finite and an E in S_β, there are measures ν_1, ν_2 such that $\mu = \nu_1 + \nu_2$, $\nu_1(x + E) \equiv 0$ and for some $\{x_n\}_{n \in \mathbb{N}}$,

$$\nu_2\left(\mathbb{R} \setminus \bigcup_{n \in \mathbb{N}} (x_n + E)\right) = 0.$$

5.56. Let $Q(a_1, a_2)$ in $\mathbb{R}^2$ be $\left\{ (x_1, x_2) : |x_1 - a_1| + |x_2 - a_2| \le \frac{1}{2} \right\}$. For $(\mathbb{R}^2, \mathsf{S}_\beta, \mu)$, if horizontal and vertical lines are null sets (μ) then

$$f : \mathbb{R}^2 \ni (a_1, a_2) \mapsto \mu(Q(a_1 a_2)) \in \mathbb{R}$$

is continuous.

5.57. For $(\mathbb{R}^2, \mathsf{S}_\beta, \mu)$ totally finite, if all lines are null sets, E is a bounded Borel set, and $0 < a < \mu(E)$ then there is a Borel set F such that $\mu(F) = a$ (cf. **4.24**).

5.58. For $\{(\mathbb{R}^n, \mathsf{S}_\beta, \mu_m)\}_{m=0}^{\infty}$ and each f in $C_{00}(\mathbb{R}^n, \mathbb{C})$, if

$$\lim_{m\to\infty} \int_{\mathbb{R}^n} f(\mathbf{x})\, d\mu_m(\mathbf{x}) = \int_{\mathbb{R}^n} f(\mathbf{x})\, d\mu_0(\mathbf{x}) \tag{5.6}$$

then: a) for U open, $\underline{\lim}_{m\to\infty}\mu_m(U) \geq \mu_0(U)$; b) if E is a Borel set and $\mu_0(\partial E) = 0$ then $\lim_{m\to\infty} \mu_m(E) = \mu_0(E)$ (cf. **4.19**).

5.59. $\mathsf{S}_\beta(\mathbb{R}^m) \times \mathsf{S}_\beta(\mathbb{R}^n) = \mathsf{S}_\beta(\mathbb{R}^{m+n})$, $\mathsf{S}_\lambda(\mathbb{R}^m) \times \mathsf{S}_\lambda(\mathbb{R}^n) \subsetneqq \mathsf{S}_\lambda(\mathbb{R}^{m+n})$.

5.60. For some product measure situation $\mathsf{X}_{\gamma\in\Gamma}(X_\gamma, \mathsf{S}_\gamma, \mu_\gamma)$ and some subset Γ' of Γ, $P_{\Gamma'}$ does not preserve measurability.

5.61. There are measure situations $(X_i, \mathsf{S}_i, \mu_i)$ for which $\mathsf{S}_1 \times \mathsf{S}_2$ is resp. is not the set of all empty, finite, or countable unions of pairwise disjoint rectangles $E_1 \times E_2, E_i \in \mathsf{S}_i, i = 1, 2$.

5.62. For $(\mathbb{R}^n, \mathsf{S}_\beta, \mu)$, if $\operatorname{supp}(\mu) \stackrel{\text{def}}{=} K$ is compact and for some $\mathbb{R}$-valued polynomial p and all $\mathbb{R}$-valued polynomials q, $\int_{\mathbb{R}^n} p(\mathbf{x})\,(q(\mathbf{x}))^2\, d\mu(\mathbf{x}) \geq 0$ then $p \geq 0$ on K.

5.63. If $\{f_n\}_{n\in\mathbb{N}} \subset C([0,1], \mathbb{R})$, $f_n \stackrel{\text{a.e.}}{\to} f$, and $0 \leq a < 1$ then $[0,1]$ contains a compact set K such that $\lambda(K) > a$ and f is continuous on K.

5.64.

i. The function f is in $AC([a,b], \mathbb{R})$ iff: a) f is continuous, b) $f \in BV([a,b], \mathbb{R})$, and c) $\{\lambda(E) = 0\} \Rightarrow \{\lambda(f(E)) = 0\}$.

ii. If x) is one of a),b),c) there is an f for which x) does not hold and for which the other two do hold.

iii. If $E \in \mathsf{S}_\lambda$ and $f \in AC([a,b], \mathbb{R})$ then $f(E) \in \mathsf{S}_\lambda$.

5.65. If f is not constant on $[a,b]$ and if $f'(x) \doteq 0$ then $f \notin \operatorname{Lip}(1)$ on $[a,b]$.

5.66. If $f \in \mathbb{R}^{\mathbb{R}} \cap \operatorname{Lip}(1)$ then for all E in S_λ and for some constant M, $\lambda(f(E)) \leq M\lambda(E)$.

5.67. For some f, g in $\mathbb{R}^{\mathbb{R}}$, f is continuous and open, g is measurable (λ), and $g \circ f$ is not measurable (λ).

5.68. For some E in S_λ and some ϕ in $C(\mathbb{R}, \mathbb{R})$, $\phi(E) \notin \mathsf{S}_\lambda$.

5.69. For (X, S, μ), if each singleton set $\{x\}$ is in S and $2^X \setminus \mathsf{S} \neq \emptyset$ then there is a diset Λ and a net $n : \Lambda \ni \lambda \mapsto n_\lambda \in \mathcal{M}$ such that n_λ converges in the product topology for $\mathcal{M}$ to an f in $\mathbb{R}^X \setminus \mathcal{M}$.

5.70. If $0 \leq f \in L^1(X, \mu)$ then for

$$\nu : \mathsf{S} \ni E \mapsto \nu(E) \stackrel{\text{def}}{=} \int_E f(x)\, d\mu(x),$$

the essential suprema of f with respect to μ and ν are the same.

5.71. If $E \in \mathsf{S}_\lambda$ and $\lambda(E) > 0$ then $\#(E) = \mathfrak{c}$.

5.72. If $([0,\infty), \mathsf{S}_\beta, \mu)$ complex and $\int_{[0,\infty)} e^{-nx}\, d\mu(x) = 0,\ n \in \mathbb{N}$, then $\mu(\mathsf{S}_\beta) = \{0\}$.

5.73. For $([0,\infty), \mathsf{S}_\beta, \mu)$, if $\mu([0,\infty)) = 1$ then

$$\int_{[0,\infty)} (1 - \mu([0,x)))\, dx = \int_{[0,\infty)} x\, d\mu(x).$$

5.74. For $([0,\infty), \mathsf{S}_\beta, \mu)$, if $0 \le f \in \mathcal{M}$ and f is bounded then

$$\{f \in L^1([0,\infty), \mu)\} \Rightarrow \left\{\int_{[0,\infty)} \mu(\{y : x \le f(y)\})\, dx < \infty\right\}.$$

5.75. (Banach) For f in $C([0,1], \mathbb{R})$, if $f([0,1]) \stackrel{\text{def}}{=} [m, M]$,

$$\nu : [m,M] \ni y \mapsto \nu(y) \stackrel{\text{def}}{=} \begin{cases} \#(\{x : f(x) = y\}) & \text{if } \#(\{x : f(x) = y\}) < \infty \\ \infty & \text{otherwise,} \end{cases},$$

and $0 \cdot \infty \stackrel{\text{def}}{=} 0$ then $\int_{[m,M]} \nu(y)\, dy = \operatorname{var}_{[a,b]}(f)$.

5.76. If $[a,b] \subset \bigcup_{n=1}^N [a_n, b_n]$ then $\sum_{n=1}^N (b_n - a_n) \ge b - a$.

5.77. The plane $\mathbb{R}^2$ is not the countable union of lines.

5.78. If $(0,1] \cap \mathbb{Q} \stackrel{\text{def}}{=} \{r_k\}_{k\in\mathbb{N}}$ and $f(x) \stackrel{\text{def}}{=} \sum_{k=1}^\infty \dfrac{1}{k^2 |x - r_k|^{\frac{1}{2}}}$ then $f(x) < \infty$ a.e.

5.79. If $\mathbb{Q} \stackrel{\text{def}}{=} \{r_n\}_{n\in\mathbb{N}}$ then $\mathbb{R} \setminus \bigcup_{n\in\mathbb{N}} \left(r_n - \frac{1}{n^2}, r_n + \frac{1}{n^2}\right) \neq \emptyset$. Is there an enumeration $\{s_n\}_{n\in\mathbb{N}}$ of $\mathbb{Q}$ such that $\mathbb{R} \setminus \bigcup_{n\in\mathbb{N}} \left(s_n - \frac{1}{n}, s_n + \frac{1}{n}\right) \neq \emptyset$?

5.80. True or false: for $\left([0,1]^2, \mathsf{S}_{\lambda_2}, \lambda_2\right)$ if $E \in \mathsf{S}_{\lambda_2}$ and $\lambda(E_{x_1}) \le \dfrac{1}{2}$ a.e. then $\lambda(\{x_2 : \lambda(E_{x_2}) = 1\}) \le \dfrac{1}{2}$?

5.81. The set $E \stackrel{\text{def}}{=} \mathbb{R}^2 \setminus \{(x_1, x_2) : x_1 - x_2 \in \mathbb{Q}\}$ contains no measurable rectangle $A_1 \times A_2$ such that $\lambda_2(A_1 \times A_2) > 0$.

5.82. If $f \in \mathbb{R}^{\mathbb{R}^2}$, f_{x_1} is Borel measurable for all x_1, and $f_{x_2} \in C(\mathbb{R}, \mathbb{R})$ for all x_2 then f is Borel measurable.

5.83. If E is dense in $\mathbb{R}$ and for some f in $\mathbb{R}^{\mathbb{R}^2}$ f_{x_1} is Lebesgue measurable for all x_1 in E, and $f_{x_2} \in C(\mathbb{R},\mathbb{R})$ a.e. (λ) then f is Lebesgue measurable.

5.84. If $f \in \mathbb{R}^{\mathbb{R}^2}$, $f_{x_1} \in \mathcal{M}$ for all x_1, and $f_{x_2} \in C(\mathbb{R},\mathbb{R})$ for all x_2 then for every g in $\mathbb{R}^{\mathbb{R}} \cap \mathcal{M}$, $h : \mathbb{R} \ni x \mapsto f(g(x_2), x_2) \in \mathcal{M}$.

5.85. Is $f : \mathbb{R}^2 \ni (x_1, x_2) \mapsto \dfrac{x_1 x_2}{(x_1^2 + x_2^2)}$ in $L^1([-1,1]^2, \lambda_2)$?

5.86. For

$$f : \mathbb{R}^2 \ni (x_1, x_2) \mapsto \begin{cases} \dfrac{(x_1^2 - x_2^2)}{(x_1^2 + x_2^2)} & \text{if } x_1^2 + x_2^2 > 0\,, \\ 0 & \text{otherwise} \end{cases}$$

how are $\int_{[0,1]} \left(\int_{0,1]} f(x_1, x_2)\, dx_1 \right) dx_2$ and $\int_{[0,1]} \left(\int_{[0,1]} f(x_1, x_2)\, dx_2 \right) dx_1$ related?

5.87. For some f in $C((0,1)^2, \mathbb{R})$, $\int_{(0,1)^2} |f(x_1, x_2)|\, d\lambda_2(x_1, x_2) < \infty$ and for some x_1 in $(0,1)$, $\int_{(0,1)} |f(x_1, x_2)|\, dx_2 = \infty$.

5.88. True or false: if (X, S, μ_i), $i = 1, 2$, are totally finite then

$$\left\{ \{\mu_i \ll \mu_j,\ i \neq j\} \wedge \left\{ \frac{d\mu_1}{d\mu_2} \in L^\infty(X, \mu_2) \right\} \right\} \Rightarrow \left\{ \frac{d\mu_2}{d\mu_1} \in L^\infty(X, \mu_1) \right\}?$$

5.89. If $([0,1], \mathsf{S}_\beta, \mu)$ is totally finite and $\mu \ll \lambda$ then

$$\lim_{a \downarrow 0} \frac{\mu([0,1] \cap (x - a, x + a))}{\lambda([0,1] \cap (x - a, x + a))}$$

exists a.e. (λ).

5.90. True or false: for some E in $\mathsf{S}_\lambda([0,1])$,

$$\{0 \le a < b \le 1\} \Rightarrow \left\{ \lambda(E \cap [a,b]) = \frac{(b-a)}{2} \right\}?$$

5.91. If $\{J_n\}_{n \in \mathbb{N}}$ is a set of open intervals in $\mathbb{R}$ and $\{C_n\}_{1 \le n < N}$ is the set of components of $A \stackrel{\text{def}}{=} \bigcup_{n \in \mathbb{N}} J_n$, then: a) $\sum_{1 \le n < N} \lambda(C_n) \le \sum_{1 \le n < N} \lambda(J_n)$; b) for M in $\mathbb{N}$, there is a subset $\{J_{n_k}\}_{1 \le k < K}$ of pairwise disjoint intervals such that $\lambda\left(\bigcup_{n=1}^{M} J_n\right) \le 2\lambda\left(\bigcup_{1 \le k < K} J_{n_k}\right)$.

5.92. True or false: there is a signed $([0,1], \mathsf{S}, \mu)$ such that $\mu \not\equiv 0$, $\mu \ll \lambda$, and for all a in $[0,1]$, $\mu([0,a]) = 0$?

5.93. Lebesgue measure λ_n is rotation-invariant.

5.94. There is in $\mathbb{R}$ a Lebesgue null set of the second category.

5.95. If $0 < \lambda(E) < \infty$ and $0 \le f \in \mathcal{M}$ then $g : \mathbb{R} \ni x \mapsto \int_E f(x-t)\,dt$ is in $L^1(\mathbb{R},\lambda)$ iff $f \in L^1(\mathbb{R},\lambda)$.

5.96. If $\mu(\mathbb{R}) = 1$ then $\mu * \lambda = \lambda$, i.e., $\lambda(E) \equiv \int_{\mathbb{R}} \mu(E+x)\,dx$.

5.97. If $T \in (\mathbb{R}^m)^{\mathbb{R}^n}$, then T is continuous if $T(\mathbf{x}+\mathbf{y}) \equiv T(\mathbf{x}) + T(\mathbf{y})$ and $T^{-1}(\mathsf{O}(\mathbb{R}^n)) \subset \mathsf{S}_\beta(\mathbb{R}^n)$.

5.98. For x in $[0,1]$, let $a_n(x)$ be the nth decimal marker for x. If $k \ne 0,1$ and

$$f : [0,1] \ni x \mapsto \begin{cases} k & \text{if } a_n(x) \ne 0,\ n \in \mathbb{N} \\ 1 & \text{if } \min\{\, n \ :\ a_n(x) = 0 \,\} \text{ is even} \\ 0 & \text{otherwise} \end{cases}$$

then f is Lebesgue measurable. What is $\int_{[0,1]} f(x)\,dx$?

5.99. Let E in $[0,1]$ be the set of x such that the nth decimal marker of x is 2 or 7. True or false: a) E is closed? b) E is open? c) E is countable? d) E is dense in $[0,1]$? e) $E \in \mathsf{S}_\beta$? If $E \in \mathsf{S}_\lambda$, find $\lambda(E)$.

5.100. For x in $[0,1]$ and m in $\mathbb{N} \setminus \{1\}$, let $\sum_{n=1}^\infty \frac{\epsilon_n(x)}{m^n}$ be the m-ary representation of x. If $0 \le k < m$ and

$$A_N(x,k,m) \stackrel{\text{def}}{=} \#(\{\, n \ :\ \epsilon_n(x) = k,\ 1 \le n \le N \,\})$$

then

$$E \stackrel{\text{def}}{=} \left\{ x \ :\ \lim_{N\to\infty} \frac{A_N(x,k,m)}{N} \text{ does not exist} \right\} \in \mathsf{S}_\lambda \setminus (\mathsf{O} \cup \mathsf{F}).$$

5.101. For some E in $\mathsf{S}_\lambda(\mathbb{R})$, $\lambda(E) < \infty$, and if $a < b$ then

$$0 < \lambda(E \cap [a,b]) < b - a.$$

5.102. For some E in S_λ, if $a < b$ then $\lambda(E \cap [a,b]) \cdot \lambda([a,b] \setminus E) > 0$.

5.103. True or false: if $\{E_n\}_{n\in\mathbb{N}} \subset \mathsf{S}_\lambda([0,1])$, the E_n are pairwise different, and for some positive a and all n in $\mathbb{N}$, $\lambda(E_n) \ge a$ then there is an infinite subsequence $\{E_{n_k}\}_{k\in\mathbb{N}}$ such that $\lambda\left(\bigcap_{k\in\mathbb{N}} E_{n_k}\right) > 0$?

5.104. If $\{A_n\}_{n\in\mathbb{N}} \subset \mathsf{S}_\lambda(\mathbb{R})$, $\lambda\left(\dot{\bigcup}_{n\in\mathbb{N}} A_n\right) < \infty$, and

$$G_k \stackrel{\text{def}}{=} \{\, x \ :\ x \in A_n \text{ for exactly } k \text{ different values of } n \,\}$$

then $G_k \in \mathsf{S}_\lambda(\mathbb{R})$, $k \in \mathbb{N}$ and $\sum_{k=1}^\infty k\lambda(G_k) = \sum_{n=1}^\infty \lambda(A_n)$.

5.105. If f is integrable (λ) over every interval and for each interval J of length 1 or $\sqrt{2}$, $\int_J f(x)\,dx = 0$ then $f(x) \doteq 0$.

5.106. If $f \in L^\infty_{\mathbb{R}}(\mathbb{T},\tau)$, $\|f\|_\infty = 1$, and for some x_0, and some N, $|\sigma_N(f)(x_0)| = 1$ then $f(x)$ is constant a.e. The conclusion need not hold if f is $\mathbb{C}$-valued.

5.107. There is a bijection $f : [0,1] \mapsto [0,1]$ such that: a) E is Lebesgue measurable iff $f(E)$ is Lebesgue measurable, b) for a Lebesgue measurable E, $\lambda(E) = \lambda(f(E))$, and c) $f\left(\left(\frac{1}{4},\frac{3}{4}\right)\right) = \left[\frac{1}{4},\frac{3}{4}\right]$.

5.108. For some Lebesgue measurable function f, finite everywhere, if $a < b$ then $\int_{[a,b]} f(x)\,dx = \infty$.

5.109. If $f \in \mathbb{R}^{\mathbb{R}}$ and f is S_λ-measurable, there is a sequence $\{f_n\}_{n\in\mathbb{N}}$ of $\mathbb{Q}$-valued S_λ-measurable functions converging uniformly and monotonely to f.

5.110. If $f \in C(\mathbb{R},\mathbb{R})$, g is S_λ-measurable, and for every null set (λ) N, $f^{-1}(N)$ is S_λ-measurable then $g \circ f$ is S_λ-measurable (cf. **5.67**).

5.111. If $\lambda^*(A) > 0$ and $0 < \theta < 1$, there is an interval J such that $\lambda^*(A \cap J) > \theta \cdot \lambda^*(J)$.

5.112. For (X, S, μ), if $f \in \mathcal{M}$ for some g in $\mathcal{M}$, $g(x) \doteq f(x)$ and $\sup_x |g(x)| = \|f\|_\infty$.

5.113. If $f \in \mathbb{R}^{[0,1]} \cap \mathcal{M}$ then for a unique a_0:

$$\text{a) } \lambda(\{x : f(x) \ge a_0\}) \ge \frac{1}{2};$$

$$\text{b) } \lambda(\{x : f(x) \ge a\}) < \frac{1}{2} \text{ if } a_0 < a < \infty.$$

5.114. If $f \in L^\infty([0,1],\lambda)$ and

$$\int_0^1 x^n f(x)\,dx = 0, \text{ or } \int_0^1 e^{\pm inx} f(x)\,dx = 0, \; n \in \mathbb{Z}^+,$$

then $f(x) \doteq 0$ (cf. **2.79**).

5.115. If $f, g \in L^\infty((0,\infty),\lambda)$ and $\int_{(0,\infty)} (|f(x)| + |g(x)|)/x\,dx < \infty$ then $\int_{(0,\infty)} \left(|f(xy)g(1/y)|/y\right)\,dy < \infty$ a.e.

5.116. If $\{f_n\}_{n\in\mathbb{N}} \subset L^\infty(\mathbb{R},\lambda)$, $f_n \overset{\text{meas}}{\to} f$, and for some g in $L^1(\mathbb{R},\lambda)$, $|f_n(x)| \le |g(x)|$ a.e. then $f \in L^1(\mathbb{R},\lambda)$ and

$$\lim_{n\to\infty} \int_{\mathbb{R}} f_n(x)\,dx = \int_{\mathbb{R}} f(x)\,dx.$$

5.117. If $\lambda(E) > 0$ and $\{x, y \in E\} \Rightarrow \left\{\frac{1}{2}(x+y) \in E\right\}$ then E contains a nonempty open set.

5.118. $\#(\mathsf{S}_\lambda) = 2^{\mathfrak{c}}$.

5.119. If $E \subset \mathbb{R}^n$ and for each $\mathbf{a}$ in $\mathbb{R}^n$, there is a positive $r_{\mathbf{a}}$ such that $B(\mathbf{a}, r_{\mathbf{a}}) \cap E \in \mathsf{S}_\lambda$, then $E \in \mathsf{S}_\lambda$.

5.120. If $\lambda(E) = 0$ and $\#(E) > \aleph_0$ then E contains a subset F in $\mathsf{S}_\lambda \setminus \mathsf{S}_\beta$.

5.121. If μ^* is an outer measure on $2^{[0,1]}$, $\mathsf{S}_\beta \subset \mathsf{S}_\mu$ [the set of (Caratheodory) μ^*-measurable subsets of $[0,1]$], and $\mu \ll \lambda$ then $\mathsf{S}_\lambda \subset \mathsf{S}_\mu$.

5.122. True or false: if $\{f_n\}_{n=0}^\infty \subset \mathbb{R}^{[0,1]}$, each f_n is monotonely increasing, and $f_n \overset{\text{meas}}{\to} f_0$ then $\lim_{n\to\infty} f_n(x) = f_0(x)$ at each point of continuity x of f_0?

5.123. If $0 \le f \in \mathbb{R}^{[0,1]}$ and f is S_λ-measurable then: a)

$$A_f \overset{\text{def}}{=} \{g \ : \ g \in L^1([0,1], \lambda), \ |g(x)| \le f(x) \text{ a.e.}\}$$

is a $\| \ \|_1$-closed subset of $L^1([0,1], \lambda)$; b) if A_f is $\| \ \|_1$-compact then f is integrable.

5.124. For some E in S_{λ_2}, $E + E \notin \mathsf{S}_{\lambda_2}$.

5.125. For some E in S_λ, $E + E \notin \mathsf{S}_\lambda$.

5.126. If, for $\{([0,1], \mathsf{S}_\lambda, \mu_n)\}_{n=0}^\infty$,

$$\mu_0([0,1]) = 1 = \lim_{n\to\infty} \mu_n([0,1])$$

and for each E in S_λ, $\mu_n(E) \le \mu_0(E)$, $n \in \mathbb{N}$, then there is a sequence $\{\mu_{n_k}\}_{k\in\mathbb{N}}$ such that for each g in $L^1([0,1], \mu_0)$, $\int_{[0,1]} g(x)\, d\mu_{n_k}(x)$ exists and $\lim_{k\to\infty} \int_{[0,1]} g(x)\, d\mu_{n_k}(x) = \int_{[0,1]} g(x)\, d\mu_0(x)$.

5.127. For $([0,1], \mathsf{S}_\lambda, \lambda)$, if E is in S_λ and

$$f_n(x) \overset{\text{def}}{=} n \int_0^{\frac{1}{n}} \chi_E(x+t)\, dt, \ n \in \mathbb{N},$$

then: a) $0 \le f_n \le 1$; b) $f_n \in AC(\mathbb{R}, \mathbb{R})$; c) $f_n \overset{\text{a.e.}}{\to} \chi_E$; d) $f_n \overset{\| \ \|_1}{\to} \chi_E$.

5.128. If $f \in AC([a,b], \mathbb{R})$, f is strictly monotonely increasing, and $f([a,b]) = [c,d]$ then for every Borel set E in $[a,b]$,

$$\int_{f^{-1}(E)} f'(x)\, dx = \lambda(E).$$

5.129. For some f in $BV([0,1],\mathbb{C}) \cap C([0,1],\mathbb{C})$, and some null set A, $f(A) \in \mathsf{S}_\lambda([0,1])$ and $\lambda(f(A)) > 0$.

5.130. a) If $f \in BV([0,1],\mathbb{C})$ then f' is finite a.e. (λ) and

$$f' \in L^1([0,1],\lambda).$$

b) If $f \in AC([0,1],\mathbb{R})$ then f is the difference of two absolutely continuous monotonely increasing functions.

5.131. True or false: if $f \in \mathbb{R}^{\mathbb{R}}$ then f is continuous a.e. iff f is equal a.e. to a continuous function?

5.132. If f and g are nonnegative Lebesgue measurable functions on $\mathbb{R}$, $A_y \stackrel{\text{def}}{=} \{x : f(x) \geq y\}$, and $h(y) \stackrel{\text{def}}{=} \int_{A_y} g(x)\,dx$ then

$$\int_{\mathbb{R}} f(x)g(x)\,dx = \int_0^\infty h(y)\,dy.$$

5.133. An f is in Lip(1) on $[0,1]$ iff: a) for some g in

$$L^\infty([0,1],\lambda),\, f(x) - f(0) = \int_0^x g(t)\,dt;$$

or b) for some sequence $\{f_n\}_{n\in\mathbb{N}}$ in Lip(1),

$$\lim_{n\to\infty} f_n(0) = f(0),\ \lim_{n\to\infty} \operatorname{var}_{[0,1]}(f - f_n) = 0,$$

and the Lipschitz constants $K(f_n)$, $n \in \mathbb{N}$, form a bounded set.

When $f \in \mathbb{C}^{\mathbb{R}}$ and for all x, $f(x-0)$ exists, $f(x) = f(x-0)$, and $\lim_{x\to-\infty} f(x) = 0$, then $\mu_f : [a,b) \mapsto f(b) - f(a)$ is the interval function associated to f. When, to boot, f is a function of bounded variation, μ_f is uniquely extensible to a (complex) measure, again denoted μ_f, on S_λ.

5.134. True or false: if $f \in C([0,1],\mathbb{C})$, there is a complex $([0,1],\mathsf{S}_\beta,\mu)$ such that $\mu([a,b)) = \mu_f([a,b))$?

5.135. If f is monotonely increasing on $\mathbb{R}$ and $g = f^2$ then $\mu_g \ll \mu_f$. What is $\dfrac{d\mu_g}{d\mu_f}$?

5.136. If $f \in L^1(\mathbb{R},\lambda)$ then for some sequence $\{f_n\}_{n\in\mathbb{N}}$ in $C_{00}(\mathbb{R},\mathbb{C})$, $g(x) \stackrel{\text{def}}{=} \sum_{n=1}^\infty f_n(x)$ exists and $f(x) \doteq g(x)$.

5.137. An f is in $L^2([0,1],\lambda)$ iff $f \in L^1([0,1],\lambda)$ and there is a monotonely increasing function g such that if $0 \le a \le b \le 1$ then

$$\left|\int_a^b f(x)\,dx\right|^2 \le (g(b)-g(a)) \cdot |b-a|.$$

5.138. If $g \in L^2([0,1],\lambda) \setminus \{0\}$ and $G(x) \stackrel{\text{def}}{=} \int_0^x g(t)\,dt$ then $\|G\|_2 < \|g\|_2$.

5.139. There is a systematic enumeration $\left\{\{J_{nk}\}_{1\le k\le 2^{n-1}}\right\}_{n\in\mathbb{N}}$ of the set of open intervals deleted in the construction of the Cantor set. For $f(x) \stackrel{\text{def}}{=} \sum_{n=1}^{\infty}\left(\sum_{k=1}^{2^n-1} n\chi_{J_{nk}}\right)$, what is $\int_{[0,1]} f(x)\,dx$?

5.140. If $\left(\mathbb{R}^2, \mathsf{S}_\beta\left(\mathbb{R}^2\right), \mu\right)$ is totally finite then for the net

$$\nu : \mathsf{S}_\beta \ni A \mapsto \nu(A) \stackrel{\text{def}}{=} \mu(A \times \mathbb{R}),$$

there is a map $\mathsf{S}_\beta(\mathbb{R}) \ni B \mapsto f_B \in L^1([0,1],\nu)$ such that:

$$\{B_1 \cap B_2 = \emptyset\} \Rightarrow \{f_{B_1\cup B_2} = f_{B_1} + f_{B_2}\};$$
$$\{B \in \mathsf{S}_\beta(\mathbb{R})\} \Rightarrow \left\{\mu(A\times B) = \int_A f_B(x)\,d\nu(x)\right\}.$$

5.141. True or false: if a) $\{c_n\}_{n\in\mathbb{N}} \subset \mathbb{R}$, b) $\lambda(A) > 0$, and c) for all t in A, $\lim_{n\to\infty} e^{ic_nt}$ exists then $\lim_{n\to\infty} c_n$ exists?

5.142. If $f \in L^\infty(\mathbb{R},\lambda)$, $1 \le p < \infty$, $-\infty < a < b < \infty$, $0 < c < 1$, and $\left(\int_a^b |f(x)|\,dx\right)^p \le c(b-a)^{p-1}\int_a^b |f(x)|^p\,dx$ then $f \doteq 0$.

5.143. If $f \in C([0,1],\mathbb{R})$, $f(0) = 0$, and $f'(0)$ exists then

$$g : [0,1] \ni x \mapsto \frac{f(x)}{x^{3/2}}$$

is in $L^1([0,1],\lambda)$.

5.144. If $f \in L^1([0,1],\lambda)$, $g \in L^\infty(\mathbb{R},\lambda)$, and g is periodic with period one $[g(t+1) \equiv g(t)]$ then $\lim_{n\to\infty}\int_{[0,1]} f(t)g(nt)\,dt = \int_{[0,1]} f(t)\,dt \cdot \int_{[0,1]} g(t)\,dt$.

5.145. Each finite subset of $L^1(X,\mu)$ is uniformly integrable.

5.146. For (X,S,μ), if $\{f_n\}_{n\in\mathbb{N}} \subset L^1(X,\mu)$, $f_n \ge 0$, $f_n \stackrel{\text{a.e.}}{\to} 0$, and $\int_X \max\{f_1(x),\ldots,f_n(x)\}\,d\mu(x) \le M < \infty$ then $\int_X f_n(x)\,d\mu(x) \to 0$ as $n \to \infty$.

5.147. a) $\lim_{R\to\infty}\int_0^R \frac{\sin x}{x}\,dx = \frac{\pi}{2}$; b) $\lim_{R\to\infty}\int_0^R \left|\frac{\sin x}{x}\right|\,dx = \infty$.

5.2. Probability Theory

Conventions

A probability measure situation (Ω, S, P) is one for which $\Omega \in \mathsf{S}$, $P(\Omega) = 1$, and subsets of null sets are in S [(Ω, S, P) is complete]. In these circumstances a subset $\mathcal{S}$ of 2^Ω is independent iff for each finite subset $\{E_1, \ldots, E_n\}$ of $\mathcal{S}$, $P\left(\bigcap_{k=1}^n E_k\right) = \prod_{k=1}^n P(E_k)$. A subset $\mathcal{F}$ of $\mathcal{M}$ is independent iff for every finite subset $\{f_1, \ldots, f_n\}$ of $\mathcal{F}$ and every finite subset $\{B_1, \ldots, B_n\}$ of Borel subsets of $\mathbb{R}$, $\left\{f_1^{-1}(B_1), \ldots, f_n^{-1}(B_n)\right\}$ is an independent subset of 2^Ω. When $\mathcal{S}$ resp. $\mathcal{F}$ is independent, the elements of $\mathcal{S}$ resp. $\mathcal{F}$ are called, by abuse of language, independent.

Traditionally, elements of $\mathcal{S}$ are called events and the elements of $\mathcal{M}$ are called random variables. Furthermore, if f is integrable then $E(f) \stackrel{\text{def}}{=} \int_\Omega f(\omega)\,dP(\omega)$ is the expected value of f; if f^2 is integrable, then $\operatorname{Var}(f) \stackrel{\text{def}}{=} \sigma^2(f) \stackrel{\text{def}}{=} \int_\Omega (f(x) - E(f))^2\,dP(\omega)$ is the variance of f.

Two particularly elementary and yet important instances of a probability measure situation (Ω, S, P) are: a) for some discrete measure P, $(\mathbb{N}, 2^{\mathbb{N}}, P)$; b) $([0,1], \mathsf{S}_\lambda, \lambda)$. Every (Ω, S, P) contains at most countably many atoms. When S is metrized according to

$$d : \mathsf{S} \times \mathsf{S} \ni (A, B) \mapsto d(A, B) \stackrel{\text{def}}{=} P(A \Delta B)$$

and S is d-separable, then (Ω, S, P) is, by abuse of language, measure-isomorphic for some P to

$$(\mathbb{N}, 2^{\mathbb{N}}, P),\ ([0,1], \mathsf{S}_\lambda, \lambda),\ \text{or}\ (\mathbb{N}, 2^{\mathbb{N}}, P) \cup ([0,1], \mathsf{S}_\lambda, \lambda).$$

Unless the contrary is indicated in the current **Section**, the underlying probability measure situation is to be taken as a general (Ω, S, P).

5.148. When some of or all the members of $\mathcal{S}$ are replaced by their complements, the resulting set $\widetilde{\mathcal{S}}$ is independent iff $\mathcal{S}$ is independent.

5.149. a) The set $\{E_\lambda\}_{\lambda\in\Lambda}$ is independent iff the set $\{\chi_{E_\lambda}\}_{\lambda\in\Lambda}$ is independent. b) If $\mathcal{S} \stackrel{\text{def}}{=} \left\{\{f_{jk}\}_{k=1}^{K_j}\right\}_{j=1}^J$ is independent and if g_j is a Borel measurable function on $\mathbb{R}^{K_j}$, $1 \le j \le J$, then $\left\{h_j \stackrel{\text{def}}{=} g_j\left(f_{j1}, \ldots, f_{jK_j}\right)\right\}_{j=1}^J$ is independent.

5.150. a) If $P(A) = 0$ or $P(A) = 1$ and $B \in \mathsf{S}$ then $\{A, B\}$ is independent. b) If $f(\omega)$ is constant a.e. and $g \in \mathcal{M}$ then $\{f, g\}$ is independent.

5.151. If $\{f, g\}$ is independent and $\{f, g\} \subset L^1(\Omega, P)$ then $fg \in L^1(\Omega, P)$ and

$$\int_\Omega f(\omega)g(\omega)\,dP(\omega) = \int_\Omega f(\omega)\,dP(\omega) \cdot \int_\Omega g(\omega)\,dP(\omega). \tag{5.7}$$

5.152. If P in $\left(\mathbb{N}, 2^{\mathbb{N}}, P\right)$ is such that

$$P(n) = \begin{cases} 2^{-n!} & \text{if } n = 2, 3, \ldots \\ 1 - \sum_{n=2}^\infty 2^{-n!} & \text{if } n = 1 \end{cases}$$

then: a) if neither A nor B is $\emptyset$ nor $\mathbb{N}$, then $\{A, B\}$ is not independent; b) if neither $f(\omega)$ nor $g(\omega)$ is constant a.e. then $\{f, g\}$ is not independent.

5.153. For $([0, 1], \mathsf{S}_\lambda, \lambda)$ if $f \in \mathcal{M}$, f is strictly monotone on some nonempty subinterval (a, b) of $[0, 1]$, and $\{f, g\}$ is independent then $g(\omega)$ is constant a.e.

For $S \subset L^2(X, \mu)$, span(S) is the set of all (finite) linear combinations of elements in S while

$$S^\perp \stackrel{\text{def}}{=} \left\{ g \ : \ g \in L^2(X, \mu), \{f \in S\} \Rightarrow \left\{ \int_X f(x)\overline{g(x)}\,d\mu(x) = 0 \right\} \right\}.$$

For M a subspace of $L^2(X, \mu)$, dim(M) is the cardinality of a maximal linearly independent subset of M. A subset S of M spans M when the $\|\ \|_2$-closure of span(S) is M (cf. **Chapter 6**).

5.154. If $S \stackrel{\text{def}}{=} \{f_\gamma\}_{\gamma \in \Gamma} \subset L^2(\Omega, P)$ and S is independent then: a)

$$\left\{\dim\left(L^2(\Omega, P)\right) \geq 3\right\} \Rightarrow \left\{\dim\left(S^\perp\right) \geq 1\right\}.$$

b) $\{\#(\Gamma) \geq \aleph_0\} \Rightarrow \left\{\dim\left(S^\perp\right) \geq \aleph_0\right\}$.

5.155. If $S \stackrel{\text{def}}{=} \{A_n\}_{n \in \mathbb{N}}$ is independent then $P\left(\overline{\lim}_{n\to\infty} A_n\right)$ is 0 or 1 according as $\sum_{n=1}^\infty P(A_n)$ converges or diverges. [The preceding situation illustrates the zero-one law **[Ko]**.]

5.156. For $([0, 1], \mathsf{S}_\lambda, \lambda)$: a) the set

$$\mathcal{R} \stackrel{\text{def}}{=} \{r_n : [0, 1] \ni x \mapsto \text{sgn}\left(\sin\left(2^n \pi x\right)\right)\}_{n-0}^\infty$$

is independent; b) the set $\mathcal{W} \stackrel{\text{def}}{=} \{W_n\}_{n=0}^\infty$ of all finite products of pairwise different members of $\mathcal{R}$ (cf. **Solution 5.154**) spans $L^2([0, 1], \lambda)$.

[**Note 5.3:** The functions r_n resp. W_n are the Rademacher resp. Walsh functions [**Fin, KacS, Wal**].]

5.3. Ergodic Theory

Conventions

The context is a totally finite (X, S, μ) and an autojection $T : X \overset{\text{aut}}{\mapsto} X$ such that

$$\{E \in \mathsf{S}\} \Rightarrow \{\{T(E) \in \mathsf{S}\} \wedge \{\, T^{-1}(E) \in \mathsf{S}\} \wedge \{\mu(E) = \mu\left(T(E)\right)\}\}.$$

For f in $L^1(X, \mu)$ and n in $\mathbb{N}$,

$$s_n : X \ni x \mapsto \sum_{k=0}^{n-1} f\left(T^k(x)\right), \ \overline{F} \overset{\text{def}}{=} \overline{\lim}_{n\to\infty} \frac{s_n}{n}, \text{ and } \underline{F} \overset{\text{def}}{=} \underline{\lim}_{n\to\infty} \frac{s_n}{n}$$

are the principal objects of study.

5.157. If $E \in \mathsf{S}$ then $\mu\left(T^{-1}(E)\right) = \mu(E)$.

5.158. If

$$A \overset{\text{def}}{=} \left\{ x \ : \ \sup_{n\in\mathbb{N}} s_n(x) > 0 \right\} \text{ and } E \in \mathsf{S}$$

then

$$\int_{A\cap E} f(x)\, d\mu(x) \geq 0.$$

5.159. If

$$A_a \overset{\text{def}}{=} \left\{ x \ : \ \sup_{n\in\mathbb{N}} \frac{s_n(x)}{n} > a \geq 0 \right\} \text{ and } E \in \mathsf{S}$$

then

$$\int_{A_a\cap E} f(x)\, d\mu(x) \geq a \cdot \mu\left(A_a \cap E\right).$$

5.160. a) $\overline{F}(x) = \overline{F}(T(x))$ and $\underline{F}(x) = \underline{F}(T(x))$; b) $\overline{F}(x) \doteq \underline{F}(x)$; c) $\left\{\frac{s_n}{n}\right\}_{n\in\mathbb{N}}$ is uniformly integrable.

5.161. For

$$F(x) \overset{\text{def}}{=} \begin{cases} \overline{F}(x) & \text{if } \overline{F}(x) = \underline{F}(x) \\ 0 & \text{otherwise} \end{cases},$$

if $E \in \mathsf{S}$ then

$$\int_E f(x)\, d\mu(x) = \int_E F(x)\, d\mu(x).$$

[**Note 5.4:** The conclusions in **5.160**b) and **5.161** embody the pointwise ergodic theorem of G.D. Birkhoff.]

In $L^2(X,\mu)$ the autojection T engenders a unitary automorphism U in $\left[L^2(X,\mu)\right]$: $U : L^2(X,\mu) \ni f \overset{\text{endo}}{\mapsto} [U(f) : X \ni x \mapsto f(T(x))]$.

5.162. In the context just described, if $f \in L^2(X,\mu)$ then for some $L(f)$ in $L^2(X,\mu)$, $\dfrac{\sum_{k=0}^{n-1} U^k(f)}{n} \overset{\|\ \|_2}{\to} L(f)$.

[**Note 5.5:** The assertion in **5.162** is the mean ergodic theorem of von Neumann. It is valid in the context of any abstract Hilbert space $\mathfrak{H}$ and a unitary (automorphism) $U : \mathfrak{H} \mapsto \mathfrak{H}$.]

5.163. For z_0 in $\mathbb{T}$, f in $C(\mathbb{T},\mathbb{C})$, and the autojection

$$T \stackrel{\text{def}}{=} T_{z_0} : \mathbb{T} \ni z \mapsto z_0 \cdot z$$

of $\mathbb{T}$, $L \stackrel{\text{def}}{=} \lim_{n\to\infty} \dfrac{s_n(z_0)}{n}$ exists. There is a $(\mathbb{T}, \mathsf{S}_\beta, \mu_{z_0})$ such that for each f in $C(\mathbb{T},\mathbb{C})$, $L = \int_{\mathbb{T}} f(z)\, d\mu_{z_0}(z)$.

6

Topological Vector Spaces

6.1. The Spaces $L^p(X,\mu)$, $1 \le p \le \infty$

Conventions

The topological vector space V (over a topological field $\mathbb{K}$) is locally convex iff each element of some neighborhood base at $\mathbf{O}$, is convex. A subset B of V is bounded iff for every neighborhood U of $\mathbf{O}$ and some positive t, $B \subset tU$. When some open subset of V is bounded, V is locally bounded. A subset S of V is circled iff $\mathbb{K}$=$\mathbb{C}$ and for every complex number z such that $|z| \le 1$ and every $\mathbf{x}$ in S, $z\mathbf{x} \in S$. For a circled bounded neighborhood of U of $\mathbf{O}$, the associated Minkowski functional M_U is:

$$M_U : V \ni \mathbf{x} \mapsto \inf\{\alpha \ : \ \alpha \ge 0, \mathbf{x} \in \alpha U\}.$$

A map $p : V \ni \mathbf{x} \mapsto p(\mathbf{x}) \in [0,\infty)$ is a quasinorm iff for some k in $[1,\infty)$: a) $p(\mathbf{x}+\mathbf{y}) \le k\,(p(\mathbf{x}) + p(\mathbf{y}))$; b) $p(t\mathbf{x}) = |t|\mathbf{x}$. When $k=1$, p is a seminorm. A seminorm p is a norm iff $\{p(\mathbf{x}) = 0\} \Leftrightarrow \{\mathbf{x} = \mathbf{O}\}$. A subset R of V is radial or absorbing iff for every finite subset F of V and some real number $r(F)$, $\lambda R \supset F$ if $|\lambda| \ge r(F)$.

The dual space of V is $[V,\mathbb{C}]$ and is denoted V^*. Elements of a dual space are tagged with a superscript *.

The spaces V and V^* are paired and for $\mathbf{v}$ in V and $\mathbf{v}^*$ in V^*, the value of $\mathbf{v}^*(\mathbf{v})$ is denoted $(\mathbf{v},\mathbf{v}^*)$.

To each T in $[V,W]$ there corresponds in $[W^*,V^*]$ a unique adjoint T^* satisfying (and defined by): $(T\mathbf{x},\mathbf{y}^*) \equiv (\mathbf{x},T^*\mathbf{y}^*)$.

The weak topology for V is $\sigma(V,V^*)$; its basic neighborhoods are

$$U(\mathbf{x};\mathbf{x}_1^*,\ldots,\mathbf{x}_n^*;\epsilon) \stackrel{\text{def}}{=} \{\mathbf{y} \ : \ |(\mathbf{y}-\mathbf{x},\mathbf{x}_k^*)| < \epsilon,\ 1 \le k \le n\},$$

determined by an $\mathbf{x}$ in V, an n in $\mathbb{N}$, $\mathbf{x}_1^*,\ldots,\mathbf{x}_n^*$ in V^*, and a positive ϵ. The weak* topology for V^* is $\sigma(V^*,V)$; its basic neighborhoods are

$$U(\mathbf{x}^*;\mathbf{x}_1,\ldots,\mathbf{x}_n;\epsilon) \stackrel{\text{def}}{=} \{\mathbf{y}^* \ : \ |(\mathbf{x}_k,\mathbf{y}^*-\mathbf{x}^*)| < \epsilon,\ 1 \le k \le n\},$$

determined, mutatis mutandis. For a net $n : \Lambda \ni \lambda \mapsto n(\lambda) \in V$ resp. $n : \Lambda \ni \lambda \mapsto n(\lambda) \in V^*$, $n(\lambda) \stackrel{\text{w}}{\to} \mathbf{x}$ resp. $n(\lambda) \stackrel{\text{w}^*}{\to} \mathbf{x}^*$ indicates the convergence, in the weak resp. weak* topology of V resp. V^*, of the net to $\mathbf{x}$ resp. $\mathbf{x}^*$.

For a subset A resp. A^* of the topological vector space V resp. V^*

$$A^\perp \stackrel{\text{def}}{=} \{\mathbf{x}^* : (\mathbf{a}, \mathbf{x}^*) = 0, \ \mathbf{a} \in A\} \ (\subset V^*)$$

resp.

$$A_\perp \stackrel{\text{def}}{=} \{\mathbf{x} : (\mathbf{x}, \mathbf{a}^*) = 0, \ \mathbf{a}^* \in A^*\} \ (\subset V).$$

When $(V, \| \ \|)$ is a normed vector space, e.g., a Banach space, V is uniformly convex iff for each ϵ in $(0, 2]$ and some positive $\delta(\epsilon)$,

$$\{\{\|\mathbf{x}\|, \|\mathbf{y}\| \leq 1\} \wedge \{\|\mathbf{x} - \mathbf{y}\| \geq \epsilon\}\} \Rightarrow \left\{\left\|\frac{1}{2}(\mathbf{x} + \mathbf{y})\right\| \leq 1 - \delta(\epsilon)\right\}.$$

The set $B \stackrel{\text{def}}{=} \{\mathbf{x}_n\}_{n \in \mathbb{N}}$ in a topological vector space V is a basis for V iff for each $\mathbf{x}$ in V and for a unique sequence $\{a_n(\mathbf{x})\}_{n \in \mathbb{N}}$ in $\mathbb{K}^{\mathbb{N}}$, $\lim_{N \to \infty} \sum_{n=1}^{N} a_n(\mathbf{x})\mathbf{x}_n = \mathbf{x}$ in the topology of V. Associated with a basis are the maps

$$S_N : V \ni \mathbf{x} \mapsto \sum_{n=1}^{N} a_n(\mathbf{x})\mathbf{x}_n, \ N \in \mathbb{N}, \ P_n : V \ni \mathbf{x} \mapsto a_n(\mathbf{x})\mathbf{x}_n, \ n \in \mathbb{N}.$$

When V is a Banach space, according as the topology is norm-induced, weak, or (when V is some W^*) weak*, B is a norm-basis (a Schauder basis), a weak basis, or a weak* basis.

A $\mathbb{K}$-Hamel basis for V is a $\mathbb{K}$-maximal linearly independent subset $H \stackrel{\text{def}}{=} \{\mathbf{x}_\lambda\}_{\lambda \in \Lambda}$ of V. Thus there are unique functions

$$a_\lambda : V \ni \mathbf{x} \mapsto a_\lambda(\mathbf{x}) \in \mathbb{K}, \ \lambda \in \Lambda,$$

such that $\mathbf{x} = \sum_{\lambda \in \Lambda} a_\lambda(\mathbf{x})\mathbf{x}_\lambda$. (For each $\mathbf{x}$, only finitely many a_λ are nonzero.)

When $1 < p < \infty$, $p' \stackrel{\text{def}}{=} \dfrac{p}{p-1}$, i.e., $\dfrac{1}{p} + \dfrac{1}{p'} = 1$. When $p = 1$, $p' \stackrel{\text{def}}{=} \infty$; when $p = \infty$, $p' \stackrel{\text{def}}{=} 1$. For (X, S, μ), two measurable functions f and g in $\mathbb{C}^X$ are equivalent iff $f(x) \doteq g(x)$ (μ). When $1 \leq p < \infty$, $L^p(X, \mu)$ is the set of equivalence classes of measurable functions f such that $\int_X |f(x)|^p \, d\mu(x) \stackrel{\text{def}}{=} \|f\|_p^p < \infty$. When $p = \infty$, the condition $\|f\|_\infty < \infty$ is imposed.

When $L^p(X, \mu)$ is regarded as an $\mathbb{R}$-module, $L^p_{\mathbb{R}}(X, \mu)$ is a closed $\mathbb{R}$-submodule of $L^p(X, \mu)$.

The notation $\ell^p(X)$ is used for $L^p(X, \mu)$ when μ is counting measure ζ. Similarly, the notation $c_0(X)$ is used for $C_0(X, \mathbb{C})$ when the topology of X is discrete.

When X in (X, S, μ) is a topological space, it is assumed to be a Hausdorff space, S is S_β, for each open set U, $\mu(U) > 0$, and for each compact set K, $\mu(K) < \infty$: the situation is symbolized by $(X_{top}, \mathsf{S}_\beta, \mu)$.

For a set X, a subset N of X, and an f in $\mathbb{C}^X$,

$$\operatorname{osc}_N(f) \stackrel{\text{def}}{=} \sup_{a,b \in N} |f(a) - f(b)|.$$

For a topological space X, a set of nonnegative continuous functions $\{\phi_\gamma\}_{\gamma\in\Gamma}$ in $\mathbb{R}^X$ is a partition of unity subordinate to an open cover $\{U_\gamma\}_{\gamma\in\Gamma}$ of X iff a) each ϕ_γ is not identically zero, b) $\phi_\gamma(x) = 0$ off U_γ, c) for each x in X, only finitely many $\phi_\gamma(x)$ are different from zero, and d) $\sum_{\gamma\in\Gamma} \phi_\gamma(x) \equiv 1$.

For a locally compact group G and f, g in $L^1(G, \mu)$, their convolution $f * g$ is $G \ni x \mapsto \int_G f(y)\, g(y^{-1}x)\, d\mu(y) = \int_G f(y) g^{[x]}(y^{-1})\, d\mu(y)$.

The set $\widehat{G} \stackrel{\text{def}}{=} \{\alpha : \alpha \in [G, \mathbb{T}]\}$ of continuous homomorphisms of G into $\mathbb{T}$ is the dual group or group of characters $\alpha : G \ni x \mapsto (\alpha, x) \in \mathbb{T}$ of G. For f in $L^1(G, \mu)$, $\widehat{f} : \widehat{G} \ni \alpha \mapsto \widehat{f}(\alpha) \stackrel{\text{def}}{=} \int_G f(x)\overline{(\alpha, x)}\, d\mu(x)$ is, when G is abelian, the Gelfand-Fourier transform of f.

For vector spaces V, W and h in $[V, W]$, $\ker(h) = h^{-1}(\mathbf{O})$.

6.1. If $f \in L^1(X, \mu)$, and $\int_E f(x)\, d\mu(x) \le a$ whenever $\mu(E) < \infty$, then $\int_X f(x)\, d\mu(x) \le a$. If the hypothesis $f \in L^1(X, \mu)$ is dropped, does the conclusion follow?

6.2. For some (X, S, μ) and for some sequence $\{f_n\}_{n\in\mathbb{N}}$ in $L^1(X, \mu)$, $f_n \stackrel{u}{\to} 0$ and $\int_X f_n(x)\, d\mu(x) \uparrow \infty$.

6.3. If $\{f_n\}_{n\in\mathbb{N}} \subset L^1(X, \mu)$, $0 \le f_n \le 1$, $f_n \stackrel{\text{a.e.}}{\to} 1$, and for some E of finite measure, $f_n(x) \equiv 1$ off E then $\lim_{n\to\infty} \int_X (1 - f_n(x))\, d\mu(x) = 0$.

6.4. For (X, S, μ_i), $i = 1, 2$, such that $X \in \mathsf{S}$, $\mu_1(X) < \infty$, and $p \in [1, \infty)$, and for some f in $L^p(X, \mu_1)$,

$$\{E \in \mathsf{S}\} \Rightarrow \left\{\mu_2(E) = \int_E f(x)\, d\mu_1(x)\right\}$$

iff for some a in $\mathbb{R}$, and all measurable partitions $\{E_n\}_{n\in\mathbb{N}}$ of X,

$$\sum_{n=1}^{\infty} \frac{(\mu_2(E_n))^p}{(\mu_1(E_n))^{p-1}} \le a.$$

6.5. If (X, S, μ) is totally σ-finite then for p in $(1, \infty)$, f in $L^p(X, \mu)$, and $E_t \stackrel{\text{def}}{=} \{x : |f(x)| \ge t\}$, $\|f\|_p^p = p\int_0^\infty t^{p-1}\mu(E_t)\, dt$.

6.6. For (X, S, μ), if: a) $1 \le p < \infty$, b) g is measurable, c) $E_{\neq}(g, 0)$ is σ-finite, and d) $gh \in L^1(X, \mu)$ for all h in $L^p(\mathbb{R}, \lambda)$, then $g \in L^{p'}(X, \mu)$. Without c) the conclusion is invalid.

6.7. For $(\mathbb{R}, \mathsf{S}_\lambda, \lambda)$ and p in $[1, \infty)$, if g is measurable and for all h in $L^p(\mathbb{R}, \lambda)$, $\int_{-T}^{T} g(x)h(x)\,dx$ has a finite limit as $T \to \infty$ then $g \in L^{p'}(\mathbb{R}, \lambda)$.

6.8. For p in $(1, \infty)$, (X, S, μ), and (Y, T, ν), if f in $\mathbb{R}^{X \times Y}$ is such that a) f is $\mathsf{S} \times \mathsf{T}$ measurable, b) for almost every x in X, $f_x \in L^p(Y, \nu)$, c) for almost every y in Y, $f_y \in L^1(X, \mu)$ then

$$\left(\int_Y \left|\int_X f(x,y)\,d\mu(x)\right|^p d\nu(y)\right)^{\frac{1}{p}} \le \int_X \left(\int_Y |f(x,y)|^p\,d\nu(y)\right)^{\frac{1}{p}} d\mu(x).$$

[**Note 6.1:** The result above may be viewed as a "continuous" version of Minkowski's inequality, e.g., when $Y \stackrel{\text{def}}{=} \{1, 2, \ldots, n\}$ and $\nu = \zeta$ (counting measure).]

6.9. If $\{\mathbf{x}_n\}_{n \in \mathbb{N}} \subset \ell^1(\mathbb{N})$, then $\mathbf{x}_n \xrightarrow{\|\ \|_1} \mathbf{x}$ iff $\mathbf{x}_n \xrightarrow{\mathrm{w}} \mathbf{x}$.

6.10. For some f in $C^\infty([0, \infty), \mathbb{C}) \cap L^1([0, \infty), \lambda)$, $\sum_{n=1}^{\infty} f(n) = \infty$.

6.11. For $\{(X, \mathsf{S}, \mu_n)\}_{n=0}^{\infty}$, if $\mu_0(X) = 1 = \lim_{n\to\infty} \mu_n(X)$ and $\mu_n \le \mu_0$ then for a subsequence $\{\mu_{n_k}\}_{k \in \mathbb{N}}$, a measure μ_∞, and all g in $L^1(X, \mu_0)$, $\lim_{k\to\infty} \int_X g(x)\,d\mu_{n_k}(x) = \int_X g(x)\,d\mu_\infty(x)$.

6.12. If f is a nonnegative Lebesgue measurable function on $([0, 1]$ and $E_n \stackrel{\text{def}}{=} \{x : n - 1 \le f(x) < n\}$, $n \in \mathbb{N}$, then $f \in L^1([0, 1], \lambda)$ iff $\sum_{n=1}^{\infty} n\lambda(E_n) < \infty$.

6.13. If

$$f : \mathbb{R} \ni x \mapsto \begin{cases} x^2 \sin\left(\dfrac{1}{x^2}\right) & \text{if } x \neq 0 \\ 0 & \text{if } x = 0 \end{cases},$$

then f' exists everywhere but $f' \notin L^1(\mathbb{R}, \lambda)$.

6.14. True or false: if $f \in L^1([0, n], \infty)$, $n \in \mathbb{N}$, and

$$F(x) \stackrel{\text{def}}{=} \frac{1}{x} \int_0^x f(t)\,dt \ge f(x)$$

a.c. on $(0, \infty)$ then f is a monotonely decreasing function?

6.15. If $f \in L^1([0, 1], \lambda)$ and $\int_0^1 f(x)\,dx \neq 0$ then for some g in $\mathcal{M}$, $\int_0^1 |f(x)| \cdot |g(x)|\,dx < \infty$ and $\int_0^1 |f(x)| \cdot |g(x)|^2\,dx = \infty$.

6.16. If $f \in L^1([0,1],\lambda)$, $f(x) \geq 0$ a.e., and $\int_0^1 (f(x))^n\,dx$ is n-free then for some E in S_λ, $f(x) \doteq \chi_E(x)$. How is the conclusion altered if the assumption: $f(x) \geq 0$ a.e. is dropped?

6.17. If $f \in L^1(\mathbb{R},\lambda)$ and for every open set U such that $\lambda(U) = 1$, $\int_U f(x)\,dx = 0$ then $f(x) \doteq 0$.

6.18. True or false: if $f \in L^1(\mathbb{R},\lambda)$ and $\int_{\mathbb{R}} |x|^n\,|f(x)|\,dx \leq 1, n \in \mathbb{N}$, then $f(x) = 0$ a.e. on $M \stackrel{\text{def}}{=} \{x : |x| \geq 1\}$?

6.19. If $f \in L^1(\mathbb{R},\lambda)$ and for every open set U, $\int_U f(x)\,dx = \int_{\overline{U}} f(x)\,dx$ then $f(x) = 0$ a.e.

6.20. If f in $L^1_{\mathbb{R}}([a,b],\lambda)$ and

$$\{a < c < d < b\} \Rightarrow \left\{\lim_{|h|\downarrow 0} \frac{1}{h}\int_c^d (f(x+h) - f(x))\,dx = 0\right\}$$

then for some constant k, $f(x) \doteq k$.

6.21. If $f \in L^1(\mathbb{R},\lambda)$ and $\|f_{[t]} - f\|_1 \leq |t|^2$ then $f(x) \doteq 0$.

6.22. If $q \in L^1(\mathbb{R},\lambda)$ then for some constant C_q,

$$\{f \in C^\infty_{00}(\mathbb{R},\mathbb{C})\} \Rightarrow \left\{\left|\int_{\mathbb{R}} |q(x)| \cdot |f(x)|^2\,dx\right| \leq C_q \int_{\mathbb{R}} |f(x)|^2 + |f'(x)|^2\,dx\right\}.$$

6.23. If $f \in L^1(\mathbb{R},\lambda)$ then

$$T(f) : \mathbb{R} \ni x \mapsto \begin{cases} f\left(x - \dfrac{1}{x}\right) & \text{if } x \neq 0 \\ 0 & \text{otherwise} \end{cases}$$

is also in $L^1(\mathbb{R},\lambda)$ and $\int_{\mathbb{R}} f(x)\,dx = \int_{\mathbb{R}} T(f)(x)\,dx$.

6.24. If $f, g \in L^1([0,1],\lambda)$ then $f(x) \doteq g(x)$ iff for all h in $C^\infty([0,1],\mathbb{R})$, $\int_0^1 f(x)h(x)\,dx = \int_0^1 g(x)h(x)\,dx$.

6.25. If $f \in L^1([0,\infty),\lambda)$, $g \in \mathcal{M}$, and for t in $[1,\infty)$, $|tg(t)| \leq M < \infty$ then $\lim_{t\to\infty} \dfrac{1}{t}\displaystyle\int_1^t f(s)g(s)\,ds = 0$.

6.26. If $f \in L^1([0,1],\lambda)$ and f is continuous at 0 then for each n in $\mathbb{N}$, $f_n : [0,1] \ni x \mapsto f(x^n)$ is in $L^1([0,1],\lambda)$.

6.27. If f in $L^1(\mathbb{R},\lambda)$ and $f_n \stackrel{\text{def}}{=} f_{[n]} \cdot \chi_{[0,1]}$ then $\left\{\sum_{n=1}^N f_n\right\}_{N=1}^\infty$ is a $\|\ \|_1$-Cauchy sequence.

6.28. For (X, S, μ) totally finite,

$$\left\{\begin{array}{l}\{\{f_n\}_{n\in\mathbb{N}} \subset L^\infty(X,\mu)\} \wedge \{\{g_n\}_{n\in\mathbb{N}} \subset L^1(X,\mu)\} \\ \wedge \{\|f_n\|_\infty \le M < \infty\} \wedge \left\{f_n \overset{\|\ \|_1}{\to} f\right\} \wedge \left\{g_n \overset{\|\ \|_1}{\to} g\right\}\end{array}\right\} \Rightarrow \left\{f_n g_n \overset{\|\ \|_1}{\to} fg\right\}.$$

6.29. If $f_n \ge 0$ and $\lim_{n\to\infty} \int_X f_n(x)\,d\mu(x) = 0$ then

$$\lim_{n\to\infty} \int_X \left(1 - e^{-f_n(x)}\right) d\mu(x) = 0.$$

6.30. If $f_n \in L^1(\mathbb{R}, \lambda)$ and $f_n \overset{\text{a.e.}}{\to} f$ then $f_n \overset{\|\ \|_1}{\to} f$ iff a) for each positive ϵ and some A_ϵ in S_λ, $\lambda(A_\epsilon) < \infty$ and $\sup_{n\in\mathbb{N}} \int_{\mathbb{R}\setminus A_\epsilon} |f_n(x)|\,dx < \epsilon$ and b) $\lim_{\lambda(B)\to 0} \sup_{n\in\mathbb{N}} \int_B |f_n(x)|\,dx = 0$.

6.31. If $\alpha \in (-1,0)$, $S_N \overset{\text{def}}{=} \sum_{n=0}^N \binom{\alpha}{n} x^n$, and $f(x) \overset{\text{def}}{=} (1-x)^\alpha$ then in $L^1([0,1],\lambda)$, $\lim_{N\to\infty} \|S_n - f\|_1 = 0$ (cf. **2.72**).

6.32. If: a) $f > 0$; b) f is monotonely increasing on $\mathbb{R}$; c)

$$\lim_{x\to\infty} \frac{x}{f(x)} = 0, \text{ i.e., } x = o(f(x)),\ \{g_n\}_{n\in\mathbb{N}} \subset \mathcal{M}([0,1])$$

$$g_n \overset{\text{a.e.}}{\to} g, f\circ(|g_n|) \in \mathcal{M}, \int_0^1 f(|g_n(x)|)\,dx \le M < \infty,\ n \in \mathbb{N};$$

then $\{g_n\}_{n\in\mathbb{N}} \subset L^1([0,1],\lambda)$ and $\lim_{n\to\infty} \int_0^1 |g_n(x) - g(x)|\,dx = 0$.

6.33. If $p \in (1,\infty)$ and $\{f_n\}_{n=0}^\infty \subset L^p(\mathbb{R},\lambda)$ then a) and b) following are equivalent: a) $\|f_n\|_p \le K < \infty$ and $\lim_{n\to\infty} \int_0^x f_n(t)\,dt \equiv \int_0^x f_0(t)\,dt$; b) $f_n \overset{\text{w}}{\to} f_0$.

6.34. Is $A \overset{\text{def}}{=} \{f : f \in L^1([0,1],\lambda),\ |f(x)| \ge 1 \text{ a.e.}\}$ $\|\ \|_1$-closed? Is A $\sigma\left(L^1([0,1],\lambda), L^\infty([0,1],\lambda)\right)$-closed?

6.35. If $f \in L^1([0,1],\lambda)$ and $S \overset{\text{def}}{=} \{x : 0 \le x \le 1,\ f(x) \in \mathbb{Z}\}$ then $S \in \mathsf{S}_\lambda$ and $\lim_{n\to\infty} \int_0^1 |\cos \pi f(x)|^n\,dx = \lambda(S)$.

6.36. a) If $g \in L^1(\mathbb{T},\tau)$ then

$$T : C(\mathbb{T},\mathbb{C}) \ni f \mapsto \left\{g * f : \mathbb{T} \ni x \mapsto \int_{\mathbb{T}} g\left(y^{-1}x\right) f(y)\,d\tau(y)\right\} \in C(\mathbb{T},\mathbb{C})$$

is a compact linear endomorphism. b) For $f_n : \mathbb{T} \ni x \mapsto x^n + \dfrac{1}{x^n}$, $n \in \mathbb{N}$, $\lim_{n\to\infty} \|T(f_n) - \mathbf{O}\|_\infty = 0$.

6.37. If $f \in L^1(\mathbb{R}, \lambda)$ and for some n in $\mathbb{N}$, $f(x) \doteq 0$ off $[-n, n]$ then $\widehat{f} \in C_0^\infty(\mathbb{R}, \mathbb{C})$.

6.38. If $f \in L^1(\mathbb{R}, \lambda)$ and $\text{supp}\left(\widehat{f}\right)$ is compact then for some g not identically zero and in $L^1(\mathbb{R}, \lambda) \cap C(\mathbb{R}, \mathbb{C})$, $f * g(x) \doteq 0$.

6.39. If $f \in L^1(\mathbb{R}, \lambda)$ and $f * f = f$ then $f(x) \doteq 0$.

6.40. If $\lambda(E) > 0$, $f \in L^1(\mathbb{R}, \lambda)$, $f \geq 0$, and $\int_E f(x)\,dx = 1$ then $\left|\int_E f(x)e^{-ix}\,dx\right| < 1$.

6.41. If $O \neq f \in L^1(\mathbb{R}, \lambda)$ and $f \geq 0$, then for all nonzero t, $\left|\widehat{f}(t)\right| < \|f\|_1$.

6.42. If $(\mathbb{T}, \mathsf{S}_\beta, \nu)$ is such that $\nu(\mathbb{T}) = 1$ then $\left|\int_{\mathbb{T}} x\,d\nu(x)\right| \leq 1$ and equality obtains iff for some a in $\mathbb{T}$, $\nu(a) = 1$.

6.43. For a locally compact group G and the Haar measure situation $(G, \mathsf{S}_\beta, \mu)$, if $1 < p \leq \infty$, $f \in L^1(G, \mu)$, and $g \in L^p(G, \mu)$ then

$$f * g \in L^p(G, \mu) \text{ and } \|f * g\|_p \leq \|f\|_1 \|g\|_p.$$

For a locally compact group G with identity e, the set $\mathcal{U}(e)$ of neighborhoods of e is a diset with respect to the partial order:

$$\{\mathcal{U}(e) \ni U \prec U' \in \mathcal{U}(e)\} \Leftrightarrow \{U \subset U'\}.$$

6.44. If the net $n : \mathcal{U} \ni U \mapsto n_U \in L^1(G, \mu)$ satisfies the following conditions: a) $n_U \geq 0$ and $n_U(x) = n_U(x^{-1})$; b) for all U, $\|n_U\|_1 = 1$; c) $n_U(x) = 0$ off U; then for p in $[1, \infty)$ and f in $L^p(G, \mu)$, $n_U * f \xrightarrow{\|\ \|_p} f$. [In particular, n is an approximate identity for $L^1(G, \mu)$, regarded as a Banach algebra with respect to convolution as multiplication.]

6.45. If G is a locally compact abelian group and n is an approximate identity for $L^1(G, \mu)$ then $\widehat{n} \xrightarrow{u} 1$.

6.46. a) For t in $(0, \infty)$ what is c_t if $c_t \int_{-\infty}^{\infty} \exp\left(\left(\frac{x}{t}\right)^2\right) dx = 1$? b) For $g_t : \mathbb{R} \ni x \mapsto c_t \exp\left(\left(\frac{x}{t}\right)^2\right)$ and f in $L^p(\mathbb{R}, \lambda)$, $\lim_{t\to 0} \|g_t * f - \|_p = 0$.

6.47. If $f \in L^1(\mathbb{R}, \lambda)$, $f(x) = 0$ off $[-n, n]$, and $\text{supp}\left(\widehat{f}\right)$ is compact then $f \doteq 0$.

6.48. If $g \in L^1(\mathbb{R}, \lambda)$ then $T_g : L^2(\mathbb{R}, \lambda) \ni f \mapsto \int_{\mathbb{R}} g(x-y)f(y)\,dy \stackrel{\text{def}}{=} g * f$ is not compact unless $g \doteq 0$.

6.49. For $\mu : 2^{[1, \infty)} \ni E \mapsto \sum_{x \in E} x$, what is $\left(L^1([1, \infty), \mu)\right)^*$?

6.50. For some nonempty open interval I,

$$\{\{p \in I\} \wedge \{f \in L^1(\mathbb{R}, \lambda)\}\} \Rightarrow \left\{\int_{\mathbb{R}} f(x-y)\frac{|\sin(1/y)|}{|y|^{\frac{1}{2}}}\,dy \in L^p(\mathbb{R}, \lambda)\right\}.$$

6.51. If $p \in [1, \infty]$ and every nonempty open set in X of $(X_{top}, \mathsf{S}_\beta, \mu)$ has positive measure then each equivalence class corresponding to an element of $L^p(X, \mu)$ contains at most one continuous function. For some $(X_{top}, \mathsf{S}_\beta, \mu)$ of the kind described, some equivalence class contains no continuous function.

6.52. If $1 \le p < \infty$, the map

$$f : [0,1]^2 \ni (x,y) \mapsto \begin{cases} (xy-1)^{-1} & \text{if } xy \ne 1 \\ 0 & \text{if } xy = 1 \end{cases}$$

is in $L^p\left([0,1]^2, \lambda_2\right)$ iff $p = 1$.

6.53. If (G, S, μ) is the Haar measure situation for a locally compact group, $p \in [1, \infty]$, $g \in L^p(G, \mu)$, and $n : \Lambda \ni \lambda \mapsto n(\lambda) \in G$ is a net converging to the e in G then $\lim_{n \to e} \left\|g + g_{[n]}\right\|_p = 2\|g\|_p$.

6.54. If $f \in L^p(\mathbb{R}, \lambda)$ and $1 \le p < \infty$ then for some sequence $\{a_n\}_{n \in \mathbb{N}}$ in $(0, \infty)$, $a_n \to 0$ and if $|b_n| < a_n, n \in \mathbb{N}$, $f_{[b_n]} \overset{\text{a.e.}}{\to} f$.

6.55. If $0 \le f, g$, $f \in L^p(\mathbb{R}^n, \lambda_n)$, $g \in L^{p'}(\mathbb{R}^n, \lambda_n)$, and for all positive t, $E_t \overset{\text{def}}{=} \{\mathbf{x} : g(\mathbf{x}) > t\}$ then

$$\int_0^\infty \left(\int_{E_t} f(\mathbf{x})\,d\mathbf{x}\right) dt = \int_{\mathbb{R}^n} f(\mathbf{x})g(\mathbf{x})\,d\mathbf{x}.$$

6.56. If $1 < p < \infty$, $\{a_n\}_{n \in \mathbb{N}}, \{b_n\}_{n \in \mathbb{N}} \subset \mathbb{C}$, and for all n in $\mathbb{N}$,

$$\left|\sum_{n=0}^N a_n b_n\right|^p \le \int_0^1 \left|\sum_{n=0}^N b_n t^n\right|^p dt$$

then for some unique f in $L^{p'}([0,1], \lambda)$, $a_n \equiv \int_0^1 t^n f(t)\,dt$.

6.57. a) If $1 < p < \infty$ and for (X, S, μ), $\mu \ne 0$, what is $\mathcal{E}_p$, the set of extreme points of the unit ball $B(\mathbf{O}, 1)$ of $L^p(X, \mu)$? b) What is the situation when $p = 1$ and μ is nonatomic?

6.58. If $1 < p < \infty$ and A and B are closed subspaces of $L^p(X, \mu)$ then $A = B$ iff $A^\perp = B^\perp$.

6.59. a) For a uniformly convex normed vector space $(Y, \| \ \|)$, p in $(1,\infty)$ and ϵ positive, there is a positive $\delta_p(\epsilon)$ such that if $\|\mathbf{x}\|, \|\mathbf{y}\| \le 1$, and $\|\mathbf{x}-\mathbf{y}\| \ge \epsilon$ then

$$\left\|\frac{\mathbf{x}+\mathbf{y}}{2}\right\|^p \le (1-\delta_p(\epsilon))\left(\frac{\|\mathbf{x}\|^p + \|\mathbf{y}\|^p}{2}\right). \tag{6.1}$$

b) If $1 < p < \infty$, then $L^p(X,\mu)$ is uniformly convex.

6.60. If $1 \le p \le \infty$ and (X, S, μ) is totally finite then for $\{f_n\}_{n=0}^\infty$ in $L^p(X,\mu)$,

$$\left\{\left\{f_n \overset{\text{a.e.}}{\to} f_0\right\} \wedge \left\{\lim_{n\to\infty} \|f_n\|_p = \|f_0\|_p\right\}\right\} \Rightarrow \left\{f_n \overset{\| \|_p}{\to} f_0\right\}.$$

6.61. For (X, S, μ), if $\{E_n\}_{n\in\mathbb{N}} \subset \mathsf{S}$ and $L^1(X,\mu) \ni \chi_{E_n} \overset{\| \|_1}{\to} f$ then for some E in S, $f = \chi_E$.

6.62. If (X, S, μ) is totally finite and $\mathbf{O} \ne f \in L^\infty(X,\mu)$ then

$$\|f\|_\infty = \lim_{n\to\infty} \frac{\|f\|_{n+1}^{n+1}}{\|f\|_n^n}.$$

6.63. For (X, S, μ), $L^\infty(X,\mu) = L^1(X,\mu)$ iff $\dim\left(L^1(X,\mu)\right) < \infty$.

6.64. If (X, S, μ) is σ-finite and $f \in L^1(X,\mu)$ then

$$T_f : L^\infty(X,\mu) \ni g \mapsto \int_X f(x)g(x)\,d\mu(x)$$

is in $(L^\infty(X,\mu))^*$.

6.65. If $k \in L^\infty(\mathbb{R}, \lambda)$ and $\int_{\mathbb{R}} e^{-(x-y)^2} k(y)\,dy \equiv 0$ then $k(x) \doteq 0$.

6.66. If $1 \le p \le \infty$, $S \in [L^p(\mathbb{T}, \tau)]$, and $S\left(f_{[t]}\right) \equiv S(f)_{[t]}$ (S and translation commute), then: a) for all f and g in $L^\infty(\mathbb{T}, \tau)$,

$$S(f * g) = f * S(g) = S(f) * g;$$

b) for some sequence $\{a_n\}_{n\in\mathbb{N}}$ in $\mathbb{C}$ and all f in $L^p(\mathbb{T}, \tau)$, $(\widehat{S(f)})_n = a_n \widehat{f}_n$.

6.67. $(L^\infty([0,1], \lambda))^* \setminus L^1([0,1], \lambda) \ne \emptyset$.

6.68. If $f \in L^\infty([0,1], \lambda)$ and for each x in $[0,1]$, and some g_x in $L^\infty([0,1], \mathbb{C})$, $f(t) \doteq g_x(t)$ and $\lim_{t\to x} g_x(t) \overset{\text{def}}{=} v_x$ exists then for some g in $C([0,1], \mathbb{C})$, $f(t) \doteq g(t)$.

6.2. Hilbert Space $\mathfrak{H}$

Conventions

A vector space $\mathfrak{H} \stackrel{\text{def}}{=} \{\mathbf{x}\}$ over $\mathbb{C}$ is a Hilbert space when there is defined an inner product: $\mathfrak{H}^2 \ni \{\mathbf{x}, \mathbf{y}\} \mapsto (\mathbf{x}, \mathbf{y})$ such that

$$\begin{aligned} &(a\mathbf{x} + b\mathbf{y}, \mathbf{z}) = a(\mathbf{x}, \mathbf{z}) + b(\mathbf{y}, \mathbf{z}) \\ &(\mathbf{x}, \mathbf{y}) = \overline{(\mathbf{y}, \mathbf{x})}, \ (\mathbf{x}, \mathbf{x}) \geq 0, \\ &\left\{(\mathbf{x}, \mathbf{x}) \left(\stackrel{\text{def}}{=} \|\mathbf{x}\|^2\right) = 0\right\} \Leftrightarrow \{\mathbf{x} = \mathbf{O}\} \end{aligned}$$

[$(\cdot, \cdot)$ is conjugate bilinear and positive definite]. It is assumed that $\mathfrak{H}$ is a Banach space in the $\| \ \|$-induced metric. The paradigm for $\mathfrak{H}$ is $L^2(X, \mu)$:

$$\begin{aligned} (f, g) &\stackrel{\text{def}}{=} \int_X f(x)\overline{g(x)}\, d\mu(x); \\ \|f\| &\stackrel{\text{def}}{=} \|f\|_2; \end{aligned}$$

Because $\mathfrak{H}^*$ and $\mathfrak{H}$ are isomorphic, the adjoint T^* of a T in $[\mathfrak{H}]$ is itself in $[\mathfrak{H}]$ and satisfies $(T\mathbf{x}, \mathbf{y}) = (\mathbf{x}, T^*\mathbf{y})$. When $U \in [\mathfrak{H}]_a$ and, to boot, $(U\mathbf{x}, U\mathbf{y}) = (\mathbf{x}, \mathbf{y})$, U is unitary and $U^* = U^{-1}$.

The gradient map is $\nabla : C(\mathbb{R}^n, \mathbb{C}) \ni f \mapsto \left(\frac{\partial f}{\partial x_1}, \ldots, \frac{\partial f}{\partial x_n}\right)$ and the Laplace map is $\Delta : C^2(\mathbb{R}^n, \mathbb{C}) \ni f \mapsto \sum_{k=1}^n \frac{\partial^2 f}{\partial x_k^2}$. For h in $\mathbb{R} \setminus \{0\}$,

$$\Delta_h f(x) \stackrel{\text{def}}{=} \begin{cases} \dfrac{f(x+h) - f(x)}{h} & \text{if } f \in \mathbb{C}^{\mathbb{R}} \\ \dfrac{f\left(xe^{i2\pi h}\right) - f(x)}{h} & \text{if } f \in \mathbb{C}^{\mathbb{T}} \end{cases}.$$

6.69. The span of the image of $\gamma : \left[-\frac{1}{2}, \frac{1}{2}\right] \ni t \mapsto \left\{\frac{1}{n-t}\right\}_{n \in \mathbb{N}}$ is $\| \ \|_2$-dense in $\ell^2(\mathbb{N})$.

6.70. For $F_N : \ell^2(\mathbb{N}) \ni \{a_n\}_{n\in\mathbb{N}} \mapsto \left(\sum_{n\geq N} |a_n|^2\right)^{\frac{1}{2}}$, $F_N \stackrel{\text{u}}{\to} O$ on each $\| \ \|_2$-compact set.

6.71. If $f, g \in C^2(\mathbb{R}^2, \mathbb{C}) \cap L^2(\mathbb{R}^2, \lambda)$, $\Delta f, \Delta g \in L^2(\mathbb{R}^2, \lambda)$, and each component of $\nabla f, \nabla g$ is in $L^2(\mathbb{R}^2, \lambda)$ then

$$\int_{\mathbb{R}^2} f(\mathbf{x})\Delta g(\mathbf{x})\, d\mathbf{x} = \int_{\mathbb{R}^2} g(\mathbf{x})\Delta f(\mathbf{x})\, d\mathbf{x}.$$

6.72. For some curve $\gamma : [0,1] \ni t \mapsto \gamma(t) \in \mathfrak{H}$,

$$\{0 \le t_1 < t_2 \le t_3 < t_4 \le 1\} \Rightarrow \{(\gamma(t_2) - \gamma(t_1)) \perp (\gamma(t_4) - \gamma(t_3))\}.$$

6.73. True or false: if $\int_{[0,\infty)} |f_n'(x)|^2 \, dx \le M^2 < \infty$ and $|xf_n(x)| \le 1$ then the sequence $\{f_n\}_{n\in\mathbb{N}}$ contains: a) a pointwise convergent subsequence? b) a subsequence converging uniformly on $[0,\infty)$? c) a subsequence $\| \ \|_2$-convergent in $L^2([0,\infty),\lambda)$?

6.74. For $E \stackrel{\text{def}}{=} \{(x,y) \ : \ 0 \le |x| \le y \le 1\}$, if $f \in L^2(E,\lambda)$ then

$$A \stackrel{\text{def}}{=} \underline{\lim}_{y=0} \int_{-y}^{y} |f(x,y)| \, dx = 0.$$

6.75. If $\{f_n\}_{n\in\mathbb{N}}$ is $\| \ \|_2$-bounded in $L^2([0,1],\lambda)$ and $f_n \stackrel{\text{meas}}{\to} 0$ then $\lim_{n\to\infty} \|f_n\|_1 = 0$.

6.76. For f, g in $L^2(\mathbb{R},\lambda)$, if $\lim_{h\to 0} \|\Delta_h f - g\|_2 = 0$ then for some constant c, $f(x) \doteq \int_0^x g(t)\,dt + c$.

6.77. For $T \stackrel{\text{def}}{=} L^2([0,1],\lambda) \ni f \mapsto \left\{x \mapsto \int_0^x f(t)\,dt\right\}$ and $P \stackrel{\text{def}}{=} T + T^*$: a) $\|T\| \le 2^{-\frac{1}{2}}$; b) $P^2 = P$ and $P\left(L^2([0,1],\lambda)\right) = \mathbb{C}$; c) T is compact.

6.78. If $S \subset C([0,1],\mathbb{C})$ and S is a closed subspace of $L^2([0,1],\lambda)$ then: a) S is $\| \ \|_\infty$-closed in $C([0,1],\mathbb{C})$; b) for all f in S and some positive M, $\|f\|_2 \le \|f\|_\infty \le M\|f\|_2$; c) for all y in $[0,1]$, some k_y in $L^2([0,1],\lambda)$, and all f in S, $f(y) = \int_0^1 k_y(x)f(x)\,dx$.

6.79. If $f \in L^2(\mathbb{T},\tau)$ and for some positive C and a,

$$\int_{\mathbb{T}} |h\Delta_h f(x)|^2 \, d\tau(x) \le C|h|^{1+a}$$

then $\sum_{n=-\infty}^{\infty} \left|\widehat{f_n}\right| < \infty$.

6.80. The set $A \stackrel{\text{def}}{=} \left\{f \ : \ f \in C^\infty([0,1]), \int_0^1 \frac{f(x)}{x}\,dx = 0\right\}$ is $\| \ \|_2$-dense in $L^2([0,1],\lambda)$.

6.81. If $f \in L^2([0,1],\lambda)$ then $f(t) \equiv t$ iff $\int_0^1 t^n f(t)\,dt = \frac{1}{n+2}, n \in \mathbb{Z}^+$.

6.82. If $\{f_n\}_{n\in\mathbb{N}} \subset L^2(X,\mu)$ and $\|f_n - f_{n+1}\|_2 \le 2^{-n}$ then for some f in $L^2(X,\mu)$, $f_n \stackrel{\text{a.e.}}{\to} f$ and $f_n \stackrel{\| \ \|_2}{\to} f$.

6.83. If $\{f_n\}_{n\in\mathbb{N}}$ is orthonormal in $L^2(X,\mu)$ and $\|f_n\|_\infty \le M < \infty$ then $\sum_{n=1}^{\infty} \frac{|f_n(x)|}{n}$ converges a.e. □

6.84. For an orthonormal sequence $\{\mathbf{x}_n\}_{n\in\mathbb{N}}$ in $\mathfrak{H}$, $\mathbf{x}_n \xrightarrow{w} \mathbf{O}$.

6.85. Let M be the $\| \ \|_2$-closure of

$$S \stackrel{\text{def}}{=} \{ f \ : \ f \in C([-1,1],\mathbb{C}), \ f(x) = f(-x) \}.$$

a) What is a simple formula for the orthogonal projection of $L^2([-1,1],\lambda)$ onto M? b) There are polynomials $p_n, n \in \mathbb{N}$, constituting an orthonormal basis for M.

6.86. For an infinite orthonormal system $\{\mathbf{x}_n\}_{n\in\mathbb{N}}$ in $\mathfrak{H}$ and

$$E \stackrel{\text{def}}{=} \{ \mathbf{x}_m + m\mathbf{x}_n \ : \ n > m, \ m,n \in \mathbb{N} \} :$$

a) $\mathbf{O}$ is in the weak closure E^w of E; b) if $F \subset E$ and F is $\| \ \|$-bounded then $\mathbf{O} \notin F^w$; c) no subsequence of E converges weakly to $\mathbf{O}$.

6.87. a) If $\{\mathbf{x}_n\}_{n\in\mathbb{N}}$ is a complete orthonormal set in $\mathfrak{H}$, $0 \le c \le 2^{-\frac{1}{2}}$, and $\|\mathbf{y}_n - \mathbf{x}_n\| \le c^n$, $n \in \mathbb{N}$, then $\{\mathbf{y}_n\}_{n\in\mathbb{N}}$ is a norm-basis for $\mathfrak{H}$. b) If $\mathbf{x} = \sum_{n=1}^{\infty} a_n\mathbf{y}_n$ then $\sum_{n=1}^{\infty} |a_n|^2$ is finite.

6.88. If $\Phi \stackrel{\text{def}}{=} \{\phi_\gamma\}_{\gamma\in\Gamma}$ and $\Psi \stackrel{\text{def}}{=} \{\psi_\delta\}_{\delta\in\Delta}$ are complete orthonormal sets in $\mathfrak{H}$ then $\#(\Gamma) = \#(\Delta)$.

6.89. If $K \in L^2\left([0,1]^2, \lambda_2\right)$ then for

$$T : L^2([0,1],\lambda) \ni f \mapsto \left\{ x \mapsto \int_0^1 K(x,y)f(y)\,dy \right\} :$$

a) $T \in \left[L^2([0,1],\lambda)\right]$; b) for each positive a and some T_a in $\left[L^2([0,1],\lambda)\right]$, $\operatorname{im}(T_a)$ is finite-dimensional and $\|T - T_a\| < a$.

6.90. If $\dim(\mathfrak{H}) = \infty$ and S is the σ-ring generated by the $\| \ \|$-open sets of $\mathfrak{H}$, there is no $(\mathfrak{H}, \mathsf{S}, \mu)$ such that $0 < \mu(B(\mathbf{O},1)) < \infty$ and μ is translation-invariant.

6.91. If $\dim(\mathfrak{H}) = \infty$, $p \in (0,\infty)$, and ρ^p denotes Hausdorff measure, then $\rho^p(E) = \begin{cases} \infty & \text{if } E \ne \emptyset \\ 0 & \text{otherwise.} \end{cases}$

6.92. The $\| \ \|_1$-induced topology for $L^2([0,1],\lambda) \stackrel{\text{def}}{=} \mathfrak{H}$ is not stronger than $\sigma(\mathfrak{H}, \mathfrak{H}^*)$.

6.93. If M is a closed subspace of $\mathfrak{H} \stackrel{\text{def}}{=} \ell^2(\mathbb{N})$,

$$D \stackrel{\text{def}}{=} \{\{a_n\}_{n\in\mathbb{N}} : \{a_n\}_{n\in\mathbb{N}}, \{na_n\}_{n\in\mathbb{N}} \in \mathfrak{H}\},$$
$$A \stackrel{\text{def}}{=} \{\{a_n\}_{n\in\mathbb{N}} : \{a_n\}_{n\in\mathbb{N}} \in \mathfrak{H}, \{na_n\}_{n\in\mathbb{N}} \in M^\perp\},$$

and $M \cap D = \{\mathbf{O}\}$ then $A \cap M$ is $\| \|_2$-dense in $\mathfrak{H}$.

6.94. The set

$$S \stackrel{\text{def}}{=} \left\{ f : f(x) = \sum_{n=1}^\infty a_n \sin 2n\pi x,\ x \in [0,1],\ \sum_{n=1}^\infty |na_n| \le 1 \right\}$$

is $\| \|_2$-compact in $L^2([0,1],\lambda)$.

6.95. If S in $L^2([0,1],\lambda)$ is the set of f such that for some g in $L^2([0,1],\lambda)$ and some constant c, $f(x) = \int_0^x g(t)\,dt + c$ then $T : S \ni f \mapsto g$ is a well-defined map for which the graph $\mathcal{G} \stackrel{\text{def}}{=} \{\{f, Tf\} : f \in S\}$ is closed in the $\| \|_2$-induced product topology of $\mathfrak{H}^2$.

6.96. If (X, S, μ) is totally finite, M is a subspace of $L^2(X,\mu)$, and for all f in M, $|f(x)| \le C\|f\|_2$ then $\dim(M) \le C^2\mu(X)$ $(< \infty)$.

6.97. If $f \in L^\infty(X,\mu)$ then: a) the operator norm of the map

$$T_f : L^2(X,\mu) \ni g \mapsto fg \ (\in L^2(X,\mu))$$

is $\|f\|_\infty$ iff X contains no atoms having infinite measure, by abuse of language, no infinite atoms; b) if X contains no infinite atoms and $\|k\|_\infty = \infty$ then $k \cdot L^2(X,\mu) \not\subset L^2(X,\mu)$.

6.98. What are useful necessary and sufficient conditions that T_f in **6.97** be surjective?

6.99. In **6.97** when $X = [0,1]$ and $\mu = \lambda$: a) if $f(t) \equiv t$, T_f has no eigenvalues; b) if $S \in [\mathfrak{H}]$ and for all f in $L^\infty([0,1],\lambda)$, $ST_f = T_fS$ then for some g in $L^\infty([0,1],\lambda)$, $S = T_g$.

6.100. For $\mathbf{a} \stackrel{\text{def}}{=} \{a_n\}_{n\in\mathbb{N}}$ in $\ell^\infty(\mathbb{N})$ and

$$T_{\mathbf{a}} : \ell^2(\mathbb{N}) \ni \mathbf{x} \stackrel{\text{def}}{=} \{x_n\}_{n\in\mathbb{N}} \mapsto \{a_nx_n\}_{n\in\mathbb{N}} \stackrel{\text{def}}{=} T_{\mathbf{a}}\mathbf{x} :$$

a) $T_{\mathbf{a}} \in [L^2(\mathbb{N})]$, $\|T_{\mathbf{a}}\| = \|\mathbf{a}\|_\infty$; b) λ is an eigenvalue of $T_{\mathbf{a}}$ iff for some n_0, $\lambda = a_{n_0}$; c) if $a_n \not\to 0$ then $T_{\mathbf{a}}(B(0,1))$ is not norm-compact.

6.101. If $\|\mathbf{x}_n\| \to \|\mathbf{x}_0\|$ as $n \to \infty$ and $\mathbf{x}_n \stackrel{\text{w}}{\to} \mathbf{x}_0$ then $\mathbf{x}_n \stackrel{\| \|}{\to} \mathbf{x}_0$.

6.102. If $\{\mathbf{x}_n, \mathbf{y}_n\}_{n\in\mathbb{N}}$ is a subset of the unit ball in $\mathfrak{H}$ then

$$\left\{\lim_{n\to\infty}(\mathbf{x}_n, \mathbf{y}_n) = 1\right\} \Rightarrow \left\{\lim_{n\to\infty}\|\mathbf{x}_n - \mathbf{y}_n\| = 0\right\}.$$

6.103. If $\mathsf{B} : \mathfrak{H}^2 \ni \{\mathbf{x}, \mathbf{y}\} \mapsto \mathsf{B}(\mathbf{x}, \mathbf{y}) \in \mathbb{C}$ is a conjugate bilinear form on $\mathfrak{H}^2$ and $|\mathsf{B}(\mathbf{x}, \mathbf{y})| \le \|\mathbf{x}\|\,\|\mathbf{y}\|$ then for some T in $[\mathfrak{H}]$, $\mathsf{B}(\mathbf{x}, \mathbf{y}) = (\mathbf{x}, T\mathbf{y})$ and $\|T\| \le 1$.

6.104. The topology $\sigma(\mathfrak{H}, \mathfrak{H}^*)$ is a Hausdorff topology.

6.105. For the topology $\sigma(\mathfrak{H}, \mathfrak{H}^*)$, $\mathfrak{H}$ is of the second category iff $\dim(\mathfrak{H})$ is finite.

6.106. If M is a closed subspace of $\mathfrak{H}$ and $\mathbf{x} \in \mathfrak{H}$, then

$$\inf\{\|\mathbf{y} - \mathbf{x}\| \;:\; \mathbf{y} \in M\} = \sup\{|(\mathbf{z}, \mathbf{x})| \;:\; \mathbf{z} \in M^\perp \cap \partial B(\mathbf{O}, 1)\}.$$

6.107. If $f \in L^2(\mathbb{T}, \tau)$ and $\widehat{f}_n \ne 0$, $n \in \mathbb{Z}$, then

$$M \stackrel{\text{def}}{=} \overline{\text{span}\left(\{f_{[t]}\}_{t\in\mathbb{T}}\right)} = L^2(\mathbb{T}, \tau).$$

6.108. For $L^2([0,1], \lambda) \stackrel{\text{def}}{=} \mathfrak{H}$, the domain of the differentiation operator D is, by definition, the set of functions f such that $f \in \mathfrak{H}$, Df exists a.e., and $Df \in \mathfrak{H}$. The domain S of D is dense and the graph of D is closed.

6.109. For $\mathbf{f} : \mathbb{R} \ni x \mapsto \mathbf{f}(x) \in \mathfrak{H}$, if $F(x) \stackrel{\text{def}}{=} (\mathbf{f}(x), \mathbf{y})$ is, for each $\mathbf{y}$, differentiable with respect to x then for all x_0 in $\mathbb{R}$, $\lim_{x\to x_0}\|\mathbf{f}(x) - \mathbf{f}(x_0)\| = 0$.

6.3. Abstract Topological Vector Spaces

Conventions

In a topological vector space V over a topological field $\mathbb{K}$ (usually $\mathbb{C}$, occasionally $\mathbb{R}$), the maps $\mathbb{K} \times V \ni \{a\mathbf{x}\} \mapsto a\mathbf{x}$, $V \times V \ni \{\mathbf{x}, \mathbf{y}\} \mapsto \mathbf{x} + \mathbf{y}$ are assumed to be continuous.

For a set $\{V_\lambda\}_{\lambda\in\Lambda}$ of topological vector spaces, the direct sum $\bigoplus_{\lambda\in\Lambda} V_\lambda$ is the set $\mathsf{X}_{\lambda\in\Lambda} V_\lambda$ endowed with the weakest topology for which each map

$$P_\mu : \mathsf{X}_{\lambda\in\Lambda} V_\lambda \ni \{\mathbf{x}_\lambda\}_{\lambda\in\Lambda} \mapsto \mathbf{x}_\mu \in V_\mu$$

is continuous. Vector addition and multiplication by scalars are performed componentwise.

When W is a closed subspace of V, the quotient topology for V/W is the strongest topology for which the map $V \ni \mathbf{x} \mapsto \mathbf{x}/W$ is continuous.

When V is a normed space or a Banach space, so are W and V/W, and the quotient topology for V/W is induced by the quotient norm

$$\| \ \|_Q : V/W \ni \mathbf{x}/W \mapsto \inf \{ \|\mathbf{y}\| \ : \ \mathbf{y} - \mathbf{x} \in W \}.$$

When V is a Banach space, so is V^*, $F_{\mathbf{x}} : V^* \ni \mathbf{x}^* \mapsto (\mathbf{x}, \mathbf{x}^*)$ is in $(V^*)^* \stackrel{\text{def}}{=} V^{**}$, and $\Theta : V \ni \mathbf{x} \mapsto F_{\mathbf{x}}$ is the isometry mapping V into V^{**}; V is reflexive iff Θ is surjective.

For a map $\mathbf{f} : V \mapsto W$ between normed vector spaces and an $\mathbf{x}$ in V, when for some $\mathbf{df}(\mathbf{x})$ in $[V, W]$,

$$\lim_{\substack{\mathbf{h} \neq \mathbf{O} \\ \|\mathbf{h}\| \to 0}} \frac{\|\mathbf{f}(\mathbf{x} + \mathbf{h}) - \mathbf{f}(\mathbf{x}) - \mathbf{df}(\mathbf{x})(\mathbf{h})\|}{\|\mathbf{h}\|} = 0$$

then $\mathbf{df}(\mathbf{x})$ is the differential of $\mathbf{f}$ at $\mathbf{x}$.

6.110. For the Banach space X let $\mathcal{K}(X)$ be the set of all compact elements of $[X]$. If $TK = KT$ for all K in $\mathcal{K}(\mathfrak{H})$, then for some constant c, $T = c \cdot \text{id}$.

6.111. For Banach spaces V, W and $\{T_n\}_{n \in \mathbb{N}}$ in $[V, W]$, if, for all $\{\mathbf{v}, \mathbf{w}^*\}$ in $V \times W^*$, $\sup_{n \in \mathbb{N}} |(T_n \mathbf{v}, \mathbf{w}^*)| < \infty$ then $\sup_{n \in \mathbb{N}} \|T_n\| < \infty$.

6.112. For a Banach space V if $T \in V^V$ resp. $S \in (V^*)^{V^*}$ and for all $\{\mathbf{v}, \mathbf{v}^*\}$ in $V \times V^*$, $(T(\mathbf{v}), \mathbf{v}^*) = (\mathbf{v}, S(\mathbf{v}^*))$ then $T \in [V]$ resp. $S \in [V^*]$ and $S = T^*$.

6.113. Any two $\mathbb{C}$-Hamel bases for a vector space V are of the same cardinality.

6.114. If V is an infinite-dimensional Banach space and H is a $\mathbb{C}$-Hamel basis for V then $\#(H) \geq \mathfrak{c}$.

6.115. If $H \stackrel{\text{def}}{=} \{\mathbf{x}_\lambda\}_{\lambda \in \Lambda}$ is an $\mathbb{R}$-Hamel basis for an infinite-dimensional Banach space V then at least one of the (linear) maps a_λ is discontinuous.

6.116. If T is a not necessarily continuous homomorphism of the Banach space V into the Banach space W and $\ker(T) \stackrel{\text{def}}{=} T^{-1}(\mathbf{O})$ is closed then true or false: a) T is continuous? b) T is open?

6.117. Let W be a closed subspace of the Banach space V. If $\mathbf{x} \in V \setminus W$ and $a > 0$ then for some $\mathbf{z}^*$ in V^*, $\|\mathbf{z}^*\| \leq 1$ and

$$(\mathbf{x}, \mathbf{z}^*) > \inf \{ \|\mathbf{z} - \mathbf{w}\| \ : \ \mathbf{w} \in W \} - a \stackrel{\text{def}}{=} d(\mathbf{x}, W) - a.$$

6.118. If V is a Banach space, there is no involution

$$\# : V \ni \mathbf{x} \mapsto \mathbf{x}^{\#} \in V$$

such that: a) $\left(\mathbf{x}^{\#}\right)^{\#} = \mathbf{x}$ (# has period two); b) $(a\mathbf{x} + b\mathbf{x})^{\#} = \overline{a}\mathbf{x}^{\#} + \overline{b}\mathbf{y}^{\#}$ (# is conjugate linear); c) for all $\{\mathbf{x}, \mathbf{x}^*\}$ in $V \times V^*$, $\left(\mathbf{x}^{\#}, \mathbf{x}^*\right) = \overline{(\mathbf{x}, \mathbf{x}^*)}$.

6.119. If V is a Banach space and $M \subset E^*$ then $(M_\perp)^\perp$ is the weak* closure of span(M).

6.120. If M is a closed subspace of a Banach space V then $\left(M^\perp\right)_\perp = M$.

6.121. For some Banach space V, V^* contains a $\| \ \|$-closed subspace M such that $M \subsetneqq (M_\perp)^\perp$.

6.122. a) If K is a convex subset of a Banach space and K is $\| \ \|$-closed then K is weakly closed. b) If K is a convex subset of a Banach space then the $\| \ \|$-closure and weak closure of K are the same.

6.123. If W is a finite-dimensional subspace of a normed vector space V then W is closed and $[V, W]_e$ contains an idempotent P.

6.124. If f is a linear but not necessarily continuous map of a Banach space V into $\mathbb{C}$ then either $\ker(f) \stackrel{\text{def}}{=} f^{-1}(0)$ is closed or $\ker(f)$ is a dense proper subset of V (cf. **6.116**).

6.125. If a) V and W are Banach spaces, b) M is a subspace of W, c) $\dim(W/M) < \infty$, and d) for some T in $[V, W]$, $T(V) = M$ then M is closed. The conclusion can fail if d) is dropped.

6.126. If V is a Banach space then:

$$\begin{aligned}&\text{a) } \{V^* \text{ is separable}\} \Rightarrow \{V \text{ is separable}\};\\ &\text{b) } \{V \text{ is separable}\} \not\Rightarrow \{V^* \text{ is separable}\}.\end{aligned}$$

6.127. If V is a Banach space and V^* is separable then the σ-algebra generated in 2^{V^*} by the $\| \ \|$-open sets of V^* is the same as the σ-algebra generated by the weak* open sets of V^*.

6.128. If V is a separable infinite-dimensional Banach space then V contains a dense linearly independent subset.

6.129. Let V be a Banach space containing a sequence $\{\mathbf{v}_n\}_{n\in\mathbb{N}}$ such that for all $\mathbf{v}^*$ in V^*, $\sum_{n=1}^\infty |(\mathbf{v}_n, \mathbf{v}^*)| < \infty$. True or false: if $\{a_n\}_{n\in\mathbb{N}}$ is in $c_0(\mathbb{N})$ then $\left\{\sum_{n=1}^N a_n\mathbf{v}_n\right\}_{N\in\mathbb{N}}$ is a $\| \ \|$-convergent?

6.130. If $\{\mathbf{x}_n\}_{n\in\mathbb{N}}$ is a dense subset of $B(\mathbf{O}, 1)$ in the Banach space V then: a) $T : \ell^1(\mathbb{N}) \ni \{b_n\}_{n\in\mathbb{N}} \mapsto \sum_{n=1}^\infty b_n\mathbf{x}_n$ is in $\left[\ell^1(\mathbb{N}), V\right]_e$; b) $\ell^1(\mathbb{N})/\ker(T)$ endowed with its quotient norm is isometrically isomorphic to V.

6.131. If V is a Banach space, $T \in [V]$, $\|T\| < 1$, and $\mathbf{y} \in V$ then the equation $\mathbf{x} = T\mathbf{x} + \mathbf{y}$ has a solution.

6.132. What is the adjoint S^* of

$$S : \ell^2(\mathbb{N}) \ni \mathbf{x} \stackrel{\text{def}}{=} (x_1, x_2, \ldots) \mapsto (0, x_1, x_2, \ldots)?$$

Although $\|(S^*)^n\| \equiv 1$, for each $\mathbf{x}$, $(S^*)^n \mathbf{x} \stackrel{\| \|_2}{\to} \mathbf{O}$.

6.133. If $\{M_n\}_{n\in\mathbb{N}}$ is a sequence of closed subspaces of a Banach space V and $M \stackrel{\text{def}}{=} \bigcup_{n\in\mathbb{N}} M_n$ is a closed subspace of V then for some n_0, $M = M_{n_0}$.

6.134. If V is a normed vector space over $\mathbb{C}$ then V is a Banach space iff

$$\left\{\sum_{n=1}^{\infty} \|\mathbf{x}_n\| < \infty\right\} \Rightarrow \left\{\lim_{N\to\infty} \sum_{n=1}^{N} \mathbf{x}_n \text{ exists}\right\}. \tag{6.2}$$

6.135. For a $\| \|$-compact subset A of a Banach space V and K the $\| \|$-closure of the convex hull $\text{Conv}(A)$ of A: a) K is $\| \|$-compact; b) for $\mathbf{x}^*$ in V^*, $|(\mathbf{x}, \mathbf{x}^*)| \,\big|_K$ achieves its maximum value on K; c) for each $\mathbf{x}$ in K there is defined on $\mathsf{S}(A)$ a Borel measure $\mu_{\mathbf{x}}$, such that $\|\mu_{\mathbf{x}}\| = 1$ and for all $\mathbf{x}^*$ in V^*, $(\mathbf{x}, \mathbf{x}^*) = \int_A (\mathbf{y}, \mathbf{x}^*)\, d\mu_{\mathbf{x}}(\mathbf{y})$.

6.136. For Banach spaces V, W, Z and a bilinear map $\mathsf{B} : V \times W \mapsto Z$, B is (jointly) continuous: a) iff for some constant C, $\|\mathsf{B}(\mathbf{v}, \mathbf{w})\| \le C\|\mathbf{v}\|\,\|\mathbf{w}\|$; b) iff B is continuous separately in each argument.

6.137. If V and W are Banach spaces, $\mathbf{f} \in W^V$, and $\mathbf{df} \equiv \mathbf{O}$ then $\mathbf{f}$ is a constant.

6.138. For a Banach space V, a linear map $T : V \mapsto V$ is continuous iff

$$\left\{\mathbf{x}_n \stackrel{\text{w}}{\to} \mathbf{O}\right\} \Rightarrow \left\{T(\mathbf{x}_n) \stackrel{\text{w}}{\to} \mathbf{O}\right\}. \tag{6.3}$$

6.139. If V is a normed vector space, $\mathbf{x}^* \in V^*$, and $M = \ker(\mathbf{x}^*)$, then $d(\mathbf{x}, M) \cdot \|\mathbf{x}^*\| \equiv |(\mathbf{x}, \mathbf{x}^*)|$.

6.140. If $B \stackrel{\text{def}}{=} \{\mathbf{x}_n\}_{n\in\mathbb{N}}$ is a Schauder basis for a Banach space V then: a) the maps $\mathbf{x}_n^* : V \ni \mathbf{x} \mapsto a_n(\mathbf{x})$ are in V^*; b) $\{\mathbf{x}_n, \mathbf{x}_n^*\}_{n\in\mathbb{N}}$ is a maximal biorthogonal set in $V \times V^*$; c) $S_N \in [V]$, $S_N^2 = S_N$; d) $P_n \in [V]$, $P_n^2 = P_n$, $n \in \mathbb{N}$; e) there are constants S and P such that $\|S_N\| \le S$, $N \in \mathbb{N}$, $\|P_n\| \le P, n \in \mathbb{N}$.

6.141. If $\{\mathbf{x}_n, \mathbf{x}_n^*\}_{n\in\mathbb{N}}$ is a biorthogonal set then for N in $\mathbb{N}$, the maps $S_N : V \ni \mathbf{x} \mapsto \sum_{n=1}^{N} (\mathbf{x}, \mathbf{x}_n^*)\, \mathbf{x}_n$ are in $[V]$. Furthermore, $X \stackrel{\text{def}}{=} \{\mathbf{x}_n\}_{n\subset\mathbb{N}}$ is a $\| \|$-basis iff $L \stackrel{\text{def}}{=} \text{span}\left(\{\mathbf{x}_n\}_{n\in\mathbb{N}}\right)$ is dense in V and for some M,

$$\|S_N\| \le M < \infty. \tag{6.4}$$

6.142. The set $X \stackrel{\text{def}}{=} \{\mathbf{x}_n\}_{n\in\mathbb{N}}$ is a basis for a Banach space V iff: a) no $\mathbf{x}_n$ is $\mathbf{O}$; b) $L \stackrel{\text{def}}{=} \operatorname{span}(X)$ is dense in V; c) for some finite K

$$\{\{m \le n\} \wedge \{\{a_k\}_{k=1}^n \subset \mathbb{C}\}\} \Rightarrow \left\{\left\|\sum_{k=1}^m a_k\mathbf{x}_k\right\| \le K\left\|\sum_{k=1}^n a_k\mathbf{x}_k\right\|\right\}. \tag{6.5}$$

6.143. A $\sigma(V, V^*)$-basis $X \stackrel{\text{def}}{=} \{\mathbf{x}_n\}_{n\in\mathbb{N}}$ for a Banach space V is a $\| \ \|$-basis for V.

6.144. If $X \stackrel{\text{def}}{=} \{\mathbf{x}_n\}_{n\in\mathbb{N}}$ is a basis for a Banach space V then for some sequence $\{\epsilon_n\}_{n\in\mathbb{N}}$ in $(0,\infty)$,

$$\{\|\mathbf{y}_n - \mathbf{x}_n\| \le \epsilon_n,\ n \in \mathbb{N}\} \Rightarrow \left\{\{\mathbf{y}_n\}_{n\in\mathbb{N}} \text{ is a } \| \ \|\text{-basis for } V\right\}.$$

6.145. For some infinite-dimensional normed vector space V, there is a countable $\mathbb{R}$-Hamel basis.

6.146. Let $\{x_n\}_{n\in\mathbb{N}}$ be a Schauder basis for a Banach space V. For $t \stackrel{\text{def}}{=} \sum_{n=1}^\infty \epsilon_n(t)2^{-n}$ (the binary representation of t in $[0,1]$) and

$$C(\mathbf{x}) \stackrel{\text{def}}{=} \left\{t \ :\ t \in [0,1],\ \sum_{n=1}^\infty \epsilon_n(t)\,(\mathbf{x},\mathbf{x}_n^*)\,\mathbf{x}_n \ \| \ \|\text{-converges}\right\}:$$

a) $C(\mathbf{x})$ is Lebesgue measurable; b) $C \stackrel{\text{def}}{=} \bigcap_{\mathbf{x}\in V} C(\mathbf{x})$ is Lebesgue measurable and dense in $[0,1]$; c) $\lambda(C) = 0$ or $\lambda(C) = 1$; d) $\lambda(C) = 0$ or $C = [0,1]$.

6.147. Let k be a measurable function in $\mathbb{R}^{\mathbb{R}^n}$ and for a in $(0,n)$, some c, and all $\mathbf{x}$, assume $|k(\mathbf{x})| \le c\|\mathbf{x}\|^{a-n}, n = 1,2$. For the map

$$K : C_{00}(\mathbb{R}^n,\mathbb{R}) \ni f \mapsto k * f,$$

b positive, and $\mathbf{y}$ fixed: a) if $1 - a/n < 1/q$ then

$$\int_{B(\mathbf{O},b)} |k(\mathbf{x}-\mathbf{y})|^q\, d\lambda_n(\mathbf{x}) \stackrel{\text{def}}{=} M_q < \infty;$$

b) if $1 - a/n < 1/q < 1$ then K has a unique extension to $L^1(B(\mathbf{O},b),\lambda_n)$, and the extension is in $\left[L^1(B(\mathbf{O},b),\lambda_n), L^q(B(\mathbf{O},b),\lambda_n)\right]$.

6.148. For a positive let E_a be

$$\left\{f \ :\ f \in \mathbb{C}^{\mathbb{C}}, f(0) = 0, \|(f)\| \stackrel{\text{def}}{=} \sup_{s\ne t} \frac{|f(s)-f(t)|}{|s-t|^a} < \infty\right\}.$$

As a function space, $(E_a, \| \ \|)$ is a Banach space.

6.149. For

$$F \stackrel{\text{def}}{=} \left\{ f \ : \ f \in \mathbb{C}^{[0,1]},\ f(0) = 0,\ \|f\| \stackrel{\text{def}}{=} \sup_{s \neq t} \frac{|f(s) - f(t)|}{|s - t|} < \infty \right\}$$

and $\|f\|' \stackrel{\text{def}}{=} \|f\|_\infty + \|f\|$, there is no constant K satisfying $\|f\|' \leq K\|f\|$ for all f in F.

6.150. The set $K \stackrel{\text{def}}{=} \left\{ f \ : \ f \in \mathbb{C}[x], \deg(f) \leq k, \int_0^1 |f(x)|\,dx \leq 1 \right\}$ is a $\| \ \|$-compact subset of $L^1([0,1], \lambda)$.

6.151. Both $L^1(\mathbb{R}, \lambda) \cap C_0(\mathbb{R}, \mathbb{C})$ and $\left(L^1(\mathbb{R}, \lambda)\right)^\frown$ are $\| \ \|_\infty$-dense in $C_0(\mathbb{R}, \mathbb{C})$.

6.152. For m in $\mathbb{N}$, let P_m be $\{ p \ : \ p \in \mathbb{R}[x],\ \deg(p) \leq m \}$. If

$$L : P_{2n+1} \mapsto \mathbb{R}$$

is a linear functional such that $L(p^2) > 0$ when $p \neq 0$ and $\deg(p) \leq n$, then for some p_{n+1} in P_{n+1}: a) for all q in P_n, $L(p_{n+1} \cdot q) = 0$; b) p_{n+1} has $n+1$ different real zeros; c) if a) and b) obtain for some $\widetilde{p_{n+1}}$ then $\widetilde{p_{n+1}}$ is a constant multiple of p_{n+1}.

6.153. By abuse of notation, $V \stackrel{\text{def}}{=} L^{\frac{1}{2}}([0,1], \lambda)$ is a topological vector space with respect to the weakest topology making the map

$$V \ni f \mapsto \int_0^1 |f(x)|^{\frac{1}{2}}\,dx \stackrel{\text{def}}{=} \|f\|_{\frac{1}{2}}^{\frac{1}{2}}$$

continuous. It follows that $V^* = \{\mathbf{O}\}$.

6.154. For $(X, \mathsf{S}_\beta, \mu)$, $\left(L^2(X, \mu)\right)^n$ normed according to

$$\|\{f_1, \ldots, f_n\}\| \stackrel{\text{def}}{=} \left(\sum_{k=1}^n \|f_k\|_2^4 \right)^{\frac{1}{4}}$$

is a Banach space V. What is V^*?

6.155. If V is a Banach space and $T \in \left[V, \ell^1(\mathbb{N})\right]_e$ then: a) for some P in $[V]$, $P^2 = P$ and $P(V) = \ker(T)$; b) for the product topology of the direct sum, $F : \ker(T) \oplus \{(\mathrm{id} - P)(V)\} \ni \{\mathbf{x}, \mathbf{y}\} \mapsto \mathbf{x} + \mathbf{y} \in V$ is a continuous open map.

6.156. If V is a normed vector space, $\{\mathbf{y}_k\}_{k=1}^K$ is a finite subset of V, and M is a $\| \ \|$-closed subspace of V then $\operatorname{span}\left(\{\mathbf{y}_k\}_{k=1}^K, M\right)$ is closed.

6.157. a) If $\{a_n\}_{n\in\mathbb{N}} \in \ell^1(\mathbb{N})$, for (X, S, μ), $\{f_n\}_{n\in\mathbb{N}} \subset L^1(X,\mu)$, and $\|f_n\|_1 \le M < \infty$ then $F : [0,1] \ni x \mapsto \sum_{n=1}^{\infty} a_n f_n(x)$ is in $L^1([0,1],\lambda)$.

b) For what pairs $\left(\{a_n\}_{n\in\mathbb{N}}, \{f_n\}_{n\in\mathbb{N}}\right)$ in $\mathbb{C}^{\mathbb{N}} \times \left(L^1([0,1]\lambda)\right)^{\mathbb{N}}$ is

$$\|F\|_1 = \sum_{n=1}^{\infty} |a_n| \cdot \|f_n\|_1 \tag{6.6}$$

valid?

c) For what $\{a_n\}_{n\in\mathbb{N}}$ does (6.6) hold for each sequence $\{f_n\}_{n\in\mathbb{N}}$ contained in $L^1([0,1],\lambda)$?

d) For what sequences $\{f_n\}_{n\in\mathbb{N}}$ does (6.6) hold for each $\{a_n\}_{n\in\mathbb{N}}$ in $(\ell^1(\mathbb{N}))$?

6.158. If V and W are separable infinite-dimensional Banach spaces, there is an isomorphism $T : V \mapsto W$. [Absent further hypotheses, T need not be continuous. If V and W are Hilbert spaces, some such T is continuous.]

6.159. The set

$$V \stackrel{\text{def}}{=} \left\{ f \ : \ f \in C([0,1],\mathbb{R}),\ \int_0^{\frac{1}{2}} f(x)\,dx - \int_{\frac{1}{2}}^1 f(x)\,dx = 1 \right\}$$

is a $\| \ \|_\infty$-closed convex subset of $C([0,1],\mathbb{R})$ and no element in V is of minimal norm.

6.160. If E is a locally convex topological vector space and U is a convex circled neighborhood of $\mathbf{O}$ then the Minkowski functional M_U is a seminorm.

6.161. If p is a seminorm for a topological vector space E then the following statements are equivalent: a) p is continuous at $\mathbf{O}$; b) p is uniformly continuous; c) $U \stackrel{\text{def}}{=} \{\mathbf{x} : p(\mathbf{x}) < 1\}$ is a convex, open, and circled neighborhood of $\mathbf{O}$; d) p is the Minkowski functional of U.

6.162. (Kolmogorov) A topological vector space E has a norm-induced topology iff E contains a bounded and convex neighborhood of $\mathbf{O}$.

6.4. Banach Algebras

Conventions

A Banach algebra is a Banach space $(A, \| \ \|)$ and a $\mathbb{C}$-algebra in which multiplication $A^2 \ni \{\mathbf{x},\mathbf{y}\} \mapsto \mathbf{xy} \in A$ is separately continuous. In consequence, it may be assumed that $\|\mathbf{xy}\| \le \|\mathbf{x}\|\,\|\mathbf{y}\|$.

An ideal I in A is a proper subalgebra that is A-invariant: $AI \cup IA \subset I$. A regular or modular ideal I is one such that the quotient algebra A/I contains an identity e (thus when $u/I = e$, u is identity modulo I). Ideals in the category of Banach algebras are closed unless the contrary is indicated.

When A is commutative, $\mathcal{M}$ is the set of regular maximal ideals of A and the radical $\mathcal{R}(A)$ of A is the intersection of all its regular maximal ideals: $\mathcal{R}(A) = \bigcap_{M\in\mathcal{M}} M$.

A generalized nilpotent in a Banach algebra A is an $\mathbf{x}$ for which $\lim_{n\to\infty} \|\mathbf{x}^n\|^{\frac{1}{n}} = 0$. When A is commutative, $\mathcal{R}(A)$ is the set of generalized nilpotents.

By definition, $[A, \mathbb{C}]$ consists of the continuous nonzero algebra $\mathbb{C}$-homomorphisms.

When G is a locally compact group, the Banach space $L^1(G, \mu)$ is a Banach algebra $A(G)$ with respect to convolution as multiplication. When X is a locally compact Hausdorff space, the Banach space $C_0(X, \mathbb{C})$ is a Banach algebra $A(X)$ with respect to pointwise multiplication.

A derivation $D : A \ni \mathbf{x} \mapsto D\mathbf{x}$ is an element of $[A]$ that satisfies the product rule for derivatives: $D(\mathbf{xy}) = D(\mathbf{x})\mathbf{y} + \mathbf{x}D(\mathbf{y})$.

The disc $\{ z : |z - a| \le r \}$ in $\mathbb{C}$ is denoted $D(a, r)$. For a region Ω in $\mathbb{C}$, $H(\Omega)$ is the set of functions f in $\mathbb{C}^\Omega$ and holomorphic in Ω.

6.163. If $\mathbf{n}$ is a bounded approximate identity for a Banach algebra A then for each k in $\mathbb{N}$, $\mathbf{n}^k$ is also an approximate identity.

6.164. For the norm $\| \ \|_\infty$, the set $A(U)$ of functions holomorphic in $U \stackrel{\text{def}}{=} \{ z : z \in \mathbb{C}, |z| < 1 \}$ and continuous on $\{ z : z \in \mathbb{C}, |z| \le 1 \}$ is a Banach algebra with respect to pointwise multiplication of functions. If $f \in A(U)$ and $g \stackrel{\text{def}}{=} f|_{\mathbb{T}}$ then: a) $\widehat{g}_n = 0$ if $n < 0$; b) for some sequence $\{p_n\}_{n\in\mathbb{N}}$ of polynomials, $p_n \stackrel{\text{u}}{\to} f$ on $D(0, 1)$.

6.165. For $A(D(0,1)^\circ)$ as in **6.164**, if $h \in [A, \mathbb{C}]$ then for some z_h in $D(0, 1)$, $h(f) \equiv f(z_h)$.

6.166. Let $A^{(*)}$ be the same *set* as $A(U)$ in **6.165** but as a Banach algebra with respect to convolution as multiplication:

$$f * g(z) \stackrel{\text{def}}{=} \int_{[0,z]} f(z - w)g(w)\, dz,$$

and with norm $\| \ \|_\infty$. Each f in $A^{(*)}$ is a generalized nilpotent.

6.167. If X is a compact Hausdorff space then for each h in $[A(X), \mathbb{C}]$ and some x_h in X, $h(f) \equiv f(x_h)$.

6.168. a) If G is abelian and $\alpha \in \widehat{G}$ then

$$h : A(G) \ni f \mapsto \int_G f(x)\overline{(\alpha, x)}\, d\mu(x) \in \mathbb{C} \stackrel{\text{def}}{=} \widehat{f}(\alpha)$$

is in $[A(G), \mathbb{C}]$.

b) If $h \in [A(G), \mathbb{C}]$ for some α in $\widehat{G}$, $h(f) \equiv \widehat{f}(\alpha)$.

6.169. The algebra $A(G)$ has an identity iff G is discrete.

6.170. If I is a regular ideal in A then the closure $\overline{I}$ of I is also an ideal.

6.171. For G the locally compact group $\mathbb{R}$, what is an example of an ideal I dense (whence not closed) in $A(G)$ $\left(= L^1(\mathbb{R}, \lambda)\right)$?

6.172. Let A be a commutative Banach algebra with identity. If D is a continuous derivation on A then $D(A) \subset \mathcal{R}(A)$.

COMPLEX ANALYSIS:

PROBLEMS

7 Elementary Theory

7.1. Geometry in $\mathbb{C}$

Conventions

For each subset S of $\mathbb{C}$, the set of complex conjugates of elements of S is denoted $\overline{S}$. On the other hand, the topological closure of S, i.e., the intersection of all closed sets containing S, is denoted S^c.

The symbol Ω, with or without subscripts, denotes a region, i.e., a nonempty connected open subset of $\mathbb{C}$. Concordantly, $H(\Omega)$ denotes the set of functions mapping Ω into $\mathbb{C}$ and holomorphic in Ω; in particular, $H(\mathbb{C}) \stackrel{\text{def}}{=} \mathcal{E}$, the set of entire functions; when Ω^c is compact,

$$A(\Omega) \stackrel{\text{def}}{=} H(\Omega) \cap C(\Omega^c, \mathbb{C}).$$

For a in $\mathbb{C}$ and r in $[0, \infty)$, $D(a,r) \stackrel{\text{def}}{=} \{z : z \in \mathbb{C}, |z-a| \le r\}$ and $C_a(r) \stackrel{\text{def}}{=} \partial D(a,r)$; in particular, $U \stackrel{\text{def}}{=} D(0,1)^\circ$ and $\mathbb{T} = C_0(1)$.

The symbol z denotes both a complex number and the geometric object that z represents as an element of the Gaußian plane. If $z_1, z_2 \in \mathbb{C}$ are two points, then $\{z_1, z_2\} \stackrel{\text{def}}{=} \{z : z = z_1 + t(z_2 - z_1), t \in \mathbb{R}\}$ is the straight line they determine and $[z_1, z_2] \stackrel{\text{def}}{=} \{tz_1 + (1-t)z_2 : 0 \le t \le 1\}$ is their convex hull.

The lexicographic partial order $\preceq$ of $\mathbb{C}$ is defined according to

$$\{z_1 \preceq z_2\} \Leftrightarrow \{\{\Re(z_1) \le \Re(z_2)\} \wedge \{\Im(z_1) \le \Im(z_2)\}\}.$$

A topological space X is simply connected iff every closed curve γ such that $\gamma^* \subset X$ is homotopic to a constant in X. Although a simply connected space X need not be connected, e.g., when $X = U \cup (2+U)$, unless the contrary is stated, only simply connected regions in $\mathbb{C}$ are considered.

When $f \stackrel{\text{def}}{=} u + iv \in \mathbb{C}^{\mathbb{C}}$ and $\gamma(t) = p(t) + iq(t), 0 \le t \le 1$, the symbol $\int_\gamma f(z)\,dz$ means that the Riemann-Stieltjes integrals

$$\int_0^1 (u(p(t), q(t)) + iv(p(t), q(t)))\,dp(t) \stackrel{\text{def}}{=} A$$

$$\int_0^1 (u(p(t), q(t)) + iv(p(t), q(t)))\,dq(t) \stackrel{\text{def}}{=} B$$

exist and $\int_\gamma f(z)\,dz \stackrel{\text{def}}{=} \int_0^1 f(\gamma(t))\,d\gamma(t) \stackrel{\text{def}}{=} A + iB$.

The extended complex plane $\mathbb{C}_\infty$, intuitively the construct $\mathbb{C}\cup\{\infty\}$, is the set $\Sigma \stackrel{\text{def}}{=} \{(\xi,\eta,\zeta) : \xi,\eta,\zeta \in \mathbb{R}, \xi^2+\eta^2+\zeta^2 = 1\}$. Furthermore the set $\Sigma \setminus \{(0,0,1)\}$ may be identified with $\mathbb{C}$ according to the stereographic projection: $\Theta : \Sigma \setminus \{(0,0,1)\} \ni (\xi,\eta,\zeta) \mapsto \left(\frac{\xi}{1-\zeta}, \frac{\eta}{1-\zeta}\right) \in \mathbb{C}$ while $(0,0,1)$ is identified with ∞. The center of the stereographic projection is $(0,0,1)$ and Θ maps (ξ,η,ζ) in $\Sigma \setminus \{(0,0,1)\}$ into the intersection of the line through $(0,0,1)$ and (ξ,η,ζ) and the plane

$$\{(a,b,0) : a,b \in \mathbb{R}\} \stackrel{\text{def}}{=} \mathbb{C}.$$

If $z_i, 1 \le i \le 4$, are four elements of $\mathbb{C}$, the number

$$X(z_1,z_2,z_3,z_4) \stackrel{\text{def}}{=} \frac{(z_1-z_2)(z_3-z_4)}{(z_1-z_3)(z_2-z_4)}$$

is their cross ratio or anharmonic ratio. For z_2, z_3, z_4 fixed and pairwise different, $X(z,z_2,z_3,z_4) \stackrel{\text{def}}{=} M(z)$ is a function on $\mathbb{C}_\infty \setminus \{z_3\}$ and may be extended to $\mathbb{C}_\infty$ by defining $M(z_3)$ to be ∞. There are constants a,b,c,d such that $M(z) = \frac{az+b}{cz+d} \stackrel{\text{def}}{=} T_{abcd}(z)$ and

$$ad - bc = (z_1-z_3)(z_2-z_3)(z_2-z_1) \neq 0.$$

More generally, when $ad-bc \stackrel{\text{def}}{=} \Delta \neq 0$, the map $T_{abcd} : \mathbb{C} \ni z \mapsto \frac{az+b}{cz+d}$ is a Möbius transformation. By definition,

$$T_{abcd}\left(-\frac{d}{c}\right) = \infty \text{ and } T_{abcd}(\infty) = \frac{a}{c}.$$

Correspondingly, there is $\widetilde{T}_{abcd} \stackrel{\text{def}}{=} \Theta^{-1}T_{abcd}\Theta : \Sigma \setminus \{(0,0,1)\} \mapsto \Sigma$, which may be extended by continuity to a self-map of Σ. When ambiguity is unlikely, the subscript $_{abcd}$ is dropped. Note that if $\alpha \neq 0$ then

$$T_{(\alpha a)(\alpha b)(\alpha c)(\alpha d)}(z) = T_{abcd}(z)$$

whence if $\alpha = \frac{1}{\sqrt{\Delta}}$ then $(\alpha a)(\alpha d) - (\alpha b)(\alpha c) = 1$, and as the need arises the value of Δ may be taken to be 1.

Each Möbius tranformation T is invertible. The inverse of T_{abcd} is $T_{abcd}^{-1} : \mathbb{C} \ni z \mapsto \frac{-dz+b}{cz-a} = T_{(-d)bc(-a)}$. Thus each T is one-one:

$$\{T(z) = T(z')\} \Leftrightarrow \{z = z'\}.$$

The set of all Möbius transformations is denoted **Mö**.

When $r > 0$, $z \neq a$, and $k \stackrel{\text{def}}{=} \frac{r^2}{|z-a|^2}$, the point $z^\rho \stackrel{\text{def}}{=} a + k(z-a)$ is the reflection of z in $C_a(r)$ and z is the reflection of z^ρ in $C_a(r)$: $z = (z^\rho)^\rho$.

If $\theta \in [0, 2\pi)$, the image γ^* of $\gamma : \mathbb{R} \ni t \mapsto a + te^{i\theta}$ is a straight line $L_a(\theta)$ in $\mathbb{C}$. If $z \notin L_a(\theta)$ the reflection of z in $L_a(\theta)$ is $z^\rho \stackrel{\text{def}}{=} a + e^{i\theta}\overline{(z-a)}$ and z is the reflection of z^ρ in $L_a(\theta)$.

In **Figure 7.1** there is a depiction of the geometry of the situation for the circle $C_a(r)$.

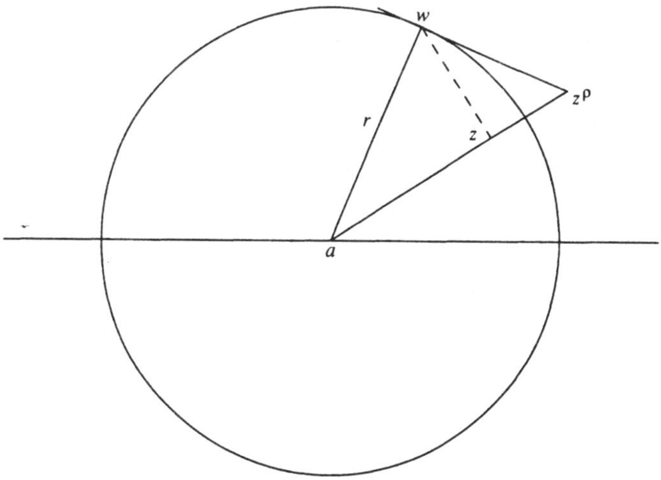

$$\begin{aligned} |w-z|^2 &= |z-a| \cdot |z^\rho - z| - |z-a| \cdot |z^\rho - a| - |z-a|^2 \\ &= |z-a| \cdot |z^\rho - a| - \left(r^2 - |w-z|^2\right) \\ |z-a| \cdot |z^\rho - a| &= r^2. \end{aligned}$$

Figure 7.1.

For the line $L_0(0)$, the reflection z^ρ of z is $\overline{z}$ and for $L_a(\theta)$ regarded as a mirror, z^ρ is the mirror image of z.

Thus the superscript $^\rho$ is used as a generic notation for a map $z \mapsto z^\rho$ performed with respect to some circle or line.

For $f \in H\left(D(a,R)^\circ\right)$ and $0 < r < R$, $M(r;f)) \stackrel{\text{def}}{=} \max_{|z|=r} |f(z)|$ and $m(r;f) \stackrel{\text{def}}{=} \min_{|z|=r} |f(z)|$.

7.1. If K is a circle lying on Σ then $\Theta(K \setminus \{(0,0,1)\})$ is a circle or a straight line.

7.2. a) What are the algebraic descriptions of the boundary and interior of the triangle $\Delta(pqr)$ having as vertices the noncollinear points p,q,r? b) If $p,q, \in D(0,1)^\circ$, $p,q,1$ are noncollinear, and s lies in the interior of the triangle $\Delta(pq1)$ then $R(s) \stackrel{\text{def}}{=} \dfrac{|1-s|}{1-|s|} \le K(p,q) < \infty$.

7.3. If $T_1, T_2 \in$ **Mö** and neither is the identity id then: a) $T_1(z) = z$ has at most two solutions in $\mathbb{C}$ (each such z is a fixed point of T_1); b) $\widetilde{T_1}$, the extension by continuity of T_1 to all Σ, has at least one fixed point; c) if z_1 is the only point fixed by T_1 and z_2 and z_3 are two points fixed by T_2 then $T_1T_2 \neq T_2T_1$; d) if T_1 and T_2 have the same fixed point(s), they commute: $T_1T_2 = T_2T_1$; e) when $a_1d_1 - b_1c_1 = 1$ and $c_1 \neq 0$, $|T_1(z)| = |z|$ iff $|c_1z + d_1| = 1$ (the equation of the isometric circle of T_1). What are the center and radius of the isometric circle of T_1?

7.4. If $\{z_1, z_2, z_3\}$ resp. $\{w_1, w_2, w_3\}$ are two sets of three points in $\mathbb{C}_\infty$ then for precisely one T in **Mö**, $T(z_i) = w_i, 1 \le i \le 3$.

7.5. If $T \in$ **Mö** then for any four p,q,r,s,

$$X(T(p), T(q), T(r), T(s)) = X(p,q,r,s).$$

7.6. For the family $\mathcal{F}$ of all lines and circles in $\mathbb{C}$ and for $T \in$ **Mö**: a) if K is in $\mathcal{F}$ then $T(K) \in \mathcal{F}$; b) if z^ρ is the reflection of z in the circle $C_a(r)$ and $T \in$ **Mö** then $T(z^\rho)$ is the reflection of $T(z)$ in $T(C_a(r))$.

7.7. Which T in **Mö** map U onto U?

7.8. If $z \neq a$ then the reflection z^ρ of z in $C_a(r)$ is $a + \dfrac{r^2}{\overline{z} - \overline{a}}$.

7.9. The elements of (z, z') in $\mathbb{C}^2_\infty$ are a pair of mutual reflections in C resp. L iff $X(z', z_1, z_2, z_3) = \overline{X(z, z_1, z_2, z_3)}$ and the validity of the equation is independent of the choice of the triple z_1, z_2, z_3.

7.10. Three points p,q,r are collinear iff $\dfrac{p-q}{q-r}$ is real.

7.11. a) If $S \in$ **Mö**, and $S^{-1}(0, \infty, 1) = (z_2, z_3, z_4)$ then

$$X(z_1, z_2, z_3, z_4) = S(z_1).$$

b) Four points p,q,r,s are cocircular or collinear iff $X(p,q,r,s)$ is real.

7.12. With respect to composition of transformations: a) **Mö** is a group generated by the subset

$$T_0 : \mathbb{C} \setminus \{0\} \ni z \mapsto \frac{1}{z}, \; T_{ab} : \mathbb{C} \ni z \mapsto az + b, \; a, b \in \mathbb{C};$$

b) as a group, **Mö** is isomorphic to $SL(2,\mathbb{C})$, the multiplicative group of all 2×2 matrices M with entries from $\mathbb{C}$ and for which $\det(M) = 1$; c) each T in **Mö** is the composition of an even number of reflections.

7.13. If $a, b, c, d \in \mathbb{R}$ and $ad - bc > 0$ then T_{abcd} leaves $\{ z : \Im(z) > 0 \}$ invariant.

7.14. If $C_a(r)$ and $C_b(s)$ are two nonintersecting circles then for some T in **Mö**, $T(C_a(r))$ and $T(C_b(s))$ are concentric.

7.15. For what values of θ in $[0, 2\pi)$ does $\lim_{r\uparrow\infty} e^{re^{i\theta}}$ exist?

7.2. Polynomials

7.16. If $P_n(z) \stackrel{\text{def}}{=} \sum_{k=0}^{n-1}(k+1)z^k$ and $0 < r < 1$ then for some n_0, if $n > n_0$ then P_n has no zeros in $D(0, r)$.

7.17. If $a \in \mathbb{R}$, $b > 0$, and for n in $\mathbb{N}$, $f_n(z) = z^{2n} + az^{2n-1} + b$ then f_n has exactly n zeros with positive imaginary parts.

7.18. If $f \in H(U)$ and for each z in U and some n_z in $\mathbb{N}$, $f^{(n_z)} = 0$ then f is a polynomial.

7.19. If $a_0 > a_1 > \cdots > a_n > 0$ then $\sum_{k=0}^{n} a_k z^k \stackrel{\text{def}}{=} f(z) \neq 0$ in $D(0,1)$.

7.20. For a polynomial p, an a in $\mathbb{C}$, and $E(a) \stackrel{\text{def}}{=} \{ z : p(z) = a \}$, the number of components of $E(a)$ does not exceed $\deg(p)$.

7.21. If $p, q \in \mathbb{C}[z]$ and $\deg(p) = n$ then: a) $\Omega \stackrel{\text{def}}{=} \{ z : \#\left(p^{-1}(z)\right) = n \}$ is open; b) $\tilde{q}(z) \stackrel{\text{def}}{=} \sum_{p(w)=z} q(w)$ is holomorphic in Ω.

7.3. Power Series

Conventions

The radius of convergence of the power series $\sum_{n=0}^{\infty} c_n(z-a)^n$ is $R_{\mathbf{c}}$.

For f in $\mathbb{C}^{\mathbb{C}}$, $Z(f)$ is the set of zeros of f, $S(f)$ is the set of singularities, $P(f)$ is the set of poles, and $ES(f)$ is the set of isolated essential singularities of f. The statements $Z(f) = Z(g)$ or $P(f) = P(g)$ are to be interpreted as meaning not only that the sets are equal but that the corresponding multiplicities or orders are the same as well.

When $w = re^{i\theta}$, $r > 0$, and $0 \leq \theta < 2\pi$ then $w^z \stackrel{\text{def}}{=} e^{z(\ln r + i\theta)}$.

7.22. If $\sum_{n=0}^{\infty} |c_n - c_{n+1}| < \infty$, $\lim_{n\to\infty} c_n = 0$, and $0 < \delta < 2$ then $\sum_{n=0}^{\infty} c_n z^n$ converges absolutely and uniformly in the compact set

$$K \stackrel{\text{def}}{=} \{ z : |z| \leq 1, |z-1| \geq \delta \}.$$

7.23. If $\{f_m\}_{m\in\mathbb{N}} \subset \mathcal{E}$, $f_m(z) = \sum_{n=0}^{\infty} c_{mn} z^n$, $\{|f_m(z)|\}_{m\in\mathbb{N}}$ is bounded on every compact set, and $\lim_{m\to\infty} c_{mn} \stackrel{\text{def}}{=} c_n$ then $\sum_{n=0}^{\infty} c_n z^n$ converges everywhere and represents an entire function f such that $f_m \stackrel{\text{u}}{\to} f$ on every compact set.

7.24. For $\sum_{n=0}^{\infty} c_{mn} z^n, m = 1, 2$, if $\sum_{n=0}^{\infty} c_{1n} c_{2n} z^n \stackrel{\text{def}}{=} \sum_{n=0}^{\infty} c_n z^n$ then $R_{\mathbf{c}} \ge R_{\mathbf{c}_1} R_{\mathbf{c}_2}$. What is an example for which $R_{\mathbf{c}} > R_{\mathbf{c}_1} R_{\mathbf{c}_2}$?

7.25. a) What are the c_k such that for z near 2,

$$\zeta(z) \stackrel{\text{def}}{=} \sum_{n=1}^{\infty} \frac{1}{n^z} = \sum_{k=0}^{\infty} c_k (z-2)^k?$$

b) What is $R_{\mathbf{c}}$?

7.26. If $f(z) = \sum_{n=0}^{\infty} c_n z^n$ in $D(0,R)^\circ$ and for some k and some r in $(0,R)$, $|c_k| = \dfrac{M(r;f))}{r^k}$ then $f(z) = c_k z^k$.

7.27. If $\Re(c_n) \ge 0, n \in \mathbb{N}$, and $R_{\mathbf{c}} = 1$ then 1 is a singularity of $f(z) \stackrel{\text{def}}{=} \sum_{n=1}^{\infty} c_n z^n$.

7.28. If $f(z) = \sum_{n=0}^{\infty} a_n z^n$, $a_n \ge 0, |z| < 1$, $f'(1)$ exists, and in $D(1,t)^\circ$, $f(z) = \sum_{n=0}^{\infty} b_n (z-1)^n$ then $R_{\mathbf{a}} \ge 1 + t$.

7.29. If $f(z) = \sum_{n=0}^{\infty} c_n z^n$, $R_{\mathbf{c}} = 1$, and $S(f) \cap \mathbb{T} = P(f)$ then $\sup_n |c_n|$ is finite.

7.30. If, for k in $\mathbb{N}$ and z in U, $P_k(z) \stackrel{\text{def}}{=} (1-z)^{k+1} \sum_{n=1}^{\infty} n^k z^n$ then $P_k \in \mathbb{N}[z]$ and $\deg(P_k) = k$.

7.31. If $K \stackrel{\text{def}}{=} \{ f \ : \ f(z) = \sum_{n=0}^{\infty} a_n z^n,\ |a_n| \le n+1 \}$ then K contains a sequence $\{f_n\}_{n\in\mathbb{N}}$ converging uniformly on every compact subset of U.

7.32. If $L(z,n) \stackrel{\text{def}}{=} -\sum_{k=1}^{n} \dfrac{z^k}{n+1-k} + \sum_{k=1}^{n} \dfrac{z^{n+k}}{k}$, $n \in \mathbb{N}$ then: a) there is a constant K such that $|L(z,n)|\ |_{D(0,1)} \le K$; b) $\sum_{n=1}^{\infty} \dfrac{z^{2^{n^2}} L(z, 2^n)}{n^2}$ represents a function f in $A(U)$ although the power series for f does not converge absolutely at any point of $\mathbb{T}$.

7.33. If $(X, \mathcal{S}, \mu)$ is a complex measure situation, Ω is a region in $\mathbb{C}$, and g is a measurable function such that $g(X) \subset \mathbb{C} \setminus \Omega$ then

$$f : \Omega \ni z \mapsto \stackrel{\text{def}}{=} \int_X \frac{d\mu(x)}{g(x) - z}$$

is in $H(\Omega)$.

7.34. (Inverse function theorem) If $f \in H(U)$, $f(0) = 0 \neq f'(0)$, $r \in (0,1)$, $f(z) \neq 0$ for z in $A(0,0:r)$, and

$$g(w) \stackrel{\text{def}}{=} \frac{1}{2\pi i} \int_{|z|=r} \frac{zf'(z)}{f(z) - w}\, dz \tag{7.1}$$

then for some s in $(0,\infty)$, $g \in H\left(D(0,s)^\circ\right)$ and for all w in $D(0,s)^\circ$,

$$f \circ g(w) = w.$$

8
Functions Holomorphic in a Disc

8.1. General Results

Conventions

For f in $H(U)$ and t in $[0, 2\pi)$,

$$f^*(t) \stackrel{\text{def}}{=} \begin{cases} \lim_{r\uparrow 1} f\left(re^{it}\right) & \text{when the limit exists} \\ 0 & \text{otherwise} \end{cases}$$

is the radial limit function. The (Hardy class)

$$\left\{f \,:\, f \in H(U), \sup_{z\in U} |f(z)| < \infty\right\}$$

is H^∞.

8.1. In $H(U)$ there is no $f, g,$ or h such that:

$$\left\{f\left(\frac{1}{n}\right)\right\}_{n\in\mathbb{N}} = \left\{1, 0, \frac{1}{2}, \frac{1}{3}, 0, \frac{1}{4}, \frac{1}{5}, \frac{1}{6}, 0, \ldots\right\};$$

$$\left\{g\left(\frac{1}{n}\right)\right\}_{n\in\mathbb{N}} = \left\{1, -\frac{1}{2}, \frac{1}{3}, -\frac{1}{4}, \ldots\right\};$$

$$\left\{h\left(\frac{1}{n}\right)\right\}_{n\in\mathbb{N}} = \left\{\frac{1}{2}, \frac{1}{2}, \frac{1}{4}, \frac{1}{4}, \frac{1}{6}, \frac{1}{6}, \ldots\right\}.$$

8.2. If $f \in H(U)$ and a) $\left|f\left(z^2\right)\right| \geq |f(z)|$ or b) $\Re(f) = (\Im(f))^2$ then f is a constant.

8.3. If $f \in C(U, \mathbb{C})$, $\{f_n\}_{n\in\mathbb{N}} \subset H(U)$, and for all r in $(0, 1)$,

$$\lim_{n\to\infty} \int_{-\pi}^{\pi} \left|f_n\left(re^{i\theta}\right) - f\left(re^{i\theta}\right)\right| d\theta = 0$$

then $f \in H(U)$.

8.4. If $f \in H\left(D(0, r)^\circ\right)$ and $n \in \mathbb{N}$ then for some s in $(0, \infty)$ and some g in $H\left(D(0, s)^\circ\right)$,6

$$\{f(z) = (g(z))^n\} \vee \{f(z) = z(g(z))^n\} \vee \cdots \vee \left\{f(z) = z^{n-1}(g(z))^n\right\}.$$

8.5. If $f(z) = \sum_{n=0}^{\infty} c_n z^n$, $\underline{\lim}_{n\to\infty} |c_n| > 0$, $R_c = R < \infty$, and the only singularity of f in $\mathbb{C}$ is a pole a of order 1 then $\lim_{n\to\infty} \frac{c_n}{c_{n+1}} = a$.

8.6. If $f \in H(U)$ and $f'(U) \subset D(0, M)$ then for some F in $C(D(0,1), \mathbb{C})$, $F|_U = f$.

8.7. For $A \stackrel{\text{def}}{=} \{e^{i\theta} : 0 \le a \le \theta < b \le 2\pi\}$, if $f \in H(U) \cap C(U \cup A, \mathbb{C})$ and $f(A) = \{0\}$ then $f(z) \equiv 0$.

8.8. If $f, g \in H(U)$, f is injective, $f(0) = g(0) = 0$, and $g(U) \subset f(U)$ then for r in $[0,1)$, $g(D(0,r)) \subset f(D(0,r))$.

8.9. If $z \in U$ then $\frac{1}{1-z} = \prod_{n=0}^{\infty} \left(1 + z^{2^n}\right)$.

8.10. If, for z in U,

$$f(z) \stackrel{\text{def}}{=} \prod_{n\in\mathbb{N}} \left(1 + z^{2n}\right) \text{ and } g(z) \stackrel{\text{def}}{=} \prod_{n\in\mathbb{N}} \left(1 - z^{2(2n-1)}\right)$$

then $f, g \in H(U)$ and $f(z)g(z) \equiv 1$ in U.

8.11. For no f in $A(U)$, $f(z)|_{\mathbb{T}} = \frac{1}{z}$.

8.12. If $f \in A(U)$ and $|f(z)|\,|_{\mathbb{T}} \equiv K$ then f is a rational function.

8.13. If $f \in A(U)$ and $0 \notin f(\mathbb{T})$ then for some r in $(0,1)$, the number of zeros N of f in U is $\frac{1}{2\pi}\int_0^{2\pi} \frac{f'\left(re^{i\theta}\right)}{f\left(re^{i\theta}\right)} re^{i\theta}\, d\theta$.

8.14. For $f(z) \stackrel{\text{def}}{=} \sum_{n=0}^{\infty} c_n z^n \in A(U)$: a) $\sum_{n=0}^{\infty} |c_n|^2 < \infty$; b) if $z = re^{i\theta}, 0 \le \theta < 2\pi, r < 1$ then:

$$\begin{aligned} f(z) &= \frac{1}{2\pi}\int_0^{2\pi} \Re\left(\frac{e^{it}+z}{e^{it}-z}\right) f\left(e^{it}\right) d; \\ &= \int_0^{2\pi} \frac{1}{2\pi} \frac{1-r^2}{1 - 2r\cos(\theta - t) + r^2} f\left(e^{it}\right) dt; \\ &\stackrel{\text{def}}{=} \int_0^{2\pi} P_r(\theta - t) f\left(e^{it}\right) dt; \end{aligned}$$

c) for P_r, Poisson's kernel, there obtain:

i. $P_r(t) = \frac{1}{2\pi} \sum_{n=-\infty}^{\infty} r^{|n|} e^{int} \ge 0$;

ii. $\int_0^{2\pi} P_r(t)\, dt \equiv 1$;

iii. $P_r(t) \stackrel{\text{u}}{\to} 0$ if $0 < \delta$, $t \in [\delta, \pi - \delta]$, and $r \uparrow 1$.

[**Note 8.1:** Because $P_r(t)$ is a periodic function with period 2π, i – iii above may be re-interpreted for $P_r(t+\pi) \stackrel{\text{def}}{=} \Pi_r(t)$. The interval of integration $[0, 2\pi]$ is replaced by $[-\pi, \pi]$ and iii becomes: iii'. if $0 < \epsilon < \pi$ and $t \notin (-\pi + \epsilon, \pi - \epsilon)$ then $\Pi_r(t) \stackrel{\text{u}}{\to} 0$ as $r \to 1$. In sum, $\{\Pi_r\}_{r \in [0,1)}$, is, like the set $\{F_N\}_{N \in \mathbb{N}}$ of Fejér's kernels, an approximate identity in the sense of **6.44**.]

8.15. If $f \in H(D(a,R)^\circ)$ and $0 < r < R$ then

$$f(a) = \frac{1}{\pi r^2} \int_0^{2\pi} \left(\int_0^r f\left(a + se^{it}\right) s\, ds \right) dt.$$

8.16. If $f \in H(U)$ and $f(U) \subset U$ then $|f'(0)| \le 1$.

8.17. If $f \stackrel{\text{def}}{=} u + iv \in H(U)$ then $f^{(k)}(0) = k! \dfrac{1}{\pi r^k} \displaystyle\int_0^{2\pi} u\left(re^{i\theta}\right) e^{-ik\theta}\, d\theta$.

8.18. If $f(z) = \sum_{n=0}^\infty c_n z^n$, $0 < R < R_c$, and

$$\Re(f(z)) \stackrel{\text{def}}{=} u(r,\theta) \le M < \infty$$

in $D(0,R)^\circ$ then: a) $\alpha \stackrel{\text{def}}{=} \Re(c_0) = \dfrac{1}{2\pi} \displaystyle\int_0^{2\pi} u(r,\theta)\, d\theta$; b) $|c_n| \le \dfrac{2(M-\alpha)}{R_c^n}$.

8.19. For $f(z) \stackrel{\text{def}}{=} \sum_{n=0}^\infty c_n z^n$ in $H(D(0,R)^\circ)$ and injective on $D(0,R)^\circ$, when r in $(0,R)$, $L(r)$ resp. $A(r)$ denotes the length of $f(C_0(r))$ resp. the area of $f(D(0,r))$. Then:

$$\text{a) } L(r) \ge 2\pi r\,|f'(0)|; \text{ b) } A(r) = \pi \sum_{n=1}^\infty n\,|c_n|^2 r^{2n}; \text{ c) } A(r) \ge \pi r^2 |f'(0)|^2.$$

8.20. If $f \in H(U)$ then $I_2(r) \stackrel{\text{def}}{=} \int_0^{2\pi} \left|f\left(re^{i\theta}\right)\right|^2 d\theta$ is a monotonely increasing function on $[0,1)$ (cf. **15.9**).

8.21. If $f \in H(U)$, $f(U) \subset D(0,1)$, and $f(z) = z^n \left(\sum_{k=n}^\infty c_k z^{k-n}\right)$ then for z in U, $|f(z)| \le |z|^n$.

8.22. If $f, f^{-1} \in H(U)$, $f(0) = 0$, and $n \in \mathbb{N}$ then for exactly n functions $g_k, 1 \le k \le n$, in $H(U)$, $f(z^n) = (g_k(z))^n$, $1 \le k \le n$.

8.23. If $f \in H(U)$, $x_n \downarrow 0$, and $f(x_n) \in \mathbb{R}, n = 2, 3, \ldots$, then for each ν in $\mathbb{Z}^+$, $f^{(\nu)}(0) \in \mathbb{R}$.

8.24. For $S \stackrel{\text{def}}{=} \{f_n\}_{n \in \mathbb{N}} \subset H(U)$ and r in $[0,1)$, if $\int_0^{2\pi} \left|f_n\left(re^{i\theta}\right)\right| d\theta \le 1$ then S contains an $L^1(U, \lambda_2)$-Cauchy sequence.

8.25. For no function f in $H(U)$, $|f(z_n)| \to \infty$ whenever $|z_n| \to 1$.

8.2. Applications of Möbius Transformations

8.26. If $f(z) = 1 + \sum_{n=1}^{\infty} c_n z^n$, $R_c = 1$, and $\Re(f(z)) > 0$ for z in U then: a) in U, $\frac{1-|z|}{1+|z|} \le |f(z)| \le \frac{1+|z|}{1-|z|}$; b) $\lim_{n\to\infty} c_n = 0$.

8.27. If $f \in H(U)$, $Z(f) \cap U = \{z_k\}_{k=1}^{K}$, and $f(U) \subset U$ then: a) when $K < \infty$ and $z \in U$, $|f(z)| \le \prod_{k=1}^{K} \left|\frac{z - z_k}{1 - \overline{z_k} z}\right| \stackrel{\text{def}}{=} \prod_{k=1}^{K} |\Phi_{z_k}(z)|$; b) if $K = \infty$ and $\lim_{k\to\infty} |z_k| = 1$ then

$$|f(z)| \le \prod_{k\in\mathbb{N}} |\Phi_{z_k}(z)| \tag{8.1}$$

and either $\sum_{k=1}^{\infty} (1 - |z_k|) < \infty$ or $f \equiv 0$.

8.28. (Blaschke) If $\{z_n\}_{n\in\mathbb{N}} \subset U$ and $\sum_{n=1}^{\infty} (1 - |z_n|) < \infty$ then

$$B(z) \stackrel{\text{def}}{=} \prod_{n=1}^{\infty} \operatorname{sgn}(z_n)\left(-\Phi_{z_n}(z)\right) \in H^{\infty}. \tag{8.2}$$

8.29. If $f \in H(U)$, $f(U) \subset U$, and for two points a, b in U, $f(a) = a$ and $f(b) = b$, then $f(z) \equiv z$: U contains at most one f-fixed point.

8.30. a) If $f \in H(U)$ and $f(U) \subset U$ then

$$\frac{|f'(z)|}{1 - |f(z)|^2} \le \frac{1}{1 - |z|^2}.$$

b) If $g(z) \stackrel{\text{def}}{=} \sum_{n=0}^{\infty} c_n z^n \in H(U)$ and $g(U) \subset D(0, R)$ then

$$R|c_1| \le R^2 - |c_0|^2.$$

8.31. If $f \in H(U)$, $f(U) = U$, and f is injective then f is a Möbius transformation (cf. **7.7**).

9
Functions Holomorphic in a Region

9.1. General Regions

Conventions

The sets $\left\{ z : \Im(z) \gtrless 0 \right\}$ are denoted $\Pi_\pm$.

9.1. a) If $f \in H(\Omega)$ and $a \in \Omega$ for two points b, c in Ω, $f'(a) = \dfrac{f(b) - f(c)}{b - c}$.

b) For γ a Jordan curve and Ω a region containing γ^* together with the bounded component Ω_1 of $\mathbb{C} \setminus \gamma^*$, if $f \in H(\Omega)$ and on γ^*, $|f(z)| \equiv M$, a constant, then f is a constant or for some a in Ω_1, $f(a) = 0$.

9.2. For a region Ω, if $f, g \in H(\Omega)$ and

$$\{f(a) = f(b)\} \Rightarrow \{g(a) = g(b)\}$$

then for some h in $H(f(\Omega))$, $g = h \circ f$.

9.3. For some region Ω and some f in $H(\Omega)$, no sequence $\{p_n\}_{n \in \mathbb{N}}$ of polynomials converges to f uniformly on compact subsets of Ω.

9.4. If Ω is simply connected, $f, g \in H(\Omega)$, and $k \in C\left(f'(\Omega) \times g'(\Omega), \mathbb{C}\right)$ then for some h in $H(\Omega)$, $h' = k\left(f', g'\right)$.

9.2. Regions Ω Containing $D(0,1)$

Conventions

For the items in this **Section** it is to be understood that the region Ω contains $D(0,1)$.

9.5. If $f \in H(\Omega)$ and $f(\mathbb{T}) \subset \mathbb{R}$ then f is a constant.

9.6. If $f \in H(\Omega)$ and f has n simple zeros in U then $\Re(f)$ has at least $2n$ zeros on $\mathbb{T}$.

9.7. If $f, g \in H(\Omega) \cap C\left(\Omega^c, \mathbb{C}\right)$, $f\left(\Omega^c\right) \cup g\left(\Omega^c\right) \subset U^c$, $g(\partial\Omega) = \mathbb{T}$, and $\dfrac{f}{g}$ is bounded on $\Omega \setminus g^{-1}(0)$ then $|f(z)| \leq |g(z)|$ on Ω.

9.8. If f in $H(\Omega)$, $f_n(z) \stackrel{\text{def}}{=} f(z^n), n \in \mathbb{N}$, and $f(0) = 0$ then: a) $\lim_{N\to\infty} F_N \stackrel{\text{def}}{=} \lim_{N\to\infty} \sum_{n=1}^{N} f_n \stackrel{\text{def}}{=} F$ exists in $D(0,1)$; b) $F' = \sum_{n=1}^{\infty} f_n'$.

9.9. If $f \in H(\Omega)$ and $f(D(0,1)) \subset U$ then U contains precisely one a such that $f(a) = a$.

9.10. If $f \in H(\Omega)$ and $f(U) = U$ then $f(\mathbb{T}) = \mathbb{T}$.

9.3. Other Special Regions

Conventions

When $0 \le r < R < \infty$, the symbol $A(a, r : R)$ denotes the annulus $\{ z : r \le |z - a| \le R \}$.

9.11. If $\Omega = \{ z : \Re(z) > 0 \}$, $f \in H(\Omega)$, $|f| \le R$, $\lim_{r\downarrow 0} f(r) = a$, and $b \in \Omega$ then $\lim_{r\downarrow 0} f(rb) = a$.

9.12. If $f \in H(A(0, 0 : R)^\circ) \stackrel{\text{def}}{=} H(\Omega)$ and $\Re(f(z)) \le M$ on Ω then zero is a removable singularity of f.

9.13. If $f \in H\left(A(0, r_1 : r_2)^\circ\right)$ and $f(z) \not\equiv 0$ then for some r in (r_1, r_2), $f(z) \ne 0$ on $C_0(r)$.

9.14. If $f \in H(A(0, r : R)^\circ) \stackrel{\text{def}}{=} H(\Omega)$, $\{p_n\}_{n\in\mathbb{N}} \subset \mathbb{C}[z]$, and $p_n \stackrel{\text{u}}{\to} f$ on every compact subset of Ω then for some F in $H(D(0, R)^\circ)$, $F(z)\big|_\Omega = f(z)$.

9.15. If

$$f \in H(A(0, 1 : 2)^\circ) \stackrel{\text{def}}{=} H(\Omega),\ f \in C(\Omega^c \setminus \{1\}, \mathbb{C}),$$
$$\varlimsup_{r\uparrow 2} \left|f\left(re^{i\theta}\right)\right| \le 1,\ 0 \le \theta < 2\pi, \text{ and } \varlimsup_{r\downarrow 1} \left|f\left(re^{i\theta}\right)\right| \le 1,\ 0 \le \theta < 2\pi,$$

then $f(\Omega) \subset U^c$.

9.16. If $\{L_k\}_{1\le k\le n}$ is a set of straight lines in $\mathbb{C}$, $\Sigma \stackrel{\text{def}}{=} \mathbb{C} \setminus \bigcup_{k=1}^{n} L_k$, and $f \in H(\Sigma) \cap C(\mathbb{C}, \mathbb{C})$ then there is an entire function F such that $F\big|_\Sigma = f_\Sigma$.

9.17. For $\Omega \stackrel{\text{def}}{=} A(0, r : 1)^\circ$, if $f \in H(\Omega)$ then for some M, Ω contains a sequence $\{z_n\}_{n\in\mathbb{N}}$ such that $\lim_{n\to\infty} |z_n| = 1$ and $|f(z_n)| \le M$ (cf **8.25**).

9.18. If Ω is convex, $f \in H(\Omega)$, f is not a constant, and

$$\Re(f'(z)) \cdot \Im(f'(z)) \ne 0$$

in Ω then f is injective on Ω.

9.19. If $\Omega \stackrel{\text{def}}{=} \mathbb{C} \setminus \{0\}$, $g \in H(\Omega)$, g is injective, and $g(\Omega) \subset \Omega$, then for some α, $g(z) = \alpha z$ or $g(z) = \dfrac{\alpha}{z}$.

9.20. If $\{f_n\}_{n\in\mathbb{N}} \subset H(\Omega)$, $f_n \xrightarrow{\mathrm{u}} f$ on each compact subset of Ω, and each f_n is never zero in Ω then $f \equiv 0$ or f is never zero in Ω. There are examples for each situation.

9.21. Assume $\{f_n\}_{n\in\mathbb{N}} \subset H(\Omega)$, for each a in Ω, $f_n \xrightarrow{\mathrm{u}} f$ on compact subsets of some neighborhood $V(a)$, and $f \not\equiv 0$. If $b \in \Omega$ and $f(b) = 0$ then for each neighborhood $V(b)$ contained in Ω, and some $n(V(b))$ in $\mathbb{N}$, $\{n > n(V(b))\} \Rightarrow \{Z(f_n) \cap V(b) \neq \emptyset\}$.

9.22. If $\Omega \stackrel{\text{def}}{=} \{z : \Re(z) > 0\}$, $f \in H(\Omega)$, and $f(z+1) \equiv 2f(z)$ then there is an entire function F such that $F(z)|_\Omega = f(z)$.

9.23. Assume for $\Omega \stackrel{\text{def}}{=} \{z : \Im(z) > 0\}$, f in $H(\Omega) \cap C(\Omega^c, \mathbb{C})$, z in $\Omega^c \setminus \{0\}$, and M, r positive, that $|f(z)| \le M|z|^{-r}$. True or false: a) for z in Ω, $f(z) = \dfrac{1}{2\pi i}\displaystyle\int_{-\infty}^{\infty} \frac{f(t)}{(t-z)}\,dt$? b) if f is not a constant then

$$f\left(i\mathbb{R}^+\right) \subset \mathbb{C} \setminus \mathbb{R}?$$

9.24. For f, g in $H\left(A(0, r : \infty)^\circ\right) \stackrel{\text{def}}{=} H(\Omega)$, if $r < R \le \dfrac{|z|}{2}$ then: a) $\int_{C_0(R)} f(z-w)g(w)\,dw \stackrel{\text{def}}{=} f * g(z)$ exists and its value is independent of R; b) $f * g \in H(\Omega)$; c) if f, g are rational so is $f * g$.

9.25. If $f \in H(\Omega)$, a is a zero of order k of f, and g has a pole of order m at zero then $h \stackrel{\text{def}}{=} g \circ f$ has a pole of order km at a.

9.26. True or false: if $f \in H(D(a,r) \setminus \{a\})$ then for some unique A in $\mathbb{C}$, $\int_\gamma \left(f(z) - \dfrac{A}{z-a}\right) dz = 0$ for every closed rectifiable curve-image γ^* in $A(a, 0 : r)$?

9.27. If $a \in U$, $f \in H(U \setminus \{a\}) \cap C(D(0,1) \setminus \{a\}, \mathbb{C})$, and $f(\mathbb{T}) \subset \mathbb{R}$ then for some A in $\mathbb{C}$, some B in $\mathbb{R}$, and some entire function F,

$$f(z) = F\left(\frac{Az^2 + Bz + \overline{A}}{(z+a)(1+\overline{a}z)}\right).$$

9.28. If $\{f_n\}_{n\in\mathbb{N}} \subset H(\Omega)$, for each z in Ω, $\{|f_n|\}_{n\in\mathbb{N}}$ is bounded in some $N(z)$, $\{a_k\}_{k\in\mathbb{N}} \subset \Omega$, $\lim_{k\to\infty} a_k = a \in \Omega$, and $\lim_{n\to\infty} f_n(a_k) = 0, k \in \mathbb{N}$, then $f_n \xrightarrow{\mathrm{u}} 0$ on every compact subset of Ω.

9.29. True or false: for $\Sigma \stackrel{\text{def}}{=} U \setminus (-1, 1)$ and f in $H(\Sigma)$: a) if $|f(z)|$ is bounded in Σ then for some F in $H(U)$, $F(z)|_\Sigma = f(z)$? b) if for some F in $C(U, \mathbb{C})$, $F(z)|_\Omega = f(z)$ then $F \in H(U)$?

9.30. For some simply connected region Ω and some f in $H(\Omega)$, $f(\Omega)$ is not simply connected.

9.31. For $\Omega \overset{\text{def}}{=} \{ z : |\Im(z)| < 1 \}$ and some f in $H(\Omega)$,

$$\lim_{\Re(z)\to\infty} f(\Re(z)) = 0$$

while

$$\lim_{\Re(z)\to\infty} f(z)$$

does not exist.

In one form, the maximum modulus theorem states that if V is a bounded open set (not necessarily a region) and $f \in H(V)$ then $|f(z)|$ has no local maximum in V. Hence if $f \in A(V)$ and for each a in ∂V, $\overline{\lim}_{z\to a}|f(z)| \le M$ then $\max_{z\in V^c}|f(z)| \le M$.

For a subset S of $\mathbb{C}$, $\partial_\infty G$ denotes the boundary of S in $\mathbb{C}_\infty$.

9.32. If $f \in H(\Omega)$ and for each a in $\partial_\infty\Omega \overset{\text{def}}{=} F$, $\overline{\lim}_{z\to a}|f(z)| \le M$ then $\sup_{z\in\Omega}|f(z)| \le M$.

9.33. The open mapping theorem implies the maximum modulus theorem.

10
Entire Functions

10.1. Elementary Theory

Conventions

Unless the contrary is indicated, every function introduced in this **Chapter** is entire. An entire function that is not a polynomial is transcendental.

When f in $H(\Omega)$ is injective, f is conformal [and $f^{-1} \in H(f(\Omega))$].

10.1. If $\lim_{|z|\to\infty} \dfrac{|f(z)|}{|z|} = 0$ then f is a constant.

10.2. If $g \in (0,\infty)^{[0,\infty)}$, $\lim_{x\to\infty} |g(x)| = 0$, and

$$|f(z)| \le |z| \cdot |g(|z|)| \text{ in } \mathbb{C} \setminus \{0\}$$

then $f \equiv 0$.

10.3. If $m \in \mathbb{N}$ and $\lim_{r\to\infty} \dfrac{M(r;f)}{r^m} = 0$ then $f \in \mathbb{C}[z]$ and $\deg(f) \le m-1$ (cf. **7.26**).

10.4. If, for some k in $\mathbb{Z}^+$, some g, and all large $|z|$, $|f(z)| \le \left|z^k g(z)\right|$ then for some rational h, $f = hg$.

10.5. If $\alpha, A > 0$ and $|f(z)| \le A|z|^\alpha$ then $f \in \mathbb{C}[z]$.

10.6. For a sequence $\{z_n\}_{n\in\mathbb{N}}$, what are necessary and sufficient conditions that for each sequence $\mathbf{w} \stackrel{\text{def}}{=} \{w_n\}_{n\in\mathbb{N}}$, and some f (depending on $\mathbf{w}$), $f(z_n) = w_n, n \in \mathbb{N}$?

10.7. If $f(\mathbb{R}) \subset \mathbb{R}$ and $f(i\mathbb{R}) \subset i\mathbb{R}$ then $f(-z) = -f(z)$.

10.8. For some g, h, $g(\mathbb{R}) \cup h(\mathbb{R}) \subset \mathbb{R}$ and $f = g + ih$.

10.2. General Theory

10.9. True or false: a) for some transcendental f, $f(\mathbb{C}) = \mathbb{C}$? b) for some f, $f(\mathbb{C}) = \{ z : \Re(z) > 0 \}$? c) for some injective f, $f(\mathbb{C}) \subsetneqq \mathbb{C}$.

10.10. An f is transcendental iff for some sequence $\{z_n\}_{n\in\mathbb{N}}$, $z_n \to \infty$ and $\sup_{n\in\mathbb{N}} |f(z_n)| < \infty$.

10.11. If f is transcendental then for some sequence $\{z_n\}_{n\in\mathbb{N}}$, $|z_n| \uparrow \infty$ and for each p in $\mathbb{C}[z]$, $\lim_{n\to\infty} f(z_n)p(z_n) = 0$.

10.12. If $g \in \mathbb{C}[z]$, $F \in C(\mathbb{C},\mathbb{C})$ and $F(g) = f$ then $F \in \mathcal{E}$.

10.13. If $p, q \in \mathbb{C}[z]$ and $e^{f(z)} + p(z) = e^{g(z)} + q(z)$ then $p = q$.

10.14. If f is not a constant then $(f^3 + f^2)(\mathbb{C}) = \mathbb{C}$.

10.15. If $a \in (0,\infty)$ and $E_a \overset{\text{def}}{=} \{ z : |f(z)| < a \}$ then: a)

$$E_a^c = \{ z : |f(z)| \le a \};$$

b) every bounded component C of E_a meets $Z(f)$.

10.16. If $f(z) = f(z+1)$ then for some sequence $\{c_n\}_{n\in\mathbb{Z}}$,

$$f(z) = \sum_{n=-\infty}^{\infty} c_n e^{2\pi i n z},$$

the series converging to $f(z)$ uniformly on compact sets.

10.17. If $f(z) = f(z+1) = f\left(z+\sqrt{2}\right)$ then f is a constant.

10.18. If f is not a constant then for some curve $\gamma : [0,1] \ni t \mapsto \gamma(t) \in \mathbb{C}_\infty$, $\gamma(1) = (0,0,1)$ and $|f(\gamma(t))| \uparrow \infty$ as $t \uparrow 1$.

10.19. If $m \in \mathbb{N}$ and $f(z) \overset{\text{def}}{=} \int_{[0,z]} \exp\left(w^{2m}\right) dw$ then f is not injective and $f(\mathbb{C}) = \mathbb{C}$.

10.20. If $e^{f(z)} + e^{g(z)} \equiv 1$ then f and g are constants.

10.21. If $Z(f) = \emptyset$ then either $f(z) = ce^z$ or $f(z) + e^z$ has infinitely many zeros.

10.22. If a, b are two points in $\mathbb{C}$ and both $f^{-1}(a)$ and $f^{-1}(b)$ are finite nonempty sets then $f \in \mathbb{C}[z]$.

10.23. If f is injective then $f \in \mathbb{C}[z]$ and $\deg(f) = 1$.

10.24. If $f(\mathbb{C}) = \mathbb{C}$ and $f' \neq 0$ then f is injective.

[**Note 10.1:** When $f(z) = e^z$ then f is not injective although $f' \neq 0$.]

10.25. If $\#(Z(f))$ is finite then for some p in $\mathbb{C}[z]$ and some g, $f = pg$.

10.26. True or false: a) if $|\Re(f)|$ is bounded then $|f|$ is a constant? b) if $f \in H(U)$ and $|\Re(f)|$ is bounded then $|f|$ is bounded?

10.27. Let L_1 and L_2 be lines in $\mathbb{C}$ and assume that $f^{-1}(L_1) \subset L_2$. If A_i, B_i are the two components (half-spaces) of $\mathbb{C}\setminus L_i, i = 1, 2$ then: a) for an appropriate choice of labels, $f(A_2) = A_1$ and $f(B_2) = B_1$; b) $f(L_2) = L_1$; c) for some constants a, b, $f(z) = az + b$.

10.28. Is there a nonconstant f such that for all z, $\Re(f(z)) \neq (\Im(f(z)))^2$?

10.29. If $Z(f) \cap Z(g) = \emptyset$ then for any h and some α, β, $\alpha f + \beta g = h$.

10.3. Order of Growth

Conventions

When f is entire and not a constant, $\rho(f) \stackrel{\text{def}}{=} \overline{\lim}_{r\to\infty} \dfrac{\ln\ln M(r;f)}{\ln r}$ is the order (of growth) of f. When $0 < \rho(f) < \infty$,

$$\tau(f) \stackrel{\text{def}}{=} \overline{\lim}_{r\to\infty} r^{-\rho(f)} \ln M(r;f)$$

is the type of its order. Thus $0 \leq \rho(f), \tau(f) \leq \infty$.

When $0 < |a_n| \uparrow \infty$ and $\sum_{n=1}^{\infty} \dfrac{1}{|a_n|^r} < \infty$,

$$v(\mathbf{a}) \stackrel{\text{def}}{=} \inf\left\{ r \;:\; r \in \mathbb{R}, \sum_{n=1}^{\infty} \left|\frac{1}{a_n}\right|^r < \infty \right\}$$

resp.

$$\delta(\mathbf{a}) \stackrel{\text{def}}{=} \sup\left\{ m \;:\; m \in \mathbb{Z}^+, \sum_{n=1}^{\infty} \left|\frac{1}{a_n}\right|^m = \infty \right\}$$

is the exponent of convergence resp. the exponent of divergence.

The function

$$E(z,p) \stackrel{\text{def}}{=} \begin{cases} 1 - z & \text{if } p = 0 \\ (1-z)\exp\left[\sum_{k=1}^{p} \dfrac{z^k}{k}\right] & \text{if } p \in \mathbb{N} \\ , p \in \mathbb{N} & \end{cases}$$

is useful in the discussion of Weierstraß factorization theorem, viz.:

If $|z_n| \uparrow \infty$, $f \in \mathcal{E}$, and $Z(f) \setminus \{0\} = \{z_n\}_{n\in\mathbb{N}}$ and $|z_n| \uparrow \infty$ then for some g in $\mathcal{E}$, some sequence $\{p_n\}_{n\in\mathbb{N}}$ contained in $\mathbb{N}$, and some λ in $\mathbb{Z}^+$, $f(z) = z^\lambda e^{g(z)} \prod_{n=1}^{\infty} E\left(\dfrac{z}{z_n}, p_n\right)$.

10.30. a) If $f(z) \stackrel{\text{def}}{=} \sum_{n=0}^{\infty} c_n z^n$ and $0 < r < R_c$ then for some $\nu_f(r)$ (the central index) in $\mathbb{Z}^+$, $\left|c_{\nu_f(r)} r^{\nu_f(r)}\right| \stackrel{\text{def}}{=} \mu_f(r) \geq |c_n| r^n, n \in \mathbb{Z}^+$, and if $k > \nu_f(r)$ then $\left|c_{\nu_f(r)}\right| r^{\nu_f(r)} > |c_k| r^k$. b) $\nu_f(r)$ is a right-continuous monotonely increasing $\mathbb{Z}^+$-valued step-function. c) If f is transcendental then $\nu_f(r) \uparrow \infty$ as $r \uparrow \infty$. d) If $a > 0$ and $g(z) \stackrel{\text{def}}{=} f(az)$ then $\rho(g) = \rho(f)$.

10.31. For the situation in **10.30c**): a)

$$\rho(f) = \inf\left\{\lambda \ : \ |f(z)| < \exp\left(|z|^\lambda\right)\right\} \stackrel{\text{def}}{=} \omega(f);$$

b) if $0 < \rho(f) < \infty$ and

$$\mathsf{K}(f) \stackrel{\text{def}}{=} \{k \ : \ k > 0 \text{ and for large } r,\ M(r;f) < \exp\left(kr^\rho\right)\} \neq \emptyset$$

then $\tau(f) = \inf \mathsf{K}(f) \stackrel{\text{def}}{=} \upsilon(f)$. c) $\rho(f) = \overline{\lim}_{r\to\infty} \dfrac{\ln\ln \mu_f(r)}{\ln r} \stackrel{\text{def}}{=} \zeta(f)$.

10.32. a) If $\rho, \tau > 0$ then $e^{\tau z^\rho}$ is of order ρ and of type τ of that order. b) e^{e^z} is of order ∞. c) If $|q| < 1$ then $f(z) \stackrel{\text{def}}{=} \sum_{n=0}^\infty q^{n^2} z^n$ is of order zero. d) If $\alpha > 0$, $F_\alpha(z) \stackrel{\text{def}}{=} \displaystyle\sum_{n=1}^\infty \left(\frac{n}{\alpha e}\right)^{-\frac{n}{\alpha}} z^n$ is of order α and of type one of that order.

10.33. What is $\nu_f(r)$ for: a) $f(z) \stackrel{\text{def}}{=} \sum_{n=0}^\infty z^n$? b) $f(z) \stackrel{\text{def}}{=} \sum_{n=0}^\infty \dfrac{z^n}{(n!)^\alpha}$ when $\alpha > 0$?

10.34. a) $\rho(f) = \sigma \stackrel{\text{def}}{=} \inf\left\{a \ : \ a \geq 0, \overline{\lim}_{r\to\infty} |f(z)|\,\big|_{|z|=r} \leq e^{|z|^a}\right\}$. b) If $f(z) \stackrel{\text{def}}{=} \sum_{n=0}^\infty c_n z^n$ then, by abuse of notation when $c_n = 0$,

$$\rho(f) = \overline{\lim}_{n\to\infty} \frac{n \ln n}{-\ln|c_n|} \stackrel{\text{def}}{=} \xi;$$

c) If $\rho(f) < \infty$ then $\tau(f) = \dfrac{1}{e\rho(f)} \overline{\lim}_{n\to\infty} n\,|c_n|^{\frac{\rho(f)}{n}}$.

10.35. When $f(z) = \sum_{n=1}^\infty \dfrac{z^n}{(n!)^\alpha}$ what are $\rho(f)$ and $\tau(f)$?

10.36. If $k \in \mathbb{N}$ and $g(z) = z^k f(z)$ then: a) $\rho(g) = \rho(f)$; b) $\rho(f) = \rho(f')$; c) $\tau(f') = e^{\rho(f)}\tau(f)$.

10.37. How are a) $\rho(f+g)$, b) $\rho(fg)$, and c) $\rho(f \circ g)$ related to $\rho(f)$ and $\rho(g)$?

10.38. If $f(z) \stackrel{\text{def}}{=} \sum_{n=0}^\infty c_n z^n$ is entire then, by abuse of notation when $c_n = 0$, $\overline{\lim}_{n\to\infty} \dfrac{n}{-\ln|c_n|} = 0$.

10.39. If $f_n(z) \stackrel{\text{def}}{=} f(z^n), n \in \mathbb{N}$, what is $\rho(f_n)$ resp. $\tau(f_n)$ in terms of $\rho(f)$ resp. $\tau(f)$?

10.40. If $\{\alpha_n\}_{n\in\mathbb{N}} \subset (0,\infty)$ then for some f, $\lim_{n\to\infty} \dfrac{M(n;f)}{\alpha_n} = \infty$.

10.41. If $\alpha > 0$, $f(z) = \sum_{n=0}^{\infty} c_n z^n$, and $|f(z)| \leq e^{|z|^\alpha}$ then $|c_n| \leq n^{-\frac{n}{\alpha}}$.

10.42. If $Z(f) = \{a_n\}_{n\in\mathbb{N}} \stackrel{\text{def}}{=} \mathbf{a}$ and $\epsilon > 0$ then $v(\mathbf{a}) \leq \rho(f)$.

10.43. For $f(z) \stackrel{\text{def}}{=} \prod_{n=2}^{\infty} \left(1 + \dfrac{z}{n(\ln n)^2}\right)$, $\rho(f) = 1$ and $\tau(f) = 0$.

10.44. If $p < s \leq p+1$ then for some $A(s)$, $\|E(z,p)\| < \exp\left(A(s)|z|^s\right)$.

10.45. For $\gamma \stackrel{\text{def}}{=} \lim_{n\to\infty} \left[\left(\sum_{k=1}^{n} \dfrac{1}{k}\right) - \ln n\right]$ (Euler's constant) and

$$\frac{1}{\Gamma(z)} \stackrel{\text{def}}{=} e^{\gamma z} z \prod_{n\in\mathbb{N}} \left[\left(1+\frac{z}{n}\right)\exp\left(-\frac{z}{n}\right)\right],$$

$\rho\left(\dfrac{1}{\Gamma}\right) = 1$ and $\tau\left(\dfrac{1}{\Gamma}\right) = \infty$.

10.46. In the context and notation of the Weierstraß factorization theorem: a) if $|z| \leq 1$ then $|E(z,p) - 1| \leq |z|^{p+1}$; b) if the exponent of divergence $\delta(\mathbf{a})$ of $\mathbf{a} \stackrel{\text{def}}{=} \{a_n\}_{n\in\mathbb{N}}$ is positive then, λ denoting the order of the zero of f at 0, $f(z) = z^\lambda e^{g(z)} \prod_{n=1}^{\infty} E\left(\dfrac{z}{a_n}, \delta(\mathbf{a})\right)$.

10.47. (Hadamard) If the order of f in **10.42** is ρ then: a) if $\delta(\mathbf{a}) = 0$ the exponential factors in the $E\left(\dfrac{z}{a_n}, 0\right)$ can be omitted; b) in any event, the function g may be chosen in $\mathbb{C}[z]$ and so that $\deg(p) \leq \lfloor\rho\rfloor$.

10.48. For $f(z) \stackrel{\text{def}}{=} \sin z$: a) $\rho(f) = 1$ and $\tau(f) = \dfrac{1}{e}$; b)

$$\sin \pi z = \pi z \prod_{n\in(\mathbb{Z}\setminus\{0\})} E\left(\frac{z}{n}, 1\right) \left(= \pi z \prod_{n=1}^{\infty}\left(1 - \frac{z^2}{n^2}\right)\right) \stackrel{\text{def}}{=} \tilde{f}(z).$$

10.49. True or false: if $g \in C(\mathbb{R}, \mathbb{R})$ then for some f,

$$|f(\Re(z))| > |g(\Re(z))|?$$

11
Analytic Continuation

11.1. Analytic Continuation of Series

Conventions

When Ω_1, Ω_2 are two regions such that $\Omega_3 \stackrel{\text{def}}{=} \Omega_1 \cap \Omega_2 \neq \emptyset$ and $f_i \in H(\Omega_i), i = 1, 2$, while $f_1(z) = f_2(z)$ in Ω_3 then the function element (f_2, Ω_2) resp. (f_1, Ω_1) is an immediate analytic continuation of the function element (f_1, Ω_1) resp. (f_2, Ω_2). When $\Omega_i, 1 \leq i \leq n$, is a finite sequence of regions such that

$$\Omega_i \cap \Omega_{i+1} \neq \emptyset, 1 \leq i \leq n-1, \; f_i \in H(\Omega_i), 1 \leq i \leq n,$$

and each (f_{i+1}, Ω_{i+1}) is an immediate analytic continuation of (f_i, Ω_i) then (f_n, Ω_n) is an analytic continuation of (f_1, Ω_1).

[**Note 11.1:** If $l - k > 1$ and $z \in \Omega_k \cap \Omega_l$ then $f_k(z)$ and $f_l(z)$ can be different.]

When $\gamma : [0, 1] \ni t \mapsto \gamma(t) \in \bigcup_{i=1}^{n} \Omega_i$ is a curve and, in the context above, $\gamma(0) \in \Omega_1, \gamma(t_i) \in (\Omega_i \cap \Omega_{i+1}), 1 \leq i \leq n-1, \gamma(1) \in \Omega_n$ then f_n is an analytic continuation of f_1 from Ω_1 to Ω_n along γ, by abuse of language, along γ^*, by further abuse of language, an analytic continuation from f_1 (to f_n).

If Ω is a simply connected region, $f \in H(\Omega)$, and $0 \notin f(\Omega)$ then $k \stackrel{\text{def}}{=} \frac{f'}{f} \in H(\Omega)$. If $a, z \in \Omega$ then for any rectifiable curve γ such that

$$\gamma^* \subset \Omega, \; \gamma(0) = a, \; \gamma(1) = z,$$

$\int_\gamma k(w)\,dw \stackrel{\text{def}}{=} K(z)$ is independent of the choice of γ and therefore defines a K in $H(\Omega)$. It follows that if $e^K = G$ then $\frac{(e^K)'}{e^K} = \frac{f'}{f}$ whence $f = f(a)e^K$. For some unique θ in $[0, 2\pi)$, $f(a) = |f(a)|e^{i\theta}$ and thus $f = e^{K + \ln|f(a)| + i\theta}$. Any of the functions $K + \ln|f(a)| + i(2m\pi + \theta), m \in \mathbb{Z}$, are branches of $\ln f$ in Ω. In particular, if Ω is a simply connected region not containing $\{0\}$, there are infinitely many branches of $\ln z$ in Ω.

When Ω is a region, $a \in \partial\Omega$, $r > 0$, and f in $H(\Omega)$ admits an immediate analytic continuation g from Ω to $D(a, r)^\circ$ then a is a regular point of $\partial\Omega$.

Otherwise a is a singular point of $\partial\Omega$. If each point of $\partial\Omega$ is a singular point then $\partial\Omega$ is a natural boundary for f.

11.1. a) For $f(z) \stackrel{\text{def}}{=} \sum_{n=1}^{\infty} nz^n$ and $N(a)$ an open neighborhood of a in $\mathbb{C} \setminus \{1\}$, there is a unique function element $(f_a, N(a))$ that is an analytic continuation of f from U. b) When $a \neq 1$ what are the $c_n(a)$ and $R_{c(a)}$ in $f_{[-a]}(z) = \sum_{n=0}^{\infty} c_n(a)(z-a)^n$?

11.2. For $f(z) \stackrel{\text{def}}{=} \sum_{n=1}^{\infty} \frac{z^n}{n^2}$, if $\Omega \cap U \neq \emptyset$, Ω, $U \cup \Omega$ and $U \cap \Omega$ are simply connected, and $\{1\} \not\subset \Omega$, there is an analytic continuation of f from U along any curve in $U \cup \Omega$.

11.3. The natural boundary for $f(z) \stackrel{\text{def}}{=} \sum_{n=1}^{\infty} z^{n!} \in H(U)$ is $\mathbb{T}$.

11.4. a) (Ostrowski) Assume $\mathbb{N} \ni p_1 < p_2 < \cdots$, $0 < \lambda \in \mathbb{R}$, $\{q_n\}_{n\in\mathbb{N}} \in \mathbb{N}$, and $q_n > \left(1 + \frac{1}{\lambda}\right) p_n, n \in \mathbb{N}$. If

$$f(z) \stackrel{\text{def}}{=} \sum_{n=1}^{\infty} c_n z^n, \ R_c = 1, \ c_n = 0, p_k < n < q_k, \ 1k \in \mathbb{N},$$

and 1 is a regular point of f then the sequence $\{s_{p_k}\}_{k\in\mathbb{N}}$ of partial sums of the series converges in some nonempty neighborhood of 1.

b) (Hadamard) In the notations and conditions of a), if

$$p_{k+1} > \left(1 + \frac{1}{\lambda}\right) p_k, c_k \neq 0, k \in \mathbb{N}, f(z) = \sum_{k=1}^{\infty} c_k z^{p_k}, \text{ and } R_c = 1$$

then $\mathbb{T}$ is a natural boundary for f.

[**Note 11.2:** In a) although the series representing f diverges for each z outside $D(0,1)$, a sequence of partial sums of the series does converge in some nonempty neighborhood of 1: there is overconvergence. By contrast, any sequence of partial sums of the series in **11.3** diverges at each point of $\mathbb{T}$.

In a), the sizes of the gaps [the sequences of successive zero coefficients] increase rapidly while the sizes of the nongaps [the sequences of successive nonzero coefficients] may increase as well; in b), the length of each nongap is 1.]

c) For a given series $\sum_{n=0}^{\infty} c_n z^n$ for which $R_c = 1$, $\mathbb{N}$ contains infinitely many pairwise disjoint sequences $\left\{\{m(p,q)\}_{q\in\mathbb{Z}^+}\right\}_{p\in\mathbb{N}}$ such that $\mathbb{T}$ is a natural boundary for each subseries $\sum_{q=0}^{\infty} c_{m(p,q)} z^{m(p,q)}, p \in \mathbb{N}$.

d) Let $f(z)$ be represented by $\sum_{n=1}^{\infty} c_n z^n$, for which $R_c = 1$. For each t in $[0,1]$, let $\{\epsilon_n(t)\}_{n\in\mathbb{T}}$ be **the** sequence of binary markers for t. For all but at most countably many of the functions,

$$f_t(z) \stackrel{\text{def}}{=} \sum_{n=1}^{\infty} \epsilon_n(t) c_n z^n, t \in [0,1],$$

$\mathbb{T}$ is the natural boundary.

11.5. a) If $a \in \mathbb{I}$ then $f(z) \stackrel{\text{def}}{=} \sum_{n=1}^{\infty} \frac{1}{2^n \left(z - e^{2\pi i n a}\right)}$ is in both

$$H(U) \stackrel{\text{def}}{=} H(\Omega_1) \text{ and } H(\mathbb{C} \setminus D(0,1)) \stackrel{\text{def}}{=} H(\Omega_2).$$

b) The natural boundary for $f|_{\Omega_i}, i = 1, 2$, is $\mathbb{T}$.

11.6. a) The domain of convergence Ω of the series $\sum_{n=1}^{\infty} \frac{z}{n(z-n)}$ is $\mathbb{C} \setminus \mathbb{N}$. b) For some F, $S(F) \cap \mathbb{C} = P(F)$ (F is meromorphic in $\mathbb{C}$) and $F(z)|_{\Omega} = \sum_{n=1}^{\infty} \frac{z}{n(z-n)}$. c) When $r \in \mathbb{R}^+ \setminus \mathbb{N}$, $\frac{1}{2\pi i}\int_{|z|=r} F(z)\,dz = [r]$.

11.7. a) The domain of convergence Ω of the series $\sum_{n=0}^{\infty} \frac{z}{z^2 - n^2}$ is $\mathbb{C} \setminus (\mathbb{Z} \setminus \{0\})$. b) The series in a) defines a meromorphic function f. c) When $r \in \mathbb{R}^+ \setminus \mathbb{N}$, $\frac{1}{2\pi i}\int_{|z|=r} f(z)\,dz = \frac{[r]}{2}$.

11.2. General Theory

11.8. If $f \in H(\Omega)$, $a \in \Omega$, and for some M,

$$\left|f^{(n)}(a)\right| \le M^n, n \in \mathbb{N},$$

then for some F in $\mathcal{E}$, $F|_{\Omega} = f$.

11.9. Assume $\Omega \supset \mathbb{T}$, $f \in H(\Omega)$, $c_n \stackrel{\text{def}}{=} \frac{1}{2\pi}\int_0^{2\pi} f\left(e^{it}\right) e^{-int}\,dt, n \in \mathbb{Z}$, and $r \stackrel{\text{def}}{=} \overline{\lim}_{n\to-\infty} |c_n|^{\frac{1}{n}}, R^{-1} = \overline{\lim}_{n\to\infty} |c_n|^{\frac{1}{n}}$. For some F in $H\left(A(0, r : R)^\circ\right)$, $F|_{\Omega} = f$.

11.10. If $\Omega \supset \mathbb{T}$ and $f \in H(\Omega)$ for some F in $H(\Omega \cup U)$, $F|_{\Omega} = f$ iff for each n in $\mathbb{Z}^+$, $\int_{\mathbb{T}} f(z) z^n\,dx = 0$.

11.11. Let Ω be a region containing the compact set K and let $\Omega \setminus K$ be a region Ω_1. If $\{f_n\}_{n\in\mathbb{N}} \subset H(\Omega)$ and $f_n \stackrel{\text{u}}{\to} f$ on every compact subset of Ω_1 then for some F in $H(\Omega)$, $F|_{\Omega_1} = f$.

11.12. If $f \in C(D(0,2),\mathbb{C}) \cap H(D(0,2)^\circ \setminus [0,1])$ then f is in $H(D(0,2)^\circ)$.

11.13. If $\Omega \supset Q \stackrel{\text{def}}{=} \{z : 0 \le \Re(z), \Im(z) \le 1\}, f \in H(\Omega), f(\partial Q) \subset \mathbb{R}$ then for some F in $\mathcal{E}$, $F|_\Omega = f$.

11.14. For $\Omega \stackrel{\text{def}}{=} \left\{z : -\frac{\pi}{2} < \Re(z) < \frac{\pi}{2}, \Im(z) > 0\right\}$, if

$$f \in H(\Omega) \cap C(\Omega^c, \mathbb{C}) \text{ and } f(\partial\Omega) \subset \mathbb{R}$$

then for some F in $\mathcal{E}$, $F|_\Omega = f$ and $F(\pi - z) = F(z + 2\pi) = F(z)$.

11.15. Assume: a) $f \in H(U)$, $\sup_{z\in U} |f(z)| < \infty$, and for some positive r, there is an analytic continuation f_r of f from U along any curve in $U \cup D(1,r)^\circ$; b) for some positive δ, $|1 - z| + |f(z)| \ge \delta$ on U. For some g, h in H^∞, $(1 - z)g(z) + f(z)h(z) \equiv 1$ on U.

11.16. Assume $f \in H\left(D\left(1, \frac{1}{2}\right)^\circ\right)$ and that f may be continued analytically along every half-line $L \stackrel{\text{def}}{=} \langle 1, \infty \rangle$ not passing through zero. For any branch β of $\ln z$ in $\Omega \stackrel{\text{def}}{=} \mathbb{C} \setminus (-\infty, 0]$ and some G in $\mathcal{E}$, the resulting analytic continuation F of f along any curve in Ω satisfies: $F(z) = G(\beta(z))$ in Ω.

11.17. If $f \in H(U \setminus \{0\}) \stackrel{\text{def}}{=} H(\Omega)$ then for some F in

$$H(\{z : \Re(z) < 0\}) \stackrel{\text{def}}{=} H(\Omega_1),$$

each $D(a,r)^\circ$ contained in Ω, and some branch $\beta(z)$ of $\ln z$, $F(\beta(z)) = f(z)$ in $D(a,r)^\circ$.

The **Problems 11.18 – 11.20** lead to some interesting properties of the Gamma function Γ first mentioned in **10.45**. Furthermore, they simplify a derivation of the connection between Γ and the Zeta function

$$\zeta(z) \stackrel{\text{def}}{=} \sum_{n=1}^{\infty} e^{-z \ln n}$$

and they lead to a method for the analytic continuation of ζ from its domain $\Omega_1 \stackrel{\text{def}}{=} \{z : \Re(z) > 1\}$ along curves in $\Omega_2 \stackrel{\text{def}}{=} \mathbb{C} \setminus \{1\}$.

11.18. a) Γ is meromorphic and $P(\Gamma) = \mathbb{Z} \setminus \mathbb{N}$; b) for z in $\Omega \stackrel{\text{def}}{=} \mathbb{C} \setminus P(\Gamma)$,

$$\Gamma(z) = \frac{1}{z} \prod_{n=1}^{\infty} \left\{ \left(1 + \frac{1}{n}\right)^z \left(1 + \frac{z}{n}\right)^{-1} \right\};$$

c) for z in Ω, Γ satisfies the functional equation

$$F(z+1) = zF(z); \tag{11.1}$$

d) if g is meromorphic, $P(g) = P(\Gamma)$, and in Ω, $g(z+1) = g(z)$ then $h \stackrel{\text{def}}{=} g\Gamma$ also satisfies (11.1); e) for each n in $\mathbb{N}$, $\Gamma(z) = \dfrac{\Gamma(z+n+1)}{\prod_{k=0}^{n-1}(z+k)}$; f) for z in Ω, $\Gamma(z) = \lim_{n\to\infty} \dfrac{n!n^z}{z(z+1)\cdots(z+n)}$; g) $\Gamma(z)\Gamma(1-z) = \dfrac{\pi}{\sin \pi z}$.

11.19. Show: a) if $\Re(z) > 1$ then

$$\Gamma(z) = \int_0^\infty e^{-t}t^{z-1}\,dt; \tag{11.2}$$

b) if $f_i, i = 1, 2$, in $(0,\infty)^{(a,b)}$ are log convex then $f_1 + g_1$ is also log convex; c) for x in $(0,\infty)$, $\ln \Gamma(x)$ is a convex function.

11.20. Show: a) for $\Omega_1 \stackrel{\text{def}}{=} \{\, z : \Re(z) > 1 \,\}$, $\zeta \in H(\Omega_1)$ and for z in Ω_1,

$$\zeta(z) = \frac{1}{\Gamma(z)} \int_0^\infty \frac{t^{z-1}}{e^z - 1}\,dt. \tag{11.3}$$

b) there is an analytic continuation of ζ along any curve in $\Omega_2 \stackrel{\text{def}}{=} \mathbb{C} \setminus \{1\}$.

11.21. If $f \in H(U)$ and for z in $\frac{1}{2}U$, $f(2z) = 2f(z)f'(z)$ then for some F in $\mathcal{E}$, $F|_U = f$.

11.22. If $f \in H(U)$ and for z in $\frac{1}{2}U$, $f(2z) = \dfrac{2f(z)}{1-(f(z))^2}$, there is a meromorphic function F such that $F|_U = f$.

11.23. a) For some power series $\sum_{n=0}^\infty c_n(z-1)^n \stackrel{\text{def}}{=} f(z)$, $e^{f(z)} = z$ in $D(1,1)^\circ$. b) How is f continued analytically along any curve in $\mathbb{C} \setminus \{0\}$?

11.24. a) For some power series $\sum_{n=0}^\infty c_n(z-1)^n \stackrel{\text{def}}{=} g(z)$, $g(z)^2 = z$ in $D(1,1)^\circ$. b) How is g continued analytically along any curve in $\mathbb{C} \setminus \{0\}$?

11.25. (Poincaré) If $f \in H(U)$, $a \in \mathbb{C}$, and $N(a)$ is an open neighborhood of a, there are at most countably many pairwise different analytic continuations $(g, N(a))$ of (f, U).

11.26. Let Ω be a region. For each a in Ω let there be at least one function element $\left(f_a, D\left(a, r_a\right)^\circ\right)$ and assume that each function element is an analytic continuation of every other one. Assume further that for each c in $\partial_\infty\Omega$, $\overline{\lim}_{z\to c} |f_a(z)| \le M < \infty$. If $a \in \Omega$, $\sup_{z\in D(a,r_a)^\circ} |f_a(z)| \le M$.

[**Note 11.3:** Informally one describes a collection of function elements conforming to the conditions in **11.26** as a regular function F: F is not single-valued (not a function) but for each value w of F at a point a, there is a function element $\left(f_a, D(a, r_a)^\circ\right)$ such that $w = f_a(a)$.

Thus **11.26** may be paraphrased as the maximum modulus theorem for regular functions.]

11.27. For $p_k, 1 \le k \le n$, in $(0, \infty)$, $\{f_k\}_{1\le k\le n}$ a set of functions holomorphic in a region Ω, assume that the hypotheses of **11.26** obtain for the regular functions $f_k(z)^{p_k}, 1 \le k \le n$. If $\phi(z) \stackrel{\text{def}}{=} \sum_{k=1}^{n} |f_k(z)|^{p_k}$ then $\sup_{z\in\Omega} \phi(z) \le Mn$.

12
Singularities

12.1. General Theory

Conventions

Each function introduced in this **Chapter** is assumed to be differentiable at all points except those specified as singularities. A singularity need not be isolated, e.g., the singularity of $\frac{1}{\sin\left(\frac{1}{z}\right)}$ at zero.

When f is defined in a region Ω and $S(f) \cap \Omega = P(f) \cap \Omega$, f is meromorphic in Ω. The set of functions meromorphic in Ω is denoted $M(\Omega)$.

When $f \in M(\Omega)$ and $a \in P(f) \cap \Omega$, the Laurent expansion for f near a is, for some N in $\mathbb{N}$, $f(z) = \sum_{n=-N}^{\infty} c_n(z-a)^n$. The sum

$$\sum_{n=-N}^{-1} c_n(z-a)^n \stackrel{\text{def}}{=} \mathcal{P}_a(f)$$

is the principal part or singular part of f near a. The residue at a is $\text{Res}(f,a) \stackrel{\text{def}}{=} c_{-1}$.

12.1. If a is a pole or an isolated essential singularity of f then a is not a pole of e^f.

12.2. If $f \in H(\mathbb{C} \setminus \{0\})$, zero is a simple pole of f, and $f(\mathbb{T}) \subset \mathbb{R}$ then for some a in $\mathbb{C} \setminus \{0\}$ and some b in $\mathbb{R}$, $f(z) = az + \overline{a}\frac{1}{z} + b$.

12.3. If $\{|a_n|\}_{n\in\mathbb{N}} \subset (0,1)$, $\lim_{n\to\infty} a_n = 0$, $f \in H\left(U \setminus \left(\{a_n\}_{n\in\mathbb{N}} \cup \{0\}\right)\right)$, and each a_n is a pole of f then $\mathbb{C} = \left(f\left(U \setminus \{a_n\}_{n\in\mathbb{N}}\right)\right)^c \stackrel{\text{def}}{=} E$.

12.4. If a is an isolated singularity of f and near a and for some m in $\mathbb{N}$, $\Re(f(z)) \le -m \ln|z-a|$ then a is removable.

12.5. Is there in $H(\mathbb{C} \setminus \{0\})$ an f such that for each positive r,

$$J_r(f) \stackrel{\text{def}}{=} \frac{1}{2\pi i}\int_{C_0(r)} f(z)\,dz = 0$$

and zero is: a) a pole of order not less than two? b) an essential singularity? c) a pole of order one?

12.6. If $f \in H(U \setminus \{0\})$ and $\int_U |f(x+iy)|^2 \, d\lambda_2(x,y) < \infty$ then 0 is a removable singularity.

12.7. a) For α in $\mathbb{C}$ and b in $\mathbb{C} \setminus \{0\}$,

$$f_b(z) \stackrel{\text{def}}{=} \sum_{n=0}^{\infty} \binom{\alpha}{n} (z-b)^n \in H\left(D(b,|b|)^\circ\right).$$

b) If $\alpha \neq 0$, $k \in \mathbb{Z}$, $\theta_b(z) \in [0, 2\pi)$, and $f_b(z) \stackrel{\text{def}}{=} |f_b(z)| e^{i\theta_b(z)}$ then for z in $D(b,|b|)^\circ$, $\exp\left(\frac{\ln|f_b(z)| + i(\theta_b(z) + 2k\pi)}{\alpha}\right) = z$. c)

$$S(f) = \begin{cases} \emptyset & \text{if } \alpha \in \mathbb{Z}^+ \\ \{0\} & \text{otherwise} \end{cases}.$$

d) Iff $\alpha \in \mathbb{Z}$, there is a (Laurent) series $\sum_{n=-\infty}^{\infty} c_n z^n \stackrel{\text{def}}{=} g(z)$ valid in $A(0, 0 : \infty)$ and such that $g(z) = f_b(z)$ in $A(0, 0 : \infty) \cap D(b, |b|)^\circ$. e) If $\alpha \notin \mathbb{Z}$, $S(f)$ $(= \{0\})$ is neither a pole nor an isolated essential singularity of f_b.

12.2. Meromorphic Functions

12.8. If $D(0,1) \subset \Omega$, $f \in M(\Omega)$, and $f(\mathbb{T}) \subset \mathbb{T}$ then f is a rational function.

12.9. If $f \in M(\mathbb{C}_\infty)$ then f is a rational function.

12.10. If $\{a_n, b_n\}_{n \in \mathbb{N}} \subset \mathbb{C}$ and $\sum_{n=1}^{\infty} |a_n - b_n| < \infty$ then $\prod_{n \in \mathbb{N}} \frac{z - a_n}{z - b_n}$ converges in $\Omega \stackrel{\text{def}}{=} \mathbb{C} \setminus \left(\{b_n\}_{n \in \mathbb{N}}\right)^c$ and represents a function in $M(\Omega)$.

12.11. For $\Omega \stackrel{\text{def}}{=} \mathbb{C} \setminus (\mathbb{Z} \setminus \mathbb{N})$, if $f \in H(\Omega) \cap M(\mathbb{C})$ and for $z+1$ in A, $f(z+1) = zf(z)$ then: a) every pole of f is simple; b) $P(f) = \mathbb{Z} \setminus \mathbb{N}$; c) $\text{Res}(f, -n) = \frac{(-1)^n f(1)}{n!}$.

12.12. If $A \subset U$ and $S \stackrel{\text{def}}{=} A^\bullet \cap U$ is empty or finite then: a) $U \setminus A$ is a region; b) if $h \in H(U \setminus A)$, $g \in M(U)$, and $|h(z)| \leq |g(z)|$ on $U \setminus A$ then for some k in $M(U)$, $k(z)\big|_{U \setminus A} = h(z)$.

12.13. If $P(f) \cap U \stackrel{\text{def}}{=} \{p_n\}_{n \in \mathbb{N}}$, $\lim_{n \to \infty} p_n \stackrel{\text{def}}{=} p_0 \in U$, $b \in \mathbb{C}$, and $f \in H\left(U \setminus \{p_n\}_{n \in \mathbb{Z}^+}\right) \stackrel{\text{def}}{=} H(\Omega)$ then U contains a sequence $\{z_n\}_{n \in \mathbb{N}}$ such that $\lim_{n \to \infty} z_n = p_0$ and $\lim_{n \in \mathbb{N}} f(z_n) = b$.

12.3. Mittag-Leffler, Runge, and Weierstraß Theorems

If Ω is a nonempty region, $D(a,r) \subset \Omega$, and $f \in H(\Omega)$ then for some sequence $\{p_n\}_{n\in\mathbb{N}}$ in $\mathbb{C}[z]$, $p_n \stackrel{u}{\to} f$ on $D(a,r)$. The previous sentence is a special case of Runge's theorem to which the following **Problems** lead and from which Mittag-Leffler's and Weierstraß's representations flow.

12.14. If F is a closed subset of $\mathbb{C}$ then $\mathbb{C}\setminus F$ consists of at most countably many components.

12.15. If K is a compact subset of $\mathbb{C}$ and $P \stackrel{\text{def}}{=} \{p_n\}_{1\le n<N\le\infty}$ meets each component of $\mathbb{C}\setminus K$ in one point then R, the set of rational functions F such that $P(f) \subset S$, is dense in $C(K,\mathbb{C})$.

12.16. For some region Ω in $\mathbb{C}_\infty$, $\mathbb{C}_\infty \setminus \Omega$ consists of uncountably many components.

12.17. If $\Omega \in \mathsf{O}(\mathbb{C})$ for some sequence $\{K_n\}_{n\in\mathbb{N}}$ in $\mathsf{K}(\mathbb{C})$: a)

$$K_n \subset K_{n+1}^{\circ} \subset \Omega;$$

b) each compact subset K of Ω is a subset of some K_n; c) each component of $\mathbb{C}_\infty \setminus K_n$ contains a component of $\mathbb{C}_\infty \setminus \Omega$.

12.18. (Runge) If Ω is an open subset of $\mathbb{C}_\infty$, and S meets each component of $\mathbb{C}_\infty \setminus \Omega$ in exactly one point, then for each f in $H(\Omega)$ and some sequence $\{R_n\}_{n\in\mathbb{N}}$ of rational functions, $P(R_n) \subset S$ and $R_n(z) \stackrel{u}{\to} f(z)$ on every compact subset of Ω.

12.19. (Mittag-Leffler) For an open subset Ω of $\mathbb{C}_\infty$, if $S \subset \Omega$, $S^\bullet \cap \Omega = \emptyset$, and for each a in S, there is a rational function $r_a(z) \stackrel{\text{def}}{=} \sum_{n=1}^{N(a)} \dfrac{c_n(a)}{(z-a)^n}$ then for some f in $M(\Omega)$, $P(f) = S$ and $\mathcal{P}_a(f) = r_a$.

12.20. (Weierstraß) If Ω is an open subset of $\mathbb{C}$, $S \subset \Omega$, $S^\bullet \cap \Omega = \emptyset$ then for some f in $H(\Omega)$, $Z(f) = S$.

[**Note 12.1:** The open sets Ω in **12.19** – **12.20** need not be regions.]

12.21. For z in $\mathbb{C}\setminus\mathbb{Z}$, $f(z) \stackrel{\text{def}}{=} \dfrac{\pi^2}{\sin^2 \pi z} = \displaystyle\sum_{n\in\mathbb{Z}} \frac{1}{(z-n)^2}$.

12.22. For z in $\mathbb{C}\setminus\mathbb{Z}$, $\pi \cot \pi z = \dfrac{1}{z} + \displaystyle\sum_{n\notin\mathbb{Z}} \left(\frac{1}{z-n} + \frac{1}{n}\right)$.

13
Harmonic Functions

13.1. Basic Properties

Conventions

When Ω is an open subset of $\mathbb{C}$, $h(x+iy)$ in $C^2(\Omega,\mathbb{R})$ is harmonic iff $\Delta h \stackrel{\text{def}}{=} h_{xx}+h_{yy}=0$ in Ω. The operator Δ is the Laplacian and the set of all functions harmonic in Ω is denoted $L(\Omega)$. When Ω is a region and h is $\Re(f)$ for some $f \stackrel{\text{def}}{=} u+iv$ in $H(\Omega)$ then v is an harmonic conjugate of h.

When Ω is open, an h in $C^2(\Omega,\mathbb{R})$ enjoys the mean value property iff $\{\{a\in\Omega\}\wedge\{D(a,r)\subset\Omega\}\}\Rightarrow\left\{h(a)=\frac{1}{2\pi}\int_0^{2\pi}h\left(a+re^{it}\right)dt\right\}$. The set of all functions enjoying the mean value property (in Ω) is $MVP(\Omega)$.

13.1. If Ω is open, $g\in H(\Omega)$, and $h\in L(g(\Omega))$ then $h\circ g\in L(\Omega)$.

13.2. If Ω is open and $h\in L(\Omega)$ then for each a in Ω, some open $N(a)$, and some f_a in $H(N(a))$, $h=\Re(f_a)$ on $N(a)$. ("Each harmonic function h is locally the real part of a holomorphic function.")

13.3. If Ω is a simply connected region and $h\in L(\Omega)$ then $L(\Omega)$ contains an harmonic conjugate of h.

13.4. If Ω is a region and $h\in L(\Omega)$, then any two harmonic conjugates of h differ by a constant.

13.5. If Ω is a simply connected region, $h\in L(\Omega)$, and $0\notin h(\Omega)$ then $L(\Omega)$ contains a pair $\{p,q\}$ such that $h=p^2-q^2$.

13.6. If $R>0$, $\Xi \stackrel{\text{def}}{=} \partial D(a,R)$, and $0\le r<R$ then for the complex measure situation $(\Xi,\mathsf{S}_\beta(\Xi),\mu)$,

$$h\left(a+re^{i\theta}\right)\stackrel{\text{def}}{=}\int_\Xi P_{r/R}(\theta-t)\,d\mu\left(a+Re^{it}\right)\in L\left(D(a,R)^\circ\right).$$

13.7. a) $L(\Omega)\subset MVP(\Omega)$. b) If Ω is a region and $f\in MVP(\Omega)$ then the following principles of the maximum and the minimum obtain:

i. (Maximum Principle) If for some a in Ω and all z in Ω, $f(a)\ge f(z)$ then f is a constant in Ω;

ii. (Minimum Principle) If for some a in Ω and all z in Ω, $f(a) \le f(z)$ then f is constant in Ω.

c) If Ω is a region then $h \in MVP(\Omega)$ iff $h \in L(\Omega)$.

13.8. If $a \in U$ then for some measure situation $(\mathbb{T}, \mathsf{S}_\beta(\mathbb{T}), \mu_a)$ and for each h in $L(U) \cap C(D(0,1), \mathbb{C})$, $h(a) = \int_{\mathbb{T}} h(z)\, d\mu_a(z)$.

13.9. If Ω is a region and $h, h^2 \in L(\Omega)$ then h or $\overline{h}$ is in $H(\Omega)$.

13.10. If $f \stackrel{\text{def}}{=} u + iv \in H(\Omega)$ and $0 \notin f(\Omega)$ then $\ln|f(z)| \stackrel{\text{def}}{=} h(z) \in L(\Omega)$.

13.11. If Ω is a region, $f \in H(\Omega)$, and $|f| \in L(\Omega)$ then f is a constant.

13.2. Developments

13.12. (Schwarz) If $h \in L\left(D(0,R)^\circ\right) \cap C(D(0,R), \mathbb{C})$ then

$$f(z) \stackrel{\text{def}}{=} \int_{|w|=R} \frac{w+z}{w-z} \frac{h(w)}{w}\, dw \in H\left(D(0,R)^\circ\right)$$

and $\Re(f) = h$.

13.13. Assume $D(0,1) \subset \Omega$, $\left\{f_n \stackrel{\text{def}}{=} u_n + iv_n\right\}_{n\in\mathbb{N}} \subset H(\Omega)$, $\{u_n\}_{n\in\mathbb{N}}$ converges uniformly in $D(0,1)$, and $\{v_n(0,0)\}_{n\in\mathbb{N}}$ converges. The sequence $\{f_n(z)\}_{n\in\mathbb{N}}$ converges uniformly on each compact subset of $D(0,1)^\circ$.

13.14. a) Assume h is harmonic in $D(0,R)^\circ$, $h(D(0,R)) \subset D(0,M)$, and v, an harmonic conjugate of h, is such that $v(0) = 0$. If $|z| \le r < R$ then $|v(z)| \le \dfrac{2M}{\pi} \ln \dfrac{R+r}{R-r}$. b) For some region Ω and some h in $L(\Omega)$, $\sup_{z\in\Omega} |h(z| < \infty$ while no harmonic conjugate of h is bounded in Ω.

13.15. (Harnack) If $h \in L\left(D(a,R)^\circ\right) \cap C(D(a,r), \mathbb{R})$ and $h(D(a,R)) \subset \mathbb{R}^+$ then for z in $D(a,R)^\circ$, $h(a) \cdot \dfrac{R-|z|}{R+|z|} \le h(z) \le h(a) \cdot \dfrac{R+|z|}{R-|z|}$.

13.16. $\displaystyle\int_0^{2\pi} \ln\left|1 - e^{it}\right| dt = 0.$

13.17. (Jensen) If $R > 0$, $\Omega \supset D(0,R)$, $f \in H(\Omega)$, $f(0) \ne 0$, and $Z(f) \cap D(0,R) = \{z_n\}_{1\le n\le N}$ (listed according to multiplicities) then

$$\ln|f(0)| + \sum_{n=1}^{N} \ln\left|\frac{R}{|z_n|}\right| = \frac{1}{2\pi} \int_0^{2\pi} \ln\left|f\left(Re^{it}\right)\right| dt.$$

13.18. (Poisson-Jensen) In the context of **13.17**, if $z \in D(0,R)$, then

$$\ln|f(z)| + \sum_{n=1}^{N} \ln\left|\frac{R^2 - \overline{z_n}z}{R(z - z_n)}\right| = \frac{1}{2\pi}\int_0^{2\pi} \Re\left(\frac{Re^{i\theta}+z}{Re^{i\theta}-z}\right) \ln\left|f\left(Re^{i\theta}\right)\right| d\theta.$$

13.19. a) If $f \in H^\infty$ and

$$\ln^+ t \stackrel{\text{def}}{=} \begin{cases} \ln t & \text{if } t \geq 1 \\ 0 & \text{if } t < 1 \end{cases}$$

then

$$\sup_{r\in[0,1)} \int_0^{2\pi} \ln^+\left|f\left(re^{i\theta}\right)\right| d\theta < \infty.$$

b) If $f \in H(U)$, $Z(f) \cap U = \{a_n\}_{n\in\mathbb{N}}$ (each zero of f is listed according to its multiplicity), and $\sup_{r\in[0,1)} \int_0^{2\pi} \ln^+\left|f\left(re^{i\theta}\right)\right| d\theta < \infty$ then

$$\sum_{n=1}^{\infty} (1 - |a_n|) < \infty \text{ or } f \equiv 0.$$

13.20. (Harnack) For a region Ω and a sequence $\{h_n\}_{n\in\mathbb{N}}$ contained in $L(\Omega)$: a) if $h_n \xrightarrow{\text{a.e.}} h$, $h \in C(\Omega, \mathbb{C})$, and for every $D(a,r)$ contained in Ω, there is a function $f_{a,r}$ integrable on $\partial D(a,r)$ and dominating $|h_n(z)|\,|_{\partial D(a,r)}$ then $h \in L(\Omega)$; b) if $h_n \xrightarrow{\text{u}} h$ on every compact subset of Ω then $h \in L(\Omega)$; c) if $h_n \uparrow h$ then either for some h, $h_n \xrightarrow{\text{u}} h$ on every compact subset of Ω (in which case $h \in L(\Omega)$) or $h \equiv \infty$.

14
Families of Functions

14.1. Sequences of Functions

Conventions

A set $\mathcal{F}$ of functions in $H(\Omega)$ is normal iff every sequence in $\mathcal{F}$ contains a subsequence converging uniformly on compact sets of Ω to a function [not necessarily in $H(\Omega)$]. When the ranges of the elements of $\mathcal{F}$ are regarded in $\mathbb{C}_\infty$ and convergence is regarded as taking place in $\mathbb{C}_\infty$, where ∞ is a legitimate limit, there is a corresponding notion of spherical normality.

For a set S of maps $f : X \mapsto Y$ between topological spaces, a base for the compact-open topology for S is the set of all sets of the form $\{f : f(K) \subset O, K \in \mathsf{K}(X), O \in \mathsf{O}(Y)\}$.

When $f \in H(U)$, $r \in [0,1)$, and $1 \le p \le \infty$, then $f_r(t) \stackrel{\text{def}}{=} f\left(re^{it}\right)$ and

$$M_p(r;f) \stackrel{\text{def}}{=} \begin{cases} \left(\frac{1}{2\pi}\int_0^{2\pi} |f_r(t)|^p \, dt\right)^{\frac{1}{p}} & \text{if } 1 \le p < \infty \\ M(r;f) & \text{if } p = \infty. \end{cases}$$

The set of f such that, by abuse of notation,

$$\|f\|_p \stackrel{\text{def}}{=} \sup_{0 \le r < 1} M_p(r;f) < \infty$$

is the Hardy class H^p. [The Hardy class H^∞ is the natural counterpart of $L^\infty(X,\mu)$.]

For a region Ω and a positive δ,,

$$\Omega_\delta \stackrel{\text{def}}{=} \left\{ z : z \in \Omega, \inf_{w \in \partial\Omega} |z - w| \ge \delta \right\}.$$

14.1. If $\{f_n\}_{n\in\mathbb{N}} \subset H(\Omega)$ is normal then for some sequence $\{m_n\}_{n\in\mathbb{N}}$, $m_n \uparrow \infty$ and $\{f_n^{m_n}\}_{n\in\mathbb{N}}$ is normal.

14.2. For some $\{f_n\}_{n\in\mathbb{N}}$ in $H(U)$, each f_n maps U conformally onto U and $f_n \stackrel{u}{\to} 1$ on compact subsets of U.

14.3. For which regions Ω is $\mathcal{F}_\Omega \stackrel{\text{def}}{=} \bigcup_{n\in\mathbb{N}} \{ f : f(z)|_\Omega = z^n \}$ normal?

14.4. For which regions Ω is $\mathcal{F}_\Omega \stackrel{\text{def}}{=} \left\{\sin nz\big|_\Omega\right\}_{n\in\mathbb{N}}$ normal?

14.5. If $2 \le q < \infty$ and $\lim_{n\to\infty} \int_U |f_n(z)|^q \, d\lambda_2 = 0$ then $f_n \stackrel{\text{u}}{\to} 0$ on compact subsets of U.

14.2. General Families

14.6. For any Ω, $H(\Omega)$ is a Banach space with respect to the norm $\| \ \|_\infty$ and the unit ball $B(\mathbf{O},1)$ is a normal family.

14.7. If $1 \le p < \infty$ then: a) $\| \ \|_p$ is a norm for the vector space H^p; b) with respect to $\| \ \|_p$, H^p is a Banach space; c) the unit ball $B(\mathbf{0},1)$ of H^p is a normal family.

14.8. If $f(z) \stackrel{\text{def}}{=} \sum_{n=0}^\infty c_n z^n \in H^2$ then for some f^* in $L^2([0,2\pi],\lambda)$: a) $\|f^* - f_r\|_2 \to 0$ as $r \uparrow 1$; b) $f_r \stackrel{\text{a.e.}}{\to} f^*$ as $r \uparrow 1$.

14.9. a) If $f \in H^2$ and f^* is the function defined in **14.8** then

$$\widehat{f^*_n} = \begin{cases} c_n & \text{if } n \in \mathbb{Z}^+ \\ 0 & \text{otherwise.} \end{cases}$$

b) If $g \in L^2([0,2\pi],\lambda)$ and, for n in $\mathbb{Z} \setminus \mathbb{Z}^+$, $\widehat{g}_n = 0$ then for some $\widetilde{g}$ in H^2, $\widetilde{g}^* \doteq g$.

14.10. If $\sum_{n=0}^\infty b_n z^n$ converges in U and $b_n \ge 0, n \in \mathbb{Z}^+$ then

$$\mathcal{F} \stackrel{\text{def}}{=} \left\{ \sum_{n=0}^\infty a_n z^n \ : \ |a_n| \le b_n \right\}$$

is normal.

14.11. For $\{a_k\}_{1\le k\le n} \subset (0,1)$ and

$$\mathcal{F} \stackrel{\text{def}}{=} \{ f \ : \ f \in H(U), f(U) \subset U, f(0) = f(a_k) = 0, 1 \le k \le n \},$$

what is $\sup_{f\in\mathcal{F}} |f'(0)|$?

14.12. The family $\mathcal{F} \stackrel{\text{def}}{=} \left\{ f \ : \ f = \sum_{k=0}^\infty c_k(f) z^k, f \in H(U) \right\}$ is normal iff for $K_k \stackrel{\text{def}}{=} \sup_{f\in\mathcal{F}} |c_k(f)|$, $\overline{\lim}_{k\to\infty} K_k^{\frac{1}{k}} \stackrel{\text{def}}{=} K \le 1$.

14.13. Assume $0 \in \Omega$, $f \in C(\Omega \times D(0,1), \mathbb{C})$, for each w in Ω,

$$f(w,z) \in H(\Omega), \frac{\partial f(w,z)}{\partial z} \in C(\Omega \times D(0,1), \mathbb{C}),$$

and for some n in $\mathbb{N}$, $f(0,z) \equiv z^n$. For each w in Ω and for some points $g_{k,w}, 1 \le k \le n$, not necessarily pairwise different, in U, $f(w, g_{k,w}) = 0$.

14.14. The family

$$\mathcal{F} \stackrel{\text{def}}{=} \{ f \ : \ f \in H(U), f(U) \subset U, f(0) = f'(0) = f''(0) = 0 \}$$

contains one and only one F such that $F\left(\frac{1}{2}\right) = \sup_{f\in\mathcal{F}}\left|f\left(\frac{1}{2}\right)\right| \stackrel{\text{def}}{=} \alpha$.

14.15. Let $S \stackrel{\text{def}}{=} \{z_n\}_{n\in\mathbb{N}}$ be a sequence of pairwise different points in $\mathbb{C}$ and assume $\{q_n\}_{n\in\mathbb{N}} \subset \mathbb{N}$. For some h in $\mathcal{I}$, the set of entire functions having at z_n a zero of order at least q_n, $\mathcal{I} = h \cdot \mathcal{E}$. (Thus $\mathcal{I}$ is a principal ideal in the algebra $\mathcal{E}$.)

14.16. If $\mathcal{F} \stackrel{\text{def}}{=} \{ f \ : \ f \in H(U), f(U) \subset U \}$ then for some f_0 in $\mathcal{F}$, $\left|\frac{\partial \Re\left(f_0\right)(0)}{\partial x}\right| = \sup_{f\in\mathcal{F}}\left|\frac{\partial \Re(f)(0)}{\partial x}\right|$.

14.17. The family $\mathcal{F} \stackrel{\text{def}}{=} \left\{ f \ : \ f \in A(U), \int_{\mathbb{T}} |f(z)|\,d\tau \le 1 \right\}$ is a normal subset of $H(U)$.

14.18. For a region Ω, if $\mathcal{F} \stackrel{\text{def}}{=} \left\{ f \ : \ f \in H(\Omega), \int_\Omega |f(z)|\,d\lambda_2 \le 1 \right\}$ then: a) for each a in Ω for some $M(a)$ in $\mathbb{R}$, and for each f in $\mathcal{F}$,

$$|f(a)| \le M(a) \int_\Omega |f(z)|\,d\lambda_2;$$

b) $\mathcal{F}$ is normal; c) $\mathcal{F}$ is compact in the compact-open topology for $H(\Omega)$.

14.19. Assume $\Omega \supset D(0,1)$, $f \in C(\Omega \times [0,1], \mathbb{C})$, $0 \notin f(\mathbb{T} \times [0,1])$, for each t in $[0,1]$, $f(z,t) \in H(\Omega)$, and $\frac{\partial f(z,t)}{\partial z} \in C(\Omega \times D(0,1), \mathbb{C})$. a)

$$I(t) \stackrel{\text{def}}{=} \frac{1}{2\pi i} \int_{\mathbb{T}} \left(\frac{\partial f(z,t)}{\partial z} \Big/ f(z,t) \right) dz$$

is t-free; b) If $g \in C([0,1], \mathbb{C})$ then $F_g(z) \stackrel{\text{def}}{=} \int_{[0,1]} f(z,t) g(t)\,dt \in H(U)$.

14.3. Defective Functions

Conventions

For a region Ω and a subset S of $\mathbb{C}$, a function f in $M(\Omega)$ is S-defective in Ω iff $f(\Omega) \subset (\mathbb{C} \setminus S)$. A family $\mathcal{F}$ of functions in $M(\Omega)$ is S-defective in Ω iff $\bigcup_{f\in\mathcal{F}} f(\Omega) \subset (\mathbb{C} \setminus S)$. The symbol $\mathcal{F}_S$ is used to denote such a family.

The items below show in what (minimal) circumstances a defective family is necessarily normal. The development is due to Landau **[La]** who created a strikingly direct proof of Bloch's and Schottky's theorems. From these Landau derived the (great) Picard theorem, which may be phrased as follows:

> If $r > 0$, a is an isolated essential singularity of f, and f is S-defective in $D(a,r)^\circ \setminus \{a\}$ then $\#(S) \le 1$.

14.20. If Ω is simply connected, $f \in H(\Omega)$, and f is $\{0,1\}$-defective, there is in $H(\Omega)$: a) a p that is $\mathbb{Z}$-defective; b) a q such that $p = q^2$; c) an r such that $p - 1 = r^2$; d) an s such that $q - r = e^s$.

14.21. For $S \stackrel{\text{def}}{=} \left\{ z \ : \ z = \pm \ln\left(\sqrt{m} + \sqrt{m-1}\right) + \frac{n\pi i}{2}, m \in \mathbb{N}, n \in \mathbb{Z} \right\}$, and in the context of **14.20**d), s is S-defective.

14.22. For S in **14.21**, $\mathbb{C} \setminus S$ contains no translate of U.

14.23. For $f \in H(\Omega)$ and $\Omega \supset D(0,R)$: a) $f(0) = 0$; b) $|f'(0)| \ge a > 0$; c) $|f'(z)| \le M$ in $D(0,R)$; d) f is $\{\gamma\}$-defective in $D(0,R)^\circ$ then $|\gamma| \ge \dfrac{a^2R}{4M}$.

14.24. (Bloch) If $\Omega \supset D(0,1)$, $f \in H(\Omega)$, and $|f'(0| \ge 1$ then for some a, $f(U) \supset D\left(a, \frac{1}{16}\right)$.

14.25. (Schottky) Assume $\Omega \supset D(0,1)$ and $f \in H(\Omega)$. For r in $[0,1)$ and some positive $\Phi(f(0),r)$, if f is $\{0,1\}$-defective then $|f(z)| < \Phi(f(0),\theta)$ in $D(0,r)$.

14.26. If a, b are two numbers and Ω is a region in $\mathbb{C}$ then $\mathcal{F}_{a,b}$ is spherically normal.

14.27. If the region S is a simply connected proper subset of $\mathbb{C}$ then for any region Ω, $\mathcal{F} \stackrel{\text{def}}{=} \{ f \ : \ f \in H(\Omega), f(\Omega) \subset S \}$ is spherically normal.

14.28. For a region Ω and $\mathcal{G} \stackrel{\text{def}}{=} \{ f \ : \ f \in H(\Omega), f(\Omega) \subset \Pi_+ \}$, the set $\mathcal{F} \stackrel{\text{def}}{=} \left\{ e^f \ : \ f \in \mathcal{G} \right\}$ is spherically normal.

14.4. Bergman's Kernel Functions

Problems 14.29 – 14.34 below lead to Bergman's kernel functions, which have practical applications in conformal mapping [**Ber1, Ber2, Hi**]. The overriding hypothesis is that $\#(\partial\Omega) \ge 2$.

14.29. For $\mathfrak{H} \stackrel{\text{def}}{=} L^2(\Omega, \lambda_2) \cap H(\Omega)$: a) $\mathfrak{H}$ is $\| \ \|_2$-complete; b) if K is a compact subset of Ω_δ, and for f in $\mathfrak{H}$, $\|f\|_K \stackrel{\text{def}}{=} \sup_{z \in K} |f(z)|$ then for some constant M_K, $\|f\|_K \le M_K \|f\|_2$; c) the unit ball of $\mathfrak{H}$ is normal.

14.30. If $\{\phi_n\}_{n\in\mathbb{N}}$ is a complete orthonormal system in $\mathfrak{H}$ and $f \in \mathfrak{H}$ then $\sum_{n=1}^{\infty}(f,\phi_n)\,\phi_n(z) \stackrel{\mathrm{u}}{\to} f(z)$ in Ω_δ.

14.31. If $\{\phi_n\}_{n\in\mathbb{N}}$ is a complete orthonormal system in $\mathfrak{H}$ and $t \in \Omega$ what is $\mu \stackrel{\text{def}}{=} \inf\left\{ \|f\|_2 \ : \ f \in \overline{\operatorname{span}\left(\{\phi_n\}_{1\le n\le N}\right)} \stackrel{\text{def}}{=} M_N, f(t) = 1 \right\}$?

14.32. For Bergman's kernel function $K(z,w) \stackrel{\text{def}}{=} \sum_{n=1}^{\infty}\phi_n(z)\phi_n(w)$: a) $|K(z,w)|^2 \le K(z,z)K(w,w)$; b) $\{w \in \Omega\} \Rightarrow \{K(w,w) > 0\}$; c)

$$\{f \in \mathfrak{H}\} \Rightarrow \left\{ f(z) = \int_\Omega K(z,w) f(w)\, d\lambda_2(w) \right\}.$$

14.33. If $w \in \Omega$, $f \in \mathfrak{H}$, and $f(w) = 1$ then $\|f\|_2 \ge \left\| \frac{K(z,w)}{K(w,w)} \right\|_2$.

14.34. If Ω is simply connected, $f(\cdot,w)$ is a conformal map of Ω onto $D\left(0, \frac{1}{\sqrt{\pi K(w,w)}}\right)^\circ \stackrel{\text{def}}{=} D(0,R)^\circ$, and $\frac{\partial f(z,w)}{\partial z}\Big|_{z=w} = 1$ then

$$f(z,w) = \frac{1}{K(w,w)} \int_w^z K(s,w)\, ds.$$

(The integral is path-independent!)

15
Convexity Theorems

15.1. Thorin's Theorem

In such disparate contexts as the Hahn-Banach theorem, seminorms, Kolmogorov's theorem (cf. **6.162**), von Neumann's theory of almost periodic functions on groups (cf. **2.93**), his theory of games [NM] and linear programming, Jensen's inequality, Hadamard's three-lines/three-circles theorems, the Hausdorff-Young theorem, probability theory, etc., the rôle played by convexity is central. The following discussion attempts to illustrate that rôle in complex analysis.

Conventions

When $\{V_k\}_{1\le k\le n}$ is a set of vector spaces, $F_k \in \mathbb{R}^{[0,\infty)}, 1 \le k \le n$, and $G \in \mathbb{R}^{\mathbb{R}}$, ϕ in $\mathbb{R}^{V_1\times\cdots\times V_n}$ is $(G; F_1,\cdots,F_n))$-convex iff whenever $0 \le t_k$, $\sum_{k=1}^n t_k = 1$, and $G\circ\phi$ is defined then $G\circ\phi(\mathbf{v}_1,\ldots,\mathbf{v}_n) \le \sum_{k=1}^n t_k F_k(\mathbf{v}_k)$.

[**Note 15.1:** In the context of Jensen's inequality as discussed in **Chapter 4**, a function ϕ convex in the ordinary sense is (id; id, id)-convex.

When

$$\begin{aligned}
&V_1 \stackrel{\text{def}}{=} L^p(X,\mu), V_2 \stackrel{\text{def}}{=} L^{p'}(X,\mu),\\
&G(x) \stackrel{\text{def}}{=} \ln x,\ F_1 \stackrel{\text{def}}{=} \ln \|\ \|_p,\ F_2 \stackrel{\text{def}}{=} \ln \|\ \|_{p'},\\
&\mathbf{x}_i \in V_i, i = 1,2,\ f(\mathbf{x}_1,\mathbf{x}_2) \stackrel{\text{def}}{=} \left|\int_X \mathbf{x}_1(x)\overline{\mathbf{x}_2(x)}\,d\mu(x)\right|
\end{aligned}$$

then Hölder's inequality states that $\ln f$ is $(G; F_1, F_2)$-convex:

$$\ln\left|\int_X \mathbf{x}_1(x)\overline{\mathbf{x}_2(x)}\,d\mu(x)\right| \le \frac{1}{p}\ln\|\mathbf{x}_1\|_p + \frac{1}{p'}\ln\|\mathbf{x}_2\|_{p'}.$$

If $h \in MVP(D(a,r)^\circ)$ then for n in $\mathbb{N}$ and s in $[0,r)$,

$$\begin{aligned}
h(a) &= h\left(\sum_{k=0}^{n-1}\frac{1}{n}\left(a+se^{\frac{2\pi ki}{n}}\right)\right)\\
&= \int_{\mathbb{T}} h(a+st)\,d\tau(t) \approx \sum_{k=0}^{n-1}\frac{1}{n}h\left(a+se^{\frac{2\pi ki}{n}}\right). \quad (15.1)
\end{aligned}$$

Thus when "=" and "≈" are read "≤," (15.1) may be viewed as a convexity property of h. By abuse of language, subharmonic functions, i.e., upper semicontinuous $\mathbb{R}$-valued functions f for which

$$\{\{a \in \Omega\} \wedge \{D(a,r) \subset \Omega\}\} \Rightarrow \left\{h(a) \leq \frac{1}{2\pi}\int_0^{2\pi} h\left(a+re^{it}\right) dt\right\},$$

may be viewed as a class of convex functions.]

When X is a set, $B, D \subset X$, and $f \in \mathbb{R}^X$ then the maximum principle in D relative to B obtains for f iff $\sup_{x\in D} f(x) \leq \sup_{x\in B} f(x)$.

For example, the maximum modulus theorem asserts, i.a., that if

$$D(a,r) \subset \Omega \text{ and } f \in H(\Omega)$$

then the maximum principle in $D \stackrel{\text{def}}{=} D(a,r)^\circ$ relative to $B \stackrel{\text{def}}{=} \partial D(a,r)$ obtains for $|f|$. For more details cf. **Chapter 9**.

Note that if $D \subset B$, the maximum principle in D relative to B obtains for all f in $\mathbb{R}^X$.

15.1. True or false: a) if a function f in $\mathbb{R}^{\mathbb{R}}$ is monotone or convex then for every finite interval $[a,b]$, the maximum principle in $[a,b]$ relative to $\{a,b\}$ obtains for f? b) if, for every finite interval $[a,b]$, the maximum principle in $[a,b]$ relative to $\{a,b\}$ obtains for f in $\mathbb{R}^{\mathbb{R}}$, then f is convex?

15.2. For a vector space V, a ϕ in $\mathbb{R}^V$ is convex iff for each λ in $\mathbb{R}$, each pair $\{\mathbf{x},\mathbf{y}\}$ in V^2, and each t in $[0,1]$, the maximum principle in the interval $[\mathbf{x},\mathbf{y}] \stackrel{\text{def}}{=} \{\mathbf{z} : \mathbf{z} = t\mathbf{x} + (1-t)\mathbf{y}, t \in [0,1]\}$ relative to $\{\mathbf{x},\mathbf{y}\}$ obtains for $\phi(t\mathbf{x}+(1-t)\mathbf{y}) - \lambda t$.

15.3. If each element of $\{\phi_\lambda\}_{\lambda\in\Lambda}$ is convex then $\phi \stackrel{\text{def}}{=} \sup_{\lambda\in\Lambda}\phi_\lambda$ is convex.

15.4. If g in $\mathbb{R}^{\mathbb{R}}$ is a monotonely increasing function continuous on the right and the maximum principle in D relative to B obtains for f then the maximum principle in D relative to B obtains as well for $g \circ f$.

15.5. (Thorin) Assume X is a vector space, C is a convex subset of X, Y is a set, and $f \in \mathbb{R}^{X\times Y}$. If, for every line segment

$$[\mathbf{x}_0,\mathbf{x}_1] \stackrel{\text{def}}{=} \{\mathbf{x}_0 + t(\mathbf{x}_1 - \mathbf{x}_0) \ : \ 0 \leq t \leq 1\}$$

contained in C and every λ in $\mathbb{R}$, the maximum principle in

$$[\mathbf{x}_0,\mathbf{x}_1] \times Y \text{ relative to } (\mathbf{x}_0 \times Y) \cup (\mathbf{x}_1 \times Y)$$

obtains for $F_{\mathbf{x}_0,\mathbf{x}_1,\lambda} : [0,1] \times Y \ni (t,y) \mapsto f(t\mathbf{x}_1 + (1-t)\mathbf{x}_0, y) - \lambda t$ then $M(\mathbf{x}) \stackrel{\text{def}}{=} \sup_{y\in Y} f(\mathbf{x},y)$ is a convex function of $\mathbf{x}$.

15.6. If Z is a vector space, $K \subset [Z, X]$, C is a convex subset of X, and $\mathbf{z} \in \bigcap_{L\in K} L^{-1}(C) \stackrel{\text{def}}{=} \Gamma$ then, in the context of **15.5**,

$$M(\mathbf{z}; K) \stackrel{\text{def}}{=} \sup\{ F(L(\mathbf{z}), y) \ : \ L \in K, y \in Y \}$$

is convex on Γ.

[**Note 15.2:** If $\Gamma = \emptyset$ the conclusion above is automatic. In any event, since each L is linear and C is convex, Γ is convex.

The result asserts in particular that if $M(\mathbf{z}; K)$ is finite at both endpoints of a line segment J lying in Γ then $M(\mathbf{z}; K)$ is bounded above on J.]

15.2. Applications

15.7. (Hadamard's Three-Lines Theorem) If

$$\Omega \stackrel{\text{def}}{=} (a, b) \times \mathbb{R}, f \in H(\Omega) \cap C(\Omega^c, \mathbb{C}), \text{ and } f(\Omega) \subset D(0, K)$$

then $\ln\left(\widetilde{M}(x; f) \stackrel{\text{def}}{=} \sup\{ |f(x + iy)| \ : \ -\infty < y < \infty \}\right)$ is convex.

15.8. (Hadamard's Three-Circles Theorem) If

$$f \in H\left(A(a, r : R)^\circ\right) \cap C\left(A(a, r : R), \mathbb{C}\right)$$

then for t in $[r, R]$, $\ln M(t; f)$ is a convex function of $\ln t$.

[**Note 15.3:** The result **3.12** combined with **15.7** implies that $\widetilde{M}(x; f)$ is itself a convex function of x; by the same token $M(r; f)$ is a convex function of $\ln r$. The greater strength of **15.7** is emphasized by **3.10**: $\{f \text{ is convex }\} \not\Rightarrow \{\ln f \text{ is convex }\}$.]

15.9. If $f \in H(U)$, $r \in (0, 1)$, $p \in (0, \infty)$, and

$$I_p(r) \stackrel{\text{def}}{=} \frac{1}{2\pi} \int_0^{2\pi} \left| f\left(re^{it}\right)\right|^p dt$$

then: a) on $(0, 1)$, $I_p(r)$ is a monotonely increasing function of r; b) $\ln I_p(r)$ is a convex function of $\ln r$.

15.10. (M. Riesz) For α, β positive,

$$\mathsf{B} : \mathbb{C}^m \times \mathbb{C}^n \ni (\mathbf{x}, \mathbf{y}) \mapsto \sum_{j=1}^{m} \sum_{k=1}^{n} a_{jk} x_j y_k \stackrel{\text{def}}{=} \mathsf{B}(\mathbf{x}, \mathbf{y}),$$

and

$$\rho_j > 0,\ \sigma_k > 0, 1 \le j \le m,\ 1 \le k \le n,$$
$$\boldsymbol{\rho} \stackrel{\text{def}}{=} (\rho_1, \ldots, \rho_m),\ \boldsymbol{\sigma} \stackrel{\text{def}}{=} (\sigma_1, \ldots, \sigma_n),$$

let S be the set of all $(\mathbf{x}, \mathbf{y}, \boldsymbol{\rho}, \boldsymbol{\sigma})$ such that

$$\sum_{j=1}^{m} \rho_j |x_j|^{\frac{1}{\alpha}} \le 1, \text{ and } \sum_{k=1}^{n} \sigma_k |y_k|^{\frac{1}{\beta}} \le 1.$$

Then the logarithm of $M(\alpha, \beta) \stackrel{\text{def}}{=} \sup_{(\mathbf{x},\mathbf{y},\boldsymbol{\rho},\boldsymbol{\sigma})\in S} \mathsf{B}(\mathbf{x}, \mathbf{y})$ is a convex function on the quadrant $Q \stackrel{\text{def}}{=} \{(\alpha, \beta) : \alpha, \beta > 0\}$.

[**Note 15.4:** The M. Riesz convexity theorem has many consequences. In particular it gives rise to the famous theorems of F. Hausdorff, G.C. and W.H. Young, and F. Riesz. These theorems apply in the context of a locally compact abelian group G and for any p in $(1, 2]$ [**We2, Zy**]. They relate the norm of a function f in $L^p(G, \mathbb{C})$ to the norm of its Fourier transform $\widehat{f}$, viz., for appropriately normalized Haar measures on G resp. $\widehat{G}$ and $p' \stackrel{\text{def}}{=} \frac{p}{p-1}$:

$$\left\|\widehat{f}\right\|_{p'} \le \|f\|_p . \tag{15.2}$$

It is important to note that the condition $p \in (1, 2]$ is essential. When $p > 2$, there are examples for which (15.2) fails.]

REAL ANALYSIS:

SOLUTIONS

1
Set Algebra and Function Lattices

1.1. Set Algebra

Conventions

References to items in **SOLUTIONS** are tagged with the prefix s to distinguish them from items in **PROBLEMS**.

1.1. If $\{A_n\}_{n\in\mathsf{N}} \subset \mathsf{M}$ then $B_N \stackrel{\text{def}}{=} \bigcup_{n=1}^{N} A_n \in \mathsf{M}$ and $B_N \subset B_{N+1}$ whence

$$\lim_{N\to\infty} B_N \left(= \bigcup_{n=1}^{\infty} A_n \right) \in \mathsf{M}.$$

Similarly, $C_N \stackrel{\text{def}}{=} \bigcap_{n=1}^{N} A_n \in \mathsf{M}$ and $C_{N+1} \subset C_N$ whence

$$\lim_{N\to\infty} C_N \left(= \bigcap_{n=1}^{\infty} A_n \right) \in \mathsf{M}.$$

□

1.2. As a monotone class, $\sigma\mathsf{R}(\mathsf{R}) \supset \mathsf{M}(\mathsf{R})$. Hence it suffices to show $\mathsf{M}(\mathsf{R})$ is a σ-ring.

For C in M and $\mathsf{B}(C) \stackrel{\text{def}}{=} \{D : C\setminus D,\ D\setminus C,\ C\cup D \in \mathsf{M}\}$:
a) $\{C \in \mathsf{B}(D)\} \Leftrightarrow \{D \in \mathsf{B}(C)\}$;
b) $\mathsf{B}(C)$ is a monotone class;
c) $\{D \in \mathsf{R}\} \Rightarrow \{\mathsf{R} \subset \mathsf{B}(D)\}$;
d) $\{D \in \mathsf{R}\} \Rightarrow \{\mathsf{M}(\mathsf{R}) \subset \mathsf{B}(D)\}$;
e) $\{A \in \mathsf{M}(\mathsf{R})\} \Rightarrow \{\mathsf{M}(\mathsf{R}) \subset \mathsf{B}(A)\}$;
f) $\mathsf{M}(\mathsf{R})$ is a ring (cf. **1.1**). □

1.3. For F in $\mathsf{F}(X)$, if $N_n \stackrel{\text{def}}{=} \bigcup_{y\in F} \left\{ x : d(x,y) < \frac{1}{n} \right\}$, then for n in $\mathbb{N}$, $\mathsf{M} \supset \mathsf{O}(X) \ni N_n \supset N_{n+1}$ and $F = \bigcap_{n\in\mathbb{N}} N_n \in \mathsf{M}$ whence $\mathsf{F}(X) \subset \mathsf{M}$. □

1.4. From **1.3** it follows that $\mathsf{M} \supset \mathsf{F}(\mathbb{R})$ whence if $a \le b$, $[a,b] \in \mathsf{M}$. Because M is monotone, $[a,b)$ $(= \lim_{n\to\infty} [a, b - \frac{1}{n}]) \in \mathsf{M}$. Since a finite union of closed intervals is closed, hence in M, any finite union of half-open intervals $[a,b)$ is in M, i.e., M contains the ring R consisting of all finite unions of half-open intervals $[a,b)$. From **1.2** it follows that $\mathsf{M} \supset \sigma\mathsf{R}(\mathsf{R})$. If $a < b$ then $(a,b) = \lim_{n\to\infty} \left[a + \frac{1}{n}, b\right) \in \sigma\mathsf{R}(\mathsf{R}) \subset \mathsf{M}$, whence $\mathsf{M} \supset \sigma\mathsf{R}(\mathsf{O})$.

Because $\mathbb{R} \in \sigma\mathsf{R}(\mathsf{O}) \cap \sigma\mathsf{R}(\mathsf{F})$, it follows that $\sigma\mathsf{R}(\mathsf{F}) = \sigma\mathsf{R}(\mathsf{O})$. The argument thus far shows that the σ-ring generated by the set of all bounded closed intervals $[a, b]$ is $\sigma\mathsf{R}(\mathsf{O})$ whence $\sigma\mathsf{R}(\mathsf{O}) = \sigma\mathsf{R}(\mathsf{F}) \supset \sigma\mathsf{R}(\mathsf{K}) \supset \sigma\mathsf{R}(\mathsf{O})$. □

1.5. If S is neither empty nor a finite set of sets, and if S is a sequence of pairwise disjoint members of S then the set Σ of all countable unions of members of S is an uncountable subset of S.

First proof. By induction, define a sequence $\{E_n\}_{n\in\mathbb{N}}$ as follows:

i. E_1, an arbitrary member of S;
ii. if $S_n \stackrel{\text{def}}{=} \{E_1, \ldots, E_n\}$ is a finite set of pairwise disjoint sets in S, there is some set E in $\mathsf{S} \setminus S_n$ and among $E_k \setminus E, E \setminus E_k, E_k$ $1 \le k \le n$, there are at least $n+1$ pairwise disjoint sets, each in S; for one of them, say E_{n+1} not in S_n, $S_{n+1} \stackrel{\text{def}}{=} \{E_1, \ldots, E_{n+1}\}$.

The sequence $S \stackrel{\text{def}}{=} \{E_n\}_{n\in\mathbb{N}}$ is an infinite subset of S and consists of pairwise disjoint sets.

Second proof. Call a set E in S indivisible iff

$$\{\mathsf{S} \ni F \subset E\} \Rightarrow \{\{F = \emptyset\} \vee \{F = E\}\}.$$

Pairwise different elements of the set I of indivisible elements of S are disjoint. If I is infinite then Σ is uncountable. If I is finite or empty let U be the union of the elements of I. Direct calculation shows that $\widetilde{\mathsf{S}} \stackrel{\text{def}}{=} \{ A \setminus U \ : \ A \in \mathsf{S} \}$ is a σ-ring and $\widetilde{\mathsf{S}} \setminus \emptyset$ is free of indivisible elements.

If $\widetilde{\mathsf{S}} = \{\emptyset\}$ then $\#(\mathsf{S}) < \aleph_0$. If $\widetilde{\mathsf{S}} \neq \{\emptyset\}$ choose a nonempty F_1 in $\widetilde{\mathsf{S}}$. Hence for some F_2 in $\widetilde{\mathsf{S}}$, $\emptyset \neq F_2 \subsetneqq F_1$ and, by induction, for n in $\mathbb{N}$, for some F_{n+1} in $\widetilde{\mathsf{S}}$, $\emptyset \neq F_{n+1} \subsetneqq F_n$. The set

$$S \stackrel{\text{def}}{=} \left\{ E_n \stackrel{\text{def}}{=} F_n \setminus F_{n+1} \ : \ n \in \mathbb{N} \right\}$$

is infinite and its members are pairwise disjoint.

Third proof. If $2^X \supset \mathsf{S} \stackrel{\text{def}}{=} \{E_n\}_{n\in\mathbb{N}}$ and S is infinite then for each x in X, $D_x \stackrel{\text{def}}{=} \left(\bigcap_{x\in E_n} E_n\right) \setminus \left(\bigcup_{x\notin E_n} E_n\right) \in \mathsf{S}$. Furthermore, if $D_x \cap D_y \neq \emptyset$, then $D_x = D_y$. Hence the pairwise different elements of $\{ D_x \ : \ x \in X \}$ are pairwise disjoint and constitute a finite or infinite subset S of S. If $x \in E_n$ then $D_x \subset E_n$. i.e., $E_n = \bigcup_{x\in E_n} D_x$. Hence S is infinite. □

1.6. a) Let E_0 be E. If $0 < \beta < \Omega$, let E_β be the set of all countable unions of differences of sets drawn from $\bigcup_{\alpha<\beta} \mathsf{E}_\alpha$.

If $A, B \in \mathsf{S}$, for some α, $A, B \in \mathsf{E}_\alpha$ whence $A \setminus B \in \mathsf{E}_{\alpha+1} \subset \mathsf{S}$. If $\{A_n\}_{n\in\mathbb{N}} \subset \mathsf{S}$, for some α in $[0, \Omega)$, $\{A_n\}_{n\in\mathbb{N}} \subset \mathsf{E}_\alpha$ whence $\bigcup_{n\in\mathbb{N}} A_n \in \mathsf{E}_{\alpha+1}$. It follows that $\bigcup_{\alpha<\Omega} \mathsf{E}_\alpha \stackrel{\text{def}}{=} \mathsf{S}$ is a σ-ring and $\sigma\mathsf{R}(\mathsf{E}) \subset \mathsf{S}$. Transfinite induction implies that for all α in $[0,\Omega)$, $\mathsf{E}_\alpha \subset \sigma\mathsf{R}(\mathsf{E})$, and there emerges the required conclusion: $\sigma\mathsf{R}(\mathsf{E}) = \mathsf{S}$.

Because $\#(\mathsf{E}) \geq 2$, $\#(\mathsf{E}_0) < (\#(\mathsf{E}))^{\aleph_0}$. If $0 \leq \beta < \Omega$ and, for all α in $[0,\beta)$, $\#(\mathsf{E}_\alpha) \leq (\#(\mathsf{E}))^{\aleph_0}$ then

$$\#\left(\bigcup_{0\leq\alpha<\beta} \mathsf{E}_\alpha\right) \leq \aleph_0 (\#(\mathsf{E}))^{\aleph_0} = (\#(\mathsf{E}))^{\aleph_0}$$

$$\#(\mathsf{E}_\beta) \leq (\#(\mathsf{E}))^{\aleph_0^2} = (\#(\mathsf{E}))^{\aleph_0}.$$

Finally, $\#([0,\Omega)) \leq \mathfrak{c}$ and $(\#(\mathsf{E}))^{\aleph_0} \geq \mathfrak{c}$ whence

$$\#(\sigma\mathsf{R}(\mathsf{E})) \leq \mathfrak{c}\cdot(\#(\mathsf{E}))^{\aleph_0} \leq \left(\#(\mathsf{E})^{\aleph_0}\right)^2 = \#(\mathsf{E})^{\aleph_0}.$$

b) For

$$\mathsf{E} \stackrel{\text{def}}{=} \left\{ \bigtimes_{k=1}^n (a_k, b_k) \;:\; -\infty < a_k < b_k < \infty, a_k, b_k \in \mathbb{Q} \right\},$$

$\sigma\mathsf{R}(\mathsf{E}) = \mathsf{S}_\beta(\mathbb{R}^n)$ whence, by virtue of a), $\#(\mathsf{S}_\beta(\mathbb{R}^n)) \leq \aleph_0^{\aleph_0} = \mathfrak{c}$. For all $\mathbf{x}$ in $\mathbb{R}^n$, $\{\mathbf{x}\} \in \mathsf{S}_\beta(\mathbb{R}^n)$ whence $\#(\mathsf{S}_\beta(\mathbb{R}^n)) = \mathfrak{c}$.

c) Since $\mathsf{M} \subset [0,\infty]^{\mathsf{S}_\beta(\mathbb{R})}$ and $\#(\mathsf{S}_\beta(\mathbb{R})) = \mathfrak{c}$ [cf. b)], it follows that

$$\#(\mathsf{M}) \leq \mathfrak{c}^{\mathfrak{c}} = \left(2^{\aleph_0}\right)^{\mathfrak{c}} = 2^{\aleph_0\mathfrak{c}} = 2^{\mathfrak{c}}.$$

On the other hand, if

$$\mathcal{B} \stackrel{\text{def}}{=} \{\{A,B\} \;:\; A,B \in \mathsf{S}_\beta(\mathbb{R}), A\cap B = \emptyset, \#(A) = \#(B) = \mathfrak{c}\}$$

then $\#(\mathcal{B}) = \mathfrak{c}$, and if $\{A,B\} \in \mathcal{B}$ and $0 < s,t < \infty$, there is a nonatomic measure μ living on $A\cup B$ and such that $\mu(A) = s, \mu(B) = t$. Hence

$$\#(\mathsf{M}) \geq \#\left(\left((0,\infty)^2\right)^{\mathcal{B}}\right) = \mathfrak{c}^{\mathfrak{c}} = 2^{\mathfrak{c}}. \qquad \square$$

1.7. For

$$\mathsf{D} \stackrel{\text{def}}{=} \{D \;:\; D \in \sigma\mathsf{R}(\mathsf{E}_0), \mathsf{E}_0 \subset \mathsf{E}, \#(\mathsf{E}_0) \leq \aleph_0\},$$

$\sigma\mathsf{R}(\mathsf{E}) \supset \mathsf{D} \supset \mathsf{E}$. If $\{A_n\}_{n\in\mathbb{N}} \subset \mathsf{D}$ then for each n in $\mathbb{N}$, there is a countable set E_{0n} such that $A_n \in \sigma\mathsf{R}(\mathsf{E}_{0n})$; $\bigcup_{n\in\mathbb{N}} A_n$ and $A_1\setminus A_2$ are in $\sigma\mathsf{R}\left(\bigcup_{n\in\mathbb{N}} \mathsf{E}_{0n}\right)$ whence D is a σ-ring and so $\sigma\mathsf{R}(\mathsf{E}) \supset \mathsf{D} \supset \sigma\mathsf{R}(\mathsf{E})$. $\square$

1.8. For each sequence $\{p,q,r,\ldots\}$ in $\mathbb{N}$, consider sequences

$$\{B_p, A_{pq}, B_{pqr}, \ldots\} \text{ resp. } \{A_p, B_{pq}, A_{pqr}, \ldots\}$$

such that

$$B_p = \bigcup_{q\in\mathsf{N}} A_{pq}, A_{pq} = \bigcap_{r\in\mathsf{N}} B_{pqr}, \ldots \tag{s1.1}$$

resp.

$$A_p = \bigcup_{q\in\mathsf{N}} B_{pq}, B_{pq} = \bigcap_{r\in\mathsf{N}} A_{pqr}, \ldots . \tag{s1.2}$$

If B is the set of all countable unions of sets B_p then $\mathsf{B} \subset \mathsf{A}$ and, symmetrically, $\mathsf{A} \subset \mathsf{B}$: $\mathsf{A} = \mathsf{B}$, i.e., A is closed with respect to the formation of countable unions and intersections of its members.

If $A \stackrel{\text{def}}{=} \bigcup_{p\in\mathsf{N}} A_p$ then

$$X \setminus A = \bigcap_{p\in\mathsf{N}} (X \setminus A_p), \ X \setminus A_p = \bigcup_{q\in\mathsf{N}} (X \setminus B_{pq}), \ldots .$$

If *iii* obtains as well and $A \in \mathsf{A}$ then $X \setminus A \in \mathsf{A}$ whence A is a σ-algebra and $\mathsf{A} \supset \sigma\mathsf{A}(\mathsf{E})$. If the last inclusion is proper and $A \in (\mathsf{A} \setminus \sigma\mathsf{A}(\mathsf{E}))$ then $A = \bigcup_{p\in\mathsf{N}} A_p$, some $A_p \notin \sigma\mathsf{A}(\mathsf{E})$, some $B_{pq} \notin \sigma\mathsf{A}(\mathsf{E}), \ldots$. Consequently (s1.1) and (s1.2) are denied. □

1.9. Items a) and b) are translations of the symbols used to define $\overline{\mathcal{E}}$ resp. $\underline{\mathcal{E}}$.

c) If $x \in \overline{\mathcal{E}}$ then $\chi_{\overline{\mathcal{E}}}(x) = 1$. Because x is in infinitely many E_n,

$$\sup_{n\in\mathsf{N}} \chi_{E_n}(x) = 1, \overline{\lim}_{n\to\infty} \chi_{E_n}(x) = 1.$$

If $x \notin \overline{\mathcal{E}}$, $\chi_{\overline{\mathcal{E}}}(x) = 0$. Since x is in at most finitely many E_n, there is an n_0 such that $\sup_{n>n_0} \chi_{E_n}(x) = 0, \overline{\lim}_{n\to\infty} \chi_{E_n}(x) = 0$.

If $x \in \underline{\mathcal{E}}$ then $\chi_{\underline{\mathcal{E}}}(x) = 1$. Because x is in all but finitely many E_n, there is an n_0 such that $\inf_{n>n_0} \chi_{E_n}(x) = 1, \underline{\lim}_{n\to\infty} \chi_{E_n}(x) = 1$. If $x \notin \underline{\mathcal{E}}$ then $\chi_{\underline{\mathcal{E}}}(x) = 0$. Since x is not in all but finitely many E_n, for infinitely many n, $x \notin E_n$, i.e., $\chi_{E_n}(x) = 0$ for infinitely many n, whence for each n_0, $\inf_{n>n_0} \chi_{E_n}(x) = 0, \underline{\lim}_{n\to\infty} \chi_{E_n}(x) = 0$. □

1.10. a) The formula $\chi_{E\cap F} = \chi_E \cdot \chi_F$ shows $\mathcal{F}$ is intersection-closed. Because $\mathbf{1} \in \mathcal{U}$ it follows that for any E in $\mathcal{F}$, $\chi_{X\setminus E} \ (= \mathbf{1} - \chi_E) \in \mathcal{U}$ whence $\mathcal{F}$ is closed with respect to the formation of complements, hence also with respect to the formation of set differences. If $\{E_n\}_{n\in\mathsf{N}} \subset 2^X$, $E \stackrel{\text{def}}{=} \bigcup_{n\in\mathsf{N}} E_n$, and $S_N \stackrel{\text{def}}{=} \bigcup_{n=1}^N E_n$ then

$$E = \dot\bigcup_{N\in\mathsf{N}} (S_{N+1} \setminus S_N) \stackrel{\text{def}}{=} \dot\bigcup_{N\in\mathsf{N}} T_N \ (T_N \in \mathcal{F}!)$$

$$\chi_E = \sum_{N=1}^{\infty} \chi_{T_N} \in \mathcal{U} \text{ whence } E \in \mathcal{F}.$$

b) If K is a compact subset of X and $N_n \stackrel{\text{def}}{=} \left\{ x \,:\, d(x,K) < \frac{1}{n} \right\}$ then $K = \bigcap_{n\in\mathbb{N}} N_n$. The function

$$f_n : X \ni x \mapsto \frac{d(x, X \setminus N_n)}{d(x, X\setminus N_n) + d(x,K)} = \begin{cases} 1 & \text{if } x \in K \\ 0 & \text{if } x \notin N_n \end{cases}$$

is in $C_{00}(X,\mathbb{C})$ while for all x in X, $0 \le f_n(x) \le 1$. Hence $f_n \in \mathcal{U}$, $n \in \mathbb{N}$. Because $\chi_K = \lim_{n\to\infty} f_n$ it follows that $\chi_K \in \mathcal{U}$ and thus $K \in \mathcal{F}$. Since $\mathcal{F}$ is a σ-ring by virtue of a), it follows that $\sigma\mathsf{R}(\mathsf{K}(X)) \subset \mathcal{F}$. □

1.11. There is a bijection $F : \mathbb{N}^{\mathbb{N}} \ni \nu \overset{\text{bij}}{\mapsto} F(\nu) \stackrel{\text{def}}{=} (\alpha,\beta) \in \left(\mathbb{N}^{\mathbb{N}}\right)^2$. If $g \in \mathcal{A}^2(\mathsf{E})$ then $g : \mathbb{N}^{\mathbb{N}} \ni \alpha \mapsto g(\alpha) \in (\mathcal{A}(\mathsf{E}))^{\mathbb{N}}$, $g(\alpha) \stackrel{\text{def}}{=} \{g(\alpha)_m\}_{m\in\mathbb{N}}$, and for each pair $\{m,\alpha\}$, some $f_{m\alpha} : \mathbb{N}^{\mathbb{N}} \ni \beta \mapsto f_{m\alpha}(\beta) \in \mathsf{E}^{\mathbb{N}}$ is in E so that $g(\alpha)_m = \bigcup_{\beta\in\mathbb{N}^{\mathbb{N}}} \bigcap_{k\in\mathbb{N}} f_{m\alpha}(\beta)_k$. For each pair $\{\alpha,\beta\}$ let $\{f_{m\alpha}(\beta)_k\}_{m,k=1}^{\infty}$ be enumerated as $\{s(\alpha,\beta)_p\}_{p\in\mathbb{N}} \stackrel{\text{def}}{=} \left\{ s(F(\nu))_p \right\}_{p\in\mathbb{N}} \stackrel{\text{def}}{=} \{r(\nu)_p\}_{p\in\mathbb{N}}$. Then

$$\begin{aligned} N_g &\stackrel{\text{def}}{=} \bigcup_{\alpha\in\mathbb{N}^{\mathbb{N}}} \bigcap_{m\in\mathbb{N}} g(\alpha)_m = \bigcup_{\alpha\in\mathbb{N}^{\mathbb{N}}} \bigcap_{m\in\mathbb{N}} \bigcup_{\beta\in\mathbb{N}^{\mathbb{N}}} \bigcap_{k\in\mathbb{N}} f_{m\alpha}(\beta)_k \\ &= \bigcup_{(\alpha,\beta)\in((\mathbb{N})^{\mathbb{N}})^2} \bigcap_{m,k\in\mathbb{N}^2} f_{m\alpha}(\beta)_k = \bigcup_{(\alpha,\beta)\in((\mathbb{N})^{\mathbb{N}})^2} \bigcap_{p\in\mathbb{N}} s(\alpha,\beta)_p \\ &= \bigcup_{\nu\in\mathbb{N}^{\mathbb{N}}} \bigcap_{p\in\mathbb{N}} r(\nu)_p \in \mathcal{A}(\mathsf{E}). \end{aligned}$$ □

1.12. a) For $\nu \stackrel{\text{def}}{=} \{n_1, n_2, \ldots\}$, let $f(\nu)_k$ be M_{n_1}, $k \in \mathbb{N}$. It follows that

$$\bigcap_{k\in\mathbb{N}} f(\nu)_k = M_{n_1} \text{ and } M_f = \bigcup_{\nu\in\mathbb{N}^{\mathbb{N}}} \bigcap_{k\in\mathbb{N}} M_{n1} \bigcup_{\nu\in\mathbb{N}^{\mathbb{N}}} M_{n1} = \bigcup_{n\in\mathbb{N}} M_n.$$

b) If f is the constant map $f : \mathbb{N}^{\mathbb{N}} \ni \nu \mapsto (M_1, M_2, \ldots)$, i.e., for all ν, k, $f(\nu)_k = M_k$, then $M_f = \bigcup_{\nu\in\mathbb{N}^{\mathbb{N}}} \bigcap_{k\in\mathbb{N}} M_k = \bigcap_{k\in\mathbb{N}} M_k$. □

1.13. For each ν in $\mathbb{N}^{\mathbb{N}}$, if $g(\nu)_k \stackrel{\text{def}}{=} \bigcap_{m=1}^{k} f(\nu)_m$ (in E) then g is regular, $M_g = \bigcap_{k\in\mathbb{N}} g(\nu)_k = \bigcap_{m\in\mathbb{N}} f(\nu)_m = M_f$. □

1.14. a) If x is in the left member of (1.1), then for some m and some $\nu \stackrel{\text{def}}{=} \{n_1, n_2, \ldots\}$, $x \in \bigcap_{k\in\mathbb{N}} M_{n_1,n_2,\ldots,n_i,m,n_{i+1},\ldots,n_{i+k}}$. However, for the sequence $\widetilde{\nu} \stackrel{\text{def}}{=} \{n_1, n_2, \ldots, n_i, m, n_{i+1}, \ldots\}$,

$$\bigcap_{k=i+2}^{\infty} f(\widetilde{\nu})_k = \bigcap_{k\in\mathbb{N}} M_{n_1,n_2,\ldots,n_i,m,n_{i+1},\ldots,n_k}.$$

Since f is regular, $\bigcap_{k=i+2}^{\infty} f(\widetilde{\nu})_k = \bigcap_{k\in\mathbb{N}} f(\widetilde{\nu})_k$, whence x is in the right member.

If x is in the right member of (1.1) then for some $\nu \stackrel{\text{def}}{=} \{n_1, n_2, \ldots\}$, $x = \bigcap_{k\in\mathbb{N}} f(\nu)_k$. Because f is regular, $x \in \bigcap_{k=i+2}^{\infty} f(\nu)_k$ which, if $m = n_{i+1}$, is $\bigcap_{k\in\mathbb{N}} M_{n_1,n_2,\ldots,n_i,m,n_{i+2},\ldots,n_{i+k}}$.

b) Because f is regular,

$$\bigcup_{k\in\mathbb{N}} M_{n_1,n_2,\ldots,n_k} = M_{n_1}$$

$$\bigcup_{\nu\in\mathbb{N}^{\mathbb{N}}}\bigcup_{k\in\mathbb{N}} M_{n_1,n_2,\ldots,n_k} = \bigcup_{\nu\in\mathbb{N}^{\mathbb{N}}} M_{n_1} = M_1 \cup M_2 \cup \cdots = \bigcup_{n\in\mathbb{N}} M_n.$$

c) If x is not in the right member of (1.2) and $x \in M$ then

$$x \in \bigcap_{\nu\in\mathbb{N}^{\mathbb{N}}}\bigcap_{k=0}^{\infty}\left((X \setminus M_{n_1,n_2,\ldots,n_k}) \cup \bigcup_{m\in\mathbb{N}} M_{n_1,n_2,\ldots,n_k,m}\right).$$

Hence if $x \in M_{n_1,n_2,\ldots,n_k}$ then for some m in $\mathbb{N}$, $x \in M_{n_1,n_2,\ldots,n_k,m}$. By hypothesis, $x \in M$ ($\stackrel{\text{def}}{=} M_{n_1,n_2,\ldots,n_k}$ when $k = 0$) whence for an $m \stackrel{\text{def}}{=} m_1$, $x \in M_{m_1}$. Thus, for an $m \stackrel{\text{def}}{=} m_2$, $x \in M_{m_1,m_2}$, etc. It follows that

$$x \in \bigcap_{k\in\mathbb{N}} M_{m_1,m_2,\ldots,m_k}, x \in \bigcup_{\nu\in\mathbb{N}^{\mathbb{N}}}\bigcap_{k\in\mathbb{N}} M_{m_1,m_2,\ldots,m_k} = M_f,$$

i.e., x is not in the left member of (1.2). □

1.15. Since every open or half-open subinterval of $[0,1]$ is a countable union of closed intervals, **1.12** implies that all relatively open subsets of $[0,1]$ are in $\mathcal{A}(\mathsf{E})$. Furthermore, $\mathcal{A}(\mathsf{E})$ is a monotone class. In **1.4**, $\mathbb{R}$ may be replaced by $[0,1]$ without affecting the conclusion. □

1.16. Because all elements of E are compact, $h(\mathsf{E})$ consists of compact sets. Since **1.13** is applicable, each E in $\mathcal{A}(\mathsf{E})$ is, for some regular g,

$$\bigcup_{\nu\in\mathbb{N}^{\mathbb{N}}}\bigcap_{k\in\mathbb{N}} g(\nu)_k.$$

Each set $g(\nu)_k \stackrel{\text{def}}{=} K_k$ is compact and $K_k \supset K_{k+1}$.

The relations $f \circ \bigcup = \bigcup \circ f$, $f^{-1} \circ \bigcap = \bigcap \circ f^{-1}$, and $f \circ \bigcap \subset \bigcap \circ f$ obtain for any map $f : X \mapsto Y$.

Moreover, if $y \in \bigcap_{k\in\mathbb{N}} h(K_k)$ then for k in $\mathbb{N}$, and some x_k in K_k, $y = h(x_k)$. Because $\{x_k\}_{k\in\mathbb{N}} \subset K_1$, for some point x in $\bigcap_{k\in\mathbb{N}} K_k$ and some subsequence $\{x_{k_m}\}_{m\in\mathbb{N}}$, $\lim_{m\to\infty} x_{k_m} = x$, $y = h(x_{k_m}) \to h(x)$ as $m \to \infty$, i.e., $y \in h\left(\bigcap_{k\in\mathbb{N}} K_k\right)$.

Thus, for the sequence $\{K_k\}_{k\in\mathsf{N}}$ of compact sets and the continuous map h, not only does $h\circ\bigcap\subset\bigcap\circ h$ hold but $h\left(\bigcap_{k\in\mathsf{N}}K_k\right)=\bigcap_{k\in\mathsf{N}}h\left(K_k\right)$ is true as well, i.e.,

$$h(E)=\bigcup_{\nu\in\mathsf{N}^{\mathsf{N}}}h\left(\bigcap_{k\in\mathsf{N}}K_k\right)=\bigcup_{\nu\in\mathsf{N}^{\mathsf{N}}}\bigcap_{k\in\mathsf{N}}h\left(K_k\right)\in\mathcal{A}(\mathsf{E}). \qquad \square$$

1.17. When $\mathsf{E}=\mathsf{S}_\lambda$, **1.13** implies an f in $\mathcal{F}$ may be assumed to be regular. Any subset E of $[0,1]$ is contained in a Borel set A such that $\lambda(A)=\lambda^*(E)$. Because $\{M_1\supset M_2\supset\cdots\}\Rightarrow\left\{\bigcap_{k\in\mathsf{N}}M_k=\bigcap_{N=1}^{\infty}\bigcup_{k=N}^{\infty}M_k\right\}$, for any sequence $\{m_1,\ldots,m_i\}$ there is a Borel set $A_{m_1,\ldots,m_i}$ such that

$$\begin{aligned}A_{m_1,\ldots,m_i}&\supset\bigcup_{\nu\in\mathsf{N}^{\mathsf{N}}}\bigcap_{k\in\mathsf{N}}M_{m_1,\ldots,m_i,n_1,\ldots,n_k}\\&=\bigcup_{\nu\in\mathsf{N}^{\mathsf{N}}}\bigcap_{k\in\mathsf{N}}M_{n_1,\ldots,n_k}\stackrel{\text{def}}{=}M_f\\ \lambda\left(A_{m_1,\ldots,m_i}\right)&=\lambda^*\left(\bigcup_{\nu\in\mathsf{N}^{\mathsf{N}}}\bigcap_{k\in\mathsf{N}}M_{m_1,\ldots,m_i,n_1,\ldots,n_k}\right).\end{aligned}\tag{s1.3}$$

Since $A_{m_1,\ldots,m_i}\cap M_{m_1,\ldots,m_i}$ serves as well as $A_{m_1,\ldots,m_i}$ in (s1.3), it may be assumed that $M_{m_1,\ldots,m_i}\supset A_{m_1,\ldots,m_i}$. Since $M\setminus(M\setminus M_f)=M_f$, to show that $M_f\in\mathsf{S}_\lambda$ it suffices to show that $\lambda^*\left(M\setminus M_f\right)=0$. However, from **1.14c)** it follows that

$$\begin{aligned}A\setminus M_f&=A\setminus\left(\bigcup_{\nu\in\mathsf{N}^{\mathsf{N}}}\bigcap_{k\in\mathsf{N}}A_{n_1,\ldots,n_k}\right)\\&\subset\bigcup_{\nu\in\mathsf{N}^{\mathsf{N}}}\bigcup_{k=0}^{\infty}\left(A_{n_1,\ldots,n_k}\setminus\bigcup_{m=1}^{\infty}A_{n_1,\ldots,n_k,m}\right).\end{aligned}\tag{s1.4}$$

The set of all finite subsets $\{n_1,\ldots,n_k\}$ of $\mathbb{N}$ is countable. Each $\{n_1,\ldots,n_k\}$ determines $A_{n_1,\ldots,n_k}\setminus\bigcup_{m=1}^{\infty}A_{n_1,\ldots,n_k,m}$. Hence there are at most countably many different summands in (s1.4). They may be enumerated, say as $\{B_p\}_{1\le p<P\le\infty}$, and $A\setminus M_f\subset\bigcup_{1\le p<P\le\infty}B_p$.

However, from **1.14c)**,

$$\begin{aligned}\bigcup_{q\in\mathsf{N}}A_{m_1,\ldots,m_i,q}&\supset\bigcup_{q\in\mathsf{N}}\bigcup_{\nu\in\mathsf{N}^{\mathsf{N}}}\bigcap_{k\in\mathsf{N}}M_{m_1,\ldots,m_i,q,n_1,\ldots,n_k}\\&=\bigcup_{\nu\in\mathsf{N}^{\mathsf{N}}}\bigcap_{k\in\mathsf{N}}M_{m_1,\ldots,m_i,n_1,\ldots,n_k}\end{aligned}$$

whence

$$A_{m_1,\ldots,m_i} \setminus \bigcup_{q\in\mathbb{N}} A_{m_1,\ldots,m_i,q} \subset A_{m_1,\ldots,m_i} \setminus \bigcup_{\nu\in\mathbb{N}^{\mathbb{N}}} \bigcap_{k\in\mathbb{N}} M_{m_1,\ldots,m_i,n_1,\ldots,n_k}.$$

The inner measure of the right member above is zero whence the left member above is a null (Borel) set, i.e., each B_p is a null set. Hence $\lambda^*(A\setminus M_f)=0$ and $M_f \in \mathsf{S}_\lambda$. □

[**Note s1.1:** An examination of the proof shows that the following generalization of Szpilrajn [**Sz**] is valid.

Let $\mathcal{B}$ be a σ-algebra contained in 2^X. Assume further that when $E\subset X$ then for some A in $\mathcal{B}$, a) $A\supset E$ and b) if $E\subset B\in\mathcal{B}$ and $F\subset A\setminus B$ it follows that $F\in\mathcal{B}$. Then for B in $\mathcal{B}$ and $f:\mathbb{N}^{\mathbb{N}}\mapsto\mathcal{B}^{\mathbb{N}}$, $B_f\in\mathcal{B}$.]

1.18. Let M be a set such that $\#(M)=\mathfrak{a}$ and let E be $\{M\}$. Then $\#(\mathcal{A}(M))=\mathfrak{a}$. □

1.19. If $\mathsf{E}\stackrel{\text{def}}{=}\mathsf{F}([0,1])$ then $\mathsf{S}_\beta([0,1])\subset\mathcal{A}(\mathsf{E})$ (**1.15**) whence

$$h(\mathsf{S}_\beta([0,1]))\subset h(\mathcal{A}(\mathsf{E}))\subset\mathcal{A}(\mathsf{F}(\mathbb{R}))\subset\mathcal{A}(\mathsf{S}_\lambda(\mathbb{R}))$$

(cf. **1.16**). Because $\mathbb{R}$ is the countable union of intervals, the argument for **1.16** may be extended to show that $\mathcal{A}(\mathsf{S}_\lambda(\mathbb{R}))=\mathsf{S}_\lambda(\mathbb{R})$. □

1.2. Function Lattices

1.20. a) It suffices to assume that $a\le b\le c$. Then

$$a\vee b=b,\ a\vee c=c,\ b\vee c=c$$
$$(a\vee b)\wedge(a\vee c)\wedge(b\vee c)=b\wedge c\wedge c=b\wedge c=c.$$

b) See a).
c) If $f(x)\equiv -1$, $g(x)\equiv x$, and $h(x)\equiv 1$ then

$$\operatorname{mid}(f,g,h)(x)=\begin{cases}-1 & \text{if } x<-1\\ x & \text{if } -1\le x<1\\ 1 & \text{if } 1\le x,\end{cases}$$

whence $\operatorname{mid}(f,g,h)\notin\{f,g,h\}$.

d) $f(x)\vee g(x)+f(x)\wedge g(x)=\begin{cases}g(x)+f(x) & \text{if } f(x)\le g(x)\\ f(x)+g(x) & \text{if } f(x)>g(x)\end{cases}.$ □

1.21. Because V is a lattice, if $f,g,h\in V$ then $f\wedge g, f\wedge h, g\wedge h\in V$. It follows that $\operatorname{mid}(f,g,h)=(f\wedge g)\vee(f\wedge h)\vee(g\wedge h)\in V$. □

1.22. For (1.3)–(1.8), honest toil succeeds when applied to the sorting out of cases as they are governed by the definition of mid. □

1.23. a) Since V is a vector space, $-f \in V$. Direct calculation shows $|f| = f \vee 0 - f \wedge 0$.

b) Because V is a lattice (1.6) implies $\mathcal{M}$ is a lattice. If $f_1, f_2 \in \mathcal{M}$, if $f_1 + f_2$ is meaningful, and $V \ni g \leq h \in V$ then for n in $\mathbb{N}$,

$$\begin{aligned}
&\phi_{in} \stackrel{\text{def}}{=} \operatorname{mid}\left(f_i, -n(|g| + |h|), n(|g| + |h|)\right) \in V,\ i = 1, 2,\\
&\phi_n \stackrel{\text{def}}{=} \operatorname{mid}\left(\phi_{1n} + \phi_{2n}, g, h\right) \in V,\\
&g \leq \phi_n \leq h, \phi_n \uparrow \operatorname{mid}\left(f_1 + f_2, g, h\right).
\end{aligned}$$

The hypotheses imply the validity of the Lebesgue monotone convergence theorem, hence the validity of the dominated convergence theorem for J. Because $g \leq \phi_n \leq h$, it follows from (1.7) and the hypotheses that $\operatorname{mid}(f_1 + f_2, g, h) = \lim_{n\to\infty} \phi_n \in \mathcal{M}$.

c) If $V \ni g \leq h \in$V and $\mathcal{M} \ni f_n \uparrow f$ then $m_n \stackrel{\text{def}}{=} \operatorname{mid}(f_n, g, h) \in V$ and $-\infty < J(g) \leq J(m_n) \leq J(h) < \infty$. The conclusion follows from (1.7). If $\mathcal{M} \ni f_n \downarrow f$ the same proof, mutatis mutandis, applies.

d) See c) and the definitions of $\overline{\lim}$, $\underline{\lim}$, and lim.

e) The result follows from the formula: $\operatorname{mid}(f, -|f|, |f|) = f$.

f) The implication $\Rightarrow$ is valid because V is a vector space. On the other hand, if $V \ni p \leq q \in V$ then (1.4) implies

$$\frac{p+q}{2} + \operatorname{mid}\left(f, -\frac{q-p}{2}, \frac{q-p}{2}\right) = \operatorname{mid}(f, p, q).$$

Because V is a function lattice, the implication $\Leftarrow$ is valid. □

1.24. For $C \stackrel{\text{def}}{=} \bigcup_{n\in\mathbb{N}} A_n$, from the equations

$$\chi_{A\cup B} = \chi_A \vee \chi_B, \chi_{A\setminus B} = \chi_{A\cup B} - \chi_B, \chi_C = \bigvee_{n\in\mathbb{N}} \chi_{A_n},$$

and the properties **1.23**b) – d), it follows that A is a σ-ring. When $\mathbf{1} \in \mathcal{M}$, the formula $\chi_{X\setminus A} = \mathbf{1} - \chi_A$ implies A is a σ-algebra. □

1.25. a) Since each $E_\circ(f, \alpha) \in f^{-1}(\mathsf{S}_\beta(\mathbb{R}))$, it suffices to note that when $\alpha_1 \leq \alpha_2$,

$$\begin{aligned}
f^{-1}([\alpha_1, \alpha_2)) &= E_<(f, \alpha_2) \setminus E_<(f, \alpha_1) = E_\geq(f, \alpha_1) \setminus E_\geq(f, \alpha_2),\\
f^{-1}((\alpha_1, \alpha_2]) &= E_\leq(f, \alpha_2) \setminus E_\leq(f, \alpha_1) = E_>(f, \alpha_1) \setminus E_>(f, \alpha_2),
\end{aligned}$$

and that $\mathsf{S}_\beta(\mathbb{R})$ is e.g., $\sigma\mathsf{R}(\{[a, b) : a \leq b, a, b \in \mathbb{Q}\})$.

b) For $(\mathbb{R}, \mathsf{S}_\lambda, \lambda)$, there is a set S in $2^{\mathbb{R}} \setminus \mathsf{S}_\lambda$ (cf. **4.39**). For all α, if $f(x) \stackrel{\text{def}}{=} x\chi_S(x) - x\chi_{\mathbb{R}\setminus S}(x)$ then

$$E_=(f, \alpha) = \pm\alpha \quad \text{while} \quad f^{-1}([0, \infty) = S \notin \mathsf{S}_\lambda. \qquad \square$$

1.26. If $f \in \mathcal{M}$ then, since $\mathbf{1} \in \mathcal{M}$, it follows from **1.23** that for n in $\mathbb{N}$,

$$n(\alpha - f) \in \mathcal{M}, \ \mathbf{1} \wedge n(\alpha - f) \in \mathcal{M}, \text{ and } (\mathbf{1} \wedge n(\alpha - f)) \vee 0 \in \mathcal{M}.$$

Furthermore, $\chi_{E_<(f,\alpha)} = \lim_{n\to\infty} (\mathbf{1} \wedge n(\alpha - f)) \vee 0$ whence again, **1.23** implies $E_<(f, \alpha) \in \mathsf{A}$.

When $f \in \mathbb{R}^X$ and $(k, n) \in \mathbb{Z} \times \mathbb{N}$, let f_{nk} be the characteristic function of $E_\le (f, k2^{-n}) \setminus E_\le (f, (k-1)2^{-n})$. Then $f = \lim_{n\to\infty} \sum_{k=-n2^n}^{n} \frac{k-1}{2^n} f_{nk}$ and if for every α, $E_\le(f, \alpha) \in \mathsf{A}$, then f is in $\mathcal{M}$. $\square$

1.27. Because $f_n \le \lim_{m\to\infty} g_m$,

$$h_m \stackrel{\text{def}}{=} (f_n - g_m) \vee 0 \downarrow_m 0 \text{ and } I(h_m) \downarrow 0.$$

Moreover, $I(h_m) \ge I(f_n - g_m)$ whence

$$0 \ge \lim_{m\to\infty} I(f_n - g_m) = I(f_n) - \lim_{m\to\infty} I(g_m).$$

Hence, $\lim_{m\to\infty} I(g_m) \ge \lim_{n\to\infty} I(f_n)$. The argument is symmetrical in the pair $\{\{f_n\}_{n\in\mathbb{N}}, \{g_m\}_{m\in\mathbb{N}}\}$. $\square$

1.28. Multiplication by nonnegative numbers preserves order among functions. Addition is a continuous operation. $\square$

1.29. For each n and some sequence $\{g_{mn}\}_{m\in\mathbb{N}}$ in L, $g_{mn} \uparrow_m f_n$. Hence if $k_n \stackrel{\text{def}}{=} g_{1n} \vee \cdots \vee g_{nn}$ then

$$L \ni k_n \le f_n, \ g_{mn} \le k_n \le f_n \text{ if } m \le n, \ g_{mn} \le \lim_{n\to\infty} k_n \stackrel{\text{def}}{=} k,$$
$$f_n = \lim_{m\to\infty} g_{mn} \le k \le f, \ k \in L_u, \ f = \lim_{n\to\infty} f_n \le k \le f.$$

Hence $f = k \in L_u$ and similarly the inequalities

$$I(f_n) = \lim_{m\to\infty} I(g_{mn}) \le I(k) \le \lim_{n\to\infty} I(f_n)$$

imply $I(f_n) \uparrow I(f)$. $\square$

1.30. For some $\{f_n\}_{n\in\mathbb{N}}$ in L, $f_n \uparrow f$. If $h \stackrel{\text{def}}{=} f_1$ and $p = \sum_{n=2}^{\infty} (f_n - f_{n-1})$ then $h \in L$, $p \in L_u$, $p \ge 0$, and $f = p + h$. $\square$

1.31. Trivialities aside, if $f \le h \in L_u$, $g \le k \in L_u$ then

$$\overline{I}(f+g) \le I(h+k) = I(h) + I(k),$$
$$\overline{I}(f+g) \le \overline{I}(f) + \overline{I}(g).$$

a) The definitions imply the conclusion. In particular, $\overline{I}$ is positive homogeneous: $\{a \ge 0\} \Rightarrow \{\overline{I}(af) = a\overline{I}(f)\}$.

b) $0 = \overline{I}(f + (-f)) \le \overline{I}(f) + \overline{I}(-f)$.

c) Since $f \le f$, if $f \in L_u$ then $\overline{I}(f) \le I(f)$. If $f \le g \in L_u$ then $I(f) \le I(g)$, whence $I(f) \le \overline{I}f)$.

d) It may be assumed that $\overline{I}(f_n) < \infty, n \in \mathbb{N}$. If $\epsilon > 0$ then for some g_n in L_u, $g_n \ge f_n$ and $I(g_n) < \overline{I}(f_n) + \frac{\epsilon}{2^n}, n \in \mathbb{N}$. From **1.29** it follows that $g \stackrel{\text{def}}{=} \sum_{n=1}^{\infty} g_n \in L_u$ and $I(g) = \sum_{n=1}^{\infty} I(g_n) \le \sum_{n=1}^{\infty} \overline{I}(f_n) + \epsilon$. Furthermore, $g \ge f$ and hence $\overline{I}(f) \le \overline{I}(g) \le \sum_{n=1}^{\infty} \overline{I}(f_n) + \epsilon$.

[**Note s1.2:** From $\left\{f \ge \sum_{n=1}^{N} f_n\right\} \Rightarrow \left\{\overline{I}f) \ge \overline{I}\left(\sum_{n=1}^{N} f_n\right)\right\}$, and the subadditivity of $\overline{I}$ it does not follow that $\overline{I}$ can be distributed validly over the sum.]

e) Again it may be assumed that $\overline{I}(f_n) < \infty$, $n \in \mathbb{N}$. From a) it follows that $\lim_{n\to\infty} \overline{I}(f_n) \le \overline{I}(f)$. If $\epsilon > 0$ for each n then for some h_n in L_u, $h_n \ge f_n$ and $I(h_n) < \overline{I}(f_n) + \frac{\epsilon}{2^n}$. Then

$$\lim_{n\to\infty} (h_1 \vee \cdots \vee h_n) \stackrel{\text{def}}{=} \lim_{n\to\infty} g_n \stackrel{\text{def}}{=} g \in L_u,$$

and $I(g_n) \uparrow I(g)$, (cf. **1.29**).

Since $g_1 = h_1$, **1.29** and mathematical induction show

$$\begin{aligned} I(g_n) + I(h_n \wedge g_{n-1}) &= I(h_n + g_{n-1}) = I(h_n) + I(g_{n-1}) \\ &\le \overline{I}(f_n) + \frac{\epsilon}{2^n} + \overline{I}(f_{n-1}) + \epsilon - \frac{\epsilon}{2^{n-1}} = \overline{I}(f_n) + \overline{I}(f_{n-1}) + \epsilon - \frac{\epsilon}{2^n}. \end{aligned}$$

Because $f_{n-1} \le h_n \wedge g_{n-1}$ it follows that $\overline{I}(f_{n-1}) \le \overline{I}(h_n \wedge g_{n-1})$ and so

$$\begin{aligned} I(h_n \vee g_{n-1}) = I(g_n) &\le \overline{I}(f_n) + \epsilon - \frac{\epsilon}{2^n} + \overline{I}(f_{n-1}) - I(h_n \wedge g_{n-1}) \\ &\le \overline{I}(f_n) + \epsilon - \frac{\epsilon}{2^n}. \end{aligned}$$

Because $\overline{I}(f_n) \le I(g_n)$ it follows that $\overline{I}(f) \le I(g) \le \overline{I}(f_n) + \epsilon$. □

1.32. a) The first inclusion follows from the fact that $I\big|_L = I$ and the second inclusion is a matter of definition.

b) Since I is additive on L, I is additive on L_u and on L^1. If $f \in L^1$ and $\epsilon > 0$ then for some g in L_u, $I(g) < I(f) + \frac{\epsilon}{2}$. Similarly, for some $-h$ in L_u, $-f \leq -h$ and $I(-h) < I(-f) + \frac{\epsilon}{2}$. Hence $h \leq f \leq g$ and $I(g + (-h)) = I(g) + I(-h) \leq I(f) + I(-f) + \epsilon = I(0) + \epsilon = \epsilon$.

Conversely, if there are a g and a $-h$ as described then

$$0 \leq \overline{I}(f) - (-\overline{I}(-f)) \leq I(g) + I(-h) = I(g + (-h)) < \epsilon.$$

c) If $\circ$ stands for any of $+, \vee, \wedge$, but the same one throughout each string of relations, and if $h_i \leq f_i \leq g_i$, $i = 1, 2$, then

$$h_1 \circ h_2 \leq f_1 \circ f_2 \leq g_1 \circ g_2,\ g_1 \circ g_2 - h_1 \circ h_2 \leq (g_1 - h_1) + (g_2 - h_2).$$

When $f_1, f_2 \in L^1$ and g_i and h_i, $i = 1, 2$, are chosen as in b), it follows that $I(g_1 \circ g_2) + I(-(h_1 \circ h_2)) < \epsilon$ whence $f_1 \circ f_2 \in L^1$; similarly, if $c \in \mathbb{R}$ then $cf_1 \in L^1$.

d) The definitions imply the statements.

e) If $f \in L^1$ then $J(f_n) \leq J(f) < \infty$. For the converse, it may be assumed that $J(f_n) < \infty$, $n \in \mathbb{N}$. If $g \stackrel{\text{def}}{=} f - f_1$ then $g \geq 0$ and $g = \sum_{n=1}^{\infty} (f_{n+1} - f_n)$. The subadditivity of $\overline{I}$ and d) in **1.31** imply

$$\overline{I}(g) \leq \sum_{n=1}^{\infty} J(f_{n+1} - f_n) = \lim_{n\to\infty} J(f_n) - J(f_1)$$
$$\overline{I}(f) = \overline{I}(f_1 + g) \leq J(f_1) + \overline{I}(g) \leq \lim_{n\to\infty} J(f_n).$$

Since $\underline{I}(f) \geq \underline{I}(f_n)$, it follows that $\underline{I}(f) \geq \lim_{n\to\infty} J(f_n)$ which, in light of the previous inequality, shows $f \in L^1$ and $J(f) = \lim_{n\to\infty} J(f_n)$.

f) The countable additivity of μ is a consequence of e).

g) If $f \in L^1$ ($= V$!) then, since L^1 is a function lattice, it follows that $f \in \mathcal{M}$ and $|f| = f \vee 0 - f \wedge 0 \in L^1$. Conversely, if $f \in \mathcal{M}$ and $|f| \in L^1$ then $-|f| \leq f \leq |f|$ and $f = \text{mid}(f, -|f|, |f|)| \in L^1$.

h) If $E \in \mathcal{M}$ and $f \stackrel{\text{def}}{=} \chi_E \in L^1$, the result follows by definition. In general, when $f \in L^1$, for some sequence $\{f_n\}_{n\in\mathbb{N}}$ in L^1, $f_n \uparrow f$ and then e) yields the conclusion. □

1.33. If $f = g - p$ as described, then **1.32a)** shows $f \in L^1$. If $f \in L^1$ then $\overline{I}(f) \in \mathbb{R}$ and for n in $\mathbb{N}$, and some f_n in L_u, $f_n \geq f$ and $I(f_n) < \overline{I}(f) + \frac{1}{n}$. Thus $\left\{g_n \stackrel{\text{def}}{=} f_1 \wedge \cdots \wedge f_n\right\}_{n\in\mathbb{N}}$ is a sequence such that for some g in L_{ul}, $g_n \downarrow g$. Moreover, $\overline{I}(g) = J(f)$, $0 \leq g - f \stackrel{\text{def}}{=} p \in L^1$, $J(p) = 0$, and $f = g - p$. □

1.34. a) Any g in L^1 is in $\mathcal{M}$ whence if $0 \le f \in \mathcal{M}$ then $f \wedge g \in \mathcal{M}$ and $|f \wedge g| \le |g| \in L^1$. It follows from **1.23e** that $f \wedge g \in L^1$.

Conversely, if $0 \le f$ and for all g in L^1, $f \wedge g \in L^1$, assume $0 \le k$ and $k \in L^1$. Then $\text{mid}(f, -k, k) = f \wedge k \in L^1$ and **1.23f** implies $f \in \mathcal{M}$.

b) If $0 \le f \in \mathcal{M}$ assume $h \in L$. Then $h \in \mathcal{M} \cap L^1$ whence a) implies $f \wedge h \in L^1$.

Conversely, assume $0 \le f$ and for all h in L, $f \wedge h \in L^1$. The lattice and monotone convergence properties of L^1 imply that if $k \in L_u$ and $I(k) < \infty$ then $f \wedge k \in L^1$ and that if $q \in L_{ul}$ then $f \wedge q \in L^1$. If $g \in L^1$ then (**1.33**) $g = q - r$, $q \in L_{ul}$, $0 \le r$, $J(r) = 0$. Because $g \le q$,

$$f \wedge q(x) - f \wedge g(x)$$
$$= \begin{cases} q(x) - g(x) = r(x) \ge 0 & \text{if } q(x), g(x) \le f(x) \\ f(x) - g(x) \ge 0 & \text{if } g(x) \le f(x) \le q(x) \\ 0 & \text{if } f(x) \le q(x), g(x) \end{cases}. \quad \text{(s1.5)}$$

Note that in (s1.5), $0 \le f(x) - g(x) \le q(x) - g(x) = r(x)$. It follows that $0 \le f \wedge q - f \wedge g = r \vee 0$ and since, $J(r) = 0$, $J(r \vee 0) = 0$. Hence, owing to **1.33**, $f \wedge g \in L^1$ and so a) implies $f \in \mathcal{M}$. □

1.35. If $f \in L^1$ then for some g in L_{ul},

$$J(|f - g|) \left(= \int_X |f(x) - g(x)| \, d\mu(x) \right) = 0$$

(cf. **1.33**). For some h in L_u, $|I(h) - J(g)| < \frac{\epsilon}{2}$ and for some f in L, $|I(h) - I(f)| < \frac{\epsilon}{2}$. Note $J|_{L_u} = I$ and $J|_L = I$. □

1.36. The definition of μ^* and the properties of $\overline{I}$ imply that μ^* is an outer measure, whence the set of Caratheodory measurable sets is a σ-algebra.

If D is Daniell measurable and $\mu^*(A) = \infty$, the subadditivity of $\overline{I}$ implies $\mu^*(A) = \mu^*(A \cap E) + \mu^*(A \setminus E)$.

If $\mu^*(A) < \infty$ and $\epsilon > 0$ then for some g in L_u, $g \ge \chi_A$ and

$$\begin{aligned} \overline{I}(\chi_A) + \epsilon \ge I(g) \ge I(g \wedge \mathbf{1}) &= I\left(g \wedge \chi_D + g \wedge \chi_{X \setminus D}\right) \\ &\ge I(g \wedge \chi_D) + I\left(g \wedge \chi_{X \setminus D}\right) \\ &\ge \overline{I}(\chi_{A \cap D}) + \overline{I}\left(\chi_{A \setminus D}\right). \end{aligned}$$

Hence Daniell measurability implies Caratheodory measurability.

When C is Caratheodory measurable, its Daniell measurability is established by showing that $\chi_C \in \mathcal{M}$, i.e., that for all g in L, $\chi_C \wedge g \in L^1$. Since the range of χ_C is the set $\{0, 1\}$, it suffices to show:

$$\{\{g \in L\} \wedge \{0 \le g \le 1\}\} \Rightarrow \{\chi_C \wedge g \in L^1\}.$$

Case 1: $\overline{I}(\chi_C) < \infty$. If $\epsilon > 0$ then for some g in L_u, $g \geq \chi_C$ and $\overline{I}(g) < I(g) + \epsilon$. Because $\mathbf{1} \in \mathcal{M}$, $L^1 \ni \mathbf{1} \wedge g \geq \chi_C$. Each of a) – e) below follows from its predecessors or from the statements made thus far: a) $\mathbf{1} \wedge g \in \mathcal{M}$; b) $B \stackrel{\text{def}}{=} E_{\geq}(\mathbf{1} \wedge g, 1) \in \mathsf{A}$; c) $g \geq \chi_B \in \mathcal{M}$; d) $\overline{I}(\chi_B) \leq I(g) < \infty$, whence $\chi_B \in L^1$; e) $\chi_B \geq \chi_C$.

Because $\chi_B \in L^1$, B is Daniell measurable and hence Caratheodory measurable. The following equations and inequalities are validated, i.a., because *both* B and C are Caratheodory measurable, $\chi_B \in L^1$, $\underline{I}$ is superadditive, and for all sets A, $\overline{I}(\chi_A) = \mu^*(A)$ (by definition):

$$\overline{I}(\chi_B) = \overline{I}(\chi_B \wedge \chi_C) + \overline{I}(\chi_{B \setminus C}) = \overline{I}(\chi_C) + \overline{I}(\chi_B - \chi_C)$$
$$\overline{I}(\chi_B - \chi_C) = \overline{I}(\chi_B) - \overline{I}(\chi_C) = \underline{I}(\chi_B) + \underline{I}(-\chi_C) \leq \underline{I}(\chi_B - \chi_C).$$

Thus $\chi_B - \chi_C \in L^1$, $\chi_C = \chi_B - (\chi_B - \chi_C) \in L^1$, i.e., $C \in \mathsf{A}$.

Case 2: $\overline{I}(\chi_C) = \infty$. Let A be $E_{>}(g, 0)$ which, owing to **1.24**, is in A.

Case 2a: $\chi_A \in L^1$. Then A is Daniell measurable, hence Caratheodory measurable, and $\chi_C \wedge g = \chi_C \wedge (g \wedge \chi_A) = (\chi_C \wedge \chi_A) \wedge g = (\chi_{C \cap A}) \wedge g$. Thus $C \cap A$ is Caratheodory measurable and, since $0 \leq \chi_{C \cap A} \leq \chi_A \in L^1$, $\overline{I}(\chi_{C \cap A}) < \infty$. The conclusion in Case 1 shows that $\chi_{C \cap A} \wedge g \in L^1$, whence $\chi_C \wedge g \in L^1$.

Case 2b: $\chi_A \in \mathcal{M} \setminus L^1$. For $\epsilon > 0$ let $E_{>}(g, \epsilon)$ be A_ϵ. Then $\chi_{A_\epsilon} \in \mathcal{M}$ and $\overline{I}(\chi_{A_\epsilon}) \leq \overline{I}\left(\frac{1}{\epsilon} \cdot g\right) = \frac{1}{\epsilon} I(g) < \infty$ whence $\chi_{A_\epsilon} \in L^1$. The conclusion in Case 2a implies that $C \cap A \in \mathsf{A}$. Setting ϵ at $\frac{1}{n}$, $n \in \mathbb{N}$, yields the result. □

[**Note s1.3:** Because the set of Caratheodory measurable sets is complete (subsets of null sets are Caratheodory measurable), it follows that A is complete if $\mathbf{1} \in \mathcal{M}$.]

2
Topology, Limits, and Continuity

2.1. Topology

2.1. a) If A contains the range of the net n converging to a then for each for each $N(a)$, n is eventually in $N(a)$ whence $N(a) \cap A \neq \emptyset$ and so $a \in \overline{A}$. If Λ is the diset $\mathcal{N}(a)$ then the range of the net $n : \Lambda \ni N \mapsto n(N) \in N \cap A$ is contained in A and n converges to a.

The proofs for b) – f) are similar exegeses of the definitions. □

2.2. For $\mathcal{N}(x)$ and $\Lambda_N \stackrel{\text{def}}{=} \{\lambda : \lambda \in \Lambda,\ n(\lambda) \in N \in \mathcal{N}(x)\}$, since n is frequently in each N of $\mathcal{N}(x)$, $\Gamma \stackrel{\text{def}}{=} \bigcup_{N \in \mathcal{N}(x)} \Lambda_N$ is a subdiset of Λ. Then $m \stackrel{\text{def}}{=} n\big|_\Gamma$ converges to x. □

2.3. If $\mathcal{F}$ converges to x, and n is a net corresponding to $\mathcal{F}$, then every $N(x)$ contains an element of $\mathcal{F}$ and hence n is eventually in $N(x)$.

Conversely, if every net corresponding to $\mathcal{F}$ converges to x, and if $\mathcal{F}$ does not converge to x then some $N(x)$ fails to contain any element of $\mathcal{F}$. Hence some net n corresponding to $\mathcal{F}$ is such that for each member F of $\mathcal{F}$, $n(F) \in F \setminus N(x)$. Thus n fails to converge to x, a contradiction. □

2.4. The net n converges to x iff n is eventually in each $N(x)$, i.e., iff for some λ_0, if $\lambda \succ \lambda_0$ then $B_\lambda \subset N$, i.e., iff $\mathcal{F}$ converges to x. □

2.5. a) $\overline{\lim}_{x=a} f(x) = -\underline{\lim}_{x=a}\left(-f(x)\right)$. If f is usc and $x \in E_<(f,\alpha)$ then $\overline{\lim}_{y=x} f(y) = f(x) < \alpha$, whence for some open $N(x)$, $\sup_{y \in N(x)} f(y) < \alpha$, i.e., $N(x) \subset E_<(f,\alpha)$.

Conversely, if each $E_<(f,\alpha)$ is open and $f(x) \neq \overline{\lim}_{y=x} f(y) \stackrel{\text{def}}{=} \alpha$ then $f(x) \stackrel{\text{def}}{=} \beta < \alpha$ and so for some $N(x)$ and all y in $N(x)$, $f(y) < \dfrac{\alpha+\beta}{2}$. It follows that $\sup_{y \in N(x)} f(y) < \dfrac{\alpha+\beta}{2} < \alpha$ and so $\overline{\lim}_{y=x} f(y) < \dfrac{\alpha+\beta}{2} < \alpha$, a contradiction.

For lsc, the dualities among lsc/usc, sup / inf, and +/– apply. □

2.6. For some $\widetilde{N(a)}$ and some L in $[0,\infty)$, $\sup_{x \in \widetilde{N(a)}} f(x) = M + L$. The set $\Lambda \stackrel{\text{def}}{=} \left\{ N : N = N(a) \subset \widetilde{N(a)} \right\}$, partially ordered according to: $N' \succ N$ iff $N' \subset N$, is a diset. For N in Λ and some ϵ_N in $[0,\infty)$, $\sup_{x \in N} f(x) = M + \epsilon_N$ and the net $\epsilon : \Lambda \ni N \mapsto \epsilon_N \in [0,\infty)$ converges to

0. Furthermore, for some x_N in N, $f(x_N) \geq M + \frac{1}{2}\epsilon_N$. Finally, the net $n : \Lambda \ni N \mapsto x_N$ converges to a and $f \circ n$ converges to L.

For $\underline{\lim}_{x=a} f(x)$, the duality $\overline{\lim}/\underline{\lim}$ applies. □

2.7. Because $\#\left(\{0,1\}^{\mathbb{N}}\right) = \mathfrak{c}$, and since each element of S is representable by a sequence in that element, it follows that $\#(s) \leq \mathfrak{c}$. on the other hand, if $s \in s$ and σ represents s, the sequence σ' arising by replacing each term in an infinite subsequence of terms in σ by its opposite counterpart (0 by 1, 1 by 0) is not equivalent to σ: $\#(s) \geq \mathfrak{c}$. □

2.8. If x adheres to $\mathcal{F}$ then every $n(x)$ meets every element of $\mathcal{F}$ whence the filter $\mathcal{F}' \stackrel{\text{def}}{=} \{n(x) \cap f \ : \ f \in \mathcal{F}\}$ is finer than $\mathcal{F}$ and converges to x.

If $\mathcal{F}$ converges to x then each $N(x)$ contains some element E of $\mathcal{F}$. if $F \in \mathcal{F}$ then $F \cap E \in \mathcal{F}$ whence $N(x)$ contains every $F \cap E$ and thus, for each F in $\mathcal{F}$, $N(x) \cap F \neq \emptyset$, i.e., x adheres to $\mathcal{F}$. □

2.9. Assume X is compact and $\mathcal{U} \stackrel{\text{def}}{=} \{F\}$ is an ultrafilter in 2^X. each $\overline{F}$ is closed and if $\bigcap_{F \in \mathcal{U}} \overline{f} = \emptyset$ then for some finite subset $\{F_1, \ldots, F_n\}$ of $\mathcal{U}$, $\bigcap_{k=1}^n F_k \subset \bigcap_{k=1}^n \overline{F_k} = \emptyset$, which is impossible for elements of a filter. Hence some x is in $\bigcap_{F \in \mathcal{U}} \overline{F}$. For the filter $\mathcal{N}(x)$,

$$\{N(x) \cap F \ : \ N(x) \in \mathcal{N}(x), f \in \mathcal{U}\}$$

generates a filter $\mathcal{V}$ that converges to x. Furthermore, $\mathcal{V} \succ \mathcal{U}$ and, since $\mathcal{U}$ is an ultrafilter, $\mathcal{V} = \mathcal{U}$.

Conversely, if every ultrafilter converges, let $\mathcal{E} \stackrel{\text{def}}{=} \{E\}$ be a set of closed sets such that every finite subset of $\mathcal{E}$ has a nonempty intersection. then $\mathcal{E}$ generates a filter $\mathcal{F}$ contained in an ultrafilter $\mathcal{U}$ that converges, say to x. For $N(x)$ and E in $\mathcal{E}$, there is in $\mathcal{U}$ a U that is contained in $N(x)$ and $N(x) \supset U \cap E \neq \emptyset$. since E is closed, $x \in E$: $x \in \bigcap_{E \in \mathcal{E}} E$. In other words, if a set $\mathcal{E}$ of closed sets enjoys the finite intersection property then the intersection of all the sets of $\mathcal{E}$ is nonempty. Hence X is compact.

If X is compact and $n : \lambda \mapsto X$ is a net, the corresponding filter $\mathcal{F} \stackrel{\text{def}}{=} \left\{B_\lambda \stackrel{\text{def}}{=} \{n(\mu) \ : \ \mu \succ \lambda\}\right\}$ is a subset of an ultrafilter $\mathcal{U}$ converging to some x in X. Hence x is in the closure of each element U of $\mathcal{U}$; in particular x is in the closure of each B_λ. Thus for each $N(x)$ and each λ in Λ, there is a U contained in $N(x)$, and in the nonempty set $B_\lambda \cap U$, there is an $n(\lambda')$. Note that $\lambda' \succ \lambda$. The set λ' of all such λ' is cofinal in Λ and the cofinal net $n|_{\lambda'}$ converges to x.

Conversely, if $\mathcal{U}$ is an ultrafilter in 2^X, and n is a net corresponding to $\mathcal{U}$, there is a cofinal net converging to some x in X. Hence $\mathcal{U}$ converges to x whence every ultrafilter converges and x is compact. □

[**Note s2.1:** When X is not compact a point x may lie in the closure of the range $n(\Lambda)$ of a net $n : \Lambda \mapsto X$ and yet no cofinal net $n|_{\Lambda'}$ converges to x [**Ar, GeO, Rin**].]

2.10. The map $f:[0,1)\ni x\mapsto \frac{1}{1-x}\sin\left(\frac{1}{1-x}\right)\in\mathbb{R}$ is continuous and $f([0,1))=\mathbb{R}$. □

2.11. Define the equivalence relation $\sim$ on $\mathbb{R}$ according to $x\sim y$ iff $x-y\in\mathbb{Q}$. Then $\sim$ decomposes $[0,1]$ into equivalence classes and, via the Axiom of Choice, there is a set E consisting of one element from each equivalence class. Thus $E-E\setminus\{0\}\subset \mathbb{I}_{\mathbb{R}}$ and $(E-E)^\circ=\emptyset$. □

2.12. The function $D:[0,1)^2\ni\{x,y\}\mapsto\left|\frac{x}{1-x}-\frac{y}{1-y}\right|$ is a new metric. Furthermore,

$$D(x,y)=\frac{|x-y|}{|(1-x)(1-y)|}\geq|x-y|=d(x,y) \tag{s2.1}$$

and $D(x,y)=d(x,y)$ iff $x=y$. Therefore, $x_n\xrightarrow{d}x$ iff $x_n\xrightarrow{D}x$, i.e., the d-topology and the D-topology are the same. A d-Cauchy sequence $\{x_n\}_{n\in\mathbb{N}}$ fails to d-converge iff $\lim_{n\to\infty}x_n=1$. For such a sequence $\{x_n\}_{n\in\mathbb{N}}$, $\{D(x_m,x_n)\}_{m,n\in\mathbb{N}}$ is unbounded and thus $\{x_n\}_{n\in\mathbb{N}}$ is not a D-Cauchy sequence. On the other hand, if $\{y_n\}_{n\in\mathbb{N}}$ is a D-Cauchy sequence, for some positive ϵ, $d(y_n,1)\geq\epsilon$. The inequality (s2.1) implies $\{y_n\}_{n\in\mathbb{N}}$ is a d-Cauchy sequence that fails to d-converge to 1 and thus d-converges to a y in $[0,1)$. Finally, $D(y_n,y)\leq\epsilon^{-2}d(y_n,y)$, i.e., $y_n\xrightarrow{D}y$. □

2.13. Because $[0,1]$ is compact and connected, each of $A\times B$, A, and B is compact and connected. If a_1 and a_2 resp. b_1 and b_2 are two elements of A resp. B then, because f^{-1} is a homeomorphism, each of the four sets

$$f^{-1}(\{a_i\}\times B)\stackrel{\text{def}}{=}I_{a_i},\ f^{-1}(A\times b_i)\stackrel{\text{def}}{=}I_{b_i},\ i=1,2,$$

is compact and connected, i.e., a closed interval. Since $I_{a_i}\cap I_{b_j}=f^{-1}(a_i,b_j)$ and $I_{a_1}\cap I_{a_2}=I_{b_1}\cap I_{b_2}=\emptyset$, the four intervals constitute an impossible configuration of subintervals of $\mathbb{R}$. □

2.14. a) If an a in C_0 has two different ternary represenur2.tex tations in which each marker is 0 or 2 let the representations differ for the first time at the nth marker. Then in one representation the nth marker is 2 and in the other it is 0. Thus $a-a$ has a ternary representation in which the first $n-1$ markers are zeros and the nth marker is not zero, i.e., $|a-a|=0\geq 3^{-n}$.

b) The map $\phi:\mathcal{D}^{\mathbb{N}}\ni\epsilon\stackrel{\text{def}}{=}\{\epsilon_n\}_{n\in\mathbb{N}}\mapsto\phi(\epsilon)\stackrel{\text{def}}{=}\sum_{n=1}^\infty 2\epsilon_n 3^{-n}\in C_0$ is a homeomorphism.

c) If $x,y\in C_0$ and $|x-y|<3^{-n}$ then $|f_n(x)-f_n(y)|<2^{-n}$.

d) If $x,y\in C_0$ and $|x-y|<3^{-n}$ then $|\phi_0(x)-\phi_0(y)|<2^{-n}$. Furthermore, if $[0,1]\ni t\stackrel{\text{def}}{=}\sum_{n=1}^\infty\epsilon_n 2^{-n}$, $\epsilon_n=0$ or 1 then $\phi_0\left(\sum_{n=1}^\infty 2\epsilon_n 3^{-n}\right)=t$.

e) If some ternary representation of a t in $[0,1]$ is given, successive ternary markers for elements x and y of C_0 can be determined by mathematical induction so that $x+y=t$. A similar argument applies for a t in $[-1,1]$.

f) $\phi_0\left(\frac{1}{3}\right)=\phi_0\left(\frac{2}{3}\right)$. □

2.15. a) For $\boldsymbol{\xi} \stackrel{\text{def}}{=} \{\xi_m\}_{m\in M} \in \mathcal{D}^M$, a minimal base $\mathcal{B} \stackrel{\text{def}}{=} \{N_m\}_{m\in M}$ for X, and each m in M, there is a dyad $A_m^i \stackrel{\text{def}}{=} \begin{cases} \overline{N_m} & \text{if } i=0 \\ X\setminus N_m & \text{if } i=1 \end{cases}$ of closed subsets of X. If $A_{\boldsymbol{\xi}} \stackrel{\text{def}}{=} \bigcap_{m\in M} A_m^{\xi_m} \neq \emptyset$ then, since X is a Hausdorff space, $A_{\boldsymbol{\xi}}$ is a single point $x_{\boldsymbol{\xi}}$.

If $A_{\boldsymbol{\xi}}=\emptyset$, the compactness of X implies that there is in M finite subset μ such that $\bigcap_{m\in\mu} A_m^{\xi_m}=\emptyset$. Then $\{\boldsymbol{\xi}' : \xi'_m=\xi_m,\ m\in\mu\}$ is an open subset of $\mathcal{D}^M$. Hence the set Ξ of those $\boldsymbol{\xi}$ such that $A_{\boldsymbol{\xi}}\neq\emptyset$ is closed in $\mathcal{D}^M$.

Furthermore, if $x\in X$, let M' be $\left\{m : x\in\overline{N_m}\right\}$. If

$$\xi'_m=\begin{cases} 0 & \text{if } m\in M' \\ 1 & \text{otherwise} \end{cases}$$

and $\boldsymbol{\xi}' \stackrel{\text{def}}{=} \{\xi'_m\}_{m\in M}$ then $x_{\boldsymbol{\xi}'}=x$. Hence the map $F:\Xi\ni\boldsymbol{\xi}\mapsto x_{\boldsymbol{\xi}}\in X$ is surjective. If N_m is a neighborhood of $x_{\boldsymbol{\xi}}$ then $V \stackrel{\text{def}}{=} \{\boldsymbol{\xi} : \xi_m=0\}\cap\Xi$ is an open subset of Ξ and $F(V)\subset N_m$. Since $\mathcal{B}$ is a base for the topology of X, F is continuous.

b) If $x\in B\cap C_0$ let $h(x)$ be x. If $x\in C_0\setminus B$ and $\sum_{n=1}^{\infty}\frac{x_n}{3^n}$ is the ternary representation of x, define a sequence $\{z_n\}_{n\in\mathbb{N}}$ by induction as follows:

> When $\alpha\in\{0,2\}$, let α' denote $\alpha+2 \bmod 4$. If B contains a b for which the first ternary marker b_1 is x_1, let z_1 be x_1; otherwise all $b_1\neq x_1$, in which case let z_1 be x'_1. Thus B contains a b such that the first ternary marker b_1 is z_1. Having defined $z_1,\ldots,z_n$ so that B contains a b for which the first, ..., nth ternary markers $b_1,\ldots,b_n$ are $z_1,\ldots,z_n$, let z_{n+1} be x_{n+1} if B contains a b such that the ternary markers $b_1,\ldots,b_n,b_{n+1}$ for b are $b_1,\ldots,b_n,x_{n+1}$; otherwise, all $b_1,\ldots,b_n,b_{n+1}$ are $b_1,\ldots,b_n,x'_{n+1}$, in which case let z_{n+1} be x'_{n+1}. Thus B contains a z such that the ternary markers for b are $z_1,z_2,\ldots$.

For z as just defined let $h(x)$ be z. If $x,y\in C_0$ then $|x-y|<3^{-N}$ iff the ternary markers $x_n,y_n,1\le n\le N$, are equal, in which case the (corresponding) ternary markers for $h(x)$ and $h(y)$ are the same. Hence h is a continuous surjection of C_0 on B.

c) If $\mathsf{W}=\aleph_0$ then $\mathcal{D}^M$ is homeomorphic to C_0 while Ξ is a closed subset B of C_0. Thus $F\circ h\left(C_0\right)=X$. □

2.16. a) Since $\{\emptyset, X\} \subset \mathcal{T}$ and $\mathcal{T}$ is closed with respect to the formation of arbitrary unions and finite intersections, $(X, \mathcal{T})$ is a topological space.

If $\mathcal{U} \stackrel{\text{def}}{=} \{U_\lambda\}_{\lambda\in\Lambda}$ is an open cover of X then x_0 is in some U_λ. Hence $X \setminus U_\lambda$ is finite and thus can be covered by finitely many elements of $\mathcal{U}$: they, together with U_λ, constitute a finite subcover of $\mathcal{U}$.

If $x \neq y$ and neither is x_0 then each is open; if $x = x_0$ then $X \setminus \{y\}$ is a neighborhood of x_0 and y is open.

b) Because $x_0 \notin \{x\}$, $\{x\} \in \mathcal{T}$.

c) $\mathcal{B} \stackrel{\text{def}}{=} \{\, \{x\} \ : \ x \neq x_0 \,\} \cup (X \setminus \{x_0\})$ is a base for $\mathcal{T}$ and $\#(\mathcal{B}) = \#(X)$. Because each x other than x_0 is open, the weight of X is at least $\#(X)$. □

2.17. Each nonempty O_γ contains a basic neighborhood

$$N(\sigma, \mathsf{E}) = \bigtimes_{i=1}^{n} \{x_{m_i}\} \times \mathcal{D}^{M\setminus\sigma}$$

specified by: a) a finite subset $\sigma \stackrel{\text{def}}{=} \{m_1, \ldots, m_n\}$ of M; b) some finite set $\mathsf{E} \stackrel{\text{def}}{=} \{x_{m_1}, \ldots, x_{m_n}\}$ of binary markers.

It suffices to show that any set $\mathcal{N}$ of pairwise disjoint basic neighborhoods is empty, finite, or countable.

i. If two basic neighborhoods $N(\sigma, \mathsf{E})$ and $N(\sigma', \mathsf{E}')$ are disjoint then $\sigma \cap \sigma' \neq \emptyset$ and for some m in $\sigma \cap \sigma'$, the coordinates x_m and x'_m are each other's opposite: $x_m + x'_m = 1$. The index m is a hit-index for the pair $\{N(\sigma, \mathsf{E}), N(\sigma', \mathsf{E})\}$.

ii. If, for k in $\mathbb{N}$, $\Sigma_k \stackrel{\text{def}}{=} \{\, N(\sigma, \mathsf{E}) \ : \ N(\sigma, \mathsf{E}) \in \mathcal{N}, \#(\sigma) = k \,\}$ then $\mathcal{N} = \bigcup_{k\in\mathbb{N}} \Sigma_k$.

iii. If $\Sigma_1 \neq \emptyset$ then *i* and *ii* imply $\#(\Sigma_1) = 1$ or 2.

iv. If each of $\Sigma_1, \ldots, \Sigma_{k-1}$ is empty or finite and $\#(\Sigma_k) \geq 2$, fix $N \stackrel{\text{def}}{=} N(\sigma, \mathsf{E})$ in Σ_k. Owing to *i* and the existence of another N' in Σ_k, some hit-index m_1 is in σ. Let T_1 in Σ_k be the set of all N' such that $x'_{m_1} = 1 - x_{m_1}$. Replace each N' in T_1 by a basic neighborhood $\widetilde{N'}$ for which the restriction at index m_1 is removed. Since the N' are pairwise disjoint and $x'_{m_1} = 1 - x_{m_1}$, any pair of the N' has a hit-index different from m_1. Each $\widetilde{N'}$ is determined by exactly $k - 1$ indices and, owing to the conclusion in the last sentence, the set $\widetilde{T_1}$ of the $\widetilde{N'}$ from T_1 consists of pairwise disjoint basic neighborhoods. By inductive assumption $\#\left(\widetilde{T_1}\right)$ $(= \#(T_1))$ is finite.

If $\Sigma_k \setminus T_1 \neq \emptyset$, then some hit-index m_2 is in $\sigma \setminus m_1$ and $\Sigma_k \setminus T_1$ contains a subset T_2 consisting of an N' such that $x'_{m_2} = 1 - x_{m_2}$. The corresponding set $\widetilde{T_2}$ consists of pairwise disjoint basic neighborhoods, each determined by precisely $k - 1$ indices. By inductive assumption, $\#\left(\widetilde{T_2}\right)$ $[= \#(T_2)]$ is finite.

For some K in $\mathbb{N}$, $K \leq \#(\sigma)$ and in finitely many repetitions of the procedure just described, there are generated

$$T_1, T_2 \ (\subset \Sigma_k \setminus T_1), \ldots, T_K \ (\subset ((\Sigma_k \setminus T_1) \setminus T_2) \cdots).$$

Each T_i is a finite set of basic neighborhoods, and the union of the T_i is Σ_k: $\#(\Sigma_k)$ is finite. Owing to *ii*, $\mathcal{N}$ is empty, finite, or countable. □

2.18. Let $\mathcal{B}_\lambda \stackrel{\text{def}}{=} \{U_{\lambda,n}\}_{n\in\mathbb{N}}$ be a countable base for X_λ. Since each set in $\mathcal{O}$ contains a basic neighborhood constructed with elements from finitely many bases, it suffices to show that a set $\mathcal{N} \stackrel{\text{def}}{=} \{N(\sigma, \mathsf{E})\}$ of pairwise disjoint basic neighborhoods is empty, finite, or countable.

A basic neighborhood $N(\sigma, \mathsf{E})$ is determined by a finite set σ of indices from Λ, and a finite set E of sets $U_{\lambda,n_\lambda}, \lambda \in \sigma, n_\lambda \in \mathbb{N}$, from $\mathcal{B}_\lambda$.

The finiteness of each index set σ permits the exhaustion process described in **2.17** to be carried out, mutatis mutandis. The following should be noted.

i. Disjointness arises at a hit-index because of inequality of the base elements used at a hit-index.

ii. The set T_1 is here a set T_{11} consisting of all N' having at the hit-index λ_1 a fixed base element different from the base element for N at the hit-index λ_1; T_{12} consists of all N' having at the hit-index λ_1 a fixed base element different from the base element for N and from the base element used in the construction of T_{11}; ... . Once the first hit-index λ_1 has produced its countable string $T_{11}, T_{12}, \ldots$, the process is repeated for a second hit-index λ_2, etc. The refinement T_{ij} permits the conclusion that the elements of $\widetilde{T_{ij}}$ are pairwise disjoint.

iii. Each Σ_k may be infinite but cannot be uncountable. □

[**Note s2.2:** If basic neighborhoods are determined by not more than countably many indices, the argument remains valid. If each space X_λ has weight W and if neighborhoods are determined by not more than countably many indices, the cardinality of a set of pairwise disjoint open sets in X cannot exceed W. At the heart of the reasoning is the fact that if $\mathfrak{a}$ is an infinite cardinal then $\aleph_0 \mathfrak{a} = \mathfrak{a}$.]

2.19. If $f : \mathcal{D}^M \stackrel{\text{sur}}{\mapsto} X$ is continuous then for each of the uncountably many points x other than x_0, the set $\{x\}$ is open and $f^{-1}(\{x\})$ is an open subset of $\mathcal{D}^M$. The sets $\{x\}$ are pairwise disjoint, whence the sets $f^{-1}(\{x\})$ are pairwise disjoint and constitute an uncountable set, in contradiction of the conclusion in **2.17**. □

2.20. Because $V \stackrel{\text{def}}{=} \bigcup_{\lambda\in\Lambda} (a_\lambda, b_\lambda)$ is an open subset of $\mathbb{R}$, which is separable, V is the countable union of open intervals, each contained in some

(a_λ, b_λ): for some countable subset $\{(a_{\lambda_n}, b_{\lambda_n})\}_{n\in\mathbb{N}}$, $V = \bigcup_{n\in\mathbb{N}} (a_{\lambda_n}, b_{\lambda_n})$. Let Λ_0 be $\{\lambda : b_\lambda \notin V\}$. If $(a_{\lambda'}, b_{\lambda'}] \subset (a_\lambda, b_\lambda]$ then $(a_{\lambda'}, b_{\lambda'}]$ may be cast out of the union. If $\lambda, \lambda' \in \Lambda_0$ and $\lambda \neq \lambda'$ then $(a_\lambda, b_\lambda] \cap (a_{\lambda'}, b_{\lambda'}] = \emptyset$ since otherwise, b_λ or $b_{\lambda'}$ is in V. Because there are at most countably many pairwise disjoint intervals $(a_\lambda, b_\lambda]$ in $\mathbb{R}$, Λ_0 is countable. It follows that

$$F = \bigcup_{\lambda\in\Lambda_0} (a_\lambda, b_\lambda] \cup \bigcup_{n\in\mathbb{N}} (a_{\lambda_n}, b_{\lambda_n}].$$ □

2.21. Because $\lim_{m,n\to\infty} d(m,n) = 0$, $\mathbb{N}$ is a d-Cauchy sequence but $\mathbb{R}$ contains no x such that $\lim_{n\to\infty} d(n,x) = 0$. □

2.22. If $u \notin S$ then for some sequence $\{x_n\}_{n\in\mathbb{N}}$ in S, $x_n \uparrow u$. Furthermore, $\frac{x_n}{x_{n+1}} \in S$, whence $\frac{x_n}{x_{n+1}} < u$ and thus if $\{p,n\} \subset \mathbb{N}$ then $0 < x_n < u^p x_{n+p}$. Since $u < 1$ it follows that $x_n \equiv 0$, a contradiction. □

2.23. The map $f : \mathbb{R} \setminus \mathbb{Q} \ni t \mapsto \begin{cases} \frac{1}{2} + \frac{t}{2t+2} & \text{if } t > 0 \\ \frac{1}{2} + \frac{t}{2-2t} & \text{if } t < 0 \end{cases}$ illustrated in **Figure s2.1** maps $\mathbb{R} \setminus \mathbb{Q}$ homeomorphically onto $(0,1) \setminus \mathbb{Q}$.

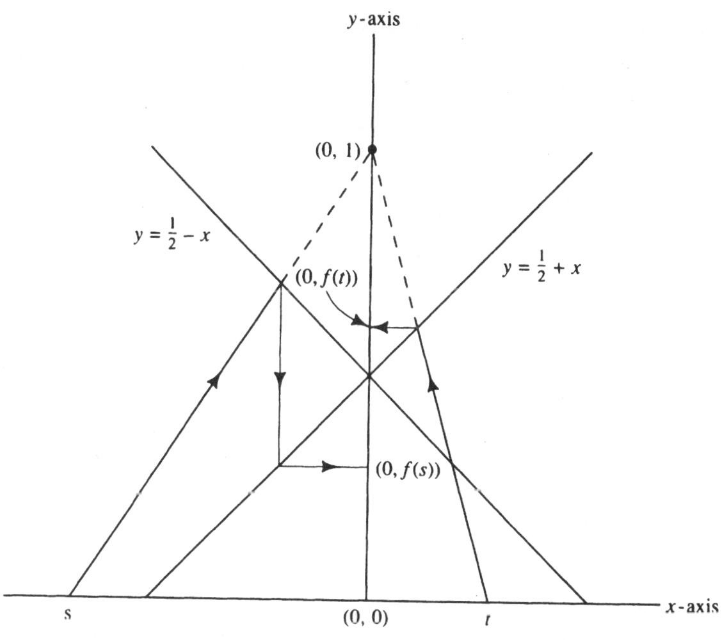

Figure s2.1. The homeomorphism $f : \mathbb{R} \setminus \mathbb{Q} \overset{\text{sur}}{\mapsto} (0,1) \setminus \mathbb{Q}$.

□

2.24. If $p_0 \in U$ then for some positive r_0, $B(p_0, r_0)^\circ \subset U$. Because there are no isolated points in X, there are in $B(p_0, r_0)$ two different points p_{00}, p_{01} and a positive r_1 such that

$$B(p_{00}, r_1) \cap B(p_{01}, r_1) = \emptyset \text{ and } B(p_{00}, r_1) \cup B(p_{01}, r_1) \subset B(p_0, r_0).$$

By induction, for each dyadic rational number $\sum_{k=1}^{K} \epsilon_k 2^{-k}$, there can be defined a point $p_{\epsilon_1,\ldots,\epsilon_K}$ and for each K a positive r_K so that $r_K \downarrow 0$,

$$B\left(p_{\epsilon_1,\ldots,\epsilon_{K-1},0}, r_K\right) \cap B\left(p_{\epsilon_1,\ldots,\epsilon_{K-1},1}, r_K\right) = \emptyset, \text{ and}$$
$$B\left(p_{\epsilon_1,\ldots,\epsilon_{K-1},0}, r_K\right) \cup B\left(p_{\epsilon_1,\ldots,\epsilon_{K-1},1}, r_K\right) \subset B\left(p_{\epsilon_1,\ldots,\epsilon_{K-1}}, r_{K-1}\right).$$

The cardinality of the closure of the set of all points $p_{\ldots}$ is $\mathfrak{c}$. □

2.25. If no point of K is isolated, each point of K is a limit point and since K is compact, K is perfect. Because X is a complete metric space, $\#(K) \geq \mathfrak{c}$ (cf. **2.24**), a contradiction. □

2.26. If $\mathbb{R}^n \neq \overline{\bigcup_{n\in\mathbb{N}} F_n^\circ}$ then $\mathbb{R}^n \setminus \overline{\bigcup_{n\in\mathbb{N}} F_n^\circ}$ contains a nonempty compact ball B and $B = \bigcup_{n\in\mathbb{N}} B \cap (F_n \setminus F_n^\circ)$. Since each summand is closed and its interior is empty, each summand is nowhere dense. The complete metric space B is the countable union of nowhere dense sets, a contradiction. □

2.27. a) Since $\mathbb{R}^n$ is separable, it has a countable base $\{N_n\}_{n\in\mathbb{N}}$. Every $N(a)$ contains an N_{n_a} containing a. If every $A \cap N(a)$ is at most countable, then $A \subset \bigcup_{n\in\mathbb{N}} A \cap N_{n_a}$, which is countable.

b) If $s \in \overline{S}$ then $s = \bigcap_{n\in\mathbb{N}} B\left(s, \frac{1}{n}\right)^\circ$ and $B\left(s, \frac{1}{n}\right)^\circ \cap S \stackrel{\text{def}}{=} S_n \neq \emptyset$. For t in S_n and some m in $\mathbb{N}$, $B\left(t, \frac{1}{m}\right) \subset B\left(s, \frac{1}{n}\right)$ whence $s \in S$.

c) $\#(S) \geq \#(S_n) > \aleph_0$.

d) If $n = 1$ and $A \cap N(a)$ is uncountable then $\underline{\lim}\{x : x \in A \cap N(a)\}$ serves for b and $\overline{\lim}\{y : y \in A \cap N(a)\}$ serves for c. □

2.28. Let P be the self-dense kernel of X. If $x \in P^\bullet$ then $\{x\} \cup P$ is self-dense whence $x \in P$, i.e., P is closed and hence P is perfect.

If $x \in C \stackrel{\text{def}}{=} X \setminus P$ then x is isolated: C consists of isolated points. Because D is dense, $x \in D$ and since D is countable, $\#(C) \leq \aleph_0$. □

2.29. Because $\mathcal{A}(\mathsf{F}([0,1])) \subset \mathcal{A}(\mathsf{S}_\lambda([0,1]))$ and $\mathcal{A}(\mathsf{S}_\lambda([0,1])) = \mathsf{S}_\lambda([0,1])$ (cf. **1.16**), it suffices to prove that every Lebesgue measurable subset of $[0,1]$ is empty, finite, countable, or of cardinality $\mathfrak{c}$.

The Cantor-Bendixson theorem implies that every closed subset of $[0,1]$ is empty, finite, countable, or of cardinality $\mathfrak{c}$. If $E \in \mathsf{S}_\lambda([0,1])$ it follows that $\lambda(E) = \sup\{\lambda(F) : F \in \mathsf{F}([0,1]), F \subset E\}$. Hence $\#(E) \leq \aleph_0$ or $\#(E) = \mathfrak{c}$. □

2.30. a) Let Y be $X \times X$ metrized according to the formula

$$D : Y \times Y \ni ((a_1, b_1), (a_2, b_2)) \mapsto d(a_1, a_2) + d(b_1, b_2).$$

Since (X, d) is compact, so is (Y, D). The map f serves to define a map $F : Y \ni y \stackrel{\text{def}}{=} (x_1, x_2) \mapsto (f(x_1), f(x_2)) \stackrel{\text{def}}{=} F(y) \in Y$. If, for some y in Y, some positive ϵ, and all n, $D(F^n(y), y) \geq \epsilon$ then, since f is expansive, for each k in $\mathbb{N}$, $D\left(F^{n+k}(y), F^k(y)\right) \geq \epsilon$. In other words, for any two p, q in $\mathbb{N}$, $D(F^p(y), F^q(y)) \geq \epsilon$, i.e., $\{F^m(y)\}_{m \in \mathbb{N}}$ contains no Cauchy sequence, a contradiction of the compactness of Y.

Hence if $\epsilon > 0$ then for each y in Y, and some $n \stackrel{\text{def}}{=} n(y)$ in $\mathbb{N}$, $D(F^n(y), y) < \epsilon$. If $y = (a, b)$ then:

$$d(a,b) \leq d(f^n(a), f^n(b)) \leq d(f^n(a), a) + d(a,b) + d(b, f^n(b));$$
$$d(f^n(a), a) + d(b, f^n(b)) = D(F^n(y), y) < \epsilon;$$
$$d(a,b) \leq d(f(a), f(b)) \leq d(f^n(a), f^n(b)) < d(a,b) + \epsilon.$$

b) Because f is expansive, f is injective. Since f is an isometry, f is continuous and, because X is compact, f^{-1} is continuous on $f(X)$. If $f(X) \subsetneqq X$ then $Z \stackrel{\text{def}}{=} X \setminus f(X)$ is open and nonempty whence it contains some closed ball $B(a, r)$ of positive radius. But then $f(B(a,r)) \subset f(X)$ and it follows that the sequence $\{f^n(B(a,r))\}_{n \in \mathbb{N}}$ consists of pairwise disjoint closed balls, each of radius r. Hence $\{f^n(a)\}_{n \in \mathbb{N}}$ contains no Cauchy sequence, again in contradiction of the compactness of X.

c) For each x in X, $f^{-1}(x) \neq \emptyset$. The Axiom of Choice implies there is a map $g : X \mapsto X$ such that $f \circ g(x) = x$ and, since f is contractive, g is expansive. From a) and b) it follows that g is an isometric auteomorphism. Since $f \circ g = \text{id}$, it follows that

$$f = f \circ \text{id} = f \circ \left(g \circ g^{-1}\right) = (f \circ g) \circ g^{-1} = \text{id} \circ g^{-1} = g^{-1}. \qquad \square$$

2.31. For $T : \mathbb{T} \ni e^{2\pi i\theta} \mapsto e^{2\pi i(2\theta/(1+\theta))} \in \mathbb{T}$ and all x, $\lim_{n\to\infty} T^n(x) = 1$. If d is any T-invariant metric compatible with the topology of X then $d(T^n(x), T^n(y)) \equiv d(x, y)$, whence $d(x,y) \equiv 0$, i.e., $d(T(x), T(y)) = d(x, y)$ iff $x = y$. $\square$

2.32. It suffices to show that every sequence $\{T_n\}_{n \in \mathbb{N}}$ in E contains a convergent subsequence. The proof follows the lines of the proof of the Arzelà-Ascoli theorem. Because X is a compact metric space, X contains a dense sequence $S \stackrel{\text{def}}{=} \{x_n\}_{n \in \mathbb{N}}$. Because each T_n maps X into itself, there is a convergent subsequence $T_{n_m}(x_1)$, a convergent subsubsequence $T_{n_{m_j}}(x_2)$, etc. The diagonal sequence $\left\{T_{n_1}, T_{n_{m_2}}, \ldots\right\} \stackrel{\text{def}}{=} \{G_1, G_2, \ldots\}$ is such that for each m, $\{G_p(x_m)\}_{p \in \mathbb{N}}$ converges.

Owing to the equicontinuity of $\{G_p\}_{p\in\mathbb{N}}$, if $\epsilon > 0$ there is a positive η such that $\{d(x,y) < \eta\} \Rightarrow \{\forall(p)\,\{d(G_p(x), G_p(y)) < \epsilon\}\}$. Furthermore $\left\{N_m \stackrel{\text{def}}{=} \{x : d(x, x_m) < \eta\}\right\}_{m\in\mathbb{N}}$ is an open cover of X and thus contains a finite subcover $\{N_m\}_{1\le m\le M}$. Furthermore, there is a $k(\epsilon)$ such that if $p, q > k(\epsilon)$ and $1 \le m \le M$ then $d(G_p(x_m), G_q(x_m)) < \epsilon$. It follows that if if $x \in X$ then for some m in $[1, M]$, $x \in N_m$ and if $p, q > k(\epsilon)$ then

$$\begin{aligned} d(G_p(x), G_q(x)) \le d(G_p(x), G_p(x_m)) &+ d(G_p(x_m), G_q(x_m)) \\ &+ d(G_q(x_m), G_q(x)) < 3\epsilon. \end{aligned}$$ □

2.33. If f is continuous then

$$F : X \ni x \mapsto (x, f(x)) \in X \times Y$$

is also continuous. Because X is compact, $F(X)$, which is $\mathcal{G}(f)$, is compact and thus closed.

Conversely, if $\mathcal{G}(f)$ is closed and f is not continuous, there is a sequence $\{x_n\}_{n=0}^{\infty}$ such that $\lim_{n\to\infty} x_n = x_0$ and for some positive ϵ and all n, $\delta(f(x_n), f(x_0)) \ge \epsilon$. Owing to the compactness of Y, there is a subsequence $\{x_{n_k}\}_{k\in\mathbb{N}}$ such that $\lim_{n\to\infty} f(x_{n_k}) \stackrel{\text{def}}{=} y$ exists and is not $f(x_0)$. Since $\mathcal{G}(f)$ is closed, (x_0, y) is in $\mathcal{G}(f)$. On the other hand, for some subsequence $\{x_{m_l}\}_{l\in\mathbb{N}}$, $\lim_{l\to\infty} x_{m_l} = x_0$ and $\lim_{l\to\infty} f(x_{m_l}) \stackrel{\text{def}}{=} z$ exists, whence $(x_0, z) \in \mathcal{G}(f)$. In sum, the two different points (x_0, y) and (x_0, z) belong to $\mathcal{G}(f)$, a contradiction. □

2.34. Let $\mathcal{B} \stackrel{\text{def}}{=} \{U_n\}_{n\in\mathbb{N}}$ be a base for X. Then every open set in X is a countable union of members of $\mathcal{B}$, whence $\#(\mathsf{O}) \le \mathfrak{c}$. Furthermore, $\theta : \mathsf{O} \ni U \mapsto X \setminus U \in \mathsf{F}$ is a bijection. □

2.35. a) The maps $f_n : [0,1] \ni x \mapsto \dfrac{x}{n}$ are in A_i whereas $f_n \stackrel{\text{u}}{\to} 0$ and $0 \notin A_i$: A_i is not closed.

b) If $\{f_n\}_{n\in\mathbb{N}} \subset A_s$ and $f_n \stackrel{\text{u}}{\to} f$ then for y in $[0,1]$ and n in $\mathbb{N}$, there is an x_n such that $f_n(x_n) = y$. If $x_{n_k} \to x_0$ as $k \to \infty$ then $f(x_{n_k}) \to f(x_0)$ as $k \to \infty$. Because $|y - f(x_{n_k})| = |f_{n_k}(x_{n_k}) - f(x_{n_k})| \le \|f_{n_k} - f\|_\infty$, it follows that $f(x_0) = y$ and so A_s is closed.

c) As depicted in **Figure s2.2**,

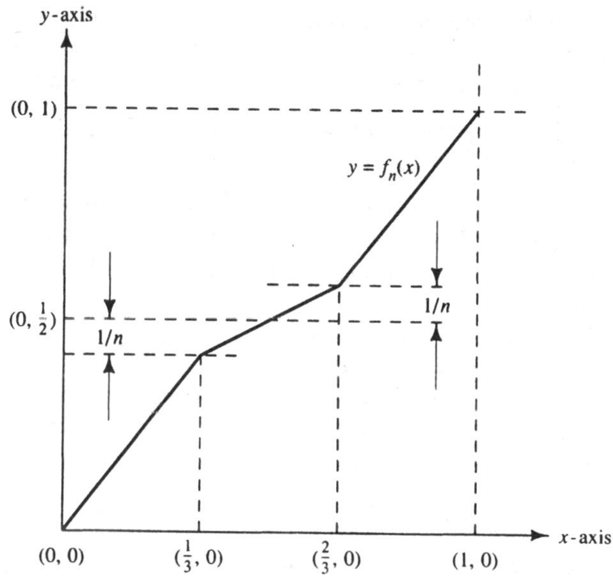

Figure s2.2. Graph of the equation $y = f_n(x)$.

when $n = 3, 4, \ldots$, the maps

$$f_n; [0,1] \ni x \mapsto \begin{cases} \dfrac{3(n-2)x}{2n} & \text{if } 0 \le x \le \frac{1}{3} \\ 6\dfrac{x-\frac{1}{2}}{n} + \dfrac{1}{2} & \text{if } \frac{1}{3} < x \le \frac{2}{3} \\ f_n\left(x - \frac{2}{3}\right) + \dfrac{1}{2} & \text{if } \frac{2}{3} < x \le 1 \end{cases}$$

are in A_b. On the other hand,

$$f_n \xrightarrow{\text{u}} f : [0,1] \ni x \mapsto \begin{cases} \dfrac{3x}{2} & \text{if } 0 \le x \le \frac{1}{3} \\ \dfrac{1}{2} & \text{if } \frac{1}{3} < x \le \frac{2}{3} \\ \dfrac{3\left(x - \frac{2}{3}\right)}{2} + \dfrac{1}{2} & \text{if } \frac{2}{3} < x \le 1 \end{cases}$$

and $f \notin A_b$ whence A_b is not closed.

d) Because A is convex, it is connected.

e) The maps $f_n : [0,1] \ni x \mapsto x^n$, $n \in \mathbb{N}$, are in A, whereas no subsequence of $\{f_n\}_{n\in\mathbb{N}}$ converges uniformly: A is not compact. □

2.36. If B is not connected there are in X two closed sets F_1 and F_2 such that $(F_1 \cap B) \cup (F_2 \cap B) = B$ and $F_i \cap B \neq \emptyset$, $i = 1, 2$.

If $x_i \in F_i \cap B$, $i = 1, 2$, and $f(x_1) = f(x_2) \stackrel{\text{def}}{=} y$ then $y \in A$, x_1 and x_2 are in the connected set $C \stackrel{\text{def}}{=} f^{-1}(y)$, $C \subset B$, $C = (C \cap F_1) \cup (C \cap F_2)$, and $x_i \in C \cap F_i$, $i = 1, 2$, contradicting the connectedness of C. Hence $f(F_1 \cap B) \cup f(F_2 \cap B) = A$.

Since A is connected and neither of $f(F_1 \cap B), f(F_2 \cap B)$ is empty, it suffices to show that each is closed in A.

Thus let $n : \Lambda \to f(F_1 \cap B)$ be a net converging to some y in A. For each $n(\lambda)$ and for some x_λ in $F_1 \cap B$, $f(x_\lambda) = n(\lambda)$, whence for the net $m : \Lambda \ni \lambda \mapsto x_\lambda$, $f(m)$ converges to y. Owing to the compactness of X, for the net m, there is a cofinal diset Λ' such that $m\big|_{\Lambda'} \stackrel{\text{def}}{=} m'$ converges to some x and thus $f(m)$ converges to $f(x)$. Because $f(m') \subset f(m)$ and $f(m)$ converges to y it follows that $f(x) = y$ whence $x \in B$. Since F_1 is closed, $x \in F_1$, i.e., $x \in (F_1 \cap B)$, $y = f(x) \in f(F_1 \cap B)$. The same argument shows $f(F_2 \cap B)$ is closed in A.

If $X = Y = \{0, 1\}$ and the topology of X is 2^X while that of Y is $\{\emptyset, Y\}$, then for $f : X \ni x \mapsto x \in Y$ the hypothesis above is fulfilled but Y is connected whereas $f^{-1}(Y)$ $(= X)$ is not. □

2.37. If $\epsilon > 0$, $\mathbb{R}$ contains open sets $\{U_n\}_{n\in\mathbb{N}}$ such that $\operatorname{diam}(U_n) < \epsilon$ and $\bigcup_{n\in\mathbb{N}} U_n = \mathbb{R}$. Since X is compact, for some N, $X = \bigcup_{n=1}^{N} f^{-1}(U_n)$. Each open set $f^{-1}(U_n)$ is a union of basic neighborhoods whence for finitely many basic neighborhoods, say $V_1, \ldots, V_P$, $X = \bigcup_{p=1}^{P} V_p$. Let Z_1 be V_1, and Z_p be $(V_1 \cup \cdots \cup V_p) \setminus (V_1 \cup \cdots \cup V_{p-1})$, $p = 2, \ldots, P$. Then $X = \bigcup_{p=1}^{P} Z_p$. If $Z_p \neq \emptyset$, fix an $\mathbf{x}_p$ in Z_p. Each $\mathbf{x}$ in X is in precisely one Z_p. Thus $g : X \ni \mathbf{x} \mapsto f(\mathbf{x}_p)$ if $\mathbf{x} \in Z_p$ is well-defined.

Each basic neighborhood V_p is determined by finitely many indices $\gamma_{i,p}$, $1 \le i \le N_p$. Hence if $\mathbf{x}, \mathbf{y} \in X$ and $x_{\gamma_{i,p}} = y_{\gamma_{i,p}}$ then $\mathbf{x}$ and $\mathbf{y}$ are in the same Z_p and $g(\mathbf{x}) = g(\mathbf{y}) = f(\mathbf{x}_p)$: g is finitely determined. If $\mathbf{x} \in Z_p$ then $|f(\mathbf{x}) - g(\mathbf{x})| = |f(\mathbf{x}) - f(\mathbf{x}_p)| < \epsilon$.

For n in $\mathbb{N}$, there is a finitely determined g_n with a corresponding finite index set Γ_n and $\sup_{x\in X} |f(\mathbf{x}) - g_n(\mathbf{x})| < \frac{1}{n}$. If $\mathbf{x}$ and $\mathbf{y}$ share coordinates indexed by the countable set $\Gamma \stackrel{\text{def}}{=} \bigcup_{n\in\mathbb{N}} \Gamma_n$ then

$$
\begin{aligned}
|f(\mathbf{x}) - f(\mathbf{y})| &\le |f(\mathbf{x}) - g_n(\mathbf{x})| + |g_n(\mathbf{x}) - g_n(\mathbf{y})| + |g_n(\mathbf{y}) - f(\mathbf{y})| \\
&\le \frac{1}{n} + 0 + \frac{1}{n} = \frac{2}{n}, \ n \in \mathbb{N},
\end{aligned}
$$

whence f is countably determined. □

2.38. The function f is bounded above. Otherwise, for each n in $\mathbb{N}$, there is an x_n such that $f(x_n) > n$ whence the set $\{f^{-1}([n,\infty)) : n \in \mathbb{N}\}$ of closed sets has the finite intersection property. Hence there is a y in their intersection: for all n in $\mathbb{N}$, $f(y) \geq n$, a contradiction. It follows that $\sup_{x\in\mathbb{R}} f(x) \stackrel{\text{def}}{=} M < \infty$.

The sequence $\left\{F_n \stackrel{\text{def}}{=} \left\{x : f(x) \geq M - \frac{1}{n}\right\}\right\}$ consists of closed sets and $F_n \supset F_{n+1}$, $n \in \mathbb{N}$. Hence there is an x_0 in $\bigcap_{n\in\mathbb{N}} F_n$ and $f(x_0) = M$. □

2.39. The compact sets $L_i \stackrel{\text{def}}{=} K \setminus U_i, i = 1, 2$, are disjoint, whence there are disjoint open sets V_i such that $L_1 \subset V_1 \subset U_2$ and $L_2 \subset V_2 \subset U_1$. Furthermore, $K_i \stackrel{\text{def}}{=} K \setminus V_i$ are compact, $K_i \subset U_i, i = 1, 2$, and $K_1 \cup K_2 = K$. □

2.40. For each finite subset $\{\gamma_1, \ldots, \gamma_n\}$ of Γ, each finite subset $\{t_1, \ldots, t_n\}$ of $[0,1] \cap \mathbb{Q}$, and each m in $\mathbb{N}$,

$$U(\gamma_1, \ldots, \gamma_n; t_1, \ldots, t_n; m) \stackrel{\text{def}}{=} \left\{f : f \in [0,1]^\Gamma, |f(\gamma_i) - t_i| < \frac{1}{m}\right\}$$

is a neighborhood in Ξ. The set of all such $U(\cdots)$ is countable. Any neighborhood in Ξ is, for some $\{s_1, \ldots, s_n\}$ contained in $[0,1]$ and some positive ϵ, of the form

$$V(\gamma_1, \ldots, \gamma_n; s_1, \ldots, s_n; \epsilon) \stackrel{\text{def}}{=} \{g : g \in [0,1]^\Gamma, |g(\gamma_i) - s_i| < \epsilon\}.$$

If $|t_i - s_i| < \frac{\epsilon}{4}, 1 \leq i \leq n$, and $\frac{1}{m} < \frac{\epsilon}{8}$ then

$$U(\gamma_1, \ldots, \gamma_n; t_1, \ldots, t_n; m) \subset V(\gamma_1, \ldots, \gamma_n; s_1, \ldots, s_n; \epsilon).$$ □

2.41. As the product of compact spaces, X is compact. If X is metrizable, X is separable and hence (cf. **2.34**) $\#(X) \leq \mathfrak{c}$. However $\#(X) = 2^{\mathfrak{c}}$ $(> \mathfrak{c})$.

As a function space, X may be endowed with the metric

$$d : X \times X \ni \{f, g\} \mapsto \sup_{x\in[0,1]} |f(x) - g(x)| \stackrel{\text{def}}{=} d(f, g) \in [0,1].$$

(The reader is urged to use the language of cylinders to describe the topology induced by d. The previous paragraph implies that any metric induces a topology in which X is not compact.) □

2.42. If $Y \stackrel{\text{def}}{=} \mathbb{N}$ in the discrete topology then for all n in $\mathbb{N}$, Y and Y^n are homeomorphic (and countable). On the other hand, $\#\left(Y^{\mathbb{N}}\right) = \mathfrak{c} > \aleph_0$. □

2.43. If $\{F_n\}_{n\in\mathbb{N}}$ is a sequence of closed sets and $[0,1] = \bigcup_{n\in\mathbb{N}} F_n$, there is an open set U_2 such that $F_2 \subset U_2$ and $U_2 \cap F_1 = \emptyset$. Some component C_2 (necessarily a closed interval) of $\overline{U_2}$ meets F_2. On the other hand, if $C_2 \subset U_2$ then for some open interval (a,b), $C_2 \subset (a,b) \subset U_2$, which implies that C_2 is not a component of $\overline{U_2}$. Thus $U_2 \setminus C_2 \neq \emptyset$ and so $C_2 \setminus F_2 \neq \emptyset$. It follows that $C_2 \subset \bigcup_{n=3}^{\infty} (C_2 \cap F_n)$, whence for some n in $\mathbb{N} \setminus \{1,2\}$, $C_2 \cap F_n \neq \emptyset$. The argument just given and applied to the closed interval C_2 and then continued by induction, leads to a sequence $\{C_m\}_{m=2}^{\infty}$ of closed intervals such that $C_m \supset C_{m+1}$, $C_m \neq \emptyset$, $C_m \cap F_{m-1} = \emptyset$, $m = 2,3,\ldots$. Thus $\bigcap_{m=2}^{\infty} C_m \neq \emptyset$ while

$$\bigcap_{m=2}^{\infty} = \left(\bigcap_{m=2}^{\infty} C_m\right) \cap \left(\bigcup_{n\in\mathbb{N}} F_n\right) = \bigcup_{n\in\mathbb{N}} \left(F_n \cap \bigcap_{m=2}^{\infty} C_m\right) = \emptyset,$$

a contradiction. □

2.44. Because $F \neq \mathbb{R}^n$, the Cantor-Bendixson theorem implies ∂F is finite or countable. Thus for some M in $\mathbb{N} \cup \aleph_0$, there are in $\mathbb{R}^n$ points $\mathbf{x}_m, 1 \le m < M$, such that $B = \{\mathbf{x}_m\}_{1\le m<M}$. Assume that $\mathbf{y} \in \mathbb{R}^n \setminus F$, $\mathbf{z} \in F^\circ$, and $N(\mathbf{z}) \subset F^\circ$. Because $n \ge 2$, the cardinality of the set L of line segments joining points of $N(\mathbf{z})$ to $\mathbf{y}$ is $\mathfrak{c}$ and hence infinitely many of the segments in L fail to meet the countable set ∂F. Any such line segment ℓ is a connected set meeting both F and $\mathbb{R}^n \setminus F$ whence it meets ∂F, a contradiction. Thus $F^\circ = \emptyset$ and hence $F = \partial F$, i.e., F is countable or finite. □

2.45. Since γ is rectifiable, if $\epsilon > 0$ for some N in $\mathbb{N}$:

i. if $m = 0,\ldots,N-1$ and $t \in \left[\frac{m}{N}, \frac{m+1}{N}\right]$ the distance of $\gamma(t)$ from the line segment $\overline{\gamma\left(\frac{m}{N}\right)\gamma\left(\frac{m+1}{N}\right)}$ connecting $\gamma\left(\frac{m}{N}\right)$ and $\gamma\left(\frac{m+1}{N}\right)$ does not exceed ϵ;

ii. the sum L of the lengths of the segments

$$\overline{\gamma\left(\frac{m}{N}\right)\gamma\left(\frac{m+1}{N}\right)},\ m = 0,\ldots,N-1,$$

does not exceed $\ell(\gamma)$.

It follows that for some constant K_n, γ^* is contained in a finite union of n-dimensional boxes the total volume of which does not exceed $K_n\epsilon^{n-1}\ell(\gamma)$. □

2.46. If $-\infty \le a < \infty$ then $(a,b) = \bigcup_{n\in\mathbb{N}} \left(a, b - \frac{1}{n}\right]$ whence $\mathcal{T}$ is at least as strong as $\mathcal{E}$: $\mathcal{T} \supset \mathcal{E}$. Because $(a,b] \in \mathcal{T} \setminus \mathcal{E}$, $\mathcal{T}$ is (strictly) stronger than $\mathcal{E}$.

If two basic intervals intersect, their union is again a basic interval. Hence if $U \in \mathcal{T}$ then U is a union of a countable set of pairwise disjoint basic intervals. It follows that if $\mathcal{T}$ is separable there is a base consisting of a sequence $\{(a_m, b_m]\}_{m\in\mathbb{N}}$ of basic intervals. If $b \notin \{b_n\}_{n\in\mathbb{N}}$ and $a < b$ then for some sequence $\{n_k\}_{k\in\mathbb{N}}$, $(a,b] = \bigcup_{k\in\mathbb{N}} (a_{n_k}, b_{n_k}]$. Since $b \notin \{b_n\}_{n\in\mathbb{N}}$, for some k_0, $a_{n_{k_0}} < b < b_{n_{k_0}}$, whereas $\sup_{k\in\mathbb{N}} b_{n_k} \le b$, a contradiction: $\mathcal{T}$ is not separable.

□

[**Note s2.3:** The discussion above illustrates the distinction between the notion of a space containing a countable dense set and that of a separable space. Although $\mathbb{R}$ is not separable with respect to the topology $\mathcal{T}$, $\mathbb{Q}$ is nevertheless a countable subset of $\mathbb{R}$ and is dense with respect to the topology $\mathcal{T}$. If X is separable and S consists of one point from each element of a countable base of neighborhoods, then S is a copuntable dense set: each separable space contains a countable dense set. For a metric space (X, d) the notions coincide: X is separable iff X contains a countable dense subset (cf. **Solution 2.110**).]

2.47. If U is an open set containing $\mathbb{Q}$ and $x \in \mathbb{R}$, then for any $N(x)$, $N(x) \cap U \neq \emptyset$ and $N(x) \cap U$ is an open set not meeting $A \stackrel{\text{def}}{=} \mathbb{R} \setminus U$: A is nowhere dense. Hence if $\mathbb{Q}$ is a G_δ then $\mathbb{I}_{\mathbb{R}}$ is a set of the first category, a contradiction.

□

2.48. a) Because for all U in U, $\Delta \subset U$, it follows that $x \in V_x(U)$. Since U is a filter base, $\mathcal{V}(x)$ is a filter base.

b) If $\Delta = \bigcap_{U\in\mathsf{U}} U$ and x and y are two points in X then $\{x, y\} \notin \Delta$ and thus for some U in U, $(x, y) \notin U$ and for some W, $(y, x) \notin W$. Some Z in U is contained in $W \cap U$. The diagram in **Figure s2.3** shows why $V_x(Z) \cap V_y(Z) = \emptyset$.

Conversely, if $\mathcal{T}$ is Hausdorff topology and x and y are two points in X there are in U a U and a W such that $V_x(U) \cap V_y(W) = \emptyset$. If Z in U is contained in $U \cap W$ then $\{(x, y)\} \cup \{(y, x)\} \not\subset Z$, cf. **Figure s2.3**.

c) If $n : \Lambda \mapsto X$ is a net then (cf. **2.9**) there is a cofinal net $n|_{\Lambda'}$ converging to some x in X. Hence if $U \in \mathsf{U}$ then for some λ_0' in Λ', $n(\lambda') \in V_x(U)$ if $\lambda' \succ \lambda_0'$. Choose in U a W such that $W \circ W \subset U$. If n is a Cauchy net, then for some λ_0 in λ, $(n(\lambda), n(\mu)) \in W$ if $\lambda, \mu \succ \lambda_0$. It follows that $n(\lambda) \in V_x(U)$ if $\lambda \succ \lambda_0, \lambda_0'$.

□

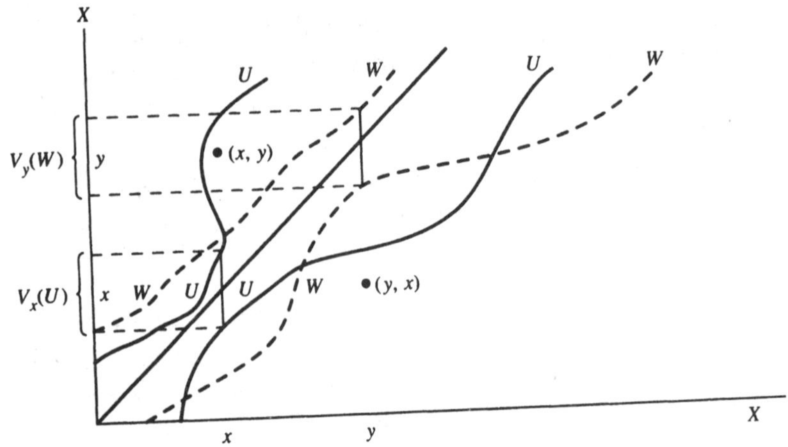

Figure s2.3.

2.49. If $U, V \in \mathsf{U}$ then some W in U is contained in $U \cap V$. It follows that $\mathcal{O}_W \succ \mathcal{O}_U, \mathcal{O}_V$. □

2.50. If $\mathcal{O}$ is a cover of X then each x in X is in some $U(x)$ belonging to $\mathcal{O}$. The set $\mathcal{O}' \stackrel{\text{def}}{=} \{U(x)\}_{x\in X}$ is an open cover of X and $\#(\mathcal{O}') \le \#(X)$. □

2.51. If $\#(D) > 1$, Z is a connected subset of D, and $\#(Z) > 1$ then, since D is closed, it may be assumed that Z is closed. If $z \in Z$ then $z \notin \bigcup_{N\in\mathcal{N}} V_N$ and for some N in $\mathcal{N}$, $z \in N$. Choose x and y in Z and an $N(x)$ in $\mathcal{N}$ so that $y \notin N(x)$. Thus $Z = (Z \setminus N(x)) \cup (Z \cap N(x))$ and neither summand is empty. Because $N(x)$ is open, $Z \setminus N(x)$ is closed.

Furthermore, because $V_N \supset \partial N$, $\overline{N(x)} \setminus V_{N(x)} = N(x) \setminus V_{N(x)}$ and thus $Z \cap N(x) = Z \cap \left(\overline{N(x)} \setminus V_{N(x)}\right)$, i.e., $Z \cap N(x)$ is closed and so Z is not connected, a contradiction. □

2.52. If $0 < a < b$ then for some m_0 in $\mathbb{N}$,

$$\{m > m_0\} \Rightarrow \{(m+1)a < mb\}$$

(otherwise $a \ge b$). Hence $\bigcup_{m>m_0}(ma, mb) = ((m_0+1)\,a, \infty)$. Since G is unbounded and open, it contains a nonempty interval (c, d) contained in $((m_0+1)\,a, \infty)$, i.e., for some m_1 in $(m_0+1, \infty)\cap\mathbb{N}$, $(m_1 a, m_1 b) \cap (c, d) \ne \emptyset$, i.e., (a, b) contains an x_1 such that $m_1 x_1 \in (c, d)$ $(\subset G)$. Thus if

$$A_m \stackrel{\text{def}}{=} \{x : mx \in G\} \text{ and } D_n \stackrel{\text{def}}{=} \bigcup_{m\ge n} A_m$$

then each D_n is dense in $[0,\infty)$. Because G is open, each A_m is open and thus each D_n is a dense open subset of $[0,\infty)$. However, $D = \bigcap_{n\in\mathbb{N}} D_n$ and the Baire category theorem implies that D is a G_δ, dense in $[0,\infty)$. □

2.2. Limits

2.53. The inequalities

$$\sum_{n=1}^{2^{N+1}} a_n \geq 2a_2 + \sum_{n=1}^{N} 2^{N-1} a_{2^N}, \quad \sum_{n=1}^{\infty} (a_{2^n+1} + \cdots a_{2^{n+1}}) \leq \sum_{n=1}^{\infty} 2^n a_{2^n}$$

imply the assertion. □

2.54. For n, K in $\mathbb{N}$,

$$\sum_{k=1}^{K} \frac{a_{n+k}}{A_{n+k}} \geq \sum_{k=1}^{K} \frac{A_{n+k} - A_{n+k-1}}{A_{n+K}} = 1 - \frac{A_n}{A_{n+K}}. \tag{s2.2}$$

For large K, the last members in (s2.2) are near one, whence the leftmost members of (s2.2) do not constitute a null sequence.

On the other hand, if $0 < d, x < 1$, the mean value theorem implies $1 - x^d > d(1-x)$ whence if $2 > r = 1 + d > 1$

$$\begin{aligned} \sum_{n=1}^{\infty} \left(\frac{1}{A_n^d} - \frac{1}{A_{n+1}^d} \right) &= \sum_{n=1}^{\infty} \frac{A_{n+1}^d - A_n^d}{A_n^d A_{n+1}^d} \\ &\geq d \sum_{n=1}^{\infty} \frac{A_{n+1}^{d-1} (A_{n+1} - A_n)}{A_{n+1}^d A_n^d} \geq d \sum_{n=1}^{\infty} \frac{a_{n+1}}{A_{n+1}^{d+1}}. \end{aligned} \tag{s2.3}$$

The first member of (s2.3) converges. □

2.55. It may be assumed that $b_1 > 1$. If $\epsilon_n \stackrel{\text{def}}{=} \dfrac{(-1)^n}{\sum_{k=1}^n b_k}$ then

$$\sum_{n=1}^{\infty} (-1)^n \epsilon_n b_n = \infty \quad (\mathbf{2.54}),$$

$$\sum_{n=1}^{\infty} (-1)^n b_n \text{ converges,}$$

whence $\sum_{n=1}^{\infty} (-1)^n b_n (1 + \epsilon_n) \stackrel{\text{def}}{=} \sum_{n=1}^{\infty} (-1)^n a_n$ diverges. □

2.56. For large n, $\left| \dfrac{a_n}{b_n} \right| < 1$ and

$$\frac{a_n}{a_n + b_n} = \frac{a_n}{b_n} \frac{1 - \dfrac{a_n}{b_n}}{1 - \left(\dfrac{a_n}{b_n} \right)^2} = \frac{a_n}{b_n} \left(1 - \frac{a_n}{b_n} \right) \left(1 + \sum_{k=1}^{\infty} \left(\frac{a_n}{b_n} \right)^{2k} \right).$$

Furthermore, $\delta_n \stackrel{\text{def}}{=} \sum_{k=1}^{\infty} \left(\frac{a_n}{b_n}\right)^{2k}$ is small and positive and

$$\sum_{n=1}^{\infty} \frac{a_n}{a_n + b_n} = \sum_{n=1}^{\infty} \left(\frac{a_n}{b_n} - \left(\frac{a_n}{b_n}\right)^2\right)(1+\delta_n).$$

For some N in $\mathbf{N}$, if $n > N$ then

$$\left(\frac{a_n}{b_n} - \left(\frac{a_n}{b_n}\right)^2\right)(1+\delta_n) \geq 0,\ 1+\delta_n \leq 2, \text{ and}$$

$$\sum_{n=N+1}^{\infty} \frac{a_n}{a_n + b_n} \leq 2 \sum_{n=N+1}^{\infty} \left(\frac{a_n}{b_n} - \left(\frac{a_n}{b_n}\right)^2\right) < \infty. \qquad \square$$

2.57. *i.* Because $\sum_{n=1}^{\infty} a_n$ converges absolutely, there is an autojection $\pi : \mathbf{N} \mapsto \mathbf{N}$ such that $a_{\pi(n)} \downarrow 0$, $\sum_{n=1}^{\infty} a_{\pi(n)} = 1$, and S for the latter series is the same as that for the former. Thus it may be assumed that $a_n \downarrow 0$.

If, for some N, $a_N > \sum_{k=N+1}^{\infty} a_k$ then S fails to contain the interval $\left(\sum_{k=N+1}^{\infty} a_n, a_N\right)$.

On the other hand, if $a_n \leq \sum_{k=n+1}^{\infty} a_k$, $n \in \mathbf{N}$, and $0 < t < 1$ then for some n_1 in $\mathbf{N}$, $S_1 \stackrel{\text{def}}{=} a_{n_1} \leq t < a_{n_1-1} \leq \sum_{k=n_1}^{\infty} a_k$. By induction, there can be defined in $\mathbf{N}$ a strictly increasing sequence $\{n_p\}_{1\leq p<P\leq\infty}$ such that $S_{p-1} \leq S_p \stackrel{\text{def}}{=} a_{n_1} + \cdots + a_{n_p} \leq t \leq S_{p-1} + a_{n_p-1} \leq S_{p-1} + \sum_{k=n_p}^{\infty} a_k$. Either $P < \infty$ and $S_P = t$ or $P = \infty$ and $S_p \uparrow \sum_{p=1}^{\infty} a_{n_p} = t$.

If $t = 0$ then $t = \sum_{k=1}^{\infty} 0 \cdot a_k$ and if $t = 1$ then $t = \sum_{k=1}^{\infty} 1 \cdot a_k$.

ia. The argument in *i* remains valid owing to the added hypothesis $\lim_{n\to\infty} a_n = 0$.

ii. For the map

$$\theta : C_0 \ni x \stackrel{\text{def}}{=} \sum_{n=1}^{\infty} \frac{2\epsilon_n(x)}{3^n} \mapsto \theta(x) \stackrel{\text{def}}{=} \sum_{n=1}^{\infty} \epsilon_n(x) a_n,$$

if $\eta > 0$, there is an n_0 such that $\sum_{n=n_0+1}^{\infty} a_n < \eta$. If $|x - x'| < 3^{-(n_0+1)}$ then $|\theta(x) - \theta(x')| < \sum_{n=n_0+1}^{\infty} a_n < \eta$ whence θ is continuous.

For a given $a \stackrel{\text{def}}{=} \sum_{n=1}^{\infty} \epsilon_n a_n$, if $x = \sum_{n=1}^{\infty} \frac{2\epsilon_n}{3^n}$ then $\theta(x) = a$, i.e., θ is a continuous surjection of the Cantor set C_0 onto S.

If $x > x'$, there is an n_0 such that

$$\epsilon_n(x) - \epsilon_n(x') = \begin{cases} 0 & \text{if } n \leq n_0 \\ 1 & \text{if } n = n_0 + 1 \end{cases}$$

$$\theta(x) - \theta(x') \geq a_{n_0+1} > 0$$

whence θ is a strictly monotonely increasing function. Since C_0 is compact, $\theta : C_0 \mapsto S$ is a homeomorphism. (Thus S is a perfect nowhere dense subset of $[0,1]$).

iia. If $1 > a \stackrel{\text{def}}{=} 1 - r > 0$ then $ar^{n-1} \leq \sum_{n=k}^{\infty} ar^k = \frac{ar^n}{1-r}$ iff $r \geq \frac{1}{2}$ (whence $ar^{n-1} > \sum_{n=k}^{\infty} ar^k = \frac{ar^n}{1-r}$ iff $r < \frac{1}{2}$).

iii. If $r \geq \frac{1}{2}$ then $S = [0,1]$ and $\lambda(S) = 1$. If $0 \leq r < \frac{1}{2}$, $[0,1] \setminus S$ is the union of pairwise disjoint open intervals, the first of length $1 - 2r$, each of the next two of length $(1-2r)r, \ldots,$ each of the next 2^{n-1} of length $(1-2r)r^{n-1}, \ldots,$ and $\lambda(S) = 1 - \sum_{n=1}^{\infty}(1-2r)(2r)^{n-1} = 0$.

iv. By induction, there can be defined $a_{11} \stackrel{\text{def}}{=} a_1$, $a_{12} \stackrel{\text{def}}{=}$ the first a_i less than a_{11}, $a_{13} \stackrel{\text{def}}{=}$ the first a_i such that $a_{12} > a_i$ and $a_{11} > a_{12} + a_i$, etc. The process is repeated on the sequence $\{a_n\}_{n\in\mathbb{N}} \setminus \{a_{1q}\}_{q\in\mathbb{N}}$, to produce a sequence $\{a_{2q}\}_{q\in\mathbb{N}}, \ldots$.

v. The condition implies that for every subsequence $\{a_{pq}\}_{q\in\mathbb{N}}$ as described, $a_{p1} > \sum_{q=2}^{\infty} a_{pq}$. □

2.58. The series converges resp. diverges according as $\underline{\lim}_{n\to\infty} q_n > 1$ resp. $\overline{\lim}_{n\to\infty} q_n < 1$. However, if P is the set of primes in $\mathbb{N}$ then

$$\sum_{n\in P} \frac{1}{n^2} \text{ converges although } \underline{\lim}_{n\to\infty} q_n = 0,$$

$$\sum_{n\in P} \frac{1}{n} \text{ diverges although } \overline{\lim}_{n\to\infty} q_n = 1.$$ □

2.59. If $f_n : \mathbb{R} \ni x \mapsto \sum_{k=0}^{n-1}(1-x)^{n-1}$ then $xf_n(x) = 1 - (1-x)^n$ and

$$M \stackrel{\text{def}}{=} \sum_{k=1}^{\infty}\left(1 - \left(1 - 2^{-k}\right)^n\right) = \sum_{k=1}^{\infty} 2^{-k} f_n\left(2^{-k}\right).$$

Direct integration and graphical considerations imply

$$\int_0^1 f_n(x)\,dx = \sum_{p=1}^{n} \frac{1}{p}, \ \ln(n+1) < \sum_{p=1}^{n} \frac{1}{p} < 1 + \ln n,$$

$$\frac{1}{2}\sum_{k=0}^{\infty} 2^{-k} f_n\left(2^{-k}\right) \leq \int_0^1 f_n(x)\,dx \leq \sum_{k=1}^{\infty} 2^{-k} f_n\left(2^{-k}\right).$$

Because $n \geq 2$, $1 \leq \frac{\ln n}{\ln 2}$ and so

$$\ln(n+1) < \sum_{p=1}^{n} \frac{1}{p} = \int_0^1 f_n(x)\,dx \leq M,$$

$$\frac{1}{2}+\frac{1}{2}M \le \int_0^1 f_n(x)\,dx = \sum_{p=1}^{n}\frac{1}{p} < 1+\ln n,$$

$$\ln(n+1) \le M \le 1+2\ln n,\ \ln n \le M \le \left(2+\frac{1}{\ln 2}\right)\ln n. \qquad \square$$

2.60. If $p_n(x) \stackrel{\text{def}}{=} \sum_{k=1}^{M} a_{nk}x^k$, $n \in \mathbb{N}$, and the $M+1$ points at which convergence takes place are $\alpha_1,\ldots,\alpha_{M+1}$, the determinant of the Vandermonde matrix

$$V \stackrel{\text{def}}{=} \begin{pmatrix} 1 & \alpha_1 & \cdots & \alpha_1^M \\ \vdots & \vdots & \ddots & \vdots \\ 1 & \alpha_M & \cdots & \alpha_{M+1}^M \end{pmatrix}$$

is $\prod_{1\le k<l\le M+1}(\alpha_l-\alpha_k)$ which is not zero because the α_m are pairwise different. Thus, if $\lim_{n\to\infty} p_n(\alpha_m) \stackrel{\text{def}}{=} A_m$ then

$$\lim_{n\to\infty} V\begin{pmatrix} a_{n1} \\ \vdots \\ a_{nM}\end{pmatrix} = \begin{pmatrix} A_1 \\ \vdots \\ A_M\end{pmatrix}$$

$$\lim_{n\to\infty}\begin{pmatrix} a_{n1} \\ \vdots \\ a_{nM}\end{pmatrix} = V^{-1}\cdot\begin{pmatrix} A_1 \\ \vdots \\ A_M\end{pmatrix} \stackrel{\text{def}}{=} \begin{pmatrix} a_1 \\ \vdots \\ a_M\end{pmatrix}$$

and $f(x) = \lim_{n\to\infty} p_n(x) = \sum_{k=1}^{M} a_k x^k$.

On the other hand, the Weierstraß approximation theorem implies that if $M=\infty$ then f need not be a polynomial. $\square$

2.61. a) If $x\in\mathbb{Q}$, then for large all large m, $m!x\in 2\mathbb{N}$ whence $s_m=1$ and the right member is one. If $x\notin\mathbb{Q}$ then for each m, $|\cos(m!x\pi)|<1$ whence $s_m=0$ and the right member is zero.

b) If $x=0$ then $\arctan(nx)=0$; if $x>0$, $\lim_{n\to\infty}\arctan(nx)=\frac{\pi}{2}$; if $x<0$, $\lim_{n\to\infty}\arctan(nx)=-\frac{\pi}{2}$.

c) If $\epsilon_n \stackrel{\text{def}}{=} \sum_{k=n+1}^{\infty}\frac{1}{k!}$ then $(n+1)!\epsilon_n - 1 \stackrel{\text{def}}{=} \delta_n \downarrow 0$. Thus

$$n\sin(2\pi e n!) = n\sin(2\pi\epsilon_n n!) = \frac{\frac{n(1+\delta_n)}{n+1}\sin\left(2\pi\left(\frac{1+\delta_n}{n+1}\right)\right)}{\frac{1+\delta_n}{n+1}} \to 2\pi$$

as $n\to\infty$.

If $e\in\mathbb{Q}$ then for some n, $en!\in\mathbb{N}$ and if $m>n$ then $m\sin(2\pi e m!)=0$, a contradiction.

d) If $x \in \mathbb{Q}$, for all large m, $\sin^2(m!\pi x) = 0$; if $x \notin \mathbb{Q}$, $\sin^2(m!\pi x) > 0$. □

2.62. For x in $[0,\infty)$,

$$\begin{aligned} 0 \le F(x) &= \int_x^\infty e^{-\frac{(t-x)(t+x)}{2}}\, dt = \int_0^\infty e^{-\frac{u(u+2x)}{2}}\, du \\ &\le \int_0^1 e^{-\frac{u(u+2x)}{2}}\, du + \int_1^\infty e^{-\frac{u+2x}{2}}\, du \stackrel{\text{def}}{=} I + II. \end{aligned}$$

As $x \uparrow \infty$ the integrands in I and II decrease monotonely to zero. The dominated convergence theorem applies. □

2.63. For a in $(0,1)$, each

$$E_{kn}(a) \stackrel{\text{def}}{=} \left\{ x \,:\, x \in [2k, 2(k+1)\pi], \left|\sin\left(x + \frac{r_n\pi}{n}\right)\right| > a \right\},\ k,n \in \mathbb{N},$$

is the union of two disjoint intervals of total length $\ell(a)$ that is independent of both k and n and decreases monotonely to zero as $a \uparrow 1$. Thus

$$\begin{aligned} &\int_{2k\pi}^{2(k+1)\pi} \left| e^{-x}\left(\sin\left(x + \frac{r_n\pi}{n}\right)\right)^n \right| dx \\ &\quad \le \int_{E_{kn}(a)} + \int_{[2k\pi,2(k+1)\pi]\setminus E_{kn}(a)} \\ &\quad \le 2e^{-2k\pi}\left(1 - e^{-\ell(a)}\right) + 2a^n\left(2 + e^{-2\pi}\right)e^{-2k\pi}, \end{aligned}$$

$$\begin{aligned} &\left|\int_0^\infty e^{-x}\left(\sin\left(x + \frac{r_n\pi}{n}\right)\right)^n dx\right| \\ &\quad \le \sum_{k=0}^\infty \int_{2k\pi}^{2(k+1)\pi} \left| e^{-x}\left(\sin\left(x + \frac{r_n\pi}{n}\right)\right)^n \right| dx \\ &\quad \le 2\left(1 - e^{-\ell(a)}\right)\sum_{k=0}^\infty e^{-2k\pi} + 2a^n\left(2 + e^{-2\pi}\right)\sum_{k=0}^\infty e^{-2k\pi} \\ &\quad \le \frac{2}{1 - e^{-2\pi}}\left(\left(1 - e^{-\ell(a)}\right) + a^n\left(2 + e^{-2\pi}\right)\right) \stackrel{\text{def}}{=} \frac{2}{1 - e^{-2\pi}}(I + II). \end{aligned}$$

If a in $(0,1)$ is near 1, I is small and for such an a, II is small if n is large. □

2.64. If $\epsilon \in (0,1)$ then

$$\left|1 - e^{(\epsilon x)^2}\right| \le 2\epsilon x e^{(\epsilon x)^2} \text{ and } \left|1 - e^{(\epsilon x)^2}\right| e^{-x^3} \le 2\epsilon x e^{-x^2(x-\epsilon^2)}.$$

The integrand converges to zero on $[0,\infty)$, is bounded on $[0,2]$, and is bounded on $[2,\infty)$ by an integrable function. The dominated convergence theorem applies. □

2.65. The dominated convergence theorem implies the limit is zero. □

2.66. If $0 < \epsilon < 1 - \delta < 1$, the change of variable $x \to 1 - y$ yields

$$\begin{aligned}\int_{\epsilon}^{1-\delta} f(x)\,dx &= \int_{1-\epsilon}^{\delta} \frac{(1-y)\ln(1-y)}{-y}\,dy \\ &= \left\{-y + \sum_{n=2}^{\infty} \frac{y^n}{n^2(n-1)}\right\}\Bigg|_{1-\epsilon}^{\delta} \\ &= 1 - \delta - \epsilon + \sum_{n=2}^{\infty} \frac{\delta^n - (1-\epsilon)^n}{n^2(n-1)} < \infty.\end{aligned}$$

If $\epsilon = \frac{1}{p}, \delta = \frac{1}{q}, p, q \in \mathbb{N}$, the dominated convergence theorem permits a passage to the limit as $p, q \to \infty$. □

2.67. Because $\sum_{n=1}^{\infty} |t_{mn}| \le M$, the infinite product $\prod_{n=1}^{\infty}(1 + t_{mn}z)$ converges for all z in $\mathbb{C}$ and represents an entire function f_m.

If $x \in (-1, 1)$ then

$$\begin{aligned}\ln(1+x) &= \sum_{n=1}^{\infty} \frac{(-1)^{n+1}x^n}{n} \stackrel{\text{def}}{=} x - \frac{x^2}{2}(\alpha(x)), \\ 1 + x &= \exp\left(x - \frac{x^2}{2}\alpha(x)\right).\end{aligned}$$

Hence there are functions $\alpha_{mn}(z)$ such that if $|z| < \frac{1}{2}$ then for some constant K, $|\alpha_{mn}(z)| \le K$ and $1 + t_{mn}z = \exp\left(t_{mn}z - \frac{t_{mn}^2 z^2}{2}\alpha_{mn}(z)\right)$. Thus if $|z| \le \frac{1}{2}$ then

$$\begin{aligned}\sum_{n=1}^{\infty} \left|t_{mn}^2 \alpha_{mn}(z)\right| &< \infty, \\ s_m(z) \stackrel{\text{def}}{=} \sum_{n=1}^{\infty} t_{mn}^2 \alpha_{mn}(z) &\text{ exists}, \\ |s_m(z)| \le K \sum_{n=1}^{\infty} t_{mn}^2 &\quad (\to 0 \text{ as } m \to \infty), \\ f_m(z) \stackrel{\text{def}}{=} \prod_{n \in \mathbb{N}} (1 + t_{mn}z) = \exp\left(\sum_{n=1}^{\infty} t_{mn}z - \frac{s_m(z)}{2}\right) &\to e^z \text{ as } m \to \infty.\end{aligned}$$

Since f_m and exp are entire functions, $\lim_{m\to\infty} f_m(z) = e^z$ for all z in $\mathbb{C}$. □

2.68. If $s_n \uparrow s < \infty$, the convergence-preserving properties of Toeplitz matrices imply that $\lim_{m\to\infty} \sigma_m = s$.

If $s_n \uparrow \infty$ then for each positive M there is and N_M such that $s_n > M$ if $n > N_M$. Thus

$$\begin{aligned}\sigma_m &= \sum_{n=1}^{N_M} t_{mn}s_n + \sum_{n=N_M+1}^{\infty} t_{mn}s_n \\ &\geq s_1\left(\sum_{n=1}^{N_M} t_{mn}\right) + M\left(\sum_{n=N_M+1}^{\infty} t_{mn}\right) \stackrel{\text{def}}{=} S_1 + S_2.\end{aligned}$$

As $m \to \infty$ and N_M is fixed, $S_1 \to 0$ and $S_2 \to M$, whence $\sigma_m \to \infty$. □

[**Note s2.4:** Absent the monotonicity of $\{s_n\}_{n\in\mathbb{N}}$, the conclusion $\lim_{n\to\infty} \sigma_n = s$ can fail to obtain. If, e.g., $s_n = (-1)^{n+1}n, n \in \mathbb{N}$, then $\sigma_{2n} = -\dfrac{1}{2}$, $\sigma_{2n-1} > 0$, and $\lim_{n\to\infty} \sigma_n$ does not exist.]

2.69. For $\sum_{n=1}^{N} \frac{a_n}{n} \stackrel{\text{def}}{=} s_N$, $\left\{\frac{\sum_{n=1}^{N} s_n}{N} \stackrel{\text{def}}{=} \sigma_N\right\}_{N\in\mathbb{N}}$ is the result of applying the special Toeplitz matrix $T_1 \stackrel{\text{def}}{=} \{t_{mn}\}_{m,n=1}^{\infty}$ for which

$$t_{mn} = \begin{cases}\dfrac{1}{m} & \text{if } 1 \leq n \leq m \\ 0 & \text{otherwise}\end{cases}$$

to the sequence $\{s_n\}_{n\in\mathbb{N}}$. Hence $\lim_{N\to\infty} \sigma_N = \lim_{n\to\infty} s_n \stackrel{\text{def}}{=} s$. Furthermore, $\dfrac{1}{N}\sum_{n=1}^{N} a_n = \dfrac{1}{N}\sum_{n=1}^{N} n\dfrac{a_n}{n} = \dfrac{1}{N}\left(\sum_{n=2}^{N} n(s_n - s_{n-1}) + s_1\right)$ and Abel summation yields

$$A_N = -\frac{1}{N-1}\left(\sum_{n=1}^{N-1} s_n\right)\cdot\left(\frac{N-1}{N}\right) + s_N \to -s\cdot 1 + s = 0$$

as $N \to \infty$. □

2.70. For all real x, $|\sin x| \leq |x|$ whence $|u_{n+1}| \leq |u_n|$ (≤ 1), $n \in \mathbb{N}$. Thus $\lim_{n\to\infty} |u_n| \stackrel{\text{def}}{=} A$ exists. Let $\{p_m\}_{m\in\mathbb{N}}$ resp. $\{q_m\}_{m\in\mathbb{N}}$ be the sequence of positive resp. negative terms in $\{u_n\}_{n\in\mathbb{N}}$. Then $p_m \downarrow p \geq 0$ and $q_m \uparrow q \leq 0$. Owing to the continuity of sin, $\sin p = p$ and $\sin q = q$, whence $p = q = 0$. □

2.71. If $x > 1$ then $f_{n+1}(x) > f_n(x)$ and if α as described exists and is finite then $1 < \alpha = x^\alpha$, $x = \alpha^{\frac{1}{\alpha}}$ whence $1 < x \leq \sup_{\alpha>1} \alpha^{1/\alpha} = e^{1/e}$.

If $0 < x \leq 1$ then $1 \geq f_{2(n+1)}(x) > f_{2n}(x) > 0$ whence $\lim_{n\to\infty} f_n(x)$ exists and is finite. □

2.72. The ratio test implies that for any real α, the series converges absolutely and uniformly in every closed subinterval of $(-1,1)$; if $\alpha = 0$ or $\alpha \in \mathbb{N}$, the same conclusion obtains on $[-1,1]$ itself.

If $\alpha \in (0,\infty) \setminus \mathbb{N}$ and $a_n \stackrel{\text{def}}{=} \left|\binom{\alpha}{n}\right|$ then for n in $([\alpha]+1,\infty)$,

$$na_n - (n+1)a_{n+1} = \alpha a_n > 0$$

whence for some γ in $[0,\infty)$, $na_n \downarrow \gamma$, and

$$\sum_{n=[\alpha]+1}^{\infty} \left|\binom{\alpha}{n}\right| = \frac{1}{\alpha}\sum_{n=1}^{\infty}(na_n - (n+1)a_{n+1}) = \frac{1}{\alpha}(a_1 - \gamma).$$

The Weierstraß M-test implies the uniform and absolute convergence of $\sum_{n=0}^{\infty}\binom{\alpha}{n}x^n$ on $[-1,1]$.

For $f_\alpha(x) \stackrel{\text{def}}{=} \sum_{n=0}^{\infty}\binom{\alpha}{n}x^n$, α in $\mathbb{R}$, and x in $(-1,1)$, the following equations may be derived by valid term-by-term differentiation and multiplication by $1+x$ of the convergent power series representing $f_{\alpha-1}(x)$ and $f'_\alpha(x)$: $f'_\alpha = \alpha f_{\alpha-1}$, $(1+x)f_{\alpha-1}(x) = f_\alpha(x)$, $(1+x)f'_\alpha(x) = \alpha f_\alpha(x)$. Hence the derivative of $F(x) \stackrel{\text{def}}{=} \dfrac{f_\alpha(x)}{(1+x)^\alpha}$ is zero, i.e., F is a constant. Furthermore, $F(0) = 1$. □

[**Note s2.5:** If $L \in C([-1,1],\mathbb{C})$ and L is piecewise linear then L is a linear combination of translates of $p : t \mapsto t$ and $q : t \mapsto |t|$. Because $|t| = \left(1 + (t^2 - 1)\right)^{\frac{1}{2}}$, q is uniformly approximable by polynomials. Since every g in $C([-1,1],\mathbb{C})$ is uniformly approximable on $[-1,1]$ by functions like L, there emerges the Weierstraß approximation theorem:

> Every g in $C([-1,1],\mathbb{C})$ is uniformly approximable on $[-1,1]$ by polynomials.

The Weierstraß approximation theorem is a vital ingredient in the proof of the Stone-Weierstraß theorem.]

2.3. Continuity

2.73. Since $A = \bigcup_{n\in\mathbb{Z}} A \cap [n,n+1]$, it follows that A is the union of *compact* summands K_n: $A = \bigcup_{n\in\mathbb{N}} K_n$. Since f is continuous, each $f(K_n)$ is compact, whence closed, and $f(A) = \bigcup_{n\in\mathbb{N}} f(K_n)$, an F_σ. □

2.74. Because V is a union of open intervals (a_m, b_m), $m \in \mathbb{N}$, and each (a_m, b_m) is a union of closed intervals $\left[a_m + \frac{1}{n}, b_m - \frac{1}{n}\right]$, $n \in \mathbb{N}$, it follows that V is an F_σ and thus $f(V)$ is also an F_σ (cf. **2.73**). □

2.75. The function $g : \mathbb{T} \ni z \mapsto f(z) - f(-z)$ is such that $g(z) = -g(-z)$. Hence $g(\mathbb{T})$ is a connected set containing meeting both $[0, \infty)$ and $(-\infty, 0]$, whence $0 \in g(\mathbb{T})$.

□

2.76. If $x \sim y$ iff $f(x) = f(y)$ then $\sim$ is an equivalence relation and if $\Xi \stackrel{\text{def}}{=} [0,1]/\sim$, there is a map $\theta : [0,1] \ni x \mapsto x/\sim \stackrel{\text{def}}{=} \xi \in \Xi$. For $\mathcal{T} \stackrel{\text{def}}{=} \{\theta^{-1}(U) : U \text{ open in } [0,1]\}$, $(\Xi, \mathcal{T})$ is a topological space,

$$\widetilde{g} : \Xi \ni \xi \mapsto g \circ \theta^{-1}(\xi) \text{ and } \widetilde{f} : \Xi \ni \xi \mapsto f \circ \theta^{-1}(\xi)$$

are well-defined and continuous, and the diagram below

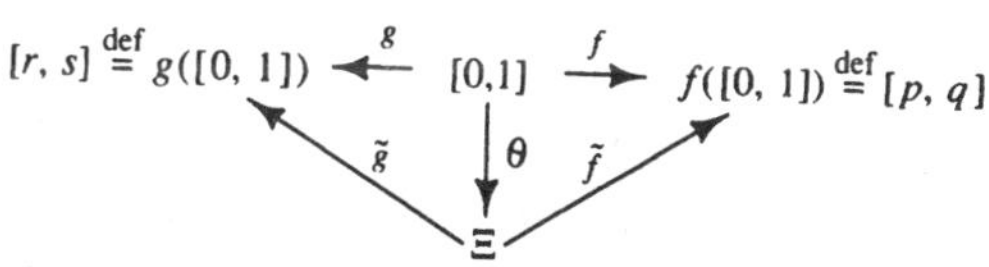

Figure s2.4.

is commutative. Furthermore, $\widetilde{f}$ is injective and, owing to the definition of $\mathcal{T}$, $\widetilde{f}^{-1}$ is continuous, i.e., $\widetilde{f}$ is a homeomorphism. For $\widetilde{g} \circ \widetilde{f}^{-1} \stackrel{\text{def}}{=} G$, if U is open in $[r, s]$ then $\widetilde{g}^{-1}(U)$ is open in Ξ and $\widetilde{f}(U)$ is open in $[p, q]$: $G \stackrel{\text{def}}{=} \widetilde{g} \circ \widetilde{f}^{-1}$ is a continuous map of $[p, q]$ onto $[r, s]$.

Stripped of its disguise, the discussion above shows that for y in $[p, q]$, the equation $G(y) = g(x)$ defines $G(y)$ uniquely and that G is continuous on $[p, q]$. The Weierstraß approximation theorem implies that there is a sequence $\{p_n\}_{n \in \mathbb{N}}$ of polynomials such that $p_n \stackrel{\text{u}}{\to} G$ on $[p, q]$, i.e., $p_n \circ f \stackrel{\text{u}}{\to} g$ on $[0, 1]$.

□

2.77. *i.* Let G be the set of all g as described. From **2.76** it follows that there is a sequence $\{p_n\}_{n \in \mathbb{N}}$ polynomials such that $p_n \circ f \stackrel{\text{u}}{\to} g$. Since L is linear and continuous, $0 = L(p_n \circ f) \to L(g)$ as $n \to \infty$.

ii. Let H be the set of all h as described. For each n, $f^{\{n\}}$ is a continuous, odd, and piecewise linear function. The zeros of $f^{\{n\}}$ are $\pm x_n$ and $\lim_{n \to \infty} \pm x_n = \pm 1$. The graph of $y = f^{\{n\}}(x)$ is a polygon consisting of three segments: $\langle(-1, -1)(-x_n, 0)\rangle$, $[-x_n, x_n]$, and $\langle(x_n, 0), (1, 1)\rangle$. **Figure s2.5** is offered to provide a graphical presentation of the form of $f^{\{n\}}$. Note that $\lim_{n \to \infty} x_n = 1$.

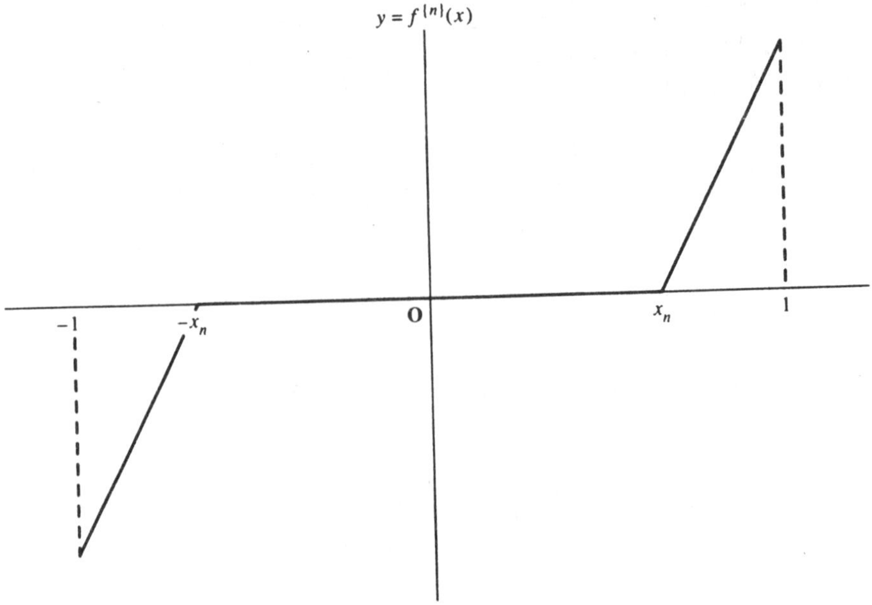

Figure s2.5.

The linear span S of $\{f^{\{n\}}\}_{n\in\mathbb{N}}$ is a set of continuous, odd, and piecewise linear functions, each vanishing on $\left[-\frac{1}{3},\frac{1}{3}\right]$. The graph of each h in H is uniformly approximable by polygonal graphs corresponding to functions in S: S is $\|\ \|_\infty$-dense in H. Because L is continuous, for each such h in H, $L(h)=0$. □

2.78. Since the $\mathbb{R}$-span R_N of A_N is a separating algebra of functions in $C([0,1],\mathbb{R})$, the Stone-Weierstraß theorem implies that R_N is $\|\ \|_\infty$-dense in $C([0,1],\mathbb{R})$. Consideration of the real and imaginary parts of functions in $C([0,1],\mathbb{C})$ shows that S_N is $\|\ \|_\infty$-dense in $C([0,1],\mathbb{C})$. □

2.79. Fejér's theorem for Fourier series implies the Weierstraß approximation theorem and thus implies both conclusions. □

2.80. In $A \stackrel{\text{def}}{=} \{f : f\in C([0,1],\mathbb{C}), f(0)=0\}$ there is a $\|\ \|_\infty$-dense sequence $\{g_n\}_{n\in\mathbb{N}}$. For n in $\mathbb{N}$, there are polynomials $p_1,p_2,\ldots$, so that a) $\deg(p_1)\stackrel{\text{def}}{=} d_1$, $p_1(0)=0$, and $\|g_1-p_1\|_\infty < 2^{-1}$, and b) after $p_1,\ldots,p_n$ have been chosen so that $\deg(p_k)\stackrel{\text{def}}{=} d_k$, $1\le k\le n$, and

$$\sup_{0\le x\le 1}\left|g_k(x)-p_1(x)-p_2\left(x^{d_1+1}\right)-\cdots-p_k\left(x^{d_{k-1}+1}\right)\right| < 2^{-k},\ 1\le k\le n,$$

there is a polynomial p_{n+1} such that

$$\sup_{0\le x\le 1}\left|g_{n+1}(x)-p_1(x)-\cdots-p_{n+1}\left(x^{d_n+1}\right)\right|<2^{-(n+1)}.$$

Thus $\sum_{n=1}^{\infty} p_n\left(x^{d_{n-1}+1}\right)$ is a power series $\sum_{n=1}^{\infty} a_n x^n$. If $f \in A$ then there is a subsequence $\{g_{n_k}\}_{k\in\mathbb{N}}$ such that $g_{n_k}(x) \stackrel{\text{u}}{\to} f(x)$ whence there are m_k such that $\sum_{n=1}^{m_k} a_n x^n \stackrel{\text{u}}{\to} f(x)$ as $k \to \infty$. □

2.81. If $f(x) \not\to 0$ as $x \to \infty$ there is a sequence $\{x_n\}_{n\in\mathbb{N}}$ and a positive δ such that $x_n \uparrow \infty$ and $|f(x_n)| \ge 2\delta$. Because f is continuous, each x_n belongs to an interval (a_n, b_n) where $|f(x)| > \delta$. Hence $G \stackrel{\text{def}}{=} \bigcup_{n\in\mathbb{N}} (a_n, b_n)$ is an open set unbounded above. Hence for G, D as in **2.52** is dense in $[0, \infty)$ and for some h in $(a, b) \cap D$ and infinitely many n in $\mathbb{N}$, $nh \in G$. Thus infinitely often, $|f(nh)| > \delta$, whereas $f(nh) \to 0$ as $n \to \infty$. □

2.82. a) For n in $\mathbb{N}$, let E_n be $f_n^{-1}(0)$. Then each E_n is closed while the hypothesis implies $\bigcup_{n\in\mathbb{N}} E_n = [0, 1]$. The Baire category theorem implies that some E_{n_0} is not nowhere dense and hence contains a nonempty subinterval (a, b). For x in $[a, b]$, $0 = f_{n_0}(x) = \int_0^x f_{n_0-1}(t)\,dt$. Furthermore, $f_n' = f_{n-1}$ whence on $[a, b]$, $f_{n_0-1}(x) = f_{n_0-2}(x) = \cdots = f_0(x) = 0$ and so $(a, b) \subset [a, b] \subset E_0$.

b) If $|f_0(0)| = \delta > 0$ then for some positive b, $|f_0(x)| \ge \frac{\delta}{2}$ if $x \in [0, b]$. Thus for each n in $\mathbb{N}$, $|f_n(b)| \ge \frac{\delta}{2} b^n > 0$, whereas for some n_b, $f_{n_b}(b) = 0$, a contradiction. Thus $f_0(0) = 0$.

If $0 < b \le 1$, $f_0(x) \not\equiv 0$ in $[0, b]$, and $f_0(x)$ does not change sign in $[0, b]$ then for each n in $\mathbb{N}$, $f_n(x) \ne 0$ on $(0, b]$, a contradiction when $n = n_b$. It follows that if $0 < b \le 1$ then at some b_1 in $(0, b)$, f_0 changes sign, at some b_2 in $(0, b_1)$, f_0 changes sign, ..., whence f_0 changes sign infinitely often in $(0, b]$ and thus $f_0(x) = 0$ infinitely often in $(0, b)$. □

2.83. For each n in $\mathbb{N}$, let E_n be $\left\{x : f^{(n)}(x) = 0\right\}$. If $x \in (0, 1)$ then $x \in E_{n_x}$ whence $\bigcup_{n=0}^{\infty} E_n = (0, 1)$. Because each E_n is closed, the Baire category theorem implies that some E_N contains a nonempty open interval (a, b), i.e., $f^{(N)}(x) \equiv 0$ on (a, b). By successive integration it follows that there are constants $c_0, c_1, \ldots, c_{N-1}$ such that on (a, b), $f^{(N-1)}(x) = c_0$, $f^{(N-2)}(x) = c_0 x + c_1, \ldots$, and

$$f(x) = \frac{c_0}{(N-1)!}x^{N-1} + \frac{c_1}{(N-2)!}x^{N-2} + \cdots + c_{N-1}.$$ □

[**Note s2.6:** Compare the argument above with that for **2.82a**).]

2.84. If $0 \le x < y \le 1$ and $g(x) = g(y)$ then for m as described, $x = g^{\{m\}}(x) = g^{\{m\}}(y) = y$, a contradiction. Thus g is strictly monotone and since $g(0) = 1-g(1) = 0$, g is strictly monotonely increasing. If $g(x) \gtrless x$ for some x then $g^{\{m\}}(x) \gtrless g^{\{m-1\}} \gtrless \cdots \gtrless g(x) \gtrless x$, a contradiction. □

2.85. a) If $f(x) = \begin{cases} \frac{1}{x} & \text{if } 0 < x \\ 0 & \text{if } x \le 0 \end{cases}$, then f is left-continuous and unbounded on any open interval containing 0.

b) If $f(x)$ is unbounded in $[0,1]$ then for some sequence $\{x_k\}_{k\in\mathbb{N}}$ in $[0,1]$, $\lim_{k\to\infty} x_k \stackrel{\text{def}}{=} x_0$ exists and $|f(x_k)| > k$. Thus there emerges the contradiction: $\overline{\lim}_{x=x_0} f(x) = \infty \ne f(x_0)$. □

2.86. If $\|f\|_\infty \stackrel{\text{def}}{=} \delta \ge 1$ then for some a in $(0,1]$, $|f(a)| = 1$ and if $0 \le x < a$ then $|f(x)| < 1$. If $0 < \epsilon < 1$ and if S is equicontinuous, there is a positive δ such that if $|x-a| < \delta$ then for all n in $\mathbb{N}$,

$$\epsilon > |f^n(a) - f^n(x)| \ge |f^n(a)| - |f^n(x)| = 1 - |f^n(x)|.$$

However, for large n and x in $(0,a)$, $|f^n(x)| < 1-\epsilon$ whence there emerges the contradiction $1-\epsilon > |f^n(x)| > 1-\epsilon$.

Conversely, if $\|f\|_\infty \stackrel{\text{def}}{=} \epsilon < 1$ then for x, y in $[0,1]$,

$$|f^n(x) - f^n(y)| < 2\epsilon^n \downarrow 0.$$ □

2.87. If, for n in $\mathbb{N}$ and f in $C([0,1],\mathbb{C})$, $L_n(f) \stackrel{\text{def}}{=} \dfrac{\int_0^1 x^n f(x)\,dx}{\int_0^1 x^n\,dx}$ then $L_n \in C([0,1],\mathbb{C})^*$ and $\|L_n\| \le 1$. If p is a polynomial and $p \ne 0$, then $\lim_{n\to\infty} L_n(p) = p(1)$. If $f \in C([0,1],\mathbb{C})$ then

$$|L_n(f) - L_n(p)| = |L_n(f-p)| \le \|f-p\|.$$

The Weierstraß approximation theorem implies $\lim_{n\to\infty} L_n(f) = f(1)$. □

2.88. If $\sup_{f\in A} \|f''\| \stackrel{\text{def}}{=} K < \infty$ then for f in A and all x in $[0,1]$,

$$|f'(x)| \le |f'(0)| + K\int_0^x 1\,dt \le |f'(0)| + K,$$

$$f'(0) = f(1) - f(0) - \int_0^1 \left(\int_0^x f''(t)\,dt\right) dx,$$

$$|f'(0)| \le 2 + \frac{K}{2},\quad M \le 2 + \frac{3K}{2} < \infty.$$

On the other hand, if $K = \infty$ then for each n in $\mathbb{N}$, and for some f_n and some x_n in $[0,1]$, $|f_n''(x_n)| > n$. If for some x_0 and some n, $|f_n''(x_0)| \le n-1$ then, since $\|f_n'''\|_\infty \le 1$, for any x, $|f_n''(x)| \le |f_n''(x_0)| + \int_{x_0}^x |f_n'''(t)|\,dt \le n$, a contradiction. Hence for all n in $\mathbb{N}$ and all x in $[0,1]$, $|f_n''(x)| > n-1$ and, since f_n'' is continuous, it must be of one sign, say $f_n''(x) > n-1$, throughout $[0,1]$.

If $f_n(x) = f_n(0) + f_n'(0)x + \frac{f_n''(0)}{2}x^2 + R_2(x)$ then $|R_2(x)| \le \frac{1}{6}$ in $[0,1]$. The hypotheses imply

$$\left|f_n''(0)x^2 + 2f_n'(0)x\right| \le 2\left(|f_n(x) - f_n(0)| + R_2(x)\right) \le \frac{13}{3} < 5.$$

Hence if $f_n''(0) \stackrel{\text{def}}{=} a_n$, and $2f_n'(0) \stackrel{\text{def}}{=} b_n$ the discussion turns on behavior of the quadratic polynomial $a_nx^2 + b_nx$ on the interval $[0,1]$. Since $a_n \ge n-1$ the graph of the equation $y = a_nx^2 + b_nx$ has one of the four forms indicated in **Figure s2.6**.

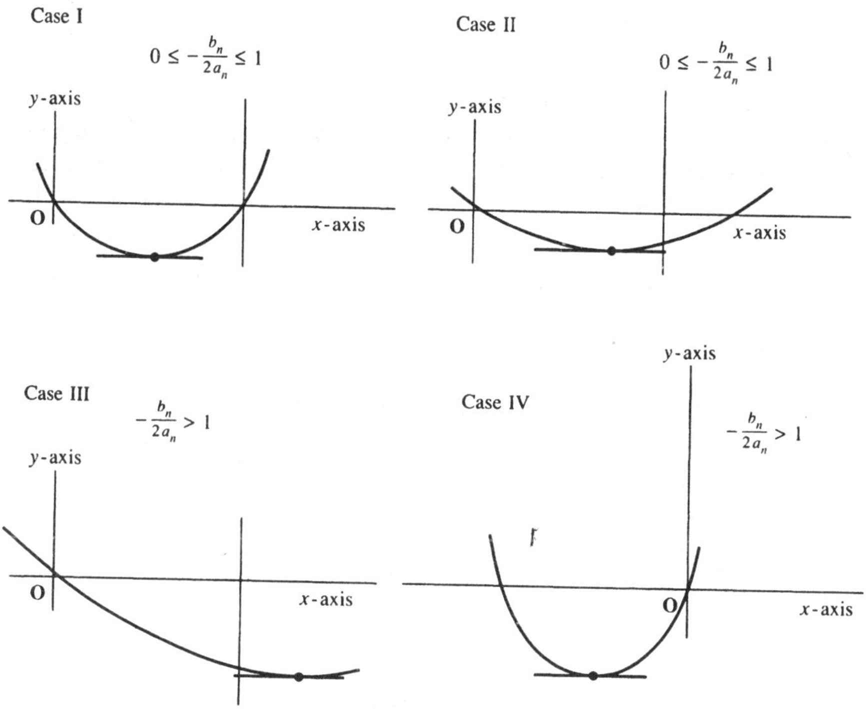

Figure s2.6.

In Cases I and II,

$$-5 < -\frac{b_n^2}{4a_n}, \; -2\sqrt{5a_n} < b_n < 0, \; a_n - 2\sqrt{5a_n} < a_n + b_n \le |a_n + b_n| \le 5.$$

Upon division by $\sqrt{a_n}$, there emerges the inequality $n - 1 \le a_n \le 20$, a contradiction if $n > 21$.

In Case III,

$$-\frac{b_n}{2a_n} > 1,\ b_n < -2a_n,\ -5 \le a_n + b_n < -a_n,\ 5 \ge a_n \ge n-1,$$

a contradiction if $n > 7$.

In Case IV, $b_n > 0$ whence

$$5 \ge |a_n + b_n| = a_n + b_n \ge a_n \ge n-1,$$

a contradiction if $n > 7$. Hence $K = \infty$ is an impossibility. A similar argument applies if it is assumed that $f_n''(x) \le -(n-1)$. □

2.89. Note that for a) and c), $C_0(X, \cdots) = C(X, \cdots)$.

a) The Weierstraß approximation theorem implies that P is true.

b) For $f : [0,\infty) \ni x \mapsto e^{-x^2} \in \mathbb{C}$, $\lim_{x\to\infty} f(x) = 0$ whereas for any nonconstant polynomial p, $\lim_{x\to\infty} |p(x)| = \infty$ and, for a constant polynomial p, $\sup_{0\le x<\infty} |f(x) - p(x)| \ge \frac{1}{2}$: P is false.

c) The function $f : X \ni z \mapsto \overline{z}$ is continuous on X but not holomorphic in X°, whereas any uniform limit of a sequence of polynomials is holomorphic in X°: P is false. □

2.90. The Stone-Weierstraß theorem implies that the compactness of F is sufficient while **Solution 2.89**b) shows that the compactness of F is also necessary. □

2.91. There is a positive δ such that if $|x-y| < \delta$ then $|f(x) - f(y)| < 1$. If $x \in (a,b)$ then $\left|x - \frac{a+b}{2}\right| \le \frac{b-a}{2} = \left[\frac{b-a}{\delta}\right]\frac{\delta}{2}$. If $\mathbb{N} \ni n_0 > \frac{b-a}{\delta}$ then for any x in (a,b) and for some m in $\mathbb{N} \cap [1, n_0]$, between $\frac{a+b}{2} \overset{\text{def}}{=} x_1$ and $x \overset{\text{def}}{=} x_m$, there are $x_2, \ldots, x_{m-1}$ for which $|x_{k+1} - x_k| < \delta$, $1 \le k \le m-1$. Hence $\left|f(x) - f\left(\frac{a+b}{2}\right)\right| \le \sum_{k=1}^m |f(x_k) - f(x_{k-1})| \le n_0$. □

2.92. For x in $\mathbb{R}$, let $F(x)$ resp. $G(x)$ be $f\left(e^{2\pi i x}\right)$ resp. $g\left(e^{2\pi i x}\right)$. Then

$$\begin{aligned}\int_{\mathbb{T}} f(z) g(z^n)\, d\tau(z) &= \int_0^1 F(x)G(nx)\, dx \\ &\approx \frac{1}{n}\sum_{k=0}^{n-1} \int_k^{k+1} F\left(\frac{k}{n}\right) G(x)\, dx \\ &\approx \left(\frac{1}{n}\sum_{k=0}^{n-1} F\left(\frac{k}{n}\right)\right) \int_0^1 G(x)\, dx.\end{aligned}$$

The last expression converges to $\int_0^1 F(x)\, dx \int_0^1 G(x)\, dx$. □

2.93. The Arzelà-Ascoli theorem implies that the closure $\overline{\text{Conv}(A)}$ of the convex hull of A is an equicontinuous subset of $B(\mathbf{O}, \|f\|_\infty)$.

If $F(x) \stackrel{\text{def}}{=} \int_{\mathbb{T}} f(xy)\,d\tau(y)$ then, since $\tau(\mathbb{T}) = 1$ and τ is a nonnegative measure on $\mathbb{T}$, $F \in \overline{\text{Conv}(A)}$. Because τ is translation-invariant, for all z in $\mathbb{T}$, $F(zx) = F(x)$, i.e., F is constant and $C(f) = \int_{\mathbb{T}} f(y)\,d\tau(y)$.

If G is constant and in $\overline{\text{Conv}(A)}$ then G is uniformly approximable by linear combinations of translates of f. Again the fact that $\tau(\mathbb{T}) = 1$ and integration show that for any positive δ, $|G - C(f)| < \delta$. □

2.94. Because $\bigcup_{n\in\mathbb{N}} \{x : \sup_{f\in\mathcal{F}} |f(x)| \le n\} \stackrel{\text{def}}{=} \bigcup_{n\in\mathbb{N}} U_n = \mathbb{R}$, at least one $\overline{U_n}$, say $\overline{U_M}$, contains a nonempty open set V. If $x \in V$ then for some $\{x_n\}_{n\in\mathbb{N}}$ contained in U_M, $\lim_{n\to\infty} x_n = x$ whence

$$\{f \in \mathcal{F}\} \Rightarrow \left\{\lim_{n\to\infty} f(x_n) = f(x)\right\}.$$

However, for f in $\mathcal{F}$, $|f(x_n)| \le M$. □

2.95. For L in $C(X, \mathbb{C})^*$, the Riesz representation theorem implies that for some complex Borel measure μ and all f in $C(X, \mathbb{C})$, $L(f) = \int_X f(x)\,d\mu(x)$. The bounded convergence theorem applies. □

2.96. Since $A_{\mathbb{R}}$, consisting of the elements of A that are formed with the use of $\mathbb{R}$-valued functions f_i, g_i, is a separating subalgebra of $C(X, \mathbb{R})$, it follows from the Stone-Weierstraß theorem that $A_{\mathbb{R}}$ is $\|\ \|_\infty$-dense in $C(X, \mathbb{R})$. Moreover, $A = A_{\mathbb{R}} + iA_{\mathbb{R}}$. □

2.97. a) If $x \in \text{Cont}(f)$ then for each n in $\mathbb{N}$, there is an $N_n(x)$ such that $d(f(y), f(x)) < \frac{1}{n}$ if $y \in N_n(x)$. For each n in $\mathbb{N}$,

$$\text{Cont}(f) \subset \Omega_n \stackrel{\text{def}}{=} \bigcup_{x\in\text{Cont}(f)} N_n(x)$$

whence $\text{Cont}(f) \subset \bigcap_{n\in\mathbb{N}} \Omega_n \stackrel{\text{def}}{=} \Omega$.

On the other hand, if $y \notin \text{Cont}(f)$, there is a positive ϵ such that in each neighborhood $N(y)$ there is an x such that $d(f(x), f(y)) \ge \epsilon$. Choose N in $\mathbb{N}$ so that $\frac{1}{N} < \frac{\epsilon}{2}$. If $y \in \Omega$ then $y \in \Omega_N$ and for some z in $\text{Cont}(f)$, $y \in N_N(z)$. There is a neighborhood $N(y)$ contained in $N_N(z)$ and in $N(y)$ there is an x such that $d(f(x), f(y)) \ge \epsilon$. Hence,

$$\epsilon \le d(f(x), f(y)) \le d(f(x), f(z)) + d(f(z), f(y)) < \frac{1}{N} + \frac{1}{N} < \epsilon,$$

a contradiction. Thus $\{y \notin \text{Cont}(f)\} \Rightarrow \{y \notin \Omega\}$, $\text{Cont}(f) = \Omega$, a G_δ.

b) The sets

$$F_{km} \stackrel{\text{def}}{=} \left[\bigcap_{m\le n}\left\{x \,:\, |f_m(x)-f_n(x)| \le \frac{1}{k}\right\}\right],$$

$$F_k \stackrel{\text{def}}{=} \bigcup_{m\in\mathbb{N}} F_{km},\ G_{km} \stackrel{\text{def}}{=} F^{\circ}_{km}, \text{ and}$$

$$G_k \stackrel{\text{def}}{=} \bigcup_{m\in\mathbb{N}} G_{km},\ G \stackrel{\text{def}}{=} \bigcap_{k\in\mathbb{N}} G_k$$

are such that:

- ▷ each F_{km} is closed because the f_n are continuous;
- ▷ each F_k is X because the f_n converge everywhere;
- ▷ if $F^{\circ}_{km} = \emptyset$ then F_{km} is nowhere dense because F_{km} is closed;
- ▷ the set $R_{km} \stackrel{\text{def}}{=} F_{km} \setminus F^{\circ}_{km}$ is closed and its interior is empty, whence R_{km} is nowhere dense and $R_k \stackrel{\text{def}}{=} \bigcup_{m\in\mathbb{N}} R_{km}$ is of the first category;
- ▷ each G_k, as a union of open sets is open, and G, as the intersection of a countable set of open sets, is a G_δ.

If $G = \emptyset$ then $G \subset \text{Cont}(f)$. If $x \in G$ then for each k in $\mathbb{N}$ and some m_k in $\mathbb{N}$, $x \in F^{\circ}_{km_k}$, whence $|f_{m_k}(x) - f_n(x)| \le \frac{1}{k}$, $m_k \le n$. Because $F^{\circ}_{km_k}$ is open (and nonempty), it contains an $N(x)$ and for every z in $N(x)$, $|f_{m_k}(z) - f_n(z)| \le \frac{1}{k}$, $m_k \le n$, whence $|f_{m_k}(z) - f(z)| \le \frac{1}{k}$ and so

$$|f(z)-f(x)| \le |f(z)-f_{m_k}(z)| + |f_{m_k}(z)-f_{m_k}(x)| + |f_{m_k}(x)-f(x)|. \tag{s2.4}$$

The first and third terms in the right member of (s2.4) do not exceed $\frac{1}{k}$. Owing to the continuity of f_{m_k}, $N(x)$ contains a neighborhood $W(x)$ such that if $z \in W(x)$ then the second term is less than $\frac{1}{k}$. Hence at each point x of G, the limit function f is continuous: $\text{Cont}(f) \supset G$.

c) When X is a complete metric space then, since R_k is nowhere dense and $G_k = X \setminus R_k$, G_k is dense. The set G_k, as a union of open sets, is open and so G_k is a dense open set. Baire's Theorem implies that G is dense.

It follows that $\text{Cont}(f)$ is both dense and a G_δ. □

2.98. a) If f is an extreme point in $B(\mathbf{O},1)$ then $\|f\|_\infty = 1$.

If $|f(x)| \equiv 1$ then $\text{sgn}(f)f = |f| = 1$. If $g, h \in B(\mathbf{O},1) \setminus \{f\}$, $g \ne h$, $t \in (0,1)$, and $|f| = tg + (1-t)h$ then for some a, $g(a) \ne h(a)$ and thus near a, $h(x) \ne g(x)$. Because every point on $\mathbb{T}$ is an extreme point of $D(0,1)$, $f(a) \ne tg(a) + (1-t)h(a)$. Hence f is an extreme point of $B(\mathbf{O},1)$.

If f is an extreme point of $B(\mathbf{O},1)$ and $|f(x)| \not\equiv 1$ then for some a,b,c in $(0,1)$, $a < |f(b)| < c$ and for some positive ϵ, if $|x-b| < \epsilon$ then $x \in (0,1)$ and $a < |f(x)| < c$. For some δ in $B(\mathbf{O},1)$, $\delta \geq 0$ and

$$\delta(x) = \begin{cases} 1 & \text{if } |x-b| < \dfrac{\epsilon}{2} \\ 0 & \text{if } |x-b| > \epsilon \end{cases}.$$

Then for small η,

$$g_\pm \stackrel{\text{def}}{=} f \pm \eta\delta\overline{\text{sgn}(f)} \in B(\mathbf{O},1) \setminus \{f\},\ g_+ \neq g_-, \text{ and } f = \frac{1}{2}(g_+ + g_-),$$

a contradiction.

In sum, f is an extreme point iff $|f(x)| \equiv 1$.

b) As in a), $\|f\|_\infty = 1$. If $|f(x)| \equiv 1$ then X is compact and the argument in a) shows that f is an extreme point.

If $|f(x)| \not\equiv 1$ then $\max_{x\in X}|f(x)| - \min_{x\in X}|f(x)| \stackrel{\text{def}}{=} 3\Delta > 0$. The sets

$$E_M \stackrel{\text{def}}{=} \left\{ x \ :\ |f(x)| \geq \max_{x\in X}|f(x)| - \Delta \right\},$$
$$E_m \stackrel{\text{def}}{=} \left\{ x \ :\ |f(x)| \leq \min_{x\in X}|f(x)| + \Delta \right\},$$

are disjoint, nonempty, compact, and contained in disjoint compact neighborhoods N_M resp. N_m. Since X is connected, $X \setminus (N_M \cup N_m) \neq \emptyset$. In $C_0(X,\mathbb{C})$ there is an ϵ such that $\epsilon(x) \geq 0$, $\epsilon\big|_{N_M\cup N_m} = 0$, $\epsilon \neq 0$, and $\|\epsilon\|_\infty < \dfrac{\Delta}{6}$. The functions $g_\pm \stackrel{\text{def}}{=} f \pm \epsilon$ are in $B(\mathbf{O},1)$, neither is f, and $f = \frac{1}{2}g_+ + \frac{1}{2}g_-$.

In sum, if X is a connected, locally compact and not compact Hausdorff space then there is no extreme point in $B(\mathbf{O},1)$. □

[**Note s2.7:** For $C([0,1],\mathbb{C})$, the cardinality of the set of extreme points of $B(\mathbf{O},1)$ is $\mathfrak{c}$. However, in $C([0,1],\mathbb{R})$, the set of extreme points of $B(\mathbf{O},1)$ consists of the two constants ± 1. By contrast, if $X = \mathbb{N}$, then $C_0(X,\mathbb{R}) = \mathbf{c}_0$ and each vector $\mathbf{e}_n \stackrel{\text{def}}{=} \{\delta_{mn}\}_{m\in\mathbb{N}}$ is an extreme point of B_1.]

2.99. a) If $X = C_0$ (**the** Cantor set) then, since C_0 is totally disconnected, for each x in C_0, $f_{C_0}^{-1}(x)$ contains a component. It follows that

$$\mathfrak{c} \geq \#(\mathcal{C}) \geq \#(C_0) = \mathfrak{c}.$$

b) Each component of A is either a point or a nondegenerate closed interval. There can be no more than countably many components of the form $[a_n, b_n], 0 \le a_n < b_n \le 1, 1 \le n < N \le \infty$, since components are pairwise disjoint and each of the form described meets $\mathbb{Q}$. Thus

$$B \stackrel{\text{def}}{=} A \setminus A^\circ = A \setminus \biguplus_n (a_n, b_n)$$

is closed and nowhere dense and a) implies $\#(B) = \mathfrak{c}$. The Cantor-Bendixson theorem implies that B is the union of a set P homeomorphic to C_0 and a countable set D: $B = P \uplus D$. Furthermore, the self-dense kernel of D is empty. If $U \stackrel{\text{def}}{=} \biguplus_{1 \le n < N} (a_n, b_n)$ then P, D, and U are pairwise disjoint and $A = P \uplus D \uplus U$. If $1 \le n < N$ then $f_{C_0}([a_n, b_n])$ is a single point and thus $f_{C_0}(B) = f_{C_0}(A)$.

c) Because every compact metric space X is the continuous image of C_0 (cf. **2.15**), there is a continuous surjection $h_X : C_0 \stackrel{\text{sur}}{\mapsto} X$. Hence if $g_X \stackrel{\text{def}}{=} h_X \circ f_{C_0}\big|_B$ then g_X is a continuous surjection of B on X. □

2.100. If $f_0 = 0$ and for n in $\mathbb{N}$,

$$f_n : [0,1] \ni x \mapsto \begin{cases} nx & \text{if } 0 \le x \le \frac{1}{n} \\ n\left(-x + \frac{2}{n}\right) & \text{if } \frac{1}{n} < x \le 2n \\ 0 & \text{if } \frac{2}{n} < x \le 1 \end{cases},$$

then $\|f_n\|_\infty \le 1$ and $\lim_{n\to\infty} f_n(x) \equiv 0$. If $L \in C([0,1], \mathbb{C})^*$ then for some complex measure μ on $[0,1]$,

$$\{f \in C([0,1], \mathbb{C})\} \Rightarrow \left\{ L(f) = \int_{[0,1]} f(x)\, d\mu(x) \right\}.$$

The bounded convergence theorem implies $\lim_{n\to\infty} L(f_n) = 0$ whereas $f_n\left(\frac{1}{n}\right) \equiv 1$. □

2.101. For $L_n : C([0,1], \mathbb{C}) \ni f \mapsto f\left(1 - \frac{1}{n}\right)$ and $L \stackrel{\text{def}}{=} \sum_{n=1}^\infty \frac{(-1)^n L_n}{2^n}$, if $\|f\|_\infty \le 1$ then $|L(f)| \le \sum_{n=1}^\infty 2^{-n} = 1$, i.e., $\|L\| \le 1$. For each k in $\mathbb{N}$ and some f_k in $C([0,1], \mathbb{C})$, $\|f_k\|_\infty = 1$ and

$$f_k\left(1 - \frac{1}{n}\right) = \begin{cases} (-1)^n & \text{if } n = 1, 2, \ldots, k \\ 0 & \text{if } n = k+1, k+2, \ldots \end{cases}.$$

Then $L(f_k) = 1 - 2^{-k}$ whence $\|L\| = 1$.

On the other hand, if $\|f\|_\infty \le 1$ and $|L(f)| = 1$ then for some θ in $[0, 2\pi]$, $L(f) = e^{i\theta}$, $L\left(e^{-i\theta} f\right) = 1$. For all n in $\mathbb{N}$,

$$e^{-i\theta}(-1)^n L_n(f) \in D(0,1).$$

Assume that for some n_0, $e^{-i\theta}(-1)^{n_0}L_{n_0}(f) \stackrel{\text{def}}{=} a \neq 1$. Then

$$b \stackrel{\text{def}}{=} \frac{\sum_{n\neq n_0} L_n(f)}{1-\dfrac{1}{2^{n_0}}} \in D(0,1),\ 1 = L\left(e^{-i\theta}f\right) = \frac{1}{2^{n_0}}a + \left(1-\frac{1}{2^{n_0}}\right)b,$$

a contradiction since 1 is an extreme point of $D(0,1)$. It follows that $e^{-i\theta}(-1)^n L_n(f) \equiv 1$ and so $\lim_{x\to 1} f(x)$ does not exist, a contradiction. Hence there is in $C([0,1],\mathbb{C})$ no f such that $\|f\|_\infty \leq 1$ and $L(f) = 1$. $\square$

2.102. a) Let Q be the set of all polynomials with rational complex coefficients. Then $A \supset Q$ and the Weierstraß approximation theorem implies A is $\| \ \|_\infty$-dense in $C([0,1],\mathbb{C})$.

If $\{r_k\}_{k\in\mathbb{N}}$ is an enumeration of $\mathbb{Q}$ then for $\{k,n\} \subset \mathbb{N}$,

$$E_{kn} \stackrel{\text{def}}{=} \{f \ : \ f \in C([0,1],\mathbb{C}), f(r_k) \in \{r_1,\ldots,r_n\}\}$$

is closed in $C([0,1],\mathbb{C})$ and $A \subset \bigcup_{\{k,n\}\subset\mathbb{N}} E_{kn} \stackrel{\text{def}}{=} B$. If $f \in B$ and $k \in \mathbb{N}$, then $f(r_k) \in \mathbb{Q}$ whence $f \in A$: $A = B$ and B is an F_σ.

b) The set of all polynomials with exactly one irrational coefficient is also $\| \ \|_\infty$-dense in $C([0,1],\mathbb{C})$ and no such function maps $\mathbb{Q}$ into $\mathbb{Q}$: $A^\circ = \emptyset$. $\square$

2.103. As a closed subspace of a Banach space, A is itself a Banach space. The distributive character of multiplication implies M_g is linear. For x in $[0,1]$ and $T_x : A \ni f \mapsto g(x)f(x) \in \mathbb{C}$, it follows that $|T_x(f)| \leq |g(x)| \cdot \|f\|_\infty$ whence $T_x \in C([0,1],\mathbb{C})^*$. Because $M_g(A) \subset A$, $\sup_{x\in[0,1]} |T_x(f)| = \|gf\|_\infty$. The uniform boundedness principle applies to the set $\{T_x\}_{x\in[0,1]}$ of linear functionals: $\sup_{x\in[0,1]} \|T_x\| \stackrel{\text{def}}{=} M < \infty$. Hence

$$\|gf\|_\infty = \sup_{x\in[0,1]} |g(x)f(x)| \leq M\|f\|_\infty. \qquad \square$$

2.104. a) For g in $C([0,1],\mathbb{C})$ and f in A, $g\cdot f\left(\frac{1}{2}\right) = 0$ whence A is an ideal. If A is a principal ideal then for some k in $C([0,1],\mathbb{C})$, $k \cdot C([0,1],\mathbb{C}) = A$. Hence $k(1/2)\,\mathbf{1}(1/2) = k(1/2)\cdot 1 = 0$ and so $k(1/2) = 0$. If $|k| \stackrel{\text{def}}{=} r$ then both r and $\sqrt{r}$ are in A whence for some g in $C([0,1],\mathbb{C})$, $kg = \sqrt{r}$. Since $\sqrt{r} \geq 0$, it follows that $\sqrt{r} = kg = |kg| = |k|\cdot|g| = r|g|$ whence $|g| = r^{-\frac{1}{2}}$. However, since $r\left(\frac{1}{2}\right) = 0$, $|g(x)|$ is unbounded and hence $g \notin C([0,1],\mathbb{C})$, a contradiction: A is not a principal ideal.

b) If $g_0(x) \stackrel{\text{def}}{=} x - \dfrac{1}{2}$ then $g_0 \cdot C([0,1],\mathbb{C})$ contains the set of all polynomials vanishing at $\dfrac{1}{2}$. From the Stone-Weierstraß theorem it follows that $\overline{g_0 \cdot C([0,1],\mathbb{C})} = A$. $\square$

2.105. a) Because $B_K \subset A_K$ and A_K is closed, it follows that $\overline{B_K} \subset A_K$. There is a sequence $\{V_n\}_{n\in\mathbb{N}}$ of open sets such that $V_n \supset V_{n+1}, n \in \mathbb{N}$, and $\bigcap_{n\in\mathbb{N}} V_n = K$. For n in $\mathbb{N}$ and some f_n in A_K, $K \prec f_n \prec V_n$.

For f in A_K, m in $\mathbb{N}$, and some open set U_m containing K, $|f(x)| < \frac{1}{m}$ if $x \in U_m$. If $\epsilon > 0$ and $\frac{1}{\epsilon} < m_0 \in \mathbb{N}$ then for some n_0 greater than m_0, $\overline{V_n} \subset U_m$ if $n > n_0$. If $m > n_0$ then $f_m \cdot f \in B_K$ and $\|f_m \cdot f - f\|_\infty < \epsilon$. Hence $\overline{B_K} = A_K$.

b) As in **2.104** it follows that if $K \neq \emptyset$ then A_K is an ideal. The argument showing that A in **2.104** is not a principal ideal may be repeated, mutatis mutandis, to prove that A_K is not a principal ideal. If $K = \emptyset$ then $A_K = C([0,1], \mathbb{C})$, which by definition is not an ideal. □

2.106. a) If $a \in X$ and N is a neighborhood of a then for x in N and f in S, $f(x) \leq h(x)$, whence

$$\inf_{x\in N} f(x) \leq \inf_{x\in N} h(x)$$

$$\underline{\lim}_{x=a} f(x) = \sup_{N\ni a}\left(\inf_{x\in N} f(x)\right) \leq \sup_{N\ni a}\left(\inf_{x\in N} h(x)\right) = \underline{\lim}_{x=a} h(x).$$

Because f is continuous, $\underline{\lim}_{x=a} f(x) = \lim_{x\to a} f(x) = f(a)$ and since $f \in S$, $f(a) \leq h(a)$, whence $h(a) \stackrel{\text{def}}{=} \bigvee_{f\in S} f(a) \leq \underline{\lim}_{x=a} h(x)$. By definition, $h(a) \geq \underline{\lim}_{x=a} h(x)$. A dual argument leads to the dual conclusion for $\bigwedge$, etc.

b) If X is locally compact, h is lsc, $h \geq 0$, and $\alpha < h(a)$, there is an $N(a)$ such that $\alpha < \inf(h(N(a))$. Urysohn's lemma implies that for some f in $C_{00}(X, \mathbb{R})$, $f(a) = \alpha$, $0 \leq f \leq \alpha$ and $f = 0$ off $N(a)$. Hence $f < h$ and $h = \bigvee\{f : f \in C_{00}(X, \mathbb{R})\}$. The lsc/usc duality applies for upper semicontinuity. □

2.107. a) Because $\underline{\lim}_{x=a}(f(x) + g(x)) \geq \underline{\lim}_{x=a} f(x) + \underline{\lim}_{x=a} g(x)$, if f and g are lsc then

$$f(a) + g(a) \geq \underline{\lim}_{x=a}(f(x) + g(x)) \geq f(a) + g(a).$$

Hence, via lsc/usc duality, the sum of two lsc resp. usc functions is lsc resp. usc. If f is lsc resp. usc then $-f$ is usc resp. lsc. If f is lsc and g is usc then $f + g$ may be lsc (e.g., if g is continuous) or usc (e.g., if f is continuous) or neither lsc nor usc (e.g., if $f = \chi_{[0,1]}$ and $g = \chi_{(2,3)}$). All the previous remarks and the lsc/usc duality resolve the question.

b) For similar reasons, the product of two nonnegative lsc resp. usc functions is lsc resp. usc. If f is lsc and g is usc then fg may be lsc (e.g., if g is continuous) or usc (e.g., if f is continuous) or neither lsc nor usc (e.g., if $f = \chi_{[0,2]}$ and $g = \chi_{(1,3)}$). For the last f and g described, f and $-g$ are usc whereas $f \cdot (-g)$ is neither lsc nor usc. Hence the product of two usc functions may fail to be usc. The other implications can be resolved by lsc/usc duality.

c) The implicit assumption here is that f is defined on the range of g. If $g = \chi_{[0,1]}$ and $f(x) = 1 - x$ then f is continuous and $0 = f(1) = 1 - f(0)$. For any neighborhood N of 1, $g(N) = \{0, 1\}$,

$$\sup_{x \in N} f \circ g(x) = \max\{f(0), f(1)\} = 1 > 0 = f(1) = f \circ g(1).$$

Hence $\overline{\lim}_{x=1} f \circ g(x) = 1 > f \circ g(1)$: $f \circ g$ is not usc even though f is continuous and g is usc. Similar examples for combinations of lsc and usc show that there are no instances of valid implications. □

2.108. a) If f is continuous at x then for every neighborhood N of $f(x)$, there is a neighborhood V of x such that $f(V) \subset N$. If $\mathcal{F}$ converges to x then $\mathcal{F}$ refines the neighborhood filter at x and thus $f(\mathcal{F})$ refines the neighborhood filter at $f(x)$, i.e., $f(\mathcal{F})$ converges to $f(x)$. The argument for the converse proceeds similarly.

b) If f is continuous at x and n converges to x then n is eventually in each neighborhood of x and thus $f(n)$ is eventually in each neighborhood of $f(x)$. The argument for the converse proceeds similarly. □

2.109. Let W be a vicinity in W. If $x \in X$ and $y = f(x)$ then for $V_y(W)$ (cf. **2.48**) and some $U(x)$ in U, $f(V_x(U(x))) \subset V_y(W)$. Since $X = \bigcup_{x \in X} V_x(U(x))$ and X is compact, it follows that for some finite set $\{x_1, \ldots, x_N\}$, $X = \bigcup_{n=1}^{N} V_{x_n}(U(x_n))$. If $(x, x') \in \bigcap_{n=1}^{N} U(x_n)$ then $(f(x), f(x')) \in W$. □

2.110. *i.* If G is a group, H is a subgroup, and $G/H \stackrel{\text{def}}{=} K$ is the coset space then $\#(G) = \#(H)\#(K)$. Hence if H and K are countable, so is G: countability is a QL property in $\mathcal{G}$.

ii. The group S_3 of permutations of the set $\{1, 2, 3\}$ contains the normal subgroup A_3 of even permutations, the cyclic group of order three, and the quotient group S_3/A_3 is S_2, the cyclic group of order two. Both A_3 and S_2 are abelian while S_3 is not: abelianity is not a QL property in $\mathcal{G}$.

iii. If, in the context of *i*, the groups are topological and both H and K are compact then every open cover $\{U_\lambda\}_{\lambda \in \Lambda}$ of G has a refinement $\mathcal{V} \stackrel{\text{def}}{=} \{V_\gamma\}_{\gamma \in \Gamma}$ such that each V_γ is open and $\overline{V_\gamma}$ is compact.

Let $h : G \mapsto G/H$ be the quotient map taking elements of G to their cosets. Then $h(\mathcal{V})$ is an open cover of K. Because K is compact, it is covered by finitely many $h(V_{\gamma_i}), 1 \le i \le n$, and $G \subset \bigcup_{i=1}^{n} \overline{V_{\gamma_i}} \cdot H$. Each summand, as the continuous image under the map $G \times G \ni \{x, y\} \mapsto xy \in G$ of the compact set $\overline{V_{\gamma_i}} \times H$ in $G \times G$, is compact. As the finite union of compact sets, G is compact: compactness is a QL property in $\mathcal{LCG}$.

iv. For a metric topological group G, a closed subgroup H, and the coset space $K \stackrel{\text{def}}{=} G/H$, let $\{k_n\}_{n\in\mathbb{N}}$ resp. $\{h_m\}_{m\in\mathbb{N}}$ be a dense subset of K resp. H. In G, there is a set $\{g_n\}_{n\in\mathbb{N}}$ such that $g_n/H = k_n$.

If U is open in G then $U/H \stackrel{\text{def}}{=} V$ is open in K and for some k_n in V and some z in U, $z/H = k_n$. Because $z/H = g_n/H$, it follows that $zg_n^{-1} \in H$ and $Ug_n^{-1} \cap H$ is open in H. Thus there is an h_m in $Ug_n^{-1} \cap H$, i.e., $Ug_n^{-1} \ni h_m$, and $h_m g_n \in U$: $\{h_m g_n\}_{m,n\in\mathbb{N}}$ is dense in G: separability is a QL property in $\mathcal{MTG}$.

v. If, for the topological group G and closed subgroup H, both H and $K \stackrel{\text{def}}{=} G/H$ are separable, their associated uniformities have countable bases and so both H and K are metric and contain countable dense subsets whence, as the argument in *iv* shows, G contains a countable dense subset $\{g_n\}_{n\in\mathbb{N}}$.

Furthermore: a) for some sequence $\{V_n\}_{n\in\mathbb{N}}$ of open neighborhoods containing the identity of G,

$$V_n = V_n^{-1},\ V_{n+1}^2 \subset V_n,\ n \in \mathbb{N},$$

and $\{V_n \cap H\}_{n\in\mathbb{N}}$ is a base of neighborhoods of the identity in H; b) for some sequence $\{U_n\}_{n\in\mathbb{N}}$ of open neighborhoods containing the identity of G, $\{U_n/H\}_{n\in\mathbb{N}}$ is a base of neighborhoods of the identity coset of K.

If U is an open neighborhood of the identity in G, there is an open neighborhood W and a V_n such that $W^2 \subset U$ and $V_n \cap H \subset W \cap H$. For

$$\begin{aligned} P_n &\stackrel{\text{def}}{=} \left(G \setminus \overline{V_{n+1} \cdot (H \setminus V_n)}\right) \cap (U_n H), \\ Q_n &\stackrel{\text{def}}{=} P_1 \cap \cdots \cap P_n, n \in \mathbb{N}, \end{aligned}$$

and some k greater than n,

$$\begin{aligned} & U_k/H \subset (W \cap V_{n+1})/H, \\ & Q_k \subset P_k \cap P_n \subset (U_k H) \cap \left(G \setminus \overline{(V_{n+1} \cdot (H \setminus V_n))}\right), \\ & \quad \subset (W \cap V_{n+1}) H \cap [(W \cap V_{n+1}) \cdot (G \setminus (H \setminus V_n))], \\ & \quad \subset (W \cap V_{n+1}) \cdot (H \cap V_n), \\ & \quad \subset (W \cap V_{n+1}) \cdot (W \cap H) \subset W^2 \subset U. \end{aligned}$$

Hence G is metric and contains a countable dense subset $\{x_n\}_{n\in\mathbb{N}}$. The set $\{B(x_n, r)\}_{\substack{r>0 \\ r\in\mathbb{Q}}}$ is a countable base for G. □

[**Note s2.8:** The argument in *v* can be extended to show that if K contains a dense subset of cardinality $\mathfrak{k}$ and H contains a dense subset of cardinality $\mathfrak{h}$ then G contains a dense subset of cardinality $\mathfrak{h} \cdot \mathfrak{k}$.

The metrizability of a topological group for which there is a countable base at the identity is a theorem of Kakutani [**Kak**]. Thus v is a consequence of two propositions: a) if H and G/H have countable bases at their identities then G has a countable base at its identity; b) iv.]

2.111. a) If $\lim_{x\to a} f(x)$ exists and $V(\epsilon)$ is a typical vicinity, then a is in a neighborhood U such that if $y \in U$ then $|f(y) - L| < \frac{\epsilon}{2}$. It follows that $|n(U') - L| < \epsilon$ if $U' \subset U$.

Conversely, if n as described is a Cauchy net then, since $\mathbb{R}$ is complete, n converges to some L and $L = \lim_{x\to a} f(x)$.

b) By definition, f is continuous at a iff $f(a) = \lim_{x\to a} f(x)$. □

2.112. The conclusion is a consequence of the definitions. □

2.113. The argument is, mutatis mutandis, the argument for the case of sequences. □

2.114. If X is not compact, then some sequence $S \stackrel{\text{def}}{=} \{x_n\}_{n\in\mathbb{N}}$ in X contains no Cauchy subsequence. For each n and some positive r_n, the open balls $U_n \stackrel{\text{def}}{=} B(x_n, r_n)^\circ$ are pairwise disjoint. It may be assumed that $r_n \downarrow 0$. Because X contains no isolated points, for some y_n in U_n, $y_n \neq x_n$, i.e., $d(x_n, y_n) \stackrel{\text{def}}{=} 2\delta_n > 0$. If

$$f_n(x) \stackrel{\text{def}}{=} \frac{d\left(x, \left(X \setminus B(x_n, 2\delta_n)^\circ\right)\right)}{d(x, B(x_n, \delta_n)) + d\left(x, X \setminus B(x_n, 2\delta_n)^\circ\right)}$$

then f_n is continuous and

$$f_n(x) = \begin{cases} 1 & \text{if } x \in B(x_n, \delta_n) \\ 0 & \text{if } x \in X \setminus B(x_n, 2\delta_n)^\circ \end{cases}.$$

The series $\sum_{n=1}^{\infty} f_n(x)$ converges throughout X since for each x, there is an $n(x)$ such that $f_n(x) = 0$ if $n > n(x)$. For similar reasons, $f \stackrel{\text{def}}{=} \sum_{n=1}^{\infty} f_n$ is continuous. However, contrary to the hypothesis, f is not uniformly continuous: if $\epsilon = 0.5$, $|f(x_n) - f(y_n)| = |1 - 0| = 1 > \epsilon$ while $\lim_{n\to\infty} d(x_n, y_n) = 0$. □

3

Real- and Complex-valued Functions

3.1. Real-valued Functions

3.1. For each n in D and some index m greater than n, $s_m > s_{n-1}$. Let n' be the least such m. If $n < p < n'$ then

$$s_{n'} - s_{n-1} = s_{n'} - s_{p-1} + s_{p-1} - s_{n-1} > 0,$$

whereas $s_{p-1} - s_{n-1} \leq 0$, and so $s_{n'} - s_{p-1} > 0$. In other words, each p in (n, n') is distinguished, i.e., the numbers $n, n+1, \ldots, n'-1$ constitute a block or a part of a block.

Thus a monotonely increasing enumeration of the elements of D begins with a block: $\overline{n_1} \stackrel{\text{def}}{=} n_1, \overline{n_1}+1, \ldots, \overline{n_1}'-1$. Let $\overline{n_2}$ be the first distinguished index after $\overline{n_1}'-1$ and then continue the enumeration: $\overline{n_2}, \overline{n_2}+1, \ldots, \overline{n_2}'-1$, etc. In this way D is completely enumerated and if $n^{\#}$ is the last member of a block to which n belongs and if $n^{\#} > n$, then $s_{n^{\#}} > s_{n-1}$. □

3.2. If $D = \emptyset$ then D is open and no further discussion is required. If $D \neq \emptyset$, $x \in D$, $\overline{\lim}_{y=x} f(y) \stackrel{\text{def}}{=} L_x$, $x' > x$, and $f(x') > L_x$, for some $N(x)$, $x \in N(x), x' \notin N(x)$ and $\sup_{y \in N(x)} f(y) < f(x')$. Hence $N(x) \subset D$, D is open, and uniquely expressible as follows: $D = \bigcup_n (a_n, b_n)$.

Assume $x \in (a_n, b_n)$ and $f(x) > L_{b_n}$. Because $b_n \notin D$, if $x'' > b_n$ then $L_{b_n} \geq f(x'')$ whence $L_x \geq f(x) > f(x'')$. Thus if $f(x') > L_x$ then $x' \in (x, b_n]$. Let c be the supremum of all such x'. Then $x < c \leq b_n$. If $c = b_n$ then $f(c) \geq L_x$ as claimed. If $c < b_n$ there is a c' such that $c' > c$ and $f(c') > L_c$ $(\geq f(c) \geq L_x)$ and so $c' \in (x, b_n]$, $c \leq c'$, a contradiction. Hence $c = b_n$. □

[**Note s3.1:** **Figure s3.1**, showing the graph of $y = f(x)$ for a continuous f, suggests the origin of the name running water lemma for the result in **3.2**. The situations in **3.1** and **3.2** are analogous. Both discoveries are due to F. Riesz who used them to give perspicuous proofs of a) the differentiability a.e. of a monotone function and b) the Birkhoff pointwise ergodic theorem, (cf. **5.160, 5.161**).]

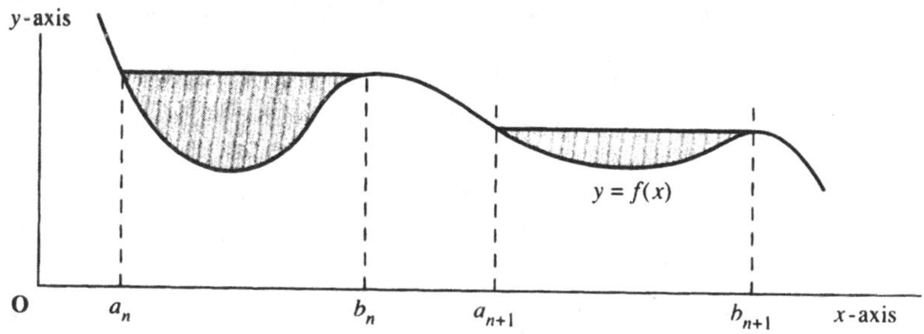

Figure s3.1.

3.3. If $E \stackrel{\text{def}}{=} \{p : p \in \mathbb{R}[x], \deg(p) \le R+S-1\}$ then E is isomorphic to $\mathbb{C}^{R+S}$. For $T : E \ni p \mapsto \left(p(1), \dots, p^{(R-1)}(1), p(2), \dots, p^{(S-1)}(2)\right) \in \mathbb{C}^{R+S}$, if $T(p) = \mathbf{O}$ and $p(x) = \sum_{k=0}^{R+S-1} A_k (x-1)^k$ then

$$A_0 = A_1 = \cdots = A_{R-1} = 0.$$

If $1 \le s \le S$ then

$$p^{(s-1)}(x) = \sum_{k=R}^{R+S-1} \frac{k!}{(k-s+1)!} A_k (x-1)^{k-s+1}$$

whence $\sum_{k=R}^{R+S-1} \dfrac{1}{(k-s+1)!} k! A_k = 0,\ 1 \le s \le S$, a system of S homogeneous linear equations in the S variables, $k!A_k, R \le k \le R+S-1$. For $N \stackrel{\text{def}}{=} R+S-1$, the matrix of coefficients is, modulo column/row permutations,

$$M(N,S) \stackrel{\text{def}}{=} \begin{pmatrix} \dfrac{1}{N!} & \dfrac{1}{(N-1)!} & \cdots & \dfrac{1}{(N-S+1)!} \\ \vdots & \vdots & \ddots & \vdots \\ \dfrac{1}{(N-S+1)!} & \dfrac{1}{(N-S)!} & \cdots & \dfrac{1}{(N-2S+2)!} \end{pmatrix}.$$

Mathematical induction and the row operations for reducing $M(N,S)$ to a manageable form show there are positive constants $K(N,S)$ such that $\det(M(N,S)) = K(N,S)\det(M(N-1,S-1))$. Hence $\det(M(N,S)) \neq 0$, $A_k = 0$, i.e., T is surjective. □

3.4. If

$$\delta_n : \mathbb{R} \ni x \mapsto \stackrel{\text{def}}{=} \begin{cases} 0 & \text{if } x < b_n \\ 1 & \text{if } b_n \le x \end{cases}, \ n \in \mathbb{N},$$

$f : \mathbb{R} \ni x \mapsto \stackrel{\text{def}}{=} \sum_{n=1}^{\infty} a_n \delta_n(x)$, and $\epsilon > 0$, there is an N such that $N > n$ and $\sum_{m=N+1}^{\infty} a_m < \epsilon$. There is a positive δ such that if $y < b_n \le x$ and

$x - y < \delta$ then, among $b_1, \ldots, b_N$, only b_n lies in $(y, x]$. Hence

$$a_n + \epsilon \geq f(x) - f(y) = \sum_{b_m \in (y,x]} a_m = a_n + \sum_{\substack{m \neq n \\ b_m \in (y,x]}} a_m \geq a_n,$$

whence $f(b_n + 0) - f(b_n - 0) = a_n$ and so $\text{Cont}(f) \subset S$.

Furthermore, if $x \in S$ and $\epsilon > 0$ for some N, $\sum_{n=N+1}^{\infty} a_n < \epsilon$. There is a positive δ such that

$$(x - \delta, x + \delta) \cap \{b_n\}_{n \in \mathbb{N}} \subset \{b_n\}_{N+1 \leq n < \infty},$$

whence if $y \in (x - \delta, x + \delta)$ then $|f(x) - f(y)| < \epsilon$. Thus $S \subset \text{Cont}(f)$. □

3.5. Direct calculations reveal the truth of each of the following statements.

i. On each of the intervals $(-\infty, x_1), (x_1, x_2), \ldots, (x_{n-1}, x_n), (x_n, \infty)$, s is a linear function.

ii. On $\mathbb{R}$, s is continuous.

iii. If $x_k \stackrel{\text{def}}{=} x_1 + \delta_k, 1 \leq k \leq n$, then $0 = \delta_1 < \delta_2 < \cdots < \delta_n$ and

$$s(\theta) = \begin{cases} \sum_{k=1}^{n} x_k - n\theta & \text{on } (-\infty, x_1] \\ (2p - n)\theta + (n - 2p)x_1 - \sum_{k=1}^{p} \delta_k + \sum_{k=p+1}^{n} \delta_k & \text{on } [x_p, x_{p+1}) \\ n\theta - \sum_{k=1}^{n} x_k & \text{on } [x_n, \infty) \end{cases}.$$

iv. On $(-\infty, x_1]$, s is monotonely decreasing, on $[x_n, \infty)$ s is monotonely increasing, and on $[x_k, x_{k+1}]$

$$s \text{ is } \begin{cases} \text{monotonely decreasing} & \text{if } 2k - n < 0 \\ \text{constant} & \text{if } 2k - n = 0 \\ \text{monotonely increasing} & \text{if } 2k - n > 0 \end{cases}.$$

Hence if n is odd, $s(\theta)$ reaches its minimal value when $\theta = x_{\frac{n+1}{2}}$. If n is even, $s(\theta)$ reaches its minimal value throughout the interval $\left[x_{\frac{n}{2}}, x_{\frac{n}{2}+1}\right]$. □

3.6. The Heine-Borel theorem implies there is a finite set $\{x_k\}_{1 \leq k \leq K}$ in $\mathbb{R}$ and a set $\{\delta_k\}_{1 \leq k \leq K}$ of positive numbers such that:

$$x_1 - \delta_1 < p < x_1 < x_2 - \delta_2 < x_1 + \delta_1 < x_2 < \cdots$$
$$< x_K - \delta_K < x_K < x_{K-1} + \delta_{K_1} < q < x_K + \delta_K;$$

$$\bigcup_{1 \leq k \leq K} (x_k - \delta_k, x_k + \delta_x) \supset [p, q];$$

$$\{x_{k-1} - \delta_{k-1} < a_k < x_{k-1} < b_k < x_{k-1} + \delta_{k-1}\}$$
$$\Rightarrow \{f(a_k) \leq f(b_k)\}, k = 2, 3, \ldots, K + 1.$$

Hence if $x_2 - \delta_2 < y_2 < x_1 + \delta_1, \ldots, x_K - \delta_K < y_K < x_{K-1} + \delta_{K-1}$ then $f(p) \leq f(y_2) \leq f(y_3) \leq \cdots \leq f(y_{K-1}) \leq f(q)$. □

3.7. If ϕ is convex,

$$u < a < c < b < v,\ \alpha \stackrel{\text{def}}{=} \frac{b-c}{b-a},\ \text{and}\ \beta \stackrel{\text{def}}{=} \frac{c-a}{b-a}$$

then $\alpha, \beta > 0, \alpha + \beta = 1$, and $c = \alpha a + \beta b$. Hence

$$\frac{\phi(b) - \phi(a)}{b-a} = \frac{\phi(b) - \alpha\phi(a) - \beta\phi(b)}{b - \alpha a - \beta b} \leq \frac{\phi(b) - \phi(c)}{b-c}$$

and, by a similar calculation, $\dfrac{\phi(c) - \phi(a)}{c-a} \leq \dfrac{\phi(b) - \phi(a)}{b-a}$. The inequalities $u < x \leq x' < y \leq y' < v$ can be applied first when $x = a$, $x' = c$, and $y = b$ and then when $x' = a$, $y = c$, and $y' = b$. The argument is reversible.

The sketch in **Figure s3.2** shows the geometry of the situation. □

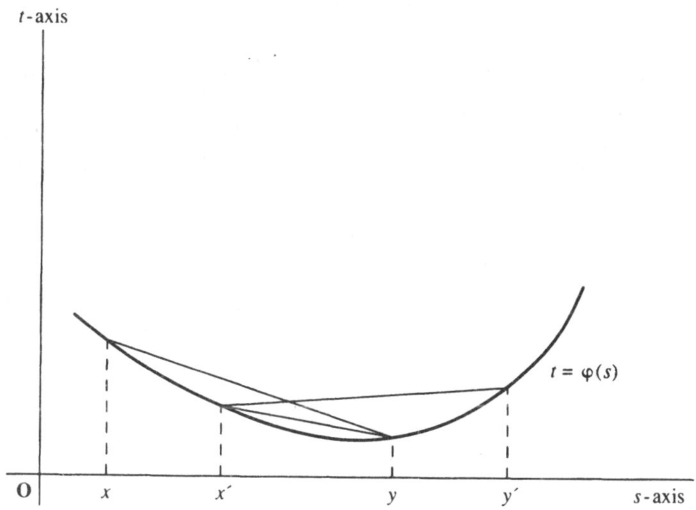

Figure s3.2.

3.8. If ϕ is neither monotonely decreasing nor monotonely decreasing, there are numbers a, b, c such that $a < b < c$ and either a) $\phi(b) > \max\{\phi(a), \phi(c)\}$ or b) $\phi(b) < \min\{\phi(a) < \phi(c)\}$. The convexity of ϕ rules out a) and implies for b) that if $a'' < a' < a$ then $\phi(a') \geq \phi(a)$ [whence $\phi(a'') \geq \phi(a')$] and if $b'' > b' > b$ then $\phi(b') \geq \phi(b)$ [whence $\phi(b'') \geq \phi(b')$].

Hence ϕ is monotonely decreasing on $(-\infty, a)$ and ϕ is monotonely increasing on (b, ∞).

If $p \stackrel{\text{def}}{=} \sup\{a : \phi\downarrow \text{ on } (-\infty, a)\}$, $q \stackrel{\text{def}}{=} \inf\{b : \phi\uparrow \text{ on } (b,\infty)\}$ then $p \leq q$ and $[p,q]$ is the required interval.

Examples: a) $\phi(x) = e^x$; b) $\phi(x) = x + |x|$; c) $\phi(x) \equiv 0$; d) $\phi(x) = e^{-x}$; e)

$$\phi(x) = \begin{cases} (x+a)^2 & \text{if } -\infty < x \leq -a < 0 \\ 0 & \text{if } -a < x \leq a \\ (x-a)^2 & \text{if } a < x, \infty \end{cases};$$

f) $\phi(x) = |x| - x$. □

3.9. If $u < c \leq x < y \leq d < v$ then

$$\frac{\phi(c)-\phi(a)}{c-a} \leq \frac{\phi(y)-\phi(x)}{y-x} \leq \frac{\phi(b)-\phi(d)}{b-d}$$

and so $\phi \in \text{Lip}(1)$.

The criterional inequality of **3.7** shows that the difference quotient

$$\frac{\phi(x+h)-\phi(x)}{h}$$

is a monotonely increasing function of h. It follows that the right- and left-hand derivatives $D_{\pm}\phi$ of ϕ exist everywhere.

In particular, at each point a, $D_-\phi(a) \leq D_+\phi(a)$. [Any line through $(a, \phi(a))$ and with a slope m satisfying $D_-\phi(a) \leq m \leq D_+\phi(a)$ is a supporting line.]

From **3.9** and the differentiability a.e. of a monotone function it follows that ϕ is differentiable a.e. □

[**Note s3.2:** The fact that (u,v) is open is essential to the continuity and differentiability properties of ϕ. For example, if (u,v) is replaced by $(0,1]$ and

$$\phi(x) = \begin{cases} 0 & \text{if } 0 < x < 1 \\ 1 & \text{if } x = 1 \end{cases}$$

then ϕ is convex but is discontinuous when $x = 1$. Similar observations apply when (u,v) is replaced by any nonopen interval.]

3.10. The functions $\psi : x \mapsto x^2$ and $\phi : x \mapsto e^{-x}$ are convex, but $\phi \circ \psi : x \mapsto e^{-x^2}$ is not convex. □

3.11. a) If ϕ'' exists, $\phi'' > 0$, and ϕ is not convex then there is a triple p, q, r such that $u < p < q < r < v$ and $\phi(q) > \frac{r-q}{r-p}\phi(p) + \frac{q-p}{r-p}\phi(r)$. The map $f : x \mapsto \phi(x) - \frac{r-x}{r-p}\phi(p) + \frac{x-p}{r-p}\phi(r)$ is continuous and positive at q, whence f has a positive maximum on $[p,r]$. Because $f(p) = f(r) = 0$, this maximum occurs at some s in (p, r) and thus $0 < \phi''(s) = f''(s) \le 0$, a contradiction.

b) If $p < r$ and $\phi'(p) > \phi'(r)$, the convexity of ϕ implies

$$\phi(r) > \phi(p) + \phi'(p)(r - p),$$

whence

$$\begin{aligned}\phi(r) + (p - r)\phi'(r) &> \phi(r) + (p - r)\phi'(p)\\ &> \phi(p) + \phi'(p)(r - p) + \phi'(p)(p - r) = \phi(p).\end{aligned}$$

In geometric terms, the inequalities above say that the point $(p, \phi(p))$ does not lie above the supporting line through $(r, \phi(r))$, a contradiction. Hence

$$\{p \le r\} \Rightarrow \{\phi'(p) \le \phi'(r)\},$$

i.e., ϕ' is monotonely increasing and so $\phi'' \ge 0$.

An alternative proof uses a truncated Taylor series. If $x \in (s, t)$ then for suitable ξ and η,

$$\begin{aligned}\phi(s) &= \phi(x) + \phi'(x)(s - x) + \phi''(\xi)\frac{(s - x)^2}{2!},\\ \phi(t) &= \phi(x) + \phi'(x)(t - x) + \phi''(\eta)\frac{(t - x)^2}{2!}.\end{aligned}$$

For some λ in $(0, 1)$, $x = \lambda s + (1 - \lambda)t$ and, owing to the nonnegativity of ϕ'', for some nonnegative P,

$$\begin{aligned}\lambda\phi(s) + (1 - \lambda)\phi(t) &= \phi(x) + \phi'(x)\left(\lambda s - \lambda x + (1 - \lambda)t - (1 - \lambda)x\right) + P\\ &\ge \phi(x) + 0 = \phi(\lambda s + (1 - \lambda)t).\end{aligned}$$

Note that if $\phi(x) = x$ then ϕ is convex and yet $\phi''(x) \equiv 0$: the implication $\{\phi'' \text{ exists and } \phi \text{ is convex}\} \Rightarrow \{\phi'' > 0\}$ is invalid. □

3.12. a) If $\ln(\phi)$ is convex, $x, y \in \mathbb{R}, 0 \le \alpha, \beta, \alpha + \beta = 1$ then

$$\begin{aligned}\ln(\phi)(\alpha x + \beta y) &\le \alpha \ln(\phi)(x) + \beta \ln(\phi)(y)\\ \phi(\alpha x + \beta y) &\le (\phi(x))^\alpha \cdot (\phi(y))^\beta.\end{aligned}$$

However for u, v positive, $u^\alpha v^\beta \le \alpha u + \beta v$. (Proof. Because

$$(t^\alpha)'\Big|_{t=c} = \begin{cases}\alpha & \text{if } c = 1\\ \alpha c^{\alpha-1} \le \alpha & \text{if } c > 1\\ \alpha c^{\alpha-1} \ge \alpha & \text{if } 0 < c < 1\end{cases},$$

$$(\alpha t + \beta)' \equiv \alpha,$$

$$t^\alpha\Big|_{t=1} = \alpha t + \beta\Big|_{t=1} = 1,$$

it follows that for all positive t, $t^\alpha \le \alpha t + \beta$. Hence if $t = \frac{u}{v}$ then

$$\frac{u^\alpha}{v^\alpha} v \le \left(\alpha \frac{u}{v} + \beta\right) v,$$
$$u^\alpha v^\beta \le \alpha u + \beta v.]$$

By induction it follows that if

$$\alpha_k, u_k \ge 0,\ 1 \le k \le K, \text{ and } \sum_{k=1}^{K} \alpha_k = 1$$

then, ambiguous cases excluded,

$$\prod_{k=1}^{K} u_k^{\alpha_k} \le \sum_{k=1}^{K} \alpha_k u_k,$$

as required.

b) Graphical considerations show that the map $\phi : x \mapsto 1 + |x|$ is convex. However, for $\psi \stackrel{\text{def}}{=} \ln \phi$,

$$\psi'(x) = \begin{cases} \dfrac{1}{1+x} & \text{if } x > 0 \\ -\dfrac{1}{1-x} & \text{if } x < 0 \end{cases},$$

whence

$$\{x \ne 0\} \Rightarrow \left\{\psi''(x) = -\frac{1}{(1-x)^2} < 0\right\}.$$

Hence **3.11** implies ψ is not convex. □

3.13. *i.* True: There is a sequence $\{x_n\}_{n\in\mathbb{N}}$ such that $x_n < \frac{x_{n+1}}{2}$ and $g(x) \ge n$ if $x \le x_n$. If $\phi(x_n) = n - 1, n \in \mathbb{N}$, and if ϕ is piecewise linear, continuous, and nonnegative then ϕ is also convex and monotonely decreasing, $\phi \le g$, and $\phi(x) \to \infty$ as $x \to 0$.

ii. False: For $g : [0, \infty) \ni x \mapsto \ln(1 + x)$, if the convex function ϕ is such that $\phi \le g$ then (cf. **Figure s3.3**), since $-g$ is convex, g is concave and

$$\phi\left(\frac{e^n}{n}\right) = \phi\left(\left(1 - \frac{1}{n}\right) \cdot 0 + \frac{1}{n} e^n\right)$$
$$\le \frac{\ln(1 + e^n)}{e^n} \cdot \frac{e^n}{n} = 1 + \frac{\ln(1 + e^{-n})}{n}.$$

As $n \to \infty$, $\frac{e^n}{n} \to \infty$ and $1 + \frac{\ln(1 + e^{-n})}{n} \to 1$ whence $\phi(x) \not\to \infty$ as $x \to \infty$. □

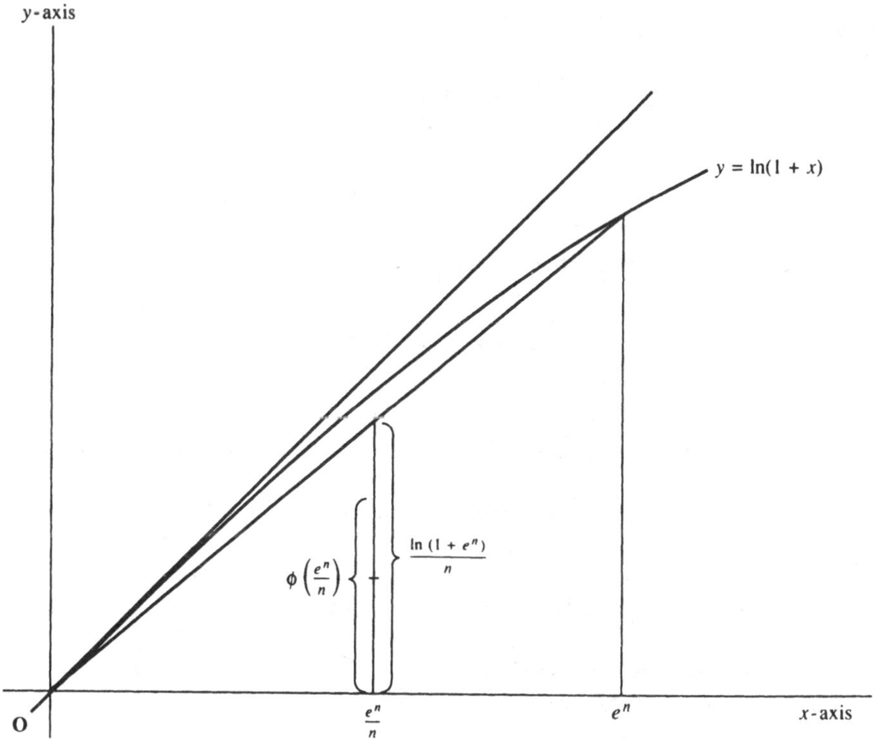

Figure s3.3. The graph of $y = \ln(1+x)$

[**Note s3.3:** Associated with the notion of convexity is that of midpoint convexity, viz.:

A function f in $\mathbb{R}^{(a,b)}$ is midpoint convex iff for all x, y in (a,b), $f\left(\frac{x}{2}+\frac{y}{2}\right) \le \frac{1}{2}f(x)+\frac{1}{2}f(y)$.

Hence if $\alpha, \beta \in \{t : t = k2^{-n}, k, n \in \mathbb{N}\}$, and $\alpha + \beta = 1$, then $f(\alpha x + \beta y) \le \alpha f(x) + \beta f(y)$. If, to boot, f is continuous, then f is convex. The existence of a discontinuous linear (hence convex) function ϕ in $\mathbb{R}^{\mathbb{R}}$ (cf. **[GeO]** and, for generalizations, **6.115**) shows that the converse is false.]

3.14. For the maps

$$k : [0,\infty) \ni s \mapsto \int_0^s b(r)\,dr,$$
$$g : [0,\infty) \ni s \mapsto k(s) + c(s),$$
$$T : C([0,\infty),\mathbb{R}) \ni f \mapsto \left(F : [0,\infty) \ni s \mapsto \int_0^s a(r)f(r)\,dr\right),$$

the following obtain:

$$\frac{d\left(T(y)(s)\right)}{ds} - a(s)T(y)(s) \le a(s)g(s);$$

$$\frac{d\left(e^{-\int_t^s a(r)\,dr}T(y)(s)\right)}{ds} \le e^{\int_s^t a(r)\,dr}a(s)g(s);$$

$$T(y)(t) \le \int_0^t a(s)g(s)e^{\int_s^t a(r)\,dr}\,ds;$$

$$\int_0^t a(s)\left[y(s) - g(s)e^{\int_s^t a(r)\,dr}\right]ds \le 0;$$

from which the result follows.

Owing to the importance of fixed-point theory, the alternative solution below is of interest.

If M_0 is an endomorphism of a Banach space X and if, for all $\mathbf{x}$ in X, $\sum_{n=1}^{\infty} M_0^n(\mathbf{x})$ converges then for each $\mathbf{y}$ in X, the map

$$M_{\mathbf{y}} : X \ni \mathbf{x} \mapsto \mathbf{y} + M_0(\mathbf{x})$$

has a unique fixed point, namely $\mathbf{P} \stackrel{\text{def}}{=} \mathbf{y} + \sum_{n=1}^{\infty} M_0^n(\mathbf{x})$. If, to boot, X is a function space and M_0 preserves positivity then

$$\{\mathbf{x} \le \mathbf{y} + M_0(\mathbf{x})\} \Rightarrow \{\mathbf{x} \le \mathbf{P}\}.$$

More particularly, if $X = C([0,a],\mathbb{R})$, $0 \le t \le a$, and $M_0 = T$, the results just cited are applicable: for $\mathbf{y}$ read g, for $\mathbf{x}$ read y, and for $\mathbf{P}$ read

$$t \mapsto g(t) + \int_0^t a(s)\left(e^{\int_s^t a(r)\,dr}\right)g(s)\,ds. \qquad \square$$

3.15. The solution involves two steps: a) showing that for a, b in $\mathbb{R}$, the differential equation problem:

$$y'' + (1+q)y = 0, y(0) = a, \quad y'(0) = b \tag{s3.1}$$

has a unique solution; b) the solution is bounded.

a) Because $q \in L^1([0,\infty))$, the map

$$T : C([0,R],\mathbb{R}) \ni f \mapsto \left([0,\infty) \ni t \mapsto \int_0^t \sin(t-s)q(s)f(s)\,ds\right)$$

is one for which $\sum_{n=1}^{\infty} \|T^n\|_\infty$ converges if $0 \le t \le R$. Hence (cf. **3.14**) if $w \in C([0,\infty),\mathbb{R})$, the map $M_w : C([0,\infty),\mathbb{R}) \ni f \mapsto w - T(f)$ has a unique fixed point v. If w'' is continuous and $w(0) = a, w'(0) = b$ then

$v(0) = a, v'(0) = b$ and direct calculation shows $v'' + (1+q)v = 0$: (s3.1) has a solution.

Since the differential equation is homogeneous, the question of uniqueness of solution for the problem is resolved by examining the case in which $a = b = 0$. The original equation is equivalent to the system

$$y_1' = +1 \cdot y_2,\ y_2' = -(1+q)y_1 + 0 \cdot y_2. \tag{s3.2}$$

Any solutions y_1, y_2 of the system must satisfy

$$y_1(t) = \int_0^t y_2(s)\,ds,\ y_2(t) = \int_0^t -(1+q(s))y_1(s)\,ds.$$

Because

$$y_1(t) \le \int_0^t y_2(s)\,ds,\ -y_2(t) \le -\int_0^t (-y_2(s))\,ds,$$

the second proof in **Solution 3.14** applied to (s3.2) shows $y_1 = 0$ (whence $y_2 = 0$): the problem has at most one solution.

b) If $w'' + w = 0, w(0) = a$,and $w'(0) = b$ then

$$|v(t)| \le \|w\|_\infty + \int_0^t |q(s)|\,|v(s)|\,ds;$$

3.14 applied again yields

$$|v(t)| \le \|w\|_\infty \exp\left(\int_0^t |q(s)|\,ds\right) \le \|w\|_\infty \exp\left(\|q\|_1\right),$$

whence v is bounded. □

3.16. a) Since f is continuous, $E_{\neq}(f, 0)$ is the union of a null set N and an open subset U of $(0, 1)$. Furthermore,

$$U = (U \cap E_{>}(f,0)) \uplus (U \cap E_{<}(f,0)) \stackrel{\text{def}}{=} U_+ \uplus U_-$$

and $U_\pm$ are uniquely of the form $\biguplus_{n\in\mathbb{N}} (a_{n\pm}, b_{n\pm})$. On $U_\pm$, $|f| = \pm f$ and $|f|''$, as the limit of second order difference quotients for $\pm f$, is measurable. Because f is in $C^2([0,1], \mathbb{R})$,

$$h : [0,1] \ni x \mapsto \begin{cases} 0 & \text{if } x \in [0,1] \setminus U \\ |f|'' & \text{otherwise} \end{cases}$$

is in $L^\infty_{\mathbb{R}}([0,1], \lambda)$.

b) Since

$$\int_0^1 g(x)h(x)\,dx = \sum_{n=1}^{\infty} \int_{a_{n\pm}}^{b_{n\pm}} g(x)h(x)\,dx,$$

it suffices to consider a single term of the sum. Then, (a, b) denoting any one of $(a_{n\pm}, b_{n\pm})$, integration by parts twice yields

$$\int_a^b g(x)h(x)\,dx = g(x)|f|'(x)\big|_a^b - g''(x)|f(x)|\big|_a^b + \int_a^b g''(x)|f(x)|\,dx.$$

Since $g(b)\,|f|'(b) \leq 0$ and $g(a)\,|f|'(a) \geq 0$, the first summand in the right member above is nonpositive. Because $|f|(a) = |f|(b) = 0$, the second summand is zero. Hence

$$\int_a^b g(x)h(x)\,dx \leq \int_a^b g''(x)|f(x)|\,dx.$$

□

3.17. *i.* The map

$$f : (0,1) \ni x \mapsto \frac{1}{2}\sin\left(\frac{1}{x}\right)$$

is continuous and

$$\left\{\frac{2}{n\pi}\right\}_{n\in\mathbb{N}}$$

is a Cauchy sequence, whereas its f-image is not.

ii. If f is not continuous then for some x_0 in $(0,1)$ and some sequence $\{x_n\}_{n\in\mathbb{N}}$,

$$\lim_{n\to\infty} x_n = x_0 \text{ and } f(x_n) \not\to f(x_0)$$

as $n \to \infty$. If

$$y_{2n} = x_n,\ y_{2n-1} = x_0,\ n \in \mathbb{N},$$

then $\{y_n\}_{n\in\mathbb{N}}$ is a Cauchy sequence and its f-image is not, a contradiction. □

3.18. The hypothesis implies that if $x_n \to 0$ as $n \to \infty$ then $\{f(x_n)\}_{n\in\mathbb{N}}$ is a Cauchy sequence. Thus the argument in **Soluition 3.17*ii*** implies $\lim_{x\to 0} f(x)$ exists. □

3.19. The map $f : \mathbb{R} \ni r \mapsto \lambda\{\, x \,:\, \sin x > r, x \in [0, 4\pi]\,\}$ is monotonely decreasing, $f([-1,1]) = f(\mathbb{R}) = [0, 4\pi]$, and on $[-1,1]$, f is injective. Thus $g \stackrel{\text{def}}{=} f^{-1}\big|_{[0,4\pi]}$ is monotonely decreasing and $g(x) > r$ iff $x < f(r)$. □

3.20. a) If f_k is the function for which the graph of $y = f_k(x)$ is that in **Figure s3.4** then $\lim_{k\to\infty} f_k(x) \equiv 1$ and $f_k'\left(\frac{1}{2}\right) = k, k \in \mathbb{N}$.

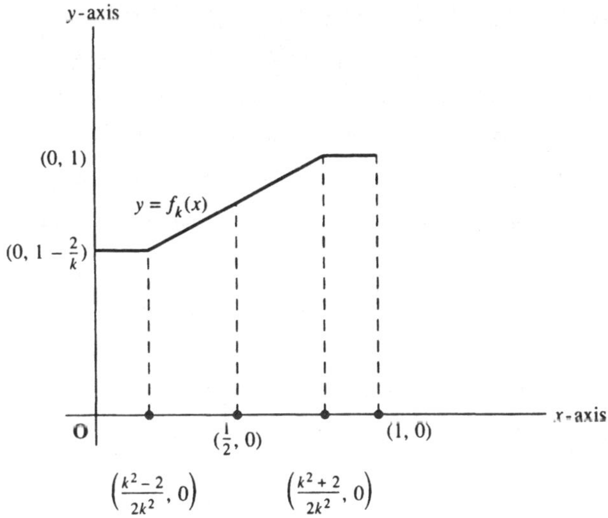

Figure s3.4.

b) Because each f_k is monotone, each f_k' exists a.e.,

$$E \stackrel{\text{def}}{=} \{x \ : \ f_k'(x) \text{ exists and is finite, } k \in \mathbb{N}\}$$

is measurable, and $\lambda(E) = 1$. The Fundamental Theorem of Calculus for Lebesgue integration implies $f_k(1) - f_k(0) \geq \int_E f_k'(x)\,dx, k \in \mathbb{N}$, and Fatou's lemma (cf. **4.3**) permits the conclusion:

$$\overline{\lim}_{k\to\infty}\left(f_k(1) - f_k(0)\right) = 0 \geq \underline{\lim}_{k\to\infty} \int_E f_k'(x)\,dx \geq \int_E \underline{\lim}_{k\to\infty} f_k'(x)\,dx.$$

However, $f_k'(x) \geq 0$ if $x \in E$. □

3.21. Since $\mathbf{A_R}$ is countable and contains $\mathbb{Q}$, $\mathbf{A_R}$ is a dense F_σ. If, for some f, $\mathbf{A_R} = \text{Cont}(f)$ it is also a (dense) G_δ (cf. **2.97**) and therefore a set of the second category, a contradiction.

As an F_σ, $\mathbf{A_R}$ is the union of countably many closed sets F_n such that $F_n \subset F_{n+1}$. If $B_n \stackrel{\text{def}}{=} (F_n \setminus F_{n-1}) \setminus (F_n \setminus F_{n-1})^\circ$ and

$$f(x) = \begin{cases} 2^{-n} & \text{if } x \in B_n \\ 0 & \text{if } x \notin \bigcup_{n\in\mathbb{N}} B_n \end{cases}$$

then $\text{Discont}(f) = \mathbf{A_R}$ whence $\text{Cont}(f) = \mathbb{R} \setminus \mathbf{A_R}$ [GeO]. □

3.22. Because $f_n'(x) - f_n'(y) = \int_y^x f_n''(t)\,dt$, the sequence $\{f_n'\}_{n\in\mathbb{N}}$ is equicontinuous. If M' as described does not exist then, via passage to subsequences as needed, it may be assumed that $a_n \stackrel{\text{def}}{=} \|f_n\|_\infty \downarrow 0$ and that $M_n' \stackrel{\text{def}}{=} \|f_n'\|_\infty \uparrow \infty$. Hence for some positive ϵ and for each n in $\mathbb{N}$, there is an x_n such that $\{|x - x_n| < \epsilon\} \Rightarrow \left\{|f_n'(x)| > \frac{1}{2}M_n'\right\}$. For some c_n,

$$f_n(x) = \int_{x_n}^x f_n'(t)\,dt + c_n, n \in \mathbb{N},$$

and since $f_n(x_n) = c_n$, it follows that $|c_n| \le a_n, n \in \mathbb{N}$. If $|x - x_n| < \epsilon$ then $|f_n(x)| \ge \frac{1}{2}M_n'\epsilon - a_n$. Hence $\|f_n\|_\infty \ge \frac{1}{4}M_n'\epsilon - a_n \to \infty$ as $n \to \infty$, a contradiction. □

3.23. If U is open in $\mathbb{R}$, $f^{-1}(U)$ is the union of two disjoint sets:

$$C_U \stackrel{\text{def}}{=} \text{Cont}(f) \cap f^{-1}(U) \text{ and } D_U \stackrel{\text{def}}{=} \text{Discont}(f) \cap f^{-1}(U).$$

By hypothesis, $\lambda(D_U) = 0$.

If $x \in C_U$, for some positive $\epsilon(x)$, $\{y : |f(x) - y| < \epsilon(x)\} \subset U$ and there is a positive $\delta(x)$ such that

$$f(\{x' : |x' - x| < \delta(x)\}) \subset \{y : |f(x) - y| < \epsilon(x)\} \subset U.$$

Hence $f^{-1}(U) = \left(\bigcup_{x\in C_U} \{x' : |x' - x| < \delta(x)\}\right) \cup D_U$, a union of an open (hence measurable) set and a null set. □

3.24. If $f'(x) = \chi_{(-\infty,0]} \circ f(x)$ is valid on an open set U containing $\{0\}$ then

$$f'(x) = \begin{cases} 0 & \text{if } f(x) > 0 \\ 1 & \text{if } f(x) \le 0 \end{cases}, \tag{s3.3}$$

and, by hypothesis, $f(0) = 0$.

i. If $\epsilon > 0$ and $f(x) = 0$ on $[0, \epsilon)$ then $f'(x) = 0$ on $(0, \epsilon)$ while (s3.3) requires that $f'(x) = 1$.

ii. If $f(x) > 0$ throughout $(0, \epsilon)$ then $f'(x) = 0$ throughout $(0, \epsilon)$ whence for some positive constant p, $f(x) = p$ throughout $(0, \epsilon)$ and hence f is not continuous at zero.

iii. If $f(x) < 0$ throughout $(0, \epsilon)$ then $f'(x) = 1$ throughout $(0, \epsilon)$ and for some constant k, $f(x) = x + k$ throughout $(0, \epsilon)$. Because $f(0) = 0$, $k = 0$; the hypothesis implies $f'(x) = 0$ throughout $(0, \epsilon)$, a contradiction.

Hence if $\epsilon > 0$ then f:

iv. cannot be zero throughout $[0, \epsilon)$;

v. can be neither only positive nor only negative throughout $(0, \epsilon)$.

Thus, owing to (s3.3), if $\epsilon > 0$, f' assumes both values zero and one on $(0, \epsilon)$ and no others. As a derivative, f' enjoys the intermediate value property and a contradiction emerges. □

3.25. If, for some x, y, $x < y$ and $f(x) > f(y)$ then for some positive p and all ϵ in $(0, p)$, $f(y) < f(x) - \epsilon(y - x)$.

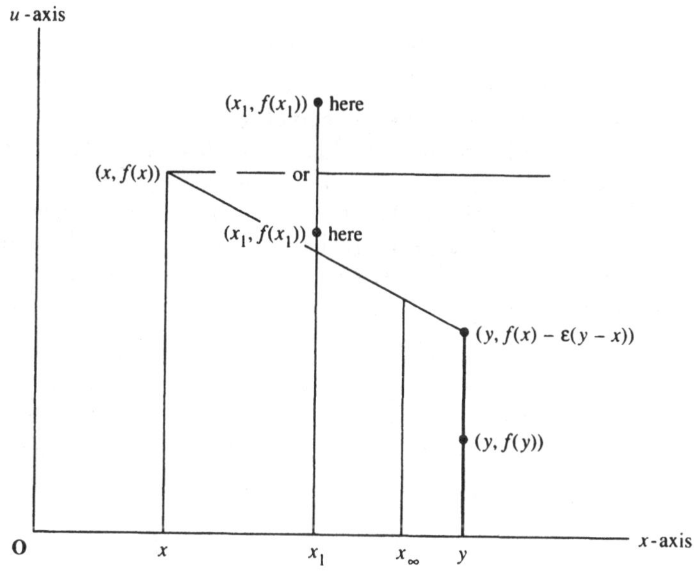

Figure s3.5. The situation: $x < y, f(x) > f(y)$, and $x_\infty < y$.

As indicated in **Figure s3.5**, by hypothesis, for some x_1 in (x, y),

$$f(x_1) \geq f(x) - \epsilon(x_1 - x).$$

If

$$y > x_\infty \stackrel{\text{def}}{=} \sup\{\, x_1 \;:\; x < x_1 < y, f(x_1) \ge f(x) - \epsilon\,(x_1 - x) \,\}$$

then $f(x_\infty) \ge f(x) - \epsilon\,(x_\infty - x)$ and, by hypothesis, for some z in (x_∞, y),

$$\begin{aligned} f(z) &\ge f(x_\infty) - \epsilon\,(z - x_\infty) \\ &\ge f(x) - \epsilon\,(x_\infty - x) - \epsilon\,(z - x_\infty) = f(x) - \epsilon(z - x), \end{aligned}$$

a contradiction. Thus

$$x_\infty = y \text{ and } f(y) \ge f(x) - \epsilon(y - x).$$

Since the last inequality obtains for all ϵ in $(0, p)$, $f(y) \ge f(x)$, another contradiction. □

3.26. For the Cantor function ϕ_0, let f_1 be ϕ_0.

i. On the closure of each interval $J_{1n}, n \in \mathbb{N}$, deleted in the construction of the Cantor set C_0, define the homolog f_{1n} of f_1.
ii. On the closure of each interval $J_{2n}, n \in \mathbb{N}$, deleted in the construction of the sequence $\{f_{1n}\}_{n\in\mathbb{N}}$, define the homolog f_{2n} of f_1.
iii. By induction, for each k in $\mathbb{N}$, there can be defined a sequence $\{f_{kn}\}_{n\in\mathbb{N}}$ such that: a) the domain of f_{kn} is the closure of one of the intervals deleted in the construction of $\{f_{k-1,n}\}_{n\in\mathbb{N}}$; b) each f_{kn} on its domain is the homolog of f_1.

The sequence $\{f_{1n}\}_{n\in\mathbb{N}}, \{f_{2n}\}_{n\in\mathbb{N}}, \ldots$ can be rearranged so that there emerges a single sequence $\{h_n\}_{n\in\mathbb{N}}$, each member of which, on its original domain, is homologous to f_1 on $[0,1]$. Furthermore, each h_n may be extended to be constantly zero to the left of its domain and constantly one to the right of its domain. Then $h \stackrel{\text{def}}{=} \sum_{n=1}^{\infty} \frac{h_n}{2^n}$ is continuous and monotonely increasing while $1 - h \stackrel{\text{def}}{=} f$ is monotonely decreasing. Owing to Fubini's theorem on derivatives of sequences of functions, $h' \doteq \sum_{n=1}^{\infty} \frac{h_n'}{2^n}$ and, since $h_n' \doteq 0, n \in \mathbb{N}$, $f' \doteq 0$. In sum, the function f meets all the requirements: f is a monotonely decreasing function, hence monotonely increasing on no interval and yet $f' \doteq 0 \ge 0$. □

3.27. For x, y in $(0,1)$, the formula

$$d : (0,1)^2 \ni \{x, y\} \mapsto \max\{|f(x) - f(y)|, |x - y|\}$$

defines a metric and furthermore, $d(x,y) \geq |x-y|$. Because f is continuous, if $\lim_{n\to\infty} |x_n - x| = 0$ then $\lim_{n\to\infty} |f(x_n) - f(x)| = 0$, whence $\lim_{n\to\infty} d(x_n, x) = 0$: d and $|\ |$ endow $(0,1)$ with the same topology.

Since $\{d(x,y) < \epsilon\} \Rightarrow \{|f(x) - f(y)| < \epsilon\}$, f is uniformly continuous on $((0,1), d)$. □

3.28. If $n = 1$, f may be regarded as the restriction to $(-r, r)$ of a function F holomorphic in $D(0,r)^\circ$. If $f \not\equiv 0$ then $F^{-1}(0)$ is countable and thus $\lambda\left(f^{-1}(0)\right) = 0$.

Assume the result is valid when $n = 1, 2, \ldots, N-1$. When $n = N$, let A be $f^{-1}(0) \cap B(\mathbf{O}, r)^\circ$. Fubini's theorem for product measures implies that if $A_{x_N} \stackrel{\text{def}}{=} \{(x_1, \ldots, x_{N-1}) : (x_1, \ldots, x_{N-1}, x_N) \in A\}$ then

$$\lambda_N(A) = \int_{-r}^{r} \lambda_{N-1}\left(A_{x_N}\right) dx_N.$$

For fixed x_N in $(-r, r)$, A_{x_N} in $\mathbb{R}^{N-1}$ is the possibly empty subset on which f, regarded as a function on $\mathbb{R}^{N-1}$ when x_n is fixed, is zero. By inductive assumption, $\lambda_{N-1}\left(A_{x_N}\right) = 0$ whence $\lambda_N(A) = 0$. □

[**Note s3.4:** The last result implies, when Mat_{nn} is regarded as a subset of $\mathbb{R}^{n^2}$, Nondiag_{nn} is the set of nondiagonable $n \times n$ matrices, and Sing_{nn} is the set of singular $n \times n$ matrices, that

$$\lambda_{n^2}\left(\text{Nondiag}_{nn}\right) = \lambda_{n^2}\left(\text{Sing}_{nn}\right) = 0.$$

Thus "almost all" $n \times n$ matrices are invertible and diagonable.

In the same vein, every algebraic or analytic variety in $\mathbb{R}^n$ is a null set.]

3.29. For f, g in V, $M_f \vee M_g$ resp. $k_f \vee k_g$ serves for $M_{f\vee g}$ resp. $k_{f\vee g}$ and $M_f \wedge M_g$ resp. $k_f \wedge k_g$ serves for $M_{f\wedge g}$ resp. $k_{f\wedge g}$. The vector space character of V flows from the definitions.

The Daniell construction in **Chapter 1** shows that there is a measure ν such that $P(f) = \int_{\mathbb{R}^n} f(\mathbf{x})\, d\nu(\mathbf{x})$ whence $\nu = \mu$. Each nonnegative f in V is the limit of a monotonely increasing sequence in $C_0(\mathbb{R}^n, \mathbb{R})$ and the monotone convergence theorem implies the validity of (3.1) for f. For an arbitrary f in V, $f = f^+ + f^-$. □

3.30. For t_0 fixed in $\mathbb{R}$, the map $g : \Sigma \ni \mathbf{x} \mapsto (f(\mathbf{x}, t_0))^2 + \left(\frac{\partial f(\mathbf{x}, t_0)}{\partial t}\right)^2$ is in $C(\Sigma, \mathbb{R})$ and $g > 0$. Because Σ is compact, $m \stackrel{\text{def}}{=} \min_{\mathbf{x}\in\Sigma} g(\mathbf{x}, t_0) > 0$ and for some $\mathbf{x}_{t_0}$ in Σ, $g(\mathbf{x}_{t_0}, t_0) = m$.

If, arbitrarily close to t_0, there is a t_k and in Σ a corresponding $\mathbf{x}_{t_k}$ such that $g(\mathbf{x}_{t_k}, t_k) = 0$ then it may be assumed that for some $\mathbf{x}_\infty$ in Σ,

$\lim_{k\to\infty} \|\mathbf{x}_{t_k} - \mathbf{x}_\infty\| = 0$ and thus $g(\mathbf{x}_\infty, t_0) = 0$, a contradiction. Hence, for some open set $N(t_0)$, if $t \in N(t_0)$ then $\min_{\mathbf{x}\in\Sigma} g(\mathbf{x}, t) > 0$.

If, for each k in $\mathbb{N}$, there are in $N(t_0)$ two points, t_{k1}, t_{k2} such that $|t_{k1} - t_0| + |t_{k2} - t_0| < \frac{1}{k}$ and for some $\mathbf{x}_k$ in Σ, $f(\mathbf{x}_k, t_{ki}) = 0$, $i = 1, 2$, then Rolle's theorem implies for some t_{k3} between t_{k1} and t_{k2}, $\frac{\partial f(\mathbf{x}_k, t_{k3})}{\partial t} = 0$. It may be assumed that for some $\overline{\mathbf{x}}$ in Σ, $\lim_{k\to\infty} \|\mathbf{x}_k - \overline{\mathbf{x}}\| = 0$. Hence $\frac{\partial f(\overline{\mathbf{x}}, t_0)}{\partial t} = 0$. Since $\min_{\mathbf{x}\in\Sigma} g(\mathbf{x}, t) > 0$, $f(\overline{\mathbf{x}}, t_0) \neq 0$. Thus for large k, $f(\mathbf{x}_k, t_{k1}) \cdot f(\mathbf{x}_k, t_{k2}) \neq 0$, a contradiction. It follows that for t_1, t_2 close enough to t_0 and different from each other (one may be t_0), and any $\mathbf{x}$ in Σ, $(f(\mathbf{x}, t_1))^2 + (f(\mathbf{x}, t_2))^2 > 0$. □

3.31. For Σ as in **3.30**, let K be $\Sigma \cap f^{-1}(0)$. Then K is compact and $g(\mathbf{x}) > 0$ on K. Hence for some positive ϵ and some open U containing K, $g(\mathbf{x}) \geq \epsilon$ on U. For some positive δ, $f(\mathbf{x}) \geq \delta$ on the compact set $\Sigma \setminus U$. Because U may be chosen so that $\mathbf{O} \notin U$ and since if $\mathbf{x} \neq \mathbf{O}$,

$$f(\mathbf{x}) = \|\mathbf{x}\|^m f\left(\frac{\mathbf{x}}{\|\mathbf{x}\|}\right), \quad g(x) = \|\mathbf{x}\|^m g\left(\frac{\mathbf{x}}{\|\mathbf{x}\|}\right),$$

it follows that

$$\left\{\frac{\mathbf{x}}{\|\mathbf{x}\|} \in \Sigma \setminus U\right\} \Rightarrow \left\{\frac{f(\mathbf{x})}{\|\mathbf{x}\|^m} \geq \delta\right\} \text{ and } \left\{\frac{\mathbf{x}}{\|\mathbf{x}\|} \in U\right\} \Rightarrow \left\{\frac{g(\mathbf{x})}{\|\mathbf{x}\|^m} \geq \epsilon\right\}.$$

The required inequality obtains if $C = \frac{1}{\delta}$ and $D = \frac{1}{\epsilon}$. □

3.32. For $\mathbf{a} \stackrel{\text{def}}{=} (a_1, \ldots, a_n)$ in $B_1(\mathbb{R}^n)$ let $\mathbf{f}(\mathbf{a})$ be $(f_1(\mathbf{a}), \ldots, f_n(\mathbf{a}))$. If $\mathbf{b} \stackrel{\text{def}}{=} (b_1, \ldots, b_n) \in B_1(\mathbb{R}^n)$ then for some c_{ij} between a_j and b_j,

$$\begin{aligned} f_i(\mathbf{a}) - f_i(\mathbf{b}) &= (f_i(a_1, \ldots, a_n) - f_i(b_1, a_2, \ldots, a_n)) + \cdots \\ &\quad + (f_i(b_1, a_2, \ldots, a_n) - f_i(b_1, b_2, a_3, \ldots, a_n)) + \cdots \\ &\quad + (f_i(b_1, \ldots, b_{n-1}, a_n) - f_i(b_1, \ldots, b_n)) \\ &= \sum_{j=2}^{n-1} \frac{\partial f_i(b_1, \ldots, b_{j-1}, c_{ij}, b_{j+1}, \ldots, a_n)}{\partial x_j}(a_j - b_j) \\ &\quad + \frac{\partial f_i(c_{i1}, a_2, \ldots, a_n)}{\partial x_1}(a_1 - b_1) \\ &\quad + \frac{\partial f_i(b_1, \ldots, b_{n-1}, c_{in})}{\partial x_n}(a_n - b_n). \end{aligned}$$

There is a positive δ such that

$$\left\{\left\|\frac{\partial f_i}{\partial x_j}\right\|_\infty < \delta, i, j = 1, \ldots, n\right\} \Rightarrow \left\{\|\mathbf{f}(\mathbf{a}) - \mathbf{f}(\mathbf{b})\| < \frac{1}{2}\|\mathbf{a} - \mathbf{b}\|\right\}.$$

Because $\mathbf{d}(\mathrm{id}) = \mathrm{id}$, it follows that if $\mathbf{g} = \mathbf{f} - \mathrm{id}$ then $\mathbf{d}(\mathbf{g}) = \mathbf{d}(\mathbf{f}) - \mathrm{id}$.

There is a positive η such that

$$\left\{\sup_{\mathbf{x}} \|\mathbf{d}(\mathbf{f})(\mathbf{x}) - \mathrm{id}\| < \eta\right\} \Rightarrow \left\{\left\|\frac{\partial f_i}{\partial x_j} - \delta_{ij}\right\|_\infty < \delta, i, j = 1, \ldots, n\right\},$$

whence $\|\mathbf{a}-\mathbf{b}\| - \|\mathbf{f}(\mathbf{a}) - \mathbf{f}(\mathbf{b})\| \le \|\mathbf{g}(\mathbf{a}) - \mathbf{g}(\mathbf{b})\| < \frac{1}{2}\|\mathbf{a}-\mathbf{b}\|$. Consequently, $\|\mathbf{f}(\mathbf{a}) - \mathbf{f}(\mathbf{b})\| \ge \frac{1}{2}\|\mathbf{a} - \mathbf{b}\|$ and thus $\mathbf{f}$ is injective. □

3.33. For $\mathbf{x}$ in $\mathbb{R}^n$, $\mathbf{d}(f)(\mathbf{x}) \in [\mathbb{R}^n, \mathbb{R}]$ and, since $[\mathbb{R}^n, \mathbb{R}]$ and $\mathbb{R}^n$ are naturally isomorphic as vector spaces, if $f \in C^2(\mathbb{R}^n, \mathbb{R})$ then $\mathbf{d}^2(f)$ is in $[\mathbb{R}^n, [\mathbb{R}^n]]$. [For each $\mathbf{x}$, $\mathbf{d}^2(f)(\mathbf{x})$ may be represented by the Hessian matrix $\left(\frac{\partial^2 f(\mathbf{x})}{\partial x_i \partial x_j}\right)_{i,j=1}^n$.]

By definition, there is a map $\epsilon : [-1, 1] \ni t \mapsto (0, \infty)$ such that $\lim_{t\to 0} \epsilon(t) = 0$ and for some positive δ,

$$\{\|\mathbf{u}\| < \delta\} \Rightarrow \left\{\left\|\mathbf{d}f(\mathbf{x}_0 + t\mathbf{u}) - \mathbf{d}f(\mathbf{x}_0) - \mathbf{d}^2 f(\mathbf{x}_0)(t\mathbf{u})\right\| = \epsilon(t)|t| \cdot \|\mathbf{u}\|\right\}.$$

Since $\mathbf{d}f(\mathbf{x}_0) = \mathbf{O}$, $\left\|\mathbf{d}f(\mathbf{x}_0 + t\mathbf{u}) - \mathbf{d}^2 f(\mathbf{x}_0) t\mathbf{u}\right\| = \epsilon(t)|t| \cdot \|\mathbf{u}\|$.

If $\mathbf{d}f(\mathbf{x}_o + t\mathbf{u}) = \mathbf{O}$ for some nonzero t in $[-1, 1]$ and some nonzero $\mathbf{u}$, then, since $\mathbf{d}^2 f(\mathbf{x}_0)^{-1}$ exists, $\left\|\mathbf{d}^2 f(\mathbf{x}_0)^{-1}\right\| \stackrel{\text{def}}{=} K > 0$. If $\mathbf{y} \stackrel{\text{def}}{=} \mathbf{d}^2 f(\mathbf{x}_0)$ then $t\mathbf{u} = \mathbf{d}^2 f(\mathbf{x}_0)^{-1}(\mathbf{y}) \le K\|\mathbf{y}\|$, whence $\mathbf{y} \ne \mathbf{O}$ and $\|\mathbf{y}\| \le \epsilon(t)K\|\mathbf{y}\|$, i.e., $1 \le \epsilon(t)K$.

If $\|\mathbf{x}_n - \mathbf{x}_0\| < \frac{1}{n}$, $\mathbf{d}f(\mathbf{x}_n) = 0, n \in \mathbb{N}$, and $\frac{1}{n_0} < \frac{\delta}{2}$ then when $n > n_0$, there is a $\mathbf{u}_n$ such that $\|\mathbf{u}_n\| = \frac{\delta}{2}$ and a t_n such that $\mathbf{x}_n = \mathbf{x}_0 + t_n\mathbf{u}_n$ and $|t_n| < \frac{1}{n}$. But then $1 \le \epsilon(t_n) K$, a contradiction, since $\lim_{n\to\infty} \epsilon(t_n) = 0$. Hence for some open set $N(x_0)$, $\mathbf{d}f(\mathbf{y}) \ne 0$ for all $\mathbf{y}$ in $N(x_0) \setminus \{\mathbf{x}_0\}$. □

3.34. If $W = \mathrm{var}_{[0,1]}(f) - \frac{\epsilon}{4}$, there is a partition P_0 determined by $n+2$ partition points for which $\mathrm{var}_{[0,1],P_0}(f) > W - \frac{\epsilon}{2}$. There is a positive δ such that $\{|x - y| < \delta\} \Rightarrow \left\{|f(x) - f(y)| < \frac{\epsilon}{8n}\right\}$. If $|P| < \min\{\delta, |P_0|\} \stackrel{\text{def}}{=} a$ and P_1 is the common refinement of P_0 and P, then

$$\mathrm{var}_{[0,1],P}(f) + 2n\frac{\epsilon}{8n} \ge \mathrm{var}_{[0,1],P_1}(f) \ge \mathrm{var}_{[0,1],P_0}(f) > W - \frac{\epsilon}{2}$$

$$\mathrm{var}_{[0,1],P}(f) > W - \frac{3\epsilon}{4} = \mathrm{var}_{[0,1]}(f) - \epsilon.$$

When $f = \chi_{\{\frac{1}{2}\}}$ then $\mathrm{var}_{[0,1]}(f) = 2$. However, for any partition P determined by points in $[0, 1] \setminus \left\{\frac{1}{2}\right\}$, $\mathrm{var}_{[0,1],P}(f) = 0$. □

3.35. There are functions $f_i, 1 \le i \le 4$, each monotonely increasing, nonnegative, and bounded for which $f = f_1 - f_2 + i(f_3 - f_4)$. For each f_i, $\int_{[0,1]} f_i'(x)\,dx \le f_i(1) - f_i(0) < \infty$, whence $f' \in L^1([0,1],\lambda)$. □

3.36. Only the situation for left-continuity is discussed. The other situations are treated, mutatis mutandis.

If f is of bounded variation and left-continuous at a, if $\epsilon > 0$, and if

$$0 \stackrel{\text{def}}{=} x_0 < x_1 < \cdots < x_n < x_{n+1} < a$$

$$\sum_{k=0}^{n} |f(x_{k+1}) - f(x_k)| + |f(a) - f(x_{n+1})| > \operatorname{var}_{[0,a]}(f) - \epsilon$$

then $\operatorname{var}_{[0,x_{n+1}]}(f) > \operatorname{var}_{[0,a]}(f) - \epsilon - |f(a) - f(x_{n+1})|$. As $x_{n+1} \uparrow a$, $\lim_{x_{n+1}\uparrow a} \operatorname{var}_{[0,x_{n+1}]}(f) \ge \operatorname{var}_{[0,a]}(f) - \epsilon$. Because $\operatorname{var}_{[0,x]}(f)$ increases monotonely, $\operatorname{var}_{[0,a]}(f) \ge \lim_{x_{n+1}\uparrow a} \operatorname{var}_{[0,x_{n+1}]}(f) \ge \operatorname{var}_{[0,a]}(f) - \epsilon$.

Conversely, if $\operatorname{var}_{[0,x]}(f)$ is left-continuous at a, then

$$|f(a) - f(x_{n+1})| \le \operatorname{var}_{[0,a]}(f) - \operatorname{var}_{[0,x_{n+1}]}(f)$$

and as $x_{n+1} \uparrow a$ the right member above descends to zero. □

3.37. If, for some a, $|f(a+0) - f(a-0)| = \Delta > 0$ then for all small positive δ, $f|(a+\delta) - f(a-\delta)| > \frac{\Delta}{2}$ and so on $[a-\delta, a+\delta]$, f assumes no values between $(f(a-0) \wedge f(a+0)) + \frac{\Delta}{3}$ and $(f(a-0) \vee f(a+0)) - \frac{\Delta}{3}$, a contradiction. □

[**Note s3.5:** Hence if a function f is of bounded variation and is a derivative then f is continuous.]

3.38. There is a sequence $\{f_{mn}\}_{m,n\in\mathbb{N}}$ of piecewise linear, continuous, even functions such that

$$f_{mn}(x) = \begin{cases} 1 & \text{if } |x-a| < \frac{1}{m} \text{ or } |x+a| < \frac{1}{m} \\ 0 & \text{if } x \notin \left(-a - \frac{2}{m}, -a + \frac{2}{m}\right) \cup \left(a - \frac{2}{m}, a + \frac{2}{m}\right) \end{cases}$$

$$f_{mn} \downarrow \chi_{[-a-\frac{1}{m}, -a+\frac{1}{m}]} + \chi_{[a-\frac{1}{m}, a+\frac{1}{m}]} \text{ as } n \to \infty.$$

Then $0 = \lim_{n\to\infty} \int_{-1}^{1} f_{mn}(x) g(x)\,dx = \int_{-a-\frac{1}{m}}^{-a+\frac{1}{m}} + \int_{a-\frac{1}{m}}^{a+\frac{1}{m}} g(x)\,dx$. Because g is continuous at a and $-a$, if $\epsilon > 0$ and m is large,

$$\left| m\left(\int_{-a-\frac{1}{m}}^{-a+\frac{1}{m}} + \int_{a-\frac{1}{m}}^{a+\frac{1}{m}} g(x)\,dx \right) - (g(-a) + g(a)) \right| < \epsilon.$$ □

3.39. The assertion holds if $f(x) = \chi_{\{1\}}(x)$. □

3.40. Direct calculation shows that on $[0,1]$ if $a \le 1$ then f_a is not of bounded variation and that if $a > 1$ then f_a is absolutely continuous. □

3.41. If

$$f(x) = \begin{cases} (x-1)\sin\left(\dfrac{1}{x-1}\right) & \text{if } 0 \le x < 1 \\ 0 & \text{if } x = 1 \end{cases}$$

then f has the required properties. □

3.42. If $0 < \alpha < 1$, C_α is a Cantor subset of $[0,1]$, and $\lambda(C_\alpha) = \alpha$, then

$$f(x) \stackrel{\text{def}}{=} \int_0^x (1 - \chi_{C_\alpha}(t))\, dt$$

is absolutely continuous.

If $[a,b] \subset [0,1]$ then (a,b) is either a subinterval of some interval deleted in the construction of C_α or some such interval is a subinterval of (a,b). It follows that

$$f(b) - f(a) \ge b - a.$$

Furthermore $f' \doteq 1 - \chi_{C_\alpha}$ whence $f' \doteq 0$ on C_α. □

3.43. The map θ in **Solution 2.57*ii*** resp. $\theta^{-1} \stackrel{\text{def}}{=} \xi$ is a continuous strictly monotonely increasing function on C_0 resp. S. The intervals deleted from $[0,1]$ in the construction of C_0 are in bijective order-preserving correspondence with intervals deleted from $[0,1]$ to yield S.

An analog ϕ_S of ϕ_0 should be monotonely increasing, continuous, and constant on each interval deleted in the construction of S. Owing to the denseness of $[0,1] \setminus S$, the values of ϕ_S on the deleted intervals then determine the values of ϕ_S throughout $[0,1]$. Thus there is some arbitrariness about the definition of ϕ_S.

A straightforward analogy with ϕ_0 is

$$\phi_S : S \ni x = \sum_{n=1}^{\infty} \epsilon_n(x)(1-r)r^{n-1} \mapsto \sum_{n=1}^{\infty} \frac{\epsilon_n(x)}{2^n}$$

which is a continuous and monotonely increasing map of S onto $[0,1]$. Consequently, at the ends of each interval deleted in the construction of S, the values of ϕ_S are the same and ϕ_S may be extended in one and only one way to a continuous and monotonely increasing map of $[0,1]$ onto itself. **Figure s3.6** depicts the graph of the second approximant to ϕ_S.

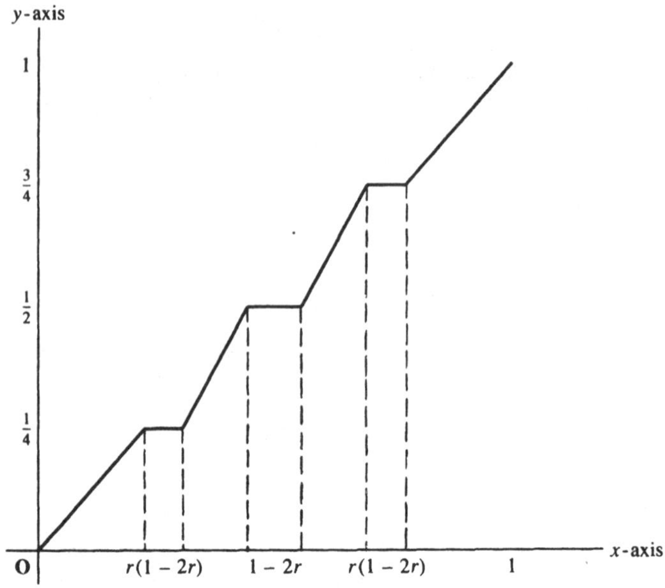

Figure s3.6.

a) The complement in the unit square of each approximant to $\mathcal{G}_S$ consists of two congruent figures. Hence $\int_{[0,1]} \phi_S(x)\,dx = \frac{1}{2}$.

b) The length of $\mathcal{G}_S$ may be approximated by the lengths of polygons that represent the successive approximants to $\mathcal{G}_S$. Induction shows that, β_n denoting the length of each interval in the nth group of deleted intervals, at the Nth stage of approximation, the length of the polygonal approximant is $\sum_{n=1}^{N} 2^{n-1}\beta_n + \sqrt{1 + \left(1 - \sum_{n=1}^{N} 2^{n-1}\beta_n\right)^2}$. Hence the length of $\mathcal{G}_S$ is 2 (!).

c) Symmetry arguments show $\int_{[0,1]} \phi_\alpha(x)\,dx = \frac{1}{2}$. On the other hand, the discussion in b) now leads to $1 - \alpha + \sqrt{1 + \alpha^2}$ as the length of Γ_α. The length of Γ_α is less than 2 if $0 < \alpha < 1$. □

[**Note s3.6:** When $0 < r < \frac{1}{2}$, a more interesting analogy with ϕ_0 (for which $r = \frac{1}{3}$) follows.

Dual to $a_n \overset{\text{def}}{=} (1-r)r^{n-1}, n \in \mathbb{N}$, is $b_n \overset{\text{def}}{=} r(1-r)^{n-1}, n \in \mathbb{N}$. Furthermore, $\sum_{n=1}^{\infty} b_n = 1$ and $b_n \le \sum_{k=n+1}^{\infty} b_k$, whence the set T of all subsums of $\sum_{n=1}^{\infty} b_n$ is $[0,1]$.

Thus the formula

$$\Phi_S : S \ni x \stackrel{\text{def}}{=} \sum_{n=1}^{\infty} \epsilon_n(x)(1-r)r^{n-1} \mapsto \sum_{n=1}^{\infty} \epsilon_n(x) r(1-r)^{n-1}$$

defines a continuous monotonely increasing map of S onto $[0,1]$. Hence the values of Φ_S at the endpoints of each interval deleted in the construction of S are the same and Φ_S may be extended in one and only one way to a continuous and monotonely increasing function mapping $[0,1]$ onto itself.

The reader might wish to calculate $\int_{[0,1]} \Phi_S(x)\,dx$ and the length of the graph of $y = \Phi_S(x)$.]

3.44. If $p \equiv 0$, $E_=(p) = \mathbb{R}$, $E_>(p) = E_<(p) = \emptyset$. The definitions imply that if $p(x) \stackrel{\text{def}}{=} a_0x^n + \cdots + a_n$ and $a_0 \neq 0$ then

$$\begin{gathered} E_{R,j}(p') = E_{R,j+1}(p), 0 \le j \le n-1 \\ E_R(p) = E_{R,0}(p) \cap E_R(p') \\ E_=(p) = \emptyset \\ E_{>,n}(p) = \begin{cases} \emptyset & \text{if } a_0 < 0 \\ \mathbb{R} & \text{if } a_0 > 0 \end{cases} \\ E_{<,n}(p) = \begin{cases} \mathbb{R} & \text{if } a_0 < 0 \\ \emptyset & \text{if } a_0 > 0 \end{cases}. \end{gathered}$$

Thus it suffices to discuss only $E_>(p)$ and to assume that $a_0 > 0$. When $n = 1$, $E_>(p) = \left(\frac{-a_1}{a_0}, \infty\right)$. If $n = N$ and, when $n = 1, \ldots, N-1$, for some α in $\mathbb{R}$, $E_>(p) = (\alpha, \infty)$, then for some β in $\mathbb{R}$, $E_>(p') = (\beta, \infty)$ and $E_{>,0}(p)$ is a union of a finite number of pairwise disjoint open intervals.

If $E_{>,0}(p) = \mathbb{R}$ then for some β in $\mathbb{R}$, $E_>(p) = (\beta, \infty)$.

If $r_k < s_k < r_{k+1}$, $E_{>,0}(p) = \bigcup_{k=1}^{K} (r_k, s_k)$, and $\beta \le s_{K-1}$, then for some γ in (s_{K-1}, ∞), $p'(\gamma) = 0$ in contradiction of the definition of β. Hence for $\alpha \stackrel{\text{def}}{=} \max\{\beta, r_K\}$, $E_>(p) = (\alpha, \infty)$ (and $s_K = \infty$). □

3.45. If f is Riemann integrable, it must be bounded since otherwise the set of approximating Riemann sums is unbounded in $\mathbb{R}$.

Since

$$\text{Discont}(f) = \bigcup_{n\in\mathbb{N}} \left\{ x : \overline{\lim}_{y=x} f(y) - \underline{\lim}_{y=x} f(y) \ge \frac{1}{n} \right\} \stackrel{\text{def}}{=} \bigcup_{n\in\mathbb{N}} E_{\frac{1}{n}},$$

if $\lambda(\text{Discont}(f)) = \delta > 0$ then for some n_0, $\lambda\left(E_{\frac{1}{n_0}}\right) \stackrel{\text{def}}{=} \epsilon > 0$. For each partition used to calculate the upper and lower Riemann sums, the sum of the lengths of the partition intervals containing $E_{\frac{1}{n_0}}$ is at least ϵ and thus the difference between the upper and lower Riemann sums is at least $\frac{\epsilon}{n_0}$, whence f is not Riemann integrable, a contradiction.

Conversely, if f is bounded and $\lambda(\text{Discont}(f)) = 0$ then for positive ϵ, E_ϵ is compact and $\lambda(E_\epsilon) = 0$. Hence E_ϵ can be covered by finitely many pairwise disjoint open intervals of small length-sum. For a partition consisting of such intervals and others (on each of which f is uniformly continuous), the difference between the upper and lower Riemann sums is small if the mesh of the partition is small. □

3.46. For $\mathbb{Q} \cap [0,1] \stackrel{\text{def}}{=} \{r_k\}_{k\in\mathbb{N}}$, $f_n \stackrel{\text{def}}{=} \sum_{k=1}^n \chi_{r_k}$, $\lim_{n\to\infty} f_n = \chi_{\mathbb{Q}\cap[0,1]}$, each f_n is Riemann integrable while $\text{Discont}\left(\chi_{\mathbb{Q}\cap[0,1]}\right) = [0,1]$ and $\chi_{\mathbb{Q}\cap[0,1]}$ is not Riemann integrable (cf. **3.45**). □

3.47. Let ϕ_0 be the Cantor function and let f be $\phi_0 \cdot \chi_{[0,1]\setminus C_0}$. Because f is bounded and $\lambda(\text{Discont}(f)) = 0$, f is Riemann integrable.

For α positive, there is a homeomorphism $H : C_\alpha \stackrel{\text{homeo}}{\mapsto} C_0$ **[GeO]** and the Tietze extension theorem implies that H has a continuous extension $g : [0,1] \mapsto [0,1]$. If $h \doteq f \circ g$ then $\lambda(\text{Discont}(h)) = \lambda(C_\alpha) > 0$ and h is not Riemann integrable. □

3.48. If f is strictly increasing, all its difference quotients are nonnegative and thus $f' \geq 0$ everywhere. If D contains a connected set containing two points, then $f'(x) = 0$ on a nonempty open interval (a,b), whence f is constant on (a,b), a contradiction.

Conversely if $f' \geq 0$ everywhere, Rolle's theorem implies f is monotonely increasing. If $a < b$ and $f(a) = f(b)$ then $f'(x) = 0$ on (a,b), an impossibility if D is totally disconnected. □

3.49. If m is a strict local maximum value of f, if $A_m \stackrel{\text{def}}{=} S \cap f^{-1}(m)$, and if $\#(A_m) > \aleph_0$ then by virtue of **2.27**, for some a in A_m, and every $N(a)$, $\#(N(a) \cap A_m) > \aleph_0$, whence arbitrarily close to a there are points where the value of f is m, i.e., a is not the site of a strict local maximum: $\#(A_m) \leq \aleph_0$.

Let M be the set of strict local maxima of f. If $\#(M) > \aleph_0$ then for some m in M, and all p in $(0,\infty)$, $\#((m, m+p) \cap M) > \aleph_0$ whence $\#\left(f^{-1}((m, m+p) \cap M)\right) > \aleph_0$. Thus in $E \stackrel{\text{def}}{=} f^{-1}((m, m+p) \cap M)$ there is a b such that for every $N(b)$, $\#(N(b) \cap E) > \aleph_0$. Because $\#(A_m) \leq \aleph_0$ for some c in $N(b) \cap E$, $f(c) \neq m$ and so $f(c)$ is in $(m, m+p)$ whence $m \notin S$: $\#(M) \leq \aleph_0$ and $\#(S) = \sum_{m\in M} \#(A_m) \leq \aleph_0$. □

3.50. If $a < b$, there is a sequence $\{[a_n, b_n)\}_{n\in\mathbb{N}}$ such that

$$a_n \stackrel{\text{def}}{=} p_n 2^{-q_n},\ b_n \stackrel{\text{def}}{=} (p_n + 1)\, 2^{-q_n},\ p_n, q_n \in \mathbb{N},\ (a,b) = \bigcup_{n\in\mathbb{N}} [a_n, b_n).$$

For some n_0, $a_* \overset{\text{def}}{=} \inf_{n\le n_0} a_n$ and $b^* \overset{\text{def}}{=} \sup_{n\le n_0} b_n$ are such that

$$|f(a)-f(a_*)|+|f(b)-f(b^*)|<b-a$$
$$|f(b)-f(a)|\le |f(b)-f(b^*)|+\sum_{n=1}^{n_0}|f(b_n)-f(a_n)|+|f(a_*)-f(a)|$$
$$<b-a+\sum_{n=1}^{n_0}2^{-q_n}\cdot M<(M+1)(b-a).$$

Hence $f\in \text{Lip}(1)$ and so $f\in AC$ whence for all x,

$$f\left(x+2^{-n}\right)-f(x)=\int_x^{x+2^{-n}} f'(t)\,dt$$
$$\lim_{n\uparrow\infty}\Delta_n(x)=\frac{\int_x^{x+2^{-n}} f'(t)\,dt}{2^{-n}}=0.$$

The Fundamental Theorem of Calculus for Lebesgue integration implies that $f'(x)\doteq 0$. Hence for each n in $\mathbb{N}$ and each k in $\mathbb{Z}$, $f(0)=f\left(k2^{-n}\right)$. Since $\{k2^{-n}\}_{\substack{k\in\mathbb{Z}\\ n\in\mathbb{N}}}$ is dense in $\mathbb{R}$ and f is continuous, $f(x)\equiv f(0)$. □

3.51. If no such a exists then, for all x, $f(x)-g(x)\ne 0$. Since f and g are continuous, $f-g$ is of one sign, say $f-g>0$ and on $[0,1]$, for some positive δ, $f(x)>g(x)+\delta$. Thus $f\circ f>g\circ f+\delta=f\circ g+\delta>g\circ g+2\delta$. By induction, it follows that $\{n\in\mathbb{N}\}\Rightarrow\left\{f^{\{n\}}>g^{\{n\}}+n\delta\right\}$, an impossibility if $n\delta>1$.

For $f:[0,\infty)\ni\mapsto \ln(1+x)$ and $g:[0,\infty)\ni x\mapsto e^{1+x}$, it follows that $f\circ g(x)=1+x=g\circ f(x)$ and yet, for all nonnegative x, $\ln(1+x)<e^{1+x}$.

□

3.52. Because f is continuous, if $\epsilon,|\delta|>0$ then for some h_δ in $\mathbb{Q}\setminus\{0\}$, $|h_\delta-\delta|\le\delta^2$ and $|f(c+\delta)-f(c+h_\delta)|<\epsilon|\delta|$. Hence

$$\left|\frac{f(c+\delta)-f(c)}{\delta}-L\right|\le\left|\frac{f(c+\delta)-f(c+h_\delta)}{\delta}\right|$$
$$+\left|\frac{f(c+h_\delta)-f(c)}{h_\delta}\cdot\frac{h_\delta}{\delta}-L\right|.$$

For small $|\delta|$, the second term in the right member above is small since $\dfrac{h_\delta}{\delta}$ is near one and the first member is small by virtue of the choice of h_δ.

If $f=\chi_{\mathbb{Q}}$ and $c=0$ then $\lim_{h\in\mathbb{Q}\setminus 0,h\to 0}\dfrac{f(c+h)-f(c)}{h}=0$ although, since f is nowhere continuous, $f'(0)$ does not exist. □

3.53. If $\text{supp}(h) \subset [a,b]$ then for $F(x) \stackrel{\text{def}}{=} \int_a^x f(t)\,dt$, integration by parts implies $\int_{\mathbb{R}} f(t)h(t)\,dt = -\int_a^b F(t)h'(t)\,dt = -\int_a^b g(t)h'(t)\,dt$ whence $F = g$. Because $F' = f$, g' exists and is f. □

3.54. If $\lim_{n\to\infty} \mathbf{f}(\mathbf{x}_n) = \mathbf{y}$ then $\lim_{n\to\infty} \mathbf{x}_n \stackrel{\text{def}}{=} \mathbf{x}$ exists and $\mathbf{f}(\mathbf{x}) = \mathbf{y}$. Hence $\mathbf{f}(\mathbb{R}^n)$ is closed.

The inequality (3.2) implies that $\mathbf{f}$ is injective and hence, confined to a closed ball in $\mathbb{R}^n$, is a homeomorphism. From Brouwer's invariance of domain theorem it follows that $\mathbf{f}$ is open whence $\mathbf{f}(\mathbb{R}^n)$ is both open and closed. Since $\mathbb{R}^n$ is connected, $\mathbf{f}$ is surjective: $\mathbf{f}$ is bijective. Finally, (3.2) also implies $\mathbf{f}^{-1}$, which exists because $\mathbf{f}$ is bijective, is continuous. □

3.55. The proof is an application of the average principle: an average lies between the supremum and infimum of the numbers averaged.

If $|f(\mathbf{O})| = \sup_{\mathbf{x}\in B(\mathbf{O},1)} |f(\mathbf{x})| \stackrel{\text{def}}{=} M$, $\|\mathbf{x}_0\| = r < 1$, and $|f(\mathbf{x}_0)| < M$ then

$$|f(\mathbf{O})| = M = \left|\frac{1}{2\pi r}\oint_{\|\mathbf{y}\|=r} f(\mathbf{y})\,d\mathbf{y}\right| \le \frac{1}{2\pi r}\oint_{\|\mathbf{y}\|=r} |f(\mathbf{y})|\,d\mathbf{y}. \tag{s3.4}$$

Because f is continuous, $|f(\mathbf{y})| < M$ near $\mathbf{x}_0$ whence the value of the last member in (s3.4) is less than M, a contradiction.

If f is constant then $|f(\mathbf{O})| = \sup_{\mathbf{x}\in B(\mathbf{O},1)} \|f(\mathbf{x})\|$. □

3.56. Because f is uniformly continuous on K, if $\epsilon > 0$ there is a positive δ such that on K, $\{|x-y| < \delta\} \Rightarrow \{|f(x) - f(y)| < \epsilon\}$. Consequently,

$$\{|x-y| < \epsilon\} \Rightarrow \left\{\left|f_{[t]}(x) - f_{[t]}(y)\right| = |f(x+t) - f(y+t)| < \epsilon\right\},$$

whence $\left\{f_{[t_n]}\right\}_{n\in\mathbb{N}}$ is uniformly bounded and equicontinuous. The Arzelà-Ascoli theorem implies the result. □

3.57. For any x in $[0,1]$ and some ξ in $(0,x)$,

$$f(x) \le |f(x)| \le \int_0^x f(t)\,dt = f(\xi)x \le \|f\|_\infty x.$$

It follows by repeated integration that for each n in $\mathbb{N}$ and each x in $[0,1]$, $|f(x)| \le \|f\|_\infty \dfrac{x^n}{n!}$. □

3.58. For each real a, there are coefficients $b_n(a)$, $n \in \mathbb{N}$, and a positive $r(a)$ such that for x in $(a-r(a), a+r(a))$, $f'(x) = \sum_{n=1}^\infty b_n(a)(x-a)^n$. The series $\sum_{n=1}^\infty b_n(a)\dfrac{(x-a)^{n+1}}{n+1}$ converges in $(a-r(a), a+r(a))$ and represents a function g_a.

On the other hand, $\int_0^x f'(t)\,dt = g_a(x)$ for x near a and, since f' is continuous, $g_a' = f'$, $f = g_a + \text{constant}$, whence f is real analytic. □

3.59. If $t \in (0,1]$, $f^{(n+1)}(tx)$ is a monotonely increasing function of x. Taylor's formula

$$f(x) = f(0) + f'(0)x + \cdots + \frac{f^{(n)}(0)}{n!}x^n + \left(\int_0^1 (1-t)^n f^{(n+1)}(tx)\,dt\right)\frac{x^{n+1}}{n!}$$
$$\stackrel{\text{def}}{=} f(0) + f'(0)x + \cdots + \frac{f^{(n)}(0)}{n!}x^n + R_n(x)$$

leads to the conclusion that if $0 \le x < c$ then

$$\begin{aligned}0 \le R_n(x) &\le \frac{x^{n+1}\int_0^1 (1-t)^{n+1} f^{(n+1)}(tc)\,dt}{n!}\\ &= \frac{x^{n+1}\left(f(c) - f(0) - f'(0)c - \cdots - \frac{f^{(n)}(0)c^n}{n!}\right)}{c^{n+1}}\\ &\le \frac{f(c)x^{n+1}}{c^{n+1}}\end{aligned}$$

whence for all x, $\lim_{n\to\infty} R_n(x) = 0$. □

3.60. a) Abel summation implies

$$\begin{aligned}&\left|\sum_{n=m}^{M} a_n \sin nx \sin\frac{x}{2}\right|\\ &\quad= \frac{1}{2}\left|a_m\cos\left(n - \frac{1}{2}\right)x\right.\\ &\qquad\left.+ \sum_{n=m}^{M}(a_{n+1} - a_n)\cos\left(n + \frac{1}{2}\right)x - a_M\cos\left(M + \frac{1}{2}\right)x\right|\\ &\quad\le \frac{1}{2}\left[a_m + \sum_{n=m}^{M-1}(a_n - a_{n+1}) + a_M\right] = a_m.\end{aligned}$$

If $x \in (0,\pi)$ and $n \stackrel{\text{def}}{=} \left[\frac{1}{x}\right]$ then when $m \le n$

$$|S_m(x)| \le \sum_{n=1}^{m} na_n x \le Kmx \le Knx \le K.$$

If $m > n$ then, since

$$\frac{1}{a_{n+1}}\sin\frac{1}{2}x \ge \frac{1}{a_{n+1}}\frac{x}{\pi} \ge \frac{1}{a_{n+1}(n+1)}\frac{1}{\pi} \ge \frac{1}{K\pi},$$

it follows that

$$|S_m(x)| \le |S_n(x)| + \left|\sum_{k=n+1}^{m} a_k \sin kx\right| \le K + \frac{a_{n+1}}{\sin\frac{x}{2}} \le K + K\pi.$$

Hence if $x \in (0,\pi]$, $|S_N(x)| \le K(1+\pi)$.

b) When $\theta \in (0, 2\pi)$, Euler's formula $e^{i\theta} = \cos\theta + \sin\theta$ and the formula for the sum of a geometric progression imply that there is a constant C_1 such that

$$\left|\sum_{n=1}^{N} \sin n\theta\right| \stackrel{\text{def}}{=} |K_N(\theta)| \le \frac{C_1}{|\sin(\theta/2)|}.$$

Abel summation implies that if $N \ge M$ and $0 < a \le \theta \le b < 2\pi$ then for some constant C_2,

$$\begin{aligned}|S_N(\theta) - S_M(\theta)| \le &\sum_{k=M+1}^{N} |a_k - a_{k+1}| \cdot |K_k(\theta)| \\ &+ |a_N| \cdot |K_N(\theta)| + |a_{M+1}| \cdot |K_{M+1}(\theta)| \\ \le &\, C_2 (|a_N| + |a_{M+1}|)\end{aligned},$$

whence convergence is uniform on $[a, b]$.

c) If $\sum_{n=1}^{N} \sin n\theta$ converges uniformly on $[0, 2\pi]$ as $N \to \infty$, $\theta = \frac{\pi}{2N}$, and $\epsilon > 0$ then for large M and all N in (M, ∞),

$$\frac{\epsilon}{2} > S_N(\theta) - S_M(\theta) \ge (N-M)a_N = \frac{N-M}{N} Na_N = \left(1 - \frac{M}{N}\right) Na_N.$$

Conversely, if $a_n \downarrow 0$ and $\lim_{n\to\infty} na_n = 0$ the conclusion in a) implies that it suffices to prove uniform convergence on $\left[0, \frac{\pi}{4}\right]$. Owing to pointwise convergence everywhere on $[0, 2\pi]$, consideration can be focussed on $\sum_{n=M+1}^{\infty} a_n \sin n\theta \stackrel{\text{def}}{=} R_M$ and $\epsilon_M \stackrel{\text{def}}{=} \max_{n \ge M}\{na_n\}$.

If $0 < \theta \le \frac{\pi}{4}$ then $\left[\frac{1}{\theta}\right] \stackrel{\text{def}}{=} \nu(\theta) \in \mathbb{N}$ and for M in $\mathbb{N}$,

$$R_M = \begin{cases} \sum_{n=M+1}^{\nu(\theta)} + \sum_{n=\nu(\theta)+1}^{\infty} a_n \sin n\theta \stackrel{\text{def}}{=} R'_M + R''_M & \text{if } M+1 \le \nu(\theta) \\ \sum_{n=M+1}^{\infty} a_n \sin n\theta \stackrel{\text{def}}{=} S_M & \text{if } M+1 > \nu(\theta) \end{cases}.$$

If $M + 1 \le \nu(\theta)$, the inequality $|\sin n\theta| \le |n\theta|$ implies

$$|R'_M| \le \theta \sum_{n=M+1}^{\nu(\theta)} na_n \le \frac{1}{\nu(\theta)} \epsilon_{M+1} (\nu(\theta) - M) \le \epsilon_{M+1}.$$

In sum, $R'_M \xrightarrow{u} 0$.

If $0 \le \theta \le \frac{\pi}{4}$ then $\sin\frac{\theta}{2} \ge \frac{\pi\theta}{8}$. Since

$$|K_n(\theta)| \le \frac{C_1}{|\sin(\theta/2)|},$$

if $M + 1 \le \nu(\theta)$ Abel summation reveals that for some constant C_3,

$$\begin{aligned}|R''_M| &\le \sum_{n=\nu(\theta)}^{\infty} (a_n - a_{n+1})\,|K_n(\theta)| + a_{\nu(\theta)}\left|K_{\nu(\theta)-1}(\theta)\right| \\ &\le 2C_1 a_{\nu(\theta)}\frac{8}{\pi\theta} \le \frac{16C_2}{\pi}\left(\nu(\theta)+1\right)a_{\nu(\theta)} < C_3\epsilon_M.\end{aligned}$$

If $M + 1 > \nu(\theta)$ the same kind of calculation reveals that for some constant C_4, $|S_M| \le C_4\epsilon_M$. Hence if $C \stackrel{\text{def}}{=} \max\{C_1, C_2, C_3, C_4\}$ then

$$\left|\sum_{n=M+1}^{\infty} a_n \sin n\theta\right| \le C\epsilon_m. \qquad \square$$

3.2. Complex-valued Functions

3.61. For n in $\mathbb{N}$ and j in $[0, n-1]$,

$$x_n\left(\frac{j+1}{n}\right) = x_n\left(\frac{j}{n}\right) + \frac{1}{n}f\left(x_n\left(\frac{j}{n}\right)\right)$$

and induction shows, since f is defined on $[0,1]$, that

$$0 \le x_n\left(\frac{j}{n}\right) \le 1, 0 \le x_n\left(\frac{j}{n}\right) + \frac{1}{n}f\left(x_n\left(\frac{j}{n}\right)\right) \le 1.$$

Direct calculation shows that

$$\left\{|t-s| < \frac{1}{2n}\right\} \Rightarrow \{|x_n(t) - x_n(s)| \le \|f\|_\infty |t-s|\},$$

whence $\{x_n\}_{n\in\mathbb{N}}$ is uniformly bounded and equicontinuous. The Arzelà-Ascoli theorem implies that there is a subsequence $\{x_{n_k}\}_{k\in\mathbb{N}}$ and a continuous function x such that $x_{n_k} \xrightarrow{u} x$.

The equicontinuity of $\{x_{n_k}\}_{n\in\mathbb{N}}$ and the uniform continuity of f imply that if $\eta > 0$, there is an n_0 and a positive δ such that if $n_k > n_0$ then

$$\{|t-s| < \delta\} \Rightarrow \{|f(x_{n_k}(t)) - f(x_{n_k}(s))| < \eta\}.$$

Hence if $n_k > \max\left\{\frac{1}{\delta}, n_0\right\}$ and $\frac{j}{n_k} \le s < t \le \frac{j+1}{n_k}$ then

$$\frac{x_{n_k}(t) - x_{n_k}(s)}{t-s} = f\left(x_{n_k}\left(\frac{j}{n_k}\right)\right)$$
$$f\left(x_{n_k}(s)\right) - \eta \le \frac{x_{n_k}(t) - x_{n_k}(s)}{t-s} \le f\left(x_{n_k}(s)\right) + \eta$$
$$f\left(x(s)\right) - 2\eta \le \frac{x(t) - x(s)}{t-s} \le f(x(s)) + 2\eta.$$

It follows that

$$f(x(s)) - 2\eta \le \underline{\lim}_{t=s} \frac{x(t)-x(s)}{t-s} \le \overline{\lim}_{t=s} \frac{x(t)-x(s)}{t-s} \le f(x(s)) + 2\eta$$

whence $x'(s)$ exists and $x'(s) = f(x(s))$.

Because $X(s) \stackrel{\text{def}}{=} \int_0^s f(x(t))\,dt$ and $x(s)$ are both solutions of

$$z' = f(z), z(0) = 0, \tag{s3.5}$$

the uniqueness theorem for differential equations implies $X = x$. If the entire sequence $\{x_n\}_{n\in\mathsf{N}}$ fails to converge to x, some subsequence converges uniformly to a different limit, say y. However both x and y are solutions of (s3.5). □

3.62. If $\{f_k\}_{k\in\mathsf{N}} \subset N_n$ and $\lim_{k\to\infty} \|f_k - f\|_\infty = 0$, then for each k and some x_k in $[0,1]$, if $h \ne 0$, $\left|\frac{f_k(x_k+h) - f_k(x_k)}{h}\right| \le n$. It may be assumed that for some x in $[0,1]$, $x - x_k \stackrel{\text{def}}{=} \delta_k \to 0$ as $k \to \infty$. Thus for large k, $\left|\frac{\delta_k + h}{h}\right|$ is near one and

$$\begin{aligned}\left|\frac{f(x+h) - f(x)}{h}\right| \le & \left|\frac{f(x+h) - f_k(x+h)}{h}\right| \\ & + \left|\frac{f_k(x_k+\delta_k+h) - f_k(x_k)}{\delta_k+h}\right| \cdot \left|\frac{\delta_k+h}{h}\right| \\ & + \left|\frac{f_k(x_k) - f(x_k)}{h}\right| + \left|\frac{f(x_k) - f(x)}{h}\right| \\ \stackrel{\text{def}}{=} & \text{ I+II+III+IV}.\end{aligned}$$

For any nonzero h and large k, I, III, and IV are small while II does not exceed a number near n. Hence $f \in N_n$ and so N_n is closed.

There is a piecewise linear continuous "sawtooth" function G for which at each point x, $\left|\frac{G(x+h) - G(x)}{h}\right| > n$ if h is sufficiently small while $\|G\|_\infty$

is arbitrarily small [GeO]. For any neighborhood $N(f)$ of an f in N_n, there is a polynomial p such that $p+G \in N(f)\setminus N_n$. Hence the interior N_n° of N_n is empty and so, as a closed set with no interior, N_n is nowhere dense. □

[**Note s3.7:** The set $A \stackrel{\text{def}}{=} C([0,1],\mathbb{C}) \setminus \left(\bigcup_{n\in\mathbb{N}} N_n\right)$ is of the second category and consists entirely of (continuous) nowhere differentiable functions.]

3.63. Write $f_n = \Re(f_n) + i\Im(f_n) \stackrel{\text{def}}{=} u_n + iv_n$. If $\|f_n\|_\infty \not\to 0$ as $n \to \infty$, it may be assumed that for some positive δ, some x_∞, and some sequence $\{x_n\}_{n\in\mathbb{N}}$, $u_n(x_n) \geq \delta$ and $\lim_{n\to\infty} x_n = x_\infty$. The mean value theorem and the hypothesis imply

$$|u_n(x) - u_n(x_\infty)| \leq |x - x_\infty|,$$

whence for large n and all x in $\left[x_\infty - \frac{\delta}{4}, x_\infty + \frac{\delta}{4}\right]$, $u_n(x) \geq \frac{\delta}{4}$.

If g is the piecewise linear function such that

$$g(x) = \begin{cases} 1 & \text{on } \left[x_\infty - \frac{\delta}{4}, x_\infty + \frac{\delta}{4}\right] \\ 0 & \text{off } \left[x_\infty - \frac{\delta}{2}, x_\infty + \frac{\delta}{2}\right] \end{cases}$$

then for large n, $\int_0^1 u_n(x)g(x)\,dx \geq \frac{\delta^2}{16}$, a contradiction. □

3.64. Because $f(0) = 0$, for some F in $C^1([-\pi,\pi],\mathbb{C})$,

$$F(x) = \begin{cases} f(x) & \text{if } x \in [0,\pi] \\ -f(-x) & \text{if } x \in [-\pi, 0]. \end{cases}$$

Integration by parts and the Schwarz inequality show that if

$$F(x) = \sum_{n=1}^{\infty} b_n \sin nx$$

[since $F \in C^1([-\pi,\pi],\mathbb{C})$, the right member converges everywhere to $F(x)$] then

$$b_n = \frac{1}{\pi n}\left(\int_{-\pi}^{0} f'(x)\cos nx\,dx + \int_0^\pi f'(x)\cos nx\,dx\right)$$

$$|b_n| \leq \frac{\|f'\|_2}{\sqrt{\pi} n}$$

$$2\int_0^\pi |f(x)|^2\,dx = \int_{-\pi}^{\pi} |F(x)|^2\,dx = \pi\sum_{n=1}^{\infty}|b_n|^2 \leq \left(\sum_{n=1}^{\infty}\frac{1}{n^2}\right)\int_0^\pi |f'(x)|^2\,dx$$

as required. □

3.65. The Fundamental Theorem of Calculus implies that if E is a bounded subset of $C^1([0,1],\mathbb{C})$ then E is equicontinuous. The Arzelà-Ascoli theorem implies $\overline{T(E)}$ is compact. □

3.66. a) Since $\| \ \|^{(1)} \geq \| \ \|_\infty$, X is closed in $C^1([0,1],\mathbb{C})$.

b) Because $\| \ \|_\infty \leq \| \ \|^{(1)}$, the inclusion map

$$T : C^1([0,1],\mathbb{C}) \hookrightarrow C([0,1],\mathbb{C})$$

restricted to X is a continuous bijection of Banach spaces and thus T is open, whence T^{-1} is continuous. Hence for some M, $\| \ \|^{(1)} \leq M\| \ \|_\infty$. Let k be $\frac{1}{M}$ and let K be 1.

c) The mean value theorem for derivatives implies that a bounded subset of $C^1([0,1],\mathbb{C})$ is equicontinuous. By virtue of **3.65**, the inclusion map T of b) above is compact, whence $\overline{T(E)}$ is compact. In particular, $\overline{T\left(B(\mathbf{O},1)^\circ\right)}$ is compact while, since T is open, $T\left(B(\mathbf{O},1)^\circ\right)$ is open. Since a Banach space is finite-dimensional iff it is locally compact, it follows that X is finite-dimensional. □

3.67. Since $f \in C^1(\mathbb{T},\mathbb{C})$, f is represented by its Fourier series:

$$f(z) = \sum_{n\in\mathbb{Z}\setminus\{0\}} a_n z^n.$$

Furthermore, the Fourier series for f' is $\sum_{n\in\mathbb{Z}\setminus\{0\}} n a_n z^{n-1}$, whence by virtue of Parseval's equation,

$$\|f'\|_2^2 = \sum_{n\in\mathbb{Z}\setminus\{0\}} n^2 |a_n|^2, \ \|f\|_2^2 = \sum_{n\in\mathbb{Z}\setminus\{0\}} |a_n|^2 \leq \|f'\|_2^2.$$

Equality obtains iff for n other than ± 1, $a_n = 0$. □

3.68. Note that $E_1 = L^2([0,1],\lambda)$, $D\left(C^1([0,1],\mathbb{C})\right) \subset E_1$, and for all f in $C^1([0,1],\mathbb{C})$, $\|D(f)\|' \leq \|f\|''$. Because $C^1([0,1],\mathbb{C})$ is dense in E_1, D may be extended to a linear continuous map $\widetilde{D} : E_2 \mapsto L^2([0,1],\lambda)$.

If $\widetilde{D}(f) = 0$ then for some sequence $\{f_n\}_{n\in\mathbb{N}}$ in $C^1([0,1],\mathbb{C})$,

$$\lim_{n\to\infty} \|f_n - f\|'' = \lim_{n\to\infty} \|D(f_n)\|' = 0.$$

The Schwarz inequality and the choice of the sequence $\{f_n\}_{n\in\mathbb{N}}$ imply

$$\lim_{n\to\infty} \int_0^1 |f_n(x)| \, dx = 0$$

$$\lim_{m,n\to\infty} \int_0^1 \left(|f_m(x) - f_n(x)|^2 + |f_m'(x) - f_n'(x)|^2\right) dx = 0.$$

It may be assumed that $\lim_{n\to\infty} f_n \doteq f$ and $\lim_{n\to\infty} f'_n \doteq 0$. The Fundamental Theorem of Calculus implies that for all x in $[0,1]$,

$$|f_m(0) - f_n(0)| \le |f_m(x) - f_n(x)| + \int_0^1 |f'_m(t) - f'_n(t)|\, dt.$$

Hence $\lim_{n\to\infty} f_n(0) \stackrel{\text{def}}{=} c$ exists and, since x may be chosen off the null set on which $\{f_n\}_{n\in\mathbb{N}}$ fails to converge, $f(x) \doteq c$ as required. □

3.69. a) The triangle inequality $|a+b| \le |a| + |b|$ for complex numbers shows that $D(0,1)$ is convex.

b) If $0 \le t_n \le t_{n+1}$ and $\sum_{n=1}^N t_i = 1$, the Euclidean structure of $\mathbb{R}^2$ yields the result if $N = 2$. If $N > 2$ and $0 < t_1 < 1$ then the formulæ

$$z' \stackrel{\text{def}}{=} \sum_{n=1}^N t_n z_n = t_1 z_1 + \left(\sum_{n=2}^N t_n\right) \sum_{n=2}^N \frac{t_n}{\sum_{n=2}^N t_n} z_n \stackrel{\text{def}}{=} t_1 z_1 + t'_2 z'_2$$

and mathematical induction reduce the argument to the case when $N = 2$. □

3.70. If $\sum_{n=1}^N a_n e^{inx} \equiv 0$ on $[a,b]$ then $f : \mathbb{C} \ni z \mapsto \sum_{n=1}^N a_n e^{inz}$ is an entire function vanishing on a compact infinite subset of $\mathbb{C}$. Hence $f(z) \equiv 0$ and thus $f(x) \equiv 0$ on $[0, 2\pi]$. The orthogonality relations

$$\int_0^{2\pi} e^{inx} e^{-imx}\, dx = \begin{cases} 0 & \text{if } m \ne n \\ 2\pi & \text{if } m = n \end{cases}$$

lead to the conclusion that $a_n \equiv 0$. □

3.71. If $A \stackrel{\text{def}}{=} (a_{ij})_{i,j=1}^{n,n}$ and $\det(A) = 0$, the result follows. Hence it may be assumed that $\det(A) \ne 0$, whence the rows of A are linearly independent vectors $\mathbf{a}_1, \ldots, \mathbf{a}_n$ in $\mathbb{C}^n$. The Gram-Schmidt orthonormalization process applied to $\{\mathbf{a}_i\}_{1\le i\le n}$ provides an orthonormal set $U \stackrel{\text{def}}{=} \{\mathbf{u}_i\}_{1\le i\le n}$. Furthermore, there is a lower triangular matrix Γ such that

$$\Gamma \begin{pmatrix} \mathbf{a}_1 \\ \vdots \\ \mathbf{a}_n \end{pmatrix} = \begin{pmatrix} \mathbf{u}_1 \\ \vdots \\ \mathbf{u}_n \end{pmatrix}. \tag{s3.6}$$

If $\mathbf{u}_i = (u_{i1}, \ldots, u_{in})$, (s3.6) may be written $\Gamma (a_{ij})_{i,j=1}^n = (u_{ij})_{i,j=1}^n$. The diagonal entries of Γ are $\gamma_{ii} \stackrel{\text{def}}{=} \left\|\mathbf{a}_i - \sum_{j=1}^{i-1} (\mathbf{a}_i, \mathbf{u}_j)\right\|^{-1} \ge \|\mathbf{a}_i\|^{-1}$. Because $(u_{ij})_{i,j=1}^{n,n}$ is a unitary matrix, $\left|\det\left[(u_{ij})_{i,j=1}^{n,n}\right]\right| = 1$, whence

$$|\det(A)| = \frac{1}{\prod_{i=1}^n |\gamma_{ii}|} \le \prod_{i=1}^n \|\mathbf{a}_i\|.$$ □

3.72. For m in $\mathbb{N}$ and z in $D(0,1)$, the functions

$$F_p(z) \stackrel{\text{def}}{=} \frac{f\left(\frac{p}{m}z\right)}{f\left(\frac{p-1}{m}z\right)},\ 1 \le p \le m,$$

are continuous and $|F_p(z)| \equiv 1, 1 \le p \le m$. For large m, and all z, because the numerator and denominator of the fraction defining F_p are near each other, $F_p(z) \neq -1, 1 \le p \le m$. Hence $F_p(D(0,1))$ is a connected compact subarc A_p of $\mathbb{T} \setminus \{-1\}$, i.e., there is a continuous function

$$\eta_p : D(0,1) \ni z \mapsto \eta_p(z) \in (-\pi, \pi)$$

such that $F_p(z) = e^{i\eta_p(z)} \in A_p$. Since

$$f(z) = f(0) \cdot \prod_{1 \le p \le m} \frac{f\left(\frac{p}{m}z\right)}{f\left(\frac{p-1}{m}z\right)},$$

it follows that if $f(0) \stackrel{\text{def}}{=} e^{i\eta_0}, -\pi < \eta_0 \le \pi$, and $\phi(z) \stackrel{\text{def}}{=} \eta_0 + \sum_{k=0}^{m} \eta_k(z)$ then $f(z) = e^{i\phi(z)}$. □

3.73. If $G(\theta) \stackrel{\text{def}}{=} g\left(e^{i(\theta+\pi)}\right) - g\left(e^{i\theta}\right)$ then $G(\pi) = g(1) - g(-1) = -G(0)$. Hence either $G(0) = 0$, $p = 1$, and $\widetilde{p} = -1$ or for some θ near π, $G(\theta)$ and $G(0)$ are of opposite sign, whence at some ψ in $[0, \pi)$, $G(\psi) = 0$. Thus if $p = e^{i\psi}$ and $\widetilde{p} = e^{i(\psi+\pi)}$ then $g(p) = g(\widetilde{p})$. □

3.74. Because $h|_{\mathbb{T}}$ maps $\mathbb{T}$ into itself, **3.73** implies that at least one of some pair of antipodal points is not left fixed. □

3.75. If no point is left fixed by f then for z in $D(0,1)$, the closed half-line determined by z and $f(z)$ and of which $f(z)$ is the endpoint meets $\mathbb{T}$ in a point $h(z)$. The map h is continuous and h leaves each point of $\mathbb{T}$ fixed, in contradiction of **3.74**. □

4
Measure and Topology

4.1. Borel Measures

4.1. a) Because $L(f) \in (u,v)$, if $y = m(x-a) + \phi(a)$ is the equation of a supporting line through $(a, \phi(a))$, for x in X,

$$m\left(f(x) - a\right) + \phi(a) \le \phi \circ f(x)$$
$$L\left(m\left(f(x) - a\right) + \phi(a)\right) = m(L(f) - L(f)) + \phi(L(f)) \le L(\phi \circ f).$$

b) If $(*)$ obtains for every simple Lebesgue measurable function f, assume $p, q \in \mathbb{R}$, $0 \le \alpha, \beta$, and $\alpha + \beta = 1$. For $f \stackrel{\text{def}}{=} p\chi_{(0,\alpha]} + q\chi_{[\alpha,1)}$,

$$\int_{[0,1]} f(x)\,dx = \alpha p + \beta q$$
$$\int_{[0,1]} \phi \circ f(x)\,dx = \int_0^\alpha \phi(p)\,dx + \int_\alpha^1 \phi(q)\,dx = \alpha\phi(p) + \beta\phi(q)$$
$$\phi\left(\int_{[0,1]} f(x)\,dx\right) = \phi\left(\alpha p + \beta q\right) \le \int_{[0,1]} \phi \circ f(x)\,dx = \alpha\phi(p) + \beta\phi(q).$$

□

[**Note s4.1:** The idea behind the original version of Jensen's inequality is that approximating Lebesgue sums for $\int_X f(x)\,d\mu(x)$ are, because $\mu(X) = 1$, convex combinations of values of f. The convexity of ϕ implies an inequality between ϕ applied to the Lebesgue sums for f and the Lebesgue sums for $\phi \circ f$. The continuity of ϕ permits a valid passage from the inequalities for the Lebesgue sums to the same inequalities for Lebesgue integrals.

The impact of b) is that Jensen's inequality when $X = [0,1]$ implies Jensen's inequality for any X as described in a).]

4.2. By abuse of language, the functional $\underline{\lim}$ resp. $\overline{\lim}$ is superadditive resp. subadditive and positive homogeneous, i.e., if $t \ge 0$ then

$$\underline{\lim}\,(f+g) \ge \underline{\lim}\, f + \underline{\lim}\, g$$
$$\overline{\lim}\,(f+g) \le \overline{\lim}\, f + \overline{\lim}\, g$$
$$t\,\underline{\lim}\, f = \underline{\lim}\, tf,\ t\,\overline{\lim}\, f = \overline{\lim}\, tf.$$

□

4.3. With respect to inclusion as a partial order, the set Λ of all finite subsets of $\mathbb{N}$ is a diset. If $\mu(X) = 1$, for $\lambda \in \Lambda$, $\mu(\lambda) \stackrel{\text{def}}{=} \max\{n : n \in \lambda\}$, and for $\{f_n\}_{n\in\mathbb{N}}$, $n_\lambda \stackrel{\text{def}}{=} f_{\mu(\lambda)}$, Jensen's inequality applies for the convex functionals $\overline{\lim}_{\lambda\in\Lambda} n_\lambda$ and $-\underline{\lim}_{\lambda\in\Lambda} n_\lambda$.

More generally, if $\{f_m\}_{m\in\mathbb{N}} \subset L^1(X,\mu)$ then $E_{\neq}(f_m, 0) \stackrel{\text{def}}{=} E_m$ is σ-finite: $0 < \mu(A_n) < \infty$ and $\dot{\bigcup}_{n\in\mathbb{N}} A_n = \bigcup_{m\in\mathbb{N}} E_m$. Then the conclusion in the previous paragraph applies and, owing to the positive homogeneity of $\overline{\lim}$ and $\underline{\lim}$, leads to:

$$\begin{matrix}\overline{\lim}_{k\to\infty} \\ \underline{\lim}_{k\to\infty}\end{matrix} \int_{A_n} f_k(x)\, d\mu(x) \begin{matrix}\le \\ \ge\end{matrix} \int_{A_n} \begin{matrix}\overline{\lim}_{k\to\infty} \\ \underline{\lim}_{k\to\infty}\end{matrix} f_k(x)\, d\mu(x).$$

Hence if $\underline{\lim}_{k\to\infty} f_k \in L^1(X,\mu)$ then

$$\begin{aligned}\underline{\lim}_{k\to\infty} \int_X f_k(x)\, dm(x) &\ge \sum_{n=1}^{\infty} \underline{\lim}_{k\to\infty} \int_{A_n} f_k(x)\, d\mu(x) \\ &\ge \sum_{n=1}^{\infty} \int_{A_n} \underline{\lim}_{k\to\infty} f_k(x)\, d\mu(x) = \int_X \underline{\lim}_{k\to\infty} f_k(x)\, d\mu(x).\end{aligned}$$

Mutatis mutandis, $\overline{\lim}_{k\to\infty} \int_X f_k(x)\, d\mu(x) \le \int_X \overline{\lim}_{k\to\infty} f_k(x)\, d\mu(x)$, [which is true as well if $\overline{\lim}_{k\to\infty} f_k \notin L^1(X,\mu)$]. □

4.4. a) If $0 < \epsilon_1 < \epsilon_2$ then $\{\text{diam}(U) < \epsilon_1\} \Rightarrow \{\text{diam}(U) < \epsilon_2\}$ whence, for fixed A, $\rho_\epsilon^p(A)$ is a monotonely increasing function of ϵ.

b) If $\delta > 0$, $\{U_{mn}\}_{m,n\in\mathbb{N}}$ is a double sequence of open sets, each of diameter less than ϵ, if $\bigcup_{m\in\mathbb{N}} U_{mn} \supset A_n, n \in \mathbb{N}$, and $\sum_{m=1}^{\infty} (\text{diam}(U_{mn}))^p < \rho_\epsilon^p(A_n) + \frac{\delta}{2^n}$, then

$$\begin{aligned}\rho_\epsilon^p\left(\bigcup_{n\in\mathbb{N}} A_n\right) &\le \sum_{m,n\in\mathbb{N}} (\text{diam}(U_{mn}))^p = \sum_{n=1}^{\infty}\left(\sum_{m=1}^{\infty} (\text{diam}(U_{mn}))^p\right) \\ &\le \sum_{n=1}^{\infty} \rho_\epsilon^p(A_n) + \delta \le \sum_{n=1}^{\infty} \rho^p(A_n) + \delta\end{aligned}$$

$$\rho^p\left(\bigcup_{n\in\mathbb{N}} A_n\right) \le \sum_{n=1}^{\infty} \rho^p(A_n).$$

c) It may be assumed that $0 < \ell(\gamma) < \infty$ and, by a change of scale, that $\ell(\gamma) = 1$. The parametrization of γ may be chosen so that $\ell(\gamma([0,t])) = t$, i.e., so that the parameter is arc-length.

In these circumstances, if $m \in \mathbb{N}$ and $1 \leq k \leq m-1$ then

$$d\left(\gamma\left(\frac{k}{m}\right), \gamma\left(\frac{k+1}{m}\right)\right) < \frac{1}{m}.$$

Hence $\bigcup_{r=1}^{m} B\left(\gamma\left(\frac{2r-1}{2m}\right), \frac{1}{2m-1}\right)^{\circ} \supset A$ and so $\rho^{1}_{\frac{1}{2m-1}}(A) \leq \frac{2m}{2m-1}$. As $m \to \infty$ there emerges the inequality $\rho^{1}(A) \leq 1$.

For some partition $Q \stackrel{\text{def}}{=} \{s_j\}_{0 \leq j \leq J}$ of $[0,1]$,

$$\sum_{j=0}^{J-1} d\left(\gamma\left(s_{j+1}\right), \gamma\left(s_j\right)\right) > 1 - \delta$$

and it may be assumed that $\min_j d\left(\gamma\left(s_{j+1}\right), \gamma\left(s_j\right)\right) \stackrel{\text{def}}{=} \Delta > 0$.

If $0 < \epsilon_1 < \max\left\{\frac{\delta}{4J}, \Delta\right\}$, then for some positive η, $|t' - t''| < \eta$ and $d\left(\gamma(t'), \gamma(t'')\right) < \epsilon_1$. If $|P| < \min\{\eta, |Q|\}$ then for some $k_1, k_2, \ldots$,

$$0 = t_0 = s_0 < t_1 < \cdot < t_{k_1} \leq s_1 < t_{k_1+1} < \cdot < t_{k_2} < s_2 < \cdot < t_K = s_J = 1.$$

Hence

$$\begin{aligned} 1 - \delta < \sum_j &\left(\sum_{k=k_j+1}^{k_{j+1}} d\left(\gamma\left(t_{k+1}\right), \gamma\left(t_k\right)\right)\right) \\ &+ \sum_j d\left(\gamma\left(s_j\right), \gamma\left(t_{k_j}\right)\right) + d\left(\gamma\left(t_{k_p+1}\right), \gamma\left(s_j\right)\right). \end{aligned}$$

The second sum in the right member above is less than $J \cdot 2\epsilon_1 < \frac{\delta}{2}$ and so the first (double) sum exceeds $1 - \frac{3\delta}{2}$. Because γ is a homeomorphism, there is a positive ϵ such that

$$\left\{d\left(\gamma\left(t'\right), \gamma\left(t''\right)\right) < \epsilon\right\} \Rightarrow \left\{|t' - t''| < \min\left\{\eta, \min_j \left|s_{j+1} - s_j\right|\right\}\right\}.$$

[In other words, if points corresponding on $\gamma([0,1])$ to successive partition points of $[0,1]$ are close to each other, the associated approximants to $\ell(\gamma)$ are close to $\ell(\gamma)$.]

For δ, ϵ as in the preceding paragraph and α positive, let $\{U_m\}_{m \in \mathrm{N}}$ be a sequence of open sets such that

$$A \subset \bigcup_{m \in \mathrm{N}} U_m, \ \operatorname{diam}\left(U_m\right) < \epsilon, \ m \in \mathbb{N}, \ \sum_{m=1}^{\infty} \operatorname{diam}\left(U_m\right) < \rho^{1}_{\epsilon}(A) + \alpha.$$

Since A is compact, for some M in $\mathbb{N}$, $A \subset \bigcup_{m=1}^{M} U_m$. Let t_0 be 0 and for some m not exceeding M, let V_0 be a U_m containing $\gamma(t_0)$. If $\{V_i\}_{0\le i\le k}$ have been defined so that each is a U_m, $m \le M$, $\gamma(t_i) \in V_i$, no two V_i are the same, and $t_k = 1$, the process stops. Otherwise, $t_{k+1} \stackrel{\text{def}}{=} \sup\{ t : \gamma(t) \in V_k \}$ and V_{k+1} is a $U_m, m \le M$, covering $\gamma(t_{k+1})$. Thus V_{k+1} is different from all the V_i previously defined. Since M is finite, there emerges a set $\{V_i\}_{1\le i\le K\le M}$ such that no two V_i are the same, $t_K = 1$, and $\max_i d(\gamma(t_{i+1}), \gamma(t_i)) < \epsilon$ whence $\sum_{i=0}^{K} d(\gamma(t_{i+1}), \gamma(t_i)) > 1-\delta$. On the other hand, $\sum_{i=0}^{K} d(\gamma(t_{i+1}), \gamma(t_i)) \le \sum_{i=0}^{K} \operatorname{diam}(V_i) < \rho_\epsilon^1(A) + \alpha$. Thus $1-\delta \le \rho^1(A) + \alpha$. □

4.5. It suffices to show $\rho_\epsilon^q(A) \le \epsilon^{q-p}\rho_\epsilon^p(A)$. If $\{U_m\}_{m\in\mathbb{N}}$ is a sequence of open sets such that $\bigcup_{m\in\mathbb{N}} U_m \supset A$, $\operatorname{diam}(U_m) < \epsilon$, $m \in \mathbb{N}$, it follows that $\rho_\epsilon^q(A) \le \sum_{m=1}^{\infty} (\operatorname{diam}(U_m))^q \le \epsilon^{q-p} \sum_{m=1}^{\infty} (\operatorname{diam}(U_m))^p$. □

[**Note s4.2:** Besicovitch **[Bes, GeO]** showed that in $\mathbb{R}^3$ there is a homeomorphic image S of $\partial B(\mathbf{O}, 1)$ and $\lambda_3(S)$ is large while the two-dimensional surface area of S is small. However it is impossible for $\rho^3(S)$ to be positive while $\rho^2(S)$ is small.]

4.6. If $E \subset \mathbb{R}^p$ then $\lambda_p^*(E) \le (\operatorname{diam}(E))^p$. Hence if $\{U_m\}_{m\in\mathbb{N}}$ is a sequence of open sets such that $\bigcup_{m\in\mathbb{N}} U_m \supset A$, $\operatorname{diam}(U_m) < \epsilon, m \in \mathbb{N}$, then $\lambda_p^*(A) \le \sum_{m=1}^{\infty} \lambda_p(U_m) \le \sum_{m=1}^{\infty} (\operatorname{diam}(U_m))^p$ whence $\lambda_p^*(A) \le \rho^p(A)$.

If $\lambda_p^*(A) = \infty$ then for any constant c_p, $\lambda_p^*(A) \ge c_p\rho^p(A)$. If $\lambda_p^*(A) < \infty$ and $\delta, \epsilon > 0$, then for some open U, $A \subset U$ and $\lambda_p^*(A) > \lambda_p(U) - \dfrac{\delta}{2}$. Furthermore, U is the countable union of pairwise disjoint half-open p-dimensional cubes K_m and so $\lambda_p(U) = \sum_{m=1}^{\infty} \lambda_p(K_m)$. For a sequence $\{\eta_m\}_{m\in\mathbb{N}}$ of positive numbers, each K_m is contained in an open cube L_m such that

$$\lambda_p(L_m) < \lambda_p(K_m) + \eta_m \text{ and } \sum_{m=1}^{\infty} \eta_m < \frac{\delta}{2}.$$

Consequently $\lambda_p(U) > \sum_{m=1}^{\infty} \lambda_p(L_m) - \dfrac{\delta}{2}$.

For an open cube L of side s,

$$\lambda_p(L) = s^p = p^{-\frac{p}{2}} \operatorname{diam}(L)^p \stackrel{\text{def}}{=} c_p \operatorname{diam}(L)^p.$$

$$\text{Hence } \lambda_p^*(A) > \lambda_p(U) - \frac{\delta}{2} > c_p \sum_{m=1}^{\infty} (\operatorname{diam}(L_m))^p - \delta \ge c_p\rho_\epsilon^p(A) - \delta. \quad \square$$

4.7. Because (cf. **4.4**) ρ^p is an outer measure, it is countably subadditive and it suffices to show $\rho^p(A \cup B) \ge \rho^p(A) + \rho^p(B)$.

If $0 < \epsilon < \frac{\delta}{3}$ and $\{U_m\}_{m\in\mathbb{N}}$ is a sequence of open sets such that

$$\bigcup_{m\in\mathbb{N}} U_m \supset A \cup B,\ \operatorname{diam}(U_m) < \epsilon,\ m \in \mathbb{N},$$

then no U_m meets both A and B. Hence there are disjoint subsequences

$$\{m_k\}_{1\le k<K\le\infty} \text{ and } \{m'_l\}_{1\le l'<L\le\infty}$$

such that $\bigcup_{1\le k<K} U_{m_k} \supset A$ and $\bigcup_{1\le l<L} U_{m'_l} \supset B$. For a positive η, the U_m can be chosen so that

$$\begin{aligned}\rho^p(A\cup B) &> \sum_{m=1}^{\infty} (\operatorname{diam}(U_m))^p - \eta\\ &\ge \sum_{1\le k<K} (\operatorname{diam}(U_{m_k}))^p + \sum_{1\le l<L} \left(\operatorname{diam}\left(U_{m'_l}\right)\right)^p - \eta\\ \rho^p(A\cup B) &\ge \rho^p_\epsilon(A) + \rho^p_\epsilon(B) - \eta.\end{aligned}$$

□

4.8. If $\rho^p(S) = \infty$ then the subadditivity of ρ^p implies

$$\rho^p(S) = \rho^p(S\cap F) + \rho^p(S\setminus F).$$

If $\rho^p(S) < \infty$, it suffices to show $\rho^p(S) \ge \rho^p(S\cap F) + \rho^p(S\setminus F)$. If, for m in $\mathbb{N}$, $U_m \stackrel{\text{def}}{=} \left\{x : d(x,F) > \frac{1}{m}\right\}$ then U_m is open, $U_m \subset U_{m+1}$, and $\bigcup_{m\in\mathbb{N}} U_m = X\setminus F$. Since $((S\setminus F)\setminus U_m) = \bigcup_{k=m+1}^{\infty} (S\cup(U_k\setminus U_{k-1}))$, it follows that

$$\rho^p((S\setminus F)\setminus U_m) \le \sum_{k=m+1}^{\infty} \rho^p(S\cap(U_k\setminus U_{k-1})). \tag{s4.1}$$

If k and k' are two integers of the same parity and greater than one, then $S\cap(U_k\setminus U_{k-1})$ and $S\cap(U_{k'}\setminus U_{k'-1})$ are a positive distance apart and thus (cf. **4.7**)

$$\begin{aligned}\sum_{k\le n,\, k \text{ odd}} \rho^p(S\cap(U_k\setminus U_{k-1})) &\le \rho^p(S)\\ \sum_{k\le n,\, k \text{ even}} \rho^p(S\cap(U_k\setminus U_{k-1})) &\le \rho^p(S)\\ \sum_{m=1}^{\infty} \rho^p(S\cap(U_m\setminus U_{m-1})) &\le 2\rho^p(S) < \infty.\end{aligned}$$

It follows from (s4.1) that

$$\lim_{m\to\infty} \rho^p\left((S\setminus F)\setminus U_m\right) = 0$$
$$\rho^p\left(S\cap U_m\right) \le \rho^p\left(S\setminus F\right) \le \rho^p\left(S\cap U_m\right) + \rho^p\left((S\setminus F)\setminus U_m\right)$$
$$\lim_{m\to\infty} \rho^p\left(S\cap U_m\right) = \rho^p\left(S\setminus F\right).$$

Because $\inf\{\, d(x,y) \;:\; x\in S\cap F, y\in S\cap U_m \,\} > 0$, from **4.7** it follows that $\rho^p(S) \ge \rho^p(S\cap F) + \rho^p(S\cap U_m)$ and $\rho^p(S) \ge \rho^p(S\cap F) + \rho^p(S\setminus F)$. □

4.9. From **4.8** it follows that all Borel sets are ρ^p-measurable and **4.6** implies λ_p and ρ^p are mutually absolutely continuous. Since ρ^p is translation-invariant, the Radon-Nikodým derivative $\dfrac{d\rho^p}{d\lambda_p}$ is a constant K_p. Hence, for any Borel set A, $\lambda_p(A) = K_p\rho^p(A)$.

For any set S,

$$\begin{aligned}\lambda_p^*(S) &= \inf\{\, \lambda_p(U) \;:\; U \text{ open},\ U\supset S\,\}\\ &= K_p \inf\{\, \rho^p(U) \;:\; U \text{ open},\ U\supset S\,\}\\ &\ge K_p\rho^p(S).\end{aligned}$$

If $\epsilon, \delta > 0$, there is a sequence $\{U_m\}_{m\in\mathsf{N}}$ of open sets such that

$$\bigcup_{m\in\mathsf{N}} U_m \supset S,\ \operatorname{diam}(U_m) < \epsilon$$
$$\begin{aligned}\rho^p(S) \ge \rho^p_\epsilon(S) &\ge \sum_{m=1}^{\infty} \left(\operatorname{diam}(U_m)\right)^p - \delta\\ &\ge \sum_{m=1}^{\infty} \rho^p_\epsilon\left(U_m\right) - \delta \ge \rho^p_\epsilon\left(\bigcup_{m\in\mathsf{N}} U_m\right) - \delta\\ &\ge \inf\{\, \rho^p_\epsilon(V) \;:\; V \text{ open},\ V\supset S\,\} - \delta.\end{aligned}$$

Hence for each n in $\mathbb{N}$, some open set V_n contains S and

$$\rho^p(S) \ge \rho^p_\epsilon\left(V_n\right) - \frac{1}{n} - \delta.$$

As first, $\epsilon\to 0$, then $\delta \to 0$, and finally $n\to\infty$, the result emerges. □

4.10. The set R of μ-regular Borel sets contains $\emptyset$: $\mathsf{R} \ne \emptyset$. Direct calculations using the formulæ of set algebra imply R is a ring.

If $\epsilon > 0, \mathsf{R} \ni A_n \subset A_{n+1} \subset U_{n+1} \in \mathsf{O}(X)$, and $\mu(U_n \setminus A_n) < \frac{\epsilon}{2^n}$ then

$$A \stackrel{\text{def}}{=} \bigcup_{n\in\mathbb{N}} A_n \subset \bigcup_{n=2}^{\infty} U_n \stackrel{\text{def}}{=} U \in \mathsf{O}(X),$$

$$U \setminus A \subset \bigcup_{n=2}^{\infty} (U_n \setminus A_n),$$

$$\mu(U \setminus A) < \epsilon.$$

If $\bigcup_{n=1}^{N} A_n \stackrel{\text{def}}{=} B_N$ then $B_N \in \mathsf{R}$, for large N, $\mu(A \setminus B_N) < \frac{\epsilon}{2}$, and B_N contains a compact K_N such that $\mu(B_N \setminus K_N) < \frac{\epsilon}{2}$. Hence

$$\mu(A \setminus K_N) \le \mu(A \setminus B_N) + \mu(B_N \setminus K_N) < \epsilon :$$

$A \in \mathsf{R}$.

If $\epsilon > 0, \mathsf{R} \ni D_n \supset D_{n+1} \supset K_{n+1} \in \mathsf{K}(X)$, and $\mu(D_n \setminus K_n) < \epsilon$, then $D \stackrel{\text{def}}{=} \bigcap_{n\in\mathbb{N}} D_n \supset \bigcap_{n=2}^{\infty} K_n \stackrel{\text{def}}{=} K \in \mathsf{K}(X)$ and $\mu(D \setminus K) < \epsilon$. If $\bigcap_{n=1}^{N} D_n \stackrel{\text{def}}{=} E_N$ then for large N, $E_N \in \mathsf{R}$, $\mu(E_N \setminus D) < \frac{\epsilon}{2}$, and E_N is contained in an open set U_N such that $\mu(U_N \setminus E_N) < \frac{\epsilon}{2}$. Hence $\mu(U_N \setminus D) \le \mu(U_N \setminus E_N) + \mu(E_N \setminus D) < \epsilon$; $D \in \mathsf{R}$.

Thus R is monotone, whence R is a σ-ring (cf. **1.1, 1.2**).

If an open ball $B(x,r)^\circ$ is σ-compact it is regular by virtue of the monotone convergence theorem. If $0 < s < r$ then $B(x,s)^\circ$ is also regular. Hence for each n in $\mathbb{N}$ and each k in K, there is a regular open ball $B(k, r_k)^\circ$ such that $0 < r_k < \frac{1}{n}$ and so for some finite set $\{k_i\}_{1\le i\le I}$ contained in K, $K \subset \bigcup_{i=1}^{I} B(k_i, r_{k_i})^\circ \stackrel{\text{def}}{=} U_n$ and U_n is regular. If $x \notin K$ then $\inf\{\, d(x,k) \,:\, k \in K\,\} \stackrel{\text{def}}{=} \delta > 0$ and for some n, $\frac{1}{n} < \delta$: $x \notin U_n$, i.e., $K = \bigcap_{n\in\mathbb{N}} U_n$. Since R is a σ-ring, K is regular: $\sigma\mathsf{R}(\mathsf{K}(X)) \subset \mathsf{R}$. □

4.11. Let $\{x_n\}_{n\in\mathbb{N}}$ be dense in X. For m in $\mathbb{N}$, $X = \bigcup_{n\in\mathbb{N}} B\left(x_n, \frac{1}{m}\right)$ and if $F_{mN} \stackrel{\text{def}}{=} \bigcup_{1\le n\le N} B\left(x_n, \frac{1}{m}\right)$ then $F_{mN} \subset F_{m,N+1}$, $\lim_{N\to\infty} F_{mN} = X$, and $\lim_{N\to\infty} \mu(F_{mN}) = \mu(X)$.

If $\epsilon \in (0, \mu(X))$ and $m \in \mathbb{N}$, for some N_m, $\mu(X \setminus F_{mN_m}) < \frac{\epsilon}{2^m}$ whence

$$\mu\left(\bigcap_{m\in\mathbb{N}} F_{mN_m}\right) \ge \mu(X) - \sum_{m=1}^{\infty} \mu(X \setminus F_{mN_m}) > \mu(X) - \epsilon.$$

If $\{y_n\}_{n\in\mathsf{N}} \subset \bigcap_{m\in\mathsf{N}} F_{mN_m} \stackrel{\text{def}}{=} K_\epsilon$ then: in some $B\left(y_{n_1}, \frac{1}{1}\right)$ there is an infinite subsequence $\{y_{n_k}\}_{k\in\mathsf{N}}$; ...; in some $B\left(y_{n_{\cdot_{\cdot_{\cdot_p}}}}, \frac{1}{p}\right)$ there is an infinite sub...subsequence $\left\{y_{n_{\cdot_{\cdot_{\cdot_{p_q}}}}}\right\}_{q\in\mathsf{N}}$; .... Furthermore,

$$\Sigma \stackrel{\text{def}}{=} \left\{y_{n_1}, y_{n_{r_2}}, \ldots, y_{n_{\cdot_{\cdot_{\cdot_{p_q}}}}}, \ldots\right\}$$

is a Cauchy sequence. Because X is complete, Σ converges and because K_ϵ is closed, Σ converges to a y in K_ϵ, whence K_ϵ is compact.

Because X is a metric space, every closed set is a G_δ. Since μ is finite, every closed set is outer regular. Furthermore, if F is closed then $F \cap K_\epsilon$ is a compact subset of F and $\mu(F \setminus K_\epsilon) < \epsilon$, i.e., every closed set is regular. The set R of regular sets is a σ-ring (cf. **4.10**) and contains all closed sets, whence every Borel set is regular, i.e., μ is regular. □

4.12. The result in **4.11** is applicable to μ restricted to any $B(\mathbf{O}, r)$, i.e., μ so restricted is regular.

For a Borel set E and a positive r, if $E_r \stackrel{\text{def}}{=} E \cap B(\mathbf{O}, r)$ then

$$E = \bigcup_{m\in\mathsf{N}} E_m = \bigcup_{m\in\mathsf{N}} \left(E_m \setminus \bigcup_{k<m} E_k\right).$$

If $\epsilon > 0$ then for each m, there is an open set U_m and a compact set K_m such that

$$U_m \supset \bigcup_{k<m} E_k \supset K_m$$

$$\mu(U_m) - \frac{\epsilon}{2^m} < \mu\left(E_m \setminus \bigcup_{k<m} E_k\right) < \mu(K_m) + \frac{\epsilon}{2^m}.$$

If $\mu(E) < \infty$ then $E_m \setminus \bigcup_{k<m} E_k$ is both inner and outer regular because μ restricted to $B(\mathbf{O}, m)$ is regular.

Hence $U \stackrel{\text{def}}{=} \bigcup_{m\in\mathsf{N}} U_m \supset E, E \supset \bigcup_{m=1}^M K_m \stackrel{\text{def}}{=} \widetilde{K}_M$, U is open, $\widetilde{K_M}$ is compact, and $\mu(U) - \epsilon < \mu(E) < \mu\left(\widetilde{K_m}\right) + 2\epsilon$, whence E is regular.

If $\mu(E) = \infty$ then E is outer regular and

$$\mu(E) = \lim_{m\to\infty} \mu(E_m) \le \lim_{m\to\infty} \mu\left(\bigcup_{k=1}^m K_k\right) + \epsilon,$$

whence E is also inner regular. □

4.13. As a compact metric space, X is separable and complete. According to **4.12**, μ is regular. Hence for each x there is a positive $\delta(x)$ and a finite set $\{x_k\}_{1\le k\le K}$ such that $\mu\left(B(x,\delta(x))^\circ\right) < \frac{\epsilon}{2}$ and $X \subset \bigcup_{k=1}^K B\left(x_k,\delta\left(x_k\right)\right)^\circ$. If $\delta = \frac{1}{2}\min_k\left\{\delta\left(x_k\right)\right\}$, if $\operatorname{diam}(E) < \delta$, and if $E \cap B\left(x_k, \frac{1}{2}\delta\left(x_k\right)\right)^\circ \neq \emptyset$ then $E \subset B\left(x_k, \frac{1}{2}\delta\left(x_k\right)\right)^\circ$ whence $\mu(E) < \epsilon$. □

4.14. For $\mathcal{U} \stackrel{\text{def}}{=} \{U : U \text{ open } \mu(U) = 0\}$ partially ordered by inclusion, let $\{U_\gamma\}_{\gamma\in\Gamma}$ be a maximal linearly ordered subset of $\mathcal{U}$ and let F be the closed set $X \setminus \bigcup_{\gamma\in\Gamma} U_\gamma$.

Because μ is regular and totally finite, if $\epsilon > 0$, there is an open set V containing F and $\mu(V) < \mu(F)+\epsilon$. Then $\{V, U_\gamma\}_{\gamma\in\Gamma}$ is an open cover of X, whence for some finite set $\{\gamma_k\}_{1\le k\le K}$, $X = V \cup \bigcup_{k=1}^K U_{\gamma_k}$. Since $\{U_\gamma\}_{\gamma\in\Gamma}$ is linearly ordered, for some k_0, $X = V \cup U_{\gamma_{k_0}}$ and

$$\mu(X) \le \mu(V) + \mu\left(U_{\gamma_{k_0}}\right) = \mu(V) + 0 < \mu(F) + \epsilon.$$

Since $F \subset X$, $\mu(X) = \mu(F)$ and $\mu\left(\bigcup_{\gamma\in\Gamma} U_\gamma\right) = 0$. If F_1 is closed and $\mu\left(X \setminus F_1\right) = 0$ then $X \setminus F_1 \stackrel{\text{def}}{=} U^* \in \mathcal{U}$.

If $F_1 \subset F$ then $U^* \supsetneqq \bigcup_{\gamma\in\Gamma} U_\gamma$ which denies the maximality of $\{U_\gamma\}_{\gamma\in\Gamma}$. Hence $F_1 \not\subset F$.

Because $f^{-1}(0)$ is closed, $f(x) \doteq 0$ iff $f^{-1}(0) \supset F$. □

4.15. If E is a Borel set and if $\epsilon > 0$, then for some compact set K_ϵ contained in E, $\nu(E) < \nu\left(K_\epsilon\right) + \epsilon$. If, furthermore, $E = \bigcup_{n\in\mathbb{N}} E_n$ and each E_n is a Borel set, for each n some open set U_n contains E_n and $\nu\left(E_n\right) > \nu\left(U_n\right) - \frac{\epsilon}{2^n}$. For all N, $\sum_{n=1}^N \nu\left(E_n\right) = \nu\left(\bigcup_{n=1}^N E_n\right) \le \nu(E)$ whence $\sum_{n=1}^\infty \nu\left(E_n\right) \le \nu(E)$. For some N_1, $\bigcup_{n=1}^{N_1} U_n \supset K_\epsilon$ and so

$$\begin{aligned}\nu(E) &\le \nu\left(K_\epsilon\right) + \epsilon \le \sum_{n=1}^{N_1} \nu\left(U_n\right) + \epsilon \\ &\le \sum_{n=1}^{\infty} \nu\left(U_n\right) + \epsilon \le \sum_{n=1}^{\infty} \nu\left(E_n\right) + 2\epsilon.\end{aligned}$$ □

4.16. A finite Borel partition π of X is a finite sequence $\{E_n\}_{n=1}^N$ of Borel sets such that $X = \bigcup_{n=1}^N E_n$. When $\sigma \stackrel{\text{def}}{=} \{x_n\}_{n=1}^N$ is finite, write $\sigma \sim \pi$ iff $x_n \in E_n, 1 \le n \le N$. If $\{\sigma_1, \pi_1\} \prec \{\sigma_2, \pi_2\}$, iff $\sigma_1 \prec \sigma_2$ in the natural partial order of sets and π_2 refines π_1, i.e., $\pi_1 \prec \pi_2$ in the natural partial order of partitions then $\{\{\sigma,\pi\} : \sigma \sim \pi\} \stackrel{\text{def}}{=} \Gamma \stackrel{\text{def}}{=} \{\gamma\}_{\gamma\in\Gamma}$ is partially ordered by $\prec$.

For two elements $\{\sigma_1,\pi_1\}$ and $\{\sigma_2,\pi_2\}$ of Γ, there is a π_3, a refinement of both π_1 and π_2, such that $\sigma_1 \cup \sigma_2 \stackrel{\text{def}}{=} \sigma_3 \sim \pi_3$ and $\{\sigma_3,\pi_3\} \succ \{\sigma_1,\pi_1\}$ and $\{\sigma_3,\pi_3\} \succ \{\sigma_2,\pi_2\}$. Thus Γ is a diset.

For each $\gamma = \left\{\{x_n\}_{n=1}^N, \{E_n\}_{n=1}^N\right\}$ and each Borel set E let $\mu_\gamma(E)$ be $\sum_{x_n\in E}\mu(E_n)$.

If $\mu(X) = 0$, the conclusion is automatic. If $\mu(X) > 0$ then for f in $C(X,\mathbb{C})$ and ϵ positive, there is a Borel partition $\{E_n\}_{n=1}^N$ of $f(X)$ and $\max_n\{\operatorname{diam}(E_n)\} < \dfrac{\epsilon}{2\mu(X)}$. If

$$\pi \stackrel{\text{def}}{=} \left\{\{f^{-1}(E_n)\}_{n=1}^N\right\},\ x_n \in f^{-1}(E_n),\ 1 \le n \le N, \text{ and } \sigma \stackrel{\text{def}}{=} \{x_n\}_{n=1}^N$$

then $\{\sigma,\pi\} \stackrel{\text{def}}{=} \gamma \in \Gamma$ and if $\{\sigma_1,\pi_1\} \stackrel{\text{def}}{=} \gamma_1 \succ \gamma$, let π_1 be $\{F_m\}_{m=1}^M$. Then

$$\begin{aligned}&\left|\int_X f(x)\,d\mu_{\gamma_1}(x) - \int_X f(x)\,d\mu(x)\right|\\ &\qquad \le \sum_{m=1}^M \left|\int_{F_m} f(x)\,d\mu_{\gamma_1}(x) - \int_{F_m} f(x)\,d\mu(x)\right|.\end{aligned}$$

Because $\gamma_1 \succ \gamma$, the mth summand in the right member above does not exceed $\dfrac{\epsilon\mu(F_m)}{2\mu(X)}$ and so the whole sum does not exceed ϵ. □

4.17. If μ_1 and μ_2 are regular complex measures, $\mu \stackrel{\text{def}}{=} \mu_1 - \mu_2$, and $\epsilon > 0$ then for some (infinite) Borel partition $\{E_n\}_{n\in\mathbb{N}}$ of a Borel set E,

$$|\mu|(E) < \sum_{n=1}^\infty |\mu_1(E_n) - \mu_2(E_n)| + \epsilon.$$

For each n and some compact set K_n contained in E_n,

$$|\mu_1|(E_n \setminus K_n) + |\mu_2|(E_n \setminus K_n) < \frac{\epsilon}{2^n}.$$

If $\sum_{n=N+1}^\infty |\mu_1(E_n) - \mu_2(E_n)| < \epsilon$ then

$$|\mu|(E) < \sum_{n=1}^N |\mu_1(K_n) - \mu_2(K_n)| + 3\epsilon \le |\mu|\left(\bigcup_{n=1}^N K_n\right) + 3\epsilon,$$

whence $|\mu|$ is inner regular.

A similar argument, mutatis mutandis, shows $|\mu|$ is outer regular.

□

4.18. If $\epsilon_1 \neq \epsilon_2$ then $\partial B(x,\epsilon_1) \cap \partial B(x,\epsilon_2) = \emptyset$ whence at most countably many $\partial B(x,\epsilon)$ have positive measure. Thus for each x, there is a sequence $\{\epsilon_n(x)\}_{n\in\mathbb{N}}$ such that $0 < \epsilon_n(x) < \dfrac{1}{n}$ and $\mu(\partial B(x,\epsilon_n(x))) = 0, n \in \mathbb{N}$.

For each n and some finite set $\{x_{np}\}_{p=1}^{P_n}$, $X = \bigcup_{p=1}^{P_n} B(x_{np},\epsilon_n(x_{np}))^\circ$. If V is open and $x \in V$ then for some positive ϵ, $B(x,\epsilon)^\circ \subset V$. If $\dfrac{2}{m} < \epsilon$ then for some p, $x \in B(x_{mp},\epsilon_m(x_{mp}))^\circ$. If $y \in B(x_{mp},\epsilon_m(x_{mp}))^\circ$ then $d(x,y) < \dfrac{2}{m}$, i.e., $B(x_{mp},\epsilon_m(x_{mp}))^\circ \subset B(x,\epsilon)^\circ \subset V$. Thus

$$\mathcal{B} \stackrel{\text{def}}{=} \left\{B(x_{np},\epsilon_n(x_{np}))^\circ\right\}_{\substack{1\le p\le P_n\\ n\in\mathbb{N}}} \stackrel{\text{def}}{=} \{U_k\}_{k\in\mathbb{N}}$$

is a base for the topology of X and $\mu(\partial(U_k)) \equiv 0$. □

4.19. If $\epsilon > 0$, $\mathsf{K}(X) \ni K \subset U$, and $\mu_1(K) > \mu_1(U) - \epsilon$ (cf. **4.11**) then for some f_K in $C(X,\mathbb{C})$, $K \prec f_K \prec U$. Thus

$$\mu_1(U) \ge \int_X f_K(x)\,d\mu_1(x) \ge \mu_1(U) - \epsilon,\ \mu_n(U) \ge \int_X f_K(x)\,d\mu_n(x)$$

$$\underline{\lim}_{n\to\infty}\mu_n(U) \ge \lim_{n\to\infty}\int_X f_K(x)\,d\mu_n(x) = \int_X f_K(x)\,d\mu_1(x) \ge \mu_1(U) - \epsilon$$

whence $\underline{\lim}_{n\to\infty}\mu_n(U) \ge \mu_1(U)$.

If $\mathsf{K}(X) \ni K \subset U$ then $\int_X f_K(x)\,d\mu_n(x) \ge \mu_n(K)$ and so

$$\overline{\lim}_{n\to\infty}\mu_n(K) \le \int_X f_K(x)\,d\mu_1(x) \le \mu_1(U).$$

Hence if V is open and $V \supset \overline{U}$ then $\mu_1(V) \ge \overline{\lim}_{n\to\infty}\mu_n(\overline{U})$. If W is open and $W \supset \partial U$ then $\mu_1(W) \ge \overline{\lim}_{n\to\infty}\mu_n(\partial U)$. Because μ_1 is regular

$$0 = \mu_1(U) \ge \overline{\lim}_{n\to\infty}\mu_n(\partial U) \ge 0,$$

i.e., $\mu_n(\partial U) \equiv 0$. Hence $\mu_n(\overline{U}) \equiv \mu_n(U)$ and $\mu_1(V) \ge \overline{\lim}_{n\to\infty}\mu_n(U)$. The regularity of μ_1 implies $\mu_1(\overline{U}) = \mu_1(U) \ge \overline{\lim}_{n\to\infty}\mu_n(U)$. □

4.20. If $\epsilon > 0$ then for each x, there is an open set $U(x)$ containing x,

$$\sup\{\,|f(y_1) - f(y_2)| \;:\; y_1, y_2 \in U(x)\,\} < \epsilon,$$

and (cf. **4.19**) $\mu(\partial U(x)) = 0$. For some finite N, $\bigcup_{n=1}^N U(x_n) = X$. If $A_n \stackrel{\text{def}}{=} U(x_n) \setminus \bigcup_{m=1}^{n-1} U(x_m)$ and $V_n \stackrel{\text{def}}{=} U(x_n) \setminus \bigcup_{m=1}^{n-1} \overline{U(x_m)}$ then

$$X = \bigcup_{n=1}^N A_n = \left(\bigcup_{n=2}^N V_n\right) \cup \left(\bigcup_{n=1}^{N-1}\left(\overline{U(x_n)} \setminus \bigcup_{m=1}^{n-1} U(x_m)\right)\right) \stackrel{\text{def}}{=} U \cup F,$$

U is open, and F is closed. Hence

$$\begin{aligned}\lim_{n\to\infty}\mu_n(U)+\lim_{n\to\infty}\mu_n(F)&=\lim_{n\to\infty}\mu_n(X)=\mu_1(X)\\&=\mu_1(U)+\mu_1(F)=\lim_{n\to\infty}\mu_n(U)+\mu_1(F)\end{aligned}$$

and $\lim_{n\to\infty}\mu_n(F)=\mu_1(F)=0$. Thus if $y_k\in V_k, 2\le k\le N$, then

$$\begin{aligned}&\left|\int_X f(x)\,d\mu_n(x)-\int_X f(x)\,d\mu_1(x)\right|\\&\quad\le\left|\int_U f(x)\,d\mu_n(x)-\int_U f(x)\,d\mu_1(x)\right|+\left|\int_F f(x)\,d\mu_n(x)\right|\\&\quad\le\sum_{k=1}^N\left|\int_{V_k} f(x)\,d\mu_n(x)-f(y_k)\,\mu_n(V_k)\right|\\&\qquad+\left|\sum_{k=1}^N f(y_k)\left(\mu_n(V_k)-\mu_1(V_k)\right)\right|\\&\qquad+\left|\sum_{k=1}^N\left(f(y_k)\,\mu_1(V_k)-\int_{V_k} f(x)\,d\mu_1(x)\right)\right|\\&\qquad+\left|\int_F f(x)\,d\mu_n(x)\right|.\end{aligned}$$

According to the choice and construction of the y_k and the V_k and since $\lim_{n\to\infty}\mu_n(F)=0$,

$$\overline{\lim}_{n\to\infty}\left|\int_X f(x)\,d\mu_n(x)-\int_X f(x)\,d\mu_1(x)\right|\le 2\epsilon\mu_1(X).\qquad\square$$

4.21. For each n, μ_n may be regarded as a continuous linear functional on $C(X,\mathbb{C})$. The uniform boundedness principle implies that there is a finite M such that for all n in $\mathbb{N}$, $\mu_n(X)\le M$, whence $\mu_1(X)\le M$. The argument in **Solution 4.20** applies. □

4.22. If $x\in E$ then x belongs to an interval of positive length and hence each component of E is a nondegenerate interval. Hence E is a Borel set that is the union of at most countably many intervals. □

4.23. a) Because $x\in\mathbb{I}_{\mathbb{R}}$, $[x]<x<[x]+1$, $\{x\}\in(0,1)$, and in $\mathbb{N}$ there is a least $n(x)$ such that $\frac{1}{n(x)}<\{x\}<\frac{2}{n(x)}$. If $\delta\stackrel{\text{def}}{=}\inf\{\,y\;:\;y\in\Delta(x)\,\}>0$ then $\delta\in(0,1)$, $\{\delta\}=\delta$, $\frac{1}{n(\delta)}<\delta<\min\left\{\frac{2}{n(\delta)},\frac{1}{n(\delta)-1}\right\}$ and

$$n(\delta)\delta-\delta<1<n(\delta)\delta<2,\ [n(\delta)\delta]=1,\ \{n(\delta)\delta\}=n(\delta)\delta-1<\delta,$$

a contradiction. Hence $\delta = 0$ and if $m, n \in \mathbb{N}$ and $K \stackrel{\text{def}}{=} \left[\frac{1}{\{mx\}}\right]$ then

$$x = [mx] + \{mx\},\ k\{mx\} < 1,\ kmx = k[mx] + k\{mx\},\ \{kmx\} = k\{mx\},$$

$k = 1, 2, \ldots, K$. Thus the K equally spaced numbers $\{mx\}, \ldots, K\{mx\}$ lie in $(0,1)$. For N in $\mathbb{N}$, there is an m such that $\{mx\} < \frac{1}{N}$ whence $K \geq N$. Hence $\overline{\Delta(x)} = [0,1]$. (For an alternative approach, cf. **Solution 4.36**.)

b) The proof is based on the following derivation of Weyl's criteria for equidistributivity.

For any sequence $S \stackrel{\text{def}}{=} \{x_n\}_{n\in\mathbb{N}}$ in $\mathbb{R}$, if $0 \leq a < b \leq 1$ then S is equidistributed iff when $f = \chi_{[a,b]}$,

$$\lim_{n\to\infty} \frac{\sum_{k=1}^{n} f(x_k)}{n} = b - a \left(= \int_0^1 f(x)\,dx \right).$$

It follows that S is equidistributed iff for any step-function f,

$$\lim_{n\to\infty} \frac{\sum_{k=1}^{n} f(x_k)}{n} = \int_0^1 f(x)\,dx.$$

If f is Riemann integrable and $\epsilon > 0$ there are step-functions f_1, f_2 such that $f_1 \leq f \leq f_2$ and $\int_0^1 f_2(x)\,dx - \int_0^1 f_1(x)\,dx < \epsilon$. Hence S is equidistributed iff for every Riemann integrable function f,

$$\lim_{n\to\infty} \frac{\sum_{k=1}^{n} f(x_k)}{n} = \int_0^1 f(x)\,dx.$$

If $\phi \in \mathcal{P} \stackrel{\text{def}}{=} \left\{\sum_{n=-N_1}^{1} + \sum_{n=1}^{N_2} a_n e^{2\pi i n x}\right\}_{N_1, N_2 \in \mathbb{N}}$ and S is equidistributed then

$$\lim_{n\to\infty} \frac{\sum_{k=1}^{n} \phi(x_k)}{n} = \int_0^1 \phi(x)\,dx = 0. \tag{s4.2}$$

If g is Riemann integrable, $\int_0^1 g(x)\,dx \stackrel{\text{def}}{=} A$, and $\epsilon > 0$ then:

i. there are step-functions u_1, u_2 such that $u_1 + \epsilon \leq g \leq u_2 - \epsilon$ and $\|u_2 - u_1\|_1 < 2\epsilon$;

ii. in $C^\infty([0,1], \mathbb{C})$ there are v_1, v_2 such that $v_1 \leq u_1, u_2 \leq v_2$, and $\|u_i - v_i\|_1 < 3\epsilon, i = 1, 2$;

iii. if $B_i \stackrel{\text{def}}{=} \int_0^1 v_i(x)\,dx, i = 1, 2$, $\mathcal{P}$ contains trigonometric polynomials ϕ_1, ϕ_2 such that $\|v_i - B_i - \phi_i\|_\infty < \epsilon$.

Hence $\left|\frac{\sum_{k=1}^n v_i(x_k)}{n} - B_i - \frac{\sum_{k=1}^n \phi_i(x_k)}{n}\right| < \epsilon$, $i = 1,2$. If (s4.2) obtains for all ϕ in $\mathcal{P}$ it follows that

$$\lim_{n\to\infty} \frac{\sum_{k=1}^n v_i(x_k)}{n} = B_i,\ i = 1,2.$$

However, $\frac{\sum_{k=1}^n v_1(x_k)}{n} \le \frac{\sum_{k=1}^n g(x_k)}{n} \le \frac{\sum_{k=1}^n v_2(x_k)}{n}$, and so

$$\int_0^1 g(x)\,dx = \lim_{n\to\infty} \frac{\sum_{k=1}^n g(x_k)}{n}.$$

In sum, S is equidistributed iff (s4.2) obtains for all ϕ in $\mathcal{P}$.

If x is irrational and if $x_k \stackrel{\text{def}}{=} \{kx\}, k \in \mathbb{N}$ then for each N in $\mathbb{N}$, the formula for the sum of a geometric series implies

$$\sum_{k=1}^n e^{2\pi i N x_k} = \sum_{k=1}^n e^{2\pi i N k x} = e^{2\pi i N x} \frac{e^{2\pi i n N x} - 1}{e^{2\pi i N x} - 1},$$

whence $\left|\frac{\sum_{k=1}^n e^{2\pi i N x_k}}{n}\right| \le \frac{2}{n\left|e^{2\pi i N x} - 1\right|} \to 0$ as $n \to \infty$. □

4.24. Let M be $\sup\{\mu(E) : E \in \mathsf{S}\}$. Then since S is a monotone class, $M \in \mu(\mathsf{S})$. If $M = 0$ then $\{M\} = \{\mu(\emptyset)\} = \mu(\mathsf{S})$. If $0 < t < M$ and

$$s \stackrel{\text{def}}{=} \inf\{\mu(E) : E \in \mathsf{S}, \mu(E) \ge t\}$$

then for some sequence $\{E_n\}_{n\in\mathbb{N}}$ in S, $E_n \supset E_{n+1}$, $\mu(E_n) \downarrow s$, and

$$\mu\left(\lim_{n\to\infty} E_n\right) \stackrel{\text{def}}{=} \mu(E) = s.$$

If $s > t$ then for some sequence $\{H_n\}_{n\in\mathbb{N}}$ in S, $H_n \subset H_{n+1} \subset E$, $\mu(H_n) \le t$, and

$$r \stackrel{\text{def}}{=} \sup\{\mu(L) : L \in \mathsf{S}, L \subset E, \mu(L) \le t\} = \mu\left(\lim_{n\to\infty} H_n\right) \stackrel{\text{def}}{=} \mu(H).$$

If $s - r > 0$, $A \in \mathsf{S}$, and $A \subset (E \setminus H)$ then $\mu(A) = 0$ or $\mu(A) = s - r$, i.e., $E \setminus H$ is an atom, in denial of the hypothesis. Hence $s = r = t$. □

4.25. Fix f_0 in S and let K be $\text{supp}(f_0)$. Since K is compact, for some g in L, $K \prec g$. Let $I(g)$ be M, which may be taken as positive.

For ϵ positive and s in S, let U_s be the open set $\{x : s(x) < \epsilon\}$. If $x \in K$ then either $f_0(x) = 0$, in which case $x \in U_{f_0}$, or $f_0(x) > 0$, in which case, by hypothesis, for some f in S, $f(x) < f_0(x) \wedge \epsilon$ and thus $x \in U_f$.

In sum, $K \subset \bigcup_{f\in S} U_f$. Because K is compact, there is a finite subset $\{f_1, \ldots, f_n\}$ of S such that $K \subset \bigcup_{i=1}^{n} U_{f_i}$. If $f \in S$ then

$$h_f \stackrel{\text{def}}{=} f \wedge f_0 \wedge \cdots \wedge f_n \in S,\ h_f\big|_{U_f\setminus K} = 0, \text{ and } h_f\big|_K < \epsilon = \epsilon g\big|_K.$$

Thus $I(h_f) < \epsilon M$ whence $\inf\{ I(f) \ :\ f \in S\} = 0$. □

4.26. Each f in L_u is lsc (cf. **2.105**). If $a \in X$, $\mathcal{N}_a$ is the set of neighborhoods of a, and $f \in S$ then

$$\begin{aligned}\underline{\lim}_{x=a} F(x) &= \sup_{N\in\mathcal{N}_a} \inf_{x\in N} F(x)\\ &\geq \sup_{N\in\mathcal{N}_a} \inf_{x\in N} f(x)\\ &= \underline{\lim}_{x=a} f(x) = f(a)\end{aligned}$$

whence $\underline{\lim}_{x=a} F(x) \geq F(a)$, F is lower semicontinuous, and thus $F \in L_u$ (cf.**2.105**).

Because I is nonnegative, $A \leq I(F)$. For

$$T \stackrel{\text{def}}{=} \{g \ :\ g \in L, \text{ for some } f \text{ in } S\ g \leq f\},$$

$F = \bigvee\{g \ :\ g \in T\}$. If $A < I(F)$ then for some h in L, $h \leq F$ and $I(h) > A$. But $h = h \wedge F = h \wedge \bigvee\{g \ :\ g\in T\} = \bigvee\{h\wedge g \ :\ g \in T\}$. Since $\{h\} \cup T \subset L$, the set $\{h - h\wedge g \ :\ g \in T\}$ is one to which **4.25** applies. Hence

$$\begin{aligned}A < I(h) &= \sup\{ I(h \wedge g) \ :\ g \in T\} \leq \sup\{ I(g) \ :\ g \in T\}\\ &= \sup\{ I(f) \ :\ f \in S\} = A,\end{aligned}$$

a contradiction. □

4.27. Since I is nonnegative, $\mu^*(U) \geq \sup\{\mu^*(K) \ :\ K \in \mathsf{K},\ K \subset U\}$. However χ_U is lsc, whence $\chi_U \in L_u$ and, according to **2.105** and **4.25**,

$$\mu^*(U) = \mu(U) = I(\chi_U) = \sup\{ I(f) \ :\ f \in L,\ f \leq \chi_u\}.$$

If $\alpha < \mu(U)$ then for some f in L, $\alpha < I(f)$ and $0 \leq f \leq \chi_U$. Furthermore, for n in $\mathbb{N}$, $K_n \stackrel{\text{def}}{=} \left\{x \ :\ f(x) \geq \frac{1}{n}\right\}$ is compact,

$$\begin{gathered}K_n \subset K_{n+1} \subset U,\ \bigcup_{n\in\mathsf{N}} K_n = \{x \ :\ f(x) > 0\} \stackrel{\text{def}}{=} W \in \mathsf{O}, \text{ and}\\ \chi_{K_n} \uparrow \chi_W,\ \mu(K_n) = I(\chi_{K_n}) \uparrow I(\chi_W) \geq I(f) > \alpha,\end{gathered}$$

whence for some n, $\mu(K_n) > \alpha$.

Finally,

$$\sup\{\mu^*(V) : V \in \mathsf{O}, \mathsf{K} \ni \overline{V} \subset U\} \le \sup\{\mu^*(K) : \mathsf{K} \ni K \subset U\}.$$

However if $\mathsf{K} \ni K \subset U$ then for some V in O, $U \supset \overline{V} \supset V \supset K$. □

4.28. Because $\mu^*(E) \le \inf\{\mu^*(U) : U \in \mathsf{O},\ E \subset U\}$, only the reversed inequality needs attention, and then only if $\mu^*(E) < \infty$. By definition, for some sequence $\{g_n\}_{n\in\mathbb{N}}$ in L_u, $g_n \ge \chi_E$ and $I(g_n) \downarrow \mu^*(E)$. Since each g_n is lsc, if $\epsilon > 0$, the set $U_n \stackrel{\text{def}}{=} \{x : g_n(x) > 1-\epsilon\}$ is an open set containing E and $I(g_n) \ge \mu(U_n) \ge \mu^*(E)$. □

4.29. a) Because for f in L, $\mathbf{1} \wedge f \in L^1$, it follows that $\mathbf{1} \in \mathcal{M}$ and hence that A is a σ-algebra. If $U \in \mathsf{O}$ then $\chi_U \in L_u$ (**2.105**). Hence for all f in L, $\chi_U \wedge f \in L \subset L^1$ whence $\chi_U \in \mathcal{M}$. In light of the equivalence of Caratheodory and Daniell measurability, $U \in \mathsf{A}$. Since A is a σ-algebra, it follows that $\mathsf{O} \cup \mathsf{F} \subset \mathsf{A}$.

b) If $E \in \mathsf{O}$, $\mu^*(E) = \mu(E)$,(cf. **4.28**).

c) If $E \in \mathsf{O}$ then a) implies E is measurable and **4.28** implies the desired conclusion. If $\mu(E) < \infty$ then $f \stackrel{\text{def}}{=} \chi_E \in L^1$ and if $\epsilon > 0$, there are h and g such that $h \le f \le g$, h is usc, g is lsc, and

$$0 \le I(g-h) = I(g) - I(h) < \epsilon$$

(cf. **1.32**). Because h is usc, so is $k \stackrel{\text{def}}{=} h \vee 0$. It follows that $0 \le k \le f \le g$, $0 \le I(g-k) = I(g) - I(k) < \epsilon$. For some sequence $\{r_n\}_{n\in\mathbb{N}}$ in L, $r_n \downarrow k$ and $L \ni r_n \wedge 0 \stackrel{\text{def}}{=} k_n \downarrow k$. If $0 < \alpha < 1$ then $K \stackrel{\text{def}}{=} \{x : k(x) \ge \alpha\}$ is closed and for n in $\mathbb{N}$, $K \subset E \cap \{x : k_n(x) \ge \alpha\}$, whence K is a compact subset of E. As $\alpha \downarrow 0$ $\mu(K) \uparrow \mu(E)$. □

4.30. If $\{K_n\}_{n\in\mathbb{N}} \subset \mathsf{K}(X)$, $X = \bigcup_{n\in\mathbb{N}} K_n$, and $E \in \mathsf{A}$ then

$$E \cap K_n \stackrel{\text{def}}{=} E_n \in \mathsf{A}, \bigcup_{n\in\mathbb{N}} E_n = E, \text{ and } \mu(E_n) < \infty.$$

From **4.29**c) it follows that when $\epsilon > 0$, E_n contains a compact set B_n such that $\mu(B_n) > \mu(E_n) - \frac{\epsilon}{2^n}$. It follows that for large N, the measure of the compact set $\bigcup_{n=1}^N B_n$ is close to the measure of E.

If (X, A, μ) is σ-finite and $E \in \mathsf{A}$ then E is the countable union of measurable sets of finite measure. Hence **4.29**c) applies. □

4.31. *i.* If $\lambda(E) = 0$ then for n in $\mathbb{N}$, E can be covered by a sequence S_n of intervals of total length 2^{-n}. The set of all intervals in all the S_n is itself a sequence of intervals, the sum of all their lengths is not more than one, and each point of E belongs to at least one interval in each S_n.

Conversely, if E can be covered in the manner described and $\epsilon > 0$ then removal of finitely many of the intervals can bring the length-sum of the others below ϵ while their union covers E.

ii. Let $\{r_n\}_{n\in\mathbb{N}}$ be an enumeration of $\mathbb{Q}\cap[0,1]$. For k in $\mathbb{N}$, cover r_n with an open interval I_{kn} contained in $(0,1)$ and of length not exceeding $2^{-(k+n)}$. Then $U_k \stackrel{\text{def}}{=} \bigcup_{n\in\mathbb{N}} I_{kn}$ is open, $U_k \supset \mathbb{Q}\cap[0,1]$, and $\lambda(U_k) \le 2^{-k}$. Hence $E \stackrel{\text{def}}{=} \bigcap_{k\in\mathbb{N}} U_k$ is a G_δ contained and dense in $[0,1]$, $\lambda(E) = 0$, and E is a set of the second category. Because the complement of a dense G_δ is a set of the first category, E must meet every dense G_δ.

On the other hand, if f is Riemann integrable then $\text{Cont}(f)$, which is a G_δ, must be dense [otherwise $\lambda(\text{Discont}(f)) > 0$]. Thus E meets $\text{Cont}(f)$.

iii. Cover E in the manner described in *i* and let the set of all intervals used be S. For each interval (a,b) in S, form the function

$$f_{ab} : \mathbb{R} \ni x \mapsto \begin{cases} 0 & \text{if } -\infty < x \le a \\ x-a & \text{if } a \le x \le b \\ b-a & \text{if } b \le x < \infty \end{cases}.$$

Then $\sum_{(a,b)\in S} f_{ab}(x) \stackrel{\text{def}}{=} f(x)$ is majorized by the sum of the lengths of all the intervals in S whence f is continuous and monotonely increasing.

If $x \in E$ let I_k be the intersection of the first k intervals containing x. If $y \in I_k \setminus \{x\}$, the ratio $\dfrac{f_{ab}(y) - f_{ab}(x)}{y-x}$ is 1 for each of the k intervals (a,b) used to create I_k and the ratio is nonnegative everywhere since each f_{ab} is monotonely increasing. Hence $\dfrac{f(y)-f(x)}{y-x} \ge k$ and so f' exists at no point x of E. □

4.32. If $a \in \text{Cont}(f)$ and $f_n(a) \not\to f(a)$ then for some positive ϵ and some subsequence $\{f_{n_k}\}_{k\in\mathbb{N}}$

$$f_{n_k}(a) \ge f(a) + \epsilon \text{ or } f_{n_k}(a) \le f(a) - \epsilon.$$

In the first instance, for some positive δ, if $a \le x < a+\delta$ then $f(x) < f(a) + \dfrac{\epsilon}{2}$ while

$$f_{n_k}(x) \ge f_{n_k}(a) \ge f(a) + \epsilon \ge f(a) + \frac{\epsilon}{2} > f(x).$$

Thus $\lambda(\{x : f(n_k(x) \ge f(x)\}) \ge \delta, k \in \mathbb{N}$, a contradiction. The second instance is handled, mutatis mutandis. □

4.2. Haar Measure

4.33. If $f, g, h \in C_{00}(G, \mathbb{C})$ Fubini's theorem, Hölder's inequality, the Hahn-Banach theorem, and the translation-invariance of μ imply

$$\begin{aligned}\int_G |f * g(x)\overline{h(x)}|\, d\mu(x) &\leq \int_G |f(y)| \left(\int_G |g(y^{-1}x)| \cdot |h(x)|\, d\mu(x)\right) d\mu(y) \\ &\leq \int_G |f(y)|\, d\mu(y) \cdot \|g\|_p \cdot \|h\|_{p'} = \|f\|_1 \cdot \|g\|_p \cdot \|h\|_{p'} \\ \|f * g\|_p &\leq \|f\|_1 \cdot \|g\|_p. \end{aligned} \tag{s4.3}$$

Since $C_{00}(G, \mathbb{C})$ is $\| \ \|_1$-dense resp. $\| \ \|_p$-dense in $L^1(G, \lambda)$ resp. $L^p(G, \lambda)$, (s4.3) is valid for f in $L^1(G, \lambda)$ and g in $L^p(G, \mu)$. □

4.34. The function

$$F \stackrel{\text{def}}{=} \chi_E * \chi_{E^{-1}} : G \ni x \mapsto \int_G \chi_E(y^{-1}x)\, \chi_{E^{-1}}(y)\, d\mu(y).$$

is continuous, $\text{supp}(F) \subset EE^{-1}$, and $F(e) = \mu(E) > 0$. Hence on some $N(e)$, $F(x) > 0$. Thus $N(e) \subset \text{supp}(F) \subset EE^{-1}$. □

4.35. If $G = \mathbb{R}$ and, for the Cantor set C_0, $E \stackrel{\text{def}}{=} \frac{1}{2}(C_0 \cup (-C_0))$ then $\mu(E) = \lambda(E) = 0$, $E = -E$, and $E + (-E) = E + E$, which consists of all real numbers for which a ternary representation is $\sum_{n=1}^{\infty} \frac{\delta_n}{3^n}$, $\delta_n = 0, \pm 1, \pm 2$. Hence $E + E = [-1, 1]$. □

4.36. Because $\mathbb{T}$ is compact, if G is not finite, G contains a sequence $\{g_n\}_{n\in\mathbb{N}}$ such that $\left\{h_n \stackrel{\text{def}}{=} g_n g_{n+1}^{-1}\right\}_{n\in\mathbb{N}}$ is an infinite sequence converging to $e\ (= 1)$: for each $N(e)$, $N(e) \cap (G \setminus \{e\}) \neq \emptyset$. The set $\{\, h_n^k \ :\ k, n \in \mathbb{N}\,\}$ is a dense subset of G. □

4.37. The formula $a^x \cdot a^y = a^{x+y}$ shows that G is a (sub)group. If $x \in \mathbb{Q}$ then for some n in $\mathbb{N}$, $nx \in \mathbb{Z}$, whence G is finite. If $x \in \mathbb{I}_{\mathbb{R}}$ then for n in $\mathbb{N}$, $nx \notin \mathbb{Z}$, whence G is infinite and (cf. **4.36**) is dense. □

4.38. Owing to **4.36**, $\overline{G} = \overline{H} = \mathbb{T}$. If $x \in G$ then $x^2 \in H$, whence $G : H = 2$. □

4.39. *i.* If $\xi \in G \setminus H$ then $H \cap \xi H = \emptyset$ and $G = H \dot{\cup} \xi H$. Furthermore, if $SS^{-1} \cap \xi H \neq \emptyset$, i.e., if $\rho_i \in R, b_i \in H, x_i = \rho_i b_i$, $i = 1, 2$, and $x_1 x_2^{-1} \in \xi H$, then $\rho_1 \rho_2^{-1} \in \xi H \subset G$ and so, owing to the nature of R, $\rho_1 = \rho_2$. Thus $x_1 x_2^{-1} = b_1 b_2^{-1} \in H$, i.e., $x_1 x_2^{-1} \in \xi H \cap H = \emptyset$, a contradiction. It follows that $SS^{-1} \cap \xi H = \emptyset$. If L is a measurable subset of S and $\mu(L) > 0$ then $SS^{-1} \supset LL^{-1}$, which contains a $\mathbb{T}$-neighborhood of e (cf. **4.34**) and thus an element of the dense set ξH, a contradiction. It follows that the inner measure of S is zero: $\tau_*(S) = 0$.

For x in $\mathbb{T}$ and some ρ in R, $x\rho^{-1} \stackrel{\text{def}}{=} a \in G$. If $x \notin S$ then $a \notin H$, whence for some b in H, $x = \rho\xi b \in R\xi H = \xi S$. Thus

$$\mathbb{T} \setminus S \stackrel{\text{def}}{=} S^c \subset \xi S$$

and so $\tau_*(S^c) = 0$. The inner measure τ_* and outer measure τ^* are set functions such that for each measurable set M,

$$\tau^*(M \cap S) + \tau_*(M \cap S^c) = \tau(M),$$

whence $\tau^*(M \cap S) = \tau(M)$, in particular, $\tau^*(S) = 1 > 0 = \tau_*(S)$.

ii. The set $w^{-1}(S) \stackrel{\text{def}}{=} \widetilde{S}$ in $\mathbb{R}$ has properties analogous to those of S.

a) The set $\widetilde{S}$ is nonmeasurable, $\lambda_*\left(\widetilde{S}\right) = 0$, and $\lambda^*\left(\widetilde{S}\right) = \infty$.

b) The set $\widetilde{S}$ is thick and for every measurable subset M of $\mathbb{R}$,

$$\lambda_*(M \cap S) = 0 \text{ while } \lambda^*(M \cap S) = \lambda(M).$$

iii. If $p \in P$ then $-p+E = -p+(p+E) = (-p+p)+E = E$, whence P is a subgroup of $\mathbb{R}$. Because $P^\bullet \neq \emptyset$, P is dense in $\mathbb{R}$. If $\lambda(E) > 0$ then for some n in $\mathbb{Z}$, $\lambda(E \cap [n, n+1)) \stackrel{\text{def}}{=} r > 0$, whence for all p in P, $\lambda(p + (E \cap [n, n+1))) = \lambda(E \cap ([p+n, p+n+1)))$ Since P is dense in $\mathbb{R}$, choose a p in $P \cap (1, 2]$. Then the sets $kp + (E \cap [kp+n, kp+n+10)$, $k \in \mathbb{N}$, are pairwise disjoint, of measure r, and contained in E. □

4.40. If $E \in 2^{\mathbb{R}} \setminus \mathsf{S}_\lambda$, $S \stackrel{\text{def}}{=} (E \times \mathbb{R}) \cup \mathbb{R}$, and $P, Q \in S$, there is a polygonal path lying in S, connecting P to Q, and consisting of no more than three line segments.

If $y \neq 0$, the section $S_y \stackrel{\text{def}}{=} \{x : \{x, y\} \in S\} = E$ and Fubini's theorem implies $S \notin \mathsf{S}_\lambda$. □

4.41. Because $\bigcup_{n\in\mathbb{N}} h_n R = G$, if $R \in \widetilde{\mathsf{S}}$ and $\mu(R) = 0$ then $\mu(G) = 0$, a contradiction. If $\mu(R) > 0$ then RR^{-1} contains some $N(e)$ (cf. **4.34**). Since H is dense in G, some h in $RR^{-1} \cap H$ is not e. But then there are in R elements r_1, r_2 such that $r_1 = hr_2$, which, owing to the nature of R, implies $r_1 = r_2$, i.e., $h = e$, a contradiction. □

[**Note s4.3:** The existence of R in **4.41** is a consequence of the Axiom of Choice. Solovay **[Sol]** showed if

Every subset of $\mathbb{R}$ *is Lebesgue measurable*

is adjoined to the Zermelo-Fraenkel axiom system (ZF) **[Me]** for set theory, the result is as consistent as ZF itself. Because the

Axiom of Choice leads to a denial of Solovay's axiom, the existence of nonmeasurable subsets of $\mathbb{R}$ cannot be deduced in ZF alone.]

4.42. If $H \in \widetilde{\mathsf{S}}$ and $\mu(H) = 0$ then G, as the countable union of cosets of H is a null set, a contradiction.

If $\mu(H) > 0$ and: a) G is compact then $\mu(G) = \infty$, a contradiction; b) G is connected then $HH^{-1} = H$ contains some $N(e)$ whence H is open, hence also closed, a contradiction. □

4.43. Since an infinite proper subgroup contains a countable infinite subgroup, the proof for **4.42** carries over, mutatis mutandis. □

4.44. If $n = 0$ then since G contains a dense, proper, infinite subgroup H, e.g., $\mathbb{Q}^m$, the quotient group $G/H \overset{\text{def}}{=} K$ is a vector space for which there is a Hamel basis $\{k_\lambda\}_{\lambda\in\Lambda}$. If $\{g_\lambda\}_{\lambda\in\Lambda}$ is a subset of G and such that $g_\lambda/H = k_\lambda$, then $\{g_\lambda\}_{\lambda\in\Lambda}$ is $\mathbb{Q}$-linearly independent and its $\mathbb{Q}$-span is a subgroup R that is a complete set of coset representatives of H. Hence **4.41** applies.

If $m = 0$ then since $\mathbb{T}^n = \mathbb{R}^n/\mathbb{Z}^n$, $K \overset{\text{def}}{=} \mathbb{Q}^n/\mathbb{Z}^n$ is a dense countable infinite subgroup of the compact group $\mathbb{T}^n$. Hence **4.40** applies.

If $mn \neq 0$, there are nonmeasurable subgroups in each factor of G and their Cartesian product is, by virtue of Fubini's theorem, a nonmeasurable subgroup of G. □

4.45. In **4.44**, when $m = 0, n = 1$ and $K \overset{\text{def}}{=} \mathbb{Q}/\mathbb{Z}$, the corresponding nonmeasurable subgroup R of $\mathbb{T}$,then $\phi^{-1}(R) \overset{\text{def}}{=} S$ is a proper subgroup of G. For $\{k_m\}_{m\in\mathbb{N}}$ an enumeration of K and $\phi(g_m) \overset{\text{def}}{=} k_m$, if $x \in G$ then for some m in $\mathbb{N}$ and some r in R, $\phi(x) = rk_m$. Hence

$$\phi\left(xg_m^{-1}\right) = rk_m k_m^{-1} = r, \text{ i.e., } x \in Sg_m$$

and so $G = \bigcup_{m\in\mathbb{N}} Sg_m$. Thus if $S \in \widetilde{\mathsf{S}}$, there emerge the contradictions $\mu(G) = 0$ or $\mu(G) = \infty$ if G is compact resp. $S = G$ if G is connected. □

[**Note s4.4:** The question of the existence in a locally compact group G of a subgroup H not in the completion $\widetilde{\mathsf{S}}$ of S is subtle. Because S is $\sigma\mathsf{R}(\mathsf{K}(G))$, the latter may fail to contain G itself. In **[HeR]**, the structure theorem for abelian groups is used to show that every infinite compact abelian group contains a nonmeasurable subgroup.

In work yet to be submitted for publication the writer establishes the same result via the use of maximal free subsets of groups. Furthermore, it is shown that if G is an infinite connected locally compact group or if G is the Cartesian product of a set (finite or infinite) of finite-dimensional unitary groups $U(n, \mathbb{C})$ then G contains a nonmeasurable subgroup.]

4.46. a) Haar measure is regular **[Halm]**, whence E contains compact sets K_n such that $\mu(E) = \mu\left(\bigcup_{n\in\mathbb{N}} K_n\right)$, i.e., $E \setminus \left(\bigcup_{n\in\mathbb{N}} K_n\right)$ is a null set.

b) The group $\mathbb{R}_d$, i.e., $\mathbb{R}$ in the discrete topology, is a locally compact group that is not σ-compact because every σ-compact set in $\mathbb{R}_d$ is countable. □

4.47. If $N \in \mathcal{N}(e)$ then the group H generated by $U \stackrel{\text{def}}{=} N \cap N^{-1}$, i.e., $\bigcup_{n\in\mathbb{N}} U^n$, is open, hence closed, whence $H = G$. □

4.48. a) If $\mu(H) > 0$ then $HH^{-1} = H$ contains some $N(e)$ whence H is open and closed, a contradiction. b) Because H is the countable union of compact sets, $H \in \mathsf{S}$. Hence a) applies.

c) From a) it follows that H is open whence e/H is open. □

4.49. a) If E is a compact subset of K then $\phi^{-1}(E)$ is closed. If $\{U_\lambda\}_{\lambda\in\Lambda}$ is an open cover of $\phi^{-1}(E)$, it may be assumed that each $\overline{U_\lambda}$ is compact. Then $\{\phi(U_\lambda)\}_{\lambda\in\Lambda}$ is an open cover of E and thus contains a finite subcover $\{\phi(U_{\lambda_n})\}_{n=1}^N$ whence $F \stackrel{\text{def}}{=} \bigcup_{n=1}^N \overline{U_{\lambda_n}}$ is a compact set and $H \stackrel{\text{def}}{=} \phi\left(F \cap \phi^{-1}(E)\right)$ is a compact set such that $\phi(H) = E$. Hence $\mathsf{T} = \sigma\mathsf{R}(\mathsf{K}(H)) \subset \phi(\mathsf{S})$.

b) Since G is σ-compact, $G \in \mathsf{S}$. If U is open in G and $\{K_n\}_{n\in\mathbb{N}}$ is a sequence of compact sets such that $G = \bigcup_{n\in\mathbb{N}} K_n$ then

$$U = \bigcap_{n\in\mathbb{N}} U \cap (G \setminus K_n)$$

whence $\mathsf{O}(G) \subset \mathsf{S}$. If V is open in H then $\phi^{-1}(V)$ is open in G and thus $\mathsf{O}(H) \subset \mathsf{T}$: $\mathsf{S} = \sigma\mathsf{A}(\mathsf{O}(G)), \mathsf{T} = \sigma\mathsf{A}(\mathsf{O}(H))$.

c) If $G = \mathbb{R}_d$ and $H = \mathbb{R}_d/\mathbb{Z}$ then neither G nor H is σ-compact but $\mathsf{T} = \mathsf{S}$.

d) If $G = \mathbb{R}^2$, $H = \mathbb{R}$, $\mu = \lambda_2$, $\nu = \lambda$, $\phi : \mathbb{R}^2 \ni \{x, y\} \mapsto x \in \mathbb{R}$, and $E \stackrel{\text{def}}{=} \{\{x, x\} \ : \ 0 \le x \le 1\}$, then $\mu(E) = 0$ and $\nu(\phi(E)) = 1$.

e) For the situation in d), if F is a nonmeasurable subset of $[0, 1]$ and E is $\phi^{-1} \cap \{\{x, y\} \ : \ x = y\}$ then $\phi(E) \notin \widetilde{\mathsf{T}}$. □

4.50. If $\phi : G \ni x \mapsto \phi(x) \stackrel{\text{def}}{=} x/H \in G/H = K$ is the quotient map then G and K are Haar sets, $E \subsetneq K$, and $F \stackrel{\text{def}}{=} \phi^{-1}(E)$ is a subgroup of G. From **4.49**c) it follows that F is not measurable. □

4.51. Because $\nu(yE) = \mu(yEx) = \mu(Ex) = \nu(E)$, ν is a Haar measure. Furthermore,

$$\begin{aligned}\mu(Euv) = \Delta(uv)\mu(E) = \Delta(v)\mu(Eu) = \Delta(u)\Delta(v)\mu(E)\\ \Delta(uv) = \Delta(v)\Delta(u) = \Delta(u)\Delta(v) \in (0,\infty),\end{aligned}$$

whence $\Delta : G \mapsto (0,\infty)$ is a homomorphism of G into the multiplicative group $\mathbb{R}^+ \stackrel{\text{def}}{=} (0,\infty)$.

If $f \in C_{00}(G,\mathbb{C})$ and $x \in G$, the change of variable $xa \to y$ implies

$$\int_G f^{[a]}(x)\,d\mu(x) = \int_G \Delta(a)f(y)\,d\mu(y) = \Delta(a)\int_G f(x)\,d\mu(x).$$

Since f is uniformly continuous it follows that Δ is continuous. □

4.52. Since $\mu(Gx) = \mu(G) = \Delta(x)\mu(G)$ it follows that $\Delta(x) \equiv 1$. □

4.53. a) If $u \in H$ then, owing to the left-invariance of ρ,

$$F(xu) = \int_H f(xuy)\,d\rho(y) = \int_H f\left(xuu^{-1}y\right)\,d\rho(y) = F(x).$$

The continuity of F follows from the continuity of f. Furthermore, because $\text{supp}(f)$ is compact, so is $\text{supp}(F)$ $[= \text{supp}(f)/H]$.

b) If $v \in G$ and $v/H \stackrel{\text{def}}{=} w$ $(\in K)$,

$$\begin{aligned}\int_K\left(\int_H f(vxy)\,d\rho(y)\right)\,d\tau(z) &= \int_K F(wz)\,d\tau(z) = \int_K F(z)\,d\tau(z)\\ &= \int_K\left(\int_H f(xy)\,d\rho(y)\right)\,d\tau(z),\end{aligned}$$

i.e., $I : C_{00}(G,\mathbb{C}) \ni f \mapsto \int_K\left(\int_H f(xy)\,d\rho(y)\right)\,d\tau(z)$ is a translation-invariant functional. Owing to Daniell theory (cf. **Section 1.2**) and the uniqueness up to proportionality of Haar measure, there is a constant k such that $I(f) = k\int_G f(x)\,d\mu(x)$. It may be assumed that $k = 1$, whence symbolically, $d\mu(x) = d\rho(y)d\tau\,(xy/H)$.

c) If $u \in H$ then u/H is the identity of K and from b) it follows that $d\mu(xu) = \Delta(u)d\mu(x) = d\rho(yu)d\tau\,(xuyu/H) = \delta(u)d\rho(y)d\tau\,(xy/H)$ and $\Delta(u) = \delta(u)$. □

4.54. Because Δ is a continuous homomorphism, as the kernel of Δ, U is a closed normal subgroup. For a unimodular normal subgroup V and its modular function η, $1 = \eta = \Delta\big|_V$ whence $V \subset U$. □

4.55. Since G is compact, it contains a limit point and so every point in G is limit point. Because G is complete, G is perfect. If G is separable then

G is metric and then $\#(G) = \mathfrak{c}$. If G is not separable, it contains infinite separable closed subgroups. $\square$

4.56. It may be assumed that $M = 1$. If μ_2 denotes product measure on S^2 and $a \in S$ then $\mu_2\left(\theta\left((Sa)^2\right)\right) = \int_S \mu(xSa)\, d\mu(x) = 1$. For $r : S^2 \ni \{x, y\} \mapsto \{y, x\} \in S^2$, it follows that $\mu_2\left(r\theta\left((Sa)^2\right)\right) = 1$ and furthermore, $\theta\left((Sa)^2\right) \cap r\theta\left((Sa)^2\right) \neq \emptyset$. Thus $(Sa)^2$ contains pairs $\{xa, ya\}$, $\{ua, va\}$ such that $\{xa, xaya\} = \{uava, ua\}$, i.e., $xa = uava$, $x = uav$, $xaya = uavaya = ua$, $a(vaya) = a$. Thus $e \stackrel{\text{def}}{=} vaya \in Sa$ and for z in Sa, $ze = z$. Furthermore, $zez = z^2$ whence $ez = z$. Thus for any a in S, Sa contains an e that is a left and right identity. If $x \in S$ then $xa \in Sa, xa = exa, x = ex$ whence e is an identity for S. Thus for some w in S, $wa = e$, i.e., S is a group **[GeK]**. $\square$

[**Note s4.5:** The existence of μ as described together with the fact that S is a group implies that S may be endowed with its Weil topology **[We2]**.]

4.57. Let $\mathcal{F}$ be $\{A : A \subset S, A \text{ is closed}, AS \subset S\}$. Because $S \in \mathcal{F}$, it follows that $\mathcal{F} \neq \emptyset$. When $\mathcal{F}$ is partially ordered by reversed inclusion: $A_1 \prec A_2$ iff $A_1 \supset A_2$., a maximal chain $\mathcal{C} \stackrel{\text{def}}{=} \{A_\gamma\}_{\gamma\in\Gamma}$ enjoys the finite intersection property. Thus $\bigcap_{\gamma\in\Gamma} A_\gamma \stackrel{\text{def}}{=} A \neq \emptyset$: A is a minimal element of $\mathcal{F}$ and $AS \subset A$. If $a \in S$ then $aS \subset A, (aS)S \subset AS$. Since A is minimal, $aS = A$. If $x \in S$ and $a \in A$ then $ax \in A$, $axS = A = aS$, and the cancellation law implies $xS = S$. Similarly, $Sx = S$ and so, for all a, b in S, the equations $ax = b$ and $xa = b$ have solutions (unique by virtue of the cancellation law). Thus S is a group.

If x is near y then xy^{-1} is near $yy^{-1} = e$, whence $x^{-1}\left(xy^{-1}\right) = y^{-1}$ is near $x^{-1}e = x^{-1}$. Hence $x \mapsto x^{-1}$ is continuous **[GKO]**. $\square$

[**Note s4.6:** The hypotheses about compactness and the cancellation law cannot be dropped. For example, $[0, 1]$ is a compact semigroup with respect to multiplication, fails to conform to the cancellation law, and is not a group; $(0, 1]$ with respect to multiplication is a topological semigroup, conforms to the cancellation law, but is not compact [even though its completion $[0, 1]$ is compact].]

5
Measure Theory

5.1. Measure and Integration

5.1. Let μ be a finite measure. If μ is positive and

$$M \stackrel{\text{def}}{=} \sup\{\,\mu(E) \;:\; E \text{ measurable}\,\}$$

then for some sequence $\{E_n\}_{n\in\mathsf{N}}$ of measurable sets, $\mu(E_n) \uparrow M$. It may be assumed that $\{E_n\}_{n\in\mathsf{N}}$ is a monotonely increasing sequence, whence $\lim_{n\to\infty} E_n \stackrel{\text{def}}{=} E$ is measurable and $M = \lim_{n\to\infty}\mu(E_n) = \mu(E) < \infty$.

If μ is complex then $|\mu|$ is finite, $|\mu(E)| \le |\mu|(E)$, and the previous argument applies. □

5.2. a) Since $\sum_{n=1}^{\infty} |a_n| < \infty$, if $E \in 2^{\mathsf{N}}$ then E is empty, a finite set, or a countably infinite set and for any measurable partition $\{E_n\}_{n\in\mathsf{N}}$ of E, $\sum_{n=1}^{\infty}\mu(E_n)$ converges. Furthermore, the triangle inequality for the addition of complex numbers implies $|\mu|(E) = \sum_{n\in E} |a_n|$.

b) The finite additivity of μ on the ring of finite subsets of $\mathbb{N}$ is a consequence of definition. If $\sum_{n=1}^{\infty} a_n$ is conditionally convergent, then for some permutation $\pi : \mathsf{N} \mapsto \mathsf{N}$, $\sum_{n=1}^{\infty} a_{\pi(n)}$ fails to converge whence $\mu(\mathsf{N})$ is not defined. □

5.3. a) If $\mathsf{M} \stackrel{\text{def}}{=} \{\,E \;:\; E \in \sigma\mathsf{R}(\mathsf{R}),\, \mu_1(E) = \mu_2(E)\,\}$ then M is a monotone class containing R whence $\mathsf{M} \supset \sigma\mathsf{R}(\mathsf{R})$.

b) If $\mathsf{R} \stackrel{\text{def}}{=} \mathsf{R}(\{\,A \;:\; A = [a,b) \cap \mathbb{Q}\,\})$ then for all A in R, $\#(A) = \aleph_0$ unless $A = \emptyset$. If $\mu_1 \stackrel{\text{def}}{=} \zeta$ (counting measure) and $\mu_2 \stackrel{\text{def}}{=} 2\mu_1$ then $\mu_1 = \mu_2$ on R. However, if $r \in \mathbb{Q}$ then $\{r\} = \bigcap_{n\in\mathsf{N}} \left[r, r + \frac{1}{n}\right)$ whence $\{r\} \in \sigma\mathsf{R}(\mathsf{R})$ and $\mu_1(\{r\}) = 1 < \mu_2(\{r\}) = 2$. □

5.4. If $a_n \stackrel{\text{def}}{=} \frac{1}{2^n} + i\frac{1}{3^n}, n \in \mathbb{N}$, $E \subset \mathbb{N}$, and $\mu(E) \stackrel{\text{def}}{=} \sum_{k\in E} a_k$ then $(\mathsf{N}, 2^{\mathsf{N}}, \mu)$ is a complex measure situation. Furthermore, $\mathbb{N} = \bigcup_{n\in\mathsf{N}}\{n\}$, while $|\mu(\mathsf{N})| < \sum_{n=1}^{\infty} |\mu(\{n\})|$. □

5.5. Decompose μ into its real an imaginary parts and those into their positive and negative parts: $\mu = \mu_1 + i\mu_2 = \mu_1^+ - \mu_1^- + i\left(\mu_2^+ - \mu_2^-\right)$. If $A \in \mathsf{S}$ and $A \subset E$ then $E = A \cup (E \setminus A)$ and $|\mu|(E) \ge |\mu(A)|$. Hence $\nu(E) \le |\mu|(E)$.

On the other hand, if $\{E_n\}_{n\in\mathbb{N}}$ is a measurable partition of E then

$$\sum_{n=1}^{\infty}|\mu(E_n)| \le \sum_{n=1}^{\infty}\left(\mu_1^+(E_n)+\mu_1^-(E_n)+\mu_2^+(E_n)+\mu_2^-(E_n)\right)$$
$$|\mu|(E) \le \mu_1^+(E)+\mu_1^-(E)+\mu_2^+(E)+\mu_2^-(E).$$

If $P_i^{\pm}$ are Hahn decompositions for $\mu_i^{\pm}$, $i=1,2$, then

$$|\mu|(E) \le \mu_1\left(E\cap P_1^+\right)+\mu_1\left(E\cap P_1^-\right)+\mu_2\left(E\cap P_2^+\right)+\mu_2\left(E\cap P_2^-\right)$$
$$\le 4\nu(E).$$

□

5.6. Since $\mu\left(\bigcup_{k=n}^{\infty}E_k\right) \le \sum_{n=k}^{\infty}\mu(E_k) \downarrow 0$, $\mu\left(\overline{\lim}_{n\to\infty}E_n\right)=0$. □

5.7. The map $\nu : \mathsf{S} \ni E \mapsto \mu\left(f^{-1}(E)\right)$ is a finite measure and $\nu \ll \mu$. For some integrable h, $\nu(E)=\int_E h(x)\,d\mu(x)$ and if g is a bounded measurable function, $\int_X g(x)\,d\nu(x)=\int_X g(x)h(x)\,d\mu(x)$.

Since $\chi_{f^{-1}(E)}(x)=\chi_E\circ f(x)$, if $E\in\mathsf{S}$ then

$$\int_X \chi_E(x)h(x)\,d\mu(x)=\int_X \chi_E(x)\,d\nu=\nu(E)=\mu\left(f^{-1}(E)\right)$$
$$=\int_X \chi_{f^{-1}(E)}(x)\,d\mu(x)=\int_X \chi_E\circ f(x)\,d\mu(x).$$

The approximation properties of simple functions and the dominated convergence theorem apply. □

5.8. **First proof.** Fatou's lemma implies $\int_X f(x)\,d\mu(x)=0$, whence $f(x)=0$ a.e.

Second proof. If $F\in\mathsf{S}$ then $\lim_{n\to\infty}\int_F f_n(x)\,d\mu(x)=0$. Hence if $f(x)>0$ on a set of positive measure, there is a positive δ and a set E of finite positive measure on which $f(x)\ge\delta$. Egorov's theorem implies that E contains a subset F such that $\mu(F)>\frac{\delta}{2}$ and $f_n \xrightarrow{u} f$ on F. But then $\lim_{n\to\infty}\int_F f_n(x)\,d\mu(x)\ne 0$, a contradiction.

Third proof. It may be assumed that $\{f_n\}_{n\in\mathbb{N}}\subset L^1(X,\mathbb{C})$. The hypothesis implies $\lim_{n\to\infty}\|f_n\|_1=0$. □

5.9. a) The hypotheses imply $\sup_{n\in\mathbb{N}}\|f_n\|_1 \stackrel{\text{def}}{=} M<\infty$. Hence if $\epsilon>0$ and $E_\epsilon \stackrel{\text{def}}{=} \{x : x\in E, |f_n(x)-f_0(x)|\ge\epsilon\}$ then

$$\int_E |f_n(x)-f_0(x)|\,dx \le \int_{E_\epsilon}+\int_{E\setminus E_\epsilon}|f_n(x)-f_0(x)|\,dx;$$

for large n, the first term in the right member above does not exceed $2\epsilon M$ and the second term does not exceed $\epsilon\lambda(E)$.

b) Since $g_n \overset{\text{def}}{=} f_n \wedge f \overset{\text{a.e.}}{\to} f \wedge f = f$, the dominated convergence theorem implies that if $E \in \mathsf{S}$ then

$$\overline{\lim}_{n\to\infty} \int_E f_n(x)\,d\mu(x) \geq \lim_{n\to\infty} \int_E g_n(x)\,d\mu(x) = \int_E f(x)\,d\mu(x).$$

If $\epsilon > 0$ and $\overline{\lim}_{n\to\infty} \int_E f_n(x)\,d\mu(x) = \int_E f(x)\,d\mu(x) + \epsilon$ then, via passage to subsequences as needed,

$$\lim_{n\to\infty} \int_E f_n(x)\,d\mu(x) = \int_E f(x)\,d\mu(x) + \epsilon \text{ and}$$

$$\lim_{n\to\infty} \int_{X\setminus E} f_n(x)\,d\mu(x) = \overline{\lim}_{n\to\infty} \int_{X\setminus E} f_n(x)\,d\mu(x).$$

Hence

$$\begin{aligned}\lim_{n\to\infty} \int_X f_n(x)\,d\mu(x) &= \lim_{n\to\infty} \int_E f_n(x)\,d\mu(x) + \lim_{n\to\infty} \int_{X\setminus E} f_n(x)\,d\mu(x)\\ &\geq \int_E f(x)\,d\mu(x) + \epsilon + \int_{X\setminus E} f(x)\,d\mu(x),\end{aligned}$$

a contradiction. □

5.10. If $1 > \epsilon > 0$, $E_n \overset{\text{def}}{=} \{x : |f_n(x)| \geq \epsilon\}$, and

$$A_n \overset{\text{def}}{=} \int_X \frac{|f_n(x)|}{1 + |f_n(x)|}\,d\mu(x)$$

then $A_n \leq \int_{E_n} \frac{|f_n(x)|}{1+|f_n(x)|}\,d\mu(x) + \epsilon\mu(X) \leq \mu(E_n) + \epsilon\mu(X)$. If $f_n \overset{\text{meas}}{\to} 0$ then $\lim_{n\to\infty} A_n \leq \epsilon\mu(X)$.

Conversely, if $A_n \to 0$ as $n \to \infty$ and $\mu(E_n) \not\to 0$ as $n \to \infty$, there are positive δ and ϵ and a sequence $\{n_k\}_{k\in\infty}$ such that $n_k \to \infty$ as $k \to \infty$ and $\mu(E_{n_k}) \geq \delta$. Furthermore,

$$E_{n_k} = \{x : 1 \geq |f_{n_k}(x)| \geq \epsilon\} \cup \{x : |f_{n_k}(x)| > 1\} \overset{\text{def}}{=} A_{n_k} \cup B_{n_k}$$

and so

$$\begin{aligned}\int_X \frac{|f_{n_k}(x)|}{1+|f_{n_k}(x)|}\,d\mu(x) &\geq \int_{A_{n_k}} + \int_{B_{n_k}}\\ &\geq \frac{\epsilon}{2}\mu(A_{n_k}) + \frac{1}{2}\mu(B_{n_k})\\ &\geq \frac{\epsilon}{2}\cdot\mu(E_{n_k}) = \frac{\delta\epsilon}{2},\end{aligned}$$

a contradiction. □

5.11. If $a_n \not\to 0$ as $n \to \infty$, the convergence a.e. of $\sum_{n=1}^{\infty} a_n\phi_n(x)$ implies $\lim_{n\to\infty} \phi_n(x) \doteq 0$. The dominated convergence theorem implies $\lim_{n\to\infty} \|\phi_n\|_2 = 0$, contrary to the orthonormality of $\{\phi_n\}_{n\in\mathbb{N}}$. □

5.12. a) If $E = \emptyset$, the conclusion is automatic. Since each ϕ_n is in $L^2(X,\mu)$, it follows that $C \stackrel{\text{def}}{=} \bigcup_{n=1}^{\infty} \{x : \phi_n(x) \neq 0\}$ is σ-finite and $\phi = 0$ on $X \setminus C$. In the argument that follows it may be assumed that $E \subset C$ and that $\mu(E) > 0$.

Egorov's theorem implies that if $\mathsf{S} \ni A \subset E$ and $0 < \epsilon < \mu(A) < \infty$ then A contains some A_ϵ such that $\mu(A \setminus A_\epsilon) < \epsilon$ and $\phi_n \xrightarrow{\text{u}} \phi$ on A_ϵ. Since $\{\phi_n\}_{n\in\mathbb{N}}$ is orthonormal, the Schwarz inequality implies

$$\begin{aligned}\int_{A_\epsilon} |\phi(x)|\,d\mu &\le \int_{A_\epsilon} |\phi(x) - \phi_n(x)|\,d\mu + \int_{A_\epsilon} |\phi_n(x)|\,d\mu\\ &\le \|(\phi - \phi_n)\chi_{A_\epsilon}\|_\infty \mu(A_\epsilon) + \sqrt{\mu(A_\epsilon)},\end{aligned}$$

whence ϕ is integrable. From Bessel's inequality it follows that for

$$c_n \stackrel{\text{def}}{=} \int_{A_\epsilon} \overline{\operatorname{sgn}(\phi(x))}\phi_n(x)\,d\mu \int_X \chi_{A_\epsilon}\overline{\operatorname{sgn}(\phi(x))}\phi_n(x)\,d\mu,$$

$\sum_{n=1}^{\infty} |c_n|^2 \le \left\|\chi_{A_\epsilon}\overline{\operatorname{sgn}(\phi(x))}\right\|_2^2 \le \mu(A) < \infty$, whence $\lim_{n\to\infty} c_n = 0$. Since $\overline{\operatorname{sgn}(\phi)}\phi_n \xrightarrow{\text{u}} |\phi|$ on A_ϵ it follows that $\int_{A_\epsilon} |\phi(x)|\,d\mu = 0$ and thus $\phi \doteq 0$ on A_ϵ. Since $\mu(A \setminus A_\epsilon)$ can be made arbitrarily small, it follows that $\phi \doteq 0$ on A. Moreover, E is σ-finite and so there are measurable sets A_n of finite measure and such that $E = \bigcup_{n\in\mathbb{N}} A_n$, whence $\phi \doteq 0$ on E.

b) If $\overline{\lim}_{n\in\mathbb{N}}\operatorname{var}(\phi_n) < \infty$ then, since $\{\phi_n\}_{n\in\mathbb{N}}$ is orthonormal, it may be assumed that $\sup_{n\in\mathbb{N}} |\phi_n(a)| < \infty$. The Helly selection theorem **[Wi]** implies the existence of a subsequence, for convenience denoted again $\{\phi_n\}_{n\in\mathbb{N}}$, such that $\lim_{n\to\infty} \phi_n(x) \stackrel{\text{def}}{=} \phi(x)$ exists everywhere. Hence $\phi(x) \doteq 0$ and from a) there emerges the contradiction

$$0 = \int_{[a,b]} |\phi(x)|^2\,dx = \lim_{n\to\infty} \int_{[a,b]} |\phi_n(x)|^2\,dx = 1.$$ □

5.13. The functions $f_n : [0,1] \ni x \mapsto \operatorname{sgn}(\sin 2\pi 2^n x), n \in \mathbb{N}$, are in $L^\infty([0,1],\lambda)$ and $\|f_n\|_\infty \equiv 1$. Morcover, $\int_0^1 f_n(x)\,d\mu \equiv 0$ while for n in $\mathbb{N}$, $|f_n(x)| \doteq 1$. □

5.14. For $(\mathbb{R}, \mathsf{S}_\lambda, \lambda)$, if $f_n = \chi_{[0,n]} + \left(2 - \frac{x}{n}\right)\chi_{(n,2n]}, n \in \mathbb{N}$, the requirements are met. □

[**Note s5.1:** For contrast consider $(X, \mathsf{S}, \mu) \stackrel{\text{def}}{=} (\mathbb{R}, \mathsf{S}_\lambda, \lambda)$, and $f_n \stackrel{\text{def}}{=} \chi_{[-n,n]}$.]

5.15. Note that $E_n \supset E_{n+1}, n \in \mathbb{N}$, and

$$\begin{aligned}\|f\|_1 &\geq \int_{E_{n+1}} |f(x)|\,d\mu(x) \geq b_{n+1}\mu(E_{n+1}) \geq b_n\mu(E_{n+1})\\ \|f\|_1 &= \sum_{n=1}^{\infty} \int_{E_n \setminus E_{n+1}} |f(x)|\,dx\\ &\geq \sum_{n=1}^{N} b_n (\mu(E_n) - \mu(E_{n+1}))\\ &= b_1\mu(E_1) + \sum_{n=2}^{N} (b_n - b_{n-1})\mu(E_n) - b_N\mu(E_{N+1})\end{aligned}$$

whence $b_1\mu(E_1) + \sum_{n=2}^{\infty}(b_n - b_{n-1})\mu(E_n) \leq 2\|f\|_1 < \infty$. It follows that if $k \in \mathbb{N}$, there is an n_k such that

$$\frac{1}{k} > \sum_{n=n_k+1}^{\infty} (b_n - b_{n-1})\mu(E_n) \geq (b_{n_k+m} - b_{n_k})\mu(E_{n_k+m}).$$

Therefore $\frac{1}{k} + b_{n_k}\mu(E_{n_k+m}) > b_{n_k+m}\mu(E_{n_k+m})$. Since $\mu(E_{n_k+m}) \downarrow 0$ as $m \to \infty$, $\overline{\lim}_{m\to\infty} b_{n_k+m}\mu(E_{n_k+m}) = \overline{\lim}_{n\to\infty} b_n\mu(E_n) \leq \frac{1}{k}$. □

5.16.

$$\begin{aligned}&\int_X |f(x)|^p\,d\mu(x)\\ &\qquad \geq \int_{\{x\,:\,|f(x)|\geq\epsilon\}} |f(x)|^p\,d\mu(x) \geq \epsilon^p\mu(\{x : |f(x)| \geq \epsilon\}).\end{aligned}$$ □

5.17. Since $\mu_{2a} \ll \mu_3$ and $\mu_{1s} \perp \mu_3$, if $\mu_{2a}(E) > 0$ then $\mu_3(E) > 0$, whence $\mu_{1s}(E) = 0$. □

5.18. a) If $\mu_i = \mu_{ia} + \mu_{is}$ is the Lebesgue decomposition of μ_i with respect to μ_j, $i \neq j$, and μ_{ia} resp. μ_{is} lives on A_{ia} resp. A_{is}, $i = 1, 2$, then (cf. **5.17**) $A_{ia} \cap A_{js} = \emptyset, i, j = 1, 2$, $A_{is} \cap A_{js} = \emptyset, i \neq j$. Furthermore, if $E = A_{1a} \cap A_{2a}$, then $\mu_i = \mu_{iE} + \mu_{i(X\setminus E)}$, $\mu_{iE} \perp \mu_{i(X\setminus E)}, i = 1, 2$, and $\mu_{i(X\setminus E)} \perp \mu_{j(X\setminus E)}, i \neq j$. If $\frac{d\mu_{ia}}{d\mu_{ja}} \stackrel{\text{def}}{=} f_{ij}, i \neq j$, denote the Lebesgue-Radon-Nikodým (LRN) derivatives then, modulo a null set,

$$E = \{x : f_{12}(x)f_{21}(x) \neq 0\}.$$

If $\mu_{1E}(A) = 0$ then $0 = \mu_1(A \cap E) = \int_{A\cap E} f_{12}(x)\,d\mu_2(x)$. Since $f_{12}(x) > 0$ a.e. (μ_2) on E, $\mu_2(A \cap E) = \mu_{2E}(A) = 0$. An appeal to symmetry completes the argument.

b) The following relations obtain either by hypothesis or by direct deductions therefrom:

$$\mu_i = \mu_{iF} + \mu_{i(X\setminus F)}, i = 1,2, \ \mu_{iF} \ll \mu_{jF}, \mu_{iF} \ll \mu_j, i \neq j$$
$$\mu_{i(X\setminus F)} \perp \mu_{j(X\setminus F)}, \mu_{i(X\setminus F)} \perp \mu_{jF}, \mu_{i(X\setminus F)} \perp \mu_j, i \neq j.$$

Thus μ_{iF} and $\mu_{i(X\setminus F)}$ are the unique LRN components of $\mu_i, i = 1,2$, whence $\mu_{iE} = \mu_{iF}, \mu_{i(X\setminus E)} = \mu_{i(X\setminus F)}, i = 1,2$. Hence

$$\begin{aligned}\mu_1(E\setminus F) &= \mu_{1E}(E\setminus F) + \mu_{1(X\setminus E)}(E\setminus F)\\ &= \mu_{1F(E\setminus F)} + \mu_{1(X\setminus E)}(E\setminus F) = 0+0.\end{aligned}$$

Symmetry implies $\mu_1(F\setminus E) = 0$ whence $\mu_1(E\Delta F) = \mu_2(E\Delta F) = 0$. □

5.19. If $\mu_1 \perp \mu_2$ then μ_1 and μ_2 live on disjoint (measurable) sets A_1 and A_2. If $a_1, a_2 \in \mathbb{C}$ and $\{E_n\}_{n\in\mathbb{N}}$ is a measurable partition of the measurable set E then

$$\begin{aligned}\sum_{n=1}^{\infty} |a_1\mu_1(E_n) + a_2\mu_2(E_n)| &= \sum_{n=1}^{\infty} |a_1\mu_1(E_n\cap A_1) + a_2\mu_2(E_n\cap A_2)|\\ &\le (|a_1|\,|\mu_1| + |a_2|\,|\mu_2|)(E),\end{aligned}$$

whence $|a_1\mu_1 + a_2\mu_2|(E) \le (|a_1|\,|\mu_1| + |a_2|\,|\mu_2|)(E)$.

However, if $\{E_{ni}\}_{n\in\mathbb{N}}$ is a partition of $E\cap A_i, i = 1,2$, then

$$\left\{\{E_{ni}\}_{n\in\mathbb{N}, i=1,2}, E\setminus(A_1\cup A_2)\right\}$$

is a measurable partition of E. Hence

$$\begin{aligned}|a_1\mu_1 + a_2\mu_2|(E) \ge &\sum_{n=1}^{\infty} |a_1\mu_1(E_{n1}) + a_2\mu_2(E_{n1})|\\ &+ \sum_{n=1}^{\infty} |a_1\mu_1(E_{n2}) + a_2\mu_2(E_{n2})|\\ &+ |a_1\mu_1(E\setminus(A_1\cup A_2)) + a_2\mu_2(E\setminus(A_1\cup A_2))|.\end{aligned}$$

Since μ_i lives on $A_i, i = 1,2$, the last term in the right member above is zero and the first two terms may be replaced by $\sum_{n=1}^{\infty} |a_1|\,|\mu_1(E_{n1})|$ resp. $\sum_{n=1}^{\infty} |a_2|\,|\mu_2(E_{n2})|$.

Conversely, the argument in **Solution 5.18**a) may be repeated, mutatis mutandis, to produce μ_{ia}, μ_{is} such that

$$\mu_i = \mu_{ia} + \mu_{is}, \mu_{ia} \perp \mu_{is}, i = 1,2,$$
$$\mu_{ia} \ll |\mu_j|, \mu_{is} \perp |\mu_j|, i \neq j.$$

If μ_{ia} lives on $A_i, i = 1, 2$, let E be $A_1 \cap A_2$. As in **5.18a)**, $\mu_i = \mu_{iE} + \mu_{i(X\setminus E)}$, $\mu_{iE} \ll \mu_{jE}$, $\mu_{i(X\setminus E)} \perp \mu_{j(X\setminus E)}, i \neq j$, and

$$\mu_{iE}(A) = \int_A \frac{d\mu_{iE}(x)}{d\,|\mu_j|}\, d\,|\mu_j|\,(x) \stackrel{\text{def}}{=} \int_A f_{ij}(x)\, d\,|\mu_j|\,(x), i \neq j.$$

In view of the hypothesis,

$$\begin{aligned}
(\mu_{1E} \pm \mu_{2E})\,(E) &= \int_E f_{12}(x)\, d\,|\mu_2|\,(x) \pm \int_E f_{21}(x)\, d\,|\mu_1|\,(x) \\
&= \int_E (f_{12}(x) f_{21}(x) \pm f_{21}(x))\; d\,|\mu_1|\,(x), \\
|\mu_{1E} \pm \mu_{2E}|\,(E) &= \int_E |f_{12}(x) f_{21}(x) \pm f_{21}(x)|\; d\,|\mu_1|\,(x) \\
&= (|\mu_{1E}| + |\mu_{2E}|)\,(E) \\
&= \int_E (|f_{12}(x)|\; |f_{12}(x)| + |f_{12}(x)|)\; d\,|\mu_1|\,(x).
\end{aligned}$$

Since A above is an arbitrary measurable set, on E,

$$|f_{12}|\; |f_{21}| + |f_{21}| \doteq |f_{12} f_{21} \pm f_{21}| \;\; (|\mu_1|)\,.$$

Since $\mu_{iE} \ll \mu_{jE}, i \neq j$, if A is measurable then both $|\mu_1|\,(A)$ and $|\mu_2|\,(A)$ are zero or neither is zero. If $|\mu_1|\,(E) = 0$ then $\mu_1 \perp \mu_2$. On the other hand, if $|\mu_1|\,(E) \neq 0$ and

$$\begin{aligned}
B &\stackrel{\text{def}}{=} \{\, x \;:\; f_{21}(x) = 0 \,\} \cap E \\
C &\stackrel{\text{def}}{=} \{\, x \;:\; |f_{12}(x)| + 1 = |f_{12}(x) + 1| \,\} \cap E \\
D &\stackrel{\text{def}}{=} \{\, x \;:\; |f_{12}(x)| + 1 = |f_{12}(x) - 1| \,\} \cap E
\end{aligned}$$

then

$$\begin{gathered}
|\mu_1|\,(E \setminus (B \cup C)) = |\mu_1|\,(E \setminus (B \cup D)) = 0 \\
|\mu_1|\,(B) = |\mu_2|\,(B) = 0 \\
|\mu_1|\,(E \triangle C) = |\mu_1|\,(E \triangle D) = 0.
\end{gathered}$$

However, if $a \in \mathbb{C}$ and $|a| + 1 = |a + 1|$ then $a \geq 0$ and if $|a| + 1 = |a - 1|$ then $a \leq 0$. Hence $f_{12}(x) = 0$ a.e. $(|\mu_1|)$ on E. In sum, $\mu_1 \perp \mu_2$, while $\mu_{iE} = 0$ and $\mu_i = \mu_{i(X\setminus E)}, i = 1, 2$. □

5.20. a) The basic idea is to show that $\bigvee_{\gamma \in G} f_\gamma$, which a priori is not measurable, can be replaced, via a properly constructed sequence $\{g_n\}_{n \in \mathbb{N}}$, by $\bigvee_{n \in \mathbb{N}} g_n$.

If

$$E(\gamma,k,n) \stackrel{\text{def}}{=} \left\{ x \; : \; \frac{k}{2^n} \le f_\gamma(x) \right\}, (\gamma,k,n) \in \Gamma \times \mathbf{Z} \times \mathbf{N} \text{ and}$$

$$\delta(k,n) \stackrel{\text{def}}{=} \sup_{\gamma\in\Gamma} \mu(E(\gamma,k,n)), \; f_{\gamma\vee\gamma'} \stackrel{\text{def}}{=} f_\gamma \vee f_{\gamma'}$$

then $E(\gamma \vee \gamma', k, n) = E(\gamma,k,n) \cup E(\gamma',k,n)$. Hence if

$$\lim_{p\to\infty} \mu\left(E\left(\gamma_{knp},k,n\right)\right) = \delta(k,n)$$

then $\bigcup_{p\in\mathbf{N}} E\left(\gamma_{knp},k,n\right) = \lim_{p\to\infty} E\left(\gamma_{kn1} \vee \cdots \vee \gamma_{knp}, k, n\right) \stackrel{\text{def}}{=} E(k,n)$ is a measurable set and $\overline{\mu}(E(k,n)) = \delta(k,n)$.

If $\gamma \in \Gamma$ and $\mu\left(E(\gamma,k,n) \setminus E(k,n)\right) \stackrel{\text{def}}{=} \epsilon > 0$, choose p so that

$$\mu\left(E\left(\gamma_{kn1} \vee \cdots \vee \gamma_{knp}, k, n\right)\right) > \delta(k,n) - \frac{\epsilon}{2}.$$

Thus

$$\begin{aligned}
&E(\gamma,k,n)\\
&\quad = \left(E(\gamma,k,n) \setminus E\left(\gamma_{kn1} \vee \cdots \vee \gamma_{knp}, k, n\right)\right) \cup E\left(\gamma_{kn1} \vee \cdots \vee \gamma_{knp}, k, n\right),\\
&\delta(k,n) \ge \mu\left(E(\gamma,k,n)\right)\\
&\qquad \ge \mu\left(E(\gamma,k,n) \setminus \left(E\left(\gamma_{kn1} \vee \cdots \vee \gamma_{knp}, k, n\right)\right)\right)\\
&\qquad\quad + \mu\left(\left(E\left(\gamma_{kn1} \vee \cdots \vee \gamma_{knp}, k, n\right)\right)\right)\\
&\qquad \ge \epsilon + \delta(k,n) - \frac{\epsilon}{2} = \delta(k,n) + \frac{\epsilon}{2},
\end{aligned}$$

and thus $\epsilon = 0$: $\mu\left(E(\gamma,k,n) \setminus E(k,n)\right) = 0$.

If $(\gamma,n) \in \Gamma \times \mathbf{N}$ then $\bigcup_{k\in\mathbf{Z}} E(\gamma,k,n) = X$ whence $\bigcup_{k\in\mathbf{Z}} E(k,n) = X$. For n fixed and k in $\mathbf{Z}$, if $A(k,n) \stackrel{\text{def}}{=} E(k,n) \setminus \bigcup_{l=k+1}^{\infty} E(l,n)$ then

$$\{k \ne k'\} \Rightarrow \{A(k,n) \cap A(k',n) = \emptyset\},$$

whence $g_n : A(k,n) \ni x \mapsto \dfrac{k+1}{2^n}$ is well-defined. Since $\bigcup_{k\in\mathbf{Z}} A(k,n) = X$, g_n is defined on X. Furthermore, if $\gamma \in \Gamma$,

$$\begin{aligned}
\{\, x \; : \; f_\gamma(x) \ge g_n(x) \,\} &= \bigcup_{k\in\mathbf{Z}} \left(\{\, x \; : \; f_\gamma(x) \ge g_n(x) \,\} \cap A(k,n)\right)\\
&\subset \bigcup_{k\in\mathbf{Z}} \left(E(\gamma,k+1,n) \setminus \bigcup_{l=k+1}^{\infty} E(l,n) \right).
\end{aligned}$$

Since $\mu\left(E(\gamma,k,n)\setminus E(k,n)\right)\equiv 0$, it follows that

$$\mu\left(\bigcup_{k\in\mathbf{Z}}\left(E(\gamma,k+1,n)\setminus\bigcup_{l=k+1}^{\infty}E(l,n)\right)\right)=0$$

and so $\mu\{\,x\ :\ f_\gamma(x)\geq g_n(x)\,\}=0$.

The formula $E\left(\gamma_{knp},k,n\right)=E\left(\gamma_{knp},2k,n+1\right)$ implies

$$E\left(\gamma_{knp},k,n\right)\doteq E\left(\gamma_{knp},k,n\right)\cap E(2k,n+1)$$
$$E\left(\gamma_{2k,n+1,q},k,n\right)\doteq E\left(\gamma_{2k,n+1,q},k,n\right)\cap E(k,n)$$

$$\begin{aligned}
E(k,n)&=\bigcup_{p\in\mathbf{N}}E\left(\gamma_{2k,n+1,q},k,n\right)\\
&\doteq\bigcup_{q\in\mathbf{N}}\bigcup_{p\in\mathbf{N}}E\left(\gamma_{knp},2k,n+1\right)\cap E\left(\gamma_{2k,n+1,q},2k,n+1\right)\\
&\doteq\bigcup_{q\in\mathbf{N}}E\left(\gamma_{2k,n+1,q},2k,n+1\right)\cap E(k,n)\\
&\doteq E(2k,n+1)\cap E(k,n).
\end{aligned}$$

Similarly, it follows that $E(2k,n+1)\doteq E(k,n)\cap E(2k,n+1)$, whence $E(2k,n+1)\doteq E(k,n)$. Consequently

$$\begin{aligned}
A(k,n)&=E(k,n)\setminus\bigcup_{l=k+1}^{\infty}E(l,n)\doteq E(2k,n+1)\setminus\bigcup_{l=k+1}^{\infty}E(2l,n+1)\\
&=E(2k,n+1)\setminus\bigcup_{l=2k+1}^{\infty}E(l,n+1)\\
&\qquad\cup\left(\left(\bigcup_{l=2k+1}^{\infty}E(l,n+1)\right)\setminus\left(\bigcup_{l=k+1}^{\infty}E(2l,n+1)\right)\right).
\end{aligned}$$

Since

$$\begin{aligned}
&\left(\bigcup_{l=2k+1}^{\infty}E(l,n+1)\right)\setminus\left(\bigcup_{l=k+1}^{\infty}E(2l,n+1)\right)\\
&\qquad=E(2k,n+1)\setminus\bigcup_{l=k+1}^{\infty}E(2l,n+1)\\
&\qquad=A(2k+1,n+1),
\end{aligned}$$

it follows that $A(k,n)\doteq A(2k,n+1)\cup A(2k+1,n+1)$. Consequently, $g_n(x)\geq g_{n+1}(x)$ a.e., the measurable function $\lim_{n\to\infty}g_n(x)\overset{\text{def}}{=}g(x)$ exists a.e., and for all γ, $f_\gamma\leq g(x)$ a.e. Where $g(x)$ is not yet defined, let $g(x)$ be zero.

b) If h is S-measurable and $\bigvee_{\gamma\in\Gamma} f_\gamma \le h$ then

$$\begin{aligned} H &\stackrel{\text{def}}{=} \bigcup_{m\in\mathbb{N}} \left\{ x \;:\; h(x) \le g(x) - \frac{1}{2^m} \right\} \\ &\subset \bigcup_{m\in\mathbb{N}} \left\{ x \;:\; h(x) \le g_{m+1}(x) - \frac{1}{2^m} \right\}. \end{aligned}$$

The argument below shows that each summand in the right member above is a null set, whence H is a null set (μ). Since, for each n in N, $\bigcup_{k\in\mathbb{Z}} A(k,n) = X$, it suffices to show that for each k in $\mathbb{Z}$,

$$\mu\left(\left\{ x \;:\; h(x) \le g_{m+1}(x) - \frac{1}{2^m} \right\} \cap A(k, m+1)\right) \stackrel{\text{def}}{=} \mu(A) = 0.$$

In A and off a null set in A,

$$h(x) \le g_{m+1}(x) - \frac{1}{2^m} = \frac{k+1}{2^{m+1}} - \frac{2}{2^{m+1}} = \frac{k-1}{2^{m+1}}.$$

Furthermore,

$$\begin{aligned} A &= \{ x \;:\; h(x) \le g_{m+1}(x) \} \cap \left(E(k, m+1) \setminus \bigcup_{l=k+1}^{\infty} E(l, m+1) \right) \\ &= \bigcup_{p,q\in\mathbb{N}} \left(\left\{ x \;:\; h(x) \le g_{m+1}(x) - \frac{1}{2^m} \right\} \right) \\ &\qquad \cap \left\{ \left(E\left(\gamma_{k,m+1,p}, k, m+1\right) \setminus E\left(\gamma_{l,m+1,q}, l, m+1\right)\right) \right\}, \end{aligned}$$

whence it suffices to show each of the last summands above is a null set. But on a typical summand,

$$\frac{k}{2^{m+1}} \le f_{\gamma_{k,m+1,p}}(x) \le g_{m+1} - \frac{1}{2^m} = \frac{k-1}{2^{m+1}} \text{ a.e.,}$$

a contradiction unless the summand is a null set. □

5.21. There is a sequence $\left\{ \{a_{mn}\}_{m\in\mathbb{N}} \right\}_{n\in\mathbb{N}}$ such that for each m,

$$0 = a_{m1} < a_{m2} < \cdots, \quad E_{mn} \stackrel{\text{def}}{=} \{ x \;:\; a_{mn} \le f(x) \},$$
$$\sum_{n=1}^{\infty} a_{mn}\mu\left(E_{mn} \setminus E_{m,n+1}\right) = \sum_{n=1}^{\infty} \left(a_{m,n+1} - a_{mn}\right) \mu\left(E_{mn}\right), \quad \text{(s5.1)}$$

and such that as $m \to \infty$, the left member resp. right member in (s5.1) approaches $\int_X f(x)\,d\mu(x)$ resp. $(\mu \times \lambda)\left(\{ \{x,y\} \;:\; 0 \le y \le f(x) \}\right)$. □

5.22. Note that $\int_X f(y)\,d\mu(y)$ is a constant. □

5.23. An application of **5.22** leads to the equality. □

5.24. Consider the measure situation $\left(\{0,1\} \overset{\text{def}}{=} X, \mathsf{S} \overset{\text{def}}{=} 2^{\{0,1\}}, \mu\right)$ such that $\mu(\{0\}) = x, \mu(\{1\}) = 1 - x$. If $f \overset{\text{def}}{=} \chi_{\{0\}}$ then

$$E(f) = x,\ E\left(f^2\right) = x, \text{ and } \operatorname{Var}(f) = x - x^2.$$

The range of F_n is $\left\{0, \frac{1}{n}, \ldots, \frac{n}{n}\right\}$ and

$$\mu^n\left(\left\{\mathbf{x} \overset{\text{def}}{=} (x_1, \ldots, x_n) \ :\ F_n(\mathbf{x}) = \frac{k}{n}\right\}\right) = \binom{n}{k} x^k (1-x)^{n-k}$$
$$\operatorname{Var}(F_n) = \sum_{k=0}^{n}\left(\frac{k}{n} - x\right)^2 \binom{n}{k} x^k (1-x)^{n-k} = \frac{x - x^2}{n} \le \frac{1}{4n}.$$ □

[**Note s5.2:** The last inequality occurs in S. Bernstein's proof of the Weierstraß approximation theorem, viz.:

For f in $C([0,1], \mathbb{R})$ and ϵ positive, there is a positive δ such that

$$\{|x - y| < \delta\} \Rightarrow \{|f(x) - f(y)| < \epsilon\}.$$

If $n > \sup\left\{\delta^{-4}, \frac{\|f\|_\infty}{\epsilon^2}\right\}$ then since

$$\sum_{k=0}^{n}\binom{n}{k} x^k (1-x)^{n-k} \equiv 1,$$

it follows that

$$\begin{aligned}|f(x) - B_n(x)| &\overset{\text{def}}{=} \left|f(x) - \sum_{k=0}^{n} f\left(\frac{k}{n}\right)\binom{n}{k} x^k (1-x)^{n-k}\right| \\ &= \left|\sum_{k=0}^{n}\left(f(x) - f\left(\frac{k}{n}\right)\right)\binom{n}{k} x^k (1-x)^{n-k}\right| \\ &\le \left|\sum\nolimits_{\left|\frac{k}{n} - x\right| < n^{-\frac{1}{4}}}\right| + \left|\sum\nolimits_{\left|\frac{k}{n} - x\right| \ge n^{-\frac{1}{4}}}\right|.\end{aligned}$$

Owing to the domain of k in the second summand of the right member above, $\frac{\left|\frac{k}{n}-x\right|^2}{n^{-\frac{1}{2}}} \geq 1$, whence

$$\begin{aligned}|f(x)-B_n(x)| &\leq \epsilon + 2\|f\|_\infty \sum_{k=0}^{n} \binom{n}{k} x^k(1-x)^{n-k} \\ &\leq \epsilon + 2\|f\|_\infty \sum_{k=0}^{n} \frac{\left(\frac{k}{n}-x\right)^2}{n^{-\frac{1}{2}}} \binom{n}{k} x^k(1-x)^{n-k} \\ &\leq \epsilon + 2\|f\|_\infty n^{\frac{1}{2}} \sum_{k=0}^{n} \left(\frac{k}{n}-x\right)^2 \binom{n}{k} x^k(1-x)^{n-k} \\ &\leq \epsilon + \frac{\|f\|_\infty}{2n^{\frac{1}{2}}}.\end{aligned}$$

Thus f is uniformly approximable by the Bernstein polynomials B_n.]

5.25. Consider $\left(\mathbb{N}, 2^{\mathbb{N}}, \zeta\right)$, $f : \mathbb{N} \ni n \mapsto \frac{1}{n}$, and $E_n \stackrel{\text{def}}{=} \{n\}, n \in \mathbb{N}$. □

5.26. Otherwise there is a sequence $\{t_n\}_{n\in\mathbb{N}}$ converging to zero and for which $\lim_{n\to\infty} \int_X f(x,t_n)\,d\mu(x) \neq \int_X \lim_{n\to\infty} f(x,t_n)\,d\mu(x)$, in contradiction of the dominated convergence theorem. □

5.27. If

$$S(n,k) \stackrel{\text{def}}{=} \bigcap_{r,s\in\left(0,\frac{1}{n}\right)} \left\{x : |f(x,r)-f(x,s)| < \frac{1}{k}\right\}$$

$$\widetilde{S}(n,k) \stackrel{\text{def}}{=} \bigcap_{p,q\in\mathbb{Q}\cap\left(0,\frac{1}{n}\right)} \left\{x : |f(x,p)-f(x,q)| < \frac{1}{k}\right\}$$

then $S(n,k) \subset \widetilde{S}(n,k)$. On the other hand, if $x \in \widetilde{S}(n,k)$ then, owing to continuity, there are r, s such that $|f(x,r)-f(x,s)| < \frac{1}{k}$. It follows that $\widetilde{S}(n,k) = S(n,k)$ and so $S(n,k)$ is measurable.

For each k, $\mu(S(n,k)) \to \mu(X)$ as $n \to \infty$. For a suitable choice of a sequence $\{n_k\}_{k\in\mathbb{N}}$, the measure of $E \stackrel{\text{def}}{=} \bigcap_{k\in\mathbb{N}} S(n_k,k)$ exceeds $\mu(X)-\epsilon$ and the convergence of $f(x,t)$ on E is uniform. □

5.28. **First example.** If $f_n(x) = x^n$ then $\lim_{n\to\infty} f_n(x) = \chi_{\{1\}}$. If $\lambda(E) = 1$ then for any ϵ in $(0,1)$,

$$\lambda(\{x : x \in E \cap (1-\epsilon, 1]\}) = \epsilon,\ \sup\{f_n(x) : x \in E\} = 1.$$

Second example. For n in $\mathbb{N}$, if $f_n = \frac{1}{n}\chi_{(\frac{1}{n},1]} + n\chi_{(0,\frac{1}{n}]}$ then $\lim_{n\to\infty} f_n = 0$, whereas if $\lambda(E) = 1$, $\sup_{x\in E} f_n(x) = n$. □

5.29. Let S be the nonmeasurable set in **4.39**. Let $\{r_n\}_{n\in\mathbb{N}}$ be an enumeration of $\mathbb{Q}\cap[0,1)$ and define S_n to be $(S+r_n)/\mathbb{Z}$. Then there is a positive δ such that $1 > \delta = \lambda^*(S)$ $[= \lambda^*(S_n),\ n \in \mathbb{N}]$.

For t in $J_n \stackrel{\text{def}}{=} [2^{-n-1}, 2^{-n})$, let f_t be defined by the equation

$$f_t(x) = \begin{cases} 1 & \text{if } x \in S_n \text{ and } x = 2^{n+1}t - 1 \\ 0 & \text{otherwise} \end{cases}.$$

Since $[0,1) = \bigcup_{n\in\mathbb{N}} S_n$ and $(0,1) = \bigcup_{n\in\mathbb{N}} J_n$, if $t \in (0,1)$ then

$$\#(\{x : f_t(x) \neq 0\}) = \begin{cases} 1 & \text{if } 2^{n+1}t - 1 \in S_n \\ 0 & \text{otherwise} \end{cases}.$$

It follows that each f_t is a bounded measurable function different from zero for at most one x in $[0,1)$ and that if $x \in (0,1)$ then $\lim_{t\to 0} f_t(x) = 0$. In short, $f_t \stackrel{\text{a.e.}}{\to} 0$ as $t \to 0$.

If $\lambda^*(D) < \delta$ then for each n in $\mathbb{N}$, $S_n \setminus D \neq \emptyset$. Choose x_n in $S_n \setminus D$. As t traverses J_n, $2^{n+1}t - 1$ traverses $[0,1)$ and there is in J_n a t_n such that $2^{n+1}t_n - 1 = x_n$, whence $f_{t_n}(x_n) = 1$. As $n \to \infty$, $t_n \to 0$ and thus off D, $f_t \stackrel{\text{u}}{\not\to} 0$:

> Although $\lim_{t\to 0} f_t(x) = 0$ for each x in $(0,1)$, there is in $(0,1)$ no set D such that $\lambda^*(D) < \delta$ and as $t \to 0$, $f_t(x) \stackrel{\text{u}}{\to} 0$ off D.

□

5.30. a) If $\epsilon > 0$ then for some K, if $k > \frac{K}{2}$,

$$K \cdot \mu(E_{\geq}(|f|, K)) \leq \int_{E_{\geq}(|f|,K)} |f(x)|\, d\mu(x) < \frac{\epsilon}{3}.$$

Hence if $\delta \stackrel{\text{def}}{=} \frac{\epsilon}{3K}$, $\mu(E) < \delta$, then

$$E = (E \cap E_{<}(|f|, K)) \cup (E \cap E_{\geq}(|f|, K)) \stackrel{\text{def}}{=} E_1 \cup E_2$$

and $\int_E |f(x)|\, d\mu(x) = \int_{E_1} + \int_{E_2} < K \cdot \frac{\epsilon}{3K} + \frac{\epsilon}{3} < \epsilon$.

b) Assume that if $f \in \mathcal{F}$ and $\epsilon > 0$, there is a positive $\delta(\epsilon)$ such that

$$\{\mu(E) < \delta(\epsilon)\} \Rightarrow \left\{\int_E |f(x)\, d\mu(x) < \epsilon\right\}.$$

If $\epsilon > 0$, $\{f_n\}_{n\in\mathbb{N}} \subset \mathcal{F}$, and $\int_{E_{\geq}(|f_n|,n)} |f_n(x)|\, d\mu(x) \geq \epsilon$ then

$$\mu(E_{\geq}(|f_n|, n)) \geq \delta(\epsilon).$$

Since (X, S, μ) is nonatomic, $E_{\geq}(|f|, n)$ contains a subset $\widetilde{E}$ such that $\mu\left(\widetilde{E}\right) = \frac{\delta(\epsilon)}{2}$ $[< \delta(\epsilon)]$ (cf. **4.24**). It follows that for all n in $\mathbb{N}$,

$$\epsilon > \int_{\widetilde{E}} |f_n(x)|\, d\mu(x) \geq \frac{n\delta(\epsilon)}{2},$$

a contradiction.

c) For $\left(\mathbb{N}, 2^{\mathbb{N}}, \zeta\right)$, $L^1(\mathbb{N}, \zeta)$ is uniformly integrable since

$$\{\zeta(E) < 1\} \Rightarrow \{\zeta(E) = 0\} \Rightarrow \left\{\int_E |f(x)|\, d\zeta(x) = 0\right\}.$$

On the other hand, if $\epsilon > 0$, then for any K, if $k > \max\{K, \epsilon\}$, it follows that $f \stackrel{\text{def}}{=} k\chi_{\{1\}} \in L^1(\mathbb{N}, \zeta)$ and $\int_{E_{\geq}(|f|,k)} = k > \epsilon$. □

5.31. a) Since $\text{Discont}(|f|) \subset \text{Discont}(f)$, $|f|$ is integrable on every bounded interval.

b) The improper Riemann integrals $\int_{\mathbb{R}} f^{\pm}(x)\, dx$ exist and so the improper Riemann integral $\int_{\mathbb{R}} f(x)\, dx$ exists. The monotone convergence theorem implies $\int_{\mathbb{R}} f^{\pm}(x)\, d\lambda(x)$ exist.

c) If S is a nonmeasurable subset of $[0, 1]$ then $g \stackrel{\text{def}}{=} \chi_S - \chi_{[0,1]\setminus S}$ meets the requirements. □

5.32. Since f is uniformly continuous, if $\epsilon > 0$, there is in $\mathbb{N}$ an n such that $\left\{|x - y| < \frac{1}{n}\right\} \Rightarrow \{|f(x) - f(y)| < \epsilon\}$. If

$$m_k \stackrel{\text{def}}{=} \min_{\frac{k}{n} \leq x \leq \frac{k+1}{n}} f(x),\ M_k \stackrel{\text{def}}{=} \max_{\frac{k}{n} \leq x \leq \frac{k+1}{n}} f(x)$$

then $M_k - m_k < \epsilon$ and $\bigcup_{k=0}^{n-1} [m_k, M_k] \times \left[\frac{k}{n}, \frac{k+1}{n}\right]$ meets the requirements.

□

5.33. The formula for the sum of a finite geometric series and Euler's formula yield the following closed expressions:

$$D_N(\theta) = \frac{1}{2\pi} \sum_{n=-N}^{N} e^{in\theta} = \frac{1}{2\pi} e^{-iN\theta} \frac{1 - e^{i(2N+1)\theta}}{1 - e^{i\theta}}$$

$$= \frac{1}{2\pi} \frac{e^{i\left(N+\frac{1}{2}\right)\theta} - e^{-i\left(N+\frac{1}{2}\right)\theta}}{e^{i\frac{1}{2}\theta} - e^{-i\frac{1}{2}\theta}} = \frac{1}{2\pi} \frac{\sin\left(N + \frac{1}{2}\right)\theta}{\sin\left(\frac{1}{2}\theta\right)},$$

$$F_N(\theta) = \frac{1}{2\pi(N+1)} \sum_{n=0}^{N} D_n(\theta) = \frac{1}{2\pi(N+1)} \left[\frac{\sin\left(\frac{(N+1)\theta}{2}\right)}{\sin\left(\frac{\theta}{2}\right)}\right]^2.$$

Since $[0, 2\pi) \ni \theta \mapsto e^{i\theta} \stackrel{\text{def}}{=} x \in \mathbb{T}$ carries $[0, 2\pi)$ bijectively onto $\mathbb{T}$, the results above may be converted to express D_N and F_N as functions on $\mathbb{T}$. □

5.34. For f in A, the series $\sum_{n=-\infty}^{\infty} n\hat{f}_n e^{inx}$ converges uniformly on $[0, 2\pi]$ and defines a function g continuous on $[0, 2\pi]$. If $G(x) \stackrel{\text{def}}{=} \int_0^x g(t)\,dt$ on $[0, 2\pi]$ then $G(x) = -i\sum_{n=-\infty}^{\infty} \hat{f}_n e^{inx}$ and **2.79** implies

$$f(x) = iG = \sum_{n=-\infty}^{\infty} \hat{f}_n e^{inx}.$$

Thus $f \in C^1([0, 2\pi], \mathbb{C})$, $f(0) = f(1)$, and $\|f\|_\infty + \|f'\|_\infty \leq 1$. Hence if $|x - y| < \epsilon$ then $|f(x) - f(y)| < \epsilon$. It follows that A is an equicontinuous set of functions contained in the unit ball of $C([0, 1], \mathbb{C})$. The Arzelà-Ascoli theorem implies that $\overline{A}$ is compact. □

5.35. For a trigonometric polynomial $p : \mathbb{T} \ni t \mapsto \sum_{n=-N}^{N} c_n t^n$, let $L_0(p)$ be $\sum_{n=-N}^{N} a_n c_n$. Then L_0 is a linear functional defined and bounded on a dense subset of $C(\mathbb{T}, \mathbb{C})$ and so L_0 has an extension to a bounded linear functional L in $(C(\mathbb{T}, \mathbb{C}))^*$. The Riesz representation theorem yields the measure μ conforming to the requirements. □

5.36. a) Let $(\mathbb{T}, \mathsf{S}_\beta, \mu)$ be such that $\mu(\{1\}) = 1$ and $\mu(\mathbb{T} \setminus \{1\}) = 0$. Then (5.3) is the Fourier-Stieltjes series for μ.

b) The inequality $\|f * g\|_1 \leq \int_{\mathbb{T}} \left(\int_{\mathbb{T}} \left|f\left(s^{-1}t\right)\right| |g(s)|\,d\tau(s)\right) d\tau(t)$, Fubini's theorem, and the translation-invariance of Haar measure τ imply $\|f * g\|_1 \leq \|f\|_1 \|g\|_1$. Fubini's theorem then implies (5.4).

c)

$$\begin{aligned}\sigma_N^*(t) &= \frac{1}{N+1} \sum_{n=0}^{N} \left(\sum_{k=-n}^{n} \lambda_k \hat{f}_k\right) t^n \\ &= \frac{1}{N+1} \sum_{n=0}^{N} (L_n * f)\hat{\ } t^n = \Lambda_N * f(t).\end{aligned}$$

d) In consonance with the formula for the averages of the partial sums of a Fourier series, let $\sigma_N(t)$ represent the average of the first $N+1$ partial sums of a Fourier-Stieltjes series. Then, F_N denoting Fejér's kernel,

$$\begin{aligned}\sigma_N(s) &= \int_{\mathbb{T}} F_N\left(s^{-1}t\right) d\mu(t) \\ \|\sigma_N\|_1 &\leq \int_{\mathbb{T}} |d\mu(t)| \int_{\mathbb{T}} F_N\left(s^{-1}t\right) d\tau(s) = |\mu|(\mathbb{T}).\end{aligned}$$

Direct calculation shows $\sigma_N(t) = \sum_{n=-N}^{N} c_n \left(1 - \frac{|n|}{N+1}\right) t^n$. Thus if, for some μ, $\sum_{n=-N}^{N} c_n t^n$ is a Fourier-Stieltjes series then (5.5) obtains with $|\mu|(\mathbb{T})$ for M.

Conversely, if (5.5) obtains for some M, then via the identification $\mathbb{T} \leftrightarrow [0,1)$, $\left\{G_n(t) \stackrel{\text{def}}{=} \int_0^t \sigma_n(s)\,ds\right\}_{n\in\mathbb{N}}$ is a sequence of functions of uniformly bounded variation. Since $G_n(0) \equiv 0$ the Helly selection theorem is applicable. Thus there is a subsequence $\{G_{n_j}\}_{j\in\mathbb{N}}$ converging uniformly on $\mathbb{T}$ to a function G. If $n_j > |k|$, integration by parts yields

$$\left(1 - \frac{|k|}{n_j+1}\right) c_k = \int_0^1 \sigma_{n_j}(t)e^{-2\pi ikt}\,dt = G_{n_j}(1) + \frac{ik}{2\pi}\int_0^1 G_{n_j}e^{-2\pi ikt}\,dt.$$

The preceding calculation shows that as $j \to \infty$, $c_k = \int_0^1 e^{-2\pi ikt}\,dG(t)$, i.e., that the Stieltjes measure generated by G may be used for μ □

5.37. The construction of the Cantor set C_α for a positive α may be carried out on any interval $[a,b]$ in place of $[0,1]$. If $[a,b] \stackrel{\text{def}}{=} \left[0, \frac{7}{9}\right]$ and if ϕ_α is the corresponding (monotonely increasing) Cantor function, let μ be the Stieltjes measure generated by ϕ_α. □

5.38. Let $\{x_n\}_{n\in\mathbb{N}}$ be a dense subset of F. For f in $C_0(\mathbb{R},\mathbb{C})$, let $L_n(f)$ be $f(x_n)$. Then $L \stackrel{\text{def}}{=} \sum_{n=1}^\infty \frac{L_n}{2^n} \in (C_0(\mathbb{R},\mathbb{C}))^*$ and $\|L\| \le 1$. The Riesz representation theorem provides a Borel measure μ such that for f in $C_0(\mathbb{R},\mathbb{C})$, $L(f) = \int_{\mathbb{R}} f(x)\,d\mu(x)$.

If U is an open subset of $\mathbb{R}\setminus F$ and if K is a compact subset of U, there is in $C_0(\mathbb{R},\mathbb{C})$ an f such that $K \prec f \prec U$ whence $L(f) = 0$ and thus $\mu(K) = 0$: $\text{supp}(\mu) \subset F$.

If V is open and $V \cap F \neq \emptyset$, there is in $V\cap F$ an x_{n_0} and in $C_0(\mathbb{R},\mathbb{C})$ a g such that $\{x_{n_0}\} \prec g \prec V$. Hence $\mu(V) \ge L(g) \ge \frac{g(x_{n_0})}{2^{n_0}} > 0$ and so $\text{supp}(\mu) = F$. □

5.39. As a compact set in $\mathbb{R}^n$, K is $f(C_0)$ for some f in $C(C_0,\mathbb{R}^n)$ (cf. **2.15**). The Cantor function ϕ_0 determines the measure

$$\nu : \mathsf{S}_\beta([0,1]) \ni E \mapsto \lambda(\phi_0(E)).$$

For any set S in $\mathsf{S}_\lambda([0,1])$, $m(t) \stackrel{\text{def}}{=} \lambda(S\cap[0,t])$ varies continuously with t. Hence if $\lambda(S) > 0$ then $m(1) = \lambda(S)$ and $m(0) = 0$. If $\alpha \in (0,\lambda(S))$ then for some t_0 in $[0,1]$, $m(t_0) = \alpha = \lambda(S\cap[0,t_0])$. It follows that ν is nonatomic and thus that $\mu : \mathsf{S}_\beta(\mathbb{R}^n) \ni E \mapsto \nu\left(f^{-1}(E)\cap C_0\right)$ is a nonatomic measure on $\mathsf{S}_\beta(\mathbb{R}^n)$.

If $U \cap K = \emptyset$ then $f^{-1}(U) = \emptyset$, whence $\mu(U) = 0$. The linear extension F of f from C_0 to $[0,1]$ is continuous. If V is open and $V\cap K \neq \emptyset$ then $F^{-1}(V) \stackrel{\text{def}}{=} W$ is open and $W \cap C_0 \neq \emptyset$, whence $\nu(W\cap C_0) > 0$ and so $\text{supp}(\mu) = K$. □

5.40. Let $\{x_n\}_{n\in\mathbb{N}}$ be a countable dense subset of $[0,1]$. For each Borel subset A of $[0,1]$, if $\mu(A) \stackrel{\text{def}}{=} \sum_{x_n\in A} \dfrac{2}{3^n}$ then $\mu(\mathsf{S}_\beta) = C_0$. □

5.41. Jensen's inequality applied to exp implies

$$\exp\left(\int_{[0,1]} f(x)\,dx\right) \le \int_{[0,1]} \exp(f(x))\,dx.$$

Rolle's theorem implies that if $a \le x$ then $e^a(x-a) + e^a \le e^x$ and, since exp is strictly monotonely increasing, equality obtains iff $x = a$. Hence if $a \stackrel{\text{def}}{=} \int_{[0,1]} f(x)\,dx$ and $f(x) \ne a$ on a set of positive measure (μ) then, on a set of positive measure (μ), $e^a(f(x)-a)+e^a < e^{f(x)}$ and integration yields $\exp\left(\int_{[0,1]} f(x)\,dx\right) < \int_{[0,1]} \exp(f(x))\,dx$. □

5.42. If $F \in \mathsf{F}([0,1])$ and $0 \notin F$ then for some $N(F)$ and some h in $C([0,1],\mathbb{R})$, $0 \notin N(F)$ and $F \prec h \prec N(F)$. Hence $0 = h(0) \ge \int_{[0,1]} h(x)\,dx$ and so $\operatorname{supp}(\mu) = \{0\}$. Thus if $0 \le \mu(\{0\}) \stackrel{\text{def}}{=} c \le 1$ then for f in $C([0,1],\mathbb{C})$, $\int_{[0,1]} f(x)\,d\mu(x) = c\cdot f(0)$. □

5.43. For

$$A \stackrel{\text{def}}{=} \{f \,:\, f\in C([0,1],\mathbb{R}), f(x) = f(1-x), f(0) = f(1) = 0\},$$

the Stone-Weierstraß theorem implies that the span of $\{x \mapsto \sin^k \pi x\}_{k\in\mathbb{N}}$ is $\|\ \|_\infty$-dense in A. Hence if $f \in A$, $\int_0^1 f(x)\,d\mu(x) = 0$. If $E \in \mathsf{S}_\beta\left(\left[0,\frac{1}{2}\right]\right)$ then $E\cap(1-E) = \emptyset$, and $\mu(E\cup(1-E)) = \int_0^1 \chi_{E\cup(1-E)}(x)\,d\mu(x)$. The integrand in the right member preceding is the $\|\ \|_1$-limit of a sequence of functions in A: $\mu(E\cup(1-E)) = 0$. □

5.44. The linear span of $\{x\mapsto \cos^k \pi x\}_{k\in\mathbb{N}}$ is $\|\ \|_\infty$-dense in

$$\left\{f \,:\, f\in C([0,1],\mathbb{C}), f\left(\frac{1}{2}\right) = 0\right\}.$$

The argument in **Solution 5.43** applies. □

5.45. For the $\|\ \|_\infty$-closure B of $\operatorname{span}\left(\{x\mapsto \cos^k \pi x\}_{k=1}^N\right)$, the map $L : B \ni \sum_{k=1}^N c_k \cos^k \pi t \mapsto \sum_{k=1}^N c_k a_k$ is extendible to an element L_1 of B^* and, by virtue of the Hahn-Banach theorem, L_1 may be extended without increase of norm to a linear functional L_2 defined on $C([0,1],\mathbb{R})$. The Riesz representation theorem implies the existence of a finite (possibly signed) Borel measure μ such that for f in $C([0,1],\mathbb{R})$, $L_2(f) = \int_{[0,1]} f(x)\,d\mu(x)$. □

5.46. For f in $\mathcal{F}$, the map $f^* : [0,1] \ni x \mapsto \int_0^x f(t)\,dt$ is in $C([0,1],\mathbb{C})$. Thus **5.30**b) implies that for some $K(1)$,

$$\begin{aligned} |f^*(x)| &\le \int_{[0,1]} |f(t)|\,dt = \int_{[0,1]\setminus E_{\ge}(f,2K(1))} + \int_{E_{\ge}(f,2K(1))} f(t)\,dt \\ &\le 2K(1)+1, \end{aligned}$$

i.e., $\{f^*\}_{f\in\mathcal{F}}$ is uniformly bounded.

If $x \le y$, $\epsilon > 0$, and $|y-x|K(\epsilon) < \epsilon$ then

$$\begin{aligned} |f^*(x) - f^*(y)| &\le \int_{[x,y]\setminus E_{\ge}(f,2K(\epsilon))} + \int_{[x,y]\cap E_{\ge}(f,2K(\epsilon))} f(t)\,dt \\ &\le (y-x)K(\epsilon) + \epsilon < 2\epsilon \end{aligned}$$

whence $\{f^*\}_{f\in\mathcal{F}}$ is equicontinuous. The Arzelà-Ascoli theorem applies.

□

5.47. If $\mu \in P$, $A_1, A_2 \in \mathsf{S}_\beta$, $A_1 \cup A_2 = [0,1]$, and $\mu(A_1)\mu(A_2) > 0$ then $\mu_i : \mathsf{S}_\beta \ni E \mapsto \dfrac{\mu(E\cap A_i)}{\mu(A_i)}$ are in P, $\mu \ne \mu_i, i = 1,2$, and

$$\mu = \mu(A_1)\mu_1 + \mu(A_2)\mu_2.$$

Hence μ is an extreme point of P iff for all E in S_β, $\mu(E)$ is zero or one.

□

5.48. The linear span A of $\{x \mapsto e^{nx}\}_{n=0}^\infty$ is $\|\ \|_\infty$-dense $C([0,1],\mathbb{R})$ and $L : A \ni \sum_{n=0}^N a_n e^{nx} \mapsto \sum_{n=0}^N a_n t_n$ is a nonnegative linear functional.

If $g_n \in A, n \in \mathbb{N}$, $g_n \downarrow 0$, then, since L is nonnegative, $L(g_n) \downarrow a \ge 0$. Dini's theorem implies $\|g_n\|_\infty \downarrow 0$. If $k \in \mathbb{N}$, there is an m_k such that $0 \le g_n \le \dfrac{a}{k}$ if $n > m_k$ and thus $a \le L(g_n) \le \dfrac{aL(1)}{k}$ whence $a = 0$.

If $\sup\{\,|L(h)| : h \in A, \|h\|_\infty = 1\,\} = \infty$ then for some $\{h_n\}_{n\in\mathbb{N}}$ contained in A, $\|h_n\|_\infty \equiv 1$ while $|L(h_n)| \ge n$. Hence $\dfrac{h_n}{\sqrt{n}} \stackrel{\text{def}}{=} H_n \stackrel{\text{u}}{\to} 0$ while $|L(H_n)| \ge \sqrt{n}$. However $0 < g_n \stackrel{\text{def}}{=} H_n + \dfrac{(n+1)\|H_n\|_\infty}{n} \in A, n \in \mathbb{N}$, while $g_n \stackrel{\text{u}}{\to} 0$. Hence for some subsequence $\{g_{n_k}\}_{k\in\mathbb{N}}$, $g_{n_k} \downarrow 0$ and $L(g_{n_k}) \downarrow 0$ while

$$L(g_{n_k}) \ge |L(H_{n_k})| - \left|\frac{(n_k+1)\|H_{n_k}\|_\infty}{n_k}\right| \cdot |L(1)|.$$

The first term in the right member is unbounded while the second converges to zero as $k \to \infty$, a contradiction. In sum, for some constant M and each h in A, $|L(h)| \le M\|h\|_\infty$. Hence L may be extended to an $\widetilde{L}$ in $C([0,1],\mathbb{R})^*$.

The Riesz representation theorem applies. □

5.49. If μ_1 is a measure and $R \stackrel{\text{def}}{=} [a,b] \times [c,d] \subset X$ then direct calculation shows that $\mu_1(R) = \int_R (1+4xy)\,dydx$. It follows that if $S \in \mathsf{S}_\beta(X)$ then $\mu_1(S) = \int_S (1+4xy)\,dydx$, a formula that fails when $0 < a \le b \le 1$ and $S \stackrel{\text{def}}{=} ([0,a] \times [0,b]) \cap X$. A similar calculation reveals a contradiction for the putative measure μ_2. □

5.50. Integration by parts shows that if $h(x) = 0$ off $[a,b]$ then, $F(x)$ denoting $\int_a^x f(t)\,dt$,

$$\begin{aligned}-\int_{\mathbb{R}} f(t)h(t)\,dt &= -\left[F(t)h(t)\Big|_a^b - \int_a^b F(t)h'(t)\,dt\right]\\ &= \int_a^b F(t)h'(t)\,dt = \int_a^b h'(t)\,d\mu(t).\end{aligned}$$

Hence $\mu([a,b]) = \int_a^b F(t)\,dt$ and if $E \in \mathsf{S}_\beta$ then $\mu(E) = \int_E F(t)\,dt$, whence $\mu \ll \lambda$ and $\frac{d\mu}{d\lambda}(x) = \int_{-\infty}^x f(t)\,dt$. □

5.51. a) The formulæ

$$\mathbb{R} \setminus (x+E) = x + (\mathbb{R} \setminus E),\ E\Delta(x+E) = (\mathbb{R} \setminus E)\,\Delta\,(x + (\mathbb{R} \setminus E))$$

lead to the conclusion.

b) If $\lambda(E) \cdot \lambda(\mathbb{R} \setminus E) > 0$, the metric density theorem and translation-invariance of λ imply that for some p resp. q in E resp. $\mathbb{R} \setminus E$ and for all sufficiently small positive a,

$$\lambda(E \cap (p-a,p+a)) > 1.8a \text{ and } \lambda((\mathbb{R} \setminus E) \cap (q-a,q+a)) > 1.8a.$$

If $x \stackrel{\text{def}}{=} p - q$ then

$$\begin{aligned}\lambda((\mathbb{R} \setminus E) \cap (q-a,q+a)) &= \lambda((x + (\mathbb{R} \setminus E)) \cap (x + (q-a,q+a)))\\ &= \lambda((x + (\mathbb{R} \setminus E)) \cap (p-a,p+a)) > 1.8a\end{aligned}$$

and if $2a < |p-q|$ then $(x + (\mathbb{R} \setminus E)) \cap (E \cap (p-a,p+a)) = \emptyset$, whence

$$\begin{aligned}(x + (\mathbb{R} \setminus E) \cap (p-a,p+a)) \cap (E \cap (p-a,p+a)) &= \emptyset,\\ 2a \ge \lambda(((x + (\mathbb{R} \setminus E)) \cap (p-a,p+a)) \cup (E \cap (p-a,p+a)))&\\ = \lambda(((x + (\mathbb{R} \setminus E)) \cap (p-a,p+a))) + \lambda((E \cap (p-a,p+a)))&\\ > 3.6a,&\end{aligned}$$

a contradiction. □

5.52. a) Fejér's theorem implies that if f is continuous and periodic with period 2π then f on $[0, 2\pi]$ is uniformly the limit of a sequence of trigonometric polynomials. Thus $\int_{\mathbb{R}} f(t)\, d\mu(t) = 0$ and the equality continues to obtain if f is a.e. periodic with period 2π and $f \in L^1(\mathbb{R}, \mu)$. Hence if $A \in \mathsf{S}_\beta$ and $A + 2\pi n = A, n \in \mathbb{Z}$ then

$$A = \bigcup_{n\in\mathbb{Z}} (A \cap [2\pi n, 2\pi(n+1))) \stackrel{\text{def}}{=} \bigcup_{n\in\mathbb{Z}} A_n$$
$$A_n + 2\pi = A_{n+1}$$
$$\mu\left(\bigcup_{n\in\mathbb{Z}}(A + 2\pi n)\right) = \mu(A) = \sum_{n\in\mathbb{Z}} \mu(A_n).$$

Since $\chi_A(x + 2\pi) = \chi_A(x)$, it follows that $\mu(A) = \int_{\mathbb{R}} \chi_A(x)\, d\mu = 0$ and so $\mu\left(\bigcup_{n\in\mathbb{Z}} (A + 2\pi n)\right) = 0$. If $E \in S_\beta$ and $B \stackrel{\text{def}}{=} \bigcup_{n\in\mathbb{Z}}(E + 2\pi n)$ then $B + 2\pi n = B, n \in \mathbb{Z}$, and so

$$\mu\left(\bigcup_{n\in\mathbb{Z}}(B + 2\pi n)\right) = \mu(B) = 0 = \mu\left(\bigcup_{n\in\mathbb{Z}}(E + 2\pi n)\right).$$

b) If $\{a_n\}_{n=0}^\infty \subset \mathbb{C}$, $\sum_{n=0}^\infty |a_n| = \infty$, and $\sum_{n=0}^\infty a_n = 0$ define μ by the conditions $\mu : \mathsf{S}_\beta \ni E \mapsto \begin{cases} 0 & \text{if } E \cap 2\pi\mathbb{Z} = \emptyset \\ \sum_{2\pi n \in E} a_n & \text{otherwise} \end{cases}$. [If the convention $\sum_{n\in\emptyset} a_n = 0$ is adopted, $\mu(E) = \sum_{2\pi n\in E} a_n$.]

If $m \in \mathbb{Z}$ then $\int_{\mathbb{R}} e^{-itm}\, d\mu(t) = \sum_{n=0}^\infty a_n = 0$. On the other hand, if $E \stackrel{\text{def}}{=} \{2\pi k\}_{k=0}^\infty$ then $\mu(E) = 0$ and $\mu(E - 2\pi n) = 0$ while

$$\mu(E + 2\pi) = -a_0, \ldots, \mu(E + 2\pi n) = -\sum_{n=0}^{n-1} a_k, n \in \mathbb{N}.$$

Thus if $N > 0$,

$$\sum_{n=-\infty}^{N} \mu(E + 2\pi n) = -Na_0 - (N-1)a_1 - \cdots - (N(N-1))a_{N-1}$$
$$= -N\left(\sum_{n=0}^{N-1} a_n\right) + \sum_{n=1}^{N-1} na_n.$$

If $\quad a_n = \begin{cases} 1 & \text{if } n = 0 \\ \dfrac{(-1)^n(2n+1)}{n(n+1)} & \text{if } n \in \mathbb{N} \end{cases} \quad$ then

$$\sum_{n=0}^{\infty} a_n = 0,\ \sum_{n=0}^{\infty} |a_n| = \infty,\ -N\left(\sum_{n=0}^{N-1} a_n\right) = (-1)^N$$
$$\sum_{n=1}^{N-1} na_n \stackrel{\text{def}}{=} S_{N-1} = \sum_{n=1}^{N-1} \frac{(-1)^n(2n+1)}{n+1}.$$

It follows that $S_{2N+1} \downarrow \underline{S} < 0$ and $S_{2N} \uparrow \overline{S} > 0$ whence $\sum_{n=0}^{N-1} \mu(E + 2\pi n)$ fails to converge as $N \to \infty$. □

5.53. a) Extend f so that $f(x) = 0$ off $[0,1]$. There is a Borel set E on which μ lives and $\lambda(E) = 0$. It may be assumed that $\{0,1\} \subset E \subset [0,1]$. For some N that is a null set (λ), if $x \notin N$ then $f(x-y)$ as a function of y is Borel measurable whence $\int_{\mathbb{R}} f(x-y)\,d\mu(y)$ exists and is finite a.e. (λ) iff $I_x \stackrel{\text{def}}{=} \int_{\mathbb{R}} |f(x-y)|\,d\mu(y) < \infty$ a.e. (λ).

If $E_n \stackrel{\text{def}}{=} \{x : x \in [0,1], |f(x)| \le n\}, n \in \mathbb{N}$, then $E_n \in \mathsf{S}_\lambda$ and furthermore, $\lambda(E_n) \uparrow 1$. Since $\{0,1\} \subset E$,

$$[0,2] \supset E_n + E \supset E_n \cup (E_n + \{1\}) \text{ and } E_n \cap (E_n + \{1\}) \doteq \emptyset\ (\lambda).$$

Hence $2 \ge \lambda^*(E_n + E)$, $\lambda_*(E_n + E) \uparrow 2$, and

$$2 \ge \lambda^*\left(\bigcup_{n\in\mathbb{N}} (E_n + E)\right) \ge \lambda_*\left(\bigcup_{n\in\mathbb{N}} (E_n + E)\right) \ge 2,$$

i.e., $A \stackrel{\text{def}}{=} \bigcup_{n\in\mathbb{N}} (E_n + E)$ is measurable and $\lambda(A) = 2$.

If $x \in A$, say $x \in E_{n_0} + E$, then for y in E, $x - y \in E_{n_0}$, $|f(x-y)| \le n_0$, and so $I_x < \infty$ a.e. (λ) in $[0,2]$. If $x \in \mathbb{R} \setminus [0,2]$ and $y \in E$ then $x - y \notin [0,1]$, $f(x-y) = 0$, and $I_x = 0$.

b) The Fubini-Tonelli theorems imply that if $f, g \in L^1(\mathbb{R}, \lambda)$ then $f * g(x) \stackrel{\text{def}}{=} \int_{\mathbb{R}} f(x-y)g(y)\,dy$ exists for all x off a null set (λ) and $f * g$ is in $L^1(\mathbb{R}, \lambda)$.

The Lebesgue-Radon-Nikodým decomposition

$$\mu = \mu_a + \mu_s,\ \mu_a \ll \lambda,\ \mu_a \perp \mu_s,\ \mu_s \perp \lambda$$

and a) reduce the problem to the study of $\int_{[0,1]} |f(x-y)|\,d\mu_a(y)$. If $\frac{d\mu_a}{d\lambda} \stackrel{\text{def}}{=} g$ and f, g are extended so that each is zero off $[0,1]$ then each is in $L^1(\mathbb{R}, \lambda)$ and $\int_{[0,1]} |f(x-y)|\,d\mu_a(y) = \int_{\mathbb{R}} |f(x-y|g(y)\,dy = |f| * g(x)$, which exists, is finite a.e. (λ). Furthermore, $|f| * g \in L^1(\mathbb{R}, \lambda)$ (cf. **4.33**). □

5.54. a) Since μ_n is complex, $0 < |\mu_n|(\mathbb{R}) \stackrel{\text{def}}{=} a_n < \infty$. If $b_n \stackrel{\text{def}}{=} \frac{1}{2^n a_n}$ then $\sum_{n=1}^{\infty} a_n b_n = 1$ and $|\mu_n| \ll \sum_{n=1}^{\infty} b_n |\mu_n| \stackrel{\text{def}}{=} \nu$.

b) If $f_n \stackrel{\text{def}}{=} \chi_{[n,n+1]}$ then $\mu_n : \mathsf{S}_\beta \ni E \mapsto \int_E f_n(x)\,dx$ is a nonnegative measure and if $\nu \ll \mu_n, n \in \mathbb{N}$, then $\nu((-\infty, n]) \equiv 0$ whence $\nu = 0$. □

5.55. For $\{y_n\}_{n\in\mathbb{N}}$ contained in $\mathbb{R}$, $\{y_n + E\}_{n\in\mathbb{N}}$ is a sequence contained in $\mathsf{S}_\beta(\mathbb{R})$. For S, the set of all such sequences,

$$a \stackrel{\text{def}}{=} \sup\left\{\mu\left(\bigcup_{n\in\mathbb{N}} (y_n + E)\right) : \{y_n + E\}_{n\in\mathbb{N}} \in S\right\} < \infty$$

and for m in $\mathbb{N}$, there is a sequence $\{y_{nm}\}_{n\in\mathbb{N}}$ such that

$$\mu\left(\bigcup_{n\in\mathbb{N}}(y_{nm}+E)\right)>a-\frac{1}{m}.$$

If $\{x_k\}_{k\in\mathbb{N}}$ is an enumeration of $\{y_{nm}\}_{m,n\in\mathbb{N}}$ and

$$\bigcup_{k\in\mathbb{N}}(x_k+E)\stackrel{\text{def}}{=}A$$

then $\mu(A)=a$. If $G\stackrel{\text{def}}{=}\mathbb{R}\setminus A$,

$$\nu_1:\mathsf{S}_\beta\ni B\mapsto\nu_1(B)\stackrel{\text{def}}{=}\mu(B\cap G),\text{ and }\nu_2\stackrel{\text{def}}{=}\mu-\nu_1$$

then for x in $\mathbb{R}$,

$$\nu_1(x+E)=\mu\left((x+E)\setminus A\right).$$

If $\nu_2(x+E)>0$ it follows that $\mu\left((x+E)\cup A\right)>a$, a contradiction. □

5.56. In **Figure s5.1** it can be seen that if (a_1,a_2) is near (b_1,b_2) then $|\mu\left(Q\left(a_1,a_2\right)\right)-\mu\left(Q\left(b_1,b_2\right)\right)|$ is the sum of the μ-measures of at most six nonoverlapping rectangles with sides that are parallel to the vertical or horizontal axes.

Owing to the results in **4.10** – **4.12**, μ confined to a bounded rectangle is regular. If $\{R_n\}_{n\in\mathbb{N}}$ is a monotonely decreasing sequence of closed rectangles with sides that are horizontal or vertical and $\lambda\left(R_n\right)\downarrow 0$ then $\lim_{n\to\infty}R_n$ is a point or a line segment, whence $\mu\left(R_n\right)\downarrow 0$.

If (b_1,b_2) approaches (a_1,a_2), the nonoverlapping rectangles noted in the first sentence constitute six or fewer families of rectangles partially ordered by inclusion and each family contains a cofinal sequence converging to a point or a line segment. Hence f is continuous. □

5.57. For t in $[0,\infty)$, if

$$S_t\stackrel{\text{def}}{=}\{\,(x_1,x_2)\ :\ |x_1|+|x_2|\le t\,\}$$

then $\mu\left(\partial S_t\right)=0$ and $\mu\left((0,0)\right)=0$. The argument in **Solution 5.56** shows, mutatis mutandis, that $f:[0,\infty)\ni t\mapsto\mu\left(S_t\cap E\right)$ is continuous. Since E is bounded, for large t, $f(t)=\mu(E)$ whereas $f(0)=0$. Hence for some t_a, $f\left(t_a\right)=a$ and $F\stackrel{\text{def}}{=}S_{t_a}\cap E$ meets the requirements. □

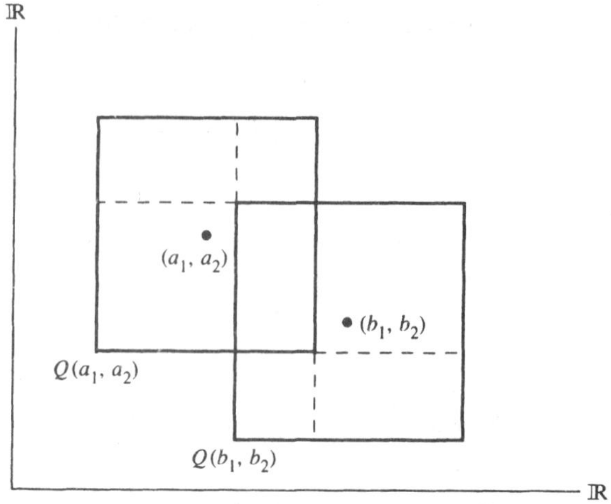

Figure s5.1.

5.58. a) If K is compact then $\mu_m(K) < \infty, m \in \mathbb{Z}^+$, whence $\mathbb{R}^n$ is σ-finite $(\mu_m), m \in \mathbb{Z}^+$.

Since each open set in $\mathbb{R}^n$ is the union of countably many compact sets, each open set is inner regular, hence regular, $(\mu_m), m \in \mathbb{Z}^+$. The set of regular sets is a σ-ring and contains all open sets whence every set in S_β is regular.

If K is compact and $\mu_0(K) > M$, (5.6) implies that for large m, $\mu_m(K) > M$.

b) The argument in **Solution 4.19** serves here as well. □

5.59. If U_m is open in $\mathbb{R}^m$ and U_n is open in $\mathbb{R}^n$ then $U_m \times U_n$ is open in $\mathbb{R}^{m+n}$, whence $\mathsf{S}_\beta(\mathbb{R}^m) \times \mathsf{S}_\beta(\mathbb{R}^n) \subset \mathsf{S}_\beta\left(\mathbb{R}^{m+n}\right)$. On the other hand, if U_{m+n} is open in $\mathbb{R}^{m+n}$ then U_{m+n} is the countable union of rectangles of the form $\bigtimes_{k=1}^{m+n} [a_k, b_k)$.

The Fubini-Tonelli theorems imply that $\mathsf{S}_\lambda \times \mathsf{S}_\lambda \subset \mathsf{S}_{\lambda_2}$. If E is a nonmeasurable subset of $\mathbb{R}$ then $A \stackrel{\text{def}}{=} E \times \{0\} \notin \mathsf{S}_\lambda \times \mathsf{S}_\lambda$ but $\lambda_2(A) = 0$ and so $A \in \mathsf{S}_{\lambda_2}$. □

5.60. Every subset of $D \stackrel{\text{def}}{=} \{(x, y) : x = y\}$ in $\mathbb{R}^2$ is a null set (λ_2) and thus is in S_{λ_2}. In particular, if E is a nonmeasurable subset of $\mathbb{R}$,

$F \stackrel{\text{def}}{=} D \cap \{(x,y) : x \in E\}$, and $\Gamma' = \{1\}$ then $F \in \mathsf{S}_{\lambda_2}$ while nevertheless $P_{\Gamma'}(F) = E \notin \mathsf{S}_\lambda$. □

5.61. Let $(X_i, \mathsf{S}_i, \mu_i)$ be $(\mathbb{R}, \mathsf{S}_\beta, \lambda)$, $i = 1, 2$. Any rectangle contained in $\Delta \stackrel{\text{def}}{=} \{(x,y) : x = y\}$ in $\mathbb{R}^2$ is a single point whence Δ is not the countable union of rectangles of the kind described.

If $(X_i, \mathsf{S}_i, \mu_i) = (\mathbb{N}, 2^{\mathbb{N}}, \zeta)$, $i = 1, 2$, then $\mathsf{S}_1 \times \mathsf{S}_2 = 2^{\mathbb{N}^2}$. □

5.62. If $A \stackrel{\text{def}}{=} p^{-1}((-\infty, 0)) \cap K \neq \emptyset$, there is in $C_{00}(\mathbb{R}^n, \mathbb{R})$ an f such that $0 \le f(\mathbf{x}) \le 1$, $f(\mathbf{x}) = 1$ at some point of A, and $f(\mathbf{x}) = 0$ off A. Since $\operatorname{supp}(\mu) = K$, $\int_{\mathbb{R}^n} f(\mathbf{x})p(\mathbf{x})\,d\mu(\mathbf{x}) \stackrel{\text{def}}{=} -c < 0$. The Stone-Weierstraß theorem implies that $f^{\frac{1}{2}}$ may be approximated by a polynomial q such that $\left\|\left(f - q^2\right) p|_K\right\|_\infty < \dfrac{c}{2\mu(K)}$, which leads to the conclusion $\int_{\mathbb{R}^n} (q(\mathbf{x}))^2 p(\mathbf{x})\,d\mu(\mathbf{x}) < -\dfrac{c}{2} < 0$, a contradiction. □

5.63. Egorov's theorem implies that if ϵ in $(0, 1-a)$, there is in S_λ an E such that $\lambda(E) > 1 - \epsilon$ and on E, $f_n \stackrel{\text{u}}{\to} f$. Hence f is continuous on E and since λ is regular, E contains a compact set K such that $\lambda(K) > 1 - \epsilon > a$. □

5.64. *i.* If $f \in AC([a,b], \mathbb{R})$ then:

a) The continuity of f follows by definition.

b) For some positive $\delta(1)$,

$$\left\{\sum_{n=1}^N |b_n - a_n| < \delta(1)\right\} \Rightarrow \left\{\sum_{n=1}^N |f(b_n) - f(a_n)| < 1\right\}.$$

Hence for a unique $N(1)$ in $\mathbb{N}$, $\dfrac{1}{N(1)+1} < \delta(1) \le \dfrac{1}{N(1)}$. If P is a partition and $|P| < \delta(1)$ then $\operatorname{var}_{[a,b],P}(f) \le N(1) + 1$.

c) If $\epsilon > 0$ then for a positive $\delta(\epsilon)$,

$$\left\{\sum_{n=1}^N |x_n - y_n| < \delta(\epsilon)\right\} \Rightarrow \left\{\sum_{n=1}^N |f(x_n) - f(y_n)| < \epsilon\right\}.$$

If $\lambda(E) = 0$ and $\epsilon > 0$, for some sequence $\{I_n\}_{n \in \mathbb{N}}$ of intervals, $E \subset \bigcup_{n \in \mathbb{N}} I_n$ and $\sum_{n=1}^\infty \lambda(I_n) < \delta(\epsilon)$. Thus $\sum_{n=1}^\infty \lambda(f(I_n)) \le \epsilon$, whence $f(E) \subset \bigcup_{n \in \mathbb{N}} f(I_n)$, $\lambda(f(E)) \le \sum_{n=1}^\infty \lambda(f(I_n)) \le \epsilon$, and so $\lambda(f(E)) = 0$.

Conversely if a), b), and c) obtain then b) implies

$$f(x) = \operatorname{var}_{[0,x]}(f) - \left(\operatorname{var}_{[0,x]}(f) - f(x)\right) \stackrel{\text{def}}{=} g(x) - h(x),$$

i.e., f is the difference of two monotonely increasing functions g and h while a) implies g and h are continuous (cf. **3.36**).

Because g is monotonely increasing, there is a possibly empty but at most countable set $\{I_m\}_{m\in S}$ of pairwise disjoint closed intervals such that if $x < y$ and x and y do not belong to one of those intervals then $g(x) < g(y)$: g is a bijection on the complement Q of the union of those intervals.

If $g \notin AC([a,b],\mathbb{R})$, for a sequence $\{a_{nk}, b_{nk}\}_{k=1}^{K_n}, n \in \mathbb{N}$, of finite sets of pairs in Q^2 and some positive ϵ,

$$a \le a_{n1} < b_{n1} \le a_{n2} < b_{n2} \le \cdots \le a_{n,K_n-1} < b_{n,K_n-1} \le a_{nK_n} < b_{nK_n} \le b$$
$$g(a_{nk}) < g(b_{nk}), 1 \le k \le K_n, n \in \mathbb{N}$$
$$\sum_{k=1}^{K_n}(b_{nk} - a_{nk}) < 2^{-n}, \sum_{k=1}^{K_n} g(b_{nk}) - g(a_{nk}) \ge \epsilon.$$

If $A_n \stackrel{\text{def}}{=} \bigcup_{k=1}^{K_n}[a_{nk}, b_{nk}]$ and $A \stackrel{\text{def}}{=} \overline{\lim}_{n\to\infty} A_n$ then $\lambda(A) = 0$. However $f(A) = g(A) - h(A)$, whence the translation-invariance of λ and c) imply $\lambda(g(A)) = 0$.

Since g is continuous and monotonely increasing, $g([a_{nk}, b_{nk}])$ are pairwise nonoverlapping compact intervals, $1 \le k \le K_n$, whence $\lambda(g(A_n)) \ge \epsilon$ and $\lambda\left(\overline{\lim}_{n\to\infty} g(A_n)\right) \ge \epsilon$. Owing to the bijective nature of g where it is applied below,

$$g^{-1}\left(\overline{\lim}_{n\to\infty} g([a_{nk}, b_{nk}])\right) = \overline{\lim}_{n\to\infty}\left(g^{-1}\left(g([a_{nk}, b_{nk}])\right)\right) = A,$$

whence $\overline{\lim}_{n\to\infty} g([a_{nk}, b_{nk}]) = g(A)$, a contradiction of c) since the null set A is carried by g to a set of positive measure. Thus g is absolutely continuous; a similar proof shows h is absolutely continuous.

ii. If $[a,b] = [0,1]$ and f is the Cantor function ϕ_0, then f is continuous and monotonely increasing. However if $E = C_0$, the Cantor set, then $\lambda(f(E)) = 1$ while $\lambda(E) = 0$: a) and b) obtain while c) fails.

If $f(x) = \chi_{(0,1]}(x) \cdot x \sin\left(\dfrac{1}{x}\right)$, $0 < a < 1$, and $x \in [a,1]$ then $|f'(x)| \le 1 + \dfrac{2}{a^2}$ whence f is absolutely continuous on $[a,1]$. Thus a) and c) obtain. If $n = 3, 5, \ldots,$ and P is a partition determined by the points $0, \dfrac{2}{n\pi}, \dfrac{2}{(n-2)\pi}, \ldots, 1$ then $\text{var}_{[0,1],P}(f) \approx \dfrac{4}{\pi}\sum_{k=1}^{n}\dfrac{1}{2k-1}$ whence f is not of bounded variation on $[0,1]$: a) and c) obtain while b) fails.

If $f = \chi_{\{\frac{1}{2}\}}$ then b) and c) obtain while a) fails.

iii. If E is Lebesgue measurable then owing to the regularity of λ, there is a sequence of $\{K_n\}_{n\in\mathbb{N}}$ of compact sets and there is a null set N such that $A = N \cup \bigcup_{n\in\mathbb{N}} K_n$. Hence $f(A) = f(N) \cup \bigcup_{n\in\mathbb{N}} f(K_n)$. From i it follows that $f(N)$ is a null set and, from the continuity of f, that $\bigcup_{n\in\mathbb{N}} f(K_n)$ is a union of compact sets. □

[**Note s5.3:** The proof for *iii* goes back to first principles. On the other hand, if E is Lebesgue measurable then E contains a Borel set S such that $E \setminus S \stackrel{\text{def}}{=} N$ is a null set (λ). Then $f(N)$ is a null set and (cf. **1.13** – **1.16**) $f(S) \in \mathsf{S}_\lambda$ whence $f(E) \in \mathsf{S}_\lambda$.]

5.65. If $f \in \text{Lip}(1)$ then f is absolutely continuous and for x in $[a,b]$, $f(x) = f(a) + \int_a^x f'(t)\,dt = f(a)$, i.e., f is a constant, a contradiction. □

5.66. If $|f(x) - f(y)| \le M|x-y|$ and E is Lebesgue measurable, the absolute continuity of f implies $\lambda(f(E)) = \int_E f'(t)\,dt \le M \int_E 1\,dt = M\lambda(E)$. □

5.67. The function $\phi : [0,1] \ni x \mapsto x + \phi_0(x)$ is strictly monotone and continuous whence $\phi^{-1} \stackrel{\text{def}}{=} f$ is strictly monotone and continuous and both ϕ and f, as homeomorphisms, are open.

Furthermore, $\lambda(\phi(C_0)) = 1$. Let E be a nonmeasurable subset of $\phi(C_0)$. Then $\mathsf{S}_\lambda \ni F \stackrel{\text{def}}{=} f(E) \subset C_0$ and $g \stackrel{\text{def}}{=} \chi_F \in \mathcal{M}$. If $g \circ f \stackrel{\text{def}}{=} h$ then $h^{-1}(1) = E$, whence $h \notin \mathcal{M}$. □

5.68. In **5.67**, $F \in \mathsf{S}_\lambda$ and $\phi(F) = E \notin \mathsf{S}\lambda$. □

5.69. The set Λ of all finite subsets λ of E is a diset with respect to the partial order: $\lambda \prec \lambda'$ iff $\lambda \subset \lambda'$. The net $n : \Lambda \ni \lambda \mapsto n_\lambda \stackrel{\text{def}}{=} \chi_\lambda$ is such that for each λ, $n_\lambda \in \mathcal{M}$, whereas n_λ converges to χ_E in $\mathbb{R}^X \setminus \mathcal{M}$. □

5.70. Note that $\nu \ll \mu$. Let N resp. M be the ν-essential supremum resp. μ-essential supremum of f. Thus $\{x : f(x) > M\}$ is a null set (μ) and hence a null set (ν). It follows that $N \le M$. On the other hand, if $N < M$ then $E \stackrel{\text{def}}{=} \left\{x : f(x) > \dfrac{N+M}{2}\right\}$ is a null set (ν), whence $\int_E f(x)\,d\mu(x) = 0$ and so $\dfrac{N+M}{2} \ge M$, a contradiction. □

5.71. See **2.27** and **2.28**. A proof that exploits topology rather than set theory follows.

Since $\lambda(E) > 0$, there is a compact subset K of E and $\lambda(K) > 0$. The Cantor-Bendixson theorem (**2.28**) implies that K is the union of a perfect set P and a countable set N. Furthermore, $P \ne \emptyset$ since otherwise, K is countable and $\lambda(K) = 0$. As a perfect subset of $\mathbb{R}$, P is of cardinality $\mathfrak{c}$. Hence $\mathfrak{c} \ge \#(E) \ge \#(P) = \mathfrak{c}$. □

[**Note s5.4:** Sierpinski [**Sier**] showed that the stronger implication $\{\lambda^*(E) > 0\} \Rightarrow \{\#(E) = \mathfrak{c}\}$ is equivalent to the Continuum Hypothesis (CH). Owing to Cohen's proof [**Coh2**] of the independence of CH from ZF, the implication just cited is not derivable in ZF.]

5.72. The Stone-Weierstraß theorem implies that the linear span A of $\{[0,\infty) \ni x \mapsto e^{-nx} : n \in \mathbb{N}\}$ is dense in $C_{00}([0,\infty),\mathbb{R})$. Moreover, since $|\mu|$ is bounded, if $f \in C_{00}([0,\infty),\mathbb{R})$ then $\int_0^\infty f(x)\,d\mu(x) = 0$ whence for every compact set K, $\mu(K) = 0$. From **4.10** it follows that $\mu(\mathsf{S}_\beta) = \{0\}$.

□

5.73. For n in $\mathbb{N}$, the sequence $\left\{\left[\frac{k}{2^n}, \frac{k+1}{2^n}\right)\right\}_{k=0}^\infty$ is a partition of $[0,\infty)$ and, since $1 - \mu([0,x))$ is monotonely decreasing and nonnegative,

$$\lim_{n\to\infty}\sum_{k=0}^\infty \left(1-\mu\left(\left[0,\frac{k}{2^n}\right)\right)\right)2^{-n} = \int_0^\infty (1-\mu([0,x)))\,dx \ (\leq \infty).$$

Abel summation, i.e., integration by parts with respect to counting measure, yields

$$\begin{aligned}\sum_{k=0}^K \left(1-\mu\left(\left[0,\frac{k}{2^n}\right)\right)\right)2^{-n} &= \sum_{k=1}^{K-1} k2^{-n}\cdot\mu\left(\left[\frac{k-1}{2^n},\frac{k}{2^n}\right)\right)\\ &\quad + K2^{-n}\cdot\mu\left(\left[\frac{K}{2^n},\infty\right)\right)\\ \sum_{k=0}^\infty \left(1-\mu\left(\left[0,\frac{k}{2^n}\right)\right)\right)2^{-n} &= \lim_{K\to\infty}\left(\sum_{k=1}^{K-1} k2^{-n}\cdot\mu\left(\left[\frac{k-1}{2^n},\frac{k}{2^n}\right)\right)\right.\\ &\quad \left.+ K2^{-n}\cdot\mu\left(\left[\frac{K}{2^n},\infty\right)\right)\right).\end{aligned}$$

If $\int_{[0,\infty)} x\,d\mu(x) < \infty$ then (cf. **5.15**) $\lim_{K\to\infty} K2^{-n}\cdot\mu\left(\left[\frac{K}{2^n},\infty\right)\right) = 0$ and so

$$\begin{aligned}\lim_{n\to\infty}\left(\sum_{k=1}^\infty k2^{-n}\cdot\mu\left(\left[\frac{k-1}{2^n},\frac{k}{2^n}\right)\right)\right) &= \int_0^\infty (1-\mu([0,x)))\,dx\\ &= \int_0^\infty x\,d\mu(x).\end{aligned}$$

If $\int_{[0,\infty)} x\,d\mu(x) = \infty$ then $\int_{[0,\infty)}(1-\mu([0,x)))\,dx = \infty$. □

5.74. The following definitions and equations facilitate the discussion of the question.

$$P \stackrel{\text{def}}{=} \left\{ \left[\frac{k}{2^n}, \frac{k+1}{2^n} \right) \right\}_{k=0}^{\infty} \tag{s5.2}$$

$$b_k \stackrel{\text{def}}{=} \mu \left(\left\{ y \ : \ \frac{k}{2^n} \le f(y) < \frac{k+1}{2^n} \right\} \right), 1 \le k \le K \tag{s5.3}$$

$$S_k \stackrel{\text{def}}{=} \sum_{p=k}^{K-1} b_p = \mu \left(\left\{ y \ : \ \frac{k}{2^n} \le f(y) < \frac{K}{2^n} \right\} \right),$$

$$0 \le k \le K - 1\&(\text{s5.4})$$

$$a_k \stackrel{\text{def}}{=} \frac{k}{2^n}$$

$$\sum_{k=2}^{K-1} a_k b_k = a_2 (S_1 - S_2) + \cdots + a_{K-1} (S_{K-1} - S_K) \tag{s5.5}$$

$$= \frac{1}{2^n} \sum_{k=2}^{K-1} S_k - S_K \frac{K-1}{2^n} + S_1 \frac{2}{2^n}. \tag{s5.6}$$

If $f \in L^1([0, \infty), \mu)$, then for large K and n, i.e., for small $|P|$ and large n, both members of (s5.5) approximate $\int_0^\infty f(y)\, dy$. Since f is bounded, say $f(x) \le M$, for fixed n, if $K > M2^n$ then the right member of (s5.4) is

$$\mu \left(\left\{ y \ : \ \frac{k}{2^n} \le f(y) \right\} \right)$$

while $S_K = 0$. Furthermore, $\lim_{n\to\infty} S_1 \frac{2}{2^n} = 0$. Thus the right member of (s5.6) approximates $\int_{[0,\infty)} \mu(\{ y \ : \ x \le f(y) \})\, dy$. □

5.75. If

$$J_{kn} \stackrel{\text{def}}{=} \left[\frac{k}{2^n}, \frac{k+1}{2^n} \right), n \in \mathbb{N}, -1z0 \le k \le 2^n - 1,$$

$$m_{kn} \stackrel{\text{def}}{=} \inf \{ f(x) \ : \ x \in J_{kn} \}, M_{kn} \stackrel{\text{def}}{=} \sup \{ f(x) \ : \ x \in J_{kn} \},$$

$$m \stackrel{\text{def}}{=} \inf \{ f(x) \ : \ 0 \le x \le 1 \}, M \stackrel{\text{def}}{=} \sup \{ f(x) \ : \ 0 \le x \le 1 \},$$

$$g_{kn} : \mathbb{R} \ni y \mapsto \begin{cases} 1 & \text{if } m_{kn} \le y \le M_{kn} \\ 0 & \text{otherwise.} \end{cases}$$

then $G_n(y) \stackrel{\text{def}}{=} \sum_k g_{kn}(y)$ is measurable and counts the number of J_{kn} on which the equation $f(x) = y$ has at least one solution. Furthermore, P_n denoting the partition created by the endpoints of the intervals J_{kn},

$$\int_m^M g_{kn}(y)\, dy = M_{kn} - m_{kn} \stackrel{\text{def}}{=} \text{osc}_{J_{kn}}(f) \text{ and}$$

$$\int_m^M G_n(y)\, dy = \sum_k \text{osc}_{J_{kn}}(f) = \text{var}_{[0,1],P_n}(f).$$

Since $G_n \le G_{n+1}$, $\lim_{n\to\infty} G_n \stackrel{\text{def}}{=} G$ exists and $\int_m^M G(y)\,dy = \text{var}_{[0,1]}(f)$.

On the other hand, $G_n \le \nu$ whence $G \le \nu$. If $p \in \mathbb{N}$ and $p \le \nu(y)$ then for some n, $G_n(y) \le p$. Since ν is $\mathbb{N}\cup\infty$-valued, $G = \nu$. □

5.76. It may be assumed that $a_1 \le a_2 \le \cdots \le a_N$, and $a_n \le b_n$. If $b_1 < a_2$ then $(b_1, a_2) \subset [a,b] \setminus \bigcup_{n=1}^N [a_n, b_n]$, a contradiction. Hence $b_1 \ge a_2$ and similarly, if $2 \le n \le N$ then $b_{n-1} \ge a_n$. In sum, $[a_n, b_n] \cap [a_{n+1}, b_{n+1}] \ne \emptyset$. Since for some n, $b_n \ge b$, it follows that $\sum_{n=1}^N (b_n - a_n) \ge b - a$. □

5.77. Every line in $\mathbb{R}^2$ is a null set (λ_2). □

5.78. The functions $f_n : [0,1] \ni x \mapsto \sum_{k=1}^n \dfrac{1}{k^2 |x - r_k|^{\frac{1}{2}}}$, $n \in \mathbb{N}$, are nonnegative, $f_n \uparrow f$, and $\displaystyle\int_{[0,1]} f_n(x)\,dx = \sum_{k=1}^n \frac{2\left(r_k^{\frac{1}{2}} + (1-r_k)^{\frac{1}{2}}\right)}{k^2}$. Since $\max_{0\le x\le 1}\left(x^{\frac{1}{2}} + (1-x)^{\frac{1}{2}}\right) = \sqrt{2}$, the monotone convergence theorem implies $\displaystyle\int_{[0,1]} f(x)\,dx \le \sqrt{2}\sum_{k=1}^{\infty} 1/k^2 = \frac{\sqrt{2}\pi^2}{6}$ whence $f(x) < \infty$ a.e. □

5.79. Since $\lambda\left(\bigcup_{n\in\mathbb{N}} (r_n - 1/n^2, r_n + 1/n^2)\right) \le 2\sum_{n=1}^\infty 1/n^2 = \pi^2/3$, it follows that $\mathbb{R} \setminus \bigcup_{n\in\mathbb{N}} (r_n - 1/n^2, r_n + 1/n^2) \ne \emptyset$.

For $A \stackrel{\text{def}}{=} \mathbb{N} \setminus \{n^2\}_{n=2}^\infty \stackrel{\text{def}}{=} \{a_n\}_{n\in\mathbb{N}}$, it may be assumed that $a_n < a_{n+1}$. For some sequence $\{t_k\}_{k\in\mathbb{N}}$ in $\mathbb{Q}$, $|t_k - 1| < \dfrac{1}{a_k}$, $k \in \mathbb{N}$. For the enumeration $\{r_p\}_{p\in\mathbb{N}}$ of $\mathbb{Q} \setminus \{t_k\}_{k\in\mathbb{N}}$, $s_1 \stackrel{\text{def}}{=} t_1$, $s_p \stackrel{\text{def}}{=} r_{\sqrt{p}-1}$ when $p \notin A$, and when $1 < p \in A$, s_p is the first t_k not yet chosen. Thus $\{s_p\}_{p\in\mathbb{N}}$ is an enumeration of $\mathbb{Q}$. Since

$$\bigcup_{p\in A}\left(s_p - \frac{1}{p}, s_p + \frac{1}{p}\right) \subset [-1,2],\ 1 + \sum_{p\notin A} \lambda\left(\left(s_p - \frac{1}{p}, s_p + \frac{1}{p}\right)\right) \le \pi^2/3,$$

it follows that $\mathbb{R} \setminus \bigcup_{p\in\mathbb{N}} \left(s_p - \dfrac{1}{p}, s_p + \dfrac{1}{p}\right) \ne \emptyset$. □

5.80. If $\lambda(\{x_2 : \lambda(E_{x_2}) = 1\}) > 1/2$ then

$$\begin{aligned}\lambda_2(E) &= \int_0^1 \left(\int_0^1 \lambda(E_{x_2})\,dx_2\right) dx_1 \\ &\ge \int_0^1 \left(\int_{\{x_2 : \lambda(E_{x_2})=1\}} 1\,dx_2\right) dx_1 > \frac{1}{2},\end{aligned}$$

whereas $\lambda_2(E) = \int_0^1 \left(\int_0^1 \lambda(E_{x_1})\,dx_1\right) dx_2 \le \frac{1}{2}$. □

5.81. By definition, if $A_1 \times A_2 \subset E$, $A_i \in \mathsf{S}_\lambda, i = 1, 2$, then $(A_1 - A_2)$ does not meet $\mathbb{Q}$. Furthermore, if $\lambda_2 (A_1 \times A_2) > 0$, it may be assumed that $\lambda(A_1) + \lambda(A_2) < \infty$. Then $f : \mathbb{R} \ni x_1 \mapsto \int_{\mathbb{R}} \chi_{A_1}(x_1 + x_2) \chi_{A_2}(x_2)\, dx_2$ is continuous, nonnegative, vanishes off $A_1 - A_2$, and

$$\int_{\mathbb{R}} f(x_1)\, dx_1 = \lambda(A_1)\lambda(A_2) = \lambda_2(A_1 \times A_2) > 0.$$

Hence f is nonzero on a nonempty open set U contained in $A_1 - A_2$ and there emerges the contradiction: $(A_1 - A_2) \cap \mathbb{Q} \neq \emptyset$. □

5.82. For m in $\mathbb{Z}$, n in $\mathbb{N}$, if

$$\begin{aligned} f_n : \mathbb{R}^2 \ni (x_1, x_2) \mapsto\ & n\left(\frac{m}{n} - x_1\right) f\left(\frac{m-1}{n}, x_2\right) \\ & + n\left(x_1 - \frac{m-1}{n}\right) f\left(\frac{m}{n}, x_2\right) \text{ if } x_1 \in \left[\frac{m-1}{n}, \frac{m}{n}\right] \end{aligned}$$

[on $\left[\frac{m-1}{n}, \frac{m}{n}\right] \stackrel{\text{def}}{=} [p_{m-1}, p_m]$ the value of f_n is a convex (linear) combination of $f(p_{m-1}, x_2)$ and $f(p_m, x_2)$] then on $[p_{m-1}, p_m]$, $f_n(x_1, x_2)$ is between $f(p_{m-1}, x_2)$ and $f(p_m, x_2)$. Because p_m depends on n and f_{x_2} is a continuous function of x_1, there emerges the following diagram in which each arrow represents convergence as $n \to \infty$ and $A\ R\ B\ R\ C$ means B is between A and C: $B = \text{mid}(A, B, C)$.

$$\begin{array}{ccccc} f(p_{m-1}, y) & R & f_n(x, y) & R & f(p_m, y) \\ & & \downarrow & & \\ f(p_{m-1}, y) & \rightarrow & f(x, y) & \leftarrow & f(p_m, y) \end{array} \qquad \text{(s5.7)}$$

The hypothesis implies that for each m, $\mathbb{R}^2 \ni (x, y) \mapsto f(p_{m-1}, y)$, regarded as a function on $\mathbb{R}^2$, is Borel measurable whence f is Borel measurable.

□

5.83. In analogy with the method used in **Solution 5.82**, for n in $\mathbb{N}$, E contains a sequence $\{a_{nm}\}_{m \in \mathbb{N}}$ such that

$$\begin{aligned} & \frac{1}{2n} < a_{n,m+1} - a_{nm} < \frac{1}{n},\ a_{nm} \leq x_1 \leq a_{n,m+1} \\ & f_n : \mathbb{R}^2 \ni (x_1, x_2) \mapsto \frac{a_{n,m+1} - x_1}{a_{n,m+1} - a_{nm}} f(a_{nm}, x_2) \\ & \qquad\qquad + \frac{x_1 - a_{nm}}{a_{n,m+1} - a_{nm}} f(a_{n,m+1}, x_2). \end{aligned}$$

A diagram similar to that in (s5.7) and an argument similar to that associated with (s5.7) but in which "Lebesgue measurable" replaces "Borel measurable" concludes the proof. □

5.84. This time the device used in **Solution 5.82** shows that

$$h_n : \mathbb{R}^2 \ni (x_1, x_2) \mapsto f_n(g(x_2), x_2)$$

(a function on $\mathbb{R}^2$) is Lebesgue measurable and $\lim_{n\to\infty} h_n = h$. □

5.85. Tonelli's theorem implies that

$$\int_{[-1,1]^2} |f(x_1, x_2)| \, d\lambda_2(x_1, x_2) = \int_{[-1,1]} \left(\int_{[-1,1]} \frac{|x_1 x_2|}{x_1^2 + x_2^2} \, dx_1 \right) dx_2.$$

If $x_2 \neq 0$, $\int_{[-1,1]} \frac{|x_1|}{x_1^2 + x_2^2} \, dx_1 = \ln \frac{1 + x_2^2}{x_2^2}$, whence the double and iterated integrals are finite. □

5.86. Since f is continuous on $[0,1]^2 \setminus \{(0,0)\}$ and $|f| \leq 1$, it follows that $f \in L^1\left([0,1]^2, \lambda_2\right)$, whence both integrals are the same as the integral of f over $[0,1]^2$. Since $f(x_1, x_2) = -f(x_2, x_1)$, the double integral is zero. □

5.87. For $f : (0,1)^2 \ni (x_1, x_2) \mapsto x_2^{|x_1 - \frac{1}{2}|^{\frac{1}{2}} - 1}$,

$$F(x_1) \stackrel{\text{def}}{=} \int_0^1 f(x_1, x_2) \, dx_2 \, dx_1$$

$$= \begin{cases} \left| x_1 - \frac{1}{2} \right|^{-\frac{1}{2}} & \text{if } x_1 \neq \frac{1}{2} \\ = \infty & \text{if } x_1 = \frac{1}{2} \end{cases},$$

$$\int_0^1 F(x_1) \, dx_1 = 2\sqrt{2}.$$

Hence $f \in C\left((0,1)^2, \mathbb{R}\right) \cap L^1\left((0,1)^2, \lambda_2\right)$ and yet

$$\int_0^1 f\left(\frac{1}{2}, x_2\right) dx_2 = \int_0^1 x_2^{-1} \, dx_2 = \infty.$$

□

5.88. For $\mu_1 : \mathsf{S}_\lambda([0,1]) \ni E \mapsto \int_E x^{\frac{1}{2}} \, dx$ and $\mu_2 = \lambda$,

$$\frac{d\mu_1}{d\mu_2}(x) = x^{\frac{1}{2}}, \; \frac{d\mu_2}{d\mu_1}(x) = x^{-\frac{1}{2}},$$

whence $\mu_i \ll \mu_j, i \neq j$ and each μ_i is totally finite.

Since $\frac{d\mu_1}{d\mu_2} \leq 1$, it follows that $\frac{d\mu_1}{d\mu_2} \in L^\infty([0,1], \mu_2)$. However if $M > 0$, then $\mu_1\left(\left\{ x : \frac{d\mu_2}{d\mu_1}(x) > M \right\}\right) = \mu_1\left((0, M^{-2})\right) = \frac{2M^{-3}}{3} > 0$, whence $\frac{d\mu_2}{d\mu_1} \notin L^\infty([0,1], \mu_1)$. □

5.89. If $f \stackrel{\text{def}}{=} \frac{d\mu}{d\lambda}$ and $0 < x - a < x + a < 1$, then

$$F(x) \stackrel{\text{def}}{=} \frac{\mu([0,1] \cap (x-a, x+a))}{\lambda([0,1] \cap (x-a, x+a))} = \frac{1}{2a} \int_{x-a}^{x+a} f(t)\, dt.$$

As $a \to 0$, $F \stackrel{\text{a.e.}}{\to} f$. □

5.90. For any Lebesgue measurable set E (in $\mathbb{R}$), the metric density of E is 1 a.e. on E. For E as described, the metric density of E is constantly one-half. Hence $E \notin \mathsf{S}_\lambda$. □

5.91. a) Since components are disjoint,

$$\sum_{1 \le n < N} \lambda(C_n) = \lambda\left(\bigcup_{1 \le n < N} C_n\right) = \lambda\left(\bigcup_{n \in \mathbb{N}} J_n\right) \le \sum_{1 \le n < N} \lambda(J_n).$$

b) It suffices to prove the inequality for each of the finitely many components $\{D_k\}_{1 \le k < K < N}$ of $\bigcup_{n=1}^{M} J_n$. Each D_k is the finite union of interlocking intervals denotable $(a_1, b_1), \ldots, (a_p, b_p)$ and such that

$$a_1 < a_2 < b_1 < a_3 < b_2 < \cdots < a_p < b_{p-1} < b_p.$$

If

$$\sum_{j \text{ odd}} (b_j - a_j) < \frac{\lambda(D_k)}{2}, \quad \sum_{j \text{ even}} (b_j - a_j) < \frac{\lambda(D_k)}{2} \tag{s5.8}$$

then $\sum_{1 \le j < K} (b_j - a_j) < \lambda(D_k) \le \sum_{1 \le j < K} (b_j - a_j)$, a contradiction. Hence one of the inequalities in (s5.8) is false. □

5.92. If μ as described exists then for some f in $L^1([0,1], \lambda)$ and all a in $[0,1]$, $\mu([0,a]) = \int_0^a f(x)\, dx$. Since $f = f^+ - f^-$, it follows that $\int_0^a f^+(x)\, dx = \int_0^a f^-(x)\, dx$ whence for each Lebesgue measurable set E, $\int_E f^+(x)\, dx = \int_E f^-(x)\, dx$, i.e., $f^+ \doteq f^-$, $f \doteq 0$, and $\mu = 0$, a contradiction. □

5.93. Let U represent a rotation in $\mathbb{R}^n$, i.e., with respect to an orthonormal basis X for $\mathbb{R}^n$, U is represented by a matrix $U_{XX} \stackrel{\text{def}}{=} (u_{ij})_{i,j=1}^{n,n}$ such that $U_{XX}^t U_{XX} = \text{id}$, and $\det(U_{XX}) = 1$. In particular,

$$U(B(\mathbf{x}, r)^\circ) = B(U(\mathbf{x}), r))^\circ, \; \lambda(U(B(\mathbf{x}, r)^\circ)) = \lambda(B(U(\mathbf{x}), r))^\circ).$$

Since λ is regular, if $E \in \mathsf{S}_{\lambda_n}$ there is a sequence $\left\{B(\mathbf{x}_{mn}, r_{mn})^\circ\right\}_{m,n \in \mathbb{N}}$ of open balls such that $E \subset V_m \stackrel{\text{def}}{=} \bigcup_{n \in \mathbb{N}} B(\mathbf{x}_{mn}, r_{mn})^\circ$, $V_{m+1} \subset V_m$, and $\lambda(V_m) \downarrow \lambda(E)$. However, $U(E) \subset U(V_m) = \bigcup_{n\in\mathbb{N}} B(U(\mathbf{x}_{mn}), r_{mn})^\circ$ whence $\lambda(U(E)) = \lambda(E)$. $\square$

5.94. Let $\{r_n\}_{n\in\mathbb{N}}$ be an enumeration of $\mathbb{Q}$ and for m,n in $\mathbb{N}$, let J_{mn} be $\left(r_n - 2^{-(m+n+1)}, r_n + 2^{-(m+n+1)}\right)$. Then $W_m \stackrel{\text{def}}{=} \bigcup_{n\in\mathbb{N}} J_{mn}$ is a dense open set containing $\mathbb{Q}$, $N_m \stackrel{\text{def}}{=} \mathbb{R}\setminus W_m$ is nowhere dense, $\lambda(W_m) = 2^{-m}$, and $W \stackrel{\text{def}}{=} \bigcap_{m\in\mathbb{N}} W_m$ is a dense G_δ that is a null set (λ). Furthermore, $\mathbb{R}\setminus W = \bigcup_{m\in\mathbb{N}} N_m$, a set of the first category. $\square$

5.95. If $f \in L^1(\mathbb{R},\lambda)$ then $g = f * \chi_E$ and since $\chi_E \in L^1(\mathbb{R},\lambda)$, it follows that $g \in L^1(\mathbb{R},\lambda)$.

Conversely, if $\|g\|_1 < \infty$ then the translation-invariance of λ and the Fubini-Tonelli theorems imply

$$\int_{\mathbb{R}} g(x)\,dx = \int_{\mathbb{R}}\left(\int_{\mathbb{R}} \chi_E(t) f(x-t)\,dt\right)dx = \left(\int_{\mathbb{R}} f(x)\,dx\right)\left(\int_{\mathbb{R}} \chi_E(t)\,dt\right)$$

whence $0 \le \int_{\mathbb{R}} f(x)\,dx = \dfrac{\|g\|_1}{\lambda(E)} < \infty$. $\square$

5.96. Since $\mu(\mathbb{R}) = 1$,

$$f_b : \mathbb{R} \ni x \mapsto \mu((-\infty,b)+x) \quad [= \mu((-\infty,b+x))]$$

is, for every b in $\mathbb{R}$, a monotonely increasing function such that

$$\lim_{x\to-\infty} f_b(x) = 0,\ \lim_{x\to\infty} f_b(x) = 1.$$

Hence for all b, $\int_{\mathbb{R}} f_b(x)\,dx = \infty = \lambda((-\infty,b))$. Furthermore,

$$\begin{aligned}\mu([a,b)+x) = \mu([a+x,b+x)) &= \mu((-\infty,b+x)) - \mu((-\infty,a+x)),\\ \lim_{a\to\infty} \mu([a,b)+x) &= \lim_{a\to\infty} \mu([a+x,b+x)) = 0,\\ \lim_{b\to-\infty} \mu([a,b)+x) &= \lim_{b\to-\infty} \mu([a+x,b+x)) = 0.\end{aligned}$$

If $E_n \stackrel{\text{def}}{=} \left[0,\dfrac{1}{n}\right)$ then $E_n = \bigcup_{k=0}^{m-1}\left(E_{m\cdot n} + \dfrac{k}{mn}\right)$.

Hence $\sum_{p\in\mathbb{Z}} \dfrac{\mu\left(E_n + \frac{p}{mn}\right)}{mn} = \displaystyle\sum_{k=0}^{m-1}\sum_{p\in\mathbb{Z}} \dfrac{\mu\left(E_{m\cdot n} + \frac{k}{mn} + \frac{p}{mn}\right)}{mn}$. For each k,

$$\sum_{p\in\mathbb{Z}} \frac{\mu\left(E_{m\cdot n} + \frac{k}{mn} + \frac{p}{mn}\right)}{mn} = \sum_{p\in\mathbb{Z}} \frac{\mu\left(E_{m\cdot n} + \frac{p}{mn}\right)}{mn} = \frac{\mu(\mathbb{R})}{mn} = \frac{1}{mn}$$

whence $\sum_{p\in\mathbb{Z}} \frac{\mu\left(E_n + \frac{p}{mn}\right)}{mn} = m\frac{1}{mn} = \frac{1}{n} \to \int_{\mathbb{R}} \mu(E_n + x)\, dx$ as $m \to \infty$.

Thus $\lambda(E_n) = \frac{1}{n} = \int_{\mathbb{R}} \mu(E_n + x)\, dx$. A similar calculation shows

$$\lambda\left(\left[a, a + \frac{1}{n}\right)\right) = \int_{\mathbb{R}} \mu\left(\left[a, a + \frac{1}{n}\right) + x\right) dx.$$

The map $\nu : \mathsf{S}_\beta \ni E \mapsto \int_{\mathbb{R}} \mu(E + x)\, dx$ is a measure such that for all n in $\mathbb{N}$, $\lambda\left(\left[a, a + \frac{1}{n}\right)\right) = \nu\left(\left[a, a + \frac{1}{n}\right)\right)$ whence $\nu = \lambda$. ☐

5.97. For k in $\mathbb{N}$, $\bigcup_{\mathbf{r}\in\mathbb{Q}^n} B\left(\mathbf{r}, \frac{1}{k}\right)^\circ = \mathbb{R}^n$ whence for each k there is an $\mathbf{r}_k$ such that $\lambda_m\left[T^{-1}\left(B\left(\mathbf{r}_k, \frac{1}{k}\right)^\circ\right)\right] > 0$. If $T(\mathbf{x}) \in B\left(\mathbf{r}_k, \frac{1}{k}\right)^\circ$ then

$$\lambda_m\left[-\mathbf{x} + T^{-1}\left(B\left(\mathbf{r}_k, \frac{1}{k}\right)^\circ\right)\right] = \lambda_m\left[T^{-1}\left(B\left(\mathbf{O}, \frac{1}{k}\right)^\circ\right)\right] > 0.$$

Hence $T^{-1}\left(B\left(\mathbf{O}, \frac{1}{k}\right)^\circ\right) - T^{-1}\left(B\left(\mathbf{O}, \frac{1}{k}\right)^\circ\right)$ contains a neighborhood of $\mathbf{O}$. However

$$B\left(\mathbf{O}, \frac{1}{3k}\right)^\circ - B\left(\mathbf{O}, \frac{1}{3k}\right)^\circ \subset B\left(\mathbf{O}, \frac{1}{k}\right)^\circ \subset 3B\left(\mathbf{O}, \frac{1}{3k}\right)^\circ$$
$$T^{-1}\left(B\left(\mathbf{O}, \frac{1}{k}\right)^\circ\right) \supset T^{-1}\left(B\left(\mathbf{O}, \frac{1}{3k}\right)^\circ\right) - T^{-1}\left(B\left(\mathbf{O}, \frac{1}{3k}\right)^\circ\right)$$

whence $T^{-1}\left(B\left(\mathbf{O}, \frac{1}{k}\right)^\circ\right)$ contains a neighborhood of $\mathbf{O}$. Correspondingly, if $T(\mathbf{y}) = \mathbf{z}$ then $T^{-1}\left(B\left(\mathbf{z}, \frac{1}{k}\right)^\circ\right) \supset \mathbf{y} + T^{-1}\left(B\left(\mathbf{O}, \frac{1}{k}\right)^\circ\right)$ which contains a neighborhood of $\mathbf{y}$. Thus if U is open in $\mathbb{R}^m$ then for some sequence $\left\{\mathbf{z}_p, \frac{1}{k_p}\right\}_{p\in\mathbb{N}}$ in $\mathbb{Q}^m \times \mathbb{Q}$, $U = \bigcup_{p\in\mathbb{N}} B\left(\mathbf{z}_p, \frac{1}{k_p}\right)^\circ$ and so $T^{-1}(U)$ is open, i.e., T is continuous.

[Since $T(\mathbf{x}+\mathbf{y}) \equiv T(\mathbf{x})+T(\mathbf{y})$, it follows that for n in $\mathbb{N}$, $T(n\mathbf{x}) = nT(\mathbf{x})$ and, more generally, for s in $\mathbb{Q}$, $T(s\mathbf{x}) = sT(\mathbf{x})$. If $a \in \mathbb{R}$, $T(a\mathbf{x}) = aT(\mathbf{x})$ because T is continuous.] ☐

5.98. Let E_m be $\{x : a_m(x) = 0\}$ and let E be $\bigcup_{m\in\mathbb{N}} E_m$. Each E_m is the union of 10^{m-1} pairwise disjoint intervals, each of length 10^{-m}. Hence each of $E_m, E, [0,1] \setminus E$ is in S_β and $[0,1] \setminus E = f^{-1}(k)$. Furthermore,

$$E = E_1 \uplus (E_2 \setminus E_1) \uplus (E_3 \setminus (E_1 \cup E_2)) \uplus \cdots$$

$$\lambda(E) = \frac{1}{10} \sum_{r=1}^{\infty} \left(\frac{9}{10}\right)^r = 1, \ \lambda\left(f^{-1}(k)\right) = 0$$

$$f^{-1}(1) = (E_1 \setminus E_2) \cup (E_1 \cap E_2 \cap E_3 \setminus E_4) \cup \cdots \in \mathsf{S}_\beta$$

$$\lambda\left(f^{-1}(1)\right) = \left(\frac{1}{10} - \frac{1}{10^2}\right) + \left(\frac{1}{10^3} - \frac{1}{10^4}\right) + \cdots = \frac{1}{11}.$$

Hence $f \in \mathcal{M}$, $f \geq 0$, and $\int_0^1 f(x)\,dx = \frac{1}{11}$ because $\lambda\left(f^{-1}(k)\right) = 0$. □

5.99. a) If $\{x_m\}_{m\in\mathbb{N}} \subset E$ and $\lim_{m\to\infty} x_m = x$, let the decimal representation for x be $\sum_{n=1}^{\infty} \frac{b_n}{10^n}$. If $|x - x_m| < \frac{1}{10^p}$ then $b_p = 2$ or 7 whence $x \in E$, i.e., E is closed.

b) Since $[0,1]$ is connected, E is closed, and $E \neq [0,1]$, E is not open.

c) $\#(E) = \#\left(\{2,7\}^{\mathbb{N}}\right) = \mathfrak{c}$.

d) Since $E \cap (.28, .7) = \emptyset$, E is not dense. (Alternatively, E is closed and $E \neq [0,1]$.)

e) Since E is closed, E is Lebesgue measurable. In fact, E is like the Cantor set C_0 and $\lambda(E) = 1 - .8\sum_{m=0}^{\infty}(.2)^n = 0$. □

5.100. Define a sequence $\{a_n\}_{n\in\mathbb{N}}$ as follows:

Each a_n is 0 or k and these occur alternately in blocks of size $1, 2, 2^2, 2^3, \ldots$, i.e., $0, k, k, 0, 0, 0, 0, k, k, k, k, k, k, k, k, \ldots$.

If $a = \sum_{n=1}^{\infty} \frac{a_n}{m^n}$ then on the one hand, $\overline{\lim}_{N\to\infty} \frac{A_N(x,k,m)}{N} = \frac{2}{3}$, whereas $\underline{\lim}_{N\to\infty} \frac{A_N(x,k,m)}{N} = \frac{1}{3}$. Hence $a \in E$.

If $p \in \mathbb{N}$, $x \stackrel{\text{def}}{=} \sum_{n=1}^{\infty} \frac{\epsilon_n(x)}{m^n}$, and $f_p(x) = \begin{cases} 1 & \text{if } \epsilon_p(x) = k \\ 0 & \text{otherwise} \end{cases}$ then $f_p \in \mathcal{M}$ and $A_N(\cdot, k, m) \in \mathcal{M}$ because $A_N(x,k,m) = \sum_{p=1}^{N} f_p(x)$. Furthermore, $E \in \mathsf{S}_\lambda$ because

$$[0,1] \setminus E = \left\{ x \ : \ \lim_{N\to\infty} \frac{A_N(x,k,m)}{N} \text{ exists} \right\} \in \mathsf{S}_\lambda.$$

For x in $[0,1]$, if $\delta > 0$ and $\sum_{n=1}^{\infty} \frac{\epsilon_n(x)}{m^n}$ is the m-ary representation for x, let p be such that $\frac{1}{m^p} < \delta$. The argument showing $E \neq \emptyset$ and used on the m-marker places following the pth, shows as well that there is in E a y such that $|x - y| < \delta$. Hence E is dense and since $\frac{km}{m-1} = \sum_{n=1}^{\infty} \frac{k}{m^n} \notin E$, E is not closed.

For x, δ, and p as above, if $y = \sum_{n=1}^{p} \frac{\epsilon_n(x)}{m^n} + \sum_{n=p+1}^{\infty} \frac{k}{m^n}$ then $y \notin E$ and $|x - y| < \delta$, whence $[0,1] \setminus E$ is dense and so E is not open. □

5.101. In each of the intervals deleted in the process of constructing a Cantor-like set, there can be constructed a Cantor-like set, and this process may be repeated without surcease. The Cantor-like sets so created are pairwise disjoint, their union is dense in the underlying interval, and the union may be constructed so that its measure is any number in $[0, 1)$.

In each interval $[n, n+1), n \in \mathbb{Z}$, construct a countable dense union S_n of Cantor-like sets such that $\lambda(S_n) \in (0,1)$ and $\sum_{n=-\infty}^{\infty} \lambda(S_n) < \infty$. If $E = \bigcup_{n\in\mathbb{Z}} S_n$ then $\lambda(E) < \infty$. Every nondegenerate interval (a, b) contains an interval deleted at some stage of the construction of the sequence $\{S_n\}_{n\in\mathbb{N}}$. Such a (deleted) interval contains a Cantor-like set of positive measure and so $0 < \lambda(E \cap [a, b]) < b - a$. □

5.102. The set E of **Solution 5.101** serves. □

5.103. If

$$F_{mn} \stackrel{\text{def}}{=} \left[\frac{m-1}{2^n}, \frac{m}{2^n}\right), \ 1 \le m \le 2^n, n \in \mathbb{N},$$
$$E_k \stackrel{\text{def}}{=} F_{1k} \cup F_{3k} \cup \cdots \cup F_{2^k-1,k}$$

then $\lambda(E_k) \equiv \frac{1}{2}$. If $l = k + p$ then $\lambda(E_k \cap E_l) \le \frac{1}{2^p}\lambda(E_k)$, whence if $n_k \le n_{k+1}, k \in \mathbb{N}$, then $\lambda\left(\bigcap_{k\in\mathbb{N}} E_{n_k}\right) = 0$. □

5.104. The G_k are pairwise disjoint. Hence $\lambda\left(\overline{\lim}_{n\to\infty} A_n\right) \stackrel{\text{def}}{=} \lambda(G_\infty) = 0$ (cf. **5.6**). Furthermore,

$$\begin{aligned} H_k &\stackrel{\text{def}}{=} \left\{ \bigcap_{p=1}^{k} A_{n_p} \ : \ 1 \le n_1 < n_2 < \cdots < n_k \right\} \\ &= \{ x \ : \ x \text{ belongs to at least } k \text{ of the } A_n \} \supset H_{k+1}, \end{aligned}$$

and $G_k = H_k \setminus H_{k+1}$ whence $H_k, G_k \in \mathsf{S}_\lambda$.

Since $A_n \setminus \left(\bigcup_{k\in\mathbb{N}} (G_k \cap A_n)\right) \subset G_\infty$ and

$$A_n = \left[A_n \setminus \left(\bigcup_{k\in\mathbb{N}} (G_k \cap A_n)\right)\right] \cup \left(\bigcup_{k\in\mathbb{N}} (G_k \cap A_n)\right),$$

it follows that

$$\lambda(A_n) = \sum_{k=1}^{\infty} \lambda(G_k \cap A_n), \ \sum_{n=1}^{\infty} \lambda(A_n) = \sum_{k,n\in\mathbb{N}} \lambda(G_k \cap A_n). \qquad \text{(s5.9)}$$

[The double series in the right member of (s5.9) converges absolutely and thus the order of summation is of no consequence.]

Since $G_k \subset \bigcup_{p_1<\cdots<p_k} A_{p_1} \cap \cdots \cap A_{p_k}$, it follows that

$$G_k = \biguplus_{p_1<\cdots<p_k} \left(G_k \cap A_{p_1} \cap \cdots \cap A_{p_k}\right).$$

Thus

$$\sum_{n=1}^{\infty} \lambda\left(G_k \cap A_n\right) = \sum_{n=1}^{\infty} \sum_{\substack{p_1<\cdots<p_{k-1} \\ n \notin \{p_1,\ldots,p_{k-1}\}}} \lambda\left(G_k \cap A_{p_1} \cap \cdots \cap A_{p_{k-1}} \cap A_n\right).$$

Each term in the right member above, e.g., $\lambda\left(G_k \cap A_{r_1} \cap \cdots \cap A_{r_k} \cap A_n\right)$, occurs exactly k times, namely when $n = r_k$ and $p_1 = r_1, \ldots, p_{k-1} = r_{k-1}$, when $n = r_{k-1}$ and $p_1 = r_1, \ldots, p_{k-1} = r_k$, etc. Hence (s5.9) implies

$$\sum_{n=1}^{\infty} \lambda\left(G_k \cap A_n\right) = k\lambda\left(G_k\right), \ \sum_{k=1}^{\infty} k\lambda\left(G_k\right) = \sum_{n=1}^{\infty} \lambda\left(A_n\right). \qquad \square$$

5.105. If $a > 0$ then for some k in $\mathbb{N}$, $0 < \left(\sqrt{2}-1\right)^k \stackrel{\text{def}}{=} m + n\sqrt{2} < a$, whence $S \stackrel{\text{def}}{=} \left\{m + n\sqrt{2} \ : \ m, n \in \mathbb{Z}\right\}$ is dense in $\mathbb{R}$.

If $a < b$ and $\epsilon > 0$ then for some α and β in S, $\alpha < a < b < \beta$ and $\beta - \alpha < b - a + \epsilon$. Hence

$$\int_a^b f(x)\,dx = \int_\alpha^\beta - \int_\alpha^a - \int_b^\beta f(x)\,dx = -\int_\alpha^a - \int_b^\beta f(x)\,dx.$$

As $\epsilon \to 0$, the last terms above approach zero, whence $\int_a^b f(x)\,dx = 0$. If $f \neq 0$ on a set of positive measure, it may be assumed that for some positive δ, $\lambda\left(E_>(f,\delta)\right) > 0$. However, $E_>(f,\delta)$ may be covered by a countable union of intervals and for at least one, say J, $\int_J f(x)\,dx > 0$, a contradiction.

$\square$

5.106. If $E \stackrel{\text{def}}{=} \{x \ : \ |f(x)| = a < 1\}$ is not a null set (τ) then

$$\begin{aligned}
\left|\sigma_N\left(f, x_0\right)\right| &\leq \int_{\mathbf{T}} F_N\left(y^{-1}x_0\right) |f(y)|\,d\tau(y) \\
&\leq \int_E + \int_{\mathbf{T}\setminus E} F_N\left(y^{-1}x_0\right) |f(y)|\,d\tau(y) \\
&< \int_E + \int_{\mathbf{T}\setminus E} F_N\left(y^{-1}x_0\right)\,d\tau(y) = 1,
\end{aligned}$$

a contradiction.

Hence $|f(x)| = 1$ a.e. If $A_\pm \stackrel{\text{def}}{=} f^{-1}(\pm 1)$ and $\lambda(A_+)\lambda(A_-) > 0$ then

$$|\sigma_N(f, x_0)| = \left| \int_{A_+} - \int_{A_-} F_N\left(y^{-1}x_0\right) d\tau(y) \right| < 1,$$

a contradiction. Thus $f(x) = 1$ a.e. or $f(x) = -1$ a.e.

When $\mathbb{T}$ is regarded as $[0, 2\pi)$ then $f : \mathbb{T} \ni x \mapsto x$ corresponds to $g : [0, 2\pi) \ni \theta \mapsto e^{i\theta}$. Then $\sigma_N(g, \theta) = g(\theta)$, $\|g\|_\infty = 1$. There is no constant c such that $g(\theta) \doteq c$. □

[**Note s5.5:** The sequence $\{F_N\}_{N\in\mathbb{N}}$ of Fejér kernels is an example of an approximate identity in that for any f in $L^1(\mathbb{T}, \tau)$, $\lim_{N\to\infty} \|f - F_N * f\|_1 = 0$.

If G is a locally compact group and μ is Haar measure, let $\mathcal{U}$ be the set of all open neighborhoods of the identity of G. Partially ordered by reversed inclusion: $U' \succ U$ iff $U' \subset U$, $\mathcal{U}$ is a diset. A general form of an approximate identity for $L^1(G, \mu)$ is a net $n : \mathcal{U} \ni U \mapsto n_U \in L^1(G, \mu)$ such that $n_U(x) = 0$ off U, $n_U \geq 0$, and $\|n_U\|_1 = 1$. For such a net n and f in $L^1(G, \mu)$, $n_U * f$ $\|\ \|_1$-converges to f.]

5.107. Let $g : \mathbb{Q} \cap (1/4, 3/4) \mapsto \mathbb{Q} \cap [1/4, 3/4]$ be a bijection. Extend g^{-1} to $[0, 1]$ to a map

$$G : [0,1] \ni x \mapsto \begin{cases} x & \text{if } x \notin \mathbb{Q} \cap [1/4, 3/4] \\ g^{-1}(x) & \text{otherwise} \end{cases}.$$

Then $G : [0, 1] \mapsto [0, 1]$ is a bijection and $f \stackrel{\text{def}}{=} G^{-1}$ meets the requirements. □

5.108. The construction in **Solution 5.101** may be modified to produce Cantor-like sets C_{α_n} in $[n, n+1)$ so that $\sum_{n\in\mathbb{Z}} \lambda(C_{\alpha_n}) = 1$. If A is a Cantor-like set in $[a, b]$, the corresponding Cantor-like function g_A is monotonely increasing and maps A onto $[0, 1]$ and $h_A : (a, b) \ni x \mapsto \tan\left(\frac{\pi}{2} g_A(x)\right)$ maps $A \cap (a, b)$ onto $(0, \infty)$. Hence if, for each n in $\mathbb{N}$ and for every Cantor-like set A constructed in $[n, n+1)$,

$$f(x) = \begin{cases} h_A(x) & \text{if } x \in A \\ 0 & \text{otherwise} \end{cases}$$

and $a < b$ then $\int_a^b f(x)\, dx = \infty$. □

5.109. If $E_{kn} \stackrel{\text{def}}{=} f^{-1}\left(\left[\frac{k}{2^n}, \frac{k+1}{2^n}\right)\right)$ and $f_n \stackrel{\text{def}}{=} \sum_{k=-\infty}^{\infty} \frac{k}{2^n} \chi_{E_{kn}}$ then $0 \leq f - f_n \leq 2^{-n}$ and $f_n \uparrow f$. □

5.110. If U is an open subset of $\mathbb{R}$ then $(g\circ f)^{-1}(U) = f^{-1}\left(g^{-1}(U)\right)$. Since $g^{-1}(U)$ is Lebesgue measurable, there is a Borel set A and two null sets N_1, N_2 (λ) such that $N_1 \subset A, A\cap N_2 = \emptyset, g^{-1}(U) = (A\setminus N_1)\cup N_2$. Since f is continuous, $f^{-1}(A) \in \mathsf{S}_\beta(\mathbb{R})$ while $f^{-1}(N_i) \in \mathsf{S}_\lambda(\mathbb{R}), i = 1, 2$. Hence $(g\circ f)^{-1}(U) = \left(f^{-1}(A)\setminus f^{-1}(N_1)\right)\cup f^{-1}(N_2) \in \mathsf{S}_\lambda$. □

[**Note s5.6:** Compare the result above with **5.67** where the hypothesis about the Lebesgue measurability of sets $f^{-1}(N)$ for null sets (λ) is dropped.]

5.111. The metric density theorem implies that if x is outside a null subset N of A then $\lim_{d>0,d\to 0} \dfrac{\lambda^*(A\cap(x-d,x+d))}{2d} = 1$. Since $\lambda^*(A) > 0$, $A\setminus N \neq \emptyset$ and for some positive d, $\lambda^*(A\cap(x-d,x+d)) > \theta\cdot 2d$. □

5.112. For the map

$$g : X \ni x \mapsto \begin{cases} f(x) & \text{if } |f(x)| \le \|f\|_\infty \\ 0 & \text{otherwise} \end{cases},$$

$f(x) \doteq g(x)$ and $\sup_x |g(x)| = \|f\|_\infty$. □

5.113. The map $h : [0,1] \ni x \mapsto \lambda(\{y : f(y) \ge x\})$ is a monotonely decreasing function and $0 \le h \le 1$. Furthermore, $\lim_{x\downarrow -\infty} h(x) \uparrow 1$ and $\lim_{x\uparrow\infty} h(x) \downarrow 0$.

If $x > a_0 \stackrel{\text{def}}{=} \sup\left\{x : h(x) \ge \dfrac{1}{2}\right\}$ then $h(x) < \dfrac{1}{2}$. On the other hand,

$$h(a_0) = \lambda\left(\bigcap_{n\in\mathbb{N}}\left\{y : f(y) > a_0 - \frac{1}{n}\right\}\right) \ge \frac{1}{2}.$$

Finally, if $a_1 < a_0$ and $h(a_1) \ge \dfrac{1}{2}$, the monotone character of h implies that for x in (a_1, a_0), $h(x) \ge \dfrac{1}{2}$, whence b) fails for a_1. □

5.114. The Stone-Weierstraß resp. Fejér theorem implies that for g in $C([0,1],\mathbb{R})$, $\int_{[0,1]} f(x)g(x)\,dx = 0$. The Daniell theory of integration then implies that if $f = u + iv$, the functionals

$$I_u : C([0,1],\mathbb{R}) \ni g \mapsto \int_{[0,1]} u(x)g(x)\,dx$$

$$I_v : C([0,1],\mathbb{R}) \ni g \mapsto \int_{[0,1]} v(x)g(x)\,dx$$

are both zero and their unique extensions are zero. Hence if $E \in \mathsf{S}_\lambda$ then $\int_{[0,1]} f(x)\chi_E(x)\,dx = 0$ whence $f(x) \doteq 0$. □

5.115. If $F(u) \overset{\text{def}}{=} f(e^u), G(v) \overset{\text{def}}{=} g(e^v)$ then $F, G \in L^1(\mathbb{R}, \lambda)$ and

$$\int_{(0,\infty)} \frac{\left|f(xy)g\left(\frac{1}{y}\right)\right|}{y}\, dy = \int_{\mathbb{R}} F(u+v)G(-v)\, dv = F * G(u). \qquad \square$$

5.116. It may be assumed that $\|g\|_1 > 0$. If $\epsilon > 0$, there is a positive A such that $\int_{\mathbb{R}\setminus[-A,A]} |g(x)|\, dx < \frac{\epsilon}{6}$. If $0 < b < \frac{\epsilon}{6A}$ then for some N in $\mathbb{N}$,

$$\{m, n > N\} \Rightarrow \left\{\lambda(\{x : |f_m(x) - f_n(x)| \geq b\}) < \frac{\epsilon}{6\|g\|_1}\right\}.$$

Since $|f_n| \leq |g|$ a.e., $\{f_n\}_{n\in\mathbb{N}} \subset L^1(\mathbb{R}, \lambda)$. If

$$E_{mn} \overset{\text{def}}{=} \{x : |f_m(x) - f_n(x)| \geq b\},$$

then

$$\begin{aligned}\|f_m - f_n\|_1 &\leq \int_{\mathbb{R}\setminus[-A,A]} + \int_{[-A,A]\cap E_{mn}} + \int_{[-A,A]\setminus E_{mn}} |f_m(x) - f_n(x)|\, dx\\ &< \frac{\epsilon}{3} + \frac{\epsilon}{3} + \frac{\epsilon}{3} = \epsilon.\end{aligned}$$

Hence $\{f_n\}_{n\in\mathbb{N}}$ is a $\|\ \|_1$-Cauchy sequence. Let h be its $\|\ \|_1$-limit. There is a subsequence $\{f_{n_k}\}_{k\in\mathbb{N}}$ such that $f_{n_k} \overset{\text{a.e.}}{\to} h$ whence $h \doteq f$. $\square$

5.117. For some n in $\mathbb{N}$, $\lambda(E \cap [-n, n]) \overset{\text{def}}{=} \lambda(F) > 0$ and so it may be assumed that $0 < \lambda(E) < \infty$ and hence that $\chi_E \in L^1(\mathbb{R}, \lambda)$. Consequently $f : \mathbb{R} \ni x \mapsto \int_{\mathbb{R}} \chi_E(2x - y)\chi_E(y)\, dy$ is continuous, nonnegative, and zero off $\frac{1}{2}(E + E)$ which is a subset of E. Furthermore, $f\left(\frac{1}{2}x\right) = \chi_E * \chi_E(x)$, whence if $f = 0$ then $\widehat{\chi_E}^2 = 0$. Plancherel's theorem implies $\|\chi_E\|_2^2 = 0$ while $\|\chi_E\|_1 > 0$, a contradiction. Thus f is positive on an open set U, which must be a subset of E (cf. **Solution 5.81**). $\square$

5.118. a) Every subset of the Cantor set C_0 is in S_λ, $\#(C_0) = \mathfrak{c}$, whence $2^{\mathfrak{c}} \geq \#(\mathsf{S}_\lambda) \geq 2^{\mathfrak{c}}$. $\square$

5.119. Since $\mathbb{R}^n$ is separable, there is a sequence $\{B(\mathbf{a}_n, r_n)\}_{n\in\mathbb{N}}$ such that $E \subset \bigcup_{n\in\mathbb{N}} B(\mathbf{a}_n, r_n)$, i.e., $E = \bigcup_{n\in\mathbb{N}} E \cap B(\mathbf{a}_n, r_n)$. $\square$

5.120. On the one hand, $\#(\mathsf{S}_\beta) = \mathfrak{c}$ (cf. **1.6**), but $\#(\mathsf{S}_\lambda) = 2^{\mathfrak{c}}\ (> \mathfrak{c})$ (cf. **5.118**). If $E \in \mathsf{S}_\lambda \setminus \mathsf{S}_\beta$, nothing need be proved. If $E \in \mathsf{S}_\beta$ then $\#(E) = \mathfrak{c}$ (cf. **2.29**). Since $\lambda(E) = 0$, every subset of E is in S_λ and since $\#(2^E) > \mathfrak{c} = \#(\mathsf{S}_\beta)$, E must contain a set that is not Borel measurable. $\square$

5.121. If $E \in \mathsf{S}_\lambda$, there are null sets (λ) N_1, N_2 and in S_β an A such that $E = (A \setminus N_1) \cup N_2$. Thus $\mu(N_1) = \mu(N_2) = 0$ and $\mathsf{S}_\beta \subset \mathsf{S}_\mu$, whence $E \in \mathsf{S}_\mu$. □

5.122. If $f_n(x) \equiv \chi_{[0,1)}(x) \cdot x + 2\chi_{\{1\}}$ and $f_0(x) = x, x \in [0,1]$ then each f_n is monotonely increasing, $f_n \overset{\text{a.e.}}{\to} f_0$, f_0 is continuous, but

$$\lim_{n\to\infty} f_n(1) \neq f_0(1). \qquad \square$$

5.123. a) If $\{g_n\}_{n\in\mathbb{N}} \subset A_f$ and $g_n \overset{\|\ \|_1}{\to} g$ then there is a subsequence $\{g_{n_k}\}_{k\in\mathbb{N}}$ such that $g_{n_k} \overset{\text{a.e.}}{\to} g$ whence $|g(x)| \le f(x)$ a.e.

b) If $g_n \overset{\text{def}}{=} f \wedge n, n \in \mathbb{N}$, then $\{g_n\}_{n\in\mathbb{N}} \subset A_f$. If A_f is $\|\ \|_1$-compact then $\{\|g_n\|_1\}_{n\in\mathbb{N}}$ is bounded whence $f \in L^1([0,1],\lambda)$. □

5.124. If $S \subset [0,1]$, $s \notin \mathsf{S}_{\lambda_2}([0,1]^2)$, and

$$E \overset{\text{def}}{=} (S \times \{0\}) \cup (\{0\} \times S)$$

then $\lambda_2(E) = 0$ and so $E \in \mathsf{S}_{\lambda_2}([0,1]^2)$. On the other hand, if $F \overset{\text{def}}{=} E + E$ and $x \in E$ then $F_x = S \notin \mathsf{S}_\lambda$. Since $\lambda^*(S) > 0$, Fubini's theorem implies $F \notin \mathsf{S}_{\lambda_2}([0,1]^2)$. □

5.125. For E the Cantor set C_0, $E + E = [0,2]$, whence C_0 contains a Hamel basis B. If $S \overset{\text{def}}{=} \bigcup_{r\in\mathbb{Q}}(r + B)$ and $S_1 \overset{\text{def}}{=} S \cup (-S)$ then $\lambda(S_1) = 0$ and $S_1 = -S_1$. Let S_{n+1} be $S_n + S_n$ $(= S_n - S_n)$, $n \in \mathbb{N}$. If

$$x \overset{\text{def}}{=} \sum_{b\in B} a_b b \in S_n,\ a_b \in \mathbb{Q},$$

then x is a sum of not more than 2^{n-1} members of S_1 and hence at most 2^{n-1} of the coefficients a_b are not zero. Since $C_0 + C_0 = [0,2]$, it follows that $\bigcup_{n\in\mathbb{N}} S_n = \mathbb{R}$. If each S_n is measurable then one of them, say S_{n_0}, has positive Lebesgue measure: $\lambda(S_{n_0}) > 0$. Hence for some M in $\mathbb{N}$,

$$S_{n_0} - S_{n_0}\ (= S_{n_0} + S_{n_0} = S_{n_0+1}) \supset \left(-2^{-M}, 2^{-M}\right) \overset{\text{def}}{=} U.$$

If $K \in \mathbb{N}$ and $2^{n_0} < K$, $K \sup_{1\le k\le K} |r_k| < 2^{-M}, \mathbb{Q} \setminus \{0\} \supset \{r_1, \ldots, r_K\}$ (a K-element set) then for each K-element subset $\{b_1, \ldots, b_K\}$ of B,

$$\sum_{k=1}^{K} r_k b_k \in U \setminus S_{n_0+1} = \emptyset,$$

a contradiction.

Thus some $S_n \notin \mathsf{S}_\lambda$ and if S_{n_1} is the first such S_n then $n_1 > 1$. Hence $S_{n_1-1} \in \mathsf{S}_\lambda$ and $S_{n_1-1}+S_{n_1-1} = S_{n_1} \notin \mathsf{S}_\lambda$. (This result was communicated to the writer by Harvey Diamond and Gregory Gellés.) □

5.126. Owing to the Riesz representation theorem, each μ_n may be regarded as an element F_n of $\left(L^1([0,1],\lambda)\right)^*$ $[= L^\infty([0,1],\lambda)]$ and a) implies that $\sup_{n\in\mathbb{N}} \|F_n\|_\infty \stackrel{\text{def}}{=} M < \infty$.

Let S be the $\mathbb{Q}$-span of the set of characteristic functions of open subintervals with rational endpoints in $[0,1]$. Then b) implies that the technique of proof of the Arzelà-Ascoli theorem leads to a subsequence $\{F_{n_k}\}_{k\in\mathbb{N}}$ such that $\lim_{k\to\infty} F_{n_k}(f)$ exists for each f in the countable $\|\ \|_1$-dense set S. If $g \in L^1([0,1],\lambda)$ then for f in S,

$$|F_{n_k}(g) - F_{n_l}(g)| \le |F_{n_k}(g) - F_{n_k}(f)| + |F_{n_k}(f) - F_{n_l}(f)| + |F_{n_l}(f) - F_{n_l}(g)|.$$

If $\epsilon > 0$ then for suitable f, the first and third terms of the left member above do not exceed $\frac{\epsilon}{3}$ whereupon for large k the second term does not exceed $\frac{\epsilon}{3}$. □

5.127. a) $0 \le f_n(x) = n\int_x^{x+\frac{1}{n}} \chi_E(t)\,dt \le n\frac{1}{n} = 1$.

b) If $a < b$ then

$$|f_n(b) - f_n(a)| = n\left|\int_{a+\frac{1}{n}}^{b+\frac{1}{n}} - \int_a^b \chi_E(t)\,dt\right| \le 2n\int_a^b \chi_E(t)\,dt \le 2n(b-a).$$

If $a_1 < b_1 \le a_2 < \cdots \le a_k < b_k < a_{k+1} < \cdots \le a_m < b_m$, $\epsilon > 0$, and $\sum_{k=1}^m (b_k - a_k) < \frac{\epsilon}{2n}$ then $\sum_{k=1}^m |f_n(b_k) - f_n(a_k)| < \epsilon$.

c) The Fundamental Theorem of Calculus (in the context of Lebesgue integration) implies that $f_n \stackrel{\text{a.e.}}{\to} \chi_E$.

d) Since $E \subset [0,1]$ and $f_n(x) = 0$ off $\left[-\frac{1}{n}, 1\right]$, it follows that

$$\int_{\mathbb{R}} |f_n(x) - \chi_E(x)|\,dx = \int_{-1}^{1} |f_n(x) - \chi_E(x)|\,dx$$

and the dominated convergence theorem implies $f_n \stackrel{\|\ \|_1}{\to} \chi_E$. □

5.128. If $[p,q] \subset [c,d]$ and $r = f^{-1}(p)$, $s = f^{-1}(q)$, then

$$\mu([p,q]) \stackrel{\text{def}}{=} \int_{f^{-1}([p,q])} f'(x)\,dx = f(s) - f(r) = q - p = \lambda([p,q]).$$

Since $f' \in L^1([a,b],\lambda)$, and μ coincides with λ on all intervals, $\mu = \lambda$. □

5.129. Let f be ϕ_0, **the** Cantor function, and let A be C_0, **the** Cantor set. □

5.130. a) Since $f \in BV$, there are two nonnegative monotonely increasing functions f_1, f_2 such that $f = f_1 - f_2$ and f_1', f_2' are finite a.e. The LRN theorem implies $f_i(x) \geq f_i(0) + \int_0^x f_i'(t)\,dt, i = 1, 2$.

b) It may be assumed that $f(0) = 0$ [otherwise the argument applies to $g \stackrel{\text{def}}{=} f - f(0)$]. If $f_1(x) \stackrel{\text{def}}{=} \int_0^x |f'(x)|\,dx$ and $f_2 \stackrel{\text{def}}{=} f_1 - f$, then both f_1 and f_2 are monotonely increasing absolutely continuous functions. □

5.131. Let $\left\{\{J_{nk}\}_{1\leq k\leq 2^{n-1}}\right\}_{n\in\mathbb{N}}$ be the systematic enumeration of the intervals deleted in the construction of **the** Cantor set C_0. If

$$f_{nk} = 2^{n-1}\chi_{J_{nk}},\ f = \sum_{n=1}^{\infty}\left(\sum_{k=1}^{2^{n-1}} f_{nk}\right)$$

then f is continuous on $[0,1]\setminus C_0$, whence a.e. Since f is unbounded, there is no function g continuous on $[0,1]$ and such that $f \doteq g$.

On the other hand, the function $f \stackrel{\text{def}}{=} \chi_{\mathbb{I}} \doteq 1$, but f is continuous nowhere. □

5.132. If $E, F \in \mathsf{S}_\lambda$, $f = \chi_E$, and $g = \chi_F$, then

$$A_y = \begin{cases} \mathbb{R} & \text{if } y \leq 0 \\ E & \text{if } 0 < y \leq 1 \\ \emptyset & \text{if } 1 < y \end{cases} \text{ and } \int_{\mathbb{R}} f(x)g(x)\,dx = \lambda(E \cap F),$$

whereas

$$h(y) = \begin{cases} \lambda(F) & \text{if } y \leq 0 \\ \lambda(E\cap F) & \text{if } 0 < y \leq 1 \\ 0 & \text{if } 1 < y \end{cases} \text{ and}$$

$$\int_0^{\infty} h(y)\,dy = \int_0^1 h(y)\,dy = \lambda(E\cap F) = \int_{\mathbb{R}} f(x)g(x)\,dx.$$

If $\{E_n\}_{1\leq n\leq N<\infty}$, $\{F_m\}_{1\leq m\leq M}$ are two finite sets of pairwise disjoint measurable sets, if $0 < a_1 < \cdots < a_N$, and $0 < b_1 < \cdots < b_M$ then for the nonnegative simple functions $f \stackrel{\text{def}}{=} \sum_{n=1}^N a_n\chi_{E_n}$ and $g \stackrel{\text{def}}{=} \sum_{m=1}^M b_m\chi_{F_m}$,

$$A_y = \begin{cases} \mathbb{R} & \text{if } y \leq 0 \\ E_1 \cup \cdots \cup E_N & \text{if } 0 < y \leq a_1 \\ E_1 \cup \cdots \cup E_{N-1} & \text{if } a_1 < y \leq a_2 \\ \vdots & \vdots \\ E_1 & \text{if } a_{n-1} < y \leq a_N \\ \emptyset & \text{if } a_N < y \end{cases}$$

and

$$h(y) = \begin{cases} \sum_{m=1}^{M} b_m \lambda(F_m) & \text{if } y \le 0 \\ \sum_{n,m=1}^{N,M} b_m \lambda(E_n \cap F_m) & \text{if } 0 < y \le a_1 \\ \sum_{n,m=1}^{N-1,M} b_m \lambda(E_n \cap F_m) & \text{if } a_1 < y \le a_2 \\ \vdots & \vdots \\ \sum_{m=1}^{M} b_m \lambda(E_1 \cap F_m) & \text{if } a_{N-1} < y \le a_N \\ 0 & \text{if } a_N < y. \end{cases}$$

Integration by parts shows $\int_0^\infty h(y)\,dy = \int_{\mathbb{R}} f(x)g(x)\,dx$.

The general case is handled via monotonely increasing sequences of simple functions that are approximants to f and g. □

5.133. a) If $M \overset{\text{def}}{=} \|g\|_\infty < \infty$ then $|f(b) - f(a)| \le M|b - a|$.

Conversely, if $\mathbb{R}^{[0,1]} \ni f \in \text{Lip}(1)$ then $f \in AC$, whence for some g in $L^1([0,1],\lambda)$ if $0 \le x < y \le 1$, $f(y) - f(x) = \int_x^y g(t)\,dt$. If g is not essentially bounded, it may be assumed that for all n in $\mathbb{N}$ and $\lambda(E_n) \overset{\text{def}}{=} \lambda(\{x : g^+(x) \ge n\}) > 0$.

Thus $n\lambda(E_n) \le \int_{E_n} g^+(x)\,dx$. If $a > 0$ then E_n is contained in an open set U_n such that $\lambda(U_n) < \lambda(E_n)(1 + a)$. As an open set in $\mathbb{R}$, U_n is the union of at most countably many pairwise disjoint open intervals (a_{nk}, b_{nk}): $U_n = \bigcup_k (a_{nk}, b_{nk})$.

Let K be the Lipschitz constant for f: $|f(y) - f(x)| \le K|y - x|$. Then, owing to the fact that $\mu : E \mapsto \int_E g^+(x)\,dx$ is a measure, and $\mu \ll \lambda$, for all sufficiently small positive a,

$$\begin{aligned} n\lambda(E_n)(1-a) &\le \int_{U_n} g^+(x)\,dx = \sum_k \int_{J_{nk}} g^+(x)\,dx \\ &= \sum_k f(b_{nk}) - f(a_{nk}) \le K\lambda(U_n) < K\lambda(E_n)(1+a) \end{aligned},$$

whence $n \le K\dfrac{1+a}{1-a}$, and a contradiction emerges as $a \to 0$.

If $f \in \mathbb{C}^{[0,1]}$, the argument above may be applied to the real and imaginary parts of f.

b) If $f \in \text{Lip}(1)$ and $f_n \overset{\text{def}}{=} f, n \in \mathbb{N}$ then $\{f_n\}_{n\in\mathbb{N}}$ serves.

Conversely, if $\lim_{n\to\infty} f_n(0) = f(0)$, $\lim_{n\to\infty} \text{var}_{[0,1]}(f - f_n) = 0$, and the Lipschitz constants $K(f_n)$ form a bounded set then for x in $[0,1]$,

$$|f(x) - f_n(x)| \le \text{var}_{[0,1]}(f - f_n) + |f_n(0) - f(0)|,$$

whence $f_n \overset{\text{u}}{\to} f$. If $\{(a_m, b_m)\}_{1\le m\le M}$ is a finite set of pairwise disjoint intervals contained in $[0,1]$, and $\overline{K} = \sup_{n\in\mathbb{N}} K(f_n)$ then for a set

$$\{(a_m, b_m)\}_{1\le m\le M}$$

of pairwise disjoint intervals,

$$\sum_{m=1}^{M} |f_n(b_m) - f_n(a_m)| \le K \sum_{m=1}^{M} |b_m - a_m|,$$

an inequality that persists as $n \to \infty$. □

[**Note s5.7:** Absent the boundedness of the set $\{K_n\}_{n\in\mathbb{N}}$, the limit of the f_n can fail to be absolutely continuous. The cause of the difficulty is the possibility that for a function g, $\mathrm{var}_{[0,1]}(g)$ can be small while the Lipschitz constant for g can be large owing to a steep but short rise at just one point.

The reader is urged to approximate **the** Cantor function ϕ_0, which is not absolutely continuous, by a sequence $\{f_n\}_{n\in\mathbb{N}}$ of continuous piecewise linear functions such that

$$\lim_{n\to\infty} \mathrm{var}_{[0,1]}(f_n - \phi_0) = 0$$

while the Lipschitz constants $K(f_n)$ form an unbounded set.]

5.134. If μ exists then $|\mu_f| < \infty$. If $f \in C([0,1],\mathbb{C}) \setminus BV([0,1],\mathbb{C})$ then for the putative μ, $|\mu| = \infty$. □

5.135. If $a \le b$ then $\mu_g([a,b)) = \mu_f([a,b)) \cdot (f(b) + f(a))$, whence $\mu_g \ll \mu_f$.

As $b \downarrow a$, $\dfrac{\mu_g([a,b))}{\mu_f([a,b))} = f(b) + f(a) \to f(a+0) + f(a) = \dfrac{d\mu_g}{d\mu_f}\Big|_{x=a}$. □

5.136. When the Daniell construction is applied to the Riemann integral in $C_{00}(\mathbb{R},\mathbb{C})$, among the results is $L^1(\mathbb{R},\lambda)$. □

5.137. The Schwarz inequality implies $L^2([0,1],\lambda) \subset L^1([0,1],\lambda)$ and that for f in $L^2([0,1],\lambda)$, $\left|\int_a^b f(x)\,dx\right|^2 \le |b-a| \int_a^b |f(x)|^2\,dx$. The monotonely increasing function $g : [0,1] \ni x \mapsto \int_a^x |f(t)|^2\,dt$ serves.

Conversely, for a g as described,

$$\left|\frac{1}{b-a}\int_a^b f(x)\,dx\right| \cdot \left|\frac{1}{b-a}\int_a^b \overline{f(y)}\,dy\right| \le \frac{(g(b)-g(a))}{b-a}.$$

For all a off a null set (λ), as $b \to a$, the limits on both sides exist whence

$$|f(a)|^2 \le g'(a), \quad \int_0^1 |f(x)|^2\,dx \le \int_0^1 g'(x)\,dx \le g(1) - g(0) < \infty$$

and so $f \in L^2([0,1],\lambda)$. □

5.138. The Schwarz inequality implies

$$\|G\|_2^2 = \int_0^1 \left|\int_0^x g(t)\,dt\right|^2 dx \le \int_0^1 \left(\int_0^x 1^2\,dt\right)\left(\int_0^x |g(t)|^2\,dt\right) dx$$
$$\le \int_0^1 x\|g\|_2^2\,dx = \frac{1}{2}\|g\|_2^2 < \|g\|_2^2. \qquad \square$$

5.139. By definition, $\int_0^1 f(x)\,dx = \frac{1}{3}\sum_{n=1}^{\infty} n\left(\frac{2}{3}\right)^{n-1}$. If $|x| < 1$

$$\frac{1}{1-x} = \sum_{n=0}^{\infty} x^n,\ \left(\frac{1}{1-x}\right)' = \frac{1}{(1-x)^2} = \sum_{n=1}^{\infty} nx^{n-1},$$

whence $\int_{[0,1]} f(x)\,dx = 3.$ $\square$

5.140. For a fixed Borel set B and $\mu_B : \mathsf{S}_\beta(\mathbb{R}) \ni A \mapsto \mu(A \times B)$, $\mu_B \ll \nu$. Thus the LRN theorem implies there is in $L^1([0,1],\nu)$ an f_B such that for each Borel set A, $\mu_B(A) = \int_A f_B(x)\,d\nu(x)$. For x fixed, the function $f_B(x)$ has the required properties. $\square$

5.141. If $t, s \in A$ and $u \stackrel{\text{def}}{=} t-s$ then $\lim_{n\to\infty} e^{ic_n u}$ exists. Since $\lambda(A) > 0$, there is a positive a such that $A - A \supset (-a,a)$ and so for all u in $(-a,a)$, $\lim_{n\to\infty} e^{ic_n u} \stackrel{\text{def}}{=} g(u)$ exists, $|g(u)| \equiv 1$, and $g\big|_{(-a,a)} \in \mathcal{M}$.

If $\{c_n\}_{n\in\mathbb{N}}$ is unbounded, it may be assumed that $|c_n| \uparrow \infty$, whence if $[p,q] \subset (-a,a)$ then as $n \to \infty$, $\int_p^q e^{ic_n u}\,du = \dfrac{e^{ic_n q} - e^{ic_n p}}{ic_n}$ approaches both zero (because $|c_n| \uparrow \infty$) and $\int_p^q g(u)\,du$ (by virtue of the bounded convergence theorem). Thus $g \doteq 0$, a contradiction since $|g(u)| \equiv 1$. Hence for some finite M, $|c_n| \le M$.

If $\lim_{n\to\infty} c_n$ does not exist, it may be assumed that for some b in $\left(0, \frac{\pi}{M}\right)$, if $m \ne n$, then $|c_m - c_n| \ge b$. If

$$\min\left(\frac{a}{2}, \frac{5\pi}{12M}\right) < |u| < \min\left(a, \frac{5\pi}{6M}\right)$$

then

$$\left.\begin{aligned} 0 < d \stackrel{\text{def}}{=} \min\left(\frac{ub}{4}, \frac{5b\pi}{24M}\right) &< |c_m - c_n| \cdot \frac{|u|}{2} < \frac{5\pi}{6}, \\ 0 < \sin d < \sin\left(\frac{(c_m - c_n)\,u}{2}\right) &< \frac{1}{2} \end{aligned}\right\},$$

whereas $\left|e^{ic_m u} - e^{ic_m u}\right| = \left|2ie^{i\frac{(c_m+c_n)u}{2}} \sin\left(\frac{(c_m - c_n)u}{2}\right)\right| \geq 2\sin d$, a contradiction of the convergence of $\left\{e^{ic_n u}\right\}_{n\in\mathbb{N}}$. □

5.142. For all a off a null set (λ),

$$\lim_{\substack{b\to a\\ b\neq a}} \frac{1}{b-a}\int_a^b f(x)\,dx = f(a) \text{ and } \lim_{\substack{b\to a\\ b\neq a}} \frac{1}{b-a}\int_a^b |f(x)|^p\,dx = |f(a)|^p.$$

Hölder's inequality implies

$$\begin{aligned}\left|\frac{1}{b-a}\int_a^b |f(x)|\,dx\right|^p &\leq c|b-a|^{p-1}\frac{\int_a^b |f(x)|^p\,dx}{|b-a|^p}\\ &= c\left|\frac{1}{b-a}\int_a^b |f(x)|^p\,dx\right|.\end{aligned}$$

Thus $|f(a)|^p \leq c|f(a)|^p$ and, since $0 < c < 1$, $f \doteq 0$. □

5.143. There is in $(0,1)$ an a such that

$$\{0 < x < a\} \Rightarrow \left\{\left|\frac{f(x)}{x}\right| \leq 2\left(|f'(0)| + 1\right) \stackrel{\text{def}}{=} K\right\}.$$

Hence

$$\left|\frac{f(x)}{x^{\frac{3}{2}}}\right| \leq \frac{K}{x^{\frac{1}{2}}} \text{ if } 0 < x < a,\ |f(x)| \leq \|f\|_\infty,\ \frac{1}{x^{\frac{3}{2}}} \leq \frac{1}{a^{\frac{3}{2}}} \text{ if } a \leq x \leq 1.$$

Thus $\displaystyle\int_{(0,1)} |g(x)|\,dx \leq K\int_{(0,a)} \frac{dx}{x^{\frac{1}{2}}} + \frac{1-a}{a^{\frac{3}{2}}}\|f\|_\infty < \infty.$ □

5.144. If $f \in C([0,1],\mathbb{C})$ then the two changes of variable

$$t \to s + \frac{k}{n},\ 0 \leq k \leq n-1 \text{ and } s \to \frac{r}{n}$$

lead to the equations

$$\begin{aligned}\int_{\mathbb{T}} f(t)g(nt)\,d\tau(t) &= \sum_{k=0}^{n-1}\int_{\frac{k}{n}}^{\frac{k+1}{n}} f(t)g(nt)\,d\tau(t)\\ &= \sum_{k=0}^{n-1}\frac{1}{n}\int_{[0,1]} f\left(\frac{r}{n}+\frac{k}{n}\right)g(r)\,dr. \qquad \text{(s5.10)}\end{aligned}$$

For large n and r in $[0,1]$, $\frac{r}{n}$ is near zero and thus

$$\left|f\left(\frac{r}{n}+\frac{k}{n}\right) - f\left(\frac{k}{n}\right)\right|$$

is small, whence the right member of (s5.10) is near

$$\int_{[0,1]}\left(\sum_{k=0}^{n-1} f\left(\frac{k}{n}\right)\frac{1}{n}\right) g(r)\,dr \approx \int_{[0,1]}\left(\int_{[0,1]} f(p)\,dp\right) g(r)\,dr.$$

If $f \in L^1([0,1],\lambda)$ there is in $C([0,1],\mathbb{C})$ an h that is $\|\ \|_1$-near to f. □

5.145. It suffices to prove the result for a singleton set $\{f\}$. Let $\{E_n\}_{n\in\mathbb{N}}$ be a sequence of measurable sets such that $\mu(E_n) < 2^{-(n+1)}$. It follows that $\bigcup_{n\ge m} E_n \stackrel{\text{def}}{=} F_m \supset F_{m+1}$, $\mu(F_m) \le 2^{-m}$, and $g_m \stackrel{\text{def}}{=} \chi_{F_m}|f| \downarrow 0$ a.e. Consequently, $\int_{E_m} |f(x)|\,d\mu(x) \le \int_X g_m(x)\,d\mu(x) \downarrow 0$. □

[**Note s5.8:** The conclusion in **5.145** may be interpreted as saying that if $f \in L^1(X,\mu)$, then the map

$$\nu : \mathsf{S} \ni E \mapsto \int_E f(x)\,d\mu(x) \stackrel{\text{def}}{=} \nu(E)$$

is a complex measure.]

5.146. Since $g_n(x) \stackrel{\text{def}}{=} \max\{f_1(x),\ldots,f_n(x)\}$ is a monotonely increasing sequence of integrable functions, $g \stackrel{\text{def}}{=} \lim_{n\to\infty} g_n \in L^1(X,\mu)$. Since $f_n \le g$ and $f_n \stackrel{\text{a.e.}}{\to} 0$, $\lim_{n\to\infty}\int_X f_n(x)\,d\mu(x) = 0$. □

5.147. a) Fubini's theorem and the equation $\dfrac{1}{x} = \displaystyle\int_0^\infty e^{-xt}\,dt$ (if $x > 0$) yields $\displaystyle\int_0^R \frac{\sin x}{x}\,dx = \int_0^\infty\left(\int_0^R e^{-xt}\sin x\,dx\right)dt$. Repeated integration by parts and the bounded convergence theorem lead to

$$\lim_{R\to\infty}\int_0^R \frac{\sin x}{x}\,dx = \int_0^\infty \frac{dt}{1+t^2} = \frac{\pi}{2}.$$

b) If $f(x) = \begin{cases} \dfrac{\sin x}{x} & \text{if } x \ne 0 \\ 1 & \text{otherwise} \end{cases}$ then $-f'' \ge 0$ in $\left[2n\pi, \left(2n+\frac{1}{2}\right)\pi\right]$, $n \in \mathbb{N}$, i.e., $-f$ is convex there, whence $f(x) = |f(x)| \ge \dfrac{2}{\pi}\left(1 - \dfrac{2n\pi}{x}\right) \ge 0$. Thus

$$\begin{aligned}\int_0^\infty |f(x)|\,dx &\ge \frac{2}{\pi}\sum_{n=1}^\infty \int_{2n\pi}^{(2n+\frac{1}{2})\pi}\left(1-\frac{2n\pi}{x}\right)dx \\ &\ge \sum_{n=1}^\infty\left(1 - 4n\ln\left(1+\frac{1}{4n}\right)\right).\end{aligned} \qquad \text{(s5.11)}$$

Since $1 - 4n \ln\left(1 + \frac{1}{4n}\right) = 1 - \sum_{k=1}^{\infty} \frac{(-1)^{k+1}}{k(4n)^{k-1}}$, it follows that if $n \geq 2$, the nth term in (s5.11) exceeds $\frac{1}{8n} - \frac{3}{(4n)^2}$ whence the last member in (s5.11) diverges. □

5.2. Probability Theory

5.148. If A and B are independent then

$$P(A \cap (\Omega \setminus B)) = P(A \cap \Omega) - P(A \cap B) = P(A)(1 - P(B))$$

whence A and $\Omega \setminus B$ are independent and conversely. For the case of n sets, mathematical induction serves. □

5.149. a) If E is a Borel set in $\mathbb{R}$ and $B \in \mathsf{S}$ then

$$\chi_B^{-1}(E) = \begin{cases} \emptyset & \text{if } \{0,1\} \subset \mathbb{R} \setminus E \\ B & \text{if } \{1\} \subset E \text{ and } \{0\} \not\subset E \\ \Omega \setminus B & \text{if } \{0\} \subset E \text{ and } \{1\} \not\subset E \\ \Omega & \text{if } \{0,1\} \subset E \end{cases}.$$

If $\{E_n\}_{1 \leq n \leq N} \subset \mathsf{S}_\beta(\mathbb{R}), \{F_n\}_{1 \leq n \leq N} \subset \mathsf{S}$, then **5.148** may be applied to $\bigcap_{1 \leq n \leq N} \chi_{F_n}(E_n)$.

b) If $\{A_j\}_{1 \leq j \leq J}$ are Borel sets in $\mathbb{R}$ then $h_j^{-1}(A_j) \overset{\text{def}}{=} C_j \subset \Omega$ and $\omega \in C_j$ iff

$$\left(f_{j1}(\omega), \ldots, f_{jK_j}(\omega)\right) \in g_j^{-1}(A_j) \overset{\text{def}}{=} B_j \left(\text{a Borel subset of } \mathbb{R}^{K_j}\right).$$

If each g_j is a linear combination of products of characteristic functions of pairwise disjoint measurable sets, the result follows from the independence of S. Any g_j is the limit of a sequence $\{g_{jn}\}_{n \in \mathbb{N}}$ of functions of the type just described. Thus if

$$h_{jn} \overset{\text{def}}{=} g_{jn}\left(f_{j1}, \ldots, f_{jK_j}\right), 1 \leq j \leq J, n \in \mathbb{N}, \text{ and } h_{jn}^{-1}(A_j) \overset{\text{def}}{=} C_{jn}$$

then $P\left(\bigcap_{j=1}^{J} C_{jn}\right) = \prod_{j=1}^{J} P(C_{jn})$. As $n \to \infty$, $\lim_{n \to \infty} C_{jn} = C_j$ and the equation remains valid when each C_{jn} is replaced by C_j. □

5.150. a) If $P(A) = 0$ then $P(A \cap B) = 0 = P(A)P(B)$. If $P(A) = 1$ then $P(A \cap B) = P(B) = P(A)P(B)$.

b) Since $f(\omega)$ is constant a.e., for every x, $P\left(f^{-1}(x)\right)$ is either one or zero and a) is applicable. □

5.151. The formula (5.7) is valid if f and g are characteristic functions of measurable sets because $\chi_A \chi_B = \chi_{A \cap B}$.

If $f \stackrel{\text{def}}{=} \sum_{i=1}^{m} a_i \chi_{A_i}$ and $g \stackrel{\text{def}}{=} \sum_{j=1}^{n} b_j \chi_{B_j}$ are independent, it may be assumed that the A_i resp. B_j are pairwise disjoint and that the a_i resp. b_j are pairwise different. Then every pair $\{A_i, B_j\}$ is independent and

$$\begin{aligned}\int_\Omega f(\omega)g(\omega)\,dP(\omega) &= \int_\Omega \sum_{i,j=1}^{m,n} a_i b_j \chi_{A_i}(\omega)\chi_{B_j}(\omega)\,dP(\omega)\\ &= \sum_{i,j=1}^{m,n} a_i b_j \int_\Omega \chi_{A_i}(\omega)\,dP(\omega) \int_\Omega \chi_{B_j}(\omega)\,dP(\omega)\\ &= \int_\Omega f(\omega)\,dP(\omega) \int_\Omega g(\omega)\,dP(\omega).\end{aligned}$$

Hence (5.7) is valid for simple functions.

When f and g are integrable, they are limits of sequences of simple functions. For $F_{kn} \stackrel{\text{def}}{=} \left\{\omega : \frac{k}{2^n} \le f(\omega) < \frac{k+1}{2^n}, -2^n \le k \le 2^n, n \in \mathbb{N}\right\}$, the paradigms for the functions in the sequence approximating f are

$$f_n \stackrel{\text{def}}{=} \sum_{k=-2^n}^{2^n} \frac{k}{2_n} \chi_{F_{kn}}, n \in \mathbb{N}.$$

Since f and g are independent, the functions f_n and g_n are independent. Hence

$$\int_\Omega f_n(\omega)g_n(\omega)\,dP(\omega) = \int_\Omega f_n(\omega)\,dP(\omega) \int_\Omega g_n(\omega)\,dP(\omega),$$

an equation that persists as $n \to \infty$. □

5.152. a) It suffices to show that if $\emptyset \subsetneqq A, B \subset \mathbb{N} \setminus \{1\}$ then A and B are not independent. Since $P(A) = \sum_{n\in A} 2^{-n!}$, $P(B) = \sum_{m\in B} 2^{-m!}$, and $P(A \cap B) = \sum_{k\in A\cap B} 2^{-k!}$, it suffices to show that if $a! + b! = c! + d!$ then $\{a, b\} = \{c, d\}$ (whence if $m, n \ge 2$, there is in $\mathbb{N}$ no p such that $p! = m!+n!$).

1) It may be assumed that $a \le b, c \le d$. If $b < d$ then $d! > 2b!$, whence $a! + b! < \frac{d!}{2} + \frac{d!}{2} = d! < c! + d!$, a contradiction. Arguments based on symmetry conclude the proof.

2) If $p! = m! + n!$ then as in 1), $m! + n! < \frac{p!}{2} + \frac{p!}{2} = p!$, again a contradiction.

b) Because neither f nor g is constant a.e., a) applies. □

5.153. Consider first the case where

$$0 \le a \le x - c < x + c \le b \le 1$$

and $f(x) \stackrel{\text{def}}{=} x\chi_{[a,b]}(x) \stackrel{\text{def}}{=} F_{[a,b]}(x)$. If g is not constant a.e. there is in $[0,1]$ a Borel set A such that $0 < \lambda\left(g^{-1}(A)\right) < 1$. It follows that $[x-c,x+c] = f^{-1}([x-c,x+c])$ and, since f and g are independent,

$$\frac{\lambda\left(g^{-1}(A) \cap [x-c,x+c]\right)}{2c} = \lambda\left(g^{-1}(A)\right).$$

The metric density theorem implies that the left member above approaches zero or one for all x off a null set. The right member is neither zero nor one, a contradiction: g must be constant a.e.

For the general case, $k(x) \stackrel{\text{def}}{=} f^{-1}(x) \cdot \chi_{f([a,b])}(x)$ is well-defined. If $h \stackrel{\text{def}}{=} k \circ f$ then $h = F_{[a,b]}$ and for any Borel set A, $h^{-1}(A) = f^{-1}\left(k^{-1}(A)\right)$. Since f and g are independent, **5.150**b) implies h and g are independent whence the conclusion of the preceding paragraph applies. □

5.154. a) The set $T \stackrel{\text{def}}{=} \left\{g_\gamma \stackrel{\text{def}}{=} f_\gamma - \int_\Omega f_\gamma(\omega)\,dP(\omega)\right\}_{\gamma\in\Gamma} \setminus \{\mathbf{O}\}$ is independent, $\text{span}(S) = \text{span}(T \cup \{1\})$, and $\int_\Omega g_\gamma(\omega)\,dP(\omega) \equiv 0$.

If $\#(T) = 1$ then $\dim(S) \le 2$. That trivial case aside, let $\mathcal{P}$ be the set of all possible products of at least two elements of T:

$$\mathcal{P} \stackrel{\text{def}}{=} \left\{ g_{\gamma_1} \cdot \cdots \cdot g_{\gamma_n} \ : \ 2 \le n < \infty, g_{\gamma_i} \ne g_{\gamma_j} \text{ if } i \ne j \right\}.$$

Then, according to **5.151**, if $p \stackrel{\text{def}}{=} \prod_{k=1}^K g_{\gamma_k} \in \mathcal{P}$ and $h \in T$,

$$\int_\Omega h(\omega)p(\omega)\,dP(\omega) = \prod_{k=1}^K \int_\Omega g_{\gamma_k}(\omega)\,dP(\omega) \int_\Omega h(\omega)\,dP(\omega) = 0.$$

b) If $g_\gamma \in T$ then pg_γ is either an element of $\mathcal{P}$ (if all the factors of p are different from g_γ) or g_γ is a factor , say g_{γ_1}, of p, in which case $pg_\gamma = g_{\gamma_1}^2 \prod_{k=2}^K g_{\gamma_k}$. In either case, pg_γ is the product of pairwise different independent functions and so $\int_\Omega pg_\gamma(\omega)\,dP(\omega) = 0$, i.e., $\mathcal{P} \subset T^\perp$. Since $\int_\Omega p(\omega)\,dP(\omega) \equiv 0$, $\mathcal{P} \subset S^\perp$. Because T is independent, $\{\mathbf{O}\} \subsetneqq \mathcal{P}$ and if Γ is infinite, so is $\mathcal{P}$. □

5.155. a) The range of r_0 is $\{0,1\}$ and if $n > 0$ the range of r_n is $\{0,\pm1\}$. It follows that

$$\lambda\left(r_n^{-1}(A)\right) = \begin{cases} 0 & \text{if } \{0,\pm1\} \cap A = \emptyset \\ \frac{1}{2} & \text{if } \{0,\pm1\} \cap A = \{1\} \\ \frac{1}{2} & \text{if } \{0,\pm1\} \cap A = \{-1\} \\ 1 & \text{if } \{0,\pm1\} \cap A \supset \{\pm1\}. \end{cases}$$

Direct calculation shows that the r_n are independent and indeed, e.g., if $0 < n_1 < n_2 < \cdots < n_k$, that $\lambda\left(r_{n_1}^{-1}(1) \cap r_{n_2}^{-1}(1) \cap \cdots \cap r_{n_k}^{-1}(1)\right) = 2^{-k}$.

b) The function $F_n(x,t) \stackrel{\text{def}}{=} \prod_{m=0}^{n}(1 + r_m(x)r_m(t))$ is a sum of terms of the form $W_p(x)W_p(t)$. In other words, $F_n(x,\cdot)$ may be regarded as a one-parameter family of elements in some finite-dimensional span of some of the Walsh functions. Mathematical induction implies that when $x \in \left[\frac{k}{2^n}, \frac{k+1}{2^n}\right) \subset [0,1]$, $F_n(x,t) == 2^{n+1}\chi_{\left[\frac{k}{2^n},\frac{k+1}{2^n}\right)}$ and

$$\int_{[0,1]} f(t)F_n(x,t)\,dt = 2^{n+1}\int_{\frac{k}{2^n}}^{\frac{k+1}{2^n}} f(t)\,dt. \tag{s5.12}$$

If $f \in (\operatorname{span}(\mathcal{W}))^{\perp}$ then for all k, n in $\mathbb{N}$, the right member in (s5.12) is zero. Thus for every dyadic subinterval $J \stackrel{\text{def}}{=} \left[\frac{k}{2^n}, \frac{k+1}{2^n}\right)$ of $[0,1]$, $\int_J f(t) = 0$, whence $f(t) \doteq 0$, i.e., $(\operatorname{span}(\mathcal{W}))^{\perp} = \{\mathbf{O}\}$. □

5.156. If $\sum_{n=1}^{\infty} P(A_n) < \infty$ then for each positive ϵ, there is an $n(\epsilon)$ such that $P\left(\bigcup_{n=n(\epsilon)}^{\infty} A_n\right) < \epsilon$. Hence

$$\left\{\sum_{n=1}^{\infty} P(A_n) < \infty\right\} \Rightarrow \left\{P\left(\overline{\lim}_{n\to\infty} A_n\right) = 0\right\}. \tag{s5.13}$$

[The implication (s5.13) obtains even if S is not independent.]

Assume the implication

$$\left\{\sum_{n=1}^{\infty} P(A_n) = \infty\right\} \Rightarrow \left\{P\left(\overline{\lim}_{n\to\infty} A_n\right) = 1\right\} \tag{s5.14}$$

is valid. Then the preceding paragraph shows

$$\left\{P\left(\overline{\lim}_{n\to\infty} A_n\right) = 1\right\} \Rightarrow \left\{\sum_{n=1}^{\infty} P(A_n) = \infty\right\} \tag{s5.15}$$

is also valid. Thus (s5.14) implies the validity of

$$\left\{P\left(\overline{\lim}_{n\to\infty} A_n\right) = 0\right\} \Rightarrow \left\{\sum_{n=1}^{\infty} P(A_n) < \infty\right\} \tag{s5.16}$$

whence, owing to (s5.15), (s5.16) yields

$$\left\{P\left(\overline{\lim}_{n\to\infty} A_n\right) = 0\right\} \Leftrightarrow \left\{\sum_{n=1}^{\infty} P(A_n) < \infty\right\}. \tag{s5.17}$$

Thus all conclusions hinge on the truth of (s5.14).

Note that

$$\begin{aligned}\overline{\lim}_{n\to\infty}A_n &= \bigcap_{n=1}^{\infty}\bigcup_{m=n}^{\infty}A_n\\ 1-P\left(\overline{\lim}_{n\to\infty}A_n\right) &= P\left(\Omega\setminus\overline{\lim}_{n\to\infty}A_n\right)\\ &= \lim_{n\to\infty}P\left(\bigcap_{m=n}^{\infty}(\Omega\setminus A_m)\right)\\ &= \lim_{n\to\infty}\lim_{M\to\infty}P\left(\bigcap_{m=n}^{M}(\Omega\setminus A_m)\right). \qquad \text{(s5.18)}\end{aligned}$$

Since S is independent,

$$\begin{aligned}\lim_{n\to\infty}P\left(\bigcap_{m=n}^{M}(\Omega\setminus A_m)\right) &= \lim_{n\to\infty}\lim_{M\to\infty}\prod_{m=n}^{M}P(\Omega\setminus A_m)\\ &= \lim_{n\to\infty}\lim_{M\to\infty}\prod_{m=n}^{M}(1-P(A_m)).\end{aligned}$$

However, if $\lim_{M\to\infty}\prod_{m=n}^{M}(1-\Pi(A_m))$ exists, it does not exceed one and it is different from zero iff $\sum_{m=n}^{\infty}P(A_m)<\infty$. In the current circumstances, (s5.18) implies the validity of (s5.14). □

5.3. Ergodic Theory

5.157. Since $E\in \mathsf{S}$, owing to the assumptions about T, it follows that $T^{-1}(E)\in\mathsf{S}$ and $\mu(E)=\mu\left(T\left(T^{-1}(E)\right)\right)=\mu\left(T^{-1}(E)\right)$. □

5.158. If $A_m \stackrel{\text{def}}{=} \{x : \max_{n\le m}s_n(x)>0\}$ then

$$A_m\subset A_{m+1},\, m\in\mathbb{N},\ \bigcup_{m\in\mathbb{N}}A_m=A. \qquad \text{(s5.19)}$$

For each x in X the Riesz lemma (cf. **3.1**) is applicable to the set

$$\left\{a_n(x)\stackrel{\text{def}}{=}f\left(T^n(x)\right)\right\}_{n=0}^{N}.$$

Let $D(x)$ be the set of distinguished indices. Then

$$\sum_{\substack{n\in D(x)\\0\le n\le N-1}}a_n(x)=\sum_{\substack{n\in D(x)\\0\le n\le N-1}}f\left(T^n(x)\right)>0.$$

Hence if $B_n \stackrel{\text{def}}{=} \{ x : n \in D(x) \}$, $0 \le n \le N-1$, then

$$\sum_{n=0}^{N-1} \int_{B_n} f(T^n(x))\, d\mu(x) > 0.$$

Since $T(E) = E$ and

$$\begin{aligned} B_n &= \left\{ x : \max_{n-1<p\le N-1} \left(f(T^n(x)) + \cdots + f(T^p(x)) \right) > 0 \right\} \\ &= T^{-(n-1)} A_{N-(n-1)} \end{aligned}$$

it follows that

$$\begin{aligned} \int_{B_n \cap E} f(T^n(x))\, d\mu(x) &= \int_{A_{N-(n-1)} \cap E} f(x)\, d\mu(x) \\ 0 \le \sum_{n=0}^{N-1} \int_{A_{N-(n-1)} \cap E} f(x)\, d\mu(x) &= \sum_{k=2}^{N+1} \int_{A_k \cap E} f(x)\, d\mu(x). \end{aligned}$$

The relations noted in (s5.19) imply

$$\lim_{k\to\infty} \int_{A_k} f(x)\, d\mu(x) \stackrel{\text{def}}{=} \lim_{k\to\infty} I_k = \int_A f(x)\, d\mu(x).$$

Hence $0 \le \dfrac{\sum_{k=0}^{N} I_k}{N+1} \to \displaystyle\int_{A\cap E} f(x)\, d\mu(x)$ as $N \to \infty$. □

5.159. The results in **5.158** apply to the situation in which $f - a$ replaces f and $\dfrac{s_n}{n} - a$ replaces $\dfrac{s_n}{n}$. □

5.160. a) The equations

$$\begin{aligned} \overline{F}(T(x)) &= \overline{\lim}_{N\to\infty} \frac{f(T(x)) + \cdots + f\left(T^N(x)\right)}{N} \\ &= \overline{\lim}_{N\to\infty} \left(\frac{N+1}{N} \right) \cdot \frac{s_{N+1}(x)}{N+1} - \frac{f(x)}{N} \\ &= \overline{F}(x) \end{aligned}$$

and analogous equations for $\underline{F}$ lead to the conclusion.

b) Since $\underline{F} \le \overline{F}$,

$$\{ x : \underline{F}(x) < \overline{F}(x) \} \subset \bigcup_{r,s\in\mathbb{Q}} \{ x : \underline{F}(x) < r < s < \overline{F}(x) \} \stackrel{\text{def}}{=} \bigcup_{r,s\in\mathbb{Q}} C_{rs}.$$

The argument in a) shows each C_{rs} is T-invariant, i.e., $T(C_{rs}) = C_{rs}$. Furthermore, $C_{rs} = C_{rs} \cap A_s$ since otherwise, $\sup_{n\in\mathsf{N}} \frac{s_n(x)}{n} \le s$ and then $\overline{\lim}_{n\to\infty} \frac{s_n(x)}{n} \le s$, a contradiction. Hence from **5.159** it follows that

$$\int_{C_{rs}} f(x)\,d\mu(x) = \int_{C_{rs}\cap A_s} f(x)\,d\mu(x) \ge s\mu(C_{rs}\cap A_s) = s\mu(C_{rs}).$$

If $-f$ replaces f above, it follows that $\int_{C_{rs}} f(x)\,d\mu(x) \le r\mu(C_{rs})$ whence $\mu(C_{rs}) = 0$ and $\underline{F}(x) \doteq \overline{F}(x)$.

c) For every E in S,

$$\begin{aligned}\int_E \left|\frac{s_n(x)}{n}\right| d\mu(x) &= \sum_{k=0}^{n-1}\int_E \left|\frac{f\left(T^k(x)\right)}{n}\right| d\mu(x)\\ &= \frac{1}{n}\sum_{k=0}^{n-1}\int_{T^{-k}(E)}\left|\frac{f(x)}{n}\right| d\mu(x).\end{aligned}$$

Since $f \in L^1(X,\mu)$, if $\epsilon > 0$ and $\mu(E)$ is small then $\mu\left(T^{-k}(E)\right)$ is also small and each term averaged in the rightmost member is less than ϵ. It follows that the average itself is small and that $\left\{\frac{s_n}{n}\right\}_{n\in\mathsf{N}}$ is uniformly integrable.

□

5.161. Since $\frac{s_n}{n} \overset{\text{a.e.}}{\to} F$ and since $\left\{\frac{s_n}{n}\right\}_{n\in\mathsf{N}}$ is uniformly integrable, Egorov's theorem implies that for each positive ϵ, there is in S an E_ϵ such that $\mu(E_\epsilon) < \epsilon$ and $\frac{s_n}{n} \overset{\text{u}}{\to} F$ on $X\setminus E_\epsilon$. Hence if ϵ is small and both m and n are large then both

$$\int_{E_\epsilon}\left|\frac{s_m(x)}{m} - \frac{s_n(x)}{n}\right| d\mu(x) \text{ and } \int_{X\setminus E_\epsilon}\left|\frac{s_m(x)}{m} - \frac{s_n(x)}{n}\right| d\mu(x)$$

are small. Hence $\left\{\frac{s_n}{n}\right\}_{n\in\mathsf{N}}$ is a $\|\ \|_1$-Cauchy sequence and for some g in $L^1(X,\mu)$, $\frac{s_n}{n} \overset{\|\ \|_1}{\to} g$ and $g \doteq F$. Thus

$$\int_X f(x)\,d\mu(x) = \int_X \frac{s_n(x)}{n}\,d\mu(x) \to \int_X F(x)\,d\mu(x)$$

as $n \to \infty$.

□

[**Note s5.9:** The approach using **3.1** in **5.158** – **5.161** is that of F. Riesz. He used **3.2** to prove the differentiability a.e. of a monotone function in $\mathbb{R}^{\mathbb{R}}$.]

5.162. For generality, assume $\mathfrak{H}$ is a Hilbert space and $U : \mathfrak{H} \mapsto \mathfrak{H}$ is unitary. Let $\mathcal{M}$ consist of all $\mathbf{x}$ such that $L(\mathbf{x})$ exists. Then $\mathbf{O} \in \mathcal{M}\ (\neq \emptyset)$ and it is shown next that the subspace $\mathcal{M}$ is $\| \ \|$-closed in $\mathfrak{H}$.

If $\mathcal{M} \supset \{\mathbf{x}_n\}_{n\in\mathbb{N}}$ and $\mathbf{x}_n \overset{\|\ \|}{\to} \mathbf{x}$ then for m in $\mathbb{N}$,

$$\left\|\frac{\sum_{n=0}^{M} U^n(\mathbf{x})}{M+1} - \frac{\sum_{n=0}^{N} U^n(\mathbf{x})}{N+1}\right\|$$
$$\leq \left\|\frac{\sum_{m=0}^{M} U^m(\mathbf{x}-\mathbf{x}_m)}{M+1}\right\| + \left\|\frac{\sum_{n=0}^{N} U^n(\mathbf{x}-\mathbf{x}_m)}{N+1}\right\|$$
$$+ \left\|\frac{\sum_{n=0}^{M} U^n(\mathbf{x}_m)}{M+1} - \frac{\sum_{n=0}^{N} U^n(\mathbf{x}_m)}{N+1}\right\|.$$

Since U is unitary, for large m, the first and second terms in the right member above are small and, by definition, the third term is also small, i.e., $\mathbf{x} \in \mathcal{M}$: $\mathcal{M}$ is $\| \ \|$-closed.

The following assertions and proofs lead to the conclusion.

i. If $\mathbf{x} = U(\mathbf{x})$ then $\mathbf{x} \in \mathcal{M}$ since $\mathbf{x} = \dfrac{\sum_{n=0}^{N} U^n(\mathbf{x})}{N+1}$.

ii. If $\mathbf{x} \in \mathfrak{H}$ then $\mathbf{x} - U(\mathbf{x}) \in \mathcal{M}$ since

$$\left\|\frac{\sum_{n=0}^{N} U^n(\mathbf{x})}{N+1} - U(\mathbf{x})\right\| = \frac{\|\mathbf{x} - U^{N+1}(\mathbf{x})\|}{N+1} \leq \frac{2\|\mathbf{x}\|}{N+1} \to 0$$

as $N \to \infty$.

iii. If $\mathbf{x} \in \mathcal{M}$ then $\mathbf{y} \overset{\text{def}}{=} U^{-1}(\mathbf{x}) \in \mathcal{M}$ since

$$\frac{\sum_{n=0}^{N} U^n(\mathbf{y})}{N+1} = \frac{\sum_{n=0}^{N-1} U^{n-1}(\mathbf{x})}{N} \cdot \frac{N}{N+1}.$$

iv. If $\mathbf{x} \in \mathcal{M}$ then $U(\mathbf{x}) \in \mathcal{M}$ since

$$\frac{\sum_{n=0}^{N} U^n\left(U^{-1}(\mathbf{x})\right)}{N+1} = \frac{N}{N+1} \cdot \frac{\sum_{n=0}^{N} U^{n-1}(\mathbf{x})}{N} + \frac{U^{-1}(\mathbf{x})}{N+1} \to L(\mathbf{x})$$

as $N \to \infty$. Thus $U(\mathcal{M}) = \mathcal{M}$.

v. If $\mathbf{x} \in \mathcal{M}^\perp$ then $U(\mathbf{x}) \in \mathcal{M}^\perp$ since if $\mathbf{y} \in \mathcal{M}\ [= U(\mathcal{M})]$ then for some $\mathbf{z}$ in $\mathcal{M}$, $\mathbf{y} = U(\mathbf{z})$ and then $(U(\mathbf{x}), \mathbf{y}) = (U(\mathbf{x}), U(\mathbf{z})) = (\mathbf{y}, \mathbf{z}) = 0$.

Hence if $\mathbf{x} \in \mathcal{M}^\perp$ then $\mathbf{x} - U(\mathbf{x}) \in \mathcal{M} \cap \mathcal{M}^\perp$, whence $\mathbf{x} = U(\mathbf{x})$, $\mathbf{x} \in \mathcal{M} \cap \mathcal{M}^\perp$, $\mathbf{x} = \mathbf{O}$. In sum, $\mathcal{M} = \mathfrak{H}$. □

5.163. For $f : \mathbb{T} \in z \mapsto z$ and for all z in $\mathbb{T}$,

$$\frac{\sum_{k=0}^{n-1} f\left(z^k\right)}{n} = \begin{cases} \dfrac{1-z^n}{n(1-z)} & \text{if } z \neq 1 \\ 1 & \text{if } z = 1 \end{cases},$$

whence

$$\lim_{n\to\infty} \frac{\sum_{k=0}^{n-1} f\left(z^k\right)}{n} = \begin{cases} 0 & \text{if } z \neq 1 \\ 1 & \text{if } z = 1 \end{cases}.$$

If f is a polynomial, it follows that

$$\lim_{n\to\infty} \frac{\sum_{k=0}^{n-1} f\left(z^k\right)}{n} = \begin{cases} f(0) & \text{if } z \neq 1 \\ 1 & \text{if } z = 1 \end{cases}. \tag{s5.20}$$

Fejér's theorem implies that (s5.20) is valid for any f in $C(\mathbb{T}, \mathbb{C})$ and the Riesz representation theorem shows that μ_{z_0} is a discrete measure concentrated at zero resp. one according as $z_0 \neq 1$ resp. $z_0 = 1$ and $\mu_{z_0}(\mathbb{T}) = 1$.

□

6
Topological Vector Spaces

6.1. The Spaces $L^p(X, \mu)$, $1 \le p \le \infty$

6.1. Because $E_{\neq}(f, 0)$ is σ-finite, $E = \lim_{n\to\infty} E_n$ for some sets E_n of finite measure. Hence

$$\int_X f(x)\, d\mu(x) = \int_E f(x)\, d\mu(x) = \lim_{n\to\infty} \int_{E_n} f(x)\, d\mu(x) \le a.$$

However, if $X \stackrel{\text{def}}{=} \mathbb{R}$, $\mathsf{S} \stackrel{\text{def}}{=} 2^{\mathbb{R}}$, $\mu(E) = \begin{cases} 0 & \text{if } E \text{ is finite} \\ \infty & \text{otherwise} \end{cases}$, $f \equiv 1$, and $\mu(E) < \infty$ then $\int_E f(x)\, d\mu(x) = 0$ while $\int_X f(x)\, d\mu(x) = \infty$. □

6.2. For $(\mathbb{R}, \mathsf{S}_\lambda, \lambda)$, if $f_n \stackrel{\text{def}}{=} \frac{1}{n}\chi_{[-n^2,n^2]}, n \in \mathbb{N}$, the requirements are met. □

6.3. Since $\int_X (1 - f_n(x))\, d\mu(x) = \int_E (1 - f_n(x))\, d\mu(x)$, the dominated convergence theorem applies. □

6.4. If f as described exists then $f \ge 0$ a.e. (μ_1). If $\{E_n\}_{n\in\mathbb{N}}$ is a measurable partition of X and if $\mu_1(E_n) \neq 0$, Hölder's inequality implies $\frac{(\mu_2(E_n))^p}{(\mu_1(E_n))^{p-1}} \le \int_{E_n} (f(x))^p\, d\mu_1(x)$ and so

$$\sum_{n=1}^{\infty} \frac{(\mu_2(E_n))^p}{(\mu_1(E_n))^{p-1}} \le \|f\|_p^p \stackrel{\text{def}}{=} a < \infty.$$

Conversely, if, for every measurable partition $\{E_n\}_{n\in\mathbb{N}}$ of X,

$$\sum_{n=1}^{\infty} \frac{(\mu_2(E_n))^p}{(\mu_1(E_n))^{p-1}} \le a < \infty$$

then $\{\mu_2(E) > 0\} \Rightarrow \{\mu_1(E) > 0\}$ whence $\mu_2 \ll \mu_1$ and for some f in $L^1(X, \mu)$, $\mu_2(E) \equiv \int_E f(x)\, d\mu_1(x)$. If $0 < b_n < b_{n+1} \uparrow \infty$ then

$$\{\{x : b_n \le f(x) < b_{n+1}\}\}_{n\in\mathbb{N}} \stackrel{\text{def}}{=} \{E_n\}_{n\in\mathbb{N}}$$

is a measurable partition of X and

$$\mu_2(E_n) \geq b_n\mu_1(E_n), \quad \frac{(\mu_2(E_n))^p}{(\mu_1(E_n))^{p-1}} \geq b_n^p\mu_1(E_n), \quad \sum_{n=1}^{\infty} b_n^p\mu_1(E_n) \leq a$$

whence $\|f\|_p^p \leq a$. □

6.5. When $F_n \stackrel{\text{def}}{=} \{x : b_n \leq |f(x)| < b_{n+1}\}$ then $\sum_{n=1}^{\infty} b_n^p\mu(F_n)$ is, for appropriate choices of the b_n, near $\|f\|_p^p$. On the other hand, $E_{b_{n+1}} \subset E_{b_n}$, and $F_n = E_{b_n} \setminus E_{b_{n+1}}$. Thus

$$\sum_{n=1}^{\infty} b_n^p\mu(F_n) = \sum_{n=1}^{\infty} b_n^p\left(\mu(E_{b_n}) - \mu(E_{b_{n+1}})\right).$$

For large N, $\sum_{n=N+1}^{\infty} b_n^p\mu(F_n)$ is small and

$$\sum_{n=1}^{N} b_n^p\mu(F_n) = \sum_{n=1}^{N-1}\left(b_{n+1}^p - b_n^p\right)\mu(E_{b_n}) + b_1^p\mu(E_{b_1}) - b_N^p\mu(E_{b_{N+1}}).$$

According to **5.15**, $\lim_{N\to\infty} b_N^p\mu(E_{b_N}) = 0$. Because b_1 may be chosen to be zero it follows that $\sum_{n=1}^{\infty}\left(b_{n+1}^p - b_n^p\right)\mu(E_{b_n})$ is a good approximant to $\|f\|_p^p$. The mean value theorem implies that for some θ_n in (b_n, b_{n+1}), $b_{n+1}^p - b_n^p = p\theta_n^{p-1}(b_{n+1} - b_n)$, i.e., that $\|f\|_p^p = p\int_0^\infty t^{p-1}\mu(E_t)\,dt$. □

6.6. For some $\{E_n\}_{n\in\mathbb{N}}$, $\mu(E_n) < \infty$ and $E_{\neq}(g,0) = \bigcup_{n\in\mathbb{N}} E_n$.

Assume $p = 1$ and that $\|g\|_\infty = \infty$. Some E_{n_0} contains pairwise disjoint subsets $F_m, m \in \mathbb{N}$, such that $\mu(F_m) > 0$ and

$$m + 1 > \sup\{\,|g(x)| : x \in F_m\,\} > \inf\{\,|g(x)| : x \in F_m\,\} \geq m.$$

If $a_m\mu(F_m) = \dfrac{1}{m^2}$ and $h(x) = \sum_{m=1}^{\infty} a_m \operatorname{sgn}(g(x))\,\chi_{F_m}(x)$ then

$$\int_X |h(x)|\,d\mu(x) = \sum_{m=1}^{\infty}\frac{1}{m^2} = \frac{\pi^2}{6} \text{ and}$$

$$\int_X g(x)h(x)\,d\mu(x) \geq \sum_{m=1}^{\infty} m a_m\mu(F_m) = \sum_{m=1}^{\infty}\frac{1}{m} = \infty,$$

a contradiction.

When $1 < p < \infty$, if $\mu(E_{\neq}(g,0)) < \infty$ and $\int_{E_{\neq}(g,0)} |g(x)|^{p'}\,d\mu(x) = \infty$ then $|g|$ is unbounded. If $\epsilon > 0$ and

$$G_m \stackrel{\text{def}}{=} \{x : m\epsilon \leq |g(x)| < (m+1)\epsilon\}, m = 0, 1, 2, \ldots,$$

then each G_m is measurable and $k \stackrel{\text{def}}{=} \sum_{m=1}^{\infty} m\epsilon\chi_{G_m} \le |g| < k+\epsilon$. The theory of extrema in the differential calculus shows that if $z > 0$ then $\frac{(1+z)^{p'}}{1+z^{p'}} \le 2^{p'-1}$, whence

$$|g|^{p'} = (k + (|g|-k))^{p'} \le 2^{p'-1}\left(k^{p'} + (|g|-k)^{p'}\right).$$

Because $\mu(E_{\neq}(g,0)) < \infty$, $\sum_{m=1}^{\infty}(m\epsilon)^{p'}\mu(G_m) \stackrel{\text{def}}{=} \sum_{m=1}^{\infty} a_m = \infty$. Abel's theorem (cf. **2.54**) implies that if $A_m \stackrel{\text{def}}{=} \sum_{i=1}^{m} a_i$ then

$$\sum_{m=1}^{\infty} \frac{a_m}{A_m} = \infty > \sum_{m=1}^{\infty} \frac{a_m}{A_m^p}.$$

Hence if $h = \sum_{m=1}^{\infty} \operatorname{sgn}(g)\frac{(m\epsilon)^{p'-1}}{A_m}\chi_{G_m}$ then

$$\int_X |h(x)|^p\,d\mu(x) = \sum_{m=1}^{\infty} \frac{(m\epsilon)^{p'}}{A_m^p} < \infty \text{ and}$$

$$\int_X g(x)h(x)\,d\mu(x) \ge \sum_{m=1}^{\infty}(m\epsilon)\frac{(m\epsilon)^{p'-1}}{A_m}\mu(G_m) = \sum_{m=1}^{\infty}\frac{a_m}{A_m} = \infty,$$

a contradiction. Hence if $\mu(E_{\neq}(g,0)) < \infty$ then $g \in L^{p'}(X,\mu)$.

If $\mu(E_{\neq}(g,0)) = \infty$, the preceding discussion implies that

$$\int_{E_n} |g(x)|^{p'}\,d\mu(x) \stackrel{\text{def}}{=} b_n < \infty.$$

If $\int_X |g(x)|^{p'}\,d\mu(x) = \infty$ and $B_m \stackrel{\text{def}}{=} \sum_{i=1}^{m} b_m$ then $\sum_{m=1}^{\infty} b_m = \infty$ and Abel's theorem implies $\sum_{m=1}^{\infty}\frac{b_m}{B_m} = \infty$. If $h \stackrel{\text{def}}{=} \sum_{m=1}^{\infty}\operatorname{sgn}(g)\frac{|g|^{p'-1}}{B_m}$ then

$$\int_X |h(x)|^p\,d\mu(x) = \sum_{m=1}^{\infty}\frac{b_m}{B_m^p} < \infty,\ \int_X g(x)h(x)\,d\mu(x) = \sum_{m=1}^{\infty}\frac{b_m}{B_m} = \infty,$$

a contradiction.

If $p > 1$, $X \stackrel{\text{def}}{=} \{0,1\}$, $\mu(\{0\}) = \infty$, $\mu(\{1\}) = 1$, and $g \equiv 1$ then $L^p(X,\mu) = L^1(X,\mu) = L^\infty(X,\mu)$. If $h \in L^p(X,\mu)$ then $h(0) = 0$, whence $h \in L^1(X,\mu)$, and $\int_X g(x)h(x)\,d\mu(x) = \int_X h(x)\,d\mu(x)$, whereas $\int_X |g(x)|^{p'}\,d\mu(x) = \infty$. □

6.7. For each finite T, **6.6** implies $\int_{-T}^{T} |g(x)|^{p'}\,dx < \infty$. Hölder's inequality implies that $L_T : L^p(\mathbb{R},\lambda) \ni h \mapsto \int_{-T}^{T} g(x)h(x)\,dx$ is a continuous linear functional on $L^p(\mathbb{R},\lambda)$ and the general theory of L^p spaces shows that $\|L_T\| = \left(\int_{-T}^{T}|g(x)|^{p'}\,dx\right)^{\frac{1}{p'}}$. For all h in $L^p(\mathbb{R},\lambda)$, $\overline{\lim}_{T\to\infty}|L_T(h)| < \infty$. The Banach-Steinhaus theorem implies that for some finite M and all positive T, $\|L_T\| \le M$. □

6.8. An informal proof can be given using approximants to integrals by finite sums and appropriate passages to the limit.

A formal proof uses Hölder's inequality and Fubini's theorem, viz.:

$$\begin{aligned}
&\int_Y \left|\int_X f(x,y)\,d\mu(x)\right|^p d\nu(y)\\
&\quad\le \int_Y \left(\left|\int_X f(x,y)\,d\mu(x)\right|^{p-1}\left|\int_X f(x,y)\,d\mu(x)\right|\right) d\nu(y)\\
&\quad\le \int_{X\times Y} |f(x,y)|\left|\int_X |f(x,y)|\,d\mu(x)\right|^{p-1} d(\mu\times\nu)(x,y)\\
&\quad\le \int_X \left(\int_Y |f(x,y)|\left|\int_X f(x,y)\,d\mu(x)\right|^{p-1} d\nu(y)\right) d\mu(x)
\end{aligned}$$

$$\begin{aligned}
&\int_Y |f(x,y)|\left|\int_X f(x,y)\,d\mu(x)\right|^{p-1} d\nu(y)\\
&\quad\le \left(\int_Y |f(x,y)|^p\,d\nu(y)\right)^{\frac{1}{p}} \cdot \left[\int_Y \left|\int_X f(x,y)\,d\mu(x)\right|^{p'(p-1)} d\nu(y)\right]^{\frac{1}{p'}}
\end{aligned}$$

$$\left(\int_Y \left|\int_X f(x,y)\,d\mu(x)\right|^p d\nu(y)\right)^{\frac{1}{p}} \le \int_X \left(\int_Y |f(x,y)|^p\,d\nu(y)\right)^{\frac{1}{p}} d\mu(x). \ \square$$

6.9. In any Banach space X, $\|\ \|$-convergence implies weak convergence because if $\mathbf{x}\in X, \mathbf{x}^*\in X^*$ then $|(\mathbf{x},\mathbf{x}^*)| \le \|\mathbf{x}^*\|\ \|\mathbf{x}\|$.

If $\mathbf{x}_n \xrightarrow{w} \mathbf{x}$ in ℓ^1, the Banach-Steinhaus theorem implies that for some M, $\|\mathbf{x}_n\| \le M < \infty$. Furthermore, the substitution $\mathbf{y}_n \stackrel{\text{def}}{=} \mathbf{x}_n - \mathbf{x}$ reduces the discussion to the case of weak convergence to $\mathbf{O}$. It follows that, x_{nk} denoting the kth component of $\mathbf{x}_n$, for each k, $\lim_{n\to\infty} x_{nk} = 0$.

If $\mathbf{x}_n \stackrel{\|\ \|_1}{\not\to} \mathbf{O}$ it may be assumed that $\|\mathbf{x}_n\|_1 \ge \epsilon > 0$. Hence for some sequences $\{n_m\}_{m\in\mathbb{N}}$ and $\{k_m\}_{m\in\mathbb{N}}$,

$$\sum_{k=1}^{k_1}|x_{nk}| < \frac{\epsilon}{5},\ \sum_{k=k_1+1}^{k_2}|x_{nk}| > \frac{3\epsilon}{5}, n > n_1$$

$$\cdots$$

$$\sum_{k=1}^{k_m}|x_{nk}| < \frac{\epsilon}{5},\ \sum_{k=k_m+1}^{k_{m+1}}|x_{nk}| > \frac{3\epsilon}{5}, n > n_m$$

$$\cdots.$$

It follows that if $x^*_{nk} = \operatorname{sgn}(x_{nk}), n_m < n \le n_{m+1}, k_m < k \le k_{m+1}$, then $\mathbf{x}^*_n \stackrel{\text{def}}{=} \{x^*_{nk}\}_{k\in\mathbb{N}} \in \left(\ell^1\right)^* = \ell^\infty$ and $|(\mathbf{x}_n, \mathbf{x}^*)| \ge \epsilon/5, n \in \mathbb{N}$, a contradiction. □

[**Note s6.1:** The reader might wish to suggest why the phrase "sliding hump" is associated with the discussion above.]

6.10. If

$$\beta(x) \stackrel{\text{def}}{=} \begin{cases} 0 & \text{if } x \le 0 \\ \exp\left[-\dfrac{\exp\left(-\dfrac{1}{(1-x^2)^2}\right)}{x^2}\right] & \text{if } 0 < x < 1 \\ 1 & \text{if } 1 \le x \end{cases}$$

and $\epsilon > 0$, then for suitable constants p, q, r, s and p', q', r', s',

$$g(x) \stackrel{\text{def}}{=} [p + g\beta(rx+s)] \cdot [p' + q'\beta(r'x+s')] = \begin{cases} 1 & \text{if } x = 0 \\ 0 & \text{if } |x| \ge 1/2 \end{cases},$$

$$0 \le g \le 1, \text{ and } \int_{\mathbb{R}} g(x)\,dx < \epsilon.$$

Since $g \in C^\infty(\mathbb{R}, \mathbb{C})$, for some nonnegative g_n in $C^\infty([0,\infty), \mathbb{C})$,

$$g_n(n) = \frac{1}{n^2}, \ \operatorname{supp}(g_n) = [n - \frac{1}{4}, n + \frac{1}{4}], \text{ and } \int_0^\infty g_n(x)\,dx < \frac{1}{n^2}, n \in \mathbb{N}.$$

Hence $f \stackrel{\text{def}}{=} \sum_{n=1}^\infty g_n$ meets the requirements. □

6.11. To each μ_n there corresponds the linear functional

$$G_n : L^1(X, \mu_0) \ni f \mapsto \int_X f(x)\,d\mu_n(x)$$

and the hypotheses imply $\Gamma \stackrel{\text{def}}{=} \{G_n\}_{n\in\mathbb{N}}$ is a subset of the unit ball $B(\mathbf{O}, 1)$ of $L^\infty(X, \mu_0)$, the dual space of $L^1(X, \mu_0)$. However, $B(\mathbf{O}, 1)$ is compact in the topology $\sigma\left(L^\infty(X, \mu_0), L^1(X, \mu_0)\right)$ whence Γ contains a subsequence weakly* convergent to some G_∞ in $B(\mathbf{O}, 1)$. The corresponding measure μ_∞ is what is required. □

[**Note s6.2:** A less perspicuous but more general solution of **6.11** is based on the natural embedding: $X \hookrightarrow X^{**}$, valid for all Banach spaces X.]

6.12. If $f \in L^1([0,1],\lambda)$ and $f \geq 0$ then $[0,1] = \bigcup_{n\in\mathbb{N}} E_n$ and

$$\int_{[0,1]} f(x)\,dx = \sum_{n=1}^{\infty} \int_{E_n} f(x)\,dx$$
$$\geq \sum_{n=2}^{\infty} (n-1)\lambda(E_n) \geq \sum_{n=2}^{\infty} \frac{n-1}{n} \cdot n\lambda(E_n) \geq \frac{1}{2} \sum_{n=2}^{\infty} n\lambda(E_n).$$

Conversely, if f is measurable, $f \geq 0$, and $\sum_{n=1}^{\infty} n\lambda(E_n) < \infty$ then

$$\int_{[0,1]} f(x)\,dx \leq \sum_{n=1}^{\infty} n\lambda(E_n).$$

□

6.13. The differentiability of f at any point other than zero follows from the holomorphy in $\mathbb{C} \setminus \{0\}$ of the factors of f. The difference quotient at zero is $x \sin\left(\frac{1}{x}\right)$ which approaches zero as $x \to 0$. Furthermore,

$$f'(x) = \begin{cases} 2x \sin\left(\frac{1}{x^2}\right) - \frac{2}{x} \cos\left(\frac{1}{x^2}\right) & \text{if } x \neq 0 \\ 0 & \text{otherwise} \end{cases}.$$

If $0 \leq x \leq \frac{\pi}{2}$ then $0 \leq \frac{2}{\pi}x \leq \sin x \leq x$, whence if $x^2 > \frac{2}{\pi}$ then $\frac{1}{x^2} < \frac{\pi}{2}$. Thus

$$\left|2x \sin\left(\frac{1}{x^2}\right)\right| \geq \left|2x \cdot \frac{2}{\pi x^2}\right| = \left|\frac{4}{\pi x}\right|,$$
$$\left|-\frac{2}{x} \cos\left(\frac{1}{x^2}\right)\right| \leq \frac{2}{|x|} \text{ and } |f'(x)| \geq \left(2 - \frac{4}{\pi}\right)\frac{1}{|x|} > \frac{0.7}{|x|},$$

whence $f' \notin L^1(\mathbb{R},\lambda)$. □

6.14. If $f = 2\chi_{[0,\frac{1}{3})} + \chi_{[\frac{1}{3},\frac{2}{3})} + 1.5\chi_{[\frac{2}{3},1]}$ then $F \geq f$ a.e. while f is not monotone. □

6.15. For some sequence $\{J_n\}_{n\in\mathbb{N}}$ of pairwise disjoint closed intervals, $\int_{J_n} |f(x)|\,dx \stackrel{\text{def}}{=} a_n > 0$ whence $\sum_{n=1}^{\infty} a_n < \infty$. It may be assumed that $\sum_{n=1}^{\infty} a_n^{\frac{1}{2}} < \infty$. Then $\sum_{n=1}^{\infty} a_n^{-\frac{1}{2}} \chi_{J_n} \stackrel{\text{def}}{=} g$ meets the requirements. □

6.16. For $E \stackrel{\text{def}}{=} E_=(f,1)$ and $F \stackrel{\text{def}}{=} [0,1] \setminus E$, if $\lambda(F) = 0$ then $f \doteq \chi_E$.
If $\lambda(F) > 0$ then for some m in $\mathbb{N}$,

$$\lambda\left(\left\{x : f(x) \geq 1 + \frac{1}{m}\right\}\right) + \lambda\left(\left\{x : f(x) < 1 - \frac{1}{m}\right\}\right) > 0.$$

If the first summand in the left member above is positive, then

$$\int_0^1 f(x)\,dx = \int_0^1 (f(x))^n\,dx \uparrow \infty,$$

a contradiction. Hence $f(x) \le 1$ a.e., the second summand in the left member above is positive, and $f^n \downarrow 0$ on F. Thus

$$\int_0^1 f(x)\,dx = \int_0^1 (f(x))^n\,dx = \int_E + \int_F (f(x))^n\,dx = \lambda(E) + \int_F (f(x))^n\,dx.$$

The last summand converges to zero as $n \to \infty$ whence $f \doteq \chi_E$, a contradiction.

If $f(x) \not\geq 0$ a.e. then, e.g., if $f(x) = e^{2\pi i x}$, then $\int_0^1 (f(x))^n\,dx = 0$.

If f must be $\mathbb{R}$-valued and yet $f(x) \not\geq 0$ a.e., then $f^2 \ge 0$ whence for some measurable E, $f^2 \doteq \chi_E$. If $E_\pm \stackrel{\text{def}}{=} E_=(f^\pm, 1)$ then $f = \chi_{E_+} - \chi_{E_-}$, which is χ_E for no E. □

6.17. It may be assumed that f is $\mathbb{R}$-valued. Thus for all real a,

$$\int_a^{a+1} f(x)\,dx = 0$$

and so if $b \in (0,1)$ then

$$\begin{aligned}\int_{a+1}^{a+b+1} f(x)\,dx &= \int_{a+b}^{a+b+1} - \int_{a+b}^{a+1} f(x)\,dx = -\int_{a+b}^{a+1} f(x)\,dx \\ &= -\left(\int_a^{a+1} - \int_a^{a+b} f(x)\,dx\right) = \int_a^{a+b} f(x)\,dx\end{aligned}$$

$$\frac{1}{b}\int_a^{a+b} f(x)\,dx = \frac{1}{b}\int_{a+1}^{a+b+1} f(x)\,dx$$

whence $f(x) \doteq f(x+1)$.

For k in $\mathbb{N}$, if

$$N_k \stackrel{\text{def}}{=} \{\, x \,:\, f(x) = f(x+1) = \cdots = f(x+k-1) \ne f(x+k) \,\}$$

then $\lambda(N_k) \equiv 0$ and on $\mathbb{R} \setminus \bigcup_{k\in\mathbb{N}} N_k \stackrel{\text{def}}{=} S$, $f(x) \equiv f(x+n), n \in \mathbb{N}$, i.e., $f(x) \doteq f(x+n)$. Off a null set N in S, $f(x) = \lim_{n\to\infty} n \int_x^{x+\frac{1}{n}} f(t)\,dt$. If $a \in S \setminus N$ and $U \stackrel{\text{def}}{=} \bigcup_{k=0}^{n} \left(a+k, a+k+\frac{1}{n}\right)$ then U is open, $\lambda(U) = 1$, and $\int_U f(x)\,dx = 0 = n\int_a^{a+\frac{1}{n}} f(x)\,dx \to f(a)$ as $n \to \infty$. Thus $f(x) \doteq 0$.

□

6.18. If $E \subset M \cap E_{\neq}(f,0)$ and $\lambda(E) \stackrel{\text{def}}{=} m > 0$ then it may be assumed that $1 \notin E$ and thus either $f(x) \doteq 0$ on E or $|x|^n|f(x)| \uparrow \infty$ a.e. on E. The latter alternative and Lebesgue's monotone convergence theorem imply $\int_E |x|^n|f(x)|\,dx \uparrow \infty$, whereas $\int_M |x|^n|f(x)|\,dx \le 1$, a contradiction. □

6.19. In every closed interval J let C_{Jn} be a Cantor-like subset such that $\lambda(C_{Jn}) > \lambda(J) - \frac{1}{n}$. Then $U_n \stackrel{\text{def}}{=} J \setminus C_{Jn}$ is open and $\overline{U_n} = J$. Furthermore, as $n \to \infty$, $\lambda(U_n) \to 0$ whence

$$\int_J f(x)\,dx = \lim_{n\to\infty} \int_{U_n} f(x)\,dx = \lim_{n\to\infty} \int_{\overline{U_n}} f(x)\,dx = 0$$

and so $f \doteq 0$. □

6.20. On the one hand $F(x) \stackrel{\text{def}}{=} \int_a^x f(t)\,dt$ is absolutely continuous and $F' \doteq f$. On the other hand,

$$\frac{1}{h}\int_c^d (f(x+h) - f(x))\,dx = \frac{F(d+h) - F(d)}{h} - \frac{F(c+h) - F(c)}{h}.$$

Hence for all c, d off a null set, $F'(d) = F'(c)$, i.e., $f(d) = f(c)$, and so f is constant a.e. □

6.21. Because $\left\|\widehat{f}\right\|_\infty \le \|f\|_1$, it follows that

$$\left\|\widehat{(f_{[t]} - f)}\right\|_\infty = \left\|\widehat{f}(s)\cdot(e^{it} - 1)\right\|_\infty \le |t|^2,$$

$$\left|\widehat{f}(s)\right| \cdot \left|e^{it} - 1\right| \le |t|^2,\ s,t \in \mathbb{R}, \text{ and } \left|\widehat{f}(s)\right| \cdot \left|\frac{e^{it}-1}{t}\right| \le |t|.$$

Thus when $|t|$ is near zero, $\left|\widehat{f}(s)\right|$ is small, whence $f \doteq 0$. □

6.22. If $f(x) = 0$ off $[a,b]$ and $Q(x) \stackrel{\text{def}}{=} \int_{-\infty}^x |q(t)|\,dt$ then, since $|f|^2$ is in $C_{00}^\infty(\mathbb{R}, \mathbb{C})$, integration by parts is applicable and leads to the equation

$$\int_{\mathbb{R}} |q(x)| \cdot |f(x)|^2\,dx = -\int_a^b \left(f'(x)\overline{f}(x) + f(x)\overline{f}'(x)\right) Q(x)\,dx.$$

Because $|f'| \cdot \left|\overline{f}\right| + |f| \cdot \left|\overline{f}'\right| \le |f|^2 + |f'|^2$ and $Q(x) \uparrow \|q\|_1$, it follows that

$$\begin{aligned}\int_{\mathbb{R}} |q(x)| \cdot |f(x)|^2\,dx &\le \int_a^b Q(x)\left(|f(x)|^2 + |f'(x)|^2\right)\,dx \\ &\le \|q\|_1 \int_a^b |f(x)|^2 + |f'(x)|^2\,dx,\end{aligned}$$

i.e., $\|q\|_1$ may serve for C_q. □

6.23. It suffices to prove the result when $-\infty < a < b < \infty$ and $f = \chi_{[a,b]}$ since the result for an arbitrary f follows by appealing to simple functions that are $\| \|_1$-approximants to f.

If $A \stackrel{\text{def}}{=} \left\{ x \,:\, a \le x - \frac{1}{x} \le b \right\}$ then $T(f)(x) = \chi_A(x)$. Since

$$A = \left[\frac{a - \sqrt{a^2+4}}{2}, \frac{b - \sqrt{b^2+4}}{2} \right] \cup \left[\frac{a + \sqrt{a^2+4}}{2}, \frac{b + \sqrt{b^2+4}}{2} \right],$$

it follows that $\int_{\mathbf{R}} T(f)(x)\,dx = \lambda(A) = b - a = \int_{\mathbf{R}} f(x)\,dx.$ □

6.24. The characteristic function of any finite interval $[a, b]$ contained in $[0, 1]$ is the $\| \|_1$-limit of bridging functions in $C^\infty([0,1], \mathbb{C})$. It follows that if $0 \le a < b \le 1$ then $\frac{1}{b-a}\int_a^b f(x)\,dx = \frac{1}{b-a}\int_a^b g(x)\,dx$. As $b \to a$, the equation becomes $f \doteq g$.

If $f(x) \doteq g(x)$ and if h in $C^\infty([0,1], \mathbb{R})$,

$$\int_0^1 f(x)h(x)\,dx = \int_0^1 g(x)h(x)\,dx.$$ □

6.25. If $F(x) = \int_1^x |f(s)|\,ds$ then

$$\begin{aligned} \left| \int_1^t f(s)g(s)\,dx \right| &\le M \int_1^t \frac{1}{s}|f(s)|\,ds = \frac{M}{s}F(s)\Big|_1^t + M\int_1^t \frac{F(s)}{s^2}\,ds \\ &\le M\frac{F(t)}{t} + M\|f\|_1\left(1 - \frac{1}{t}\right) \\ &\le \frac{M\|f\|_1}{t} + M\|f\|_1\left(1 - \frac{1}{t}\right) = M\|f\|_1. \end{aligned}$$

Division by t followed by passage to the limit as $t \to \infty$ yields the result. □

6.26. Because $f \in L^1([0,1], \lambda)$, f is the $\| \|_1$-limit of a sequence of step functions. However, if J is an interval then $\{ x \,:\, x^n \in J \}$ is the union of at most two intervals. Hence if s is a step function then $s_n(x) \stackrel{\text{def}}{=} s(x^n)$ is also a step function whence $f_n \in \mathcal{M}$.

Furthermore, if $a \in (0, 1)$ then

$$\int_0^1 |f_n(x)|\,dx = \int_0^a + \int_a^1 \frac{x^{\frac{1}{n}-1}|f(x)|}{n}\,dx \stackrel{\text{def}}{=} I + II.$$

Because f is continuous at zero, for some positive a, f is bounded in $[0, a)$, whence the inequality $1 - \frac{1}{n} < 1$ insures that $I < \infty$. On $[a, 1]$ the integrand in II is dominated by $\frac{a^{\frac{1}{n}-1}|f(x)|}{n}$. □

[**Note s6.3:** The proof above that f_n is measurable goes back to first principles. The measurability of f_n follows also from **5.109**.]

6.27. Since $\int_0^1 |f_n(x)|\,dx = \int_n^{n+1} |f(x)|\,dx$, it follows that

$$\int_0^1 \left|\sum_{n=N+1}^{N+K} f_n(x)\right| dx \le \sum_{n=N+1}^{N+K} \int_0^1 |f_n(x)|\,dx = \int_{N+1}^{N+K+1} |f(x)|\,dx \to 0$$

as $N, K \to \infty$. □

6.28. By hypothesis, $f \in L^1(X, \mu)$ and it follows that $\|f\|_\infty \le M$, whence $f_n g_n, fg \in L^1(X, \mu), n \in \mathbb{N}$. It may be assumed (via the use of subsequences as needed) that $f_n \overset{\text{a.e.}}{\to} f$, $g_n \overset{\text{a.e.}}{\to} g$ (whence $f_n g_n \overset{\text{a.e.}}{\to} fg$). Since $|f_n g - fg| \le 2M|g|$, $\|f_n g - fg\|_1 \to 0$ as $n \to \infty$. Furthermore, from $\|f_n g - f_n g_n\|_1 \le M\,\|g - g_n\|_1$, it follows that

$$\|f_n g_n - fg\|_1 \le \|f_n g_n - f_n g\|_1 + \|f_n g - fg\|_1 \to 0$$

as $n \to \infty$. □

6.29. For all t in $[0, \infty)$, $1 - e^{-t} \le t$, whence $1 - e^{-f_n} \le f_n$. □

6.30. If a) and b) obtain and $b > 0$, Egorov's theorem implies that $A_{\frac{b}{2}}$ contains a measurable subset B such that $\int_{A_{\frac{b}{2}} \setminus B} |f_n(x)|\,dx < \frac{b}{2^2}$, $n \in \mathbb{N}$, and $|f_n - f_m| \overset{\text{u}}{\to} 0$ on B. Hence

$$\begin{aligned}\|f_n - f_m\|_1 &\le \int_{\mathbb{R}\setminus A_{\frac{b}{2}}} + \int_{A_{\frac{b}{2}}\setminus B} + \int_B |f_n(x) - f_m(x)|\,dx \\ &\le \frac{b}{2} + \frac{b}{2^2} + \int_B |f_n(x) - f_m(x)|\,dx.\end{aligned}$$

Because convergence is uniform on B, for large n, m, the third term in the last line above is small whence for some g in $L^1(\mathbb{R}, \lambda)$, $\|f_n - g\|_1 \to 0$ as $n \to \infty$. Since a subsequence $\{f_{n_k}\}_{k\in\mathbb{N}}$ converges to g a.e., $f \doteq g$.

Conversely, if $\|f_n - f\|_1 \to 0$ as $n \to \infty$ and if $\epsilon > 0$ then for some measurable set E_ϵ and an n_0, if $n \ge n_0$ then

$$\int_{\mathbb{R}\setminus E_\epsilon} |f(x)|\,dx < \frac{\epsilon}{2}, \quad \int_{\mathbb{R}} |f_n(x) - f(x)|\,dx < \frac{\epsilon}{2^2}.$$

Furthermore, E_ϵ contains a measurable subset A_ϵ of finite positive measure and such that if $n \geq n_0$, $\int_{\mathbb{R}\setminus A_\epsilon} |f_n(x)|\,dx < \frac{\epsilon}{2^3}$ whence a) obtains.

Because $f_n \stackrel{\|\ \|_1}{\to} f$, for some finite M and all n, $\|f_n\|_1 \leq M$. For measurable B and $T_B : L^1(\mathbb{R}, \lambda) \ni g \mapsto \int_B g(x)\,dx$, $T_B(g) \to 0$ as $\lambda(B) \to 0$. The uniform boundedness principle implies that for some K, $\|T_B\| \leq K < \infty$ and $|T_B(|f_n|)| \leq |T_B(|f|)| + K\|f_n - f\|_1$. If $b > 0$, there is an n_0 such that if $n \geq n_0$ then

$$\|f_n - f\|_1 \leq \frac{b}{K}, \lim_{\lambda(B)\to 0} \sup_{n\geq n_0} |T_B(|f_n|)| \leq \lim_{\lambda(B)\to 0} |T_B(|f|)| + b = b.$$

Hence $\lim_{\lambda(B)\to 0} \sup_n |T_B(|f_n|)| = 0$ and b) obtains. □

6.31. Because

$$S_N(x) \uparrow f(x) \text{ if } -1 < x \leq 0$$
$$|S_N(x)| \leq S_N(-x) \uparrow f(-x) \text{ if } 0 \leq x < 1,$$

the dominated convergence theorem applies. □

6.32. For $F(K) \stackrel{\text{def}}{=} \sup_{x\geq K} \frac{x}{f(x)}$,

$$\begin{aligned}\int_{[0,1]} |g_n(x)|\,dx &= \int_{[0,1]\cap\{\,x\,:\,|g_n(x)|\geq K\,\}} + \int_{[0,1]\cap\{\,x\,:\,|g_n(x)|<K\,\}} |g_n(x)|\,dx \\ &\leq F(K)\int_{[0,1]\cap\{\,x\,:\,|g_n(x)|\geq K\,\}} f(|g_n(x)|)\,dx + K \leq F(K)M + K < \infty,\end{aligned}$$

whence $\{g_n\}_{n\in\mathbb{N}} \subset L^1([0,1], \lambda)$. If $a > 0$ then for all sufficiently large K, $F(K) < \frac{a}{8(M+1)}$. Hence

$$\begin{aligned}\|g_n - g_m\|_1 \leq &\int_{[0,1]\cap\{\,x\,:\,|g_n(x)-g_m(x)|\leq 2K\,\}} \\ &+ \int_{[0,1]\cap\{\,x\,:\,|g_n(x)-g_m(x)|>2K\,\}} 2\cdot\frac{|g_n(x) - g_m(x)|}{2}\,dx \\ &\stackrel{\text{def}}{=} I + II.\end{aligned}$$

Since $\int_0^1 f(|g_n(x)|)\,dx \leq M$,

$$\begin{aligned}\int_0^1 f(|g_n(x)| \vee |g_m(x)|)\,dx &= \int_{[0,1]\cap\{\,x\,:\,|g_n(x)|\geq|g_m(x)|\,\}} f(|g_n(x)|)\,dx \\ &\quad + \int_{[0,1]\cap\{\,x\,:\,|g_n(x)|<|g_m(x)|\,\}} f(|g_m(x)|)\,dx \\ &\leq 2M.\end{aligned}$$

Because $|g_n(x) - g_m(x)| \leq 2(|g_n(x)| \vee |g_m(x)|)$ and f is monotonely increasing, it follows that $f\left(\frac{|g_n(x) - g_m(x)|}{2}\right) \leq f(|g_n(x)| \vee |g_m(x)|)$ and so for all n, m, $2F(K)f\left(\frac{|g_n(x) - g_m(x)|}{2}\right)$ dominates the integrand in II. On the other hand, $II \leq 2 \cdot 2MF(K) < \frac{a}{2}$. Owing to bounded convergence, as $n, m \to \infty$, $I \to 0$. Hence for large n, m, $\|g_n - g_m\|_1 < a$. If $g_n \overset{\| \|_1}{\to} h$ then a subsequence $g_{n_k} \overset{\text{a.e.}}{\to} h$ whence $g \doteq h$. □

6.33. LEMMA. FOR ANY BANACH SPACE X AND A SEQUENCE $\{\mathbf{x}_n\}_{n=0}^{\infty}$ CONTAINED IN X, $\mathbf{x}_n \overset{\text{w}}{\to} \mathbf{x}_0$ IFF FOR SOME K, $\|\mathbf{x}_n\| \leq K < \infty$ AND FOR ALL $\mathbf{x}^*$ IN A DENSE SUBSET D OF X^*, $(\mathbf{x}_n, \mathbf{x}^*) \to (\mathbf{x}_0, \mathbf{x}^*)$ AS $n \to \infty$.

PROOF: If $\mathbf{x}_n \overset{\text{w}}{\to} \mathbf{x}_0$, the uniform boundedness principle implies that for some K, $\|\mathbf{x}_n\| \leq K < \infty$.

Conversely, if $\mathbf{y}^* \in X^*$, some $\mathbf{x}^*$ in D is near $\mathbf{y}^*$ and the condition $\|\mathbf{x}_n\| \leq K < \infty$ implies $|(\mathbf{x}_n, \mathbf{y}^*) - (\mathbf{x}_0, \mathbf{y}^*)|$ is small for large n. □

a) ⇒ b). In the current context, for any step function s,

$$\lim_{n \to \infty} \int_{\mathbb{R}} f_n(t)s(t)\,dt = \int_{\mathbb{R}} f_0(t)s(t)\,dt.$$

Because the set of step functions is dense in $L^{p'}(\mathbb{R}, \lambda)$, $f_n \overset{\text{w}}{\to} f_0$.

b) ⇒ a). The general lemma just proved implies the result. □

6.34. If $\{f_n\}_{n \in \mathbb{N}} \subset A$ and $\|f_n - f\|_1 \to 0$ as $n \to \infty$, then a subsequence converges a.e. to f. Hence $|f| \geq 1$ a.e. and so A is $\| \|_1$-closed.

Let f_n be the nth Walsh function (cf **5.156**). Then $A \supset \{f_n\}_{n \in \mathbb{N}}$. If $E \in \mathsf{S}_\beta$, Bessel's inequality implies $\int_0^1 \chi_E(x) f_n(x)\,dx \to 0$ as $n \to \infty$. Since the span of functions of the form χ_E is dense in $L^\infty([0,1], \lambda)$ and $\|f_n\|_1 \leq 1$, it follows that $f_n \overset{\text{w}}{\to} \mathbf{O}$, whereas $\mathbf{O} \notin A$. □

6.35. Because $\mathbb{Z}$ is a Borel set, $S \in \mathsf{S}_\lambda$. Furthermore,

$$\int_0^1 |\cos \pi f(x)|^n\,dx = \int_S + \int_{[0,1] \setminus S} |\cos \pi f(x)|^n\,dx,$$

$|\cos \pi f(x)| = 1$ iff $x \in S$, and $|\cos \pi f(x)|^n \to 0$ as $n \to \infty$ iff $x \in [0,1] \setminus S$, whence the bounded convergence theorem implies the result. □

6.36. a) If $\|f_n\|_\infty \leq M < \infty$ and $h_n \overset{\text{def}}{=} f_n * g$ then

$$|h_n(x+y) - h_n(x)| \leq M \left\|g_{[x+y]} - g_{[x]}\right\|_1.$$

Since $\left\|g_{[x+y]} - g_{[x]}\right\|_1 \overset{\text{u}}{\to} 0$ as $y \to 0$, it follows from the Arzelà-Ascoli theorem that $\{h_n\}_{n \in \mathbb{N}}$ contains a uniformly convergent subsequence.

b) For f_n as described, the map $[0, 2\pi] \ni \theta \mapsto e^{i\theta}$ permits the identification of $\mathbb{T}$ with $[0, 2\pi)$. Concordantly, the discussion is reduced to consideration of $F_n(\theta) = 2\cos n\theta, 0 \le \theta < 2\pi$. Because

$$|h_n(\theta)| \le \left|\int_0^{2\pi} g(\phi)\cos n\phi\, d\phi\right| \cdot |\cos n\theta| + \left|\int_0^{2\pi} g(\phi)\sin n\phi\, d\phi\right| \cdot |\sin n\theta|,$$

the Riemann-Lebesgue lemma implies $h_n \stackrel{\text{u}}{\to} 0$. □

6.37. The formula $\widehat{f}(t) = \frac{1}{2\pi}\int_{-n}^{n} f(x)e^{-itx}\, dx$ implies

$$\frac{\widehat{f}(t+s) - \widehat{f}(t)}{s} = \frac{1}{2\pi}\int_{-n}^{n} e^{-itx}\frac{e^{-isx} - 1}{s} f(x)\, dx.$$

The dominated convergence theorem permits the passage $s \to 0$ and shows $\widehat{f}'$ exists and $\widehat{f}'(t) = \frac{1}{2\pi}\int_{-n}^{n} -ixe^{-itx} f(x)\, dx$. Similar calculations show that for all k in $\mathbb{N}$, $\widehat{f}^{(k)}$ exists and $\widehat{f}^{(k)}(t) = \frac{1}{2\pi}\int_{-n}^{n} (-ix)^k e^{-itx} f(x)\, dx$. The Riemann-Lebesgue lemma implies the conclusion. □

6.38. For some h in $C_{00}^{\infty}(\mathbb{R}, \mathbb{C})$,

$$h \ne 0,\ h(x) = h(-x),\ \text{and } \operatorname{supp}(h) \cap \operatorname{supp}\left(\widehat{f}\right) = \emptyset.$$

Repeated integration by parts shows that if $g = \widehat{h}$ then

$$t^2 g(t) \in C_0(\mathbb{R}, \mathbb{C}),\, g \in L^1(\mathbb{R}, \mathbb{C}) \cap C_0(\mathbb{R}, \mathbb{C}),\ \widehat{g} = h$$
$$(g * f)\widehat{\ } = h\widehat{f} = 0,\ g * f = 0.$$

□

6.39. Because $\widehat{f * g} = \widehat{f} \cdot \widehat{g}$, it follows that in the present context, $\widehat{f}^2 = \widehat{f}$. Since $\widehat{f} \in C_0(\mathbb{R}, \mathbb{C})$ it follows that $\widehat{f} = 0$. The uniqueness theorem for Fourier transforms implies $f \doteq 0$. □

6.40. If $g \stackrel{\text{def}}{=} f \cdot \chi_E$ then $\|g\|_1 = 1$ whence $|\widehat{g}| \le 1$. If, for some real a, $\widehat{g}(1) = e^{ia}$ then $1 = \int_{\mathbb{R}} g(x)e^{-i(x+a)}\, dx = \int_{\mathbb{R}} g(y-a)e^{-iy}\, dy = \widehat{g_{[-a]}}(1)$. Because $\left\|g_{[-a]}\right\|_1 = 1$ and $g_{[-a]} \ge 0$, the problem is reduced to showing that if $f \ge 0$ and $\|f\|_1 = 1$ then $\widehat{f}(1) \ne 1$.

If $\widehat{f}(1) = 1$ then $\int_{\mathbb{R}} f(x)\cos x\, dx = 1$. If

$$E_n \stackrel{\text{def}}{=} \left\{x \,:\, x \in E, \cos x < 1 - \frac{1}{n}\right\}, n \in \mathbb{N},\ \text{and } \left|\int_{E_n} f(x)e^{-ix}\, dx\right| < 1$$

then $\int_{E_n} f(x)\cos x\,dx \le \left(1-\frac{1}{n}\right)\int_{E_n} f(x)\,dx$ and

$$\begin{aligned}1 = \int_{E_n} + \int_{\mathbb{R}\setminus E_n} f(x)\,dx &= \int_{\mathbb{R}} f(x)\cos x\,dx\\ &= \int_{E_n} + \int_{\mathbb{R}\setminus E_n} f(x)\cos x\,dx\\ &\le \left(1-\frac{1}{n}\right)\int_{E_n} + \int_{\mathbb{R}\setminus E_n} f(x)\,dx\end{aligned},$$

whence $\int_{E_n} f(x)\,dx \le \left(1-\frac{1}{n}\right)\int_{E_n} f(x)\,dx$ and $f(x) \doteq 0$ on E_n. Because $\mathbb{R} = \left(\bigcup_{n\ge 2} E_n\right) \cup \left(\bigcup_{k\in\mathbb{Z}}\{2\pi k\}\right)$, $f \doteq 0$, a contradiction. □

6.41. It may be assumed that $\|f\|_1 = 1$. If $\widehat{f}(t_0) = e^{ia} \neq 1$ then $t_0 \neq 0$ and thus if $b \stackrel{\text{def}}{=} \frac{a}{t_0}$ then, in a calculation like that in **6.40**, it follows that

$$\widehat{f_{[b]}}(t_0) \stackrel{\text{def}}{=} \widehat{g}(t_0) = 1.$$

Because $\|g\|_1 = \|f\|_1\ (= 1)$ and $g \ge 0$, the problem is reduced almost to that in **6.40**. The approach in this case is to replace E_n as defined in **6.40** by $F_n \stackrel{\text{def}}{=} \left\{x\ :\ \cos t_0 x \le 1 - \frac{1}{n}\right\}$. □

6.42. Because $\nu(\mathbb{T}) = 1$, $\left|\int_{\mathbb{T}} x\,d\nu(x)\right| \le \int_{\mathbb{T}} |x|\,d\nu(x) = \nu(\mathbb{T}) = 1$.

In the pattern of **6.40, 6.41**, if $\int_{\mathbb{T}} e^{-ia} x\,d\nu(x) = 1$ then instead of E_n or F_n, $G_n \stackrel{\text{def}}{=} \left\{x\ :\ \Re e^{-ia}x \le 1 - \frac{1}{n}\right\}$ serves. □

[**Note s6.4:** When ν above is a discrete measure its support is a finite or countable set. The conclusion when supp(ν) is a finite set is that the ν-center of gravity lies inside the unit ball of $\mathbb{C}$ or on $\mathbb{T}$, and in the latter case iff supp(ν) is a single point.]

6.43. If $\{f, g\} \subset C_{00}(G, \mathbb{C})$ then $f * g \in C_{00}(G, \mathbb{C})$ and Fubini's theorem implies $\|f * g\|_1 \le \|f\|_1 \cdot \|g\|_1$. If $h \in L^{p'}(G, \mu)$ Hölder's inequality and the translation-invariance of μ imply

$$\left|\int_G f * g(x)\overline{h(x)}\,d\mu(x)\right| \le \|f\|_1 \cdot \|g\|_p \cdot \|h\|_{p'},$$

whence $f * g \in L^p(G, \mu)$ and $\|f * g\|_p \le \|f\|_1 \cdot \|g\|_p$. □

6.44. If $h \in L^{p'}(G,\mu)$ then Fubini's theorem and Hölder's inequality imply

$$\left|\int_G (n_U * f(x) - f(x))\, h(x)\, d\mu(x)\right| \le \|h\|_{p'} \int_G \left\|f_{[y^{-1}]} - f\right\|_p n_U(y)\, d\mu(y)$$

$$\|n_U * f - f\|_p \le \int_G \left\|f_{[y^{-1}]} - f\right\|_p n_U(y)\, d\mu(y).$$

Because $\left\|f_{[y^{-1}]} - f\right\|_p \to 0$ as $y \to e$, the net $\mathcal{U} \ni U \mapsto \|n_U * f - f\|_p$ converges to zero. □

6.45. Because, for all f in $L^1(G,\mu)$, $\|n * f - f\|_1 \to 0$, it follows that

$$\left\|\widehat{n}\widehat{f} - \widehat{f}\right\|_\infty = \left\|\widehat{f}\right\|_\infty \cdot \|\widehat{n} - 1\|_\infty \to 0.$$ □

6.46. a) The formula $\int_{\mathbb{R}} e^{-x^2}\, dx = \sqrt{\pi}$ shows that $c_t = \dfrac{1}{\sqrt{\pi t}}$.

b) Choose c_n so that for $g_{tn} \overset{\text{def}}{=} c_n g_t \chi_{-[\frac{1}{n},\frac{1}{n}]}$, $\|g_{tn}\|_1 = 1, n \in \mathbb{N}$. Thus for n in $\mathbb{N}$, $g_{\frac{1}{n},n}$ is a particular instance of n_U in **6.44**. If $0 < t < \dfrac{1}{n}$ and $f \in L^1(\mathbb{R},\lambda)$ then

$$\begin{aligned}\|g_t * f - f\|_1 &\le \left\|\left(g_t - g_{\frac{1}{n},n}\right) * f\right\|_1 + \left\|g_{\frac{1}{n},n} * f - f\right\|_1 \\ &\le \left\|g_t - g_{\frac{1}{n},n}\right\|_1 \|f\|_1 + \left\|g_{\frac{1}{n},n} * f - f\right\|_1. \qquad \text{(s6.1)}\end{aligned}$$

For large n, the second term in the right member of (s6.1) is small and if $0 < t < \dfrac{1}{n}$, the first terms is small as well. □

6.47. As shown in **6.37**, $\widehat{f} \in C_0^\infty(\mathbb{R},\mathbb{C})$. The formulæ there show that $\left|\widehat{f}^{(k)}\right| \le 2n^k \|f\|_1$. If, for some b, $\widehat{f}(b) \ne 0$, let the Kth order Taylor formula for $\widehat{f}$ be

$$\sum_{k=0}^{K} \frac{\widehat{f}^{(k)}(b)(t-b)^k}{k!} + R_{K+1}(t).$$

Then for some a between b and t,

$$|R_{K+1}(t)| \le \left|\frac{\widehat{f}^{(K+1)}(a) \cdot |t-b|^{K+1}}{(K+1)!}\right| \le \frac{2n^{K+1}\|f\|_1 |t-b|^{K+1}}{(K+1)!} \to 0$$

as $K \to \infty$. Hence $\widehat{f}$ is entire and because $\operatorname{supp}\left(\widehat{f}\right)$ is compact, $\widehat{f} = 0$. Thus $f = 0$. □

6.48. If $g \doteq 0$ then $T_g\left(L^2(\mathbb{R},\lambda)\right) = \{\mathbf{O}\}$, which is $\|\ \|_2$-compact.

If $g \not\doteq 0$ and $f \in L^2(\mathbb{R},\lambda) \cap L^1(\mathbb{R},\lambda) \stackrel{\text{def}}{=} V$, then $T_g(f) \in V$ whence $\widehat{T_g(f)}$ is defined and $\widehat{T_g(f)} = \widehat{g}\cdot\widehat{f}$. Owing to the Plancherel theorem, the Fourier transform confined to V is a $\|\ \|_2$-isometry that is extendible as an isometry to $L^2(\mathbb{R},\lambda)$. Thus if the $\|\ \|_2$-closure $\overline{T_g(B(\mathbf{O},1))}$ of $T_g(B(\mathbf{O},1))$ is $\|\ \|_2$-compact then its image A under the Fourier transform is also $\|\ \|_2$-compact.

However, $A = \overline{\left\{\widehat{g}\cdot\widehat{f}\ :\ f \in B(\mathbf{O},1)\right\}}$. Because $g \not\doteq 0$, $\widehat{g} \neq 0$ and, since $\widehat{g}$ is continuous, for some a, b, $a < b$ and on $[a,b]$, $|\widehat{g}|^2 \geq d^2 > 0$. It may be assumed that $[a,b] = [0,1]$.

If $f_n \in C_{00}^{\infty}(\mathbb{R},\mathbb{C})$ and $f_n(x) = 0$ off $\left[2^{-(n+1)}, 2^{-n}\right]$ then $\{f_n\}_{n\in\mathbb{N}}$ is orthonormal and the Plancherel theorem implies $\left\{\widehat{f_n}\right\}_{n\in\mathbb{N}}$ is orthonormal and for some h_n in $L^2(\mathbb{R},\mathbb{C})$, $\widehat{h_n} = f_n$ and $\{h_n\}_{n\in\mathbb{N}}$ is also orthonormal. Consequently,

$$\|g*(f_n - f_m)\|_2 = \left\|\widehat{g}\cdot\left(\widehat{f_n} - \widehat{f_m}\right)\right\|_2 \geq d\sqrt{2} > 0.$$

Thus neither A nor $\overline{T_g(B(\mathbf{O},1))}$ is $\|\ \|_2$-compact. □

6.49. For x in $[1,\infty)$, the Kronecker function

$$\delta_x(y) \stackrel{\text{def}}{=} \chi_{\{x\}}(y) \in L^1([1,\infty),\mu).$$

If $F \in \left(L^1([1,\infty),\mu)\right)^*$ and $F(\delta_x) \stackrel{\text{def}}{=} a_x$ then

$$|a_x| \leq \|F\|\cdot\|\delta_x\|_1 = \|F\|\cdot x.$$

Conversely, if $M < \infty$ and $|a_x| \leq M\cdot x$ then

$$F : L^1([1,\infty),\mu) \ni f \mapsto \sum_{x\in[1,\infty)} f(x)\cdot a_x$$

is in $\left(L^1([1,\infty),\mu)\right)^*$ and $\|F\| \leq M$. Thus $F \in \left(L^1([1,\infty),\mu)\right)^*$ iff

$$\sup\left\{\frac{|F(\delta_x)|}{x}\ :\ x \in [1,\infty)\right\} < \infty.$$ □

[**Note s6.5:** In the context above, $\left(L^1([1,\infty),\mu)\right)^*$ is a proper subset of $L^\infty([1,\infty),\mu)$.]

6.50. If $p \geq 1$ and $f \in L^1(\mathbb{R}, \lambda)$ then (cf. **6.43**) $T_f : L^p(\mathbb{R}, \lambda) \ni g \mapsto f * g$ maps $L^p(\mathbb{R}, \lambda)$ into itself and $\|T_f(g)\|_p \leq \|f\|_1 \cdot \|g\|_p$. Thus if

$$g : \mathbb{R} \ni y \mapsto \begin{cases} \dfrac{\left|\sin\left(\frac{1}{y}\right)\right|}{|y|^{\frac{1}{2}}} & \text{if } y \neq 0 \\ 0 & \text{otherwise} \end{cases}$$

is in $L^p(\mathbb{R}, \lambda)$ then $f * g \in L^p(\mathbb{R}, \lambda)$.

However, the substitution $y \mapsto \dfrac{1}{z}$ shows that

$$\|g\|_p^p = \int_{-1}^{1} + \int_{\mathbb{R}\setminus[-1,1]} \frac{|\sin z|^p}{|z|^p} |z|^{3p-4/2}\, dz.$$

The first integral is finite iff $p > \dfrac{2}{3}$ and the second integral is finite iff $p < 2$. Hence $f * g \in L^p(\mathbb{R}, \lambda)$ for all f in $L^1(\mathbb{R}, \lambda)$ iff $\dfrac{2}{3} < p < 2$. □

[**Note s6.6:** Actually Cohen **[Coh1]** showed that if B is a Banach algebra containing a bounded approximate identity then each $\mathbf{x}$ in B may be written as a product $\mathbf{xy}$ of elements of B. From this **[HeR]** it follows that if V is a Banach space that is also a B-module then $B \cdot V$ is a closed subspace of V. That general result implies that if G is a locally compact group then $L^1(G, \mu) * L^p(G, \mu) = L^p(G, \mu)$.

It is a consequence of the results just cited that if $f \in L^p(G, \mu)$ and n is an approximate identity in $L^1(G, \mu)$ then $n * f \overset{\|\ \|_p}{\to} f$.]

6.51. Two continuous functions must differ on a nonempty open set, hence on a set of positive measure, whence if f and g are continuous and $f \doteq g$ then $f = g$.

On the other hand, for α in $(0, 1)$, let S in $L^p([0, 1], \lambda)$ be the equivalence class containing the characteristic function of a Cantor-like set C_α. If g is continuous and $g \in S$ then $g(x) = 0$ on a dense set whence $g = 0$ whereas $\|g\|_p = \alpha > 0$. □

6.52. For $R \overset{\text{def}}{=} \left\{ (u, v) \ : \ u < v < -\dfrac{u}{u+1}, -1 < u < 0 \right\}$ the map

$$T : R \ni (u, v) \mapsto T((u, v)) \overset{\text{def}}{=} \left(\sqrt{(u+1)(v+1)}, \sqrt{\frac{u+1}{v+1}} \right)$$

is a bijection between R and $(0,1)^2$. Note that for (x,y) in $(0,1)^2$,

$$u = xy - 1, v = \frac{x}{y} - 1.$$

The absolute value of the Jacobian determinant for T is $\left|\frac{1}{2(1+v)}\right|$. Thus the problem is to find the values of p in $[1,\infty)$ and for which

$$\int_R \frac{1}{|u|^p(1+v)}\,d\lambda_2(u,v) < \infty.$$

The integral above is finite iff

$$\begin{aligned}\lim_{\substack{\epsilon\uparrow 0\\ t\downarrow -1}} &\int_t^\epsilon \left(\int_u^{\frac{1}{1+u}-1} \frac{1}{|u|^p(1+v)}\,dv\right) du\\ &= \lim_{\substack{\epsilon\uparrow 0\\ t\downarrow -1}} \int_t^\epsilon \frac{|\ln(1+u)|}{|u|^p}\,du < \infty. \qquad \text{(s6.2)}\end{aligned}$$

If $-1 < u < 0$, $|\ln(1+u)| \le \sum_{n=1}^\infty \frac{|u|^n}{n}$. Hence if $p = 1$, the right member of (s6.2) does not exceed $\frac{\pi^2}{3}$. Because the integrand is nonnegative, $f \in L^1\left([0,1]^2, \lambda_2\right)$.

On the other hand, if $1 < p$ then on (t,ϵ), $\frac{|\ln(1+u)|}{|u|^p} \ge \frac{|\ln(1+\epsilon)|}{|u|^p}$ whence the right member of (s6.2) is unbounded as $\epsilon \uparrow 0$. □

6.53. If $f \in C_{00}(G,\mathbb{C})$ then

$$\left\|f - f_{[n]}\right\|_p \le \left\|f - f_{[n]}\right\|_\infty \cdot [\lambda(\operatorname{supp}(f))]^{\frac{1}{p}}$$

which approaches zero as $n \to e$. Because $C_{00}(G,\mathbb{C})$ is dense in $L^p(G,\mu)$, it follows that for any g in $L^p(G,\mu)$, $\left\|g - g_{[n]}\right\|_p \to 0$ as $n \to e$.

In particular,

$$\left|\left\|g + g_{[n]}\right\|_p - 2\|g\|_p\right| \le \left\|g + g_{[n]} - 2g\right\|_p = \left\|g_{[n]} - g\right\|_p. \qquad \square$$

6.54. According to **Solution 6.53** for some strictly monotonely decreasing null sequence $\{a_n\}_{n\in\mathbb{N}}$, if $|b_n| < a_n$ then $\left\|f_{[b_n]} - f\right\|_p^p < 2^{-n}$. Thus $\sum_{n=1}^\infty \left\|f_{[b_n]} - f\right\|_p^p < \infty$. The monotone convergence theorem implies

$$\sum_{n=1}^\infty |f(x+b_n) - f(x)|^p < \infty \text{ a.e.} \qquad \square$$

6.55. If $0 < \lambda_n(A) \cdot \lambda_n(B) < \infty$ and $f \stackrel{\text{def}}{=} \chi_A, g \stackrel{\text{def}}{=} \chi_B$, the result is directly verifiable.

If $f \stackrel{\text{def}}{=} \chi_A$, the B_k are pairwise disjoint, $0 < \lambda_n(A)\cdot\prod_{k=1}^K \lambda_n(B_k) < \infty$, $b_0 \stackrel{\text{def}}{=} 0 \le b_1 < b_2 < \cdots < b_K$, and $g \stackrel{\text{def}}{=} \sum_{k=1}^K b_k\chi_{B_k}$ then

$$F(t) \stackrel{\text{def}}{=} \int_{E_t} f(\mathbf{x})\,d\mathbf{x} = \begin{cases} 0 & \text{if } t > b+K \\ \sum_{k=p}^K \lambda_n(A\cap B_k) & \text{if } b_p \ge t > b_{p-1}, 2 \le p \le K \\ \sum_{k=1}^K \lambda_n(A\cap B_k) & \text{if } b_1 \ge t. \end{cases}$$

Hence, by Abel summation,

$$\begin{aligned}\int_0^\infty F(t)\,dt &= \sum_{p=1}^{K-1} \lambda_n\left(A\cap\left(\bigcup_{k=p}^K B_k\right)\right)(b_p - b_{p-1}) = \sum_{k=1}^K b_k\lambda_n(A\cap B_k) \\ &= \int_{\mathbb{R}^n} f(\mathbf{x})g(\mathbf{x})\,d\mathbf{x}.\end{aligned}$$

The general case can be handled via successive approximations by monotonely increasing sequences of simple functions in which first, $f = \chi_A$ and $g \in L^{p'}(\mathbb{R}^n, \lambda_n)$, second, f is a simple function and $g \in L^{p'}(\mathbb{R}^n, \lambda_n)$, and third, $f \in L^p(\mathbb{R}^n, \lambda_n)$ and $g \in L^{p'}(\mathbb{R}^n, \lambda_n)$. □

6.56. If $P \stackrel{\text{def}}{=} \mathbb{C}[x]$ is viewed as a subset of $C([0,1], \mathbb{C})$, the map

$$T : P \ni \sum_{n=1}^K b_k t^k \mapsto \sum_{k=1}^K a_k b_k \in \mathbb{C}$$

is, by hypothesis, a bounded linear functional defined on a dense subset of $L^p([0,1], \lambda)$. Hence T may be extended without increasing its norm to an element of $(L^p([0,1], \lambda))^*$, i.e., of $L^{p'}([0,1], \lambda)$. □

6.57. a) An extreme point must lie on $\partial B(\mathbf{O}, 1)$ because if $\|f\|_p < 1$ and $\|g\|_p = 1$ then
$$f = \begin{cases} (1 - \|f\|_p)\,\mathbf{O} + \|f\|_p \dfrac{f}{\|f\|_p} & \text{if } f \ne \mathbf{O} \\ \dfrac{1}{2}g + \dfrac{1}{2}(-g) & \text{if } f = \mathbf{O} \end{cases}.$$

When $p > 1$, $f \in L^p(X, \mu)$, $\|f\|_p = 1$, $g, h \in B(\mathbf{O}, 1)$, f, g, h are three points in $B(\mathbf{O}, 1)$, $a \in (0,1)$, and $f = ag + (1-a)h$ then

$$\|f\|_p \le a\|g\|_p + (1-a)\|h\|_p \le 1$$

and, according to the criterion for equality in the Minkowski inequality, for constants A, B, not both zero, $Ag + Bh \doteq 0$. It follows that

$$(aB - A(1-a))h = -Af \text{ if } A \ne 0 \text{ and } ((1-a)A - Ba)g = -Bf \text{ if } B \ne 0.$$

If $A \neq 0$, f is a multiple of h. Thus $\|h\|_p = 1$ and for some θ in $\mathbb{R}$, $f = e^{i\theta}h$, $\|g\|_p = 1$, and $a = \left|a + e^{i\theta} - 1\right|$. If $\cos\theta \neq 1$ then $a = 1$, a contradiction, and if $\cos\theta = 1$ then $e^{i\theta} = 1$, $f = h$, and yet another contradiction. If $B \neq 0$, a similar argument leads to a similar string of contradictions. In sum: $\mathcal{E}_p = \partial B(\mathbf{O}, 1)$.

b) When $p = 1$, (X, S, μ) is nonatomic, $f \in L^1_{\mathbb{R}}(X, \mu)$, and $\|f\|_1 = 1$, it may be assumed that, e.g., for some positive a, $A \stackrel{\text{def}}{=} \{x : f(x) \geq a\}$ has finite positive measure and thus may be decomposed into a union of disjoint measurable sets of positive measure: $A = A_1 \dot\cup A_2$. For some b, c such that $0 < b, c < a$ and $b\mu(A_1) = c\mu(A_2)$ and if

$$g_1 = \begin{cases} f + b & \text{on } A_1 \\ f - c & \text{on } A_2 \\ f & \text{elsewhere} \end{cases}, \; g_2 = \begin{cases} f - b & \text{on } A_1 \\ f + c & \text{on } A_2 \\ f & \text{elsewhere} \end{cases},$$

then $g_1 \neq g_2$, $\frac{1}{2}g_1 + \frac{1}{2}g_2 = f$, $\|g_1\|_1 = \|g_2\|_1 = 1$, and f is not an extreme point.

When $p = 1$, (X, S, μ) is nonatomic, and f is an extreme point then $|f|$ is also an extreme point. [PROOF: If $g, h \in B(\mathbf{O}, 1)$, a is in $(0, 1)$, and

$$|f| = ag + (1 - a)h$$

then $f = a \cdot \overline{\operatorname{sgn}(f)}g + (1 - a) \cdot \overline{\operatorname{sgn}(f)}h$, $\overline{\operatorname{sgn}(f)}g \doteq \overline{\operatorname{sgn}(f)}h$, and $g \doteq h$ on $E \stackrel{\text{def}}{=} E_{\neq}(f, 0)$. However,

$$1 = \|f\|_1 = \int_E |f(x)|\,d\mu(x) = a\int_E g(x)\,d\mu(x) + (1 - a)\int_E h(x)\,d\mu(x)$$

$$\int_E g(x)\,d\mu(x) = \int_E h(x)\,d\mu(x) = 1$$

$$1 \geq \|g\|_1 \geq \left|\int_E g(x)\,d\mu(x)\right| = 1, \; 1 \geq \|h\|_1 \geq \left|\int_E h(x)\,d\mu(x)\right| = 1,$$

whence $g \doteq h \doteq 0$ off E, i.e., $g \doteq h$, a contradiction.]

In sum, if f is an extreme point in $L^1(X, \mu)$, $|f|$ is an extreme point in $L^1_{\mathbb{R}}(X, \mu)$, an impossibility.

Hence $$\mathcal{E}_p = \begin{cases} \emptyset & \text{if } p = 1 \text{ and } (X, \mathsf{S}, \mu) \text{ is nonatomic} \\ \partial B(\mathbf{0}, 1) & \text{if } p > 1 \end{cases}.$$

□

[**Note s6.7:** If X contains atoms $\mathcal{E}_1$ can fail to be empty, e.g., in $\ell^1_{\mathbb{R}}(\{0, 1\})$, $\mathcal{E}_1$ consists of four points.]

6.58. If $A = B$ then $A^\perp = B^\perp$. Conversely, if $A^\perp = B^\perp$, it follows that $\left(A^\perp\right)^\perp = \left(B^\perp\right)^\perp$. Because $A \subset \left(A^\perp\right)^\perp$, if $f \in \left(A^\perp\right)^\perp \setminus A$ then for some g in $A^\perp$, $\int_X \overline{g(x)}f(x)\,d\mu(x) = 1$ whereas $\int_X \overline{f(x)}g(x)\,d\mu(x) = 0$. □

6.59. a) The function $f : [0, \infty) \ni t \mapsto \dfrac{1+t^p}{(1+t)^p}$ achieves its minimal value 2^{1-p} (< 1) when $t = 1$ and is strictly monotonely decreasing resp. increasing in $[0, 1)$ resp. $(1, \infty)$.

If either $\mathbf{x}$ or $\mathbf{y}$ is $\mathbf{O}$, the desired conclusion is automatic. If (6.1) is valid when $0 < \|\mathbf{y}\| \le \|\mathbf{x}\| \le 1$ and if $\mathbf{z} \stackrel{\text{def}}{=} \dfrac{\mathbf{y}}{\|\mathbf{x}\|}$ then

$$\left\| \frac{\frac{\mathbf{x}}{\|\mathbf{x}\|} + \mathbf{z}}{2} \right\|^p \le (1 - \delta_p(\epsilon)) \frac{\left\| \frac{\mathbf{x}}{\|\mathbf{x}\|} \right\|^p + \|\mathbf{z}\|^p}{2}$$

$$\left\| \frac{\mathbf{x}+\mathbf{y}}{2} \right\|^p \le (1 - \delta_p(\epsilon)) \frac{\|\mathbf{x}\|^p + \|\mathbf{y}\|^p}{2}.$$

If (6.1) is false then for some positive ϵ and vectors $\mathbf{x}_n, \mathbf{y}_n$,

$$\begin{gathered} \|\mathbf{x}_n\| = 1, \|\mathbf{y}_n\| \le 1, \|\mathbf{x}_n - \mathbf{y}_n\| \ge \epsilon \\ \lim_{n\to\infty} \frac{\left\|\frac{1}{2}(\mathbf{x}_n + \mathbf{y}_n)\right\|^p}{\frac{1}{2}\left(\|\mathbf{x}_n\|^p + \|\mathbf{y}_n\|^p\right)} = 1. \end{gathered} \tag{s6.3}$$

If, in (s6.3), $\|\mathbf{y}_n\| \le q < 1$ then, owing to the triangle inequality and the strict monotonicity of f in $[0, 1)$, for some ρ in $(0, 1)$,

$$\left\| \frac{1}{2}(\mathbf{x}_n + \mathbf{y}_n) \right\|^p \le \left(\frac{1}{2} (1 + \|\mathbf{y}_n\|)^p \right) \le \frac{\rho}{2} (1 + \|\mathbf{y}_n\|^p) = \frac{\rho}{2} \left(\|\mathbf{x}_n\|^p + \|\mathbf{y}_n\|^p\right),$$

contradicting (6.1).

Thus it may be assumed that $\|\mathbf{y}_n\| \to 1$ as $n \to \infty$ and hence if $\mathbf{z}_n \stackrel{\text{def}}{=} \dfrac{\mathbf{y}_n}{\|\mathbf{y}_n\|}$ then $\|\mathbf{z}_n - \mathbf{y}_n\| \to 0$ as $n \to \infty$. Consequently for large n, $\|\mathbf{x}_n - \mathbf{z}_n\| \ge \epsilon$ and, to boot, $\lim_{n\to\infty} \left\| \dfrac{(\mathbf{x}_n + \mathbf{y}_n)}{2} \right\| = 1$, contradicting the uniform convexity of $(Y, \| \ \|)$.

Homogeneity considerations like those treated above show that the conclusion in a) is equivalent to the following statement: If $\mathbf{x}, \mathbf{y} \in Y$ and $\max\{\|\mathbf{x}\|, \|\mathbf{y}\|\} > 0$ then

$$\left\| \frac{(\mathbf{x}+\mathbf{y})}{2} \right\|^p \le \left(1 - \delta_p \left(\frac{\|\mathbf{x}-\mathbf{y}\|}{\max\{\|\mathbf{x}\|, \|\mathbf{y}\|\}} \right)\right) \left(\frac{\|\mathbf{x}\|^p + \|\mathbf{y}\|^p}{2} \right). \tag{s6.4}$$

b) In Hilbert space $\mathfrak{H}$, e.g., in $\mathbb{C}$, the theorem of Apollonius, i.e., the parallelogram law,

$$\|\mathbf{x}+\mathbf{y}\|^2 + \|\mathbf{x}-\mathbf{y}\|^2 = 2\left(\|\mathbf{x}\|^2 + \|\mathbf{y}\|^2\right),$$

is valid. Thus for $\mathbf{x},\mathbf{y}$ in $\mathfrak{H}$, if $\|\mathbf{x}\|, \|\mathbf{y}\| \leq 1$, $\epsilon \in (0,2]$, and $\|\mathbf{x}-\mathbf{y}\| \geq \epsilon$ then $\left\|\frac{\mathbf{x}+\mathbf{y}}{2}\right\| \leq \sqrt{1-\frac{\epsilon^2}{4}} \stackrel{\text{def}}{=} 1-\delta(\epsilon)$. Thus $\mathfrak{H}$, e.g., $\mathbb{C}$, is uniformly convex.

For f,g in $B(\mathbf{O},1)$ if $\|f-g\|_p \geq \epsilon > 0$, let E be

$$\left\{t \,:\, |f(t)-g(t)|^p \geq \frac{\epsilon^p}{4}\left(|f(t)|^p+|g(t)|^p\right)\left(\geq \frac{\epsilon^p}{4}\max\left\{|f(t)|^p,|g(t)|^p\right\}\right)\right\}.$$

Owing to a), (s6.4) applies for each t in E. Hence on E,

$$\left|\frac{(f(t)+g(t))}{2}\right|^p \leq \left(1-\delta_p\left(\frac{|f(t)-g(t)|}{\max\{|f(t)|,|g(t)|\}}\right)\right)\left(\frac{(|f(t)|^p+|g(t)|^p)}{2}\right)$$

and so $\int_{X\setminus E} |f(t)-g(t)|^p\,d\mu(t) \leq \frac{\epsilon^p}{2}$. If

$$\max\left\{\int_E |f(t)|^p\,d\mu(t), \int_E |g(t)|^p\,d\mu(t)\right\} < \frac{\epsilon^p}{2^{p+1}}$$

then $\int_E |f(t)-g(t)|^p\,d\mu(t) < \frac{\epsilon^p}{2}$, in contradiction of the hypothesis, i.e., $\max\left\{\int_E |f(t)|^p\,d\mu(t), \int_E |g(t)|^p\,d\mu(t)\right\} \geq \frac{\epsilon^p}{2^{p+1}}$.

Another application of (s6.4) as applied in a) reveals

$$\begin{aligned}
&\int_X \left\{\frac{1}{2}\left(|f(t)|^p+|g(t)|^p\right)-\left(\frac{1}{2}|f(t)+g(t)|\right)^p\right\} d\mu(t)\\
&\geq \int_E \left\{\frac{1}{2}\left(|f(t)|^p+|g(t)|^p\right)-\left(\frac{1}{2}|f(t)+g(t)|\right)^p\right\} d\mu(t)\\
&\geq \int_E \delta_p\left(\frac{|f(t)-g(t)|}{\max\{|f(t)|,|g(t)|\}}\right)\left(\frac{(|f(t)|^p+|g(t)|^p)}{2}\right) d\mu(t)\\
&\geq \delta_p\left(\frac{|f(t)-g(t)|}{\max\{|f(t)|,|g(t)|\}}\right)\frac{1}{2}\max\left\{\int_E |f(t)|^p\,d\mu(t), \int_E |g(t)|^p\,d\mu(t)\right\}\\
&\geq \delta_p\left(\frac{|f(t)-g(t)|}{\max\{|f(t)|,|g(t)|\}}\right)\frac{\epsilon^p}{2^{p+2}}.
\end{aligned}$$

Hence if $f,g \in B(\mathbf{O},1)$ and $\|f-g\|_p \geq \epsilon$ then

$$\left\|\frac{f+g}{2}\right\|_p \leq \left(1-\delta_p\left(\frac{\epsilon}{4^{\frac{1}{p}}}\right)\frac{\epsilon^p}{2^{p+2}}\right)^{\frac{1}{p}}. \qquad \square$$

6.60. If $\|f_0\|_p = 0$, the conclusion is automatic. If $\|f_0\|_p > 0$, it may be assumed that for all n, $\|f_n\|_p > 0$. If $g_n \stackrel{\text{def}}{=} \frac{\|f_0\|_p f_n}{\|f_n\|_p}$ then $g_n \stackrel{\text{a.e.}}{\to} f_0$, $\|g_n\|_p = \|f_0\|_p$.

Egorov's theorem implies that for m in $\mathbb{N}$, and some E_m, $g_n \xrightarrow{u} f_0$ on E_m and $\mu(E_m) > \mu(X) - \frac{1}{m}$. However, because

$$\begin{aligned}\|g_n - f_0\|_p &= \left(\int_{E_m} + \int_{X\setminus E_m} |g_n(t) - f_0(t)|^p \, d\mu(t)\right)^{\frac{1}{p}} \\ &\le \left(\int_{E_m} |g_n(t) - f_0(t)|^p \, d\mu(t)\right)^{\frac{1}{p}} \\ &\qquad + 2\left(\int_{X\setminus E_m} |f_0(t)|^p \, d\mu(t)\right)^{\frac{1}{p}}, \end{aligned} \tag{s6.5}$$

it follows that for large m the second term in the right member of (s6.5) is small and then for large n the first term is small. Thus $g_n \xrightarrow{\|\ \|_p} f_0$.

However,

$$\begin{aligned}\|f_0\|_p \cdot \|f_n - f_0\|_p &\le \left\| \|f_0\|_p f_n - f_0 \|f_n\|_p \right\|_p + \left\| f_0 \|f_n\|_p - \|f_0\|_p f_0 \right\|_p \\ &= \|f_n\|_p \cdot \|g_n - f_0\|_p + \|f_0\|_p \cdot \left| \|f_n\|_p - \|f_0\|_p \right|. \qquad \square\end{aligned}$$

6.61. For a subsequence $\{E_{N_k}\}_{k\in\mathbb{N}}$, $\chi_{E_{n_k}} \xrightarrow{\text{a.e.}} f$. $\square$

6.62. Because $\|f\|_{n+1}^{n+1} \stackrel{\text{def}}{=} a_{n+1} \le \|f\|_\infty a_n$, $\overline{\lim}_{n\to\infty} \frac{a_{n+1}}{a_n} \le \|f\|_\infty$ while Hölder's inequality implies $\|f\|_n \le \mu(X)^{\frac{1}{n(n+1)}} \|f\|_{n+1}$. Hence

$$\underline{\lim}_{n\to\infty} \frac{a_{n+1}}{a_n} \ge \lim_{n\to\infty} \frac{\|f\|_n}{\mu(X)^{\frac{1}{n}}} = \|f\|_\infty. \qquad \square$$

6.63. If $L^1(X,\mu) = L^\infty(X,\mu)$ and $\inf\{\mu(E) : E \in \mathsf{S}, \mu(E) > 0\} = 0$ for some sequence $\{E_n\}_{n\in\mathbb{N}}$ in S and a sequence $\{k_n\}_{n\in\mathbb{N}}$ in $\mathbb{N}$,

$$k_{n+1} > 2^{k_n} \text{ and } 2^{-k_{n+1}} < \mu(E_n) < 2^{-k_n}.$$

If $F_n \stackrel{\text{def}}{=} \bigcup_{m=n}^\infty E_m$ then $F_n \supset F_{n+1}, \mu(F_n) > 0, \mu(F_n) \downarrow 0$. All the pairwise disjoint sets $G_n \stackrel{\text{def}}{=} F_n \setminus F_{n+1}$ have positive measure. Hence there emerges the contradiction $\sum_{n=1}^\infty 2^n \chi_{G_n} \in L^1(X,\mu) \setminus L^\infty(X,\mu)$. Thus

$$\inf\{\mu(E) : E \in \mathsf{S}, \mu(E) > 0\} > 0. \tag{s6.6}$$

If there is no atom in X then (s6.6) is denied.

If $\{A_n\}_{n\in\mathbb{N}}$ is an infinite sequence of pairwise different atoms then either $\sum_{n=1}^{\infty}\mu(A_n)=\infty$, in which case every nonzero constant function is in $L^\infty(X,\mu)\setminus L^1(X,\mu)$, a contradiction, or $\sum_{n=1}^{\infty}\mu(A_n)<\infty$, in which case for some sequence $\{n_k\}_{k\in\mathbb{N}}$,

$$\sum_{k=1}^{\infty}\sqrt{\mu(A_{n_k})}<\infty,\ \sum_{k=1}^{\infty}\frac{\chi_{A_{n_k}}}{\sqrt{\mu(A_{n_k})}}\in L^1(X,\mu)\setminus L^\infty(X,\mu),$$

yet another contradiction.

Hence for some N in $\mathbb{N}$, the finitely many atoms in X may be listed: $A_1,\dots,A_N$. Since (s6.6) implies $\mu\left(X\setminus\bigcup_{n=1}^{N}A_n\right)=0$, if $f\in L^1(X,\mu)$ then f is constant a.e. on each A_n, $f\doteq 0$ off $\bigcup_{n=1}^{N}A_n$, and

$$f=\sum_{n=1}^{N}f(A_n)\chi_{A_n},\ \text{i.e., }\dim\left(L^1(X,\mu)\right)=N.$$

Conversely, if $L^1(X,\mu)$ is finite-dimensional, there is a (necessarily finite) maximal linearly independent set $\{\chi_{E_n}\}_{n=1}^{N}$ of characteristic functions of measurable sets of positive measure. Then $X\setminus\bigcup_{n=1}^{N}E_n$ is a null set and, modulo a null set, X is the union of the pairwise disjoint sets

$$B_n\stackrel{\text{def}}{=}E_n\setminus\bigcup_{m=1}^{n-1}E_m,1\le n\le N.$$

If some $\mu(B_n)=0$ then $\chi_{E_n}\in\text{span}\left(\{A_m\}_{m=1}^{n-1}\right)$. Furthermore, each B_n is an atom, since otherwise there are least $N+1$ linearly independent characteristic functions of nonnull sets. It follows that $L^1(X,\mu)=L^\infty(X,\mu)$. □

6.64. The linearity of T_f follows from its form as an integral. Furthermore, $|T_f(g)|\le\|f\|_1\|g\|_\infty$. □

6.65. If $f\in L^1(\mathbb{R},\lambda)$ and $\widetilde{k}(z)\stackrel{\text{def}}{=}k\left(\frac{z}{\sqrt{t}}\right)$, $\widetilde{f}(v)\stackrel{\text{def}}{=}f\left(\frac{z}{\sqrt{t}}\right)$ then, in the notation of **6.46**,

$$\sqrt{\frac{\pi}{t}}\int_{\mathbb{R}}\left(\int_{\mathbb{R}}e^{-\frac{(u-v)^2}{t}}k\left(\frac{u}{\sqrt{t}}\right)du\right)f\left(\frac{v}{\sqrt{t}}\right)dv=\frac{\pi}{t}\int_{\mathbb{R}}g_t*\widetilde{f}(z)\widetilde{k}(z)\,dz=0.$$

Hence if $\widetilde{k}$ is regarded as an element $T_{\widetilde{k}}$ in $\left(L^1(\mathbb{R},\lambda)\right)^*$, i.e., $L^\infty(\mathbb{R},\lambda)$, then for all t in $(0,\infty)$,

$$T_{\widetilde{k}}\left(g_t*\widetilde{f}\right)=0. \tag{s6.7}$$

Because $\{g_t\}_{t\in(0,\infty)}$ is an approximate identity, as $t \to 0$, (s6.7) becomes $T_{\widetilde{k}}\left(\widetilde{f}\right) = 0$, i.e. $T_{\widetilde{k}}$ is the zero functional. Hence $\widetilde{k} \doteq 0$ and so $k \doteq 0$. □

6.66. a) Because $f * g = g * f$ it suffices to prove $S(f * g) = S(f) * g$. By hypothesis,

$$\begin{aligned} S(f) * g &= \int_{\mathbf{T}} S(f)\left(t^{-1}x\right) g(t)\, d\tau(t) = \int_{\mathbf{T}} \left(S(f)\right)_{[t^{-1}]}(x) g(t)\, d\tau(t) \\ &= \int_{\mathbf{T}} S\left(f_{[t^{-1}]}\right)(x) g(t)\, d\tau(t). \end{aligned}$$

The last integral is the limit of approximating sums of the form

$$\sum_{k=1}^{K} a_k \left(S\left(f_{[t_k^{-1}]}\right)(x) g(t_k)\right) = \left(S\left(\sum_{k=1}^{K} a_k f_{[t_k^{-1}]} g(t_k)\right)\right)(x),$$

which converge to $S\left(\int_{\mathbf{T}} f\left(t^{-1}x\right) g(t)\, d\tau(t)\right) = S(f * g)$.

b) Because $\widehat{S(f)}_n = \int_{\mathbf{T}} S(f)(t) t^{-n}\, d\tau(t)$, if $g_n(t) \stackrel{\text{def}}{=} t^n$, then, owing to the translation-invariance of τ,

$$\begin{aligned} \widehat{S(f)}_n &= x^{-n} S(f) * g_n(x) = \int_{\mathbf{T}} \widehat{S(f)}_n\, d\tau(x) = \int_{\mathbf{T}} x^{-n} S(f) * g_n(x)\, d\tau(x) \\ &= \int_{\mathbf{T}} x^{-n} \left(f * S\left(g_n\right)\right)(x)\, d\tau(x) = \widehat{f}_n \widehat{S\left(g_n\right)}_n \stackrel{\text{def}}{=} a_n \widehat{f}_n. \end{aligned}$$ □

6.67. For any μ, if $\left(L^\infty(X,\mu)\right)^* = L^1(X,\mu)$ then $L^1(X,\mu)$ is reflexive, $B(\mathbf{O},1)$ is $\sigma\left(L^1(X,\mu), \left(L^1(X,\mu)\right)^*\right)$-compact, and the Krein-Milman theorem implies that the weak closure of the convex hull of $\mathcal{E}_1$ is $B(\mathbf{O},1)$. In the present context, $L^1(X,\mu) = L^1([0,1],\lambda)$ and $\mathcal{E}_1 = \emptyset$ (cf. **6.57**) whence $L^1([0,1],\lambda)$ is not reflexive and so $\left(L^\infty(X,\mu)\right)^* \underset{\neq}{\supset} L^1(X,\mu)$. □

An alternative solution below makes no use of the Krein-Milman theorem.

The map $T : C([0,1],\mathbb{C}) \ni f \mapsto f(0)$ is a functional of norm one and, via the Hahn-Banach theorem, can be extended to a $\widetilde{T}$ in $\left(L^\infty([0,1],\lambda)\right)^*$. If $\left(L^\infty([0,1],\lambda)\right)^* = L^1([0,1],\lambda)$ then for some h in $L^1([0,1],\lambda)$ and all f in $C([0,1],\mathbb{C})$, $f(0) = \int_{[0,1]} \overline{h(x)} f(x)\, dx$. Hence $\int_{[a,b]} h(x)\, dx = 0$, on every closed subinterval $[a,b]$ of $[0,1]$, whence $h \doteq 0$. On the other hand, if $f \equiv 1$ then $f(0) = \int_{[0,1]} 1 \cdot h(x)\, dx = 1$, a contradiction. □

6.68. Let F be the equivalence class determined by f. If $\{f_k\}_{k=1}^K \subset F$, $E_k \stackrel{\text{def}}{=} \{x : f_k(x) = f(x)\}$, and $E \stackrel{\text{def}}{=} \bigcap_{k=1}^K E_k$, then $\lambda(E) = 1$ and on E,

$f_k(x) = f(x), 1 \le k \le K$. If $\{\phi_k\}_{k=1}^K$ is a partition of unity (subordinate to some open cover $\{N_k\}_{k=1}^K$ of $[0,1]$) then on E,

$$\sum_{k=1}^K \phi_k(x) f_k(x) = f(x) \sum_{k=1}^K \phi_k(x) = f(x),$$

i.e., $\sum_{k=1}^K \phi_k f_k \in F$. (The same kind of argument shows F is convex.)

The map

$$h_x : [0,1] \ni t \mapsto \begin{cases} v_x & \text{if } t = x \\ g_x(t) & \text{if } t \neq x \end{cases}$$

is in F and $h_x(t)$ is continuous. For n in $\mathbb{N}$ and x in $[0,1]$ let x lie in a neighborhood $N_n(x)$ such that $\operatorname{osc}_{N_n(x)}(h_x) < \frac{1}{n}$. Because $[0,1]$ is compact, for some n in $\mathbb{N}$, if $J_{nk} \stackrel{\text{def}}{=} \left(\frac{k}{n} - \frac{1}{n}, \frac{k}{n} + \frac{1}{n}\right), k = 0, 1, \ldots, n$, then $\operatorname{osc}_{J_{nk}}\left(h_{\frac{k}{n}}\right) < \frac{1}{n}$.

Let $\{\phi_{nk}\}_{k=0}^n$ be a partition of unity subordinate to $\{J_{nk}\}_{k=0}^n$ and let H_n be $\sum_{k=0}^n \phi_{nk} h_{\frac{k}{n}}$. Because each h_x is continuous and each ϕ_{nk} is continuous, each H_n is continuous.

It is shown next that: a) each x in $[0,1]$ is contained in a neighborhood $N(x)$ such that $\operatorname{osc}_{N(x)}(H_n) < \frac{2}{n}$; b) for all m, n in $\mathbb{N}$, all x in $[0,1]$, every neighborhood $U(x)$ of x, and some y in $U(x)$, $H_m(y) = H_n(x)$.

a) If $x \in [0,1]$ and $S(x) \stackrel{\text{def}}{=} \{k : x \in J_{nk}\}$ then $\#(S(x)) \le 2$ and x lies in a neighborhood $N(x)$ such that $N(x) \cap \left(\bigcup_{k' \notin S(x)} J_{nk'}\right) = \emptyset$. If $y_1, y_2 \in N(x)$ then

$$H_n(y_i) = \sum_{k \in S(x)} \phi_{nk}(y_i) h_{\frac{k}{n}}(y_i), i = 1, 2$$

$$|H_n(y_1) - H_n(y_2)| \le \sum_{k \in S(x)} \phi_{nk}(y_1) \left| h_{\frac{k}{n}}(y_1) - h_{\frac{k}{n}}(y_2) \right| + \sum_{k \in S(x)} \phi_{nk}(y_1) \phi_{nk}(y_2) \cdot \left| h_{\frac{k}{n}}(y_2) \right|.$$

If $A \stackrel{\text{def}}{=} \sup_{k \in S(x)} \left| h_{\frac{k}{n}}(y_2) \right| = 0$, the last summand above is zero. If $A > 0$ and for some positive b, $\operatorname{diam}(E) < b$ then $\sup_k \operatorname{osc}_E(\phi_{nk}) < \frac{1}{nA}$. It may be assumed that $\operatorname{diam}(N(x)) < b$, whence $|H_n(y_1) - H_n(y_2)| < \frac{1}{n} + \frac{1}{n}$.

b) If $G_n \stackrel{\text{def}}{=} \{x : H_n(x) = f(x)\}$ then $\lambda\left(\bigcap_{n \in \mathbb{N}} G_n\right) = 1$ whence it follows that $G \stackrel{\text{def}}{=} \{x : H_n(x) = f(x), n \in \mathbb{N}\}$ is dense in $[0,1]$. For a given $U(x)$, choose y in the nonempty set $G \cap U(x)$.

For x in $[0,1]$ and a positive ϵ,

$$|H_m(x) - H_n(x)| \le |H_m(x) - H_m(y)| \\ + |H_m(y) - H_n(y)| + |H_n(y) - H_n(x)|.$$

Hence if $\frac{2}{m}, \frac{2}{n} < \frac{\epsilon}{3}$ and if $U(x)$ is such that

$$\operatorname{osc}_{U(x)}(H_m), \operatorname{osc}_{U(x)}(H_n) < \frac{\epsilon}{3}$$

then for some y in $U(x)$, $H_m(y) = H_n(y)$. It follows that

$$|H_m(x) - H_n(x)| < \epsilon$$

if $m, n > \frac{6}{\epsilon}$. Thus $\{H_n\}_{n\in\mathbb{N}}$ is a $\| \|_\infty$-Cauchy sequence and so for some g, $H_m \overset{u}{\to} g$, g is continuous and $f \doteq g$. □

6.2. Hilbert Space $\mathfrak{H}$

6.69. If $|z| < \frac{1}{2}$, $\mathbf{a} \overset{\text{def}}{=} \{a_n\}_{n\in\mathbb{N}} \in \ell^2(\mathbb{N})$, and $\mathbf{b}_z \overset{\text{def}}{=} \left\{\frac{1}{n-z}\right\}_{n\in\mathbb{N}}$ then

$$\left|\sum_{n=k}^{m} \frac{a_n}{n-z}\right| \le \left(\sum_{n=k}^{m} |a_n|^2\right)^{\frac{1}{2}} \left(\sum_{n=k}^{m} \frac{1}{(n-1)^2}\right)^{\frac{1}{2}} \le \left(\sum_{n=k}^{m} |a_n|^2\right)^{\frac{1}{2}} \frac{\pi^2}{6}.$$

Thus $f_{\mathbf{a}}(z) \overset{\text{def}}{=} (\mathbf{a}, \mathbf{b}_z) = \sum_{n=1}^{\infty} \frac{a_n}{n-z}$ is holomorphic in $D\left(0, \frac{1}{2}\right)^{\circ}$. If the span of $\operatorname{im}(\gamma) \overset{\text{def}}{=} \gamma^*$ is not $\| \|_2$-dense in $\ell^2(\mathbb{N})$ then for some nonzero $\mathbf{a}$, $f_{\mathbf{a}}(z) \equiv 0$ on $\left[-\frac{1}{2}, \frac{1}{2}\right]$, whence $f_{\mathbf{a}}(z) \equiv 0$.

On the other hand,

$$f_{\mathbf{a}}(z) = \sum_{n=1}^{\infty} a_n \left(\sum_{k=0}^{\infty} \frac{z^k}{n^{k+1}}\right) = \sum_{k=0}^{\infty} \left(\sum_{n=1}^{\infty} \frac{a_n}{n^{k+1}}\right) z^k,$$

whence for all k in $\{0\} \cup \mathbb{N}$, $\sum_{n=1}^{\infty} \frac{a_n}{n^{k+1}} = 0$ and $a_1 = -\sum_{n=2}^{\infty} \frac{a_n}{n^{k+1}}$. As $k \to \infty$, there emerges: $a_1 = 0$. Induction shows that for all n in $\mathbb{N}$, $a_n = 0$, a contradiction. □

6.70. If K is a $\| \|_2$-compact subset of $\ell^2(\mathbb{N})$ and if $F_N \overset{u}{\not\to} \mathbf{O}$ on K then K contains a sequence $\left\{\mathbf{k}_m \overset{\text{def}}{=} \{k_{mn}\}_{n=1}^{\infty}\right\}_{m=1}^{\infty}$ such that for some positive a and each m in $\mathbb{N}$, $\sum_{n\ge m} |k_{mn}|^2 \ge a^2$. Because K is compact it may be assumed that for some $\mathbf{k} \stackrel{\text{def}}{=} \{k_n\}_{n=1}^{\infty}$ in K, $\|\mathbf{k}_m - \mathbf{k}\|_2 \to 0$ as $m \to \infty$, whence for each n, $k_{mn} \to k_n$ as $m \to \infty$. For large n_0, $\sum_{n \ge n_0} |k_n|^2 < \frac{a^2}{2^4}$ and for large m_0, $\|\mathbf{k}_m - \mathbf{k}\|_2^2 < \frac{a^2}{2^6}$ if $m \ge m_0$. Thus if $m \ge m_0 + n_0$,

$$a < \left(\sum_{n \ge n_0} |k_{mn}|^2\right)^{\frac{1}{2}} \le \left(\sum_{n \ge n_0} |k_{mn} - k_n|^2\right)^{\frac{1}{2}} + \left(\sum_{n \ge n_0} |k_n|^2\right)^{\frac{1}{2}}$$
$$< \|\mathbf{k}_m - \mathbf{k}\|_2 + \frac{a}{2^2} < \frac{a}{2^3} + \frac{a}{2^2} < a,$$

a contradiction. □

6.71. Because all functions considered are in $L^2(\mathbb{R}^2, \lambda)$, all integrals given as well as those introduced below exist. From Green's theorem it follows that for all positive R,

$$\int_{B(\mathbf{O},R)} (f(x,y)\Delta g(x,y) - \Delta f(x,y)g(x,y))\,dx\,dy$$
$$= \oint_{\partial B(\mathbf{O},R)} \left(f(x,y)\frac{\partial g(x,y)}{\partial n} - g(x,y)\frac{\partial f(x,y)}{\partial n}\right) ds$$
$$= \int_{\partial B(\mathbf{O},R)} \left(f(x,y)\frac{\partial g(x,y)}{\partial x} - g(x,y)\frac{\partial f(x,y)}{\partial x}\right) dy$$
$$+ \left(g(x,y)\frac{\partial f(x,y)}{\partial y}\right) - \left(f(x,y)\frac{\partial g(x,y)}{\partial y}\right) dx.$$

A typical estimate for the integrals in these equations is

$$\left|\int_{\partial B(\mathbf{O},R)} f(x,y)\frac{\partial g(x,y)}{\partial x}\,dy\right|$$
$$\le \int_0^{2\pi} |f(R\cos\theta, R\sin\theta)| \cdot \left|\frac{\partial g(R\cos\theta, R\sin\theta)}{\partial x}\right| \cdot |R\sin\theta|\,d\theta.$$

Thus, $A(R)$ denoting the right member above,

$$\int_{B(\mathbf{O},R)} \left|f(r\cos\theta, r\sin\theta)\frac{\partial g(r\cos\theta, r\sin\theta)}{\partial x}\right| r\,dr\,d\theta = \int_0^R A(r)\,dr$$
$$\le \|f\|_2 \cdot \left\|\frac{\partial g}{\partial x}\right\|_2$$

Hence for some $\{R_n\}_{n\in\mathbb{N}}$, $R_n \uparrow \infty$ while $A(R_n) \downarrow 0$. Similar estimates for the other three integrals show that

$$\left|\int_{B(\mathbf{O},R)} (f(x,y)\Delta g(x,y) - \Delta f(x,y)g(x,y))\,dx\,dy\right|$$

is dominated by the sum of four quantities, each depending on n and approaching zero as $n \to \infty$. □

6.72. For example, $\gamma : [0,1] \ni t \mapsto \chi_{[0,t]} \in L^2([0,1],\lambda)$ serves. □

6.73. a) If $0 \le a \le b < \infty$, the Schwarz inequality implies

$$|f_n(b) - f_n(a)| \le \int_a^b |f_n'(x)|\,dx \le M|b-a|^{\frac{1}{2}}$$

$$|f_n(x)| \le \begin{cases} |f_n(1)| + M|x-1|^{\frac{1}{2}} & \text{if } 0 \le x \le 1 \\ 1/x & \text{if } x > 1 \end{cases}$$

whence $f_n \in \text{Lip}\,(1/2)$ and $\|f_n\|_\infty \le 1 + M, n \in \mathbb{N}$.

For each interval $[0,k], k \in \mathbb{N}$, the Arzelà-Ascoli theorem implies that $\{f_n\}_{n\in\mathbb{N}}$ contains a subsequence $\{f_{kn}\}_{n\in\mathbb{N}}$ uniformly convergent on $[0,k]$ and such that $\{f_{kn}\}_{n\in\mathbb{N}} \supset \{f_{k+1,n}\}_{n\in\mathbb{N}}$. Thus $S \stackrel{\text{def}}{=} \left\{g_n \stackrel{\text{def}}{=} f_{nn}\right\}_{n\in\mathbb{N}}$ converges everywhere.

b) If $\epsilon > 0$ and $k\epsilon > 2$, there is an N such that for $\{g_n\}_{n\in\mathbb{N}}$ as in a),

$$|g_m(x) - g_n(x)| < \begin{cases} \epsilon/2 & \text{if } 0 \le x \le k \text{ and } m,n > N \\ \epsilon 2 & \text{if } k < x. \end{cases}$$

c) Because

$$\|g_m - g_n\|_2^2 = \int_{[0,k]} + \int_k^\infty |g_m(x) - g_n(x)|\,2\,dx$$

$$\le k\,\|g_m - g_n\|_\infty + \int_k^\infty 2/x^2\,dx,$$

for large k and all m,n, the second term in the right member above is small and for large m and n, the first term is small.

In sum, $\{g_n\}_{n\in\mathbb{N}}$ is pointwise, uniformly, and $\|\ \|_2$- convergent. □

6.74. Since $\int_{-y}^{y} |f(x,y)|\,dx \le \left(\int_{-y}^{y} |f(x,y)|^2\,dx\right)^{\frac{1}{2}} (2y)^{\frac{1}{2}}$, it follows that

$$\int_0^1 \frac{1}{2y}\left(\int_{-y}^{y} |f(x,y|\,dx\right)^2 dy \le \int_0^1 \left(\int_{-y}^{y} |f(x,y)|^2\,dx\right) dy \le \|f\|_2^2 < \infty.$$

On the other hand, if $A > 0$ then for all y near zero,

$$\frac{1}{2y}\left(\int_{-y}^{y} |f(x,y)|\,dx\right)^2 \ge \frac{A^2}{8y},$$

which implies that $\|f\|_2 = \infty$, a contradiction. □

6.75. If $\epsilon > 0$ and $E_n \stackrel{\text{def}}{=} \{x : |f_n(x)| \geq \epsilon > 0\}$ then $\lambda(E_n) \to 0$ as $n \to \infty$ and

$$\|f_n\|_1 = \int_{E_n} + \int_{[0,1]\setminus E_n} |f_n(x)|\,dx \leq \|f_n\|_2 \cdot (\lambda(E_n))^{\frac{1}{2}} + \epsilon.$$ □

6.76. It may be assumed that $f_n \stackrel{\text{def}}{=} \Delta_{\frac{1}{n}} f - g \stackrel{\text{a.e.}}{\to} 0$. Because

$$\int_0^x |g(t) - \Delta_h f(t)|\,dt \leq \left(\int_{\mathbb{R}} |g(t) - \Delta_h f(t)|^2\,dt \right)^{\frac{1}{2}} |x|^{\frac{1}{2}},$$

it follows that for a, x off a null set,

$$\begin{aligned} \int_a^x g(t)\,dt &= \lim_{n\to\infty} \int_a^x n\left(f\left(t + \frac{1}{n}\right) - f(t) \right) dt \\ &= \lim_{n\to\infty} n \int_x^{x+\frac{1}{n}} f(t)\,dt - \lim_{n\to\infty} n \int_a^{a+\frac{1}{n}} f(t)\,dt \\ &= f(x) - f(a). \end{aligned}$$ □

6.77. a) Because $\|Tf\|_2^2 = \int_0^1 \left| \int_0^x f(t)\,dt \right|^2 dx \leq \int_0^1 x\|f\|_2^2\,dx = \frac{1}{2}\|f\|_2^2$, it follows that $\|T\| \leq 2^{-\frac{1}{2}}$.

b) For $P : L^2([0,1],\lambda) \ni h \mapsto \int_0^1 h(x)\,dx$, $P \in [L^2([0,1],\lambda)]$, $P^2 = P$, and $P(L^2([0,1],\lambda)) = \mathbb{C}$. Because

$$\begin{aligned} (Tf, g) &= \int_0^1 \left(\int_0^x f(t)\,dt \right) \overline{g(x)}\,dx = \int_0^1 \left(\int_t^1 \overline{g(x)}\,dx \right) f(t)\,dt \\ &= \int_0^1 \left[\int_0^1 \overline{g(x)}\,dx - \int_0^t \overline{g(x)}\,dx \right] f(t)\,dt = (f, (P - T)g) \end{aligned}$$

it follows that $T^* = P - T$.

c) If $\{f_n\}_{n\in\mathbb{N}} \subset B(\mathbf{O}, 1)$ then

$$|Tf(x)| \leq x^{\frac{1}{2}} \leq 1 \text{ and } |Tf_n(x) - Tf_n(y)| \leq |x - y|^{\frac{1}{2}}.$$

The Arzelà-Ascoli theorem implies $\{Tf_n\}_{n\in\mathbb{N}}$ contains a uniformly convergent subsequence S and, because $\lambda([0,1]) = 1$, S is $\|\ \|_2$-convergent. □

6.78. a) Because $\lambda([0,1]) = 1$, $\|\ \|_2 \leq \|\ \|_\infty$, whence a $\|\ \|_\infty$-limit point of S is a $\|\ \|_2$-limit point of S.

b) Owing to a), the graph of the inclusion map

$$\iota : C([0,1],\mathbb{C}) \supset S \ni f \hookrightarrow f \in L^2([0,1],\mathbb{R})$$

is closed whence ι is continuous: $\|\iota\|$ exists. Hence from a) it follows that $\| \ \|_2 \le \| \ \|_\infty \le \|\iota\| \cdot \| \ \|_2$.

c) The map $L_y : S \ni f \mapsto f(y) \in \mathbb{C}$ is a $\| \ \|_\infty$-continuous linear functional and so for some K_y, $|L_y(f)| \le K_y\|f\|_\infty \le K_y M\|f\|_2$. The Hahn-Banach and Riesz representation theorems imply that for some k_y in $L^2([0,1],\lambda)$, $L_y(f) = \int_0^1 k_y(x)f(x)\,dx$. □

6.79. For $g_h \stackrel{\text{def}}{=} f_{[h]} - f_{[-h]}$, $\left(\widehat{g_h}\right)_n = 2i\widehat{f}_n \sin nh$ whence, owing to Parseval's equation, for some constant C_1,

$$\|g_h\|_2^2 = 4\sum_{n\in\mathbb{Z}} \left|\widehat{f}_n\right|^2 \sin^2 nh \stackrel{\text{def}}{=} 4\sum_{n\in\mathbb{Z}} r_n^2 \sin^2 nh \le C_1|h|^{1+a}.$$

If $k \in \mathbb{N}$, $N \stackrel{\text{def}}{=} 2^k$, $h = \dfrac{\pi}{2N}$, and $\dfrac{N}{2} \le |n| \le N$ then $\sin^2 nh \ge \dfrac{1}{2}$ and so for some constant C_2, $\sum_{2^{k-1}\le|n|\le 2^k} r_n^2 \le C_2 2^{-k(1+a)}$. Owing to the Schwarz inequality, for some constant C_3,

$$\sum_{2^{k-1}\le|n|\le 2^k} r_n \le \left(\sum_{2^{k-1}\le|n|\le 2^k} r_n^2\right)^{\frac{1}{2}} \cdot 2^{\frac{k+1}{2}} \le C_3 2^{-\frac{ka}{2}}.$$

It follows that $\sum_{n\in\mathbb{Z}} r_n \le \dfrac{2C_3}{1-2^{-\frac{a}{2}}} < \infty$. □

6.80. For all n in $\mathbb{Z}\setminus\{0\}$, the function $g_n : [0,1] \ni x \mapsto xe^{2\pi inx}$ is in the (vector) subspace A. Hence if $h \in A^\perp$, $n \in \mathbb{Z}\setminus\{0\}$, and $H(x) \stackrel{\text{def}}{=} xh(x)$ then $\widehat{H}_n = 0$, i.e., H is a constant. Unless $h = 0$, $h \notin L^2([0,1],\lambda)$, a contradiction: $A^\perp = \{\mathbf{O}\}$. □

6.81. If $f(t) \equiv t$, then $\displaystyle\int_0^1 t^n f(t)\,dt = \frac{1}{n+2}, n \in \mathbb{Z}^+$.

Conversely, the Weierstraß approximation theorem implies that if

$$\int_0^1 (f(t)-t)t^n\,dt \equiv 0, n \in \mathbb{Z}^+,$$

then for any continuous function g, $\int_0^1 (f(t)-t)g(t)\,dt = 0$. Since $C([0,1],\mathbb{C})$ is $\| \ \|_2$-dense in $L^2([0,1],\lambda)$, it follows that $f(t) \doteq t$. □

6.82. If $0 < \mu(E) < \infty$ then, owing to the monotone convergence theorem and the Schwarz inequality, a.e. on E,

$$\sum_{n=1}^{\infty} |f_n(x) - f_{n+1}(x)|^2 < \infty$$

$$(\mu(E))^{-\frac{1}{2}} \sum_{n=1}^{\infty} |f_n(x) - f_{n+1}(x)| < \infty$$

$$\sum_{n=1}^{\infty} (f_n(x) - f_{n+1}(x)) = \lim_{n\to\infty} (f_1(x) - f_n(x)).$$

Hence $\lim_{n\to\infty} f_n(x) \stackrel{\text{def}}{=} g_E(x)$ exists a.e. on E.

Because $S \stackrel{\text{def}}{=} \bigcup_{n\in\mathbb{N}} \{x : f_n(x) \neq 0\}$ is σ-finite, for some $\{E_m\}_{m\in\mathbb{N}}$, $\mu(E_m) < \infty, m \in \mathbb{N}$, and $S = \bigcup_{m\in\mathbb{N}} E_m$. Corresponding to each E_m there is a $g_{E_m} \stackrel{\text{def}}{=} h_m$ and $f_n \stackrel{\text{a.e.}}{\to} \sum_{m=1}^{\infty} h_m \stackrel{\text{def}}{=} f$. Because $\{f_n\}_{n\in\mathbb{N}}$ is a $\| \ \|_2$-Cauchy sequence, its $\| \ \|_2$-limit must be f. □

6.83. For S, E_m as in **6.82**, the fundamental inequality is

$$\sum_{n=1}^{\infty} \frac{\|f_n\|_2^2}{n^2} < \infty.$$

The reasoning used in **6.82** applies, mutatis mutandis. □

6.84. Bessel's inequality implies that for $\mathbf{x}$ in $\mathfrak{H}$, $(\mathbf{x}, \mathbf{x}_n) \to 0$ as $n \to \infty$. □

6.85. a) The map $P : C([-1,1], \mathbb{C}) \ni f \mapsto \left\{x \mapsto \frac{f(x) + f(-x)}{2}\right\}$ is such that for f in $C([-1,1], \mathbb{C})$, $\|Pf\|_2 \leq \|f\|_2$ and P is extendible to an element, again denoted P, of $[L^2([-1,1], \lambda)]$. Furthermore, $P^2 = P$ and $P(S) = S$ whence $M = P(L^2([-1,1], \lambda))$.

b) If $f \in L^2([-1,1], \lambda)$ and $f \sim \frac{1}{2}a_0 + \sum_{n=1}^{\infty} a_n \cos n\pi x + b_n \sin \sin n\pi x$ then because the sine function is odd, and the cosine function is even,

$$Pf \sim \frac{1}{2}a_0 + \sum_{n=1}^{\infty} a_n \cos n\pi x.$$

Hence the set of trigonometric polynomials of the cosine functions constitute an orthonormal basis for M. Each cosine function may be uniformly approximated on $[-1,1]$ by a p in $\mathbb{C}[x^2]$. The Gram-Schmidt process endows M with an orthonormal basis consisting of polynomials. □

6.86. a) If $\epsilon > 0$ and $U \stackrel{\text{def}}{=} \{\mathbf{x} : |(\mathbf{x}, \mathbf{y}_k)| < \epsilon, 1 \le k \le K\}$ is a weak neighborhood of $\mathbf{O}$, for some m, $|(\mathbf{x}_m, \mathbf{y}_k)| < \frac{\epsilon}{3}, 1 \le k \le K$, and for some n, $n > m$ and $|(\mathbf{x}_n, \mathbf{y}_k)| < \frac{\epsilon}{3m}$ whence $\mathbf{x}_n + m\mathbf{x}_n \in U$.

b) If $F \subset B(\mathbf{O}, R)$ and $\mathbf{x}_m + m\mathbf{x}_n \in F$ then $1 + m^2 \le R^2$, whence for some M in $\mathbb{N}$, $m < M$ and $m < n \le M - 1$ or $n > M - 1 \ge m$. If $a > 0$, $\mathbf{y} \stackrel{\text{def}}{=} 2a \sum_{m=1}^{M-1} \mathbf{x}_m$, and $U \stackrel{\text{def}}{=} \{\mathbf{x} : |(\mathbf{x}, \mathbf{y})| < a\}$, then

$$(\mathbf{x}_m + m\mathbf{x}_n, \mathbf{y}) = 2a > a$$

and so $F \cap U = \emptyset$.

c) If $\mathbf{z}_k \stackrel{\text{def}}{=} \mathbf{x}_{m_k} + m_k \mathbf{x}_{n_k} \stackrel{\text{w}}{\to} \mathbf{O}$ then for some K, $\|\mathbf{z}_k\| \le K$ whence for some M, $m_k \le M$. Hence for all large p,

$$\left(\mathbf{x}_{n_p}, \mathbf{z}_k\right) = \left(\mathbf{x}_{n_p}, \mathbf{x}_{m_k} + m_k \mathbf{x}_{n_p}\right) = m_k \not\to 0 \text{ as } k \to \infty. \qquad \square$$

6.87. a) Let $\mathbf{x}_n - \mathbf{y}_n$ be $\mathbf{z}_n$. If $\mathbf{x} \in \mathfrak{H}$ and $\|\mathbf{x}\| \le 1$,

$$\mathbf{x} \stackrel{\text{def}}{=} \sum_{n=1}^{\infty} a_{0n} \mathbf{x}_n = \sum_{n=1}^{\infty} a_{0n} (\mathbf{y}_n + \mathbf{z}_n)$$

$$\sum_{n=1}^{\infty} |a_{0n}| \cdot \|\mathbf{z}_n\| \le \left(\sum_{n=1}^{\infty} |a_{0n}|^2\right)^{\frac{1}{2}} \left(\sum_{n=1}^{\infty} \|\mathbf{z}_n\|^2\right)^{\frac{1}{2}}$$

$$\le 1 \cdot \left(\sum_{n=1}^{\infty} c^{2n}\right)^{\frac{1}{2}} = \left(\frac{c^2}{1 - c^2}\right)^{\frac{1}{2}} \stackrel{\text{def}}{=} r < 1.$$

Thus $\mathbf{w}_1 \stackrel{\text{def}}{=} \sum_{n=1}^{\infty} a_{0n} \mathbf{z}_n$ exists and $\|\mathbf{w}_1\| \le r$. Induction produces a sequence $\{\mathbf{w}_k\}_{k \in \mathbb{N}}$ such that

$$\mathbf{w}_k \stackrel{\text{def}}{=} \sum_{n=1}^{\infty} a_{kn} \mathbf{x}_n, \ \mathbf{w}_{k+1} \stackrel{\text{def}}{=} \sum_{n=1}^{\infty} a_{kn} \mathbf{z}_n \text{ exists}, \|\mathbf{w}_{k+1}\| \le r^{k+1}$$

$$\mathbf{x} - \sum_{n=1}^{\infty} \left(\sum_{p=0}^{P} a_{pn}\right) \mathbf{y}_n + \mathbf{w}_{P+1}$$

$$\sum_{n=1}^{\infty} \left(\sum_{p=0}^{\infty} |a_{pn}|^2\right) \le \sum_{p,q=0}^{\infty} \left(\sum_{n=1}^{\infty} |a_{pn}| \cdot |a_{qn}|\right) \le \sum_{p,q=0}^{\infty} \left(r^{p+q}\right)$$

$$\sum_{s=0}^{\infty} (s+1) r^s < \infty.$$

For $A_{Pn} \stackrel{\text{def}}{=} \sum_{p=0}^{P} a_{pn}$, $A_n \stackrel{\text{def}}{=} \lim_{P\to\infty} A_{Pn}$ exists, both $\mathbf{A}_P \stackrel{\text{def}}{=} \{A_{Pn}\}_{n\in\mathbb{N}}$ and $\mathbf{A} \stackrel{\text{def}}{=} \{A_n\}_{n\in\mathbb{N}}$ are in $\ell^2(\mathbb{N})$, and $\mathbf{A}_P \stackrel{\|\ \|_2}{\to} \mathbf{A}$. Thus

$$\left\|\mathbf{x} - \sum_{n=1}^{N} A_n \mathbf{y}_n\right\| \le \|\mathbf{w}_{P+1}\| + \left\|\sum_{n=N+1}^{\infty} A_{Pn}\mathbf{y}_n\right\| + \left\|\sum_{n=1}^{N}(A_{Pn} - A_n)\mathbf{y}_n\right\|. \tag{s6.8}$$

The third term in the right member of (s6.8) does not exceed

$$\begin{aligned}&\left\|\sum_{n=1}^{N}(A_{Pn}-A_n)\mathbf{x}_n\right\| + \left\|\sum_{n=1}^{N}(A_{Pn}-A_n)\mathbf{z}_n\right\| \\ &\le \left(\sum_{n=1}^{N}|A_{Pn}-A_n|^2\right)^{\frac12} + \sum_{n=1}^{N}|A_{Pn}-A_n|c^n \\ &\le \left(\sum_{n=1}^{\infty}|A_{Pn}-A_n|^2\right)^{\frac12}\cdot\left(1+\left(\sum_{n=1}^{\infty}c^{2n}\right)^{\frac12}\right).\end{aligned}$$

Similar calculations show that the second term in the right member of (s6.8) does not exceed $\left(\sum_{n=N+1}^{\infty}|A_{Pn}|^2\right)^{\frac12}\cdot\left(1+\left(\sum_{n=N+1}^{\infty}c^{2n}\right)^{\frac12}\right)$. Because $\|\mathbf{w}_{P+1}\|\to 0$ as $P\to\infty$, $\mathbf{x} = \sum_{n=1}^{\infty} A_n\mathbf{y}_n$.

If $\sum_{n=1}^{\infty} b_n\mathbf{y}_n = \mathbf{O}$ then $|b_n|\cdot\|\mathbf{y}_n\|\to 0$ as $n\to\infty$. Because $\|\mathbf{y}_n\|$ is near $\|\mathbf{x}_n\|$ $(=1)$, it follows that $b_n\to 0$ as $n\to\infty$. Hence $\sum_{n=1}^{\infty} b_n\|\mathbf{z}_n\| < \infty$ and $\sum_{n=1}^{\infty} b_n\mathbf{z}_n = \sum_{n=1}^{\infty} b_n(\mathbf{x}_n - \mathbf{z}_n) + \sum_{n=1}^{\infty} b_n\mathbf{z}_n = \sum_{n=1}^{\infty} b_n\mathbf{x}_n \stackrel{\text{def}}{=} \mathbf{z}$ exists. Hence $\|\mathbf{z}\|^2 = \sum_{n=1}^{\infty}|b_n|^2 \le \left(\sum_{n=1}^{\infty}|b_n|\,\|\mathbf{z}_n\|\right)^2 \le \left(\sum_{n=1}^{\infty}|b_n|^2\right)\cdot r^2$ and if $\sum_{n=1}^{\infty}|b_n|^2$ is positive, there emerges the contradiction: $1\le r^2<1$.

b) The argument in the last paragraph of a) applies, mutatis mutandis. □

6.88. Each ϕ_γ may be expressed in terms of the vectors in Ψ:

$$\phi_\gamma = \sum_{\delta\in\Delta} a_{\gamma\delta}\psi_\delta$$

and $\|\phi_\gamma\|^2 = \sum_{\delta\in\Delta}|a_{\gamma\delta}|^2$. If δ_0 is fixed and, as γ varies over Γ, $a_{\gamma\delta_0}\equiv 0$ then ψ_{δ_0} is $\|\ \|$-approximable by finite sums $\sum'_{\gamma,\delta} b_{\delta_0\gamma}a_{\gamma\delta}\psi_\delta$ in each of which the coefficient of ψ_{δ_0} is zero. Thus $\|\psi_{\delta_0}\|^2$ is approximable by finite sums $\sum'_{\gamma,\delta} b_{\delta_0\gamma}a_{\gamma\delta}(\psi_\delta,\psi_{\delta_0})$ each of which is zero, a contradiction.

Hence to each ψ_δ, there corresponds a ϕ_γ such that $a_{\gamma\delta}\ne 0$ whence $\#(\Delta)\le\#(\Gamma)$. A similar argument shows $\#(\Gamma)\le\#(\Delta)$. □

6.89. a) From the Schwarz inequality it follows that $T \in [L^2([0,1],\lambda)]$ and $\|T\| \le \|K\|_2$.

b) If $\{\phi_n\}_{n\in\mathsf{N}}$ is a complete orthonormal set for $L^2([0,1],\lambda)$ and

$$a_{mn} \stackrel{\text{def}}{=} \int_{[0,1]^2} K(x,y)\overline{\phi_m(x)\phi_n(y)}\, d\lambda_2(x,y)$$

then $\sum_{m,n\in\mathsf{N}} a_{mn}\phi_n\phi_m \stackrel{\|\ \|_2}{\to} K$. The maps

$$T_{MN} : L^2([0,1],\lambda) \ni f \mapsto \sum_{m,n=1}^{M,N} \int_0^1 a_{mn}\phi_m(x)\phi_n(y)f(y)\, dy$$

are such that $\|T_{MN} - T\| \le \sum_{m\ge M \text{ or } n\ge N} |a_{mn}|^2$. □

6.90. Let $\{\phi_n\}_{n\in\mathsf{N}}$ in $\mathfrak{H}$ be orthonormal. Then $\|\phi_m - \phi_n\| = \delta_{mn}\sqrt{2}$. The balls $B(\phi_n, 0.5)$ are pairwise disjoint and their union is a subset of $B(\mathbf{O}, 2)$. Arguments based on homothety show that for some positive r, $B(\mathbf{O},1)$ contains infinitely many pairwise disjoint translates of $B(\mathbf{O},r)$. Hence if μ is translation-invariant and $\mu(B(\mathbf{O},r)) < \infty$ then $\mu(B(\mathbf{O},r)) = 0$.

If $\mathfrak{H}$ is separable, every open set is the countable union of translates of open balls of radius not exceeding r whence every open set, in particular $\mathfrak{H}$ itself, is a null set, i.e., μ is trivial.

If $\mathfrak{H}$ is not separable, it contains a separable infinite-dimensional (Hilbert) subspace $\mathfrak{H}_1$. Each ball $B_{\mathfrak{H}_1}(\mathbf{x}_0, r)$ of $\mathfrak{H}_1$ is $\mathfrak{H}_1 \cap B(\mathbf{x}_0, r)$. If μ is a translation-invariant measure for $\mathfrak{H}$ and $0 < \mu(B(\mathbf{O},1)) < \infty$, the formula

$$\mu_0^*(B_{\mathfrak{H}_1}(\mathbf{x}_0, r)) \stackrel{\text{def}}{=} \mu(B(\mathbf{x}_0, r))$$

can be used to determine a translation-invariant outer measure μ^* in $\mathfrak{H}_1$ and therefrom a translation-invariant measure μ_0 in $\mathfrak{H}$. But then

$$0 < \mu_0(B_{\mathfrak{H}_1}(\mathbf{0},1)) = \mu(B(\mathbf{O},1)) < \infty,$$

a contradiction. □

[**Note s6.8:** If *nontrivial measure* is taken to mean that every ball of (finite) positive radius has (finite) positive measure, the argument used above and reduced to its first paragraph shows that $\mathfrak{H}$ cannot be equipped with a nontrivial (Borel) measure. Furthermore, the argument remains valid when *translation-invariant* is replaced by *unitarily invariant*.]

6.91. Because the diameter of a set in $\mathfrak{H}$ is translation-invariant, ρ^p is also translation-invariant. Every closed set is ρ^p-measurable (cf. **4.8**), whence each ball is ρ^p-measurable. If $\mathfrak{H}_1$ is an n-dimensional subspace of $\mathfrak{H}$ and

$r > 0$ then $\mathfrak{H}_1 \cap B(\mathbf{O}, r) \stackrel{\text{def}}{=} B_{\mathfrak{H}_1}(\mathbf{O}, r)$ is the n-dimensional ball and **4.6** implies: $\rho^n(B_{\mathfrak{H}_1}(\mathbf{O}, r)) \geq \lambda_n(B_{\mathfrak{H}_1}(\mathbf{O}, r)) > 0$. Thus from **6.90**, it follows that $\rho^n(B(\mathbf{O}, r)) = \infty, n \in \mathbb{N}$, and **4.5** implies $\rho^p(B(\mathbf{O}, r)) \equiv \infty$. □

6.92. The Schwarz inequality implies that if $f \in \mathfrak{H}$ then $\|f\|_1 \leq \|f\|_2$. If $k \in \mathbb{N}$ and

$$f_k(x) \stackrel{\text{def}}{=} \begin{cases} k^{-\frac{1}{4}} x^{-\frac{1}{2}+\frac{1}{k}} & \text{if } x \neq 0 \\ 0 & \text{otherwise} \end{cases}$$

then $\|f_k\|_2 \uparrow \infty, \|f_k\|_1 \downarrow 0$.

If $f \stackrel{\text{def}}{=} \sum_{k=1}^{\infty} k^{-\frac{11}{8}} f_k$ then $f \in \mathfrak{H}$, whereas

$$\begin{aligned}(f_k, f) &= k^{\frac{3}{4}} \sum_{n=1}^{\infty} \frac{n+k}{n^{\frac{5}{8}}} \geq k^{\frac{3}{4}} \int_1^{\infty} \frac{dx}{(x+k)^{\frac{13}{8}}} \\ &= \frac{8k^{\frac{3}{4}}}{5(1+k)^{\frac{5}{8}}} \to \infty\end{aligned}$$

as $k \to \infty$. If $r > 0$ and k is large then

$$f_k \in \{ f : f \in \mathfrak{H}, \|f\|_1 < r \} \setminus \{ g : g \in \mathfrak{H}, (g, f) < 1 \}.$$

Hence there is no nonempty $\| \|_1$-ball $B(\mathbf{O}, r)^\circ$ contained in the $\sigma(\mathfrak{H}, \mathfrak{H}^*)$-neighborhood $\{ g : g \in \mathfrak{H}, (g, f) < 1 \}$. □

6.93. For $T : \mathfrak{H} \ni \mathbf{x} \stackrel{\text{def}}{=} \{x_n\}_{n \in \mathbb{N}} \mapsto T(\mathbf{x}) \stackrel{\text{def}}{=} \left\{ \frac{x_n}{n} \right\}_{n \in \mathbb{N}}$,

$$T = T^*, \ T(\mathfrak{H}) = D, \text{ and } T(M^\perp) = A.$$

If A is not dense in $\mathfrak{H}$, for some nonzero $\mathbf{y}$ in $A^\perp$, whence for all $\mathbf{z}$ in $M^\perp$, $(T(\mathbf{z}), \mathbf{y}) = (\mathbf{z}, T(\mathbf{y})) = 0$. Thus $T(\mathbf{y}) \in (M^\perp)^\perp = M$. Since $M \cap D = \{\mathbf{O}\}$, $T(\mathbf{y}) = \mathbf{O}$, and so $\mathbf{y} = \mathbf{O}$, a contradiction. □

6.94. If $f \in S$ then

$$\|f\|_\infty \leq \sum_{n=1}^{\infty} |a_n| \leq \sum_{n=1}^{\infty} n |a_n| \leq 1, \ \|f\|_2^2 \leq \sum_{n=1}^{\infty} |a_n|^2 \leq 1.$$

If S contains the sequence $\{f_k(x) \sim \sum_{n=1}^{\infty} a_{kn} \sin 2n\pi x\}_{k \in \mathbb{N}}$ then via the standard selection algorithm and subsequent diagonalization, there emerge sequences $\{f_{k_i}\}_{i \in \mathbb{N}}$, $\left\{ f_{k_{i_j}} \right\}_{j \in \mathbb{N}}$, etc., such that

$$\lim_{i \to \infty} a_{k_i 1} \stackrel{\text{def}}{=} a_1, \ \lim_{j \to \infty} a_{k_{i_j} 2} \stackrel{\text{def}}{=} a_2,$$

etc. exist. If

$$g_1 \stackrel{\text{def}}{=} f_{k_1},\ g_2 \stackrel{\text{def}}{=} f_{k_{i_2}}, \ldots,$$

$$\left\{ g_p(x) \sim \sum_{n=1}^{\infty} b_{pn} \sin 2n\pi x \right\}_{p \in \mathsf{N}},$$

and if $p > r$, then for large N,

$$\|g_p - g_r\|_2^2 \le \sum_{n=1}^{N} |b_{pn} - b_{rn}|^2 + 2 \sum_{n=N+1}^{\infty} \frac{1}{n^2}.$$

Because $\sum_{n=N+1}^{\infty} \frac{1}{n^2} < \int_N^{\infty} \frac{dx}{x^2}$, the second term in the right member above does not exceed $\frac{2}{N}$ whence $\{g_p\}_{p \in \mathsf{N}}$ is a $\| \ \|_2$-Cauchy sequence with $\| \ \|_2$-limit say, g. Furthermore, $\lim_{p\to\infty} b_{pn} \stackrel{\text{def}}{=} b_n, n \in \mathsf{N}$, exist, and

$$g(x) \sim \sum_{n=1}^{\infty} b_n \sin 2n\pi x.$$

Because $\sum_{n=1}^{N} n\,|b_n| = \lim_{p\to\infty} \sum_{n=1}^{N} n\,|b_{pn}| \le 1$, $\sum_{n=1}^{\infty} n\,|b_n| \le 1$, whence $g \in S$ and so S is $\| \ \|_2$-compact. □

6.95. If $\int_0^x g_1(t)\,dt + c_1 = \int_0^x g_2(t)\,dt + c_2$ then setting x at zero shows $c_1 = c_2$. Thus $\int_0^x (g_1(t) - g_2(t))\,dt \equiv 0$, and so $g_1 \doteq g_2$: T is well-defined.

If $\{f, g\}$ is in the closure of $\mathcal{G}$ then for some sequence $\{f_n\}_{n\in\mathsf{N}}$ in S, $f_n(x) = \int_0^x g_n(t)\,dt + c_n$ and $\{f_n, g_n\} \to \{f, g\}$ as $n \to \infty$ whence

$$\|f_n - f\|_2 + \|g_n - g\|_2 \to 0 \text{ as } n \to \infty.$$

If $E \stackrel{\text{def}}{=} \{ (x,t) \ : \ 0 \le t \le x \le 1 \}$ and $K \stackrel{\text{def}}{=} \chi_E$ then $\|K\|_2 = 2^{-\frac{1}{2}}$, and

$$f(x) = \int_{[0,1]} K(x,t) g(t)\,dt + c.$$

As in **6.89**a), it follows that $\|f_n - c_n - (f_m - c_m)\|_2 < 2^{-\frac{1}{2}} \|g_n - g_m\|_2$. Furthermore,

$$\begin{aligned} |c_n - c_m| = \|c_n - c_m\|_2 &\le \|f_n - c_n - (f_m - c_m)\|_2 + \|f_n - f_m\|_2 \\ &\le 2^{-\frac{1}{2}} \|g_n - g_m\|_2 + \|f_n - f_m\|_2 \end{aligned}$$

whence $\lim_{n\to\infty} c_n \stackrel{\text{def}}{=} c$ exists.

If $F(x) \stackrel{\text{def}}{=} \int_0^x g(t)\,dt$ then

$$F(x) - f_n(x) + c_n = \int_0^1 K(x,t)\,(g(t) - g_n(t))\,dt$$
$$\|f - (F+c)\|_2 \le \|f - f_n\|_2 + \|f_n - (f - c_n)\|_2 + |c_n - c|$$
$$\le \|f - f_n\|_2 + 2^{-\frac{1}{2}}\,\|g - g_n\|_2 + |c_n - c|\,.$$

Hence $f = F + c, Tf = g$ and $\mathcal{G}$ is closed. □

6.96. If $\{f_n\}_{n=1}^N$ is an orthonormal subset of M then

$$E \stackrel{\text{def}}{=} \left\{ x \,:\, \sum_{n=1}^N |f_n(x)|^2 \ne 0 \right\} \ne \emptyset$$

and for any fixed x in E, $a_k(x) \stackrel{\text{def}}{=} \dfrac{\overline{f_k(x)}}{\left(\sum_{n=1}^N |f_n(x)|^2\right)^{\frac{1}{2}}}, 1 \le k \le N$, are well-defined.

Thus for t in X,

$$\left|\sum_{n=1}^N a_n(x) f_n(t)\right| \le C \left(\int_X \left(\sum_{n=1}^N a_n(x) f_n(s) \right)^2 d\mu(s) \right)^{\frac{1}{2}}$$
$$= C \left(\sum_{n=1}^N |a_n(x)|^2 \right)^{\frac{1}{2}} = C,$$

an inequality valid for each (x,t) in X^2. In particular it follows that when $x = t$, $\left|\sum_{n=1}^N a_n(x) f_n(x)\right|^2 \equiv \left(\sum_{n=1}^N |f_n(x)|^2\right) \le C^2$. Owing to the orthonormality of $\{f_n\}_{n=1}^N$, $\int_X \sum_{n=1}^N |f_n(x)|^2\,d\mu(x) = N \le C^2 \mu(X)$. □

6.97. a) Because $\|fg\|_2 \le \|f\|_\infty \|g\|_2$, $\|T_f\| \le \|f\|_\infty$.

If $\|T_f\| < \|f\|_\infty \stackrel{\text{def}}{=} M < \infty$ and $E_n \stackrel{\text{def}}{=} \left\{ x \,:\, |f(x)| \ge M - \dfrac{1}{n} \right\}$ then $\mu(E_n) > 0$. If, for each n in $\mathbb{N}$, $L^2(X,\mu) \setminus \{\mathbf{O}\}$ contains an h_n such that $h_n(x) \doteq 0$ off E_n then it may be assumed that $\|h_n\|_2 = 1$ whence $\|T_f\| \ge M - \dfrac{1}{n}$ and so $\|T_f\| = M$, a contradiction. Thus for some n_0, $L^2(X,\mu) \setminus \{\mathbf{O}\}$ contains no h such that $h(x) \doteq 0$ off E_{n_0}. Hence E_{n_0} contains no measurable subset of finite measure, i.e., E_{n_0} is an infinite atom.

Conversely, if E is an infinite atom then for any h in $L^2(X,\mu) \setminus \{\mathbf{O}\}$, $\chi_E h \doteq 0$, whence $\|\chi_E h\|_2 = 0$, i.e., $\|T_{\chi_E}\| = 0 < \|\chi_E\|_\infty = 1$.

b) If $E_n \stackrel{\text{def}}{=} \{x : |k(x)| > n\}$ then $E_n \supset E_{n+1}$ and $\mu(E_n) > 0, n \in \mathbb{N}$. If $\mu(E_n \setminus E_{n+1}) = 0$ for all but finitely many n, then $k(x) = \infty$ on a measurable set F of positive measure and because X contains no infinite atoms, F contains a set H of finite positive measure. Then $\chi_H \in L^2(X,\mu)$ and $k\chi_H \notin L^2(X,\mu)$.

Thus it may be assumed that for all n, $\mu(E_n \setminus E_{n+1}) > 0$ and each $E_n \setminus E_{n+1}$ contains a subset H_n of finite positive measure. If $a_n^2 \mu(H_n) = \frac{1}{n^2}$ and $g \stackrel{\text{def}}{=} \sum_{n=1}^{\infty} a_n \chi_{H_n}$ then $g \in L^2(X,\mu)$ and on H_n, $k(x)g(x) \geq \frac{1}{\sqrt{\mu(H_n)}}$ whence $\|kg\|_2 = \infty$. □

6.98. Intuition suggests that T_f is surjective iff for all h in $L^2(X,\mu)$, $\frac{h}{f} \in L^2(X,\mu)$. As given, the criterion is meaningless because $\frac{1}{f}$ may fail to be defined. The following paragraphs deal with this difficulty.

If $0 < \mu(E) < \infty$ and $f^{-1}(0) \doteq E$ then T_f cannot be surjective because $\mathbf{O} \neq \chi_E \in L^2(X,\mu)$ and if $fg = \chi_E$ then $f(x)g(x) \doteq 0$ both on E and on $X \setminus E$, a contradiction. On the other hand, if E is an infinite atom and $h \in L^2(X,\mu)$ then $h(x)\big|_E \doteq 0$. Thus if $f \in L^\infty(X,\mu)$ and $\frac{1}{f(x)}$ is essentially bounded off E then T_f is surjective.

In light of the preceding paragraphs the following criterion suggests itself:

> T_f is surjective iff for every σ-finite set E, $\left\|\frac{1}{f\chi_E}\right\|_\infty < \infty$, in the sense that $\mu\left((f\chi_E)^{-1}(0)\right) = 0$ and $\frac{1}{f\chi_E}\left((f\chi_E)^{-1}(0)\right) = 0$.

Indeed, if T_f is surjective, E is σ-finite, and $\left\|\frac{1}{f\chi_E}\right\|_\infty = \infty$ then as in **Solution 6.97**b), for some g in $L^2(E,\mu)$, $\frac{g}{f\chi_E} \notin L^2(E,\mu)$. Extend the domain of g to X so that $g(x) = 0$ off E. Then $g \in L^2(X,\mu)$. Because T_f is surjective, for some h in $L^2(X,\mu)$, $T_f(h) = g$. Because $g(x) = 0$ off E, $fh = g = f\chi_E h$ and $\frac{g}{f\chi_E} = h \notin L^2(X,\mu)$, a contradiction.

Conversely, if $h \in L^2(X,\mu)$ and for every σ-finite set E, $\left\|\frac{1}{f\chi_E}\right\|_\infty < \infty$ then $E \stackrel{\text{def}}{=} \{x : h(x) \neq 0\}$ is σ-finite and $g \stackrel{\text{def}}{=} \frac{h}{f} \in L^2(X,\mu)$, whence $T_f(g) = h$. □

6.99. a) If $g \neq \mathbf{O}$ and $T_f(g) = \lambda g$ then for all t, $(t - \lambda)g(t) = \mathbf{O}$, whence λ is not a constant.

b) For all f in $L^\infty([0,1],\lambda)$,

$$T_f(S(\mathbf{1})) = fS(\mathbf{1}) = S(T_f(\mathbf{1})) = S(f),$$

whence if $S(\mathbf{1}) \stackrel{\text{def}}{=} g$ (in $L^2([0,1],\lambda)$) then $S(f) = gf$. If $h \in L^2([0,1],\lambda)$ then $(h \vee 0) \wedge (n \cdot \mathbf{1}) + (h \wedge 0) \vee (-n \cdot \mathbf{1}) \stackrel{\text{def}}{=} h_n \stackrel{\text{a.e.}}{\to} h$ and $h_n \stackrel{\| \|_2}{\to} h$, whence $S(h_n) = gh_n \stackrel{\| \|_2}{\to} h$ and $gh_n \stackrel{\text{a.e.}}{\to} gh$ and so $gh \in L^2([0,1],\lambda)$.

If $g \notin L^\infty([0,1],\lambda)$ then for all n in $\mathbb{N}$,

$$\lambda(\{x : n \leq |g(x)| < n+1\}) \stackrel{\text{def}}{=} a_n > 0.$$

If $\sum_{n=1}^\infty \frac{|k_n|^2}{a_n} < \infty$, $\sum_{n=1}^\infty \frac{n^2|k_n|^2}{a_n} = \infty$, and $h \stackrel{\text{def}}{=} \sum_{n=1}^\infty \frac{k_n}{a_n}\chi_{E_n}$ then $h \in L^2([0,1],\lambda)$ while $gf \notin L^2([0,1],\lambda)$, a contradiction. It follows that g is in $L^\infty([0,1],\lambda)$ and $S = T_g$. □

6.100. a) See **Solution 6.97**.

b) For λ in $\mathbb{C}$, and some nonzero $\mathbf{x} \stackrel{\text{def}}{=} \{x_n\}_{n\in\mathbb{N}}$,

$$\{T_{\mathbf{a}}\mathbf{x} = \lambda\mathbf{x}\} \Leftrightarrow \{a_n x_n \equiv \lambda x_n\} \Leftrightarrow \{\exists n_0 \{x_{n_0} \neq 0\}\},$$

whence $\lambda = a_{n_0}$ and then $x_n = 0$ if $a_n \neq \lambda$.

c) For some subsequence $\{a_{n_k}\}_{k\in\mathbb{N}}$, $\lim_{k\to\infty} a_{n_k} \stackrel{\text{def}}{=} a$ exists and is not zero. If $\mathbf{x}_k \stackrel{\text{def}}{=} \{\delta_{nn_k}\}_{n\in\mathbb{N}}$ then $S \stackrel{\text{def}}{=} \{\mathbf{x}_k\}_{k\in\mathbb{N}} \subset B(\mathbf{O},1)$ whereas $T_{\mathbf{a}}(S)$ contains no $\| \|_2$-Cauchy sequence. □

[**Note s6.9:** Contrast the result in b) above with the result in **6.99**a).]

6.101. The conclusion follows from the equation

$$\|\mathbf{x}_n - \mathbf{x}_0\|^2 = \|\mathbf{x}_n\|^2 + \|\mathbf{x}_0\|^2 - 2\Re(\mathbf{x}_n, \mathbf{x}_0).$$ □

6.102. Because $\|\mathbf{x}_n - \mathbf{x}_m\|^2 = \|\mathbf{x}_n\|^2 + \|\mathbf{x}_m\|^2 - 2\Re(\mathbf{x}_n, \mathbf{x}_m)$, it follows that $\overline{\lim}_{n\to\infty} \|\mathbf{x}_n - \mathbf{x}_m\| \leq 1 + 1 - 2 = 0$. □

6.103. For fixed $\mathbf{y}$, $L_{\mathbf{y}} : \mathfrak{H} \ni \mathbf{x} \mapsto \mathsf{B}(\mathbf{x},\mathbf{y})$ is a continuous linear functional, whence for some $\mathbf{z}_{\mathbf{y}}$ in $\mathfrak{H}$, $L_{\mathbf{y}}(\mathbf{x}) = (\mathbf{x}, \mathbf{z}_{\mathbf{y}})$. Moreover, $T : \mathfrak{H} \ni \mathbf{y} \mapsto \mathbf{z}_{\mathbf{y}}$ is linear and because $|(\mathbf{x}, T\mathbf{y})| \leq \|\mathbf{x}\| \cdot \|\mathbf{y}\|$, it follows that $\|T\| \leq 1$. □

6.104. If $\|\mathbf{x} - \mathbf{y}\| = a > 0$ and $\mathbf{z} \stackrel{\text{def}}{=} \mathbf{x} - \mathbf{y}$ then $(\mathbf{x} - \mathbf{y}, \mathbf{z}) = a^2$. For any $\mathbf{u}$ and positive ϵ,

$$N(\mathbf{u};\epsilon) \stackrel{\text{def}}{=} \{\mathbf{w} : |(\mathbf{w},\mathbf{z}) - (\mathbf{u},\mathbf{z})| < \epsilon\}$$

is a (weak) neighborhood of $\mathbf{u}$ and

$$N\left(\mathbf{x}; \frac{a^2}{3}\right) \cap N\left(\mathbf{y}; \frac{a^2}{3}\right) = \emptyset.$$ □

[**Note s6.10:** There is a stronger result: The weak* topology of $\mathfrak{H}$ or of the dual X^* of any Banach space X is normal. The sequence of ideas in the proof is the following: a) The unit ball of X^* is $\sigma(X^*, X)$-compact; b) X^* is σ-compact in the topology $\sigma(X^*, X)$; c) X^* is $\sigma(X^*, X)$-Lindelöf; d) in Lindelöf spaces regularity and paracompactness are equivalent; e) every paracompact space is normal.]

6.105. If $\dim(\mathfrak{H}) \stackrel{\text{def}}{=} n$ (in $\mathbb{N}$) then the weak and $\|\ \|$-induced topologies are the same and $\mathfrak{H}$ is the complete metric space $\mathbb{C}^n$, a set of the second category.

If $\dim(\mathfrak{H}) = \infty$, W is weakly open, and $\mathbf{x} \in W$, then for some $\mathbf{w}$ in W, $\|\mathbf{w}\| \stackrel{\text{def}}{=} 1 + b > 1$ and for some $\mathbf{z}$ in $B(\mathbf{O}, 1)$, $(\mathbf{w}, \mathbf{z}) > 1 + \frac{b}{2}$. Then $U \stackrel{\text{def}}{=} \left\{ \mathbf{u} \; : \; |(\mathbf{u}, \mathbf{z}) - (\mathbf{w}, \mathbf{z})| < \frac{b}{2} \right\}$ is a weak neighborhood of $\mathbf{w}$, whereas if $\mathbf{v} \in B(\mathbf{O}, 1)$ then $|\mathbf{v}, \mathbf{z})| \leq 1$. Hence

$$|(\mathbf{w}, \mathbf{z}) - (\mathbf{v}, \mathbf{z})| \geq |(\mathbf{w}, \mathbf{z})| - |(\mathbf{v}, \mathbf{z})| \geq 1 + \frac{b}{2} - 1 = \frac{b}{2},$$

i.e., $U \cap B(\mathbf{O}, 1) = \emptyset$. Thus $W \cap U$ is a weak neighborhood of $\mathbf{w}$, $W \cap U \subset W$, and $(W \cap U) \cap B(\mathbf{O}, 1) = \emptyset$. In sum, $B(\mathbf{O}, 1)$ is weakly nowhere dense. A similar argument shows that for all nonnegative R, $B(\mathbf{O}, R)$ is weakly nowhere dense. Hence $\mathfrak{H} = \bigcup_{n \in \mathbb{N}} B(\mathbf{O}, n)$, a countable union of weakly nowhere dense sets. □

6.106. Let P be the orthogonal projection of $\mathfrak{H}$ onto M:

$$P = P^2, \ \|\mathbf{x}\|^2 = \|P\mathbf{x}\|^2 + \|\mathbf{x} - P\mathbf{x}\|^2, \ P\mathbf{x} \perp (\mathbf{x} - P\mathbf{x}).$$

Then the Hahn-Banach theorem and the preceding equalities imply

$$\begin{aligned} &\sup \{ |(\mathbf{z}, \mathbf{x})| \; : \; \mathbf{z} \in M^\perp \cap \partial B(\mathbf{O}, 1) \} \\ &\qquad = \sup \{ |(\mathbf{z}, \mathbf{x} - P\mathbf{x})| \; : \; \mathbf{z} \in M^\perp \cap \partial B(\mathbf{O}, 1) \} = \|\mathbf{x} - P\mathbf{x}\| \\ &\inf \{ \|\mathbf{x} - \mathbf{y}\| \; : \; \mathbf{y} \in M \} = \|\mathbf{x} - P\mathbf{x}\|. \end{aligned}$$

□

6.107. If $g \in M^\perp$ then $h(t) \stackrel{\text{def}}{=} (g, f_{[t]}) \equiv 0 \equiv \widehat{h}(n)$ while

$$\widehat{h}(n) = \int_{\mathbb{T}^2} g(x)\overline{f(xt)}t^{-n}\,dx\,dt = \int_{\mathbb{T}^2} g(x)\overline{f(z)}z^n x^{-n}\,dx\,dz = \widehat{g}(-n)\widehat{f}(-n),$$

whence $\widehat{g}(-n) \equiv 0$ and so $g = \mathbf{O}$. □

[**Note s6.11:** The result in **6.107** is the simple version of Wiener's Tauberian theorem. The general form asserts that if G is a locally compact abelian group and $f \in L^1(G,\mu)$ then

$$\overline{\operatorname{span}\left(\{ f_{[x]} \ : \ x \in G \}\right)} = L^1(G,\mu)$$

iff $\widehat{f}$ is never zero. The distinction between $L^2(G,\mu)$ and $L^1(G,\mu)$ is important and provokes a very much more complex proof. Wiener's original proof was hard analysis. The modern proofs use functional analysis, particularly the general theory of commutative Banach algebras [**Loo**].]

6.108. Because S includes the set A of all infinitely differentiable functions that together with all their derivatives vanish at both zero and one, S is dense.

If $\{f_n\}_{n\in\mathbb{N}} \subset S$, $f_n \overset{\|\ \|_2}{\to} f$, $D(f_n) \overset{\|\ \|_2}{\to} g$ and $h \in A$, then it may be assumed that $f_n \overset{\text{a.e.}}{\to} f$, $D(f_n) \overset{\text{a.e.}}{\to} g$ and thus

$$\begin{aligned}\int_0^1 D(f_n)(x)h(x)\,dx &= f_n(x)h(x)\big|_0^1 - \int_0^1 f_n(x)D(h)(x)\,dx\\ &= -\int_0^1 f_n(x)D(h)(x)\,dx\\ \int_0^1 g(x)h(x)\,dx &= f(x)h(x)\big|_0^1 - \int_0^1 f(x)D(h)(x)\,dx\\ &= -\int_0^1 f(x)D(h)(x)\,dx. \end{aligned} \tag{s6.9}$$

If $\int_0^x g(t)\,dt \overset{\text{def}}{=} G(x)$ then

$$\begin{aligned}\int_0^1 g(x)h(x)\,dx &= G(x)h(x)\big|_0^1 - \int_0^1 G(x)D(h)(x)\,dx\\ &= -\int_0^1 G(x)Dh(x)\,dx. \end{aligned} \tag{s6.10}$$

Since (s6.9) and (s6.10) obtain for all h in A, it follows that $f \doteq G$, i.e., f is differentiable a.e. and $D(f) \doteq g$. □

6.109. For x in $\mathbb{R}$, $\mathbf{g}(x) \overset{\text{def}}{=} \dfrac{\partial(\mathbf{f}(x),\mathbf{y})}{\partial x}$, and some $a(x,\mathbf{y})$,

$$T_x(\mathbf{y}) \overset{\text{def}}{=} (\mathbf{f}(x) - \mathbf{f}(\mathbf{x}_0), \mathbf{y}) = g(x_0,\mathbf{y})(x-x_0) + a(x,\mathbf{y})(x-x_0).$$

If $x \neq x_0$ then

$$\frac{|T_x(\mathbf{y})|}{|x-x_0|^{\frac12}} \le |x-x_0|^{\frac12}\cdot|g(x_0,\mathbf{y}) + a(x,\mathbf{y})| \to 0 \text{ as } x \to x_0.$$

The uniform boundedness principle implies that for x near x_0 and some M, $\dfrac{\|T_x\|}{|x-x_0|^{\frac{1}{2}}} \le M < \infty$. Thus

$$\|f(x) - f(x_0)\| = \sup_{\|\mathbf{y}\|=1} |T_x(\mathbf{y})| \le \|T_x\| \le M|x-x_0|^{\frac{1}{2}}. \qquad \square$$

6.3. Abstract Topological Vector Spaces

6.110. If $\{\mathbf{x},\mathbf{x}^*\} \in (V\setminus\{\mathbf{O}\}) \times (V^*\setminus\{\mathbf{O}^*\})$ then the map

$$S_{\mathbf{x}\mathbf{x}^*} : V \ni \mathbf{y} \mapsto (\mathbf{y},\mathbf{x}^*)\,\mathbf{x}$$

is compact because $\dim(\operatorname{im}(S_{\mathbf{x}\mathbf{x}^*})) = 1$. For T as described and all $\mathbf{y}$, $TS_{\mathbf{x}\mathbf{x}^*}(\mathbf{y}) = (\mathbf{y},\mathbf{x}^*)\,T(\mathbf{x}) = S_{\mathbf{x}\mathbf{x}^*}T(\mathbf{y}) = (T(\mathbf{y}),\mathbf{x}^*)\,\mathbf{x}$. Because $\mathbf{x}^* \ne \mathbf{O}$, For some $\mathbf{y}$, $(\mathbf{y},\mathbf{x}^*) \ne 0$, whence for all $\mathbf{x}$, $T(\mathbf{x}) = \dfrac{(T(\mathbf{y}),\mathbf{x}^*)}{(\mathbf{y},\mathbf{x}^*)}\mathbf{x} \stackrel{\text{def}}{=} c\mathbf{x}$. Since $T(\mathbf{x})$ is $\mathbf{y}$-free, so is c. $\square$

6.111. The uniform boundedness principle applies twice, once to the sequence $\{T_n(\mathbf{v})\}_{n\in\mathbb{N}}$ (in W regarded as a subset of W^{**}) and then to $\{T_n\}_{n\in\mathbb{N}}$ (in $[V,W]$). $\square$

6.112. If $\mathbf{x},\mathbf{y} \in V$ and $\mathbf{x}^* \in V^*$ then

$$\begin{aligned}(T(\mathbf{x}+\mathbf{y}),\mathbf{x}^*) &= (\mathbf{x}+\mathbf{y}, S(\mathbf{x}^*)) \\ &= (\mathbf{x}, S(\mathbf{x}^*)) + (\mathbf{y}, S(\mathbf{x}^*)) \\ &= (T(\mathbf{x}),\mathbf{x}^*) + (T(\mathbf{y}),\mathbf{x}^*) \\ &= (T(\mathbf{x}) + T(\mathbf{y}),\mathbf{x}^*).\end{aligned}$$

Because the equations above are valid for all $\mathbf{x}^*$ in V^*, T is additive. Similar calculations show T is a not necessarily continuous endomorphism of V and S is a not necessarily continuous endomorphism of V^*.

Furthermore, if

$$\|\mathbf{x}_n - \mathbf{x}\| + \|\mathbf{x}_n^* - \mathbf{x}^*\| + \|T(\mathbf{x}_n) - \mathbf{y}\| + \|S(\mathbf{x}_n^*) - \mathbf{y}^*\| \to 0$$

as $n \to \infty$ then for all $\mathbf{x}^*$ in V^*,

$$(T(\mathbf{x}_n),\mathbf{x}^*) = (\mathbf{x}_n, S(\mathbf{x}^*)) \to (\mathbf{x}, S(\mathbf{x}^*)) = (T(\mathbf{x}),\mathbf{x}^*)$$

as $n \to \infty$. Hence $\mathbf{y} = T(\mathbf{x})$. Similarly, for all $\mathbf{x}$ in V,

$$(\mathbf{x}, S(\mathbf{x}_n^*)) = (T(\mathbf{x}),\mathbf{x}_n^*) \to (T(\mathbf{x}),\mathbf{x}^*) = (\mathbf{x}, S(\mathbf{x}^*))$$

as $n \to \infty$ and so $S(\mathbf{x}^*) = \mathbf{y}^*$. The closed graph theorem applies. $\square$

6.113. If $\mathsf{X} \stackrel{\text{def}}{=} \{\mathbf{x}_\gamma\}_{\gamma\in\Gamma}$ and $\mathsf{Y} \stackrel{\text{def}}{=} \{\mathbf{y}_\delta\}_{\delta\in\Delta}$ are Hamel bases for V then for each γ, $\mathbf{x}_\gamma = \sum'_{\delta\in\Delta} a_{\gamma\delta}\mathbf{y}_\delta$ a finite sum. If δ_0 is fixed and, as γ varies over Γ, $a_{\gamma\delta_0} \equiv 0$ then $\mathbf{y}_{\delta_0} = \sum'_{\gamma,\delta} b_{\delta_0\gamma}a_{\gamma\delta}\mathbf{y}_\delta = \sum'_{\delta\in\Delta} c_{\delta_0\delta}\mathbf{y}_\delta$, a contradiction of the linear independence of Y. The rest of the argument given in **6.88** may be repeated, mutatis mutandis. $\square$

6.114. Because V is infinite-dimensional, the Gram-Schmidt biorthogonalization process leads to the construction in $V\times V^*$ of an infinite biorthogonal sequence $\{\{\mathbf{v}_n, \mathbf{v}_n^*\}\}_{n\in\mathbb{N}}$, i.e., $(\mathbf{v}_n, \mathbf{v}_m^*) = \delta_{mn}$. Furthermore,

$$\left\{\left\{\frac{\mathbf{v}_n}{2^n\|\mathbf{v}_n\|}, 2^n\|\mathbf{v}_n\|\,\mathbf{v}_n^*\right\}\right\}_{n\in\mathbb{N}}$$

is also a biorthogonal set $\{\{\mathbf{y}_n, \mathbf{y}_n^*\}\}_{n\in\mathbb{N}}$ and if $S \subset \mathbb{N}$, $\sum_{n\in S}\mathbf{y}_n$ $\|\ \|$-converges.

For $\{r_n\}_{n\in\mathbb{N}}$ an enumeration of $\mathbb{Q}$ and for each t in $\mathbb{R}$, an infinite subsequence $S_t \stackrel{\text{def}}{=} \{r_{n_k(t)}\}_{k\in\mathbb{N}}$ converges to t. If $t \neq t'$ then $S_t \cap S_{t'}$ is empty or finite. If $\{t_m\}_{m=1}^M$ is a finite set then, because S_{t_m} is infinite, $S_{t_m} \setminus \bigcup_{\substack{m'\neq m\\ 1\le m'\le M}} S_{t_{m'}} \neq \emptyset$.

For each t, there is a $\mathbf{z}_t \stackrel{\text{def}}{=} \sum_{r_n\in S_t}\mathbf{y}_n$. If $\mathbf{z} \stackrel{\text{def}}{=} \sum_{m=1}^M a_m\mathbf{z}_{t_m} = \mathbf{O}$ and $r_n \in S_{t_m} \setminus \bigcup_{\substack{m'\neq m\\ 1\le m'\le M}} S_{t_{m'}}$ then $(\mathbf{z}, \mathbf{y}_n^*) = a_m = 0, 1 \le m \le M$. Thus $Z \stackrel{\text{def}}{=} \{\mathbf{z}_t\}_{t\in\mathbb{R}}$ is $\mathbb{C}$-linearly independent and may be extended to a maximal $\mathbb{C}$-linearly independent set — a Hamel basis H for V. Furthermore, owing to **6.113**, $\mathfrak{c} \le \#(\mathsf{H}) = \#(H)$. $\square$

6.115. Let $S \stackrel{\text{def}}{=} \{\lambda_n\}_{n\in\mathbb{N}}$ be a countable infinite subset of Λ. Then $\sum_{n=1}^\infty \frac{\mathbf{x}_{\lambda_n}}{2^n\|\mathbf{x}_{\lambda_n}\|}$ converges and represents some $\mathbf{y}$ in V. Hence for some finite subset δ of Λ, $\mathbf{y} = \sum_{\lambda\in\delta} a_\lambda(\mathbf{y})\mathbf{x}_\lambda$. If $\lambda_{\overline{n}} \in S\setminus\delta$ and if $a_{\lambda_{\overline{n}}}$ is continuous then

$$\begin{aligned} a_{\lambda_{\overline{n}}}(\mathbf{y}) &= a_{\lambda_{\overline{n}}}\left(\sum_{\lambda\in\delta} a_\lambda(\mathbf{y})\mathbf{x}_\lambda\right) = 0 \\ &= a_{\lambda_{\overline{n}}}\left(\sum_{n=1}^\infty \frac{\mathbf{x}_\lambda}{2^n\|\mathbf{x}_{\lambda_n}\|}\right) = \frac{1}{2^{\overline{n}}\|\mathbf{x}_{\lambda_{\overline{n}}}\|}, \end{aligned}$$

a contradiction.

Hence for every countable infinite subset S of Λ, Λ contains a finite subset δ such that for every $\overline{\lambda}$ in $S\setminus\delta$, $a_{\overline{\lambda}}$ is discontinuous. $\square$

6.116. a) Both $c_0(\mathbb{N})$ and $\ell^1(\mathbb{N})$ are Banach spaces, each of cardinality $\mathfrak{c}$. Owing to **6.114**, each has a $\mathbb{C}$-Hamel basis of cardinality $\mathfrak{c}$. Any bijection between Hamel bases for $c_0(\mathbb{N})$ and $\ell^1(\mathbb{N})$ may be extended to an isomorphism $T : c_0(\mathbb{N}) \mapsto \ell^1(\mathbb{N})$ and $\ker(T) = T^{-1}(\mathbf{O}) = \{\mathbf{O}\}$, a closed set.

However, $(c_0(\mathbb{N}))^* = \ell^1(\mathbb{N})$ is $\| \ \|_1$-separable while $(\ell^1(\mathbb{N}))^* = \ell^\infty(\mathbb{N})$ is not $\| \ \|_\infty$-separable. Hence T cannot be a continuous isomorphism [even though $\ker(T)$ is closed].

b) By the same token, T^{-1} is an isomorphism of $\ell^1(\mathbb{N})$ onto $c_0(\mathbb{N})$. If T^{-1} were open, T would be continuous. □

6.117. For some $\mathbf{w}$ in W, $\|\mathbf{x} - \mathbf{w}\| > d(\mathbf{x}, W) - a$. The Hahn-Banach theorem assures the existence in V^* of a $\mathbf{z}^*$ as described. □

6.118. If $\#$ exists then for some $\{\mathbf{v}, \mathbf{v}^*\}$ in $V \times V^*$, $(\mathbf{v}, \mathbf{v}^*) = 1$. But then $(\mathbf{v}^\#, \mathbf{v}^*) = 1$, $(\mathbf{v}, i\mathbf{v}^*) = i$, and $(\mathbf{v}^\#, i\mathbf{v}^*) = \overline{(\mathbf{v}, i\mathbf{v}^*)} = -i$. On the other hand, $(\mathbf{v}^\#, i\mathbf{v}^*) = i\left((\mathbf{v}^\#, \mathbf{v}^*)\right) = i$. □

6.119. If $\mathbf{v}^* \in M$ and $\mathbf{v} \in M_\perp$ then $(\mathbf{v}, \mathbf{v}^*) = 0$ and the linearity of $(\ , \ )$ implies $\operatorname{span}(M) \subset (M_\perp)^\perp$. If $\mathbf{v}^* \in (\operatorname{span}(M))^{w*}$ [the weak* closure of $\operatorname{span}(M)$], $a > 0$, and $\mathbf{v} \in M_\perp$, then

$$\begin{aligned} &\{\mathbf{y}^* : |(\mathbf{v}, \mathbf{y}^* - \mathbf{v}^*)| < a\} \cap (M_\perp)^\perp \\ &\quad \supset \{\mathbf{y}^* : |(\mathbf{v}, \mathbf{y}^* - \mathbf{v}^*)| < a\} \cap \operatorname{span}(M) \neq \emptyset. \end{aligned}$$

Thus for some $\mathbf{y}^*$ in $(M_\perp)^\perp$, $|(\mathbf{v}, \mathbf{y}^* - \mathbf{v}^*)| = |(\mathbf{v}, \mathbf{v}^*)| < a$. Because a is an arbitrary positive number, $|(\mathbf{v}, \mathbf{v}^*)| = 0$. Thus $\mathbf{v}^* \in (M_\perp)^\perp$, i.e., $(\operatorname{span}(M))^{w*} \subset (M_\perp)^\perp$.

For $\sigma(V, V^*)$ resp. $\sigma(V^*, V)$, each of the dual pair $\{V, V^*\}$ is the dual of the other. If $\mathbf{v}^* \in (M_\perp)^\perp \setminus M^{w*}$, the Hahn-Banach theorem implies $(M^{w*})_\perp$ contains a $\mathbf{v}$ such that $(\mathbf{v}, \mathbf{v}^*) = 1$, a contradiction. □

[**Note s6.12:** Even when V is a Banach space and $V \neq V^{**}$, for $\sigma(V, V^*)$ resp. $\sigma(V^*, V)$, V and V^* are locally convex and a dual pair and each is the dual of the other [**Kö, Sch**].]

6.120. Because $M \subset (M^\perp)_\perp$, if $\mathbf{v} \in (M^\perp)_\perp \setminus M$, according to the Hahn-Banach theorem, for some $\mathbf{v}^*$ in $M^\perp$, $(\mathbf{v}, \mathbf{v}^*) = 1$, whence it follows that $\mathbf{v} \notin (M^\perp)_\perp$, a contradiction. □

6.121. For any nonreflexive Banach space V, e.g., $c_0(\mathbb{N})$, the canonical embedding $V \hookrightarrow V^{**}$ makes V a proper subspace of V^{**}. Thus $V_\perp = \{\mathbf{O}\}$ and $(V_\perp)^\perp = V^{**} \supsetneq V$. □

6.122. a) It may be assumed that $K \neq \emptyset$ and that $\mathbf{O} \in K$. If $\mathbf{v} \notin K$, the Hahn-Banach theorem implies that for some positive numbers a, b and some $\mathbf{v}^*$ in V^*, $(B(\mathbf{v}, a), \mathbf{v}^*) \subset (-\infty, -b)$ while $\mathbf{v}^*(K) \subset [0, \infty)$. Thus $\left\{\mathbf{y} : |(\mathbf{y} - \mathbf{v}, \mathbf{v}^*)| < \frac{b}{2}\right\} \cap K = \emptyset$, whence $\mathbf{v} \notin K^w$, the weak closure of K. Because $K \subset K^w$, it follows that $K = K^w$.

b) By definition $\overline{K} \subset K^w$. Because $\overline{K}$ is convex, weakly dense in K^w, and $\| \ \|$-closed it is, per the previous paragraph, weakly closed: $\overline{K} = K^w$.

□

6.123. For $B \stackrel{\text{def}}{=} \{\mathbf{w}_1, \ldots, \mathbf{w}_N\}$ a basis for W and $\mathbf{a} \stackrel{\text{def}}{=} (a_1, \ldots, a_N)$ in $\mathbb{C}^N$, $T : \mathbb{C}^N \ni \mathbf{a} \mapsto \sum_{n=1}^N a_n \mathbf{w}_n \in W$ is a linear bijection and

$$\|T(\mathbf{a})\| \leq \left(\sum_{n=1}^{N} |a_n|^2\right)^{\frac{1}{2}} \left(\sum_{n=1}^{N} \|\mathbf{w}_n\|^2\right)^{\frac{1}{2}}.$$

The open mapping theorem implies T is open whence T^{-1} is continuous. Hence if $\lim_{k\to\infty} \mathbf{v}_k = \mathbf{v}$ then $\lim_{k\to\infty} T^{-1}(\mathbf{v}_k) \stackrel{\text{def}}{=} \mathbf{a}$ exists and $T(\mathbf{a}) = \mathbf{v}$, i.e., $\mathbf{v} \in W$, W is closed.

If $\mathbf{w} \in W$ then $\mathbf{w} = \sum_{n=1}^N f_n(\mathbf{w})\mathbf{w}_n$. The Hahn-Banach theorem implies that each f_n (in W^*) may be extended to a $\mathbf{v}_n^*$ in V^*. The map $P : V \ni \mathbf{v} \mapsto \sum_{n=1}^N \mathbf{v}_n^*(\mathbf{v})\mathbf{w}_n$ is an idempotent in $[V, W]_e$ and

$$\|P\| \leq \left(\sup_{1\leq n\leq N} \|\mathbf{v}_n^*\|\right) \sum_{n=1}^{N} \|\mathbf{w}_n\|.$$

□

6.124. If f is continuous then $\ker(f) = f^{-1}(\mathbf{O})$ is closed. Conversely, if $\ker(f)$ is closed, it is a closed subspace M of V. If $M = V$ then $f = O$ and f is continuous. If $M \subsetneqq V$ for some $\mathbf{v}$ in $V \setminus M$, $f(\mathbf{v}) = 1$. If $\mathbf{z} \in V$ then $\mathbf{y} \stackrel{\text{def}}{=} \mathbf{z} - f(\mathbf{z})\mathbf{v} \in M$. If, for some $\mathbf{y}'$ in M and a in $\mathbb{C}$, $\mathbf{z} = \mathbf{y}' + a\mathbf{v}$ then $f(\mathbf{z}) = f(\mathbf{y}') + af(\mathbf{v}) = 0 + a \cdot 1 = a$, whence $\mathbf{y}' = \mathbf{y}$. The Hahn-Banach theorem implies that for some $\mathbf{v}^*$ in $M^\perp$, $(\mathbf{v}, \mathbf{v}^*) = 1$. Because $(\mathbf{z}, \mathbf{v}^*) \equiv f(\mathbf{z})$, $\mathbf{v}^* = f$ and f is continuous.

If $\ker(f)$ is not closed then f is not continuous and so $f \neq O$. If $\overline{\ker(f)} \stackrel{\text{def}}{=} M \neq V$ then for some $\mathbf{v}$ in $V \setminus M$, $f(\mathbf{v}) = 1$ and for some $\mathbf{v}^*$ in $M^\perp$, $(\mathbf{v}, \mathbf{v}^*) = 1$. The argument of the preceding paragraph shows that $\mathbf{v}^* = f$.

□

6.125. Let $\{\mathbf{y}_j\}_{j=1}^J$ be a basis for W/M. If $\mathbf{x}_j/M = \mathbf{y}_j$ then $\{\mathbf{x}_j\}_{j=1}^J$ is a basis for a closed subspace N in W and $M \cap N = \{\mathbf{O}\}$. The direct sum $U \stackrel{\text{def}}{=} V \oplus N$ with norm $\| \ \|' : U \ni \{\mathbf{v}, \mathbf{n}\} \mapsto \|\mathbf{v}\| + \|\mathbf{n}\|$ is a Banach space and $S \stackrel{\text{def}}{=} T \oplus \text{id}$ is in $[U, W]$. If $\mathbf{z} \in W$ and $\mathbf{z}/M \stackrel{\text{def}}{=} \sum_{j=1}^J a_j \mathbf{y}_j$ then $\mathbf{z} - \sum_{j=1}^J a_j \mathbf{x}_j \in M$, i.e., for some $\mathbf{w}$ in M, $\mathbf{z} = \mathbf{w} + \sum_{j=1}^J a_j \mathbf{x}_j$ and for some $\mathbf{v}$ in V, $T(\mathbf{v}) = \mathbf{w}$ whence $\mathbf{z} \in S(U)$. Thus $S : U \mapsto W$ is open.

If $\{\mathbf{m}_k\}_{k\in\mathbb{N}} \subset M$ and $\lim_{k\to\infty} \mathbf{m}_k = \mathbf{m}_0$ then because S is open, for some sequence $\{\{\mathbf{v}_k, \mathbf{n}_k\}\}_{k=0}^\infty$ in U, $\{\mathbf{v}_k, \mathbf{n}_k\} \to \{\mathbf{v}_0, \mathbf{n}_0\}$ as $k \to \infty$ and $S(\{\mathbf{v}_k, \mathbf{n}_k\}) = \mathbf{m}_k = T(\mathbf{v}_k) + \mathbf{n}_k, k \in \mathbb{Z}^+$. Hence $\mathbf{n}_k \in M \cap N = \{\mathbf{O}\}$ and so $\mathbf{m}_0 \in M$, i.e., M is closed.

On the other hand, if $\{\mathbf{x}_\lambda\}_{\lambda\in\Lambda}$ is a Hamel basis for the infinite-dimensional Banach space W, let λ' be such that the coefficient functional $a_{\lambda'}$ is not continuous (cf. **6.124**). Then $M \stackrel{\text{def}}{=} \ker(a_{\lambda'})$ is not closed although $\dim(W/M) = 1$. □

6.126. a) If V^* is separable, $V^*\setminus\{\mathbf{O}\}$ contains a dense set $\{\mathbf{v}_n^*\}_{n\in\mathbb{N}}$ and V contains a set $S \stackrel{\text{def}}{=} \{\mathbf{v}_n\}_{n\in\mathbb{N}}$ such that $\|\mathbf{v}_n\| \le 1$ and $|(\mathbf{v}_n, \mathbf{v}_n^*)| \ge \|\mathbf{v}_n^*\| - \frac{1}{n}$.

If $\mathbf{v} \in V\setminus\overline{\text{span}(S)}$ then for some $\mathbf{v}^*$, in $(\text{span}(S))^\perp$, $(\mathbf{v}, \mathbf{v}^*) = 1$. There is a sequence $\{n_k\}_{k\in\mathbb{N}}$ such that $\mathbf{v}_{n_k}^* \to \mathbf{v}^*$ as $k \to \infty$ whence

$$\left\|\mathbf{v}^* - \mathbf{v}_{n_k}^*\right\| \cdot \|\mathbf{v}_{n_k}\| \ge \left|(\mathbf{v}_{n_k}, \mathbf{v}^*) - \left(\mathbf{v}_{n_k}, \mathbf{v}_{n_k}^*\right)\right| = \left|\left(\mathbf{v}_{n_k}, \mathbf{v}_{n_k}^*\right)\right| > \left\|\mathbf{v}_{n_k}^*\right\| - \frac{1}{n}.$$

Thus $\left\|\mathbf{v}_{n_k}^*\right\| \to 0$ as $k \to \infty$, whereas $\mathbf{v}^* \ne \mathbf{O}$. Hence $V \setminus \overline{\text{span}(S)} = \emptyset$ and V is separable.

b) Although $\ell^1(\mathbb{N})$ is separable, $\left(\ell^1(\mathbb{N})\right)^* = \ell^\infty(\mathbb{N})$ is not. □

6.127. Let $\mathcal{T}$ be the $\| \ \|$-induced topology of V^*. Since $\sigma(V^*, V) \subset \mathcal{T}$ it follows that $\sigma\mathsf{A}(\sigma(V^*, V)) \subset \sigma\mathsf{A}(\mathcal{T})$. By virtue of Alaoglu's theorem, every $B(\mathbf{v}^*, r)$ is $\sigma(V^*, V)$-compact and because V^* is $\| \ \|$-separable, every $\| \ \|$-open subset of V^* is the countable union of balls, i.e., the countable union of $\sigma(V^*, V)$-compact sets. □

6.128. The construction depends on two observations:

a) If M is a proper subspace of a topological vector space V and U is a nonempty open subset of M, then for $\mathbf{x}$ in U, $-\mathbf{x} + U$ is an open subset containing $\mathbf{O}$. If $\mathbf{y} \in V$ then for some nonzero t, $t\mathbf{y} \in -\mathbf{x} + U \subset M$, i.e., $M = V$. Thus no such U exists: M contains no nonempty open set.

b) If M is a closed proper subspace of V and $\mathbf{v} \notin M$ there is an open set U such that $\mathbf{v} \in U \subset (V \setminus M)$. On the other hand, if $\mathbf{v} \in M$, U is open, and $\mathbf{v} \in U$ then the previous argument shows that the open set $U \setminus M$ is not empty whence M is nowhere dense. In sum:

> A proper subspace contains no nonempty open set and a proper closed subspace is nowhere dense.

Let $\{\mathbf{v}_n\}_{n\in\mathbb{N}}$ be dense in V. Then

$$\left\{B(\mathbf{v}_n, r_m)^\circ \ : \ 0 < r_m \in \mathbb{Q}, m, n \in \mathbb{N}\right\} \stackrel{\text{def}}{=} \{A_k\}_{k\in\mathbb{N}}$$

is a sequence of open sets. If $\mathbf{y}_1 \in V \setminus \{\mathbf{O}\}$ then $A_1 \setminus \text{span}(\mathbf{y}_1) \ne \emptyset$. If $\mathbf{y}_1, \dots, \mathbf{y}_n$ are linearly independent and

$$\mathbf{y}_k \in A_{k-1} \setminus \text{span}(\mathbf{y}_1, \dots, \mathbf{y}_{k-1}), k = 2, \dots, n,$$

then some $\mathbf{y}_{n+1}$ is in $A_n \setminus \text{span}(\mathbf{y}_1, \dots, \mathbf{y}_n)$ $(\ne \emptyset)$. The sequence $\{\mathbf{y}_n\}_{n\in\mathbb{N}}$ is linearly independent and because every open set is the union of some of the A_k, it follows that $\{\mathbf{y}_n\}_{n\in\mathbb{N}}$ is dense. □

6.129. For the maps

$$\begin{aligned}T_N &: V^* \ni \mathbf{v}^* \mapsto \{(\mathbf{v}_1, \mathbf{v}^*), \ldots, (\mathbf{v}_N, \mathbf{v}^*), 0, 0, \ldots\}, N \in \mathbb{N},\\ T &: V^* \ni \mathbf{v}^* \mapsto \{(\mathbf{v}_n, \mathbf{v}^*)\}_{n\in\mathbb{N}} \in \ell^1(\mathbb{N}):\end{aligned}$$

a) because $\|T_N(\mathbf{v}^*)\|_1 \le \left(\sum_{n=1}^N \|\mathbf{v}_n\|\right)\|\mathbf{v}^*\|$, $T_N \in [V^*, \ell^1(\mathbb{N})]$; b) since $\|(T_N - T)(\mathbf{v}^*)\|_1 = \sum_{n=N+1}^\infty |(\mathbf{v}_n, \mathbf{v}^*)|$, $\|(T_N - T)(\mathbf{v}^*)\|_1 \to 0$ as $N \to \infty$ for all $\mathbf{v}^*$ in V^*; $T(\mathbf{v}^*) \in \ell^1(\mathbb{N})$; c) the uniform boundedness principle implies $T \in [V^*, \ell^1(\mathbb{N})]$.

If $a > 0$, $M \overset{\text{def}}{=} \sup_N \|T_N\|$, and $\{a_n\}_{n\in\mathbb{N}} \in c_0$, then for some N', $|a_n| < \dfrac{a}{2M}$ if $n > N'$. If $\|\mathbf{v}^*\| \le 1$ and $p, q > N'$ then

$$\left|\mathbf{v}^*\left(\sum_{n=p}^q a_n\mathbf{v}_n\right)\right| < \frac{a}{2M}\sum_{n=p}^q |\mathbf{v}^*(\mathbf{v}_n)| = \frac{a}{2M}\|(T_p - T_q)(\mathbf{v}^*)\|_1 < a. \quad \square$$

6.130. a) For $\mathbf{x}$ in V and r in $(0,1)$, induction provides a sequence $\{\mathbf{x}_{n_k}\}_{k\in\mathbb{N}}$ such that $\left\|\mathbf{x} - \sum_{k=1}^K r^{k-1}\|\mathbf{x}\|\mathbf{x}_{n_k}\right\| < r^K\|\mathbf{x}\|$. It follows that $T(\ell^1(\mathbb{N})) = V$.

b) Hence if $\epsilon > 0$ and $\mathbf{x} \in V$ then for some $\mathbf{a} \overset{\text{def}}{=} \{a_n\}_{n\in\mathbb{N}}$ in $\ell^1(\mathbb{N})$, $\|\mathbf{a}\|_1 < \|\mathbf{x}\| + \epsilon$ and $T(\mathbf{a}) = \sum_{n=1}^\infty a_n\mathbf{x}_n = \mathbf{x}$, i.e.,

$$\inf\left\{ \left\|\{a_n\}_{n\in\mathbb{N}}\right\|_1 \ : \ \mathbf{a} \in T^{-1}(\mathbf{x}) \right\} \le \|\mathbf{x}\|.$$

In other words, if $\mathbf{a} \in \ell^1(\mathbb{N})$ then the quotient norm of $\mathbf{a}/\ker(T)$ does not exceed $\|T(\mathbf{a})\|$. Because $\|\mathbf{x}\| \le \sum_{n=1}^\infty |a_n| \cdot \|\mathbf{x}_n\| \le \|\mathbf{a}\|_1$, $\|T(\mathbf{a})\|$ does not exceed the quotient norm of $\mathbf{a}/\ker(T)$. $\square$

6.131. For the maps $S_N : V \ni \mathbf{v} \mapsto \mathbf{v} + T\mathbf{v} + \cdots + T^{N-1}\mathbf{v}$, if $N < M$ then $\|S_M\mathbf{v} - S_N\mathbf{v}\| \le \dfrac{\|T\|^N}{1 - \|T\|}$, whence $\lim_{N\to\infty} S_N\mathbf{v} \overset{\text{def}}{=} S\mathbf{v}$ exists. Furthermore, $S \in [V]$, $\|S\| \le \dfrac{1}{1 - \|T\|}$, and $S(\text{id} - T) = (\text{id} - T)S = \text{id}$. The desired solution is $\mathbf{x} = S\mathbf{y}$. $\square$

6.132. The map $T : \ell^2(\mathbb{N}) \ni \mathbf{y} \overset{\text{def}}{=} \{y_1, y_2, \ldots\} \mapsto \{y_2, y_3, \ldots\} \overset{\text{def}}{=} T\mathbf{y}$ is such that $(S\mathbf{x}, \mathbf{y}) \equiv (\mathbf{x}, T\mathbf{y})$ whence $T = S^*$. Furthermore, because $\|(S^*)^n\mathbf{x}\|_2^2 = \sum_{k=n+1}^\infty |x_k|^2$, it follows that $(S^*)^n\mathbf{x} \overset{\|\ \|_2}{\to} \mathbf{O}$. On the other hand, if $\mathbf{e}_n \overset{\text{def}}{=} \{\delta_{nm}\}_{m\in\mathbb{N}}$ then $\|\mathbf{e}_n\|_2 \equiv 1$ while $(S^*)^n\mathbf{e}_{n+1} = \mathbf{e}_1$, whence $\|(S^*)^n\| \equiv 1$. $\square$

6.133. As a closed subspace of a Banach space, M is itself a Banach space. As observed in **Solution 6.128**, if $M \supsetneqq M_n, n \in \mathbb{N}$, then each

M_n is nowhere dense and so M is a Banach space of the first category, a contradiction. □

[**Note s6.13:** If a vector space V over an infinite field $\mathbb{K}$ is the union of finitely many subspaces $V_1, \ldots, V_n$ then for some n_0, $V = V_{n_0}$ [**GeO**].]

6.134. If V is a Banach space and $\sum_{n=1}^{\infty} \|\mathbf{x}_n\| < \infty$ then

$$\left\{\sum_{n=1}^{N} \mathbf{x}_n \stackrel{\text{def}}{=} \mathbf{y}_N\right\}_{N\in\mathbb{N}}$$

is a Cauchy sequence and has a limit.

Conversely, if (6.2) is valid and $\{\mathbf{v}_n\}_{n\in\mathbb{N}}$ is a Cauchy sequence, for some subsequence $\{\mathbf{v}_{n_k}\}_{k\in\mathbb{N}}$, $\sum_{k=1}^{\infty} \left\|\mathbf{v}_{n_k} - \mathbf{v}_{n_{k+1}}\right\| < \infty$. It follows that $\lim_{k\to\infty} (\mathbf{v}_{n_1} - \mathbf{v}_{n_k})$ exists whence $\lim_{n\to\infty} \mathbf{v}_n$ exists. □

6.135. a) If $\mathbf{x} \in \text{Conv}(A)$ then for some vectors $\mathbf{v}_1, \ldots, \mathbf{v}_N$ in A and some numbers $t_1, \ldots, t_n$ in $[0,1]$, $\sum_{n=1}^{N} t_n = 1$ and $\mathbf{v} = \sum_{n=1}^{N} t_n\mathbf{v}_n$. If $\epsilon > 0$ then for some vectors $\mathbf{a}_1, \ldots, \mathbf{a}_M$ in A, $A \subset \bigcup_{m=1}^{M} B\left(\mathbf{a}_m, \frac{\epsilon}{2}\right)^{\circ}$. For each n, some $\mathbf{a}_{m_n}$ in $S \stackrel{\text{def}}{=} \{\mathbf{a}_m\}_{m=1}^{M}$, and some $\mathbf{b}_n$ in $B\left(\mathbf{O}, \frac{\epsilon}{2}\right)^{\circ}$, $\mathbf{v}_n = \mathbf{a}_{m_n} + \mathbf{b}_n$ whence $\mathbf{v} = \sum_{n=1}^{N} t_n\mathbf{a}_{m_n} + \sum_{n=1}^{N} t_n\mathbf{b}_n \stackrel{\text{def}}{=} \mathbf{a} + \mathbf{b}$. Furthermore, $\mathbf{a} \in \text{Conv}(S)$ and $\mathbf{b} \in B\left(\mathbf{O}, \frac{\epsilon}{2}\right)^{\circ}$. As the convex hull of a finite set, $\text{Conv}(S)$ is compact and $\text{Conv}(S)$ contains vectors $\mathbf{c}_1, \ldots, \mathbf{c}_P$ such that $\text{Conv}(S) \subset \bigcup_{p=1}^{P} B\left(\mathbf{c}_p, \frac{\epsilon}{2}\right)^{\circ}$ whence $\text{Conv}(A) \subset \bigcup_{p=1}^{P} B(\mathbf{c}_p, \epsilon)^{\circ}$, i.e., $\text{Conv}(A)$ is totally bounded and thus its closure K is compact.

b) Because K is $\| \;\|$-compact and $\mathbf{x}^*$ is $\| \;\|$-continuous, the maximum value of $|(\mathbf{x}, \mathbf{x}^*)|\,\big|_K$ is achieved on K.

c) The map $F : C(A, \mathbb{C}) \ni g \mapsto g(\mathbf{x})$ is a positive functional in $(C(A, \mathbb{C}))^*$ and the Riesz representation theorem implies that for some complex measure $\mu_{\mathbf{x}}$, $F(g) = \int_A g(\mathbf{y})\,d\mu_{\mathbf{x}}(y)$ and $\|\mu_{\mathbf{x}}\| = \|F\|$. Moreover, $\|F\| = 1$ and $F(\mathbf{x}^*) = \int_A (\mathbf{y}, \mathbf{x}^*)\,d\mu_{\mathbf{x}}(\mathbf{y})$. □

6.136. If b) obtains and $\mathbf{w} \in W$ then

$$\mathsf{B}_{\mathbf{w}} : V \ni \mathbf{v} \mapsto \mathsf{B}(\mathbf{v}, \mathbf{w}) \in Z$$

is continuous, whence for some constant $K_{\mathbf{w}}$ and all $\mathbf{v}$, $\|\mathsf{B}_{\mathbf{w}}(\mathbf{v})\| \le K_{\mathbf{w}}\|\mathbf{v}\|$. The hypothesis that B is continuous separately in each argument and the closed graph theorem imply that $T : W \ni \mathbf{w} \mapsto \mathsf{B}_{\mathbf{w}}$ is continuous. Hence for some constant C, $\|\mathsf{B}_{\mathbf{w}}\| \le C\|\mathbf{w}\|$ and so $\|\mathsf{B}(\mathbf{v}, \mathbf{w})\| \le C\|\mathbf{v}\| \cdot \|\mathbf{w}\|$.

Thus b) implies a), which implies B is jointly continuous, which implies b). □

6.137. Because $\mathbf{f}$ is differentiable, if $\mathbf{w}^* \in W^*$ and $\mathbf{v}, \mathbf{h} \in V$ then for the map

$$g : \mathbb{R} \ni t \mapsto (\mathbf{f}(\mathbf{v} + t\mathbf{h}), \mathbf{w}^*),$$

and an a depending on $\mathbf{v} + t_1\mathbf{h}$, $t_2 - t_1$ and $\mathbf{h}$, and approaching zero as $|t_2 - t_1| \to 0$,

$$|g(t_2) - g(t_1)| \le \|\mathbf{w}^*\| \cdot a \cdot \|\mathbf{h}\| \cdot |t_2 - t_1|.$$

Hence g' exists everywhere and $g' \equiv 0$, i.e., $g = g(0) = (\mathbf{f}(\mathbf{v}), \mathbf{w}^*)$. If $\mathbf{f}$ is not constant then for some $\mathbf{x}$ and $\mathbf{h}$ in V, $\mathbf{f}(\mathbf{x} + \mathbf{h}) \ne \mathbf{f}(\mathbf{x})$. The Hahn-Banach theorem implies that for some $\mathbf{w}^*$ in W^*, the corresponding g is not constant, a contradiction. □

6.138. If T is $\| \; \|$-continuous, $\mathbf{x}_n \overset{\text{w}}{\to} \mathbf{O}$, and $T(\mathbf{x}_n) \overset{\text{w}}{\not\to} \mathbf{O}$, it may be assumed that for some positive a, all n, and some $\mathbf{x}^*$ in V^*, $|(T(\mathbf{x}_n), \mathbf{x}^*)| \ge a$, whereas $(T(\mathbf{x}_n), \mathbf{x}^*) = (\mathbf{x}_n, T^*(\mathbf{x}^*)) \to 0$ as $n \to \infty$, a contradiction.

Conversely, if (6.3) is valid, the equation $(\mathbf{v}, S(\mathbf{v}^*)) \overset{\text{def}}{=} (T(\mathbf{v}), \mathbf{v}^*)$ defines S uniquely. Because $\| \; \|$-convergence implies weak convergence,

$$\left\{\mathbf{x}_n \overset{\| \|}{\to} \mathbf{O}\right\} \Rightarrow \left\{T(\mathbf{x}_n) \overset{\text{w}}{\to} \mathbf{O}\right\} \Rightarrow \{(T(\mathbf{x}_n), \mathbf{v}^*) \to 0\}$$

as $n \to \infty$. It follows that $S(\mathbf{v}^*) \in V^*$. The argument in **Solution 6.112** shows that both S and T are $\| \; \|$-continuous. □

6.139. It may be assumed that $\mathbf{x}^* \ne \mathbf{O}$. There are sequences $\{a_n\}_{n\in\mathbb{N}}$, $\{\mathbf{m}_n\}_{n\in\mathbb{N}}$, and $\{\mathbf{y}_n\}_{n\in\mathbb{N}}$ such that

$$a_n \downarrow 0,\ \mathbf{m}_n \in M,\ d(\mathbf{x}, M) > \|\mathbf{x} + \mathbf{m}_n\| - a_n$$
$$\|\mathbf{y}_n\| \le 1,\ |(\mathbf{y}_n, \mathbf{x}^*)| \ge \|\mathbf{x}^*\| - a_n > 0.$$

Then

$$\mathbf{x} - \frac{(\mathbf{x}, \mathbf{x}^*)\,\mathbf{y}_n}{(\mathbf{y}_n, \mathbf{x}^*)} \in M$$
$$d(\mathbf{x}, M) \le \frac{|(\mathbf{x}, \mathbf{x}^*)|}{|(\mathbf{y}_n, \mathbf{x}^*)|} \le \frac{|(\mathbf{x}, \mathbf{x}^*)|}{\|\mathbf{x}^*\| - a_n}$$
$$|(\mathbf{x}, \mathbf{x}^*)| = |(\mathbf{x} + \mathbf{m}_n, \mathbf{x}^*)| \le \|\mathbf{x}^*\| \cdot (d(\mathbf{x}, M) + a_n)$$
$$\frac{|(\mathbf{x}, \mathbf{x}^*)|}{\|\mathbf{x}^*\|} - a_n \le d(\mathbf{x}, M) \le \frac{|(\mathbf{x}, \mathbf{x}^*)|}{\|\mathbf{x}^*\| - a_n}.$$

□

6.140. a) The sequence space (function space)

$$V_1 \overset{\text{def}}{=} \left\{ \mathbf{a} \overset{\text{def}}{=} \{a_n\}_{n\in\mathbb{N}} \ : \ \sum_{n=1}^{\infty} a_n \mathbf{x}_n \ \| \; \|\text{-converges} \right\}$$

normed according to the formula $\|\mathbf{a}\|' \stackrel{\text{def}}{=} \sup_N \left\|\sum_{n=1}^N a_n\mathbf{x}_n\right\|$ is, owing to the uniform boundedness principle, a Banach space and

$$T : V_1 \ni \mathbf{a} \mapsto \sum_{n=1}^{\infty} a_n\mathbf{x}_n \stackrel{\text{def}}{=} \mathbf{x} \in V$$

is a continuous isomorphism between Banach spaces. Hence T^{-1} is an isomorphism and for all n, $|(\mathbf{x},\mathbf{x}_n^*)| \leq \dfrac{2\left\|T^{-1}(\mathbf{x})\right\|'}{\|\mathbf{x}_n\|}$.

b) The biorthogonality relations $(\mathbf{x}_n, \mathbf{x}_m^*) = \delta_{mn}$ flow from the uniqueness of the sequence $\{a_n\}_{n\in\mathbb{N}}$ in $\mathbf{x} = \sum_{n=1}^{\infty} a_n\mathbf{x}_n$. If $\{\mathbf{x}_n,\mathbf{x}_n^*\} \cup \{\mathbf{y},\mathbf{y}^*\}$ is biorthogonal then $(\mathbf{x}_n,\mathbf{y}^*) \equiv 0$ whence $(\mathbf{x},\mathbf{y}^*) \equiv 0 = (\mathbf{y},\mathbf{y}^*) = 1$, a contradiction. Hence B is maximal.

c) and d) The idempotency of each S_N and P_n is also a consequence of the uniqueness of the sequence $\{a_n\}_{n\in\mathbb{N}}$ associated with $\mathbf{x}$. Furthermore, $\|S_n(\mathbf{x})\|$ and $\|P_n(\mathbf{x})\| \leq 2\left\|T^{-1}(\mathbf{x})\right\|'$.

e) The uniform boundedness principle applies. □

6.141. Each S_N is a linear combination of the linear maps P_n (cf. **6.140**), each of which is continuous because each $\mathbf{x}_n^*$ is continuous.

If (6.4) obtains and $\mathbf{y} \in L$ then for some N, $\mathbf{y} = S_N(\mathbf{y}) = S_N^2(\mathbf{y})$. If $\mathbf{v} \in V$ then for some sequence $\{\mathbf{y}_k\}_{k\in\mathbb{N}}$ in L, $\mathbf{y}_k \stackrel{\|\ \|}{\to} \mathbf{y}$ whence

$$\|\mathbf{y} - S_N(\mathbf{y})\| \leq \|\mathbf{y} - \mathbf{y}_k\| + \|\mathbf{y}_k - S_N(\mathbf{y}_k)\| + \|S_N(\mathbf{y}_k - \mathbf{y})\|. \quad \text{(s6.11)}$$

The second term in the right member of (s6.11) is zero and the third term does not exceed $M\|\mathbf{y}_k - \mathbf{y}\|$, whence X is a basis.

If X is a basis then span(X) is dense and **6.140**e) obtains. □

6.142. If X is a basis then no $\mathbf{x}_n$ is $\mathbf{O}$, L is dense, and (6.5) obtains for $K = \sup_N \|S_N\|$: a), b), and c) obtain.

Conversely, if a), b), and c) obtain, let L_N be span $\left(\{\mathbf{x}_n\}_{n=1}^N\right)$. Then $\{\mathbf{x}_1\}$ is linearly independent. If $X_N \stackrel{\text{def}}{=} \{\mathbf{x}_n\}_{n=1}^N$ is linearly independent and X_{N+1} is linearly dependent, there are coefficients $a_1,\ldots,a_{N+1}$ such that

$$\sum_{n=1}^{N+1} a_n\mathbf{x}_n = \mathbf{O}, \sum_{n=1}^{N+1} |a_n| > 0$$

$$0 < \left\|\sum_{n=1}^{N} a_n\mathbf{x}_n\right\| \leq M\left\|\sum_{n=1}^{N+1} a_n\mathbf{x}_n\right\| = 0,$$

a contradiction. Hence X is linearly independent and there is a continuous linear idempotent $R_N : V \ni \mathbf{v} \mapsto R_N(\mathbf{v}) \in L_N$ (cf. **6.123**). The proof in **Solution 6.123** depends on the Hahn-Banach theorem used in the extensions of the coefficient functionals. These extensions can be chosen coherently so that $R_N R_{N+N'} = R_N$.

If $\|\mathbf{x}\| = 1$, $a \in (0,1)$, and $N \in \mathbb{N}$ then for some N' greater than N and some coefficients $a_1, \ldots, a_{N'}$,

$$\left\|\mathbf{x} - \sum_{n=1}^{N'} a_n \mathbf{x}_n\right\| < a,\ R_N\left(\sum_{n=1}^{N'} a_n \mathbf{x}_n\right) = \sum_{n=1}^{N} a_n \mathbf{x}_n$$
$$\left\|R_N\left(\sum_{n=1}^{N'} a_n \mathbf{x}_n\right)\right\| \le M\left\|\sum_{n=1}^{N'} a_n \mathbf{x}_n\right\|.$$

Furthermore, because $\|\mathbf{x}\| = 1$,

$$\begin{aligned}
\|R_N(\mathbf{x})\| &\le \left\|R_N\left(\mathbf{x} - \sum_{n=1}^{N'} a_n \mathbf{x}_n\right)\right\| + \left\|R_N\left(\sum_{n=1}^{N'} a_n \mathbf{x}_n\right)\right\| \\
&\le \|R_N\| \cdot a + M\left\|\sum_{n=1}^{N'} a_n \mathbf{x}_n\right\| \\
&\le \|R_N\| \cdot a + M\left(a + \|\mathbf{x}\|\right) \\
\|R_N\| &\le \frac{M(1+a)}{1-a}.
\end{aligned}$$

In sum, $\|R_N\| \le M, N \in \mathbb{N}$.

If $\mathbf{v} \in V$ and $a > 0$ then for some N in $\mathbb{N}$, any N' greater than N, and for some $\mathbf{y}$ in $L_{N'}$, $\|\mathbf{v} - \mathbf{y}\| < \dfrac{a}{M+1}$. Hence

$$\begin{aligned}
\|R_{N'}(\mathbf{v}) - \mathbf{v}\| &\le \|R_{N'}(\mathbf{v}) - \mathbf{y}\| + \|\mathbf{y} - \mathbf{x}\| = \|R_{N'}(\mathbf{v} - \mathbf{y})\| + \|\mathbf{y} - \mathbf{v}\| \\
&< \frac{(M+1)a}{M+1} = a,
\end{aligned}$$

i.e., $\lim_{N\to\infty} R_N(\mathbf{v}) = \mathbf{v}$. If $P_N \stackrel{\text{def}}{=} R_N - R_{N-1}, N \in \mathbb{N}$, then

$$P_N(\mathbf{v}) = b_N(\mathbf{v})\mathbf{x}_N \text{ and } |b_N(\mathbf{v})| \le 2M\|\mathbf{v}\|.$$

Hence the maps $\mathbf{x}_N^* : V \ni \mathbf{v} \mapsto b_N(\mathbf{v}), n \in \mathbb{N}$ are in V^* and $\{\mathbf{x}_n\}_{n\in\mathbb{N}}$ is a basis. □

6.143. If X is a weak basis for V then for each $\mathbf{v}$ in V, there is a unique sequence $\{a_n\}_{n\in\mathbb{N}}$ such that $\sum_{n=1}^{N} a_n \mathbf{x}_n \stackrel{\text{w}}{\to} \mathbf{v}$. The argument for **6.141** may be used to define a norm $\|\ \|'$ on the sequence space V_1. The uniform boundedness principle applies and the proof given for **6.141** may be repeated, mutatis mutandis. □

6.144. For the associated coefficient functionals $\mathbf{x}_n^*$, if $\epsilon_n \stackrel{\text{def}}{=} \dfrac{1}{2^n \|\mathbf{x}_n^*\|}$ and $\|\mathbf{y}_n - \mathbf{x}_n\| < \epsilon_n, n \in \mathbb{N}$, the following argument shows that $Y \stackrel{\text{def}}{=} \{\mathbf{y}_n\}_{n\in\mathbb{N}}$ is also a basis (cf. **6.87**).

If $\mathbf{z}_n \stackrel{\text{def}}{=} \mathbf{x}_n - \mathbf{y}_n$ and $\mathbf{v}_0 \in V$ then for some r in $(0,1)$,

$$\sum_{n=1}^{\infty} \|\mathbf{z}_n\| \cdot \|\mathbf{x}_n^*\| = r, \ \sum_{n=1}^{\infty} |(\mathbf{v}_0, \mathbf{x}_n^*)| \cdot \|\mathbf{z}_n\| < r \|\mathbf{v}_0\|,$$

i.e., $\mathbf{y} \stackrel{\text{def}}{=} \sum_{n=1}^{\infty} (\mathbf{v}_0, \mathbf{x}_n^*)\, \mathbf{z}_n$ exists.

Furthermore,

$$\begin{aligned}\mathbf{v}_0 &= \sum_{n=1}^{\infty} (\mathbf{v}_0, \mathbf{x}_n^*)\, \mathbf{x}_n = \sum_{n=1}^{\infty} (\mathbf{v}_0, \mathbf{x}_n^*)\, (\mathbf{y}_n + \mathbf{z}_n) \\ &= \sum_{n=1}^{\infty} (\mathbf{v}_0, \mathbf{x}_n^*)\, \mathbf{y}_n + \sum_{n=1}^{\infty} (\mathbf{v}_0, \mathbf{x}_n^*)\, \mathbf{z}_n \stackrel{\text{def}}{=} \mathbf{u}_1 + \mathbf{v}_1,\end{aligned}$$

whence, L denoting span(Y), $\mathbf{u}_1 \in \overline{L}$, and $\|\mathbf{v}_1\| \le r \|\mathbf{v}_0\|$. Similarly,

$$\mathbf{v}_1 = \mathbf{u}_2 + \mathbf{v}_2, \mathbf{u}_2 \in \overline{L}, \text{ and } \|\mathbf{v}_2\| \le r^2 \|\mathbf{v}_0\|.$$

By induction there can be defined a sequence $\{\mathbf{v}_k\}_{k\in\mathbb{N}}$ such that

$$\mathbf{v}_k = \mathbf{u}_{k+1} + \mathbf{v}_{k+1}, \mathbf{u}_{k+1} \in \overline{L}, \text{ and } \|\mathbf{v}_{k+1}\| \le r^{k+1} \|\mathbf{v}_0\|.$$

Thus $\lim_{k\to\infty} \|\mathbf{v}_0 - \mathbf{u}_{k+1}\| = 0$, i.e., $\overline{\text{span}(Y)} = V$.

For a finite sequence $\{a_n\}_{n=1}^{N}$, the biorthogonality of $\{\mathbf{x}_n, \mathbf{x}_n^*\}_{n\in\mathbb{N}}$ implies the fundamental inequality

$$\begin{aligned}\sum_{n=1}^{N} |a_n| \cdot \|\mathbf{z}_n\| &= \left(\sum_{n=1}^{N} \operatorname{sgn}(a_n) \|\mathbf{z}_n\| \mathbf{x}_n^* \right) \left(\sum_{n=1}^{N} a_n \mathbf{x}_n \right) \\ &\le r \left\| \sum_{n=1}^{N} a_n \mathbf{x}_n \right\|. \end{aligned} \tag{s6.12}$$

If $m \le n$ then, because

$$\left\| \sum_{k=1}^{n} a_k \mathbf{x}_k \right\| \le \left\| \sum_{k=1}^{n} \mathbf{y}_k \right\| + \sum_{k=1}^{n} |a_k| \cdot \|\mathbf{z}_k\|,$$

(s6.12) implies $B \stackrel{\text{def}}{=} \|\sum_{k=1}^n \mathbf{x}_n\| \le \frac{1}{1-r}\left\|\sum_{k=1}^n \mathbf{y}_k\right\| \stackrel{\text{def}}{=} \frac{A}{1-r}$. Because X is a basis, for some K, $\|\sum_{k=1}^m a_k\mathbf{x}_k\| \le K\|\sum_{k=1}^n a_k\mathbf{x}_k\|$, whence

$$\begin{aligned}\left\|\sum_{k=1}^m a_k\mathbf{y}_k\right\| &\le \left\|\sum_{k=1}^m a_k\mathbf{x}_k\right\| + \sum_{k=1}^n a_k\|\mathbf{z}_k\| \\ &\le \left\|\sum_{k=1}^m a_k\mathbf{x}_k\right\| + r\left\|\sum_{k=1}^n \mathbf{x}_k\right\| \\ &\le K\left\|\sum_{k=1}^n a_k\mathbf{x}_k\right\| + r\left\|\sum_{k=1}^n \mathbf{x}_k\right\| = (K+r)\left\|\sum_{k=1}^n a_k\mathbf{x}_k\right\| \\ &\le \frac{K+r}{1-r}\left\|\sum_{k=1}^n a_k\mathbf{y}_k\right\|.\end{aligned}$$

Thus Y is a basis by virtue of **6.142**. □

6.145. When $V \stackrel{\text{def}}{=} \mathbb{C}[x]$ is regarded as a subspace of the Banach space $C([0,1],\mathbb{C})$ then $\left\{p_n(x) \stackrel{\text{def}}{=} x^n\right\}_{n\in\mathbb{Z}^+}$ is a countable Hamel basis for V. □

6.146. Let G be $\{0,1\}^{\mathbb{N}}$ regarded as a compact group with respect to component-wise addition (modulo 2). For normalized Haar measure in G, the map $T : G \ni g \stackrel{\text{def}}{=} (\epsilon_1, \epsilon_2, \ldots) \mapsto \sum_{n=1}^\infty \frac{\epsilon_n}{2^n} \in [0,1]$ is measure-preserving and continuous with respect to the product topology for G. Moreover, T is bijective off a null set in G.

a) The maps $T_N : G \ni g \mapsto \sum_{n=1}^N \epsilon_n (\mathbf{x}, \mathbf{x}_n^*)\mathbf{x}_n \stackrel{\text{def}}{=} T_N(g)(\mathbf{x})$ are such that for each g in G, $T_N(g) \in [V]$, and for each $\mathbf{x}$, $T_N(g)(x)$ is a continuous function of g. Hence $T_N(g)$ is a Bochner measurable vector-valued map and both $C_1(\mathbf{x}) \stackrel{\text{def}}{=} \{g : \lim_{N,N'\to\infty} \|T_N(g)(\mathbf{x}) - T_{N'}(g)(\mathbf{x})\| = 0\}$ and its T-image $C(\mathbf{x}) \stackrel{\text{def}}{=} T(C_1(\mathbf{x}))$ are measurable subsets of G.

b) For $\mathbf{z}$ in V, $B(\mathbf{z}) \stackrel{\text{def}}{=} \{g : \lim_{N\to\infty} T_N(g)(\mathbf{z}) \text{ exists}\}$ is measurable and if $Z \stackrel{\text{def}}{=} \{\mathbf{z}_m\}_{m\in\mathbb{N}}$ is dense in V then $B \stackrel{\text{def}}{=} \bigcap_{m\in\mathbb{N}} B(\mathbf{z}_m)$ is measurable and $C_1 \stackrel{\text{def}}{=} \bigcap_{\mathbf{x}\in V} C_1(\mathbf{x}) \subset B$.

However, if $g \in B$, $\mathbf{x} \in V$, $M \stackrel{\text{def}}{=} \sup_{N\in\mathbb{N}} \|T_N(g)\|$, and $a > 0$, there is a $\mathbf{z}_m$ such that $\|\mathbf{z}_m - \mathbf{x}\| < \frac{a}{2M}$. Then

$$\|T_N(g) - T_{N'}(g)\| \le \|(T_N(g) - T_{N'}(g))(\mathbf{z}_m)\| + \frac{a}{2}.$$

Because $g \in B$, $\lim_{N,N'\to\infty} \|(T_N(g) - T_{N'}(g))(\mathbf{z}_m)\| = 0$, whence $g \in C_1$. Hence $B = C_1$ and both C_1 and $C \stackrel{\text{def}}{=} T(C_1)$ are measurable.

For g in C_1 and $\mathbf{x}$ in V let $\mathbf{x}(g)$ be $\sum_{n=1}^{\infty} \epsilon_n (\mathbf{x}, \mathbf{x}_n^*) \mathbf{x}_n$. Then for h in C_1, $\mathbf{x}(g+h) = \mathbf{x}(g)(h)$ whence C_1 is a dense subgroup of G and so C is dense in $[0,1]$.

c) If $\kappa \subset \mathbb{N}$, a κ-cylinder $\mathcal{Z}$ in G is a set determined by a set of subsets $A_k, k \in \kappa$: $\{g : \{n \in \kappa\} \Rightarrow \{\epsilon_n \in A_n\}\}$. If κ is finite $\mathcal{Z}$ is a finite cylinder and if $\kappa = \{n, n+1, \ldots\}$ then $\mathcal{Z}$ is a J_n. If E is measurable and a J_n for each n while F is an arbitrary finite cylinder then $\mu(E) = \mu(E)\mu(F)$. The σ-algebra of measurable subsets of G is generated by the finite cylinders, whence $\mu(E) = (\mu(E))^2$, whence $\mu(E) = 0$ or $\mu(E) = 1$. Thus $\mu(C_1) = 0$ or $\mu(C_1) = 1$ and $\lambda(C) = 0$ or $\lambda(C) = 1$.

d) If $\lambda(C) = 1$ then $\mu(C_1) = 1 = \mu(G)$. Because C_1 is a measurable subgroup of G, it follows that $C_1 = G$ whence $C = [0,1]$. □

6.147. a) If $\frac{a}{n} = 1 - \frac{1}{q} + r$ then $n, q, r, qn - qa, nqr \stackrel{\text{def}}{=} d > 0$ and

$$qn - qa = n - nqr = n - d, n = 1, 2,$$

$$\int_{B(\mathbf{O},b)} |k(\mathbf{x}-\mathbf{y})|^q \, d\lambda_n(\mathbf{x}) \le c^q \int_{B(\mathbf{O},b)} |\mathbf{x}-\mathbf{y}|^{qa-qn} \, d\lambda_n(\mathbf{x})$$

$$\le c^q \int_{B(\mathbf{y},1)^\circ} + c^q \int_{(B(\mathbf{O},b))\setminus B(\mathbf{y},1)^\circ} |\mathbf{x}-\mathbf{y}|^{qa-qn} \, d\lambda_n(\mathbf{x}). \quad \text{(s6.13)}$$

Because $d > 0$, if $n = 1$, the first term in the right member of (s6.13) is

$$\int_{B(\mathbf{y},1)^\circ} |x-y|^{-1+d} \, dx = \int_{-1}^{1} |u|^{-1+d} \, du = \frac{2}{d} < \infty.$$

If $n = 2$, the first term in the right member of (s6.13) is

$$\int_{B(\mathbf{y},1)^\circ} |\mathbf{x}-\mathbf{y}|^{-2+d} \, d\lambda_2(\mathbf{x}) = \int_0^{2\pi} \int_0^1 r^{-1+d} \, dr d\theta = \frac{2\pi}{d} < \infty.$$

Because $g : B(\mathbf{O}, b) \setminus B(\mathbf{y}, 1)^\circ \ni \mathbf{x} \mapsto \|\mathbf{x}-\mathbf{y}\|^{-n+d}$ is continuous and bounded it follows that the second term in the right member of (s6.13) is finite, $n = 1, 2$.

b) The argument in a) shows that $k \in L^q(\mathbb{R}^n, \lambda_n), n = 1, 2$, and since $q > 1$, from **6.43** it follows that for all f in $L^1(\mathbb{R}_n, \lambda_n)$, $k * f \in L^q(\mathbb{R}^n, \lambda_n)$ and $\|k * f\|_q \le \|f\|_1 \cdot \|k\|_q$. □

6.148. The relations $\|af\| = |a| \cdot \|f\|$ and $\|f+g\| \le \|f\| + \|g\|$ are ensured by the basic properties of sup. Furthermore, if $\|f\| = 0$ then f is a constant and because $f(0) = 0$, $f = 0$, whence $\| \ \|$ is a true norm.

If $\{f_n\}_{n\in\mathbb{N}}$ is a Cauchy sequence in E_a then

$$|f_n(s) - f_m(s)| = |f_n(s) - f_n(0) - (f_m(s) - f_m(0))| \le |s|^a \cdot \|f_n - f_m\|,$$

whence $\lim_{n\to\infty} f_n(s) \stackrel{\text{def}}{=} f(s)$ exists. Furthermore, if $s \neq t$ then

$$\frac{|f(s)-f(t)|}{|s-t|^a} = \lim_{n\to\infty} \frac{|f_n(s)-f_n(t)|}{|s-t|^a} \le \lim_{n\to\infty} \|f_n\| < \infty,$$

whence $f \in E_a$.

Finally,

$$\begin{aligned}\frac{|f(s)-f(t)-(f_n(s)-f_n(t))|}{|s-t|^a} &= \lim_{m\to\infty} \frac{|f_m(s)-f_m(t)-(f_n(s)-f_n(t))|}{|s-t|^a} \\ &\le \lim_{m\to\infty} \|f_m - f_n\|,\end{aligned}$$

whence E_a is complete. □

6.149. As in **6.148** it follows that $\| \ \|$ is a norm, whence $\| \ \|'$ is also a norm. If f_n is the function such that the graph of $y = f_n(x)$ is that shown in **Figure s6.1** then $\|f_n\| = n$, $\|f_n\|_\infty = \frac{1}{n}$, $\|f_n\|' = n + \frac{1}{n}$.

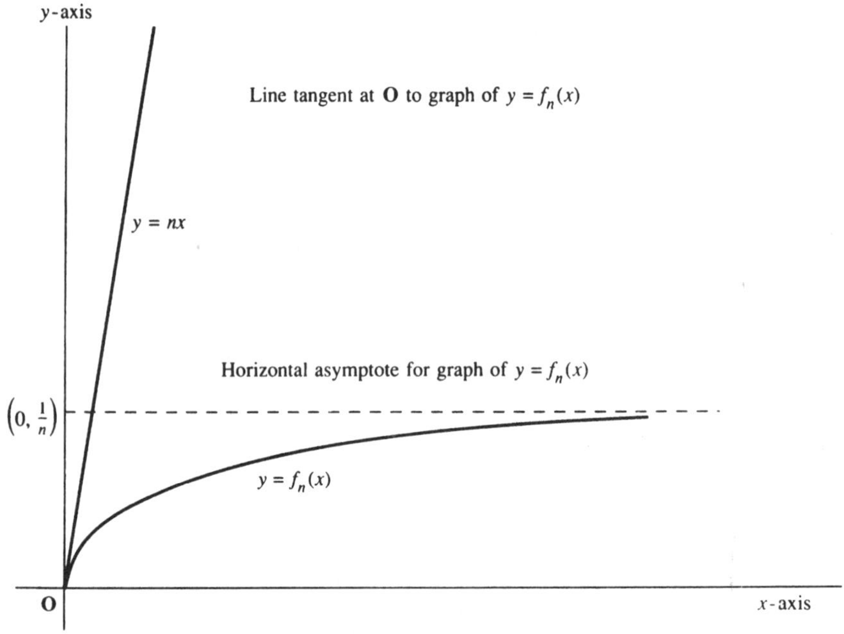

Figure s6.1.

□

6.150. Because K is a bounded closed subset of a finite-dimensional space, K is compact. □

6.151. The relation $L^1(\mathbb{R},\lambda)\cap C_0(\mathbb{R},\mathbb{C})\supset C_{00}(\mathbb{R},\mathbb{C})$ and the $\|\ \|_\infty$-denseness of $C_{00}(\mathbb{R},\mathbb{C})$ in $C_0(\mathbb{R},\mathbb{C})$ show that $L^1(\mathbb{R},\lambda)\cap C_0(\mathbb{R},\mathbb{C})$ is $\|\ \|_\infty$-dense in $C_0(\mathbb{R},\mathbb{C})$.

If $f(x)\equiv\frac{1}{\sqrt{2\pi}}\exp\left(-\frac{x^2}{2}\right)$ then $\widehat{f}(t)=\frac{1}{\sqrt{2\pi}}\exp\left(-\frac{t^2}{2}\right)$, whence $L^1(\mathbb{R},\lambda)\widehat{\ }$, which is a subalgebra of $C_0(\mathbb{R},\mathbb{C})$, contains a separating subalgebra generated by the translates of $\widehat{f}$. The Stone-Weierstraß theorem applies.

□

6.152. a) The map $\mathsf{B}: P_n^2\ni\{p,q\}\mapsto L(p\cdot q)$ defines a positive definite inner product. The Gram-Schmidt orthonormalization process applied to the sequence $\left\{g_k(x)\stackrel{\text{def}}{=}x^k\right\}_{k=0}^n$ produces for P_n a B-orthonormal basis $\{f_k\}_{k=0}^n$. If $p_{n+1}\stackrel{\text{def}}{=}g_{n+1}-\sum_{k=0}^n L(g_{n+1}\cdot f_k)f_k$ then $L(p_{n+1}\cdot P_n)=0$.

b) If p_{n+1} has ν real zeros, then $0\le\nu\le n+1$. If $\nu=0$ then $n+1$ is even, say $n+1=2m$, and there are real numbers $\{r_j,s_j\}_{j=1}^m$ such that $p_{n+1}(x)=\prod_{j=1}^m\left((x-r_j)^2+s_j^2\right)$, a sum of squares of polynomials, whence $L(p_{n+1})>0$, whereas, as found in a), $L(p_{n+1}\cdot 1)=0$.

Thus $1\le\nu$. If the ν real zeros of p_{n+1} are $\{z_j\}_{j=1}^\nu$ then $n+1-\nu$ is even, say $n+1-\nu=2m$, and $p_{n+1}(x)=\prod_{j=1}^\nu(x-z_j)\cdot\prod_{i=1}^m\left((x-r_i)^2+s_i^2\right)$. If $m\ge1$ then

$$Q(x)\stackrel{\text{def}}{=}p_{n+1}(x)\cdot\prod_{j=1}^\nu(x-z_j)=\prod_{j=1}^\nu(x-z_j)^2\cdot\prod_{i=1}^m\left((x-r_i)^2+s_i^2\right),$$

a sum of squares of polynomials, whence $L(Q)>0$. On the other hand, because $\deg\left(\prod_{j=1}^\nu(x-z_j)\right)<n$, a) implies $L(Q)=0$. Hence $m=0$.

c) If a) and b) hold for $\widetilde{p_{n+1}}$, $\deg(\widetilde{p_{n+1}})=n+1$, and $L(\widetilde{p_{n+1}}\cdot P_n)=0$, it may be assumed that the leading coefficient of $\widetilde{p_{n+1}}$ is 1. Thus

$$\begin{gathered}\deg(\widetilde{p_{n+1}}-p_{n+1})\le n\\ 0<L\left((\widetilde{p_{n+1}}-p_{n+1})^2\right)\\ =L(\widetilde{p_{n+1}}(\widetilde{p_{n+1}}-p_{n+1}))-L(p_{n+1}(\widetilde{p_{n+1}}-p_{n+1}))=0-0\end{gathered}$$

whence $\widetilde{p_{n+1}}=p_{n+1}$.

□

6.153. For n in $\mathbb{N}$ and $f_n(x)\stackrel{\text{def}}{=}n\chi_{[0,\frac{1}{n}]}$, $\|f_n\|_{\frac12}=\frac{1}{n}$. Hence for any g in $\left(L^{\frac12}([0,1],\lambda)\right)^*$, $\lim_{n\to\infty}g(f_n)=0$.

If $\sup_{f\in B(\mathbf{O},1)}|g(f)|=\infty$ then for some sequence $\{k_n\}_{n\in\mathbb{N}}$ in $B(\mathbf{O},1)$, $g(k_n)\ge n$. Owing to the continuity of g, $\sqrt{n}\le\left|g\left(\frac{k_n}{\sqrt{n}}\right)\right|\to\mathbf{O}$ as $n\to\infty$, a contradiction. Hence for some constant K and all f in V, $|g(f)| \le K\|f\|_{\frac{1}{2}}$. If $[a,b] \subset [0,1]$,

$$\left\|\chi_{[a,b]}\right\|_{\frac{1}{2}} = (b-a)^2 \text{ and } \left|g\left(\chi_{[a,b]}\right)\right| \le K(b-a)^2$$

whence $g\left(\chi_{[a,b]}\right)$ is a countably additive interval function that may be extended to a measure μ on S_β and $\mu \ll \lambda$. Consequently, for some h in $L^1([0,1],\lambda)$ and all f in V, $g(f) = \int_0^1 f(x)\overline{h(x)}\,dx$. In particular, if $t \in [0,1]$, $g\left(\left(f_n\right)_{[t]}\right) = n\int_t^{t+\frac{1}{n}} \overline{h(x)}\,dx \to 0$ as $n \to \infty$. Hence $h \doteq 0$ and $g(V) = 0$. □

6.154. Viewed as the set of all $L^2(X,\mu)$-valued functions on the set

$$Y \stackrel{\text{def}}{=} \{1,2,\ldots,n\}$$

endowed with counting measure ζ, V is a generalized $L^4(Y,\zeta)$, denoted $L^4\left(Y,\zeta,L^2(X,\mu)\right)$ to emphasize the range of the functions. The Hölder and Minkowski inequalities and the completeness of $L^2(X,\mu)$ ensure that V is a Banach space.

If $g \stackrel{\text{def}}{=} \{g_1,\ldots,g_n\} \in L^{\frac{4}{3}}\left(Y,\zeta,L^2(X,\mu)\right)$ then the map

$$F_g : L^4\left(Y,\zeta,L^2(X,\mu)\right) \ni f \stackrel{\text{def}}{=} \{f_1,\ldots,f_n\} \mapsto \sum_{k=1}^{n}(f_k,g_k)$$

is in V^*. Conversely, if $F \in V^*$ then, $(\mathbf{O},f,\mathbf{O})_k$ denoting the vector for which only the kth component f is nonzero, the map

$$h_k : L^2(X,\mu) \ni h \mapsto F(\mathbf{O},f,\mathbf{O})_k$$

is in $\left(L^2(X,\mu)\right)^*$ $\left[= L^2(X,\mu)\right]$ and $F(f_1,\ldots,f_n) = \sum_{k=1}^n (f_k,h_k)$. The Hölder and Minkowski inequalities imply that $\|F\| = \left(\sum_{k=1}^n \|h_k\|_2^{\frac{4}{3}}\right)^{\frac{3}{4}}$. □

6.155. a) By virtue of the open mapping theorem, T is open and thus for some positive r, $T(B(\mathbf{O},r)) \supset B(\mathbf{O},1)$. Hence, $\mathbf{e}_n$ denoting $\{\delta_{nm}\}_{m\in\mathbb{N}}$, for some $\mathbf{x}_n$ in $B(\mathbf{O},r)$, $T(\mathbf{x}_n) = \mathbf{e}_n$, $n \in \mathbb{N}$. If $T(\mathbf{x}) = \{a_n\}_{n\in\mathbb{N}}$ then

$$\sum_{n=1}^{\infty}|a_n|\cdot\|\mathbf{x}_n\| < \infty,\ \sum_{n=1}^{\infty} a_n\mathbf{x}_n \stackrel{\text{def}}{=} \mathbf{y} \text{ exists, } \mathbf{z} \stackrel{\text{def}}{=} \mathbf{x}-\mathbf{y} \in \ker(T)$$

$$\left\|\sum_{n=1}^{\infty} a_n\mathbf{x}_n\right\| \le \left\|\{a_n\}_{n\in\mathbb{N}}\right\|_1 \cdot r \le \|T\|\cdot\|\mathbf{x}\|\cdot r,$$

whence the map $Q : V \ni \mathbf{x} \mapsto \sum_{n=1}^{\infty} a_n \mathbf{x}_n$ is continuous and linear. Because $Q(\mathbf{x}_n) = \mathbf{x}_n$ it follows that $Q^2 = Q$, i.e., that Q is a continuous idempotent. Thus $\mathrm{id} - Q \stackrel{\text{def}}{=} P$ is also a continuous idempotent and $P(V) = \ker(T)$.

b) Because $\mathbf{x} = P(\mathbf{x}) + Q(\mathbf{x}) \stackrel{\text{def}}{=} \mathbf{u} + \mathbf{v}$ and $P(V) \cap Q(V) = \{\mathbf{O}\}$, the direct sum $Z \stackrel{\text{def}}{=} \ker(T) \oplus \mathrm{im}(Q)$ is defined. As the image of a projection, $Q(V)$ is closed, hence a Banach space, and $F : Z \ni \{\mathbf{x}, \mathbf{y}\} \mapsto \mathbf{x} + \mathbf{y} \in V$ is a continuous surjection, whence F is open. □

6.156. If $K = 1$, it may be assumed that $\mathbf{y}_1 \notin M$. The Hahn-Banach theorem implies that for some $\mathbf{y}^*$ in $M^\perp$, $(\mathbf{y}_1, \mathbf{y}^*) = 1$. By definition, each $\mathbf{y}$ in $M_1 \stackrel{\text{def}}{=} \mathrm{span}(\mathbf{y}_1, M)$ is for some a in $\mathbb{C}$ and some $\mathbf{m}$ in M, uniquely of the form $\mathbf{y} = a\mathbf{y}_1 + \mathbf{m}$. If $\{\mathbf{z}_n\}_{n\in\mathbb{N}} \subset M_1$ and $\mathbf{z}_n = a_n\mathbf{y}_1 + \mathbf{m}_n \to \mathbf{z}$ as $n \to \infty$, then $a_n = (\mathbf{z}_n, \mathbf{y}^*) \to (\mathbf{z}, \mathbf{y}^*) \stackrel{\text{def}}{=} a$ and $\mathbf{m}_n = \mathbf{z}_n - a_n\mathbf{y}_1 \to \mathbf{z} - a\mathbf{y}_1$ as $n \to \infty$. Because M is closed, $\mathbf{m} \in M$ whence M_1 is closed. Mathematical induction implies the result for any finite K. □

6.157. a) If $s_m(x) \stackrel{\text{def}}{=} \sum_{n=1}^{m} |a_n| \cdot |f_n(x)|$ then $s_m \le s_{m+1}$ and $\{s_m\}_{m\in\mathbb{N}}$ is a $\| \ \|_1$-Cauchy sequence with a limit s in $L^1(X, \mu)$. Hence it follows that $F(x) \stackrel{\text{def}}{=} \sum_{n=1}^{\infty} a_n f_n(x)$ exists a.e., $F \in L^1(X, \mu)$, and $\|F\|_1 \le \|s\|_1$.

b) If $a_n = r_n e^{i\theta_n}$ and $f_n(x) = |f_n(x)| e^{i\phi_n(x)}$, $0 \le \theta_n, \phi_n < 2\pi$ then (6.6) holds iff for all n in $\dot{\mathbb{N}}$, $\theta_n + \phi_n(x) \doteq 0 \pmod{2\pi}$.

c) The equation (6.6) holds for all sequences $\{f_n\}_{n\in\mathbb{N}}$ iff at most one a_n is different from zero.

d) The equation (6.6) holds for all sequences $\{a_n\}_{n\in\mathbb{N}}$ iff at most one f_n differs from zero a.e. □

6.158. Because V and W are separable, $\#(V) = \#(W) = \mathfrak{c}$. From **6.114** it follows that the cardinality of a Hamel basis for V or W is $\mathfrak{c}$. Any bijection between Hamel bases for V and W may be extended to an isomorphism of V and W. □

6.159. Because $\| \ \|_\infty$-convergence is uniform convergence, V is closed. The convexity of V follows from direct calculation.

Because $1 = \left| \int_0^{\frac{1}{2}} f(x)\,dx - \int_{\frac{1}{2}}^{1} f(x)\,dx \right| \le \|f\|_\infty$, each f in V is outside $B(\mathbf{O}, 1)^\circ$. Furthermore, if $\delta, \epsilon, \eta > 0$ and $\eta = \frac{\epsilon}{2} + \delta(1 + \epsilon)$, the function

$$f_\epsilon(x) = \begin{cases} 1 + \epsilon & \text{if } 0 \le x \le \frac{1}{2} - \delta \\ -\frac{1+\epsilon}{\delta}\left(x - \frac{1}{2}\right) & \text{if } \frac{1}{2} - \delta < x < \frac{1}{2} + \delta \\ -(1 + \epsilon) & \text{if } \frac{1}{2} + \delta \le x \le 1 \end{cases}$$

is in V and $\|f_\epsilon\|_\infty = 1+\epsilon$ whence $\inf\{\|f\|_\infty : f \subset V\} = 1$. However, if $g \in V$ and $\|g\|_\infty = 1$ then

$$\int_0^{\frac{1}{2}} g(x)\,dx \le \frac{1}{2},\ -\int_{\frac{1}{2}}^1 g(x)\,dx \le \frac{1}{2},\ \int_0^{\frac{1}{2}} g(x) - \int_{\frac{1}{2}}^1 g(x)\,dx \le 1$$

and equality obtains iff $g(x)\big|_{[0,\frac{1}{2}]} \equiv 1$, $g(x)\big|_{[\frac{1}{2},1]} \equiv -1$, which is impossible if g is continuous. □

6.160. By definition, $M_U(\mathbf{x}) \ge 0 = M_U(\mathbf{O})$. If $0 \ne t = |t|e^{i\theta} \in \mathbb{C}$ then $\mathbf{x} \in \alpha U$ iff $t\mathbf{x} \in |t|\alpha e^{i\theta}U$ $(= |t|\alpha U$ because U is circled). Thus $M_U(t\mathbf{x}) = |t|M_U(\mathbf{x})$.

If $\beta, \gamma > 0$, $\mathbf{x} = \beta\mathbf{u} \in \beta U$, $\mathbf{y} = \gamma\mathbf{v} \in \gamma U$ then $M_U(\mathbf{x}) = \beta$, $M_U(\mathbf{y}) = \gamma$, and, because U is convex,

$$\mathbf{x}+\mathbf{y} = (\beta+\gamma)\left(\frac{\beta}{\beta+\gamma}\mathbf{u} + \frac{\gamma}{\beta+\gamma}\mathbf{v}\right) \in (\beta+\gamma)U.$$

Hence $M_U(\mathbf{x}+\mathbf{y}) \le \beta+\gamma = M_U(\mathbf{x}) + M_U(\mathbf{y})$. □

6.161. a) ⇒ b). If $\epsilon > 0$ and for all $\mathbf{v}$ in $N(\mathbf{O})$, $p(\mathbf{v}) < \epsilon$ then when $\mathbf{x}-\mathbf{y} \in N(\mathbf{O})$, the triangle inequality implies $|p(\mathbf{x})-p(\mathbf{y})| \le |p(\mathbf{x}-\mathbf{y})| < \epsilon$.

b) ⇒ c). Because $U = p^{-1}((-\infty,1))$ and p is continuous, U is open. The basic properties of a seminorm assure that U is convex and circled.

c) ⇒ d). If $M_U(\mathbf{x}) = \beta$ and $\alpha > \beta$ then $\mathbf{x} \in \alpha U$, whence $\frac{\mathbf{x}}{\alpha} \in U$, whence $p\left(\frac{\mathbf{x}}{\alpha}\right) < 1$, whence $p(\mathbf{x}) < \alpha$, whence $p(\mathbf{x}) \le \beta = M_U(\mathbf{x})$. If $p(\mathbf{x}) = \beta$ and $\alpha > \beta$ then $p\left(\frac{\mathbf{x}}{\beta}\right) < 1$, whence $\frac{\mathbf{x}}{\beta} \in U$, whence $\mathbf{x} \in \beta U$, whence $M_U(\mathbf{x}) \le \beta = p(\mathbf{x})$. In sum: $p \le M_U \le p$.

d) ⇒ a). If $\epsilon > 0$ and $\mathbf{x} \in \frac{\epsilon}{2}U$ then $M_U(\mathbf{x}) = p(\mathbf{x}) \le \frac{\epsilon}{2} < \epsilon$ whence p is continuous at $\mathbf{O}$. □

6.162. If $\|\ \|$ induces the topology of E then $U \stackrel{\text{def}}{=} \{\mathbf{x} : \|\mathbf{x}\| \le 1\}$ is a convex circled neighborhood of $\mathbf{O}$. The set U is bounded because if $r > 0$ then $\frac{2}{r}B(\mathbf{O},r) \supset U$.

Conversely, if $N(\mathbf{O})$ is bounded and convex then for some circled neighborhood W of $\mathbf{O}$, $W+W \subset N(\mathbf{O})$ and $\operatorname{Conv}(W) \stackrel{\text{def}}{=} U \subset N(\mathbf{O})$. Hence U is also bounded. The Minkowski functional M_U is a seminorm p. If $\mathbf{x} \ne \mathbf{O}$ then, because E is a Hausdorff space, for some neighborhood Z of $\mathbf{O}$, $\mathbf{x} \notin Z$. Because U is bounded, for some positive λ, $U \subset \lambda Z$, i.e., $\frac{1}{\lambda}U \subset Z$ whence $\mathbf{x} \notin \frac{1}{\lambda}U$ and so $M_U(\mathbf{x}) = p(\mathbf{x}) > \frac{1}{\lambda} > 0$, i.e., p is a norm, say $\|\ \|$. Because U is bounded and $U = B(\mathbf{O},1)$, it follows that for any neighborhood Y

of $\mathbf{O}$, there is a positive λ such that $B\left(\mathbf{O}, \frac{1}{\lambda}\right) \subset Y$: the $\| \ \|$-topology is stronger than the given topology. On the other hand, for each n in $\mathbb{N}$, there is an $N(\mathbf{O})_n$ such that $nN(\mathbf{O})_n \subset U^\circ$, whence if $\mathbf{x} \in N(\mathbf{O})_n$ then $\|\mathbf{x}\| < \frac{1}{n}$, i.e., p is continuous at $\mathbf{O}$. Thus $B(\mathbf{O}, 1)^\circ$ is open (cf. **6.161**) and so the $\| \ \|$-topology is weaker than the given topology. □

6.4. Banach Algebras

6.163. Because $\mathbf{n}^k\mathbf{x} - \mathbf{x} = \mathbf{n}\left(\mathbf{n}^{k-1} - \mathbf{x}\right) + \mathbf{n}\mathbf{x} - \mathbf{x}$, the result follows by induction. □

6.164. a) If $z = re^{i\theta}, r < 1$ then $f(z) = \sum_{n=1}^\infty a_n z^n = \sum_{n=1}^\infty a_n r^n e^{in\theta}$. The Cauchy formula implies that for n in $\mathbb{Z}$,

$$a_n = \frac{1}{2\pi}\int_0^{2\pi} f\left(re^{i\theta}\right) e^{-in\theta} r^{-n}\, d\theta,$$

while $\widehat{g}_n = \frac{1}{\sqrt{2\pi}}\int_0^{2\pi} g\left(e^{i\theta}\right) e^{-in\theta}\, d\theta$. Because f is uniformly continuous on $D(0,1)$, it follows that $\widehat{g}_n = \frac{1}{\sqrt{2\pi}} a_n$. Note that $a_n = 0$ if $n < 0$.

b) Because g is continuous, Fejér's theorem implies that g is the uniform limit of a sequence $\{p_n\}_{n\in\mathbb{N}}$ contained in $\mathbb{C}\left[e^{i\theta}\right]$. The maximum modulus principle implies that $\|f - p_n\|_\infty \to 0$ as $n \to \infty$. □

6.165. For $p_1 : D(0,1) \ni z \mapsto z$, if $h(p_1) = z_h$ then $h\left(p_1^n\right) = z_h^n$ and so for any polynomial p, $h(p) = p(z_h)$. Hence **6.164**b) applies. □

6.166. Direct calculation shows that $\|f^{n*}\|_\infty \le \frac{\|f\|_\infty^n}{(n-1)!}$. Consideration of Maclaurin's series for e^x shows that $\lim_{n\to\infty}\left(\frac{1}{n!}\right)^{\frac{1}{n}} = 0$. □

6.167. Let M be $\ker(h)$, a closed maximal ideal. If, for each a in X and some f_a in M, $f(a)(a) \ne 0$ then in some $N(a)$, $f_a(x) \ne 0$. However, X is compact and $\bigcup_{a\in X} N(a) \supset X$, whence for some $\{a_i\}_{1\le i\le n<\infty}$, $\bigcup_{i=1}^n N(a_i) \supset X$. Then $|f_{a_i}|^2 = \overline{f_{a_i}} f_{a_i}$ and $0 < \sum_{i=1}^n |f_{a_i}|^2 \stackrel{\text{def}}{=} f \in M$, whence $\frac{1}{f} \in A(X)$ and $\mathbf{1} \in M$, a contradiction. Thus for some a_h in X and for all f in M, $f(a_h) = 0$. Furthermore, $J \stackrel{\text{def}}{=} \{k \ : \ k \in A(X), k(a_h) = 0\}$ is a proper ideal containing M, i.e., $J = M$. Thus $h' : A(X) \ni g \mapsto g(a_h) \in \mathbb{C}$ and h have the same kernel, i.e., $h = h'$. □

[**Note s6.14:** If X is locally compact the result is still valid. The argument bears on the construction of the algebra A_e obtained by adjoining a formal identity e to A:

$$A_e \stackrel{\text{def}}{=} \{\lambda e + f \; : \; \lambda \in \mathbb{C}, f \in A\}.$$

The crucial point is the existence of a bijection between $\mathcal{M}(A)$ and $\mathcal{M}(A_e) \setminus \{A\}$ [**Loo**].]

6.168. a) The translation-invariance of μ implies

$$\begin{aligned}\widehat{f * g}(\alpha) &= \int_G \left(\int_G f\left(y^{-1}x\right) g(y)\, d\mu(y)\right) \overline{(\alpha, x)}\, d\mu(x) \\ &= \int_G \left(\int_G f(z) g(y) \overline{(\alpha, yz)}\, d\mu(y)\right) d\mu(z) \\ &= \int_G f(z)\overline{(\alpha, z)} \left(\int_G g(y)\overline{(\alpha, y)}\, d\mu(y)\right) d\mu(z) = \widehat{f}(\alpha)\widehat{g}(\alpha).\end{aligned}$$

b) If $h(f) \neq 0$ and $g \in L^1(G, \mu)$ then $f * g_{[x]} = f_{[x]} * g$ and so $h(f)h\left(g_{[x]}\right) = h\left(f_{[x]}\right) h(g))$ whence $\alpha_h(x) \stackrel{\text{def}}{=} \dfrac{h\left(f_{[x]}\right)}{h(f)}$ is f-free. Furthermore, $\dfrac{h\left(f_{[xy]}\right)}{h(f)} = \left(\dfrac{h\left(f_{[x]}\right)}{h(f)}\right) \cdot \left(\dfrac{h\left(f_{[y]}\right)}{h(f)}\right)$, i.e., $\alpha_h(xy) = \alpha_h(x) \cdot \alpha_h(y)$. Because $|\alpha_h(x)| \leq \dfrac{\|f\|_1}{h(f)}$, α_h is a bounded function of x. If $|\alpha_h(x)| > 1$ then $|\alpha_h(x^n)| \uparrow \infty$, a contradiction. Hence $|\alpha_h| \leq 1$. On the other hand, $\alpha_h(x)\alpha_h\left(x^{-1}\right) = \alpha_h(e) = 1$ whence $|\alpha_h| \equiv 1$.

The Riesz representation theorem implies that for some β in $L^\infty(B, \mu)$, $h(f) = \int_G f(x)\overline{\beta(x)}\, d\mu(x)$. If n is an approximate identity for A, then $h\left(n_{[x]}\right) = h(n)\alpha_h(x)$. Because $n * g - g \stackrel{\|\ \|_1}{\to} \mathbf{O}$, it follows that

$$|h(g)\left(h(n) - 1\right)| \leq \|n * g - g\| \to 0,$$

i.e., $h(n) \to 1$ and, for h fixed, $h\left(n_{[x]}\right) \stackrel{\text{u}}{\to} \alpha_h(x)$. Hence for any f in $L^1(G, \mu)$,

$$\begin{aligned}\int_G f(x)\overline{\alpha_h(x)}\, d\mu(x) &= \lim_n \int_G f(x) h\left(n_{(x^{-1})}\right) d\mu(x) \\ &= \lim_n \int_G \left(\int_g f(x) n\left(x^{-1}y\right) \overline{\beta(y)}\, d\mu(y)\right) d\mu(x) \\ &= \lim_n \int_G (f * n)(y)\overline{\beta(y)}\, d\mu(y) \\ &= \int_G f(y)\overline{\beta(y)}\, d\mu(y) = h(f).\end{aligned}$$

□

6.169. If G is discrete then $\chi_{\{e\}}$ is the identity for $A(G)$. Conversely, if u is an identity for $A(G)$ and for each neighborhood $U(e)$, $n_U \overset{\text{def}}{=\!=} \frac{1}{\mu(U)}\chi_U$ then $U \mapsto n_U$ is an approximate identity. Thus $u * n_U = n_U \overset{\|\ \|_1}{\to} u$. Hence $u(x) = 0$ off $\{e\}$, whence $\mu(e) > 0$. Thus, since the measure of every compact set is finite, a set is compact iff it is empty or finite. Note that G is locally compact. $\square$

6.170. Let $\mathbf{u}$ be an identity modulo I. If $\mathbf{x} \in I$ and $\|\mathbf{u} - \mathbf{x}\| \overset{\text{def}}{=} \|\mathbf{z}\| < 1$ then $\mathbf{w} \overset{\text{def}}{=} -\sum_{n=1}^{\infty} \mathbf{z}^n$ exists and

$$(\mathbf{u} - \mathbf{x})\mathbf{w} - (\mathbf{u} - \mathbf{x}) - \mathbf{w} = \mathbf{z}\mathbf{w} - \mathbf{z} - \mathbf{w} = \mathbf{O}.$$

Since $\mathbf{x}, \mathbf{x}\mathbf{w}, \mathbf{u}\mathbf{w} - \mathbf{w} \in I$, it follows that $\mathbf{u} \in I$, a contradiction. Hence the distance of $\mathbf{u}$ from I is at least one and so $\mathbf{u} \notin \overline{I}$, in particular, $\overline{I} \subsetneqq A$.

The continuity of the algebraic operations implies that if $\mathbf{x}, \mathbf{y} \in \overline{I}$ and $\mathbf{z} \in A$ then $\mathbf{x} + \mathbf{y}, \mathbf{z}\mathbf{x} \in \overline{I}$ and, as a proper subalgebra of A, $\overline{I}$ is an ideal. $\square$

6.171. Let I be the set of bounded functions in $A(G)$. If $f \in A(G)$ and $b \in I$, then $g \overset{\text{def}}{=} f * b \in A(G)$ and $|g(x)| \le \|b\|_\infty \cdot \|g\|_1 < \infty$, whence I is an ideal. Because every f in $A(G)$ is the limit of simple functions and these are in I, $\overline{I} = A(G)$. $\square$

6.172. Generally, if $T, S \in [A]$ and $TS - ST$ and S commute, the endomorphism $\Delta : [A] \ni T \mapsto TS - ST$ is a derivation on $[A]$. Because $TS - ST$ and S commute, $\Delta^2(T) = O$ whence by Leibniz's rule, $\Delta^k\left(T^l\right) = O$ if $k > l$. On the other hand, $\Delta^2\left(T^2\right) = 2\left(\Delta(T)\right)^2$, whence by induction and Leibniz's rule, it follows that $\Delta^n\left(T^n\right) = n!\left(\Delta(T)\right)^n$. However if M is the norm of the operator Δ, then $\|\Delta^n\left(T^n\right)\| \le M^n\|T\|^n$ whence

$$\lim_{n\to\infty} \|\left(\Delta(T)\right)^n\|^{\frac{1}{n}} \le \lim_{n\to\infty} \frac{M\|T\|}{n!^{\frac{1}{n}}} = 0,$$

i.e., $\Delta(T)$ is a generalized nilpotent in $[A]$.

For $\mathbf{x}$ in A, let $\mathbf{x}'$ denote $D(\mathbf{x})$ and let $R_{\mathbf{x}}$ denote the endomorphism $A \ni \mathbf{z} \mapsto \mathbf{x}\mathbf{z}$.

If Δ is the map $D \mapsto DR_{\mathbf{x}} - R_{\mathbf{x}}D$, the conclusion above is that $\Delta(D)$ $(= R_{\mathbf{x}'})$ is a generalized nilpotent in $[A]$ and thus that $\mathbf{x}'$ is a generalized nilpotent in A. Since A is commutative the result follows. $\square$

[**Note s6.15:** The argument above permits the interesting conclusion that $C^\infty(\mathbb{R}, \mathbb{C})$, regarded as an algebra A with respect to the natural definitions of addition and multiplication, cannot be endowed with a norm $\|\ \|$ with respect to which A is a Banach algebra.

Indeed, if A is a Banach algebra with respect to some norm $\| \ \|$ then the Banach algebra $[A]$ of continuous endomorphisms of A consists of bounded endomorphisms $T : A \ni f \mapsto T(f) \in A$ such that $\|T\| \stackrel{\text{def}}{=} \sup\{ \|T(f)\| \ : \ \|f\| = 1 \} < \infty$. The maps

$$D : f \mapsto D(f) \stackrel{\text{def}}{=} f' \text{ and, for } g \text{ in } A, \ R_g : f \mapsto gf$$

are in $[A]$ and D is a derivation: $D(fg) = D(f)g + fD(g)$.

Because $DR_g - R_gD = R_{g'}$ and $R_gR_{g'} = R_{g'}R_g$, it follows that R_g and $DR_g - R_gD$ commute.

Fix t_0 in $\mathbb{R}$ and let ϕ_{t_0} be the functional that maps each f in A into $f'(t_0)$.

Because $f \mapsto f(t_0)$ and $f \mapsto f(t_0 + \frac{1}{n})$ are in $[A, \mathbb{C}]$, they are $\| \ \|$-continuous. Thus for n in $\mathbb{N}$,

$$L_n : A \ni f \mapsto \frac{f(t_0 + \frac{1}{n}) - f(t_0)}{\frac{1}{n}} \stackrel{\text{def}}{=} L_n(f)$$

are $\| \ \|$-continuous. For all f in A, $L_n(f) \to \phi_{t_0}(f)$ as $n \to \infty$, whence each ϕ_{t_0} is a $\| \ \|$-continuous linear functional. The closed graph theorem **[Rud]** implies that D is $\| \ \|$-continuous: $D \in [A]$.

If $t \in \mathbb{R}$, the evaluation map $E_t : A \ni f \mapsto f(t) \in \mathbb{C}$ is in $[A, \mathbb{C}]$ whence $|E_t(f)| = |f(t)| \le \|f\|$. Thus $\|f\|_\infty \le \|f\|$. Hence in the context of **6.172**

$$\|(g')^n\|^{\frac{1}{n}} \ge \| (g')^n \|_\infty^{\frac{1}{n}} = \|g'\|_\infty$$

whence $g' = 0$. Because g is an arbitrary element of A, it follows that $D = 0$, a contradiction **[SiW]**.

An alternative to the argument in the last paragraph stems from the characterization of the set $\mathcal{N}$ of generalized nilpotents of a commutative Banach algebra B as the radical $\mathcal{R}(\mathrm{B})$ of B, i.e., $\mathcal{N}$ is the intersection of the kernels of all algebraic homomorphisms of B into $\mathbb{C}$: $\mathcal{N} = \mathcal{R} = \bigcap_{h \in [B, \mathbb{C}]} \ker(h)$ **[Be, Loo]**. When $B = A$, among the elements of $[B, \mathbb{C}]$ are the evaluation maps E_t described above. Hence the generalized nilpotent g' vanishes at each t in $\mathbb{R}$, i.e., $g' = 0$, and the contradiction achieved earlier is repeated.]

COMPLEX ANALYSIS:

SOLUTIONS

7
Elementary Theory

7.1. Geometry in $\mathbb{C}$

7.1. A circle on Σ is the intersection of Σ with a plane Π for which the equation is $a\xi + b\eta + c\zeta = a^2 + b^2 + c^2$. The plane Π and Σ intersect iff $a^2 + b^2 + c^2 \le 1$. The equation $\xi^2 + \eta^2 + \zeta^2 = 1$ and the formulæ for the coordinates of $\Theta(\xi, \eta, \zeta) \stackrel{\text{def}}{=} (x, y)$ lead to the equation

$$\left(a^2 + b^2 + c^2 - c\right)\left(x^2 + y^2\right) - 2ax - 2by + a^2 + b^2 + c^2 + c = 0$$

representing a circle in $\mathbb{C}$ or, if $a^2 + b^2 + c^2 = c$, a straight line in $\mathbb{C}$. The latter circumstances imply that Π passes through $(0,0,1)$. The reasoning is reversible and leads from a circle in $\mathbb{C}$ to a circle on $\Sigma \setminus \{(0,0,1)\}$ or from a straight line in $\mathbb{C}$ to a circle passing through $(0,0,1)$ on Σ. □

7.2. a) Because $\Delta(pqr) = \{\alpha p + \beta q + \gamma r \,:\, \alpha, \beta, \gamma \ge 0, \alpha + \beta + \gamma = 1\}$, the convex hull of the set $\{p, q, r\}$, $(\Delta(pqr))^\circ$ corresponds to the supplementary conditions $\alpha\beta\gamma > 0$ and the sides $[pq]$ resp. $[qr]$, resp. $[rp]$ of $\Delta(pqr)$ correspond to the supplementary conditions $\gamma = 0$ resp. $\alpha = 0$ resp. $\beta = 0$.

b) If $1 > r > \max\{|p|, |q|\}$, choose z so that $|1 - z| = \sqrt{1 - r^2} \stackrel{\text{def}}{=} r_1$ and $|z| = r$. The lines $\{1, z\}$ resp. $\{1, \overline{z}\}$ are tangent to $C_0(r)$ at z resp. $\overline{z}$. Moreover, if $s \in D(0, r)$ then $R(s) \le \dfrac{1 + r}{1 - r} \stackrel{\text{def}}{=} K_1$.

Let $2\theta_0$ denote the size of the angle $\angle(z\,1\,\overline{z})$. Then $0 < 2\theta_0 < \pi$ and if $s \in \Delta(z\,\overline{z}\,1)$ then for some ϕ in $[\pi - \theta_0, \pi + \theta_0]$ and ρ in $(0, r_1]$, $1 - s \stackrel{\text{def}}{=} \rho e^{i\phi}$. Thus

$$|s|^2 = 1 + \rho^2 - 2\rho\cos\phi \le 1 + \rho\cos\theta_0 - 2\rho\cos\theta_0 \le \left(1 - \frac{\rho\cos\theta_0}{2}\right)^2$$

and so $1 - |s| > \dfrac{\rho\cos\theta_0}{2}$, $R(s) \le \dfrac{2}{\cos\theta_0} \stackrel{\text{def}}{=} K_2$. Because

$$\Delta(pq1) \subset D(0, r) \cup \Delta(z\,\overline{z}\,1)$$

it follows that $R(s) \le \max\{K_1, K_2\} \stackrel{\text{def}}{=} K(p, q)$. □

7.3. a) If $T_1 \stackrel{\text{def}}{=} T_{abcd}$ then $T_1(z) = z$ iff z is a solution of the quadratic equation $cz^2 + (d - a)z - b = 0$. There is no solution iff $c = d - a = 0 \ne b$. [If $c = d - a = b = 0$ then $T_1 = \text{id}$ and every z is a fixed point.]

b) If T_1 fixes no point in $\mathbb{C}$ then $\widetilde{T_1}$ fixes $(0,0,1)$ in Σ.

c) If $T_1T_2 = T_2T_1$ then $T_1T_2(z_1) = T_2(z_1)$ whence $T_2(z_1) = z_1$, e.g., $z_1 = z_2$. Furthermore, $T_2T_1(z_3) = T_1(z_3)$ whence α) $T_1(z_3) = z_1$ or β) $T_1(z_3) = z_3$. Each of α) and β) implies the contradiction $z_3 = z_1$.

[**Note s7.1:** When X is a set and G is a group, G is said to act on X iff for α, β in G and x in X, $\alpha(x) \in X$ and $\alpha(\beta(x)) = (\alpha\beta)(x)$. If 1. G acts on X, 2. α fixes exactly one point, and 3. β fixes exactly two points then $\alpha\beta \neq \beta\alpha$. The argument is in no way different from that given in c).

On the other hand, in the context of c), an alternative (elementary?) proof can be constructed on the basis of the following directly verifiable statements and references:

i. For $T \stackrel{\text{def}}{=} T_{abcd}$, $T(z) = z$ iff $cz^2 + (d-a)z - b = 0$ [**Solution 7.3**a)].

ii. $(d_1 - a_1)^2 + 4b_1c_1 = 0 \neq (d_2 - a_2)^2 + 4b_2c_2 \neq 0$.

iii. If $S = T_{pqrs}$ then $ST = TS$ iff

$$\begin{pmatrix} a & b \\ c & d \end{pmatrix}\begin{pmatrix} p & q \\ r & s \end{pmatrix} = \begin{pmatrix} p & q \\ r & s \end{pmatrix}\begin{pmatrix} a & b \\ c & d \end{pmatrix}$$

[**Solution 7.12**b)].]

d) If T_1 (and hence T_2) has no fixed point then $c_1 = c_2 = 0$ and each T_i is a translation whence $T_1T_2 = T_2T_1$. If T_1 and T_2 have fixed points and they are shared then $c_1c_2 \neq 0$. Since if $\alpha \neq 0$ then $T_{abcd} = T_{(\alpha a)(\alpha b)(\alpha c)(\alpha d)}$, it may be assumed that $c_1 = c_2 = 1$. Because a fixed point of T_i is a zero of $z^2 + (d_i - a_i)z - b_i$ it follows that $d_1 - a_1 = d_2 - a_2 \stackrel{\text{def}}{=} \delta$ and $b_1 = b_2 \stackrel{\text{def}}{=} b$. Direct calculation then shows that $T_1T_2 = T_2T_1$.

e) Since $|T'(z)| = |cz+d|^{-2}$, it follows that the equation of the isometric circle is $|cz+d| = 1$. The center is at $-\dfrac{d}{c}$ and the radius is $\dfrac{1}{|c|}$. □

7.4. Direct calculation shows that

$$\begin{aligned}\frac{(w-w_1)(w_2-w_3)}{(w-w_2)(w_1-w_3)} &\stackrel{\text{def}}{=} \frac{Pw+Q}{Rw+S} \stackrel{\text{def}}{=} M(w)\\ &= \frac{(z-z_1)(z_2-z_3)}{(z-z_2)(z_1-z_3)} \stackrel{\text{def}}{=} \frac{pz+q}{rz+s} \stackrel{\text{def}}{=} N(z)\end{aligned}$$

establishes a bijective correspondence between the extended z-plane and the extended w-plane, and that

$$w = \frac{(Sp-Qr)z + (Sq-Qs)}{(Pr-Rp)z + (Ps-Rq)} \stackrel{\text{def}}{=} \frac{az+b}{cz+d} \stackrel{\text{def}}{=} T(z).$$

It follows that $N(z_1) = 0$, resp. $\dfrac{1}{N(z_2)} = 0$ whence $M(T(z_1)) = 0$ resp. $\dfrac{1}{M(T(z_2))} = 0$ and so $w = w_1$ resp. $w = w_2$. If $z = z_3$ then $N(z) = 1$ whence $M(w) = 1$. Because the maps M, N are Möbius transformations the equation $M(w) = 1$ has the unique solution $w = w_3$.

If Z is a Möbius transformation and $Z(z_i) = w_i, 1 \le i \le 3$, then $ZT^{-1}(w_i) = w_i$ whence ZT^{-1} is a Möbius transformation with three fixed points and so $ZT^{-1} = \mathrm{id}$. In a word, T is unique. □

7.5. For Z as in **Solution 7.4**,

$$ZT^{-1}\{(T(q), T(r), T(s))\} = (Z(q), Z(r), Z(s)) = (0, \infty, 1)$$

whence

$$X(T(p), T(q), T(r), T(s)) = ZT^{-1}(T(p)) = Z(p)) = X(p, q, r, s).$$ □

7.6. a) Direct calculation shows that $|z-p| = k|z-q|$ is the equation of a circle $C_a(r)$ resp. a straight line L according as $k \neq 1$ resp. $k = 1$. When $k \neq 1$ then $r = \dfrac{k|p-q|}{|1-k^2|}$, $a = \dfrac{p-k^2q}{1-k^2}$, and $p = q^\rho$. When $k = 1$ and ϕ is the unique point in $[0, 2\pi)$ such that $p - q = |p-q|e^{i\phi}$ then L is the image γ^* of $\mathbb{R}$ under the map $\gamma : \mathbb{R} \ni t \mapsto \dfrac{p+q}{2} + te^{i(\phi - \frac{\pi}{2})}$ and again $p = q^\rho$.

For a given K and a p not in K, the preceding formulæ permit the determination of q so that $|z-p| = k|z-q|$ is the equation of K. Then $w \in T(K)$ iff $T^{-1}(w) \in K$, i.e., iff $|T^{-1}(w) - p| = |T^{-1}(w) - q| \cdot k$ or equivalently, when $(cp+d)(cq+d) \neq 0$,

$$\left|\frac{w - T(p)}{w - T(q)}\right| = \left|\frac{cq+d}{cp+d}\right| \cdot k. \qquad \text{(s7.1)}$$

When $0 = cp + d$ then $cq + d \neq 0$ and the corresponding equation is $|w - T(q)| = \dfrac{|ap+b|}{k}$. A similar formula obtains if $cq + d = 0$. Note the special case in which $k = 1$.

b) Note that (s7.1) shows that each of $T(q), T(p)$ is the reflection of the other in $T(K)$, a relationship valid as well when $(cp+d)(cq+d) = 0$. □

7.7. If $T(U) = U$ and $z \in \partial U \stackrel{\text{def}}{=} K$ then $|T(z)| \le 1$ and if $T(z) \in U$ then $z \in T^{-1}(U)$, a contradiction. Hence $T(K) \subset K$. Since the same argument applies to T^{-1}, it follows that $T(K) = K$. In particular, $T^{-1}(0)$ and $T^{-1}(\infty)$ must be reflections of each other in K. If $a = 0$ then $T^{-1}(0) \notin U$, a contradiction; if $c = 0$ then T is a translation of a dilation or contraction and cannot map U onto U unless $b = 0$ and $|a| = 1$.

Because $T^{-1}(0) = -\frac{b}{a}$ and $T^{-1}(\infty) = -\frac{d}{c}$, it follows that $U \ni -\frac{b}{a} \stackrel{\text{def}}{=} \alpha$ and $-\frac{d}{c} = \frac{1}{\overline{\alpha}}$. Thus $T(z) = -\frac{a\overline{\alpha}}{c} \cdot \frac{z-\alpha}{1-\overline{\alpha}z} \stackrel{\text{def}}{=} -\frac{a\overline{\alpha}}{c} \cdot \Phi_\alpha(z)$. Because $|T(1)| = 1$, it follows that for some θ in $[0, 2\pi)$, $T(z) = e^{i\theta}\frac{z-\alpha}{1-\overline{\alpha}z} = e^{i\theta}\Phi_\alpha(z)$. □

7.8. When $z - a = |z-a|e^{i\phi}$, $0 \le \phi < 2\pi$, reference to **Figure 7.1** shows that $k = \frac{r^2}{|z-a|^2}$ whence $z^\rho = a + \frac{r^2}{\overline{z} - \overline{a}}$. □

7.9. Since the cross-ratio is invariant with respect to Möbius transformations and since any three points on a circle or a line can be mapped via a Möbius transformation into any three points on the same circle or line, it follows that if the equation is valid, its validity is independent of the choice of z_1, z_2, z_3.

Let z_1, z_2, z_3 determine the circle $C_a(r)$. The equations below obtain by virtue of the Möbius transformations involved.

$$\begin{aligned}
\overline{X(z, z_1, z_2, z_3)} &= \overline{X(z-a, z_1-a, z_2-a, z_3-a)} \\
&= X\left(\overline{z} - \overline{a}, \frac{r^2}{z_1 - a}, \frac{r^2}{z_2 - a}, \frac{r^2}{z_3 - a}\right) \\
&= X\left(\frac{r^2}{\overline{z} - \overline{a}}, z_1 - a, z_2 - a, z_3 - a\right) \\
&= X\left(\frac{r^2}{\overline{z} - \overline{a}} + a, z_1, z_2, z_3\right).
\end{aligned}$$

Hence if $X(z', z_1, z_2, z_3) = \overline{X(z, z_1, z_2, z_3)}$ then $(z' - a)(\overline{z} - \overline{a}) = r^2$.

When the three points are on a line L, z_2 may be chosen to be ∞ whence, in the context of the problem, $\frac{z' - z_1}{z_3 - z_1} = \frac{\overline{z} - \overline{z_1}}{\overline{z_3} - \overline{z_1}}$, which implies that for any z_1 on L, $|z - z_1| = |z' - z_1|$, i.e., $L \perp [z, z']$. However, (s7.1) implies as well that $\Im\left(\frac{z' - z_1}{z_3 - z_1}\right) = -\Im\left(\frac{z - z_1}{z_3 - z_1}\right)$, i.e., that z' and z lie in opposite half-planes determined by L.

All arguments are reversible. □

7.10. If $\theta, \phi \in [0, 2\pi)$, $p - q = ae^{i\theta}$, and $q - r = be^{i\phi}$ then

$$\frac{p-q}{q-r} = \frac{a}{b}e^{i(\theta-\phi)}$$

which is real iff $\theta - \phi = 0$. □

7.11. a) See **Solution 7.5**.

b) Let R be as in **Solution 7.5**. Then

$$\{X(z_1,z_2,z_3,z_4)\in\mathbb{R}\}\Leftrightarrow\left\{R(z_1)\stackrel{\text{def}}{=}r\in\mathbb{R}\right\}$$
$$\Leftrightarrow\{z_1,z_2,z_3,z_4\in R(\mathbb{R})\}.$$

□

7.12. a) Direct calculation and the identity $\det(AB)=\det(A)\det(B)$ shows that **Mö** is a group with respect to composition. Furthermore,

$$T_{abcd}(z)=\begin{cases}\dfrac{b-\dfrac{ad}{c}}{cz+d}+\dfrac{a}{c} & \text{if } c\neq 0\\ \dfrac{a}{d}z+\dfrac{b}{d} & \text{if } c=0\end{cases}$$

$$=\begin{cases}T_{1,\frac{a}{c}}T_{b-\frac{ad}{c},0}T_0T_{cd}(z) & \text{if } c\neq 0\\ T_{ad,\frac{b}{d}}(z) & \text{if } c=0\end{cases}.$$

b) For a given T_{abcd} in **Mö** it may be assumed that $ad-bc=1$. The map $M:\mathbf{M\ddot{o}}\ni T_{abcd}\mapsto\begin{pmatrix}a&b\\c&d\end{pmatrix}\stackrel{\text{def}}{=}M(T_{abcd})$ [in $SL(2,\mathbb{C})$] satisfies $M(T_{abcd}\circ T_{pqrs})=M(T_{abcd})\cdot M(T_{pqrs})$.

c) Owing to a), it suffices to verify that each T_{ab} and T_0 is the composition of an even number of reflections.

i. If $b\neq 0$, let L_1 and L_2 be two lines perpendicular to the segment $\overline{0b}$ and such that L_1 passes through 0 while L_2 passes through $b/2$. Then for each z in L_1, reflection in L_1 leaves z unchanged and reflection in L_2 yields $z+b$. Thus for any z in L_1, $T_{0b}(z)$ is the composition of two reflections applied to z. Since $\#(L_1)>3$, **7.4** implies T_{0b} is the composition of two reflections.

ii. If $\theta>0$ then $T_{e^{i\theta},0}$ is the composition of two reflections, the first about a line L_1 through 0 and the second about a line L_2 through 0 and making an angle of size $\theta/2$ with L_1.

iii. If $a>0$ then T_{a0} is the composition of a reflection in $C_0(1)$ followed by a reflection in $C_0(\sqrt{a})$.

iv. Reflection in $\{z:\Im(z)=0\}$ followed by reflection in $C_0(1)$ is the same as T_0.

□

7.13. $\Im\left(\dfrac{a(x+iy)+b}{c(x+iy)+d}\right)=\dfrac{(ad-bc)y}{(cx+d)^2+y^2}$. □

7.14. A reflection in $C_a(r)$ followed by a reflection in a line through a reduces the problem to the case in which $C_b(s)\subset D(a,r)^\circ$. If $b=a$, there is no problem. If $b \neq a$, the composition of T_{0a} and $T_{\frac{1}{r}0}$ further reduces the problem to the cases where $C_b(s) \subset U$ and $C_a(r) = C_0(1)$. Finally, **7.4** may be used to transform $C_b(s)$ into some $C_0(s')$ while $C_0(1)$ is left invariant. □

7.15. Since $e^{re^{i\theta}} = e^{r\cos\theta}e^{ir\sin\theta}$, the limit exists iff $\cos\theta < 0$, i.e., iff $\frac{\pi}{2} < \theta < \frac{3\pi}{2}$. □

7.2. Polynomials

7.16. Because

$$\begin{aligned}P_n(z) &= \left(z + z^2 + \cdots + z^n\right)' = \left(\frac{z(1-z^n)}{1-z}\right)' \\ &= \frac{1 + z - (1+n)z^n + z^{n+1}}{1-z}\end{aligned},$$

if $|z| \leq r < 1$ then for large n, $\left|(1+n)z^n - z^{n+1}\right| < \frac{1}{r} - 1$, whence, for large n, $|P_n(z)| > 0$. □

7.17. Note that when $a = 0$ the $2n$ zeros of $f_n(z)$ are, for $k = 0, \ldots, 2n-1$, $b^{\frac{1}{2n}} e^{\frac{i(2k+1)\pi}{2n}}$. Their imaginary parts are positive iff $0 \leq k \leq n-1$.

If $a \neq 0$ then $\left|z^{2n} + b - f_n(z)\right| = \left|az^{2n-1}\right|$. For all sufficiently small positive ϵ and all sufficiently large positive R, the contour Γ consisting of the horizontal diameter and the upper half of the circumference of $D(0, R)$ is, by abuse of language, a simple closed rectifiable curve and the bounded component C of $\mathbb{C} \setminus \Gamma$ contains all the zeros of $z^{2n} + b$ that lie in the upper half-plane $\Pi_+ \stackrel{\text{def}}{=} \{z : \Im z > 0\}$. For sufficiently large positive R, if $|z| > R$ then $\left|az^{2n-1}\right| < \left|z^{2n} + b\right|$ and Rouché's theorem implies that n, the number of zeros of $z^{2n} + b$ in C is the same as the number of zeros of $f_n(z)$ in C. □

7.18. For n in $\mathbb{N}$, let A_n be $\left\{ z : n_z = n, z \in D\left(0, \frac{1}{2}\right) \right\}$. It follows that $\bigcup_{n\in\mathbb{N}} A_n = D\left(0, \frac{1}{2}\right)$, whence for some n', $\#\left(A_{n'}\right) = \mathfrak{c}$ and some z' in $D\left(0, \frac{1}{2}\right)$ is a limit point of $A_{n'}$. Hence $f^{(n')}(z) = 0$ on $A_{n'}$ and so $f^{(n')}(z) = 0$ on U. □

7.19. It may be assumed that $a_0 = 1$ whence

$$\begin{aligned}(1-z)f(z) &= 1 - \left((1-a_1)z + \cdots + (a_{n-1} - a_n)z^n + a_n z^{n+1}\right) \\ &\stackrel{\text{def}}{=} 1 - g(z).\end{aligned}$$

If $z \in D(0,1) \setminus \{1\}$ then $|g(z)| < 1$ and so $1 - g(z) \neq 0$, i.e., $f(z) \neq 0$. Furthermore, $f(1) > 1$. □

7.20. The number of components of $E(a)$ is the number of pairwise different zeros of $p - a$. □

7.21. a) If $z \in \Omega$, there are n numbers $w_i(z)$ such that $p(w_i(z)) = z$. Thus if $f(w) \stackrel{\text{def}}{=} \prod_{i=1}^{n} (w - w_i(z)) + z$ then

$$f'(w) = \sum_{i=1}^{n} \prod_{j \neq i} (w - w_j(z))$$
$$f'(w_i(z)) = \prod_{j \neq i} (w_i(z) - w_j(z)) \neq 0.$$

For each i, there is a positive r_i such that $D(w_i(z), r_i) \cap D(w_j(z), r_j) = \emptyset$ if $i \neq j$ while, by virtue of the inverse function theorem, f is injective on $W_i \stackrel{\text{def}}{=} D(w_i(z), r_i)^\circ$. Furthermore, $f(W_i) \stackrel{\text{def}}{=} V_i$ is open and

$$f(w_i(z)) = z \in V \stackrel{\text{def}}{=} \bigcap_{i=1}^{n} V_i,$$

whence Ω is open.

b) The inverse function theorem implies also that $w_i \in H(\Omega)$, whence $\tilde{q}(z) = \sum_{i=1}^{n} q(w_i(z))$ is holomorphic in Ω. □

7.3. Power Series

7.22. When $s_0(z) = 1, s_n(z) = 1 + z + \cdots + z^n, n \in \mathbb{N}$, Abel summation implies $\sum_{n=0}^{N} c_n z^n = \sum_{n=0}^{N-1} (c_n - c_{n+1}) s_n(z) + c_N s_N(z)$. If $z \in K$, $|s_n(z)| \leq \frac{2}{\delta}, n \in \mathbb{N}$, and the Weierstraß M-test applies. □

7.23. The Arzelà-Ascoli theorem and Cauchy's formula imply there is a subsequence $\{f_{m_k}\}_{k \in \mathbb{N}}$ converging uniformly on every compact set to some f in $\mathcal{E}$. From Cauchy's formula it follows that $f(z) = \sum_{n=0}^{\infty} c_n z^n$. The same argument shows that every subsequence of $\{f_m\}_{m \in \mathbb{N}}$ contains a subsubsequence converging on every compact set to f. If $f_m \stackrel{u}{\not\to} f$ on some compact set K then for some subsequence $\left\{f_{m'_k}\right\}_{k \in \mathbb{N}}$, and some positive δ, $\sup_{z \in K} \left|f_{m'_k}(z) - f(z)\right| \geq \delta$, in contradiction of the preceding sentence. □

7.24. Because $\sup_{k \geq n} |c_{1k} c_{2k}| \leq \sup_{k \geq n} |c_{1k}| \cdot \sup_{k \geq n} |c_{2k}|$, $R_{\mathbf{c}} \geq R_{\mathbf{c}_1} R_{\mathbf{c}_2}$. On the other hand, direct calculation shows that an example as required is provided when

$$c_{1n} = \begin{cases} 1 & \text{if } n \text{ is even} \\ 0 & \text{otherwise} \end{cases}, \quad c_{2n} = \begin{cases} 0 & \text{if } n \text{ is odd} \\ 1 & \text{otherwise} \end{cases},$$

in which case $R_{\mathbf{c}_1} = R_{\mathbf{c}_2} = 1 < \infty = R_{\mathbf{c}}$. □

7.25. a) When $z = x + iy$, $|n^z| = \left|e^{(x+iy)\ln n}\right| = e^{x \ln n}$. Hence if $x > 1$, $\sum_{n=1}^{\infty} |n^{-z}| = \sum_{n=1}^{\infty} e^{-x\ln n} = \sum_{n=1}^{\infty} n^{-x} < \infty$. It follows that $\zeta(z)$ is holomorphic in $\Omega \stackrel{\text{def}}{=} \{z : \Re(z) > 1\}$. Furthermore, if $y = 0$ and $x \downarrow 1$ then $\zeta(z) \uparrow \infty$ whence $1 \in S(\zeta)$. Thus the coefficients c_n may be calculated by termwise differentiation of $\sum_{n=1}^{\infty} \frac{1}{n^z}$: $c_k = \frac{\sum_{n=1}^{\infty}(-1)^k(\ln n)^k/n^2}{k!}$.

b) Since $1 \in S(\zeta)$ and $\zeta(z)$ is holomorphic in Ω, it follows that $R_{\mathbf{c}} = 1$. □

7.26. The Gutzmer coefficient estimate $\sum_{n=0}^{\infty} |c_n|^2 r^{2k} \le (M(r;f))^2$ and the hypothesis imply that $c_n = 0$ if $n \neq k$. □

7.27. If $c_n \ge 0$ and 1 is not a singularity of f then the power series $\sum_{n=0}^{\infty} \frac{f^{(n)}(1/2)}{n!}(z-1/2)^n$ converges for some z in $(1,2)$. However, since $c_n \ge 0$,

$$\sum_{n=k}^{\infty} c_n \binom{n}{k} \frac{1}{2^{n-k}} \le \sum_{n=k}^{\infty} c_n k! \binom{n}{k} \frac{1}{2^{n-k}} = f^{(k)}\left(\frac{1}{2}\right)$$

$$\sum_{k=0}^{n} \binom{n}{k} \frac{1}{2^{n-k}} \left(z - \frac{1}{2}\right)^k = z^n$$

$$\sum_{n=0}^{\infty} \left(\sum_{k=0}^{n} c_n \binom{n}{k} \frac{1}{2^{n-k}} \left(z - \frac{1}{2}\right)^k \right)$$

$$= \sum_{k=0}^{\infty} \left(\sum_{n=k}^{\infty} c_n \binom{n}{k} \frac{1}{2^{n-k}} \right) \left(z - \frac{1}{2}\right)^k \tag{s7.2}$$

$$\le \sum_{n=0}^{\infty} \frac{f^{(n)}\left(\frac{1}{2}\right)}{n!} \left(z - \frac{1}{2}\right)^n.$$

The left member of (s7.2) is $\sum_{n=0}^{\infty} c_n z^n$, which diverges when $|z| > 1$, a contradiction.

If, more generally, $\Re(c_n) \ge 0$ then for $g(z) \stackrel{\text{def}}{=} \sum_{n=0}^{\infty} \Re(c_n) z^n$, it follows that $\sum_{n=0}^{\infty} \frac{g^{(n)}(1/2)}{n!}(z-1/2)^n$ diverges when $z > 1$. On the other hand, $g^{(n)}(1/2) = \Re\left(f^{(n)}(1/2)\right)$, whence $\sum_{n=0}^{\infty} \frac{f^{(n)}(1/2)}{n!}(z-1/2)^n$ diverges. □

7.28. According to **7.27**, if $R_{\mathbf{a}} = 1$ then $1 \in S(f)$. Hence $R(\mathbf{a}) = 1+s > 1$. If $s < t$, the same argument implies $1 + s \in S(f)$, a contradiction. □

7.29. The hypothesis implies $\#\left(S(f) \cap \mathbb{T}\right) < \infty$, whence for some a on $\mathbb{T}$, some $N(a)$, and some M, if $z \in U \cap N(a)$ then $\left|\sum_{n=0}^{N} c_n z^n\right| \le M, N \in \mathbb{N}$. It follows that $|c_n z^n| \le 2M, n \in \mathbb{N}$, for all such z. □

7.30. The Maclaurin series coefficients for P_k are in $\mathbb{Z}$. Furthermore, $\sum_{n=0}^{\infty} nz^n = z\left(\frac{1}{z-1}\right)' = \frac{z}{(1-z)^2}$ and direct calculation shows that for $k = 2, 3, \ldots$, $\sum_{n=0}^{\infty} n^k z^n = \frac{z(z+1)\cdot \cdots \cdot (z+k-1)}{(1-z)^{k+1}} \stackrel{\text{def}}{=} \frac{Q_k(z)}{(1-z)^{k+1}}$ whence $Q_k \in \mathbb{N}[z]$ and $\deg(Q_k) = k$. Thus

$$\begin{aligned}\sum_{n=0}^{\infty} n^{k+1} z^n &= z\left(\frac{Q_k(z)}{(1-z)^{k+1}}\right)' \\ &= z\frac{(1-z)Q_k'(z) + (k+1)Q_k(z)}{(1-z)^{k+2}} \stackrel{\text{def}}{=} z\frac{N_k(z)}{(1-z)^{k+2}}.\end{aligned}$$

Induction shows that $N_k^{(p+1)}(z) = (1-z)Q_k^{(p+2)}(z) + (k-p)Q_k^{(p+1)}(z)$, whence if $k \ge p$ then $N_k^{(p+1)}(0) = Q_k^{(p+2)}(0) + (k-p)Q_k^{(p+1)}(0) > 0$. Since $N_k \in \mathbb{Z}[z]$, it follows that $N_k \in \mathbb{N}[z]$. □

7.31. If $0 \le r < 1$ then, as noted in **7.30**, $\sum_{n=0}^{\infty}(n+1)r^n = \frac{1}{(1-r)^2}$. Hence for m in $\mathbb{N}$, the absolute values of the functions in K are uniformly bounded in $D\left(0, 1 - \frac{1}{m}\right) \stackrel{\text{def}}{=} B_m$. The Cauchy integral formula implies they are equicontinuous on $D(0,1)$. It follows that the Arzelà-Ascoli theorem is applicable to any sequence $\left\{f_n\big|_{B_m}\right\}_{n\in\mathbb{N}}$ in K. Thus K contains a double sequence $\{g_{mn}\}_{m,n\in\mathbb{N}}$ such that $\{g_{mn}\}_{n\in\mathbb{N}} \subset \{g_{m+1,n}\}_{n\in\mathbb{N}}$ and for each m, $\{g_{mn}\}_{n\in\mathbb{N}}$ converges uniformly on B_m. The diagonal sequence $\{g_{nn}\}_{n\in\mathbb{N}}$ converges uniformly on each compact subset of U. □

7.32. a) When $z = e^{i\theta}$, $L(z,n) = 2ie^{i(2n+1)\frac{\theta}{2}} \sum_{k=n-1}^{0} \frac{\sin(n-k-1/2)\theta}{n-k}$. According to **3.60**, the sequence

$$\left\{\sin\theta + \frac{\sin 2\theta}{2} + \frac{\sin 3\theta}{2} + \cdots + \frac{\sin(2n-2)\theta}{n} + \frac{\sin(2n-1)\theta}{n}\right\}_{n\in\mathbb{N}}$$

is uniformly bounded on $[-\pi, \pi]$. It follows from the maximum modulus theorem that the sequence $\{L(z,n)\}_{n\in\mathbb{N}}$ is uniformly bounded on $D(0,1)$.

b) Because $\sum_{n=1}^{\infty} \frac{1}{n^2} = \frac{\pi^2}{6}$, $f \in A(U)$. On the other hand, if the terms of the series for f are rearranged to form a power series $\sum_{n=0}^{\infty} c_n z^n$ then

$\sum_{n=0}^{\infty}|c_n| \geq \sum_{n=1}^{\infty}\frac{2}{n^2}\left(\sum_{k=1}^{2^n}\frac{1}{k}\right)$. Since $\sum_{k=1}^{N}\frac{1}{k} \geq \ln(k+1)$, it follows that $\frac{2}{n^2}\left(\sum_{k=1}^{2^n}\frac{1}{k}\right) \geq \frac{2}{n}\ln 2$ whence $\sum_{n=1}^{\infty}|c_n| = \infty$. □

7.33. If $z \in D(a,r)^\circ \subset \Omega$ then for x in X, $\frac{1}{g(x)-z} = \sum_{n=0}^{\infty}\frac{(z-a)^n}{(g(x)-a)^{n+1}}$. The series converges uniformly on compact subsets of $D(a,r)^\circ$ and Fubini's theorem validates the equation

$$\begin{aligned}\int_X \frac{d\mu(x)}{g(x)-a} &= \int_X \sum_{n=0}^{\infty}\frac{(z-a)^n}{(g(x)-a)^{n+1}}\,d\mu(x)\\ &= \sum_{n=0}^{\infty}\int_X \frac{d\mu(x)}{(g(x)-a)^{n+1}}(z-a)^n \stackrel{\text{def}}{=} \sum_{n=0}^{\infty}c_n(z-a)^n.\end{aligned}$$ □

7.34. If $s = \min_{|z|=r}|f(z)|$ (>0) and $|w| < s$ then Rouché's theorem implies that $N(f) = N(f-w) = 1$ in $D(0,r)^\circ$. If $g(w)$ is the unique solution in $D(0,r)^\circ$ of $f(z)-w=0$, the principle of the argument implies $g(w)$ is given by the formula in (7.1). The holomorphy of g follows from the equation

$$g(w) = \sum_{n=0}^{\infty}\left(\frac{1}{2\pi i}\int_{|z|=r}\frac{zf'(z)}{(f(z))^n}\,dz\right)w^n,$$

which is valid if $w \in D(0,s)^\circ$. □

8
Functions Holomorphic in a Disc

8.1. General Results

8.1. If $f \in H(U)$ then $f'(0) = 0$ and $f'(0) = 1$ according as $f'(0)$ is calculated by using the zeros of f or the values of $1/n$ where $f(1/n) \neq 0$.

If $g \in H(U)$ then $g'(0) = \pm 1$ according as $g'(0)$ is calculated by using $g(1/n)$ for n odd or n even.

If $h \in H(U)$ then $h(z) = z$ when $z = 1/(2n), n \in \mathbb{N}$, whence $h(z) \equiv z$ although $h(z) \neq z$ when $z = 1/(2n+1), n \in \mathbb{C}$. □

8.2. a) If $0 \leq r < 1$, there is a z_r such that $M(r; f) = |f(z_r)|$. Because $0 \leq r^2 < r$, if $|z| = r$ then $M(r^2; f) = \sup_{|z|=r} |f(z^2)| \geq |f(z)|$ whence $M(r^2; f) \geq M(r; f)$, in contradiction of the maximum modulus principle unless f is a constant.

b) If $u \stackrel{\text{def}}{=} \Re(f) and v \stackrel{\text{def}}{=} \Im(f)$ then $f = v^2 + iv$. The Cauchy-Riemann equations imply $-4v^2 v_y = v_y$. If $v_y \not\equiv 0$ then $v^2 = -1/4$, a contradiction unless f is a constant. □

8.3. For r fixed in $[0,1)$, $\{f_n(re^{i\theta})\}_{n\in\mathbb{N}}$ is a $\| \ \|_1$-Cauchy sequence converging to $f(re^{i\theta})$.

If $f_n(re^{i\theta}) \stackrel{\text{def}}{=} \sum_{m=0}^{\infty} a_{nm} r^m e^{im\theta}$ for functionals F_m in $L^1([0,2\pi],\lambda)^*$, i.e., in $L^\infty([0,2\pi],\lambda)$, $a_{nm}r^m = F_m(f_n)$ and for all m, $\lim_{n\to\infty} a_{nm} \stackrel{\text{def}}{=} a_m$ exists.

Furthermore,

$$\lim_{n\to\infty} F_k(f_n) = \hat{f}_k = \begin{cases} 0 & \text{if } k < 0 \\ a_k r^k & \text{otherwise} \end{cases}.$$

Cauchy's formula implies that if $0 < r < s < 1$ then

$$\left|f_n(re^{i\theta}) - f_m(re^{i\theta})\right| \leq \frac{1}{2\pi} \int_0^{2\pi} \left| \frac{f_n(se^{i\phi}) - f_m(se^{i\phi})}{se^{i\phi} - re^{i\theta}} \right| d\phi.$$

The right member above converges to zero as $m, n \to \infty$. Thus if $0 < r < 1$ then for some g in $H(U)$, $f_n\big|_{D(0,r)} \stackrel{\text{u}}{\to} g$ and $F_k(g) = \lim_{n\to\infty} F_k(f_n) = a_k r^k$, whence $g = f$. □

8.4. For z in $D(0,r)^\circ$ and some m in $\mathbb{Z}^+$,

$$f(z) = z^m \sum_{k=0}^{\infty} a_k z^k \stackrel{\text{def}}{=} z^m h(z),\ a_0 \neq 0,$$

and $m = pn + q$, $p \in \mathbb{Z}^+$, $0 \leq q < n$. Then for some s in $(0,r)$ and g in $H\left(D(0,s)^\circ\right)$, $h = g^n$ and $f = z^q \left(z^p g(z)\right)^n$. □

8.5. Because $F(z) \stackrel{\text{def}}{=} (z-a)f(z) \stackrel{\text{def}}{=} \sum_{n=0}^{\infty} d_n z^n$ is entire,

$$\begin{aligned}\frac{F(z)}{z-a} &= \sum_{n=0}^{\infty} c_n z^n \\ &= -\frac{1}{a}\left(\sum_{k=0}^{\infty}\left(\frac{z}{a}\right)^k\right)\cdot\left(\sum_{l=0}^{\infty} d_l z^l\right) \\ &= \sum_{n=0}^{\infty} -\frac{1}{a}\left(\sum_{l=0}^{n}\frac{d_l}{a^{n-l}}\right) z^n\end{aligned},$$

$$c_n = -\frac{1}{a}\sum_{l=0}^{n}\frac{d_l}{a^{n-l}}$$

whence $\dfrac{c_n}{c_{n+1}} = a\dfrac{\sum_{l=0}^{n} d_l a^l}{\sum_{l=0}^{n+1} d_l a^l} \to a\dfrac{F(a)}{F(a)} = a$ as $n \to \infty$. □

8.6. Since $|f'(z)|$ is bounded and continuous in U, the criterion for Riemann integrability (cf. **3.45**) implies that for each θ in $[0, 2\pi)$,

$$\lim_{r\to 1}\int_{[0,re^{i\theta}]} f'(z)\,dz \stackrel{\text{def}}{=} g\left(e^{i\theta}\right)$$

exists. Because $f\left(re^{i\theta}\right) - f(0) = \int_{[0,re^{i\theta}]} f'(z)\,dz$, it follows that

$$F\left(re^{i\theta}\right) \stackrel{\text{def}}{=} \begin{cases} f\left(re^{i\theta}\right) & \text{if } 0 \leq r < 1 \\ g\left(e^{i\theta}\right) & \text{otherwise}\end{cases}$$

is in $C(D[0,1],\mathbb{C})$ and $F(z)\big|_U = f(z)$. □

8.7. The Schwarz reflection principle implies that for some function $\widetilde{f}$ holomorphic in a region Ω containing $A \cup U$, $\widetilde{f}\big|_{\Omega\cap U} = f\big|_{\Omega\cap U}$. The identity theorem for holomorphic functions implies $\widetilde{f}\big|_{\Omega\cap U} = f\big|_{\Omega\cap U} = 0$. □

8.8. The function f^{-1} is holomorphic in the open set $\Omega \stackrel{\text{def}}{=} f(U)$. Thus $h \stackrel{\text{def}}{=} f^{-1}\circ g \in H(U)$ and $h(U) \subset U$. Because $h(0) = 0$, the Schwarz lemma implies $h(D(0,r)) \subset D(0,r)$. □

8.9. Induction shows $\prod_{n=0}^{N}\left(1+z^{2^n}\right)=\sum_{k=1}^{2^{N+1}-1} z^k$. □

8.10. The convergence of each product in U is assured since

$$\sum_{n=0}^{\infty}\left|z^{2n}\right|+\sum_{n=0}^{\infty}\left|z^{2(2n-1)}\right|<\infty$$

in U. Thus $\{f,g\}\subset H(U)$.

Since $\left(1-z^{2(2n-1)}\right)=\left(1-z^{2n-1}\right)\left(1+z^{2n-1}\right)$, induction shows that a proper pairing of the factors in fg leads to partial products

$$\begin{aligned}
&\left(1-z^4\right)\\
&\left(1-z^6\right)\left(1-z^8\right)\\
&\left(1-z^{12}\right)\left(1-z^{16}\right)\left(1-z^{20}\right)\\
&\left(1-z^{24}\right)\left(1-z^{32}\right)\left(1-z^{40}\right)\left(1-z^{48}\right)\\
&\vdots
\end{aligned}$$

The estimates

$$p_N^* \stackrel{\text{def}}{=} \prod_{n=1}^{N}\left(1+|u_n|\right)\le \exp\left(\sum_{n=1}^{N}|u_n|\right)$$

$$|p_n-1| \stackrel{\text{def}}{=} \left|\prod_{n=1}^{N}\left(1+u_n\right)-1\right| \le |p_N^*-1|$$

imply that the products listed above converge to one. It follows that $f(z)g(z)=1$. □

8.11. If such an f exists then

$$f(z)=\sum_{n=0}^{\infty}c_nz^n=\sum_{n=0}^{\infty}c_nr^ne^{in\theta}, 0\le r<1.$$

Fejér's theorem applied as in **Solution 6.164** shows that $g(\theta)\stackrel{\text{def}}{=}e^{-i\theta}$ is the uniform limit on $[0,2\pi]$ of the averages $\{\sigma_N\}_{N\in\mathbb{N}}$ of the partial sums of $\sum_{n=0}^{\infty}c_ne^{in\theta}$. On the other hand, for $e_n:[0,2\pi]\ni\theta\mapsto e^{in\theta}$,

$$(g,e_n)=\int_0^{2\pi}e^{-i(n+1)\theta}\,d\theta=0.$$

Thus $(g,\sigma_N)=0, N\in\mathbb{N}$, a contradiction. □

8.12. For any arc $A\stackrel{\text{def}}{=}\left\{z\;:\;z=e^{i\theta},0\le a<\theta\le b<2\pi\right\}$, $f(A)$ is an arc contained in $\partial D(0,K)$. The Schwarz reflection principle implies that for some region Ω containing $D(0,1)$ and some F in $H(\Omega)$, $F(z)\big|_U = f(z)$. Thus for some N in $\mathbb{N}$ and $\{z_k\}_{k=1}^N$ contained in $D(0,1)$,

$$G(z) \stackrel{\text{def}}{=} F(z)\prod_{k=1}^{N}\frac{(1-\overline{z_k}z)}{z-z_k}$$

is holomorphic in some $D(0,R)^\circ$ properly containing $D(0,1)$. Hence for some h in $H\left(D(0,R)^\circ\right)$, $G = e^h$. Furthermore, $|G(z)| = K$ on $\mathbb{T}$, whence if $h \stackrel{\text{def}}{=} u + iv$ then on $\mathbb{T}$, e^u is a constant. Thus $u_x = u_y = 0$ on $\mathbb{T}$ and the Cauchy-Riemann equations imply $v_x = v_y = 0$ on $\mathbb{T}$. Hence $h' = 0$ on $\mathbb{T}$. The maximum modulus principle implies $h' \equiv 0$ and so h and hence G are constants. □

8.13. Because $0 \notin f(\mathbb{T})$, for some r in $(0,1)$, $f(z) \neq 0$ if $|z| \geq r$. Thus Rouché's theorem yields $N = \dfrac{1}{2\pi i}\displaystyle\int_0^{2\pi}\frac{f'\left(re^{i\theta}\right)}{f\left(re^{i\theta}\right)}ire^{i\theta}\,d\theta$. □

8.14. a) Since $F(\theta) \stackrel{\text{def}}{=} f\left(e^{i\theta}\right)$ is continuous, $F \in L^2\left([0,2\pi],\lambda\right)$, and

$$\widehat{F}_n = \begin{cases} c_n & \text{if } n \in \mathbb{Z}^+ \\ 0 & \text{otherwise} \end{cases}.$$

Hence $\sum_{n=0}^{\infty}|c_n|^2 = \|F\|_2^2 < \infty$.

b) Cauchy's formula and the orthogonality of $\left\{e^{in\theta}\right\}_{n\in\mathbb{Z}}$ imply that if $z \in U$ then

$$f(z) = \frac{1}{2\pi i}\int_{|z|=1}\frac{f(w)}{w-z} = \frac{1}{2\pi}\int_0^{2\pi}\frac{F(\theta)}{1-e^{-i\theta}z}\,d\theta \tag{s8.1}$$

$$\int_0^{2\pi}F(\theta)e^{in\theta}\,d\theta = 0, n \in \mathbb{N}, \text{ and}$$

$$\frac{1}{2\pi}\int_0^{2\pi}\frac{F(\theta)e^{i\theta}\overline{z}}{1-e^{i\theta}\overline{z}}\,d\theta = \frac{1}{2\pi}\sum_{n=1}^{\infty}\int_0^{2\pi}F(\theta)e^{in\theta}\overline{z}^n = 0. \tag{s8.2}$$

The result follows from (s8.1) and (s8.2), after addition and simplification, from the formula $\dfrac{1}{2\pi}\Re\left(\dfrac{e^{it}+z}{e^{it}-z}\right) = P_r(\theta - t)$.

c) *i.*

$$\begin{aligned}\sum_{n=-\infty}^{\infty} r^{|n|}e^{int} &= \sum_{n=1}^{\infty} r^n e^{in(-t)} + \sum_{n=0}^{\infty} r^n e^{int} \\ &= \frac{re^{-it}}{1-re^{-it}} + \frac{1}{1-re^{it}} = \frac{1-r^2}{1-2r\cos t + r^2} \\ &= 2\pi P_r(t)\end{aligned}$$

$$1 - 2r\cos t + r^2 = (1-r)^2 + 4r\sin^2\frac{t}{2}.$$

ii. The result a) applies when $f(z) \equiv 1$. *iii.* If $0 \le r < 1$ and $|t| \le \frac{\pi}{2}$ then

$$0 < 1 - r^2 < 2(1-r),\ 2(1-r) + 4r\sin^2\left(\frac{t}{2}\right) > (1-r)^2 + \frac{4rt^2}{\pi^2}.$$

Thus $P_r(t) < \dfrac{\pi(1-r)}{\pi^2(1-r)^2 + 4t^2}$ and

$$P_r(t) < \begin{cases} \dfrac{\pi(1-r)}{\pi^2(1-r)^2 + 4rt^2} & \text{if } |t| \le \dfrac{\pi}{2} \\ \dfrac{1-r}{\pi\,(1+r^2)} & \text{if } \dfrac{\pi}{2} < |t| < \pi \end{cases}.$$

□

8.15. Cauchy's formula implies $f(a) = \dfrac{1}{2\pi}\displaystyle\int_0^{2\pi} f\left(a + se^{it}\right) dt$ whence

$$\int_0^r f(a)s\,ds = \frac{r^2}{2} f(a).$$

□

8.16. If $r \in (0,1)$ then $f'(0) = \dfrac{1}{2\pi i}\displaystyle\int_{|z|=r} \frac{f(z)}{z^2}\,dz$, whence $|f'(0| \le \dfrac{1}{r}$.

□

8.17. If $0 \le r < 1$, $f(z) = \sum_{n=0}^{\infty} c_n z^n$, and $c_n \stackrel{\text{def}}{=} a_n + ib_n$ then

$$\begin{aligned} f\left(re^{i\theta}\right) &= \sum_{n=0}^{\infty} c_n r_n e^{in\theta} \\ &= \sum_{n=0}^{\infty} (a_n \cos n\theta - b_n \sin n\theta) + i \sum_{n=0}^{\infty} (b_n \cos n\theta + a_n \sin n\theta) \\ &= u\left(re^{i\theta}\right) + iv\left(re^{i\theta}\right), \end{aligned}$$

and $k!c_k = f^{(k)}(0)$. By appeal to Euler's formula and the observation that $\displaystyle\int_0^{2\pi} e^{im\theta}\,d\theta = 0$ when $m \ne 0$, it follows that

$$\frac{1}{2\pi}\int_0^{2\pi} u\left(re^{i\theta}\right) e^{-ik\theta}\,d\theta = \frac{c_k r^k}{2}.$$

□

8.18. a) Cauchy's formula applies.
b) From **8.17** it follows that

$$-c_n = \frac{1}{\pi R^n}\int_0^{2\pi} (M - u(r,\theta))\, e^{-in\theta}\,d\theta.$$

Because $M - u(r,\theta) \geq 0$,

$$|c_n| \leq \frac{2}{\pi r^n}\int_0^{2\pi}(M - u(r,\theta))\,d\theta = \frac{2M}{r^n}\int_0^{2\pi} - \frac{1}{\pi r^n}\int_0^{2\pi} u(r,\theta)\,d\theta. \qquad \square$$

8.19. a) The map $\gamma : [0, 2\pi] \ni t \mapsto f\left(re^{it}\right)$ is such that $\gamma^* = f\left(\partial D(0,r)\right)$ and so $L(r) = \int_0^{2\pi} |\gamma'(t)|\,dt = r\int_0^{2\pi} \left|f'\left(re^{it}\right)\right| dt$. On the other hand, Cauchy's formula implies $|f'(0)| \leq \frac{1}{2\pi}\int_0^{2\pi}\left|f'\left(re^{it}\right)\right| dt$. It follows that $L(r) \geq 2\pi r\,|f'(0)|$.

b) The Cauchy-Riemann equations and the area theorem based on Jacobian determinants show that

$$\begin{aligned}
A(r) &= \int_{D(0,r)} \left|f'(z)\right|^2 dA\\
&\stackrel{\text{def}}{=} \int_0^r \left(\int_0^{2\pi} \sum_{n,m=0}^{\infty} nmc_n\overline{c_m}s^{n+m-1}e^{i(n-m)t}\,dt\right) ds.
\end{aligned}$$

The orthogonality of the system $\left\{e^{int}\right\}_{n\in\mathbb{Z}}$ leads to the stated formula.

c) Cauchy's formula implies $(f'(0))^2 = \frac{1}{2\pi}\int_0^{2\pi}\left(f'\left(re^{it}\right)\right)^2 dt$ whence

$$\begin{aligned}
|f'(0)|^2 &\leq \frac{1}{2\pi}\int_0^{2\pi}\left|f'\left(re^{it}\right)\right|^2 dt\\
A(r) &= \int_0^r \left(\int_0^{2\pi}\left|f'\left(se^{it}\right)\right|^2 dt\right) s\,ds\\
&\geq 2\pi\int_0^r |f'(0)|^2 s\,ds = \pi r^2\,|f'(0)|^2. \qquad \square
\end{aligned}$$

8.20. If $f(z) = \sum_{n=0}^{\infty} c_n z^n$ then $I_2(r) = \sum_{n=0}^{\infty} |c_n|^2 r^{2n}$. $\square$

8.21. Because $g(z) \stackrel{\text{def}}{=} \frac{f(z)}{z^{n-1}} \in H(U)$ and

$$\left|g\left(re^{i\theta}\right)\right| \leq \frac{1}{(r+\epsilon)^{n-1}} \text{ if } r < r+\epsilon < 1,$$

it follows that $g(U) \subset D(0,1)$. Schwarz's lemma implies $|g(z)| \leq |z|$. $\square$

8.22. Since $f^{-1} \in H(U)$, for some nonzero c_1,

$$f\left(z^n\right) = z^n\left(c_1 + \cdots\right) \stackrel{\text{def}}{=} z^n g(z)$$

in U, whence $g(z) \neq 0$ in U. Thus for some h in $H(U)$, $g(z) = e^{h(z)}$ and $g_k(z) \stackrel{\text{def}}{=} ze^{\frac{h(z)+2k\pi i}{n}}$, $1 \le k \le n$, meet the requirement. The Fundamental Theorem of Algebra implies that $f(z) = (g(z))^n$ has no more than n solutions. □

8.23. Because $f(0) = \lim_{n\to\infty} f\left(\frac{1}{n}\right)$, $f(0) \in \mathbb{R}$. If $f^{(\nu)}(0) \stackrel{\text{def}}{=} c_\nu \in \mathbb{R}$ when $0 \le \nu \le m$ then $\mathbb{R} \ni \dfrac{f(x_n) - \sum_{\nu=0}^m c_\nu x_n^\nu}{x_n^{\nu+1}} \to f^{(\nu+1)}(0)$ as $n \to \infty$.

□

8.24. Cauchy's formula implies that if $0 \le r < 1$ then $|f_n(z)| \le \dfrac{1}{|r-z|}$ in $D(0,r)^\circ$. Thus on every compact subset K of U, $\left\{|f_n(z)|\,|_K\right\}_{n\in\mathbb{N}}$ is bounded and $\left\{f_n(z)|_K\right\}_{n\in\mathbb{N}}$ is equicontinuous. The Arzelà-Ascoli theorem implies that some subsequence $\{f_{n_k}\}_{k\in\mathbb{N}}$ converges uniformly on every compact subset of U. Thus

$$\|f_{n_k} - f_{n_l}\|_1 \le \int_{D(0,r)} + \int_{U\setminus D(0,r)} \left|f_{n_k}\left(re^{i\theta}\right) - f_{n_l}\left(re^{i\theta}\right)\right| r\,dr d\theta \stackrel{\text{def}}{=} I + II.$$

Owing to the uniform convergence on $D(0,r)$, I is small for large k,l. Because $\int_U \left|f_n\left(re^{ith}\right)\right| r\,dr\,d\theta \le \frac{1}{2}$, it follows that for r near one, II is small for all k,l. □

8.25. If such an f exists, $\#(Z(f)\cap U)$ is finite since otherwise, for some sequence $\{z_n\}_{n\in\mathbb{N}}$, $|z_n| \to 1$ while $f(z_n) \equiv 0$, in contradiction of the hypothesis. Hence for some r_0 in $[0,1)$, if $r_0 \le |z| < 1$ then $f(z) \neq 0$.

It follows that each singularity of $g \stackrel{\text{def}}{=} \frac{1}{f}$ in U is a pole (g is meromorphic in U) and the finite set $P(g)\cap U$ may be listed: $P(g)\cap U \stackrel{\text{def}}{=} \{p_k\}_{k=1}^n$. Then $g(z)\cdot\prod_{k=1}^n (z-p_k) \stackrel{\text{def}}{=} h(z) \in H(U)$. Furthermore, if $|z_n| \to 1$ as $n\to\infty$ then $h(z_n) \to 0$ as $n\to\infty$, whence h may be extended to a function $\widetilde{h}$ in $C(D(0,1),\mathbb{C})$ and $\widetilde{h}(\mathbb{T}) = \{0\}$. The maximum modulus theorem implies $h = 0$, a contradiction. □

[**Note s8.1:** The power series $\sum_{n=1}^\infty z^{n!}$ defines a function f in $H(U)$. For q in $\mathbb{Q}$ and t in $[0,1)$, $f\left(re^{q2\pi it}\right) \uparrow \infty$, i.e., by abuse of language, for a dense set of radii, $f^* = \infty$. The result above shows that, nevertheless, for some sequence $\{z_n\}_{n\in\mathbb{N}}$ contained in U, $|z_n| \to 1$ while $\underline{\lim}_{n\to\infty} |f(z_n)| < \infty$.]

8.2. Applications of Möbius Transformations

8.26. a) Let Ω be $\{\,w : \Re(w) > 0\,\}$. Then $\phi(w) \stackrel{\text{def}}{=} \dfrac{1-w}{1+w}$ is holomorphic in Ω and $\phi(\Omega) \subset U$. Hence $g \stackrel{\text{def}}{=} \phi\circ f \in H(U)$, $g(0) = 0$, and $g(U) \subset U$.

The Schwarz lemma implies $|g(z)| \le |z|$, i.e., $\left|\frac{1-f(z)}{1+f(z)}\right| \le |z|$. Hence $\left|\frac{1-|f(z)|}{1+|f(z)|}\right| \le |z|$ from which the conclusion follows.

b) From a) and the Cauchy formula for the coefficients it follows that if $0 < r < 1$ then $|c_n| \le \frac{1}{2\pi n! r^n}\int_0^{2\pi} \frac{1+r}{1-r}\,d\theta = \frac{1}{n!}\frac{1}{r^n}\frac{1+r}{1-r}$. The methods of the calculus show that $\min_{r\in(0,1)} \frac{1}{r^n}\frac{1+r}{1-r}$ is achieved when

$$0 < r = \frac{-1+\sqrt{1+n^2}}{n} \stackrel{\text{def}}{=} r_n < 1.$$

Because

$$r_n^n \approx \left(1-\frac{1}{n}\right)^n \approx e^{-1},$$

$$1 - r_n = 1 + \frac{1}{n} + \sqrt{1+\frac{1}{n^2}} \stackrel{\text{def}}{=} 1 + x - \sqrt{1+x^2}, \text{ and}$$

$$1 + x - \sqrt{1+x^2} \approx \left(1 - \frac{\theta}{\sqrt{1+\theta^2}}\right)x, 0 < \theta < x,$$

it follows from Stirling's formula that

$$\frac{1+r_n}{n! r_n^n (1-r_n)} \approx \frac{1}{\left(\frac{n}{e}\right)^n \sqrt{2\pi n}\cdot e^{-1}\left(1-\frac{\theta}{\sqrt{1+\theta^2}}\right)\frac{1}{n}} \to 0$$

as $n \to \infty$. □

8.27. a) If $\alpha \in U$ then $|\Phi_\alpha(z)| = 1$ on $\mathbb{T}$. Thus $\frac{f(z)}{\prod_{k=1}^n \Phi_{z_k}(z)} \stackrel{\text{def}}{=} g(z)$ is holomorphic in U and $g(U) \subset U$. Hence $|f(z)| \le \prod_{k=1}^n |\Phi_{z_k}(z)|$.

b) From a) it follows that (8.1) obtains. If $z \in U$ then $\left|\frac{z-z_k}{1-\overline{z_k}z}\right| < 1$. If $z \stackrel{\text{def}}{=} r \in (0,1)$ then direct calculation shows that

$$\begin{aligned} 1 > \mu_k &\stackrel{\text{def}}{=} 1 - \left|\frac{z-z_k}{1-\overline{z_k}z}\right| \\ &= \frac{(1-|z_k|)(1-r^2)(1+|z_k|)}{|1-rz_k|(|1-rz_k|+|r-z_k|)} > \frac{1-r^2}{6}(1-|z_k|). \end{aligned}$$

Note that $\prod_{k=1}^N (1-\mu_k) > \prod_{k=1}^{N+1}(1-\mu_k) \downarrow A \ge 0$. Furthermore, $A \le \prod_{k=1}^N (1-\mu_k) \le \exp\left(-\sum_{k=1}^N \mu_k\right)$, whence if $\sum_{k=1}^\infty \mu_k = \infty$ then

$A = 0$. Hence if $\sum_{k=1}^{\infty}(1-|z_k|) = \infty$ then for z in $(0,1)$, $\prod_{k\in\mathbb{N}}|\Phi_{z_k}| = 0$. Thus $f(z) = 0$ on $(0,1)$, whence $f \equiv 0$. □

8.28. Direct calculation shows that if $|z| \le r < 1$ then

$$|1+\operatorname{sgn}(z_n)\,\Phi_{z_n}| = \left|\frac{z_n+|z_n|}{(1-z_n z)\,z_n}\right|(1-|z_n|) \le \frac{1+r}{1-r}(1-|z_n|).$$

The argument used in **8.27**b) shows that $B \in H(U)$ and since the absolute value of each factor in (8.2) is in $[0,1)$, $B \in H^\infty$. □

8.29. It may be assumed that $a \neq 0$. If $g(z) \stackrel{\text{def}}{=} \Phi_a \circ f \circ \Phi_{-a}(z)$ and $c \stackrel{\text{def}}{=} \Phi_a(b)$ then $\Phi_{-a}(c) = b$, $c \in U$, $g(0) = 0$, and $g(U) \subset U$. Hence Schwarz's lemma implies $|g(z)| \le |z|$. Since $g(c) = c$ it follows that $g(z) \equiv z$. Hence $z = \Phi_a \circ f \circ \Phi_{-a}(z)$, i.e., $\mathrm{id} = \Phi_a \circ f \circ \Phi_{-a}$. Thus

$$f = \Phi_{-a} \circ \mathrm{id} \circ \Phi_a = \Phi_{-a} \circ \Phi_a = \mathrm{id},$$

i.e., $f(z) = z$. □

[**Note s8.2:** If $a \neq 0$, $\{\pm\operatorname{sgn}(a)\}$ is the set of Φ_a-fixed points, and each is on $\mathbb{T}$. Thus if $a \neq 0$, U contains no Φ_a-fixed point while zero (in U) is the unique Φ_0-fixed point (cf. **9.9**).

The function g may be regarded as the Φ_a-conjugate of f in the $\circ$-semigroup of self-maps of U.]

8.30. a) If $a \in U$, then $f(a) \stackrel{\text{def}}{=} b \in U$. If $g(z) \stackrel{\text{def}}{=} \Phi_b \circ f \circ \Phi_{-a}(z)$ then $g(U) \subset U, g(0) = 0$. The Schwarz lemma implies $|g(z)| \le |z|$ and $|g'(0)| \le 1$. However, $|g'(0)| = \dfrac{|f'(a)|\,(1-|a|^2)}{1-|f(a)|^2}$.

b) If $f \stackrel{\text{def}}{=} \dfrac{g}{R}$ then a) implies $|f'(0)| \le 1 - |f(0)|^2$. □

8.31. For some a in U, $f(a) = 0$, and so if $g(z) \stackrel{\text{def}}{=} f(\Phi_{-a}(z))$ then $g \in H(U)$, $g(U) = U$, g is injective, and $g(0) = 0$. Schwarz's lemma implies that for z in U, $|g(z)| \le |z|$. Furthermore, Schwarz's lemma may be applied to the inverse g^{-1} of g: for all z in U, $|g^{-1}(z)| \le |z|$. Thus

$$|z| = \left|g \circ g^{-1}(z)\right| \le \left|g^{-1}(z)\right| \le |z|.$$

Yet another application of Schwarz's lemma shows that for some θ in $[0,2\pi)$, $g^{-1}(z) = e^{i\theta}z$. Because $g^{-1}(z) = \Phi_a\left(f^{-1}(z)\right)$, it follows that

$$f^{-1}(z) = e^{i\theta}\Phi_{-ae^{i\theta}}(z),$$

i.e., $f(z) = e^{-i\theta}\Phi_{ae^{-i\theta}}(z)$. □

9
Functions Holomorphic in a Region

9.1. General Regions

9.1. a) The function $g(z) \stackrel{\text{def}}{=} f(z) - f'(a)z$ is such that $g'(a) = 0$ whence g is not injective near a, i.e., near a there are two points b, c such that $g(b) = g(c)$, whence the conclusion.

b) If f is never zero in Ω_1 then $\frac{1}{f} \in H(\Omega_1)$. The maximum modulus theorem implies $M > 0$, $|f| \leq M$, and $\left|\frac{1}{f}\right| \leq \frac{1}{M}$. Thus $M \leq |f| \leq M$: $|f(z)|$ is a constant in Ω_1. The open mapping theorem implies $f(z)$ is a constant. □

[**Note s9.1:** If $\alpha \in U \setminus \{0\}$ and $R = \left|\frac{1}{\overline{\alpha}}\right|$ then $f(z) \stackrel{\text{def}}{=} \frac{z^n - \alpha}{1 - \overline{\alpha} z^n}$ is in $H(D(0,R)^\circ)$, has n zeros in $D(0,2)^\circ$, and $|f(z)| \equiv 1$ on $\mathbb{T}$.]

9.2. For each w in $f(\Omega)$ and for some z in Ω, $w = f(z)$. The hypothesis implies that the formula $h(w) \stackrel{\text{def}}{=} g(z)$ uniquely defines h and that

$$h \circ f(z) = g(z).$$

If $f'(a) \neq 0$ then $h'(a)$ exists and is $\frac{g'(a)}{f'(a)}$. The trivial case in which f is a constant aside, if $f'(a) = 0$ and $b_n \to f(a)$ as $n \to \infty$, there is a sequence $\{a_n\}_{n \in \mathbb{N}}$ such that $a_n \to a$, $f'(a_n) \neq 0$, and $f(a_n) = b_n$. It follows that $h(b_n) = g(a_n) \to g(a)$ whence $f(a)$ is a removable singularity of h, i.e., h may be extended from a function holomorphic on $f(\Omega) \setminus \{f(f'^{-1}(0))\}$ to a function holomorphic on $f(\Omega)$. □

9.3. If $\Omega \stackrel{\text{def}}{=} D(0,2)^\circ \setminus \{0\}$, $f(z) \stackrel{\text{def}}{=} \frac{1}{z}$, $\{p_n\}_{n \in \mathbb{N}} \subset \mathbb{C}[z]$, and $p_n \stackrel{\text{u}}{\to} f$ on $\mathbb{T}$ then $0 = \int_{\mathbb{T}} p_n(z)\,dz \to \int_{\mathbb{T}} \frac{dz}{z} = 2\pi i$. □

9.4. For a fixed in Ω and z in Ω, there is a rectifiable simple arc γ such that $\gamma(0) = a$ and $\gamma(1) = z$. Because Ω is simply connected,

$$h(z) \stackrel{\text{def}}{=} \int_\gamma k\left(f'(w), g'(w)\right) dw$$

is well-defined and $h'(z) = k\left(f'(z), g'(z)\right)$. □

9.2. Regions Ω Containing $D(0,1)$

9.5. For some positive δ, if $|z| < 1+\delta$ then $f(z) = \sum_{n=0}^{\infty} c_n z^n$ and for all real θ, $\sum_{n=0}^{\infty} c_n e^{in\theta} \in \mathbb{R}$. Thus if $c_n \stackrel{\text{def}}{=} a_n + ib_n$ then

$$\sum_{n=0}^{\infty} (b_n \cos n\theta + a_n \sin n\theta) = 0$$

and the series converges uniformly on $\mathbb{R}$. Hence $a_n \equiv b_n \equiv c_n \equiv 0$. □

9.6. According to the principle of the argument, the winding number with respect to zero of the curve $\gamma : [0,1] \ni t \mapsto f\left(e^{2\pi it}\right)$ is n. Thus, as t ranges over $[0,1]$, $\theta(t) \stackrel{\text{def}}{=} \arg\left(f\left(e^{it}\right)\right)$ increases by $2n\pi$. Hence $\cos\theta(t) = 0$ at least $2n$ times. □

[**Note s9.2:** If f has nonsimple zeros in U, f is near an F having simple zeros in U and in number N equal to the sum of the multiplicities of the zeros of f. The number of zeros of $\Re(F(z))$ on $\mathbb{T}$ is $2N$ and as F approaches f the zeros of $\Re(F(z))$ on $\mathbb{T}$ approach the zeros of $\Re(f(z))$ on $\mathbb{T}$. Hence with the proper interpretation, the result above obtains when the hypothesis of the simplicity of the zeros of f is dropped.]

9.7. The zeros of g in Ω are isolated whence all the singularities of $\frac{f}{g}$ are removable: $\frac{f}{g} \stackrel{\text{def}}{=} h$ may be regarded as an element of $H(\Omega)$. Because $|f(z)| \le 1$ in Ω, $|h(z)|\,\big|_{\partial\Omega} = \frac{|f(z)|\,\big|_{\partial\Omega}}{1} \le 1$ and the maximum modulus theorem implies the result. □

9.8. a) If $r \in [0,1)$ and $|z| \le r$, for some $\epsilon(z)$, $|\epsilon(z)| < 1$ and

$$f(z) = f'(0) + \epsilon(z)z,$$
$$f_n(z) = f\left(z^n\right) = f'(0)z^n + \epsilon\left(z^n\right)z^n, \text{ and}$$
$$F_N(z) = f'(0)\left(\frac{1-z^N}{1-z}\right)z + \epsilon\left(z^n\right)\left(\frac{1-z^N}{1-z}\right)z.$$

Thus $|F_M(z) - F_N(z)| \stackrel{\text{u}}{\to} 0$ on $D(0,r)$ as $M, N \to \infty$. It follows that $\lim_{N\to\infty} F_N = \sum_{n=1}^{\infty} f_n \stackrel{\text{def}}{=} F$ exists.

b) Cauchy's formula implies $F_N' \stackrel{\text{u}}{\to} F'$. □

9.9. According to Brouwer's fixed-point theorem **3.75**, $D(0,1)$ contains an a such that $f(a) = a$. Because $f(U) \subset f(D(0,1)) \subset U$, $a \in U$. From **8.29** it follows that there is only one such a in U. □

9.10. For the curve $\gamma : [0, 2\pi] \ni t \mapsto f\left(e^{it}\right) \in D(0,1)$, if $\zeta \in (\mathbb{T} \setminus \gamma^*)$, there is a neighborhood $N \stackrel{\text{def}}{=} D(\zeta, r)^\circ$ that does not meet γ^*. For any a in the nonempty set $N \cap U$, the index of γ with respect to a is

$$\begin{aligned}\mathrm{Ind}_\gamma(a) &\stackrel{\text{def}}{=} \frac{1}{2\pi i}\int_\gamma \frac{dz}{z-a} = \frac{1}{2\pi i}\int_0^{2\pi} \frac{\gamma'(t)}{\gamma(t)-a}\,dt \\ &= \frac{1}{2\pi i}\int_0^{2\pi} \frac{f'\left(e^{it}\right) ie^{it}}{f\left(e^{it}\right)-a}\,dt = \frac{1}{2\pi i}\int_{\mathbb{T}} \frac{(f(z)-a)'}{f(z)-a}\,dz. \qquad \text{(s9.1)}\end{aligned}$$

Since a is in the unbounded component of $\mathbb{C}\setminus\gamma^*$, it follows that $\mathrm{Ind}_\gamma(a) = 0$ and thus (s9.1) implies $a \notin f(U)$, a contradiction. □

9.3. Other Special Regions

9.11. When $r_n \downarrow 0$, $f_n(z) \stackrel{\text{def}}{=} f\left(r_n z\right)$, and

$$K_m \stackrel{\text{def}}{=} \left\{ z \,:\, \frac{1}{m} \leq \Re(z) \leq \Re(b+1), |\Im(z)| \leq |\Im(b)| \leq \Re(b+1) \right\}$$

then $\{f_n\}_{n\in\mathbb{N}}$ is uniformly bounded on K_m and Vitali's theorem implies there is a sequence $\{f_{n_k}\}_{k\in\mathbb{N}}$ converging uniformly on K_m. Iteration of the (Arzelà-Ascoli) procedure used in proving Vitali's theorem leads to a subsequence, say $\{g_p\}_{p\in\mathbb{N}}$, converging uniformly on each compact set K_m. Because $|r_n b - r_n| \downarrow 0$, the result follows. □

9.12. There is a Möbius transformation $\phi : \{\, w \,:\, \Re w < M \,\} \stackrel{\text{epi}}{\mapsto} U$ and $g \stackrel{\text{def}}{=} \phi \circ f$ is bounded near zero. Hence zero is a removable singularity of g and $\phi^{-1} \circ g = f$. □

9.13. If, for each r in (r_1, r_2) and some z_r on $C_0(r)$, $f\left(z_r\right) = 0$ then $f^{-1}(0)$ contains a point of condensation z_∞ in $A(0, r_1 : r_2)^\circ$. It follows that $f(z) \equiv 0$ in $A(0, r_1 : r_2)^\circ$, a contradiction. □

9.14. If $R > R - \epsilon > r + \eta > r$, the maximum modulus theorem implies that for some F_ϵ in $H\left(D(0, R-\epsilon)^\circ\right)$, $p_n \stackrel{\text{u}}{\to} F_\epsilon$. If $z \in D(0,R)^\circ$ and $|z| > 0$ then for some positive ϵ, $|z| > r + \epsilon$. However, if $\epsilon < \delta$ and $R > R-\delta > r+\delta$ then $F_\delta(z) = F_\epsilon(z)$ in $D(0, R-\epsilon)$. Hence if $F(z) \stackrel{\text{def}}{=} F_\epsilon(z)$ then the preceding sentence assures the value of $F(z)$ is independent of the choice of ϵ. Because $p_n \stackrel{\text{u}}{\to} f$ on every compact subset of $D(0,R)^\circ$, zero is a removable singularity of F. □

9.15. Any z in Ω is contained in a subregion Ω_1 for which the maximum modulus theorem implies $f\left(\Omega_1\right) \subset U^c$. Hence $f(\Omega) \subset U^c$. □

9.16. Note that Σ is not a region. Each component C of Σ is a (convex) region and ∂C consists of finite or infinite line segments. The Schwarz reflection principle yields an entire function F_C such that $F_C\big|_C = f\big|_C$. If C_1 and C_2 are components and $C_1^c \cap C_2^c \stackrel{\text{def}}{=} B \neq \emptyset$ then $F_{C_1}(z) = F_{C_2}(z)$ on B, whence $F_{C_1} = F_{C_2}$. □

9.17. Since $f(z) = \sum_{n=-\infty}^{\infty} c_n z^n$ in $A(0, r : R)^\circ$, for $p(z) \stackrel{\text{def}}{=} \sum_{n=-\infty}^{-1} c_n z^n$ in $H\left(A(0, r : \infty)^\circ\right)$ and $q(z) \stackrel{\text{def}}{=} \sum_{n=0}^{\infty} c_n z^n$ in $H\left(D(0,1)^\circ\right)$, $f = p + q$ on $A(0, r : 1)^\circ$. Thus **8.25** applies to q and for some s in $(r, 1)$, p is bounded in $A(0, s : 1)$. □

9.18. If $f(x+iy) \stackrel{\text{def}}{=} u(x,y) + iv(x,y)$, the Cauchy-Riemann equations imply that on any nondegenerate interval $[p+iq, r+is]$ contained in Ω, one of the following four situations obtains:

$$\text{a)}\ \begin{matrix} u_x > 0,\, v_y > 0 \\ u_y < 0,\, v_x > 0 \end{matrix}, \quad \text{b)}\ \begin{matrix} u_x < 0,\, v_y < 0 \\ u_y < 0,\, v_x > 0 \end{matrix},$$

$$\text{c)}\ \begin{matrix} u_x < 0,\, v_y < 0 \\ u_y > 0,\, v_x < 0 \end{matrix}, \quad \text{d)}\ \begin{matrix} u_x > 0,\, v_y > 0 \\ u_y > 0,\, v_x < 0 \end{matrix}.$$

If $p \geq r$, $q \geq s$, and a) obtains then for $V(t) \stackrel{\text{def}}{=} v(pt + r(1-t), qt + s(1-t))$,

$$\begin{aligned} V(1) - V(0) = v(p,q) - v(r,s) &= \int_0^1 V'(t)\,dt \\ &= (p-r)\int_0^1 v_x(pt + r(1-t), qt + s(1-t))\,dt \\ &\quad + (q-s)\int_0^1 v_y(pt + r(1-t), qt + s(1-t))\,dt \\ &> 0. \end{aligned}$$

The preceding argument can be repeated in fifteen other situations, which, together with that above, exhaust all possibilities and are depicted in the following display.

u_x	v_y	$\rightarrow$	$+\ +$	$+\ +$	$-\ -$	$-\ -$
u_y	v_x		$-\ +$	$+\ -$	$-\ +$	$+\ -$
$p \geq r$	$q \geq s$		$?+$	$+?$	$-?$	$?-$
$p > r$	$q \leq s$		$+?$	$?$	$?+$	$-?$
$p \leq r$	$q \geq s$		$-?$	$?+$	$?-$	$+?$
$p \leq r$	$q \leq s$		$?-$	$-?$	$+?$	$?+$

The homolog of V is a function U and the entries $?+$, $-?$, etc. describe the circumstances of the pair $\{U(1) - U(0), V(1) - V(0)\}$: e.g., $?+$, means $U(1) - U(0) \in \mathbb{R}$ and $V(1) - V(0)\} > 0$. In each situation,

$$f(p+iq) \neq f(r+is).$$

□

9.19. There is a Laurent series representation $g(z) = \sum_{n=-\infty}^{\infty} c_n z^n$. If the series contains infinitely many nonzero terms then zero or infinity is an essential singularity of g and Picard's theorem implies g is not injective. The Fundamental Theorem of Algebra implies

$$g(z) = c_0 + c_1 z \text{ or } g(z) = \frac{c_{-1}}{z} + c_0.$$

Because g is injective, $c_{-1}c_1 \neq 0$. If $c_0 \neq 0$ then $g\left(-\frac{c_0}{c_1}\right) = 0$ or $g\left(-\frac{c_1}{c_0}\right) = 0$, each a contradiction. □

9.20. If $f(a) = 0$ and $f \not\equiv 0$ then for all z in some $A(a, 0 : r)$ contained in Ω, $f(z) \neq 0$ and thus $\int_{|z-a|=r} \frac{f'(z)}{f(z)}\, dz > 0$, whereas $\int_{|z-a|=r} \frac{f_n'(z)}{f_n(z)}\, dz \equiv 0$, a contradiction.

If $\Omega \stackrel{\text{def}}{=} U \setminus \{0\}$ and $h_n(z) \stackrel{\text{def}}{=} z^n$ then each h_n is never zero in Ω while $h_n \stackrel{\text{u}}{\to} 0$ on compact subsets of Ω.

If $\{g_n\}_{n\in\mathbb{N}}$ is a sequence of functions in $H(\Omega)$ and $g_n \stackrel{\text{u}}{\to} g$ on compact subsets of Ω, then each $f_n \stackrel{\text{def}}{=} e^{g_n}$ is never zero, $f_n \stackrel{\text{u}}{\to} f \stackrel{\text{def}}{=} e^g$ on compact subsets of Ω, and f is never zero in Ω. □

9.21. If the conclusion is false, there is some neighborhood $V(b)$ and a subsequence $\{f_{n_k}\}_{k\in\mathbb{N}}$ such that each f_{n_k} is never zero in V. From **9.20** and the equation $f(b) = 0$, it follows that $f \equiv 0$ in $V(b)$ whence also in Ω, a contradiction. □

9.22. If $a \in \Omega$ then $f(z) = \sum_{n=0}^{\infty} c_n(z-a)^n$ and $R_c \geq \Re(a)$. Hence if $f(z) = \sum_{n=0}^{\infty} b_n(z-(a+1))^n$ then $f(z+1) = \sum_{n=0}^{\infty} b_n(z-a)^n = 2f(z)$ whence $b_n = 2c_n, n \in \mathbb{N}$. Hence $\Re(a+1) \leq R_b = R_c$. It follows by induction that $R_c = \infty$. □

9.23. a) For $\Gamma_R \stackrel{\text{def}}{=} \{z : z \in \Omega, |z| = R\} \cap \Omega$, if $z \in \Omega$ and $R > |z|$, Cauchy's theorem implies $f(z) = \frac{1}{2\pi i}\left(\int_{-R}^{R} + \int_{\Gamma_R} \frac{f(t)}{t-z}\, dt\right)$. For large R and t in Γ_R, $|t-z| = |t|\cdot\left|1 - \frac{z}{t}\right| \geq \frac{R}{2}$. Thus the second integral is majorized by $2\pi M R^{-r}$, which converges to zero as $R \to \infty$.

b) If $f(z) = e^{z^2}$ then $f \in H(\Omega)$, f is not constant, and $f\left(i\mathbb{R}^+\right) \subset \mathbb{R}$. □

9.24. a) Because f and g are continuous on $C_0(R)$, $f * g(z)$ exists. Since

$$f(w), g(w) \in H\left(\left\{w : r < R_1 < |w| < R_2 \leq \frac{|z|}{2}\right\}\right),$$

Cauchy's theorem implies $f * g(z)$ is independent of R.

b) As the limit of Riemann sums, $f * g \in H(\Omega)$.
c) As rational functions, $f(z-w)$ and $g(w)$ may be written as follows:

$$f(z-w) = \sum_{k=1}^{K} \left(\sum_{m=1}^{M_k} \frac{A_{km}}{(z-w-\alpha_{km})^{\beta_{km}}} \right);$$
$$g(w) = \sum_{l=1}^{L} \left(\sum_{n=1}^{N_l} \frac{B_{ln}}{(w-\gamma_{ln})^{\eta_{ln}}} \right).$$

Hence Cauchy's formula implies that $f * g$ is rational. □

9.25. The series representations

$$f(z) = (z-a)^k (c_0 + \cdots),\ g(z) = z^{-m} (d_0 + \cdots),$$

which are locally valid, imply $h(z) = (z-a)^{-km} (e_0 + \cdots)$. □

9.26. Let A be $\mathrm{Res}(f, a)$. □

9.27. The function $g \stackrel{\text{def}}{=} f \circ \Phi_{-a}$ is in $A\left(A(0,0:1)^\circ\right)$ and $g(\mathbf{T}) \subset \mathbb{R}$. Hence if $g(z) = \sum_{n=-\infty}^{\infty} c_n z^n$ then Schwarz's reflection principle implies that for some G in $H(\mathbb{C} \setminus \{0\})$, $G(z)\big|_{U\setminus\{0\}} = g(z)$. Thus it may be assumed that $g(z) = \sum_{n=-\infty}^{\infty} c_n z^n$, a representation valid in $\mathbb{C} \setminus \{0\}$.

For each θ in $\mathbb{R}$,

$$g\left(e^{i\theta}\right) = \sum_{n=-\infty}^{\infty} c_n e^{in\theta} \in \mathbb{R},\ g\left(e^{-i\theta}\right) = \sum_{n=-\infty}^{\infty} c_n e^{-in\theta} \in \mathbb{R} \text{ and}$$
$$\overline{g\left(e^{-i\theta}\right)} = \sum_{n=-\infty}^{\infty} \overline{c_n} e^{in\theta} = g\left(e^{-i\theta}\right),$$

whence $c_n = \overline{c_{-n}}$. Thus

$$\begin{aligned} g\left(e^{i\theta}\right) + g\left(e^{-i\theta}\right) &= \sum_{n=-\infty}^{\infty} 2c_n \cos n\theta \\ &= 2c_0 + 2\sum_{n=1}^{\infty} \Re\left(c_n\right) \cos n\theta \stackrel{\text{def}}{=} k(\theta), \end{aligned}$$

an element of $C([0, 2\pi], \mathbb{R})$. The orthogonality properties of the system

$$\{\sin n\theta, \cos n\theta\}_{n \in \mathbb{Z}^+}$$

imply $c_n = \dfrac{1}{2\pi} \displaystyle\int_0^{2\pi} k(\theta) \cos n\theta \, d\theta$, i.e., each c_n is real and $c_n = c_{-n}$. In sum, if $z \neq 0$, $g(z) = 2c_0 + \sum_{n=1}^{\infty} c_n \left(z^n + z^{-n}\right)$. Induction shows that for some p_n in $\mathbb{C}[z]$, $(z^n + z^{-n}) = p_n(z + z^{-1}), n \in \mathbb{N}$. Because $\{z + z^{-1} : z \neq 0\} = \mathbb{C}$, it follows that for some entire function F, if $z \neq 0$ then $g(z) = F(z + z^{-1})$. Finally note that $f = g \circ \Phi_a$. □

9.28. Vitali's theorem applied in an $N_0(a)$ implies $f_n \stackrel{u}{\to} 0$ on compact subsets of $N_0(a)$. If $b \in \Omega$ then Ω contains a polygonal path π connecting a to b. For a set $\{N_k\}_{k=0}^K$ of open neighborhoods, their union covers π, and $N_k \cap N_{k+1} \neq \emptyset, 0 \leq k \leq K-1$. The identity theorem for holomorphic functions implies $f_n(z) \stackrel{u}{\to} 0$ on each compact subset of each N_k. Hence $\lim_{n\to\infty} f_n(b) = 0$. Vitali's theorem applies again. □

9.29. a) Note that Σ is not a region. For $\Sigma_\pm \stackrel{\text{def}}{=} \Sigma \cap \Pi_\pm$,

$$f : \Sigma_+ \ni w \mapsto (\Phi_1(w))^2$$

maps $(\Sigma_+)^\circ$ conformally on Π_+ and $g : \Pi_+ \ni z \mapsto \dfrac{1+iz}{1-iz}$ maps Π_+ conformally onto U. Hence

$$h : \Sigma \ni z \mapsto \begin{cases} f \circ g(z) & \text{if } z \in \Sigma_+ \\ f \circ g(-z) & \text{if } z \in \Sigma_- \end{cases}$$

is in $H(\Sigma)$ and $h(\Sigma) \subset U$. If $F \in H(U)$ and $F\big|_\Sigma = h$ then $\lim_{z\to\frac{1}{2}} h(z)$ exists. Direct calculations show that

$$\lim_{y\downarrow 0} h\left(\frac{1}{2} + iy\right) \neq \lim_{y\uparrow 0} h\left(\frac{1}{2} + iy\right).$$

b) Morera's and Cauchy's theorems applied to F show that $F \in H(U)$. □

9.30. The region $\Omega \stackrel{\text{def}}{=} \mathbb{C}$ is simply connected, whereas if $f(z) = e^z$ then $f(\Omega) = \mathbb{C} \setminus \{0\}$, which is not simply connected. □

9.31. If $f(z) = e^{-e^{\pi z}}$ then $\lim_{\Re(z)\to\infty} f(\Re(z)) = 0$, while $f(x+i) \uparrow \infty$ as $x \uparrow \infty$. □

9.32. If $\epsilon > 0$ then $\{z : z \in \Omega, |f(z)| > M + \epsilon\} \stackrel{\text{def}}{=} A_\epsilon$ is open.

If $a \in F \cap \mathbb{C}$ then for all z in some nonempty $D(a,r)^\circ$, $|f(z)| < M + \epsilon$. If $a \in F \setminus \mathbb{C}$, i.e., if $a = \infty$, then for some positive R and all z such that $|z| > R$ and $z \in \Omega$, $|f(z)| < M + \epsilon$. In sum, F is contained in an open set that does not meet A_ϵ whence $A_\epsilon^c \subset \Omega$. Because $\infty \notin A_\epsilon^c$ it follows that A_ϵ^c is compact. Hence the maximum modulus theorem obtains for f on A_ϵ.

However, if $z \in \partial A_\epsilon$ then

$$|f(z)| \geq M + \epsilon \text{ because } |f(z)| > M + \epsilon \text{ in } A_\epsilon \text{ and}$$
$$|f(z)| \leq M + \epsilon \text{ because } z \in \Omega,$$

i.e., $|f(z)| = M + \epsilon$, and so $|f(z)| \equiv M + \epsilon$ in A_ϵ. Thus A_ϵ is a relatively open and closed subset of Ω. Since Ω is a region, $A_\epsilon = \Omega$ or $A_\epsilon = \emptyset$. The former alternative contradicts the hypothesis re the behavior of f near $\partial_\infty \Omega$. □

9.33. If $f \in H(\Omega)$ and $f(z) \not\equiv 0$ on Ω, assume $f(a) \neq 0$. Then $f(\Omega)$ contains some closed disc $D(f(a), r)$. The geometry of the situation implies that for some w on $\partial D(f(a), r)$, $|w| > |f(a)|$. □

10
Entire Functions

10.1. Elementary Theory

10.1. Because

$$g(z) \stackrel{\text{def}}{=} \frac{f(z)-f(0)}{z} \in \mathcal{E}$$

and $|g(z)|$ is bounded, Liouville's theorem implies g is a constant. Since $\lim_{|z|\to\infty} g(z) = 0$, $f(z) \equiv f(0)$. □

10.2. Because $f(0) = 0$, **10.1** applies. □

10.3. Because $f(z) = \sum_{n=0}^{m-1} c_n z^n + z^m \sum_{n=m}^{\infty} c_n z^{n-m} \stackrel{\text{def}}{=} g(z) + z^m h(z)$, it follows that $\lim_{r\to\infty} M(r;g) = 0$ and so $\lim_{|z|\to\infty} h(z) = 0$. Since $h \in \mathcal{E}$, $h = 0$. □

10.4. It may be assumed that $g \not\equiv 0$. If $h(z) \stackrel{\text{def}}{=} \dfrac{f(z)}{g(z)}$ then for some positive r, $|h|$ is bounded in $\Omega \stackrel{\text{def}}{=} \mathbb{C} \setminus D(0,r)$. Hence every z in $Z(g) \cap \Omega$ is a removable singularity and $h \in H(\Omega)$. If $\{a_m\}_{m=1}^{M}$ is the (finite!) set of zeros of g in $D(0,r)$ then $G(z) \stackrel{\text{def}}{=} f(z)\dfrac{\prod_{m=1}^{M}(z-a_m)}{g(z)} \in \mathcal{E}$ and for some K, $|G(z)| \leq K|z|^{M+k}$. From **10.3** it follows that $G \in \mathbb{C}[z]$ whence $\dfrac{f}{g}$ is a rational function. □

10.5. Because $f \in \mathcal{E}$ and $|f(z)| \leq A|z|^{\alpha}$, it follows that $f(0) = 0$ and hence $g \stackrel{\text{def}}{=} \dfrac{f(z)}{z} \in \mathcal{E}$. If $\alpha < 1$ then $\lim_{|z|\to\infty} |g(z)| = 0$ whence $g = f = 0$.

If $\alpha \geq 1$ and $f(z) \stackrel{\text{def}}{=} \sum_{n=0}^{\infty} c_n z^n$ then $h(z) \stackrel{\text{def}}{=} \dfrac{f(z) - \sum_{n=0}^{[\alpha]-1} c_n z^n}{z^{[\alpha]}} \in \mathcal{E}$ and h is bounded, whence a constant. □

10.6. If $a \in \left(\{z_n\}_{n\in\mathbb{N}}\right)^{\bullet}$ and $w_n = (-1)^n$ then for no f, $f(z_n) = w_n$. Hence a necessary condition is that $\left(\{z_n\}_{n\in\mathbb{N}}\right)^{\bullet} = \emptyset$.

On the other hand, if $\left(\{z_n\}_{n\in\mathbb{N}}\right)^{\bullet} = \emptyset$, the Weierstraß product theorem ensures the existence of a g such that $g(z_n) \equiv 0$ and $g'(z_n) \neq 0$. Thus near z_n, $g(z) = c_{n1}(z-z_n) + c_{n2}(z-z_n)^2 + \cdots$ and $c_{n1} \neq 0$. Consequently, if $p_n(z) = \dfrac{w_n}{c_{n1}(z-z_n)}$ then near z_n,

$$g(z)p_n(z) = w_n + d_{n1}(z-z_n) + \cdots.$$

The Mittag-Leffler theorem implies the existence of a meromorphic function h such that the principal part of h at z_n is p_n. It follows that $f \stackrel{\text{def}}{=} gh \in \mathcal{E}$ and $f(z_n) = wn$. $\square$

10.7. Because $f(\mathbb{R}) \subset \mathbb{R}$, each c_n in $f(z) \stackrel{\text{def}}{=} \sum_{n=0}^{\infty} c_n z^n$ is real. Since $f(i\mathbb{R}) \subset i\mathbb{R}$, $c_{2n} \equiv 0$. $\square$

10.8. Because $f(x+iy) = \Re(f(x+iy))+i\Im(f(x+iy)) \stackrel{\text{def}}{=} u(x,y)+iv(x,y)$, it follows that $u(x,0) = \sum_{n=0}^{\infty} a_n x^n$ resp. $v(x,0) = \sum_{n=0}^{\infty} b_n x^n$, the series converging on $\mathbb{R}$. The corresponding series $\sum_{n=0}^{\infty} a_n z^n$ resp. $\sum_{n=0}^{\infty} b_n z^n$ converge on $\mathbb{C}$, represent g resp. h in $\mathcal{E}$ and $f(z) = g(z)+ih(z)$ on $\mathbb{R}$. $\square$

10.2. General Theory

10.9. a) Because $\cos\dfrac{(2n+1)\pi}{2} = 0, n \in \mathbb{Z}$, $f(z) \stackrel{\text{def}}{=} \cos z \stackrel{\text{def}}{=} \dfrac{e^{iz}+e^{-iz}}{2}$ is transcendental. If $a \in \mathbb{C}$ and then $b \stackrel{\text{def}}{=} a \pm \sqrt{a^2-1} \neq 0$ and $b + \dfrac{1}{b} = 2a$ whence if $\theta \in [0, 2\pi)$ and $b = |b|e^{i\theta}$ then

$$e^{i(-i(\ln|b|+\theta))} = b \text{ and } \cos(-i(\ln|b|+\theta)) = a.$$

b) For such an f, $f(\mathbb{C})$ is not dense, in denial of the Weierstraß-Casorati theorem.

c) Such an f is conformal, $f(\mathbb{C})$ is a simply connected proper subregion of $\mathbb{C}$ and $\mathbb{C} \setminus f(\mathbb{C})$ consists of more than one point. Thus Picard's theorem implies f is a constant function, a contradiction (cf. **9.30**). $\square$

10.10. If $f(z) \stackrel{\text{def}}{=} \sum_{n=0}^{N} c_n z^n, c_N \neq 0$, and $M > 0$ then for some positive r, if $|z| > r$ then $\left|\sum_{n=0}^{N-1} c_n z^{n-N}\right| < \dfrac{|c_N|}{2}$ and $\dfrac{|z^N c_N|}{2} > M$. It follows that $|f(z)| > M$ if $|z| > r$.

If f is transcendental then ∞ is an essential singularity of f and the Weierstraß-Casorati theorem implies that if $\epsilon > 0$ then for some z_n, $|z_n| > n$ and $|f(z_n)| < \epsilon$. $\square$

10.11. If $|z_n| \uparrow \infty$ and $f(z_n) = 0, n \in \mathbb{N}$, then for any polynomial p, $f(z_n)p(z_n) \equiv 0$.

Thus it may be assumed that $\#(Z(f)) < \infty$ and hence for some positive M, if $|z| > M$ then $f(z) \neq 0$.

If, for some sequence $\{z_n\}_{n\in\mathbb{N}}$, $|z_n| \geq n$, $|f(z_n)| < |z_n|^{-n}$, and $p \in \mathbb{C}[z]$ then for some ν in $\mathbb{N}$, some K_p, and all large $|z|$, $|p(z)| \leq K_p|z|^{\nu}$. Hence $|f(z_n)p(z_n)| \leq |z_n|^{-n} \cdot K_p|z_n|^{\nu}$, which converges to zero as $n \to \infty$.

If there is no sequence $\{z_n\}_{n\in\mathbb{N}}$ as described above then for some N in $\mathbb{N}$ and for all z in $\mathbb{C} \setminus D(0,N)^{\circ}$, $|f(z)| \geq |z|^{-N}$, i.e., if $|z| \geq N$ then

$|z^N f(z)| \geq 1$. Picard's theorem applied to the entire function $z^N f(z)$ is contradicted. □

10.12. For b in $\mathbb{C}$ for some a in $\mathbb{C}$, $g(a) = b$. If $g'(a) \neq 0$ then g is univalent near a and for some positive r, some h in $H(D(b,r)^\circ)$, and all w in $D(b,r)^\circ$, $g(h(w)) = w$. Then for k in $(0,r)$ and some function $\epsilon : D(0,s)^\circ \mapsto \mathbb{C}$ such that $\lim_{z\to 0} \epsilon(z) = 0$,

$$\begin{aligned} F(b+k) - F(b) &= g(h(b+k)) - g(h(b)) \\ &= g'(h(b))(h(b+k) - h(b)) \\ &\quad + \epsilon(h(b+k) - h(b))(h(b+k) - h(b)), \end{aligned}$$

whence $F'(b)$ exists and is $g'(a)h'(b)$.

If $g'(a) = 0$ then for some positive s, some m in $\mathbb{N}$, and some p univalent and holomorphic in $D(a,s)^\circ$, $g(z) = b + (p(z))^m$. Furthermore, for some positive t, $p(D(a,r)^\circ) \supset D(0,t)^\circ$ and in $H(D(0,t)^\circ)$ and some univalent h, $p(h(w)) = w$. The preceding argument, mutatis mutandis, applies and shows $F'(b)$ exists. □

10.13. Because $p - q = e^f\left(e^{g-f} - 1\right)$ and $e^{g-f} \in \mathcal{E}$ and is never zero, Picard's theorem implies that $e^{g(z)-f(z)} - 1 = 0$ has infinitely many solutions, whence the polynomial $p - q$ must be the constant zero. □

10.14. If the equation $w^3 + w^2 + a = 0$ fails to have at least two roots then $a \neq 0$ and there is one triple root, say r, and so $3r^2 + 2r = 0$. Since $a \neq 0$, $r \neq 0$, whence $r = -\frac{2}{3}$. But then $a = -\frac{4}{27}$ and

$$w^3 + w^2 - \frac{4}{27} = \left(w + \frac{2}{3}\right)^2 \left(w - \frac{1}{3}\right),$$

a contradiction, whence the equation always has at least two roots.

Hence if $-a \notin (f^3 + f^2)(\mathbb{C})$ then $f(z)$ omits two of the roots of

$$w^3 + w^2 + a = 0,$$

in denial of Picard's theorem. □

10.15. a) If $|f(c)| = a$ and $c \notin E_a^c$ then $D(c,r)^\circ \cap E_a = \emptyset$ for some positive r. Hence in $D(c,r)^\circ$, $|f(z)| \geq a$. The minimum modulus theorem implies f is a constant. But then $E_a = \emptyset = E_a^c$, whereas Picard's theorem implies $E_a^c = \mathbb{C}$.

Conversely, if $c \in E_a^c$ the continuity of f implies $|f(c)| \leq a$.

b) Because C is bounded, E_a is open, and C is a region, C contains a sequence $\{z_n\}_{n\in\mathbb{N}}$ converging to some c and $|f(z_n)| \downarrow \inf_{z\in C} |f(z)| \stackrel{\text{def}}{=} m$ and $m < a$. Because f is continuous, $|f(c)| = m$, i.e., $c \in C$. If $Z(f) \cap C = \emptyset$ then $m > 0$ and the minimal modulus theorem is denied. □

[**Note s10.1:** If $f(z) \stackrel{\text{def}}{=} e^z$ then $Z(f) = \emptyset$ and for each a, every component of E_a is unbounded.]

10.16. Let $f|_{[0,1]}$ be g. Because $g \in C^\infty([0,1], \mathbb{C})$, the coefficients in the Fourier series $\sum_{n=-\infty}^{\infty} c_n e^{2\pi i n t}$ for g satisfy, for each k in $\mathbb{N}$, inequalities of the form $|c_n| \le \frac{M_k}{|n|^k}$. Hence the series represents f on $\mathbb{R}$, and, owing to the cited inequalities for the coefficients, converges uniformly on each compact subset of $\mathbb{C}$: for some F, $F|_{\mathbb{R}} = f|_{\mathbb{R}}$. □

10.17. The set $\left\{ m + n\sqrt{2} \ : \ m, n \in \mathbb{Z} \right\}$ is dense in $\mathbb{R}$ (cf. **4.23**). Since

$$f\left(z + m + n\sqrt{2}\right) = f(z),$$

it follows that $f(z) \equiv f(0)$. □

10.18. Picard's theorem implies that for some sequence $\{z_n\}_{n\in\mathbb{N}}$, $|z_n| \uparrow \infty$ and $|f(z_n)| \ge n$. The broken line connecting z_1 to z_2 to $\ldots$, is γ^* for some γ. □

10.19. The substitution $w \mapsto ue^{\frac{ik\pi}{m}}$ in the integral shows that for k in $\mathbb{Z}$, $f\left(ze^{\frac{ik\pi}{m}}\right) = e^{\frac{ik\pi}{m}} f(z)$. Because $f(z) = \sum_{n=0}^{\infty} \int_{[0,z]} \frac{w^{2mn}}{n!}\, dw$, it follows that f is transcendental and so ∞ is an essential singularity of f. Thus f is not injective. Since $f(0) = 0$, if $a \notin f(\mathbb{C})$ then $a \neq 0$ and $\left\{e^{\frac{ik\pi}{m}} a\right\}_{k\in\mathbb{Z}} \not\subset f(\mathbb{C})$, in denial of Picard's theorem. □

10.20. Because $e^f = 1 - e^g$ and e^f is never zero, e^g is never one. But then g omits all the values $2n\pi i$, in contradiction of Picard's theorem unless g is a constant, in which case f is a constant. □

[**Note s10.2:** Fermat's conjecture asserts an impossibility: if $\{x, y, z\} \subset \mathbb{Z} \setminus \{0\}$ and $n \in \mathbb{N} \setminus \{1, 2\}$ then $x^n + y^n \neq z^n$:

$$\{\{\{x,y,z\} \subset (\mathbb{Z}\setminus\{0\})\} \wedge \{n \in (\mathbb{N}\setminus\{1,2\})\}\} \Rightarrow \{x^n + y^n \neq z^n\}.$$

To the writer's knowledge the conjecture has remained unresolved since 1665.

Corresponding to a triple $\{x, y, z\}$ in $(\mathbb{Z}\setminus\{0\})^3$ is a triple $\{F, G, H\}$ of entire functions that never vanish. Corresponding to $\mathbb{N}\setminus\{1,2\}$ is the larger set $\mathbb{C}$. From **10.19** it follows that

$$\{\{\{F,G,H\} \subset \mathcal{E}\} \wedge \{F,G,H \neq 0\} \wedge \{\alpha \in \mathbb{C}\}\} \Rightarrow \{F^\alpha + G^\alpha \neq H^\alpha\}.]$$

10.21. For some g, $f = e^g$ and $e^{g(z)-z}$ is transcendental unless it is a constant, which is the case iff $g(z) - z = a$, a constant. In that event, $f(z) = e^a e^z$.

If $g(z) - z$ is not a constant then $f(z) + e^z = e^z\left(e^{g(z)-z} + 1\right)$, and as in **Solution 10.19**, the second factor of f vanishes infinitely often. $\square$

10.22. If f is transcendental, ∞ is an essential singularity of f. Because $f^{-1}(a)$ is finite and nonempty, Picard's theorem implies $\#\left(f^{-1}(b)\right)$ is infinite, in denial of the hypothesis. $\square$

10.23. From **10.21** it follows that $f \in \mathbb{C}[z]$ and since f is injective, for some nonzero a and some b in $\mathbb{C}$, $f(z) = az + b$. $\square$

10.24. It may be assumed that $f(0) = 0$. The inverse function theorem implies f is locally injective. Thus for some positive r and s, some $H(D(0,s)^\circ)$, and some h, $h(D(0,s)^\circ) \subset D(0,r)^\circ$ and for z in $D(0,r)^\circ$, $h(f(z)) = z$. For any w_0 and some z_0, $f(z_0) = w_0$. Then $f([0,z_0])$ is a curve-image connecting 0 to w_0. The segment $[0,z_0]$ can be covered by finitely many open discs $D(z_n,r_n)^\circ$ in each of which f is invertible: for some $D(f(z_n),s_n)^\circ$, some h_n in $H\left(D(f(z_n),s_n)^\circ\right)$, and all z in $D(z,r_n)^\circ$, $h_n(f(z)) = z$. The monodromy theorem implies that for some F, $F(f(z)) \equiv z$. In short, f is injective. [In particular, by virtue of **10.22**, for some constant a, $f(z) = az$.] $\square$

10.25. If $p \in \mathbb{C}[z]$ and $Z(p) = Z(f)$ then $\dfrac{f}{p} \in \mathcal{E}$. $\square$

10.26. a) Because $\mathbb{C} \setminus f(\mathbb{C})$ contains a nonempty half-space, Picard's theorem implies f is a constant.

b) If $g \stackrel{\text{def}}{=} -i\Phi_{-1}$ and

$$h(w) \stackrel{\text{def}}{=} \ln|w| + i\theta,\ w \stackrel{\text{def}}{=} |w|e^{i\pi\theta},\ \theta \in (0,1)$$

then $g \circ h \stackrel{\text{def}}{=} f$ conformally maps U onto $\Omega \stackrel{\text{def}}{=} \{\, w \,:\, 0 < \Re(w) < 1 \,\}$. $\square$

10.27. a) If $f(A_2) \cap L_1 \neq \emptyset$ then $A_2 \cap f^{-1}(L_1) \neq \emptyset$ and so $A_2 \cap L_2 \neq \emptyset$, a contradiction. Hence the connected set $f(A_2)$ is contained in $\mathbb{C} \setminus L_1$, hence in one of A_1, B_1, say A_1. Similarly, the connected set $f(B_2)$ is contained in one of A_1, B_1. If the open set $f(B_2)$ is contained in A_1 then $f(\mathbb{C}) \cap B_1 = \emptyset$ in denial of the Weierstraß-Casorati theorem. Thus $f(B_2) \subset B_1$.

The Schwarz reflection principle implies that if z^{ρ_i} is the reflection of z in L_i then $(f(z))^{\rho_1} = f(z^{\rho_2})$. Hence if, e.g., $f(A_2)$ is a proper subset of A_1 then $f(B_2) \subsetneq B_1$ and thus $f(\mathbb{C})$ fails to contain two points, in contradiction of Picard's theorem.

b) If $f(L_2) \supsetneq L_1$, it may be assumed that $f(z_2) \in A_1$ for some z_2 in L_2. Thus some neighborhood $N_1(f(z_2))$ is contained in A_1. For some neighborhood $N_2(z_2)$, $f(N_2(z_2)) \subset N_1(f(z_2)) \subset A_1$. On the other hand, since $z_2 \in L_2$, $N_2(z_2) \cap B_2 \neq \emptyset$, whence $f(N_2 \cap B_2) \subset f(B_2) \subset B_1$, in contradiction of the preceding sentence.

Thus $f(A_1) = A_2$, $f(B_1) = B_2$, and $f(L_2) = L_1$.

c) There are linear transformations $z \mapsto pz+q$ resp. $w \mapsto rw+s$ that carry L_2 resp. L_1 onto $\mathbb{R}$ in the z-plane resp. w-plane. With respect to these transformations, the map f corresponds to a map F while the half-spaces A_1 resp. B_1 correspond to the half-spaces $\{\, z \,:\, \Im(w) > 0 \,\}$ resp. $\{\, w \,:\, \Im(w) < 0 \,\}$ and the half-spaces A_2 resp. B_2 correspond to the half-spaces $\{\, z \,:\, \Im(z) > 0 \,\}$ resp. $\{\, z \,:\, \Im(z) < 0 \,\}$. By this mechanism, the problem is reduced to consideration of F, which maps the upper resp. lower half of the z-plane onto the upper resp. lower half of the w-plane and $\mathbb{R}$ onto $\mathbb{R}$.

Hence for $F(x+iy) \stackrel{\text{def}}{=} U(x,y)+iV(x,y)$, $V_y(x,0) = U_x(x,0) \geq 0$, i.e., $R(x) \stackrel{\text{def}}{=} F(x+i0)$ is a monotonely increasing function. Furthermore, R is strictly increasing since otherwise F is a constant on a bounded infinite set and hence is a constant.

If either $\overline{\lim}_{x=\infty} F(x) < \infty$ or $\underline{\lim}_{x=-\infty} F(x) > -\infty$ then F omits infinitely many real values in denial of Picard's theorem. Thus F assumes each real value precisely once.

Picard's theorem implies that if F is transcendental then F assumes every real value, save at most one, infinitely often, whereas F assumes every real value precisely once. Thus $F \in \mathbb{C}[z]$: $F(z) = \sum_{n=0}^{N} c_n z^n$. Each coefficient c_n is real, and, since F is monotonely increasing on $\mathbb{R}$, $\deg(F)$ is odd. For each real r, $F(x) = r$ has precisely one real solution. If $\deg(F) \geq 3$ then F' vanishes at most finitely often, whence for some real r and a, $F(a) = r$ and $F'(a) \neq 0$: the multiplicity of the zero a is one, i.e., there are three or more zeros and only one is real. The nonreal zeros (which occur in conjugate pairs) are mapped by F onto nonreal numbers, i.e., not onto r, a contradiction. Thus for some positive α and some real β, $F(z) = \alpha z + \beta$. Correspondingly, for some a, b in $\mathbb{C}$, $f(z) = az + b$. □

10.28. The set $\left\{\, (u,v) \;:\; u \neq v^2 \,\right\}$ is an open set and the complement of an infinite set. Hence if $f \stackrel{\text{def}}{=} u + iv$ and $u \neq v^2$, Picard's theorem implies f is a constant. □

[**Note s10.3:** In several of the preceding items, the substitution $z \mapsto \dfrac{1}{w}$ reduces the discussion to the behavior of f at ∞ to the behavior of $f\left(\dfrac{1}{w}\right) \stackrel{\text{def}}{=} g(w)$ near zero, an essential singularity of g. In such instances the (weaker) Weierstraß-Casorati theorem serves instead of Picard's theorem.]

10.29. Note that $k \stackrel{\text{def}}{=} \dfrac{h}{g}$ is defined on $Z(f)$ and $k \in H(\mathbb{C} \setminus Z(g))$. For some β, β and k assume the same values with the same multiplicities on

$Z(f)$, i.e., if $a \in Z(f)$, $f(z) = (z-a)^p(a_0 + \cdots)$, and $\frac{h(a)}{g(a)} \stackrel{\text{def}}{=} A$, then near a,

$$\beta(z) - A = (z-a)^p(b_0 + \cdots) \text{ and } \frac{h(z)}{g(z)} - A = (z-a)^p(c_0 + \cdots).$$

Hence $\frac{h - \beta g}{f} \stackrel{\text{def}}{=} \alpha \in \mathcal{E}$ and $\alpha f + \beta g = h$. □

10.3. Order of Growth

10.30. a) If $|z| < R_c$, the series converges, whence $|c_n z^n| \to 0$ as $n \to \infty$. Thus $\max_{n \in \mathbb{Z}^+} |c_n z^n|$ exists and is for some n_0, $|c_{n_0} z^{n_0}|$. Furthermore,

$$\nu_f(r) \stackrel{\text{def}}{=} \max\left\{ n \ : \ n \in \mathbb{z}^+, |c_n r^n| = \max_{n \in \mathbb{Z}^+} |c_n r^n| \right\}.$$

b) If $0 \le n < \nu_f(r)$ then

$$r^n \left(\left| c_{\nu_f(r)} \right| r^{\nu_f(r) - n} - |c_n| \right) \ge 0. \tag{s10.1}$$

Because both factors in the left member of (s10.1) are nonnegative, the form of each factor implies that it increases as r increases. Thus (s10.1) remains valid as r increases, i.e., if $0 < r < r' < R_c$ and $0 \le n < \nu_f(r)$ then

$$\left| c_{\nu_f(r')} \right| (r')^{\nu_f(r)} \ge |c_n| (r')^n, \text{ whence } \nu_f(r') \ge \nu_f(r).$$

As a monotone $\mathbb{Z}^+$-valued function, ν is a step-function. If r_0 is a point of discontinuity and $r \downarrow r_0$ then (s10.1) implies $\nu_f(r) \downarrow \nu(r_0)$.

c) If $n \in \mathbb{N}$ then for some m greater than n, $c_m \ne 0$. If $r > \left(\frac{|c_n|}{|c_m|} \right)^{\frac{1}{m-n}}$ then $|c_m| r^m > |c_n| r^n$, whence for each n there is an r such that $\nu_f(r) > n$.

d) $\frac{\ln \ln M(r; g)}{\ln r} = \frac{\ln \ln M(ar; f)}{\ln ar} \cdot \frac{\ln ar}{\ln r}$. □

10.31. a) If $\epsilon > 0$ then for large r, $\frac{\ln \ln M(r; f)}{\ln r} < \rho(f) + \epsilon$, whence $M(r; f) < \exp\left(r^{\rho(f)+\epsilon}\right)$. The maximum modulus principle implies that for all z, $|f(z)| < \exp\left(|z|^{\rho(f)+\epsilon}\right)$. Hence $\omega(f) \le \rho(f)$.

On the other hand, if $\sigma < \rho(f)$ then for some $\{r_n\}_{n \in \mathbb{N}}$, $r_n \uparrow \infty$ and $M(r_n; f) > \exp(r_n^\sigma)$ whence $\sigma < \omega(f)$.

b) For large r, $\ln M(r; f) < k r^{\rho(f)}$ whence $\tau(f) \le v(f)$. If $\sigma < v(f)$ then for some $\{r_n\}_{n \in \mathbb{N}}$, $r_n \uparrow \infty$ and $M(r_n; f) > \exp\left(\sigma r_n^{\rho(f)}\right)$, whence $\sigma < \tau(f)$.

c) The Gutzmer coefficient estimate implies $\nu_f(r) \le M(r;f)$, whence $\zeta(f) \le \rho(f)$.

If $\zeta(f) = \infty$ then $\rho(f) \le \zeta(f)$. If $\zeta(f) < \beta < \infty$ then for large r, $|c_n| r^n \le \mu_f(r) < \exp\left(r^\beta\right)$ and if $r = \left(\frac{n}{\beta}\right)^{\frac{1}{\beta}}$ and n is large,

$$|c_n| r^n < \left(\frac{e\beta r^\beta}{n}\right)^{\frac{n}{\beta}}.$$

If $2e\beta \stackrel{\text{def}}{=} \delta$ then

$$\begin{aligned} M(r;f) &\le \sum_{n=0}^{[\delta r^\beta]} |c_n| r^n + \sum_{n=[\delta r^\beta]+1}^{\infty} |c_n| r^n \\ &< \left(\delta r^\beta + 1\right)\mu_f(r) + \sum_{n=[\delta r^\beta]}^{\infty} 2^{-\frac{n}{\beta}} \stackrel{\text{def}}{=} I(r) + II(r), \end{aligned}$$

$$\overline{\lim}_{r\to\infty} \frac{\ln\ln M(r;f)}{\ln r} \le \frac{\ln\ln(I + II)}{\ln r}.$$

Note that $\lim_{r\to\infty} II(r) = 0$. When $0 < x \le y$, the inequality $x + y \le 2y$ yields for large r,

$$\begin{gathered} \ln(I(r) + II(r)) \le \ln 2I(r) \text{ and} \\ \ln I(r) \le \ln\left(2\delta r^\beta\right) + \ln\mu_f(r) = \ln 2\delta + \beta\ln r + \ln\mu_f(r). \end{gathered}$$

Furthermore, if some $c_m \ne 0$ then from **10.30c)** it follows that for large r, $\mu(r) > |c_m| r^m$. If $p > 0$ then $M(r;pf) = pM(r;f)$ and $\mu_{pf}(r) = p\mu_f(r)$. It follows that $\rho(pf) = \rho(f)$ and $\zeta(pf) = \zeta(f)$. Hence it may be assumed that $|c_m| > 1$ and so $\ln\mu_f(r) > m\ln r$. If $m > \beta$, the estimate applied earlier yields

$$\begin{gathered} \ln 2\delta + \beta\ln r + \ln\mu_f(r) \le 2\ln\mu_f(r) \text{ and} \\ \frac{\ln\ln I(r)}{\ln r} \le \frac{\ln 2 + \ln\ln\mu_f(r)}{\ln r}, \end{gathered}$$

whence $\zeta(f) \ge \rho(f)$. □

10.32. In what follows, f is the symbol for the function under discussion.

a) $M(r;f) = e^{\tau r^\rho}$.

b) $\ln\ln M(r;f) = r$.

c) Note that f is entire since the Cauchy-Hadamard formula for the radius of convergence is $\lim_{n\to\infty} \frac{1}{q^n} = \infty$. Furthermore,

$$\ln\ln\mu_f(r) = 2\ln\ln r + \ln\frac{-3}{4\ln q},$$

whence $\rho(f) = 0$. d) Direct calculation shows

$$\nu_{F_\alpha}(r) = [\alpha r^\alpha], \ \mu_{F_\alpha}(r) = \exp(r^\alpha).$$

Because the coefficients in the power series representation of F_α are positive,

$$M(r; F_\alpha) = \sum_{n=1}^{\infty} \left(\frac{n}{\alpha}\right)^{-\frac{n}{\alpha}} r^n,$$

a monotonely increasing function of r, whence

$$\rho(F_\alpha) = \lim_{r\to\infty} \frac{\ln\ln M(r; F_\alpha)}{\ln r}.$$

From **10.31** it follows that $\rho(F_\alpha) = \alpha$. Furthermore, if $\epsilon > 0$ then for large r,

$$-\epsilon \ln r < \ln\ln M(r; F_\alpha) - \alpha \ln r < \epsilon \ln r \text{ and}$$
$$r^{-\epsilon} < \frac{\ln M(r; F_\alpha)}{r^\alpha} < r^\epsilon.$$

□

10.33. a) Since the series converges iff $|z| < 1$, $\nu_f(r) \equiv 0$. b) Because the factors $\frac{r}{n^\alpha}$ decrease as n increases, the absolute values of the terms increase so long as $\frac{r}{n^\alpha} \geq 1$, after which the absolute values decrease. Hence $\nu_f(r) = \left[r^{\frac{1}{\alpha}}\right]$. □

10.34. a) If $a > \sigma$ then $M(r; f) \leq e^{r^a}$, whence $\rho(f) \leq a$ and so $\rho(f) \leq \sigma$. If $\delta > 0$ and $\rho(f) = \sigma - \delta$ then $\overline{\lim}_{r\to\infty} |f(z)|\big|_{|z|=r} \leq e^{|z|^{\sigma-\delta}}$ in denial of the definition of σ.

b) To show $\xi \geq \rho(f)$, it suffices to assume $\xi < \infty$ in which case if $\epsilon > 0$ then for some $M(\epsilon)$ and all n in $\mathbb{N}$, $|c_n| < M(\epsilon) n^{-\frac{n}{\xi+\epsilon}}$. The last inequality provides a comparison between $|c_n|$ and the coefficient of z^n in the series for $F_{\xi+\epsilon} \stackrel{\text{def}}{=} F_\alpha$. Thus $M(r; f) \leq |c_0| + M(\epsilon) F_\alpha\left[(\alpha e)^{-\frac{1}{\alpha}} r\right]$. From **10.30**d) it follows that $\rho(f) \leq \alpha \ (= \xi + \epsilon)$.

On the other hand, by virtue of the Cauchy estimates, if $r > 0$,

$$|c_n| < r^{-n} M(r; f).$$

If $\rho(f) < \infty$ and $\epsilon > 0$ then for some $M_1(\epsilon)$,

$$\min_{r>0} r^{-n} M(r; f) \leq M_1(\epsilon) \min_{r>0} r^{-n} e^{r^{(\rho(f)+\epsilon)}} = M_1(\epsilon) \left(\frac{n}{e(\rho(f)+\epsilon)}\right)^{-\frac{n}{\rho(f)+\epsilon}}$$

$\left[\text{the minimum achieved when } r = \left(\frac{n}{\rho(f)+\epsilon}\right)^{\frac{1}{\rho(f)+\epsilon}}\right]$. Hence $\xi \le \rho(f)+\epsilon$, i.e., $\xi \le \rho(f)$, an inequality that is valid as well if $\rho(f) = \infty$.

c) As in b), for a finite $M(\epsilon)$, if $\eta \overset{\text{def}}{=} \frac{1}{e\rho(f)}\overline{\lim}_{n\to\infty} |c_n|^{\frac{\rho(f)}{n}} < \infty$ then for n in $\mathbb{N}$, $|c_n| \le M(\epsilon)\left(\frac{n}{e\rho(f)(\eta+\epsilon)}\right)^{-\frac{n}{\rho(f)}}$. Consequently,

$$M(r;f) \le |c_0| + M(\epsilon)F_{\rho(f)}\left((\eta+e)^{\frac{1}{\rho(f)}} r\right)$$

and, since the order of $F_{\rho(f)}$ is $\rho(f)$ and $F_{\rho(f)}$ is of type one of that order $(\rho(f))$, it follows that $\tau(f) \le \eta$.

Again, as in b), Cauchy estimates imply that if $\tau(f) < \infty$ then for any positive r and ϵ and some finite $M_1(\epsilon)$, $|c_n| \le M_1(\epsilon) r^{-n} e^{(\tau(f)+\epsilon)r^{\rho(f)}}$. The right member is minimal when $r = \left(\frac{n}{\rho(f)(\tau(f)+\epsilon)}\right)^{-\frac{n}{\rho(f)}}$. Hence $|c_n| \le M_1(\epsilon)\left(\frac{n}{e\rho(f)(\tau(f)+\epsilon)}\right)^{\frac{-n}{\rho(f)}}$ and $\eta \le \tau(f)+\epsilon$. □

10.35. Stirling's formula for $n!$ and **10.34** imply $\rho(f) = \frac{1}{\alpha}$. Similarly, $\tau(f) = \alpha$. □

10.36. a) If $f(z) \overset{\text{def}}{=} \sum_{n=0}^{\infty} c_n z^n$ then **10.34** leads to

$$\begin{aligned}\rho(g) &= \overline{\lim}_{p\to\infty} \frac{(p+k)\ln(p+k)}{-\ln|c_p|} \\ &= \overline{\lim}_{p\to\infty} \frac{p\ln p}{-\ln|c_p|}\,\frac{(p+k)\ln(p+k)}{p\ln p} = \rho(f).\end{aligned}$$

b) Because $f'(z) = \sum_{n=1}^{\infty} nc_n z^{n-1} \overset{\text{def}}{=} \sum_{n=0}^{1} d_n z^n$, **10.34** applies to show that $\frac{1}{\rho(f')} = \underline{\lim}_{n\to\infty} \frac{-\ln(n+1)-\ln|c_{n+1}|}{(n+1)\ln(n+1)} \cdot \frac{(n+1)\ln(n+1)}{n\ln n}$, from which it follows that $\rho(f') = \rho(f)$.

c) From **10.34** and a calculation similar to that in b),

$$\begin{aligned}\tau(f') &= \frac{1}{e\rho}\overline{\lim}_{n\to\infty} n\left((n+1)|c_{n+1}|\right)^{\frac{\rho}{n}} \\ &= \frac{1}{e\rho}\overline{\lim}_{n\to\infty} \frac{n}{n+1}\left((1+n)^{\frac{1}{n}}\right)^{\rho}(n+1)\left(|c_{n+1}|^{\frac{\rho}{n+1}}\right)^{\frac{n+1}{n}} \\ &= e^{\rho(f)}\tau(f).\end{aligned}$$

□

10.37. a) It may be assumed that $\rho(f) \le \rho(g)$.

If $\rho(f) < \rho(g)$ and $\epsilon > 0$ then for large r,

$$\exp\left(r^{\rho(g)-\epsilon}\right) - \exp\left(r^{\rho(f)+\epsilon}\right) < M(r;g) - M(r;f) \le M(r;f) + M(r;g),$$
$$< \exp\left(r^{\rho(f)+\epsilon}\right) + \exp\left(r^{\rho(g)+\epsilon}\right), \text{ and}$$
$$\frac{1}{2}\exp\left(r^{\rho(g)-\epsilon}\right) \le M(r;f) + M(r;g) < 2\exp\left(r^{\rho(g)+\epsilon}\right),$$

whence $\rho(f+g) = \rho(g)$.

If $\rho(f) = \rho(g)$ and $f+g$ is not a constant, then $\rho(f+g) \le \rho(f)$ $[= \rho(g)]$. However, if $f(z) = e^z, g(z) = -e^z + z$ then $\rho(f) = \rho(g) = 1$, whereas $\rho(f+g) = 0 < \rho(f)$ $[= \rho(g)]$.

b) Arguments similar to those in a) show that if $\rho(f) < \rho(g)$ then $\rho(fg) = \rho(g)$ whereas if $\rho(f) = \rho(g)$ and fg is not a constant then the sharpest general inequality is $\rho(fg) \le \rho(f)$ $[= \rho(g)]$.

c) If, for some F in $[0,\infty]^{[0,\infty]^2}$, $\rho(f \circ g) = F(\rho(f), \rho(g))$ then: a) when $f(z) = z$ and g is arbitrary, $F(\rho(f), \rho(g)) = F(0, \rho(g)) = \rho(g)$; b) when $f(z) = g(z) = e^z$ then $\infty = F(\rho(f), \rho(g)) = F(0,0)$. There is no such F. □

10.38. If, for a subsequence $\{n_k\}_{k\in\mathbb{N}}$ and a positive ϵ, $\dfrac{n_k}{-\ln|c_{n_k}|} \ge \epsilon$ then $|c_{n_k}|\left(e^{\frac{1}{\epsilon}}\right)^{p_k} \ge 1$, whence the series representing f does not converge in $\Omega \stackrel{\text{def}}{=} \left\{z : |z| > e^{\frac{1}{\epsilon}}\right\}$. □

10.39. For $f_n(z) \stackrel{\text{def}}{=} \sum_{m=0}^{\infty} c_m (z^n)^m \stackrel{\text{def}}{=} \sum_{k=0}^{\infty} a_{kn} z^{kn}$, **10.34** and **10.36** imply $\rho(f_n) = n\rho(f)$, valid even if $\rho(f) = \infty$.

Note that $\tau(f)$ is defined only iff $0 < \rho(f) < \infty$. When $0 < \rho(f) < \infty$, from **10.34** it follows that $\tau(f_n) = \tau(f)$. □

10.40. If $\overline{\lim}_{n\to\infty} \alpha_n \stackrel{\text{def}}{=} \alpha < \infty$ and $f(z) = z$, the requirements are fulfilled. Thus it may be assumed that $\alpha_n \to \infty$.

If the power series $\sum_{n=0}^{\infty} c_n z^n$ is such that $|c_n|^{\frac{1}{n}} \to 0$ as $n \to \infty$ and $\lim_{n\to\infty} \dfrac{|c_n n^n|}{\alpha_n} = \infty$ then $f(z)$ represented by the power series meets the conditions.

If $\beta_n \stackrel{\text{def}}{=} [\alpha_n]$, $K_n \stackrel{\text{def}}{=} \ln n - \dfrac{\ln \alpha_n}{\beta_n} - 1$, and

$$c_m \stackrel{\text{def}}{=} \begin{cases} 0 & \text{if } m \notin \{\beta_n\}_{n\in\mathbb{N}} \\ e^{-K_n\beta_n} & \text{if } m = \beta_n \end{cases},$$

then $$|c_m|^{\frac{1}{m}} = \begin{cases} 0 & \text{if } m \notin \{\beta_n\}_{n\in\mathbb{N}} \\ e^{-K_n} & \text{if } m = \beta_n \end{cases} \text{ and thus}$$

$$\frac{M(n;f)}{\alpha_n} \ge \frac{f(n)}{\alpha_n} \ge \frac{c_{\beta_n} n^{\beta_n}}{\alpha_n} = e^{-K_n\beta_n + \beta_n \ln n - \ln \alpha_n}$$
$$= e^{\beta_n} \uparrow \infty.$$ □

10.41. Because $M(r;f) \le e^{r^\alpha}$, it follows that $\rho(f) \le \alpha$. From **10.34** it follows that if $\epsilon > 0$ then for all large n, $\dfrac{n \ln n}{-\ln|c_n|} = \rho(f) \le \alpha + \epsilon$. Since $f \in \mathcal{E}$, for large n, $|c_n|^{\frac{1}{n}} < 1$, $-\ln|c_n| > 0$, whence $|c_n| \le n^{-\frac{n}{\alpha}}$. □

10.42. It may be assumed that $|a_n| \le |a_{n+1}|$ and that $f(0) = 1$ (whence $|a_1| > 0$). If $r > 0$, the result in **8.27**a) may be reformulated in the current context as follows:

If $f \in H(D(0,r)^\circ)$ and $Z(f) \cap D(0,r)^\circ = \{a_n\}_{n=1}^N$ then

$$|f(z) \le M(r;f) \prod_{n=1}^{N} \left| \frac{r(z-a_n)}{r^2 - \overline{a_n} z} \right|.$$

If $\epsilon > 0$ then for large r,

$$\frac{1}{|a_n|^n} \le \frac{1}{|a_1 a_2 \cdots a_n|} \le \frac{M(r;f)}{r^n} < \frac{\exp\left(r^{\rho(f)+\epsilon}\right)}{r^n}.$$

The minimum of $\dfrac{\exp\left(r^{\rho(f)+\epsilon}\right)}{r^n}$ is achieved if $r = \left(\dfrac{n}{\rho(f)+\epsilon}\right)^{\frac{1}{\rho(f)+\epsilon}}$, whence $\left|\dfrac{1}{a_n}\right| < \left(\dfrac{e(\rho(f)+\epsilon)}{n}\right)^{\frac{1}{\rho(f)+\epsilon}}$. Hence if $\delta > \epsilon$ then

$$\sum_{n=1}^{\infty} \left| \frac{1}{a_n} \right|^{\rho(f)+\delta} < \infty.$$ □

10.43. Because $M(r;f) \le \exp\left(\sum_{n=2}^{\infty} \dfrac{r}{n(\ln n)^2}\right)$, it follows that $\rho(f) \le 1$. On the other hand, $Z(f) = \left\{\dfrac{1}{n(\ln n)^2}\right\}_{n=2}^{\infty} \stackrel{\text{def}}{=} \mathbf{a}$. When $0 < s < 1$, the integral test for convergence of $\sum_{n=2}^{\infty} \dfrac{1}{n^s (\ln n)^2 s}$ leads to consideration of

$$\int 2^R \frac{dx}{x^s (\ln x)^2 s} = \left.\frac{x^{1-s}}{(\ln x)^{2s}}\right|_2^{\infty} + 2s \int_2^R \frac{x^{-s}}{(\ln x)^{2s+1}}\, dx.$$

If the integral in the left member is convergent then so is the integral in right member, whereas the first term in the right member is unbounded as $R \to \infty$. Thus $v(\mathbf{a}) = 1$ and so **10.42** implies $\rho(f) \ge 1$.

The same argument implies that for

$$f_N(z) \stackrel{\text{def}}{=} \prod_{n=N}^{\infty} \left(1 + \frac{z}{n(\ln n)^2}\right) \text{ and } \eta_N \stackrel{\text{def}}{=} \sum_{n=N}^{\infty} \frac{1}{n(\ln n)^2},$$

$M(r;f_N) \le \exp(r\eta_N)$. Thus $f(z) \stackrel{\text{def}}{=} p_N(z)f_N(z)$ and, since $\rho(p_N) = 0$, if $\epsilon > 0$ then for large r, $M(r;f) < \eta^\epsilon \exp(r\eta_N)$. Thus $\ln M(r;f) \le \epsilon + r\eta_N$ and, since $\lim_{N\to\infty} \eta_N = 0$, $\tau(f) = 0$. □

10.44. If $|z| \le \frac{1}{2}$ then

$$\begin{aligned} E(z,p) &= \exp\left(-\frac{z^{p+1}}{p+1} - \cdots\right), \\ |E(z,p)| &\le \exp\left(\frac{|z|^{p+1}}{p+1} + \cdots\right) \\ &< \exp\left(\frac{|z|^{p+1}}{1-|z|}\right) \le \exp\left(\frac{|z|^s}{1-|z|}\right), \text{ and} \\ |E(z,p)| &< \exp\left(2|z|^s\right). \end{aligned}$$

For large r,

$$|E(z,p)| < (1+r)\exp\left(r^p\left(\frac{1}{p} + \cdots + \frac{1}{r^{p-1}}\right)\right) < 2r\exp\left(\frac{2}{p}r^p\right).$$

Hence, since $p < s$, for large r, say $r \ge R$, $|E(z,p)| < \exp(2|z|^s)$. On the other hand, if $|z| \in \left[\frac{1}{2}, R\right]$ then for some K, $|E(z,p)| \le \exp(K|z|^s)$. □

10.45. Because $Z\left(\frac{1}{\Gamma}\right) = \left\{0, 1, \frac{1}{2}, \ldots\right\}$, **10.42** implies that $\rho\left(\frac{1}{\Gamma}\right) \ge 1$. The estimate in **10.44** implies that if $1 < s \le 2$ then

$$M\left(r;\frac{1}{\Gamma}\right) \le re^{\gamma r}\exp\left(A(s)\sum_{n=1}^{\infty}\left|\frac{r}{n}\right|^s\right),$$

whence $\ln M\left(r;\frac{1}{\Gamma}\right) \le \ln r + \gamma r + A(s)\frac{\pi^2 r^s}{6}$. The estimate used earlier leads to $\rho\left(\frac{1}{\Gamma}\right) \le s$: $\rho\left(\frac{1}{\Gamma}\right) = 1$.

Because $\ln r + \gamma r \le \ln M\left(r;\frac{1}{\Gamma}\right)$, it follows that $\tau\left(\frac{1}{\Gamma}\right) = \infty$. In sum,

$$\prod_{n=2}^{\infty}\left(1 + \frac{z}{n(\ln n)^2}\right) \text{ resp. } e^{az} \text{ resp. } \frac{1}{\Gamma(z)},$$

are all of order one but their types of that order are 0 resp. a resp. ∞. □

10.46. a) If $p = 0$ then $E(z,p) - 1 = z$. If $p > 0$ then as an entire function, E is representable by a power series: $E(z,p) \stackrel{\text{def}}{=} 1 + \sum_{n=1}^{\infty} c_{np} z^n$. Furthermore, $E'(z,p) = -z^p \exp\left[\sum_{n=1}^{p} \frac{z^n}{n}\right]$, whence

$$c_{np} \begin{cases} = 0 & \text{if } 1 \le n \le p \\ < 0 & \text{if } n > p \end{cases}.$$

Since $E(1,p) = 0$, it follows that

$$1 = -\sum_{n=p+1}^{\infty} c_{np} = \sum_{n=p+1}^{\infty} |c_{np}|$$

$$|(E(z,p) - 1| \le |z|^{p+1} \sum_{n=p+1}^{\infty} |c_{np}| = |z|^{p+1}.$$

b) From a), the definition of the exponent of divergence, and the general convergence properties of infinite products, it follows that

$$G(z) \stackrel{\text{def}}{=} \prod_{n=1}^{\infty} E\left(\frac{z}{a_n}, \delta(\mathbf{a})\right) \in \mathcal{E}$$

and that $\dfrac{f(z)}{z^\lambda G(z)} \in \mathcal{E}$ and is never zero, whence is of the form $e^{g(z)}$. □

10.47. a) The convergence properties of infinite products and the definition of exponent of divergence imply the result.

b) For R positive, let $n(R)$ be $\#\left(Z(f) \cap D(0,R)^\circ\right)$. Then in

$$E\left(\frac{z}{z_n}, \delta(\mathbf{z})\right) \stackrel{\text{def}}{=} \left(1 - \frac{z}{z_n}\right) \exp\left[P_n(z)\right],$$

P_n is a polynomial and $\deg(P_n) = \delta(\mathbf{c})$. Thus

$$f(z) = z^\lambda \prod_{n=1}^{n(R)} \left(1 - \frac{z}{z_n}\right) \left[\exp\left(g(z) + \sum_{k=1}^{n(R)} p_k(z)\right)\right]$$
$$\times \left[\prod_{n=n(R)+1}^{\infty} \left(1 - \frac{z}{z_n}\right) e^{P_n(z)}\right] \stackrel{\text{def}}{=} p_R(z) q_R(z).$$

Direct calculation and the maximum modulus principle imply

$$M(2R; f) \ge M\left(2R; q_R\right).$$

In $D(0,R)$, if $n > n(R)$, $\ln\left(1 - \frac{z}{z_n}\right)$ may be determined so that

$$\Im\left(\ln\left(1 - \frac{z}{z_n}\right)\right) \in [0, 2\pi)$$

and then for

$$h_R(z) \stackrel{\text{def}}{=} g(z) + \sum_{n=1}^{n(R)} P_n(z) + \sum_{n=n(R)+1}^{\infty} \left[\ln\left(1 - \frac{z}{z_n}\right) + P_n(z)\right],$$

$q_R(z) = e^{h_R(z)}$. The definition of $\delta(\mathbf{c})$ implies that for large n in the infinite series above, as in the discussion of the Weierstraß factorization theorem, the terms are dominated by the terms of $\sum_{n=1}^{\infty} \frac{2R^{\delta(\mathbf{c})+1}}{|c_n|^{\delta(\mathbf{c})+1}}$. Thus in $D(0,R)$, $\Re(h_R(z)) < \ln M(2R; f)$ and according to **8.19**, if $h_R(z) \stackrel{\text{def}}{=} \sum_{m=0}^{\infty} b_m z^m$ then $|b_m| < 2\frac{\ln M(2R;f) - \Re(b_0)}{R^m}$, $m \in \mathbb{N}$. Direct calculation shows that if $n > \rho$ then

$$b_n = \frac{g^{(n)}(0)}{n!} - \sum_{k=n(R)+1}^{\infty} \frac{1}{nc_k^n} \text{ and}$$

$$\left|\frac{g^{(n)}(0)}{n!}\right| \le 2\frac{\ln M(2R;f) - \Re(b_0)}{R^n} + \sum_{k=n(R)+1}^{\infty} \frac{1}{|c_n|^n}. \tag{s10.2}$$

If $\rho + \epsilon, n$ then $\ln M(2R; f) < (2R)^{\rho+\epsilon}$ and since $n > \kappa(\mathbf{c})$, as $R \to \infty$, both terms in the right member of (s10.2) approach zero. It follows that if $n > \rho$, $g^{(n)}(0) = 0$, i.e., g is a polynomial p and $\deg(p) \le [\rho]$. □

10.48. a) The results follow from **10.34**.

b) Because $\frac{\sin \pi z}{\widetilde{f}(z)} \stackrel{\text{def}}{=} h(z) \in \mathcal{E}$ and $Z(h) = \emptyset$, it follows that for some k in $\mathcal{E}$, $h = e^k$. From **10.43** it follows that k is a polynomial and $\deg(k) \le 1$: $k(z) = a_0 + a_1 z$. Furthermore, h is an even function whence $a_1 = 0$ and thus $e^{a_0} = 1$, i.e., $h \equiv 1$. □

10.49. For

$$M_0 \stackrel{\text{def}}{=} 2 \max_{x \in [-1,1]} |g(x)| \text{ and}$$

$$M_n \stackrel{\text{def}}{=} M_{n-1} + 2 \max_{x \in [-n-1, n+1]} |g(x)|, n \in \mathbb{N},$$

if $G(x) \stackrel{\text{def}}{=} M_1 + 2|x|$ when $0 \le x \le \frac{3}{2}$, and, according as

$$x \in \left(\frac{2n+1}{2}, \frac{2n+3}{2}\right), \text{ resp. } x \in \left(-\frac{2n+3}{2}, -\frac{2n+1}{2}\right),$$

$$G(x) \stackrel{\text{def}}{=} G\left(\frac{2n+1}{2}\right) \mp \left(M_n - G\left(\frac{2n+1}{2}\right)\right)\left(x \mp \frac{2n+1}{2}\right)$$

then $G > |g| \ge g$, $G(x) = G(-x)$, G is a monotonely decreasing resp. increasing on $(-\infty, 0]$ resp. $[0, \infty)$, continuous, and piecewise linear.

If a)

$$H_1(x) \stackrel{\text{def}}{=} \begin{cases} G(0) & \text{if } -\infty < x \le 0 \\ G(x) & \text{if } 0 < x < \infty \end{cases},$$

b) n in $\mathbb{N}$, c) k_n in $2\mathbb{N}$, d) $n < k_n < k_{n+1}$, and e) $\left(\frac{n+1}{n}\right)^{k_n} > H_1(n+1)$ then

$$\{x \in [n, n+1)\} \Rightarrow \left\{\left(\left|\frac{x}{n+1}\right|^{k_n}\right)^{\frac{1}{n}} < 1\right\},$$

whence $F_1(z) \stackrel{\text{def}}{=} \sum_{n=1}^{\infty} \left(\frac{z}{n+1}\right)^{k_n} \in \mathcal{E}$.

For

$$H_2(x) \stackrel{\text{def}}{=} \begin{cases} G(x) & \text{if } -\infty < x \le 0 \\ G(0) & \text{if } 0 < x < \infty, \end{cases}$$

a similar argument using $H_2(-x)$ in place of $H_1(x)$ yields an entire $\widetilde{F_2}$. If $F_2(z) \stackrel{\text{def}}{=} \widetilde{F_2}(-z)$ then $f \stackrel{\text{def}}{=} F_1 + F_2 \in \mathcal{E}$ and $f(\Re(z)) > g(\Re(z))$. □

11
Analytic Continuation

11.1. Analytic Continuation of Series

11.1. a) In U, $f(z) = z\left(\sum_{n=0}^{\infty} z^n\right)' = z\left(\frac{1}{1-z}\right)' = \frac{z}{(1-z)^2}$. Thus

$$F(z) \stackrel{\text{def}}{=} \frac{z}{(1-z)^2}$$

is an analytic continuation of f from U along any curve in $\mathbb{C} \setminus \{1\}$. Furthermore, F is the only analytic continuation of f by virtue of the identity theorem for holomorphic functions.

b) $c_n(a) = \frac{f^{(n)}(a)}{n!}$, $R_{c(a)} = |a-1|$. □

11.2. Note that for z in $U \cup \Omega$ if $\langle 0, z\rangle$ is a polygonal path lying in $U \cup \Omega$ (which is automatically connected) then $g(z) = \int_{\langle 0,z\rangle} \frac{1}{1-w}\,dw$ is well-defined, independent of the choice of $\langle 0, z\rangle$, and in $H(U \cup \Omega)$. If $z \in U$, then $g(z) \stackrel{\text{def}}{=} -\sum_{n=1}^{\infty} \frac{z^n}{n}$. Furthermore, $\frac{g(z)}{z} \in H(U \cup \Omega)$ and for z in U, $f(z) = \int_{\langle 0,z\rangle} \frac{g(w)}{w}\,dw$. Thus $\int_{\langle 0,z\rangle} \frac{g(w)}{w}\,dw$ is an analytic continuation of f from U along any curve in $U \cup \Omega$. □

[**Note s11.1:** For z in $U \cup \Omega$, $\frac{1}{1-z} = e^{g(z)}$, i.e., $g(z)$ is a branch of $\ln\left(\frac{1}{1-z}\right)$ or $-g(z)$ is a branch of $\ln(1-z)$.]

11.3. If $|z| < 1$ then $\sum_{n=1}^{\infty} \left|z^{n!}\right| < \frac{1}{1-|z|}$.

If $p \in \mathbb{Z}$, $q \in \mathbb{N}$, $a \stackrel{\text{def}}{=} \frac{p}{q}$, $n \geq q$, and $z = re^{2\pi i a}$ then $z^{n!} = r^{n!}$ and so $|f(z)| \to \infty$ as $r \uparrow 1$. Thus each point $e^{2\pi i a}$ is a singular point of f. Note that $\left\{e^{2\pi i a}\right\}_{a \in \mathbb{Q}}$ is dense in $\mathbb{T}$. □

11.4. a) Because 1 is a regular point of f, it follows that the result of analytically continuing f is holomorphic in some Ω containing $U \cup \{1\}$. For the holomorphic function $g(z) \stackrel{\text{def}}{=} \frac{1}{2}\left(z^\lambda + z^{\lambda+1}\right)$ note that: *i.* $g(1) = 1$; *ii.* if $z \in D(0,1) \setminus \{1\}$ then $|1+z| < 2$, whence $|g(z)| = \frac{1}{2}\left|z^\lambda\right| \cdot |(1+z)| < 1$.

Thus $g(D(0,1)) \subset U \cup \{1\} \subset \Omega$ and the Heine-Borel theorem implies that for some positive ϵ, $g(D(0,1+\epsilon)) \subset \Omega$, i.e., $D(0,1+\epsilon) \subset g^{-1}(\Omega)$.

Since $h(z) \stackrel{\text{def}}{=} f \circ g(z)$ is holomorphic in $g^{-1}(\Omega)$, it follows that for some $\{b_n\}_{n\in\mathbb{N}}$, $h(z) = \sum_{m=0}^{\infty} b_m z^m$ obtains in $D(0,1+\epsilon)^\circ$. Direct calculation shows that for k in $\mathbb{N}$,

$$s_{p_k}(z) = \sum_{n=0}^{p_k} c_n (g(z))^n = \sum_{m=0}^{(\lambda+1)p_k} b_m z^m. \tag{s11.1}$$

The right member of (s11.1) converges in $D(0,1+\epsilon)^\circ$, i.e., $\{s_{p_k}\}_{k\in\mathbb{N}}$ converges in $D(0,1+\epsilon)^\circ$ as required.

b) If $\alpha \in \mathbb{T}$ and α is a regular point of f then 1 is a regular point of $f(\alpha z)$. Hence the argument in a) applies and thus $\{s_{p_k}(z)\}_{k\in\mathbb{N}}$ converges in some $D(0,1+\epsilon)^\circ$. In the current circumstances, $\{s_{p_k}\}_{k\in\mathbb{N}} = \{s_p\}_{p\in\mathbb{N}}$. Hence if α is a regular point of f then in some $D(0,1+\epsilon)^\circ$, $\{s_p(z)\}_{p\in\mathbb{N}}$ converges, i.e., in some $D(0,1+\epsilon)^\circ$, $\sum_{n=0}^{\infty} c_n z^n$ converges. This conclusion is incompatible with the assumption that $R_c = 1$.

c) Since $\overline{\lim}_{n\to\infty} |c_n|^{\frac{1}{n}} = 1$, for some subsequence $S \stackrel{\text{def}}{=} \{c_{n_k}\}_{k\in\mathbb{N}}$, $n_{k+1} > 2n_k$ and $\overline{\lim}_{k\to\infty} |c_{n_k}|^{\frac{1}{n_k}} = 1$. By the same token, S contains pairwise disjoint subsequences $S_p \stackrel{\text{def}}{=} \{c_{m(p,q)}\}_{q\in\mathbb{Z}^+}, p \in \mathbb{N}$, such that for each p, $m(p,q+1) > 2m(p,q)$ and $\overline{\lim}_{q\to\infty} |c_{m(p,q)}|^{\frac{1}{m(p,q)}} = 1$. Then a) applies.

d) In c) let: *i.* $h_p(z)$ be $\sum_{q=0}^{\infty} c_{m(p,q)} z^{m(p,q)}$; *ii.* H_t be $\sum_{p=1}^{\infty} \epsilon_p(t) h_p$; *iii.* g be $f - H_1$; *iv.* for t in $[0,1]$, f_t be $g + H_t$. Since the sequences $\{S_p\}_{p\in\mathbb{N}}$ are pairwise disjoint, $H_t(z)$ exists for all z in U and $\mathbb{T}$ is a natural boundary for each H_t. If $\mathbb{T}$ contains a regular point for f_t, some open arc Γ_t contained in $\mathbb{T}$ consists of regular points of f_t. Furthermore, Γ_t may be chosen so that its endpoints are rational multiples of 2π. If there are uncountably many such f_t, two of their corresponding arcs, say Γ_t and Γ_s, overlap. Hence $f_t - f_s = H_t - H_s$ is holomorphic on $\Gamma_t \cap \Gamma_s$ whereas b) implies that $\mathbb{T}$ is the natural boundary for $H_t - H_s$. □

[**Note s11.2:** If $n_{k+1} > 2n_k$ then $\mathbb{T}$ is a natural boundary for $\sum_{n=0}^{\infty} z^{n_k}$. However, $\overline{\lim}_{k\to\infty} \left|e^{-\sqrt{n_k}}\right|^{\frac{1}{n_k}} = 1$ whence

$$F(z) \stackrel{\text{def}}{=} \sum_{n=0}^{\infty} e^{-\sqrt{n_k}} z^{n_k}$$

and all its derivatives are represented by power series converging uniformly on $D(0,1)$: $F \in C^\infty(D(0,1),\mathbb{C})$ and yet $\mathbb{T}$ is a natural boundary for F.]

11.5. a) If $z \notin \mathbb{T}$ then for some positive ϵ and each n in $\mathbb{N}$, $\left|z - e^{2\pi i n a}\right| > \epsilon$, whence the series converges uniformly in every compact subset of Ω_1 and in every compact subset of Ω_2.

b) For z in either of $\Omega_i, i = 1, 2$, direct calculation leads to the basic equation:

$$f\left(e^{2\pi i a} z\right) = \frac{1}{2e^{2\pi i a}}\left(\frac{1}{z-1} + f(z)\right). \tag{s11.2}$$

If $e^{2\pi i a}$ is a regular point it follows that $\lim_{z\to 1} f\left(e^{2\pi i a} z\right)$ exists whence $\lim_{z\to 1} f(z)$ cannot exist, i.e.,

$$\left\{e^{2\pi i a} \text{ is a regular point}\right\} \Rightarrow \{1 \text{ is a singular point}\}. \tag{s11.3}$$

From (s11.2) and (s11.3) it follows that if $e^{2\pi i a}$ is a regular point then $e^{-2\pi i a}$ is a singular point; in other words, at least one of $e^{\pm 2\pi i a}$ is a singular point.

However, (s11.2) and induction imply that if $e^{2\pi i a}$ is a regular point then for each k in $\mathbb{N}$, $e^{-2\pi i k a}$ is a singular point. Because $\left\{e^{-2\pi i k a}\right\}_{k\in\mathbb{N}}$ is dense in $\mathbb{T}$ (cf. **4.23**), it follows that every point of $\mathbb{T}$ is singular; in particular there emerges the contradictory conclusion that $e^{2\pi i a}$ is itself a singular point.

Thus $e^{2\pi i a}$ must be a singular point, in which case from (s11.2) it follows that the dense set $\left\{e^{2\pi i k a}\right\}_{k\in\mathbb{N}}$ consists entirely of singular points, i.e., $\mathbb{T}$ is a natural boundary for $f\big|_{\Omega_i}, i = 1, 2$. □

11.6. a) If $z \in \Omega$ and n is large, then

$$\left|\frac{z}{n(z-n)}\right| = \frac{1}{n^2}\left|\frac{z}{\left(\frac{z}{n} - 1\right)}\right| < \frac{2|z|}{n^2},$$

whence the series converges. If $z \notin \Omega$, one term of the series is undefined.

b) The series converges uniformly on every compact subset of Ω: f is holomorphic in Ω. For k in $\mathbb{N}$, $\dfrac{z}{k(z-k)} = \dfrac{1}{k} + \dfrac{1}{z-k}$ and so

$$f(z) = \frac{1}{z-k} + \frac{1}{k} + \sum_{n\neq k} \frac{z}{n(z-n)} \stackrel{\text{def}}{=} \frac{1}{z-k} + F_k(z). \tag{s11.4}$$

Because F_k is holomorphic in $D(k,1)^\circ$, it follows that $P(f) = \mathbb{N}$ and that each pole of f has multiplicity one.

c) If $k \in \mathbb{N}$, $\operatorname{Res}(f,k) = 1$. The residue theorem and the formula in (s11.4) imply the result. □

11.7. Because $\dfrac{z}{z^2-n^2} = \dfrac{1}{n^2\left(\dfrac{z^2}{n^2} - 1\right)} = \dfrac{1}{2}\left(\dfrac{1}{z-n} + \dfrac{1}{z+n}\right)$, the arguments in **Solutions 11.6a)** – **11.6c)** apply, mutatis mutandis. Note that for c), if $k \in \mathbb{Z}$, $\operatorname{Res}(f,k) = \dfrac{1}{2}$. □

11.2. General Theory

11.8. For z near a, $f(z)=\sum_{n=0}^{\infty}\frac{f^{(n)}(a)}{n!}z^n$ and the hypothesis implies that the series converges for every z. Thus the series defines an entire function F with the required properties. □

11.9. The (Laurent) expansion $\sum_{n=-\infty}^{\infty} c_n z^n$ converges in $A(0,r:R)^\circ$ and defines the function F as required. □

11.10. Cauchy's theorem implies that if F as described exists then

$$\int_{\mathbf{T}} f(z)z^n\,dz \equiv 0.$$

Conversely, if $\int_{\mathbb{T}} f(z)z^n\,dz \equiv 0$ then for z in U, define $F(z)$ to be $\frac{1}{2\pi i}\int_{\mathbf{T}}\frac{f(w)}{w-z}\,dw$. For w in $\mathbb{T}$ and z in U, $\frac{1}{w-z}=\frac{1}{w}\sum_{n=0}^{\infty}\left(\frac{z}{w}\right)^n$. As defined, $F\in C(D(0,1),\mathbb{C})$ and if $r\in[0,1]$, $\int_{|z|=r}F(z)\,dz=0$. Morera's theorem applies. □

11.11. Because $f_n \xrightarrow{\text{u}} f$ on compact subsets of Ω_1 it follows that $f\in H(\Omega_1)$. But Vitali's theorem implies that for some F in $H(\Omega)$, $f_n \xrightarrow{\text{u}} F$ on compact subsets of Ω whence $F\big|_{\Omega_1}=f$. □

11.12. If $a\in[0,1)$ then for some positive r, $D(a,r)\subset D(0,2)^\circ$.

As can be seen from consideration of **Figure s11.1** and the uniform

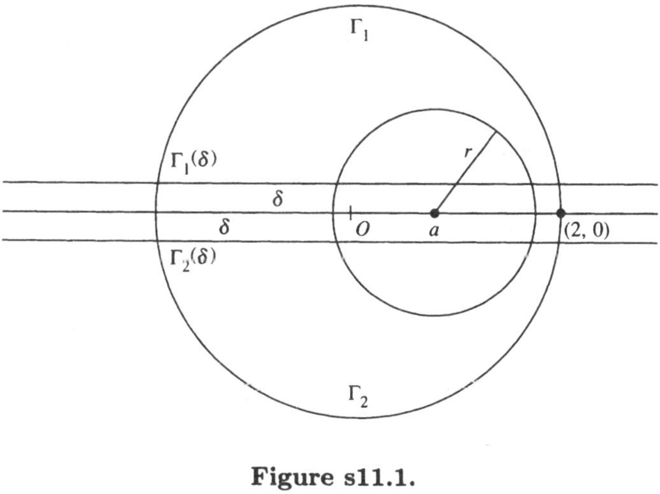

Figure s11.1.

continuity of f on $D(a,r)$,

$$\int_{|z-a|=r} f(z)\,dz = \int_{\Gamma_1} f(z)\,dz + \int_{\Gamma_2} f(z)\,dz$$
$$= \lim_{\delta\to 0}\left(\int_{\Gamma_1(\delta)} f(z)\,dz + \int_{\Gamma_2(\delta)} f(z)\,dz\right) = 0.$$

Morera's theorem implies that $f \in H\left(D(a,r)^\circ\right)$. □

11.13. Via repeated use of the Schwarz reflection principle, f may be continued analytically first from Q° to the union Ω_1 of Q and the interiors of the four squares $\pm 1+Q, \pm i+Q$, then to the union of Ω_1^c and the interiors of the eight squares adjoining the four new squares, etc. □

11.14. The Schwarz reflection principle may be applied, mutatis mutandis, as in **11.13.** □

11.15. In $D(1,r)^\circ$, $f(z) = \sum_{n=0}^\infty c_n(z-1)^n \stackrel{\text{def}}{=} f_r(z)$. The hypothesis implies $|c_0| \geq \delta$. Thus $g(z) \stackrel{\text{def}}{=} \dfrac{f(z)-c_0}{-c_0(z-1)} \in H^\infty$ and if $h(z) \equiv \dfrac{1}{c_0}$ then $(1-z)g(z) + f(z)h(z) \equiv 1$. □

11.16. Let $\sum_{n=0}^\infty c_n(z-1)^n$ be the power series representing f in $D\left(1,\frac{1}{2}\right)^\circ$. Then for each determination of $\beta(z)$ for $\ln z$ in $D\left(0,\frac{1}{2}\right)^\circ$,

$$f(z) = \sum_{n=0}^\infty c_n\left(e^{\ln z}-1\right)^n \stackrel{\text{def}}{=} g\circ\beta(z).$$

Let Ω_1 be $\beta\left(D\left(1,\frac{1}{2}\right)^\circ\right)$. Because f may be continued analytically along each line L as described, there is an analytic continuation F of f from $D\left(1,\frac{1}{2}\right)^\circ$ along any curve in Ω. The determination of $\ln z$ may be chosen so that β maps Ω bijectively (and hence biholomorphically) onto the strip $\{w : -\pi < \Im(w) < \pi, -\infty < \Re(w) < \infty\} \stackrel{\text{def}}{=} \Omega_2$. For z in Ω_1,

$$F\circ\beta(z) \stackrel{\text{def}}{=} F(w) = \sum_{n=0}^\infty c_n\left(e^{\beta(z)}-1\right)^n = \sum_{n=0}^\infty c_n\left(e^w-1\right),$$

which may be continued analytically throughout Ω_2. However, for k in $\mathbb{Z}$, $F(w) = \sum_{n=0}^\infty c_n\left(e^{\beta(z)+2k\pi i}-1\right)^n$. The Schwarz reflection principle implies that for some G in $\mathcal{E}$, $G|_{\Omega_2} = G|_{2k\pi i+\Omega_2} = F(w), k\in\mathbb{Z}$. □

11.17. Since $e^{\Omega_1} = \Omega$, for each w in Ω_1 let $F(w)$ be $f\left(e^w\right)$. It follows that $F \in H\left(\Omega_1\right)$. Note that $D(a,r)^\circ$ is simply connected and every branch of $\ln z$ is in $H\left(D(a,r)^\circ\right)$. In other words, $F(\ln z) = f(z)$ in $D(a,r)^\circ$. □

11.18. a) The formula in **10.45** shows that $\frac{1}{\Gamma} \in \mathcal{E}$ and $Z\left(\frac{1}{\Gamma}\right) = \mathbb{Z} \setminus \mathbb{N}$, whence Γ is meromorphic and $P(\Gamma) = \mathbb{Z} \setminus \mathbb{N}$.

b) Because $m^{-z} = e^{-z \ln m}$, the definitions of $\frac{1}{\Gamma(z)}$ and Euler's constant γ imply

$$\begin{aligned}
\frac{1}{\Gamma(z)} &= z \lim_{m\to\infty} \left[m^{-z} \prod_{n=1}^{m}\left(1+\frac{z}{n}\right)\right] \qquad (\text{s11.5})\\
&= z \lim_{m\to\infty} \left[\prod_{n=1}^{m-1}\left(1+\frac{1}{n}\right)^{-z} \prod_{n=1}^{m}\left(1+\frac{z}{n}\right)\right]\\
&= z \lim_{m\to\infty} \left[\prod_{n=1}^{m}\left\{\left(1+\frac{z}{n}\right)\left(1+\frac{1}{n}\right)^{-z}\right\}\left(1+\frac{1}{m}\right)^{z}\right].\\
&= z \prod_{n=1}^{\infty}\left(1+\frac{z}{n}\right)\left(1+\frac{1}{n}\right)^{-z}.
\end{aligned}$$

c)

$$\begin{aligned}
\frac{\Gamma(z+1)}{\Gamma(z)} &= \frac{1}{z+1}\left[\lim_{m\to\infty}\prod_{n=1}^{m}\frac{\left(1+\frac{1}{n}\right)^{z+1}}{1+\frac{z+1}{n}}\right] \div \left[\frac{1}{z}\lim_{m\to\infty}\prod_{n=1}^{m}\frac{\left(1+\frac{1}{n}\right)^{z}}{1+\frac{z}{n}}\right]\\
&= \frac{z}{z+1}\lim_{m\to\infty}\prod_{n=1}^{m}\left[\frac{\left(1+\frac{1}{n}\right)(z+n)}{z+n+1}\right]\\
&= z \lim_{m\to\infty}\frac{m+1}{z+m+1} = z.
\end{aligned}$$

d) $h(z+1) = g(z+1)\Gamma(z+1) = g(z)z\Gamma(z) = zh(z)$.
e) Induction and (11.1) apply.
f) If $\frac{1}{\Gamma_n(z)} \overset{\text{def}}{=} n^{-z} z \prod_{m=1}^{n}\left(1+\frac{z}{m}\right)$, then (s11.5) implies

$$\lim_{n\to\infty}\frac{1}{\Gamma_n(z)}\exp\left(-z\left(\sum_{m=1}^{n}\frac{1}{m} - \ln n\right)\right) = \frac{e^{-\gamma z}}{\Gamma(z)}$$

$$\begin{aligned}
\Gamma(z) = \lim_{n\to\infty}\Gamma_n(z) &= \lim_{n\to\infty}\frac{n!n^z}{z(z+1)\cdots(z+n)}\\
&\overset{\text{def}}{=} \lim_{n\to\infty}\Pi_n(z). \qquad (\text{s11.6})
\end{aligned}$$

g) Because $\Gamma(1-z) = -z\Gamma(-z)$ it follows that

$$\frac{1}{\Gamma(z)\Gamma(1-z)} = z \prod_{n\in\mathbb{Z}\setminus\{0\}} \left(1-\frac{z^2}{n^2}\right).$$

However the right member above is, according to **10.48**, $\dfrac{\sin \pi z}{\pi}$. □

11.19. a) Note that the dominated convergence theorem implies

$$\begin{aligned}\int_0^\infty e^{-t}t^{z-1}\,dt &= \lim_{n\to\infty}\int_0^n \left(1-\frac{t}{n}\right)^n t^{z-1}\,dt\\ &\stackrel{\text{def}}{=} \lim_{n\to\infty} Q_n(z)\end{aligned}$$

$$\int_0^n \left(1-\frac{t}{n}\right)^n t^{z-1}\,dt = n^z\int_0^1 (1-\tau)^n\tau^{z-1}\,d\tau.$$

Integration by parts and induction imply $Q_n(z) = \Pi_n(z)$ (cf. (s11..6)). Hence **11.18**f) applies.

b) By definition, $\left(f_i\left(\frac{x+y}{2}\right)\right)^2 \le f_i(x)f_i(y), i=1,2$. The Schwarz inequality in its simplest version ("the geometric mean does not exceed the arithmetic mean") implies

$$\left(f_1\left(\frac{x+y}{2}\right)+f_2\left(\frac{x+y}{2}\right)\right)^2 \le (f_1(x)+f_2(x))\,(f_1(y)+f_2(y)).$$

Owing to the continuity of the $f_i, i=1,2$, their sum is log convex.

c) The integrand $e^{-t}t^{x-1}$ in (11.2) is a log convex function of x. From b) it follows that the Riemann sums that converge to $\Gamma(x)$ are log convex whence $\Gamma(x)$ is log convex. □

11.20. a) The series representing ζ converges uniformly in every compact subset of Ω_1. For z in Ω_1, $n^{-z}\Gamma(z) = \int_0^\infty t^{z-1}e^{-nt}\,dt$. It follows by summing on n and (via Fubini's theorem) inverting the order of summation and integration that

$$\zeta(z) = \frac{1}{\Gamma(z)}\int_0^\infty \frac{t^{z-1}}{e^t-1}\,dt. \tag{s11.7}$$

b) The equation $\Gamma(z+1) = z\Gamma(z)$ permits a sequence of analytic continuations of Γ from Ω_1 to $(-1+\Omega_1)\setminus\{0,-1\}$, then to $(-2+\Omega_1)\setminus\{0,-1,-2\}$, etc., and ultimately to $\mathbb{C}\setminus(\mathbb{Z}\setminus\mathbb{Z}^+)$.

For ϵ positive, let L_ϵ be the contour described in **Figure s11.2** and for t in L_ϵ, write t as $|t|e^{-i\theta} \stackrel{\text{def}}{=} re^{-i\theta}, -\pi < \theta < \pi$.

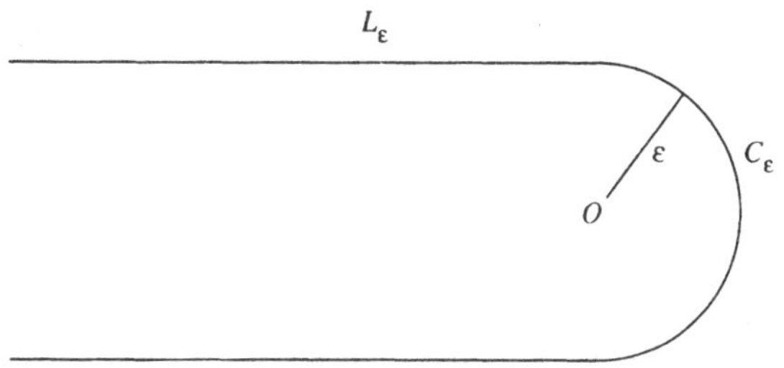

Figure s11.2.

When $\Re(t) < 0$, the integral (11.2) is related to the (contour) integral

$$\int_{L_\epsilon} e^t r^{-z} e^{i\theta t}\, dt \stackrel{\text{def}}{=} \int_{L_\epsilon} f(t;z)\, dt \stackrel{\text{def}}{=} \phi(z,\epsilon).$$

For $t \stackrel{\text{def}}{=} \sigma + i\tau$ and $z \stackrel{\text{def}}{=} x + iy$, $|f(t;z)| = e^\sigma \cdot \left(\sigma^2+\tau^2\right)^{-\frac{x}{2}} \cdot e^{-\theta y}$, and so for b, c in $\mathbb{R}$, $\lim_{a\downarrow-\infty} \int_{a+ib}^{a+ic} f(t;z)\, dt = 0$. It follows that $\phi(z,\epsilon)$ is ϵ-free. On the semicircle K_ϵ, as $\epsilon \downarrow 0$,

$$\left|\frac{1}{2\pi i}\int_{K_\epsilon} f(t;z)\, dt\right| \le \frac{e^\epsilon \cdot \epsilon^{-x} e^{\frac{y\pi}{2}} \epsilon}{2} \downarrow 0.$$

When $\Re(z) < 0$, the limit of the integral on each of the straight lines in $L_\epsilon \setminus K_\epsilon$ is, by virtue of the dominated convergence theorem,

$$e^{-i\pi z}\int_0^\infty e^{-\sigma}\sigma^{-z}\, d\sigma.$$

When $\Re(z) < 0$, $\phi(z,0) = \dfrac{\sin \pi z}{\pi}\Gamma(1-z)$, $\dfrac{1}{\Gamma(z)} = \dfrac{1}{2\pi i}\displaystyle\int_{L_0} e^t t^{-z}\, dt$. Because the last integral represents an entire function, the representation is valid in $\mathbb{C}$.

The integral in (11.3) may be treated in a manner similar to that used for the integral representing $\Gamma(z)$ when $\Re(z) > 1$. The conclusion is

$$\zeta(z) = \frac{\Gamma(1-z)}{2\pi i}\int_{L_\epsilon} \frac{e^t t^{z-1}}{1-e^t}\, dt. \tag{s11.8}$$

The integral in (s11.7) represents an entire function. Since both ζ and Γ are in $H(\Omega)$, (s11.7) provides a continuation to $\mathbb{C} \setminus \mathbb{N}$, where $\Gamma(1-z)$ is holomorphic. Because ζ is holomorphic in Ω it follows that (s11.7) provides a continuation of ζ to $\mathbb{C} \setminus \{1\}$. □

11.21. Note that f is a constant k iff $k = 0$ and that if $f(z) = \sin z$ then $f(2z) = 2f(z)f'(z)$.

If f is not constant, the formula $f(2z) = 2f(z)f'(z)$ defines F in $\bigcup_{n=1}^{\infty} 2^n U$, i.e., in $\mathbb{C}$. The product rule for derivatives shows that so defined, F is holomorphic throughout $\mathbb{C}$. The identity theorem for holomorphic functions implies that F is completely determined by the values of f in U. □

11.22. Note that f is a constant k iff $k = 0$ and if $f(z) = \tan z$ then $f(2z) = \dfrac{2f(z)}{1-(f(z))^2}$.

If F is an analytic continuation then $F(z) \neq \pm 1$ because $f(z) \neq \pm 1$. Picard's theorem implies that any nonconstant continuation F must have singularities.

Because for z in U, $\lim_{n\to\infty} f\left(2^{-n}z\right) = f(0)$, it follows that

$$f(0)((f(0))^2+1) = 0, \text{ whence } f(0) = 0 \text{ or } f(0) = \pm i.$$

The last two choices are excluded since otherwise, $f(0) = \dfrac{2f(0)}{1-1}$.

Since $f \in H(U)$, $(f(z))^2 \neq 1$ in $\left(\frac{1}{2}U\right)^c \stackrel{\text{def}}{=} K$. Extend f by iterated use of its functional equation so long as in each $2^n K, n \in \mathbb{N}$, $(f(z))^2 \neq 1$. Since there is no entire continuation of f, there is a least natural number N_1 such that $(f(z))^2 \neq 1$ in $2^{N_1}K$ and for some (at most finitely many!) z in $2^{N_1+1}K$, $(f(z))^2 = 1$. It follows that for some sequence $\{N_k\}_{k\nu\mathbb{N}}$ in $\mathbb{N}$, if $N_k < n < N_{k+1}$ then as extended, $(f(z))^2 \neq 1$ in $2^n K \setminus 2^{N_k}K$ but for some (at most finitely many!) z in $2^{N_{k+1}}K \setminus 2^{N_k}K$, $(f(z))^2 = 1$. Then F is a meromorphic extension of f. □

11.23. a) The series $-\sum_{n=1}^{\infty} \dfrac{z^n}{n}$ defines a function ϕ in $H\left(D(0,1)^\circ\right)$ and $\phi'(z) = -\dfrac{1}{1-z}$. If $g = e^\phi$ then

$$g'(z) = \frac{g(z)}{1-z} = (zg)',\ g(z) = zg + C,\ g(z) = \frac{C}{1-z}.$$

Hence $e^{-\phi(z)} = \dfrac{1-z}{C}$ and since $e^{-\phi(0)} = 1$, $C = 1$, i.e., in U, $e^{-\phi(z)} = 1 - z$ and in $D(1,1)^\circ$, $e^{-\phi(1-z)} = z$. Hence $f(z) = -\phi(1-z)$.

b) For a in $\mathbb{T} \cap \partial D(1,1)$, represent f as $\sum_{n=0}^{\infty} d_n(z-a)^n$ and thereby provide an immediate analytic continuation of f from $D(1,1)^\circ$ to $D(a,1)^\circ$.

Any b in $\mathbb{T}$ can be reached in a finite number of such immediate analytic continuations. There emerges the region $\Omega \stackrel{\text{def}}{=} D(0,2)^\circ \setminus \{0\}$ such that for any c in $\mathbb{C}\setminus\{0\}$, analytic continuation is possible from a point a in $\Omega \cap [0,c]$ along $[a,c]$. □

11.24. a) From **11.23**a) it follows that $(g(z))^2 \stackrel{\text{def}}{=} \left(e^{\frac{\ell}{2}}\right)^2 = z$. A specific form for g can be found by equating the coefficients in the equation

$$(g(z))^2 \stackrel{\text{def}}{=} \left(\sum_{n=0}^{\infty} c_n(z-1)^n\right)^2 = 1+(z-1).)$$

b) The procedure for **11.23**b) may be followed, mutatis mutandis.

□

11.25. Because $S \stackrel{\text{def}}{=} \{p+iq \ : \ p,q \in \mathbb{Q}\}$ is dense in $\mathbb{C}$, each $(g, N(a))$ may be produced via a polygonal path $\langle 0, a\rangle$ the vertices of which are in S. There are only countably many such paths. □

11.26. The argument of **Solution 9.32** may be repeated, mutatis mutandis. □

11.27. If $a \in \Omega$ and $f_k(a) \neq 0$, some branch β_k of $\ln f_k$ is holomorphic in some $D(a, r_a)$ contained in Ω. Then in $D(a,r_a)^\circ$, $f_k^{p_k}(z) \stackrel{\text{def}}{=} e^{p_k\beta_k(z)}$. Regardless of the choice of β_k, $|f_k(z)|^{p_k}$ is unambiguously defined. For each such k, the application of **11.26** shows that $\sup_{z\in\Omega}|f_k(z)|^{p_k} \leq M$ whence

$$|f(a)| \leq \sum_{f_k(a)\neq 0} \sup_{z\in\Omega} |f_k(z)|^{p_k} \leq \sum_{k=1}^{n} \sup_{z\in\Omega} |f_k(z)|^{p_k}$$

$$\sup_{z\in\Omega} |f(z)| \leq Mn.$$

[If Ω^c is compact and $f_k^{p_k}$ is continuous on Ω^c then the basic maximum modulus theorem implies $\sup_{z\in\Omega^c}|f_k(z)|^{p_k}$ is achieved on $\partial\Omega$.] □

[**Note s11.3:** The result in **11.27** is a generalized maximum modulus principle for regular functions.]

12
Singularities

12.1. General Theory

12.1. If a is a pole then for all z in some $D(a,r)^\circ$, some m in $\mathbb{N}$, and some g in $H\left(D(a,r)^\circ\right)$, $f(z) = \dfrac{g(z)}{(z-a)^m}$. Hence for some p in $\mathbb{C}[z]$ and some g in $H\left(D(a,r)^\circ\right)$, $\deg(p) < m$ and $f(z) = \dfrac{p(z)}{(z-a)^m} + g(z)$. Hence

$$e^{f(z)} = e^{\frac{p(z)}{(z-a)^m}} e^{g(z)} \stackrel{\text{def}}{=} e^{\frac{p(z)}{(z-a)^m}} G(z).$$

Because G is holomorphic near a and $e^{\frac{p(z)}{(z-a)^m}}$ has an essential singularity at a, $\left(e^{f(D(a,r)^\circ)}\right)^c = \mathbb{C}$ and so $a \notin P\left(e^f\right)$.

If a is an essential singularity of f then for any $N(a)$, $(f(N(a)))^c = \mathbb{C}$ whence $\left(e^{f(N(a))}\right)^c = \mathbb{C}$ and thus $a \notin P\left(e^f\right)$. □

12.2. In $\mathbb{C} \setminus \{0\}$,

$$f(z) = \sum_{n=-1}^{\infty} c_n z^n$$

$$g(\theta) \stackrel{\text{def}}{=} \sum_{n=-1}^{\infty} c_n e^{in\theta} \in C([0,2\pi], \mathbb{R}),$$

and for n in $\mathbb{Z}$, Cauchy's theorem implies

$$\begin{aligned}\frac{1}{2\pi}\int_0^{2\pi} g(\theta)e^{-in\theta}\,d\theta &= \overline{\frac{1}{2\pi}\int_0^{2\pi} g(\theta)e^{in\theta}\,d\theta} = c_n = \overline{c_{-n}}\\ &= 0 \text{ if } n > 1.\end{aligned}$$

□

12.3. If the assertion is false then for some positive r and some a,

$$D(a,r)^\circ \subset \mathbb{C} \setminus E$$

and so $g(z) \stackrel{\text{def}}{=} \dfrac{1}{f(z)-a} \in H\left(U \setminus \left(\{a_n\}_{n\in\mathbb{N}} \cup \{0\}\right)\right)$. Because

$$\lim_{z\to a_n} g(z) = 0, n \in \mathbb{N},$$

each a_n is a removable singularity of g, $g(a_n) = 0$, and $g \in H(U \setminus \{0\})$. If zero is a removable singularity of g then the identity theorem for holomorphic functions implies $g \equiv 0$, a contradiction. Thus zero is an isolated essential singularity of g, whence $(g(U \setminus \{0\}))^c = \mathbb{C}$. In particular, $|g|$ is unbounded in U and so $a \in E$, a contradiction. □

12.4. Since $\left|e^{f(z)}\right| = \left|e^{\Re(f(z))}\right| \le \dfrac{1}{|z-a|^m}$, it follows that $\left|(z-a)^m e^{f(z)}\right|$ is bounded near a, whence near a, $e^{f(z)} = (z-a)^{-m}\sum_{n=0}^{\infty} c_n (z-a)^n$ and $c_0 \ne 0$. If a is a pole of f then a is an isolated essential singularity of e^f, a contradiction of the preceding sentence. If a is an isolated essential singularity of f then a is an isolated essential singularity of $-f$. But near a and for some $N(a)$ and a g in $H(N(a))$, $e^{-f(z)} = (z-a)^m g(z)$, whence $\lim_{z\to a} e^{-f(z)} = 0$, in contradiction of Picard's theorem (applied to $-f$). Hence a is a removable singularity of f. □

12.5. a) When $k \ge 2$ and $f(z) = z^{-k}$, zero is a pole of order not less than two and if $r > 0$, $J_r(f) = 0$.

b) If $f(z) = e^{\frac{1}{z}} - \dfrac{1}{z}$ then zero is an essential singularity and $J_r(f) = 0$.

c) If $c_{-1} \ne 0$ and $f(z) = \dfrac{c_{-1}}{z} + \sum\limits_{n=0}^{\infty} c_n z^n$ then $J_r(f) = c_{-1}$. □

12.6. By hypothesis for some $\{c_n\}_{n\in\mathbb{Z}}$, if $z \stackrel{\text{def}}{=} re^{it} \in (U \setminus \{0\})$ then $f(z) = \sum_{n=-\infty}^{\infty} c_n z^n$. Thus

$$\begin{aligned}
&\int_U \left|f\left(re^{it}\right)\right|^2 d\lambda_2(r\cos t, r\sin t)\\
&= \lim_{\epsilon\downarrow 0} \sum_{-\infty<m,n<\infty} \int_\epsilon^{1-\epsilon} \left(\int_0^{2\pi} c_m\overline{c_n} r^{m+n+1} e^{i(m-n)t}\, dr\right) dt\\
&= \lim_{\epsilon\downarrow 0} \int_\epsilon^{1-\epsilon} \sum_{n=-\infty}^{\infty} |c_n|^2 r^{2n+1}\, dr.
\end{aligned}$$

If $n < 0$ and $c_n \ne 0$ then, as $\epsilon \downarrow 0$, $\displaystyle\int_\epsilon^{1-\epsilon} |c_n|^2 r^{2n+1}\, dr \uparrow \infty$, a contradiction.

□

12.7. a) From **2.72** it follows that the series converges in $D(b, |b|)^\circ$.

b) The argument of **Solution 2.72** applies.

c) From a) it follows that each f_b may be continued analytically along any curve in $\mathbb{C} \setminus \{0\}$. If $\alpha \in \mathbb{Z}^+$ then b) implies $f_b(z) \equiv z^n$. If $\alpha \in (\mathbb{Z} \setminus \mathbb{Z}^+)$ then $P(f_b) = \{0\}$. In all other instances, b) implies that no $(g, D(0,r)^\circ)$ is an immediate analytic continuation of $(f_b, D(b,|b|)^\circ)$.

d) See b).

e) From b) it follows that $|f_b|$ is bounded near zero. Hence zero is not a pole. Picard's theorem implies zero is not an essential singularity. □

12.2. Meromorphic Functions

12.8. The hypothesis implies that for some n in $\mathbb{N}$, $(P(f) \cap D(0,1))$ is a finite set $\{p_n\}_{n=1}^N$ whence for some region Ω_1 containing $D(0,1)$,

$$g(z) \stackrel{\text{def}}{=} f(z) \prod_{n=1}^{N} \Phi_{p_n}(z) \in H(\Omega_1).$$

Furthermore, $g(\mathbf{T}) \subset \mathbf{T}$ and thus $g(D(0,1)) \subset D(0,1)$. It follows from **8.12** that g is a rational function whence so is f. □

12.9. Since $\mathbf{C}_\infty$ is compact, $P(f)$ is finite. Enumerate $P(f)$ as $\{p_n\}_{n=1}^N$ so that each pole of multiplicity μ is listed μ times. If

$$g(z) \stackrel{\text{def}}{=} f(z) \prod_{n=1}^{N} (z - p_n)$$

is not a polynomial, ∞ is an essential singularity of g and hence of f. □

12.10. Because $\dfrac{z - a_n}{z - b_n} = 1 + \dfrac{b_n - a_n}{z - b_n}$, it follows that if $z \notin \left(\{b_n\}_{n\in\mathbb{N}}\right)^c$ then $\sum_{n=1}^{\infty} \left|\dfrac{b_n - a_n}{z - b_n}\right| < \infty$, which assures the convergence of the infinite product and the meromorphy of the function it represents. □

[**Note s12.1:** If $a_n \equiv b_n$, the conclusion is valid and vacuous. If $b_n \equiv \dfrac{1}{n}$ and $a_n \equiv \dfrac{1}{n} + \dfrac{1}{n^2}$ then $\prod_{n\in\mathbb{N}} \left(\dfrac{z - a_n}{z - b_n}\right)$ fails to converge when $z = 0$.]

12.11. Note that $0 \in A$ whence $\lim_{z\to 0} zf(z) = f(1)$. Induction shows that $f(z+n) = (z+n-1)(z+n-2)\cdots zf(z)$. Hence a), b), and c) are resolved by the equations:

$$\begin{aligned}\lim_{z\to -n} (z+n)f(z) &= \lim_{z\to -n} \frac{f(z+n+1)}{(z+n-1)(z+n-2)\cdots z} \\ &= \frac{f(1)}{(-1)(-2)\cdots(-n)}.\end{aligned}$$ □

12.12. a) The hypothesis implies $\#(A) \le \aleph_0$. If $z \in U \setminus A$ and $S = \emptyset$ then $D(0,|z|) \cap A$ is finite and there is a polygonal path $\langle 0, z\rangle$ contained in $U \setminus A$. If $S \stackrel{\text{def}}{=} \{a_1, \ldots, a_n\}$ then for some $\{r_1, \ldots, r_n\}$ contained in $(0,1)$,

$$F \stackrel{\text{def}}{=} \bigcup_{k=1}^{n} D(a_k, r_k) \subset U$$

and $\Omega \stackrel{\text{def}}{=} U \setminus F$ is a region. Now $D(0, |z|) \cap \Omega \cap A$ is finite and there is a polygonal path $\langle 0, z \rangle$ contained in $\Omega \setminus A$.

b) The function $G \stackrel{\text{def}}{=} \frac{h}{g}$ has only removable singularities in U whence $gG \stackrel{\text{def}}{=} k \in M(U)$. □

12.13. If, for some positive δ and all z near p_0, $|f(z) - b| \geq \delta$ then each p_n is a removable singularity of $g(z) \stackrel{\text{def}}{=} (f(z) - b)^{-1}$ and so g may be defined to be zero at each p_n. But then $g \equiv 0$, a contradiction. □

12.3. Mittag-Leffler, Runge, and Weierstraß Theorems

12.14. Because $\mathbb{C}$ is arcwise connected, components of open subsets of $\mathbb{C}$ are open. In each component there is a z of the form $p + iq, p, q \in \mathbb{Q}$. □

12.15. If the subspace R is not dense in $C(K, \mathbb{C})$ then for some nonzero F^* in $R^\perp$, i.e., for some nonzero complex Borel measure μ, and for every f in R, $\int_K f(z)\, d\mu(z) = 0$.

If p_n is in the component C_n of $\mathbb{C}_\infty \setminus K$ and $p_n \neq \infty$, then for some positive r, $D(p_n, r) \subset C_n$ and for z in $\mathbb{C}_\infty \setminus K$, $\sum_{k=0}^\infty \frac{(z - p_n)^k}{(w - p_n)^{k+1}} = \frac{1}{w - z}$ in K and the series converges uniformly on K. Since

$$g(z) \stackrel{\text{def}}{=} \int_K \frac{d\mu(w)}{w - z} \in H(\mathbb{C}_\infty \setminus K),$$

it follows that $g = 0$.

If $p_n = \infty$, a similar argument applies to the series $\frac{1}{w - z} = \sum_{k=0}^{\infty} \frac{w^k}{z^{k+1}}$.

In sum, $\int_K \frac{d\mu(w)}{w - z} \equiv 0$ in $\mathbb{C}_\infty \setminus K$.

If $f \in H(\Omega)$ then Cauchy's formula and Fubini's theorem imply that for any cycle Γ contained in $\Omega \setminus K$,

$$\begin{aligned}
\int_K f(z)\, d\mu(z) &= \int_K \left\{ \frac{1}{2\pi i} \int_\Gamma \frac{f(w)}{w - z}\, dw \right\} d\mu(z) \\
&= \frac{1}{2\pi i} \int_\Gamma f(w)\, dw \int_K \frac{d\mu(z)}{w - z} \\
&= 0,
\end{aligned}$$

a contradiction since μ is nonzero. □

12.16. The complement Ω of the Cantor set C_0 is a region. Its complement is totally disconnected and of cardinality $\mathfrak{c}$. □

12.17. a) For n in $\mathbb{N}$, $K_n \stackrel{\text{def}}{=} D(0,n) \cap \left\{ z \ : \ \inf_{w\notin\Omega} |z-w| \geq \frac{1}{n} \right\}$ is compact while $W_n \stackrel{\text{def}}{=} D(0,n+1)^\circ \cap \left\{ z \ : \ \inf_{w\notin\Omega} |z-w| > \frac{1}{n+1} \right\}$ is open and $K_n \subset W_n \subset K_{n+1}^\circ$.

b) If $\mathsf{K}(\mathbb{C}) \ni K \subset \Omega$ then for some n in $\mathbb{N}$, $K \subset D(0,n)$ and

$$\inf_{\substack{z\in K\\ w\notin\Omega}} |z-w| > \frac{1}{n}.$$

c) Because $U_n \stackrel{\text{def}}{=} \mathbb{C}_\infty \setminus K_n \supset \mathbb{C}_\infty \setminus \Omega \stackrel{\text{def}}{=} F$, ∞ is in each unbounded component of U_n and each unbounded component of F. On the other hand, if C is a bounded component of U_n then by definition, for some z in U_n and some w in $\mathbb{C} \setminus \Omega$, $z \in D\left(w, \frac{1}{n}\right)^\circ$ while $D\left(w, \frac{1}{n}\right)^\circ \subset \mathbb{C}_\infty \setminus K_n$. Since $D\left(w, \frac{1}{n}\right)^\circ$ is connected, z is in component of $\mathbb{C}_\infty \setminus K_n$. In sum:

Each component of U_n meets a component of F.

If a component C of F meets two components C_1, C_2 of U_n then

$$C_3 \stackrel{\text{def}}{=} C \cup C_1 \cup C_2$$

is a connected proper superset of both C_1 and C_2, is contained in U_n, and denies the maximality of C_1 and C_2. □

12.18. For $\{K_n\}_{n\in\mathbb{N}}$, as in **12.17**, **12.15** applies and provides a rational function R_n with poles in S and such that $|f(z) - R_n(z)| < \frac{1}{n}$ on K_n. If K is a compact subset of Ω then for some n, $K \subset K_n$. □

12.19. For $\{K_n\}_{n\in\mathbb{N}}$ as in **12.17**, the sets

$$S_1 \stackrel{\text{def}}{=} S \cap K_1^\circ,\ S_n \stackrel{\text{def}}{=} S \cap \left(K_n^\circ \setminus K_{n-1}\right), n \geq 2$$

are finite and $\rho_n(z) \stackrel{\text{def}}{=} \sum_{a\in S_n} r_a(z)$ is a rational function holomorphic in an open set containing K_{n-1}. Thus **12.17** applies, yields a rational function R_n such that $|\rho_n(z) - R_n(z)| < \frac{1}{2^n}$ on K_{n-1}, and $f \stackrel{\text{def}}{=} \rho_1 + \sum_{n=2}^\infty (\rho_n - R_n)$ meets the requirements. □

12.20. Each component C of Ω is a region. Hence **12.19** applies: for some $g \stackrel{\text{def}}{=} \rho_1 + \sum_{n=2}^\infty (\rho_n - R_n)$, $P(g) \cap C = S \cap C$. Since C is not necessarily simply connected, for a fixed in C, if γ^* is a rectifiable curve-image connecting a to z in $C \setminus S$, then $\int_\gamma g(w)\,dw$ is defined modulo $2\pi i$. Hence $G \stackrel{\text{def}}{=} \exp\left(\int_\gamma g(w)\,dw\right)$ is unambiguously defined and $G(C\setminus S) \subset (\mathbb{C} \setminus \{0\})$.

Furthermore, if $b \in S \cap C$, $z \in C \setminus S$ and $z \to b$ then $G(z) \to 0$. Its form shows that $G \neq 0$ in $C \setminus S$. Thus $S \cap C$ is not only a set of removable singularities for G, but if $G(S) \stackrel{\text{def}}{=} \{0\}$ then $G \in H(C)$.

The argument may be repeated for each component of Ω. □

[**Note s12.2:** If $f \in \mathcal{E}$ and $Z(f) \stackrel{\text{def}}{=} \{a_n\}_{n\in\mathbb{N}}$ then $\frac{f'}{f} \in M(\mathbb{C})$. The Mittag-Leffler theorem combined with the argument in **12.20** yields the Weierstraß product representation for f.]

12.21. If $n \in \mathbb{Z}$, the principal part of $f(z) \stackrel{\text{def}}{=} \frac{\pi^2}{\sin^2 \pi z}$ near n is $\frac{1}{(z-n)^2}$. Hence the Mittag-Leffler representation of f is, for some g in $\mathcal{E}$,

$$f(z) = g(z) + \sum_{n\in\mathbf{Z}} \frac{1}{(z-n)^2} \stackrel{\text{def}}{=} g(z) + h(z).$$

Note that both f and h are periodic with period 1. Furthermore,

$$\left|\sin^2 \pi(x+iy)\right| = \frac{1}{2}(\cosh 2\pi y - \cos 2\pi x)$$

whence $\lim_{|y|\to\infty} |f(x+iy)| = 0$. Finally, $\lim_{|y|\to\infty} |h(x+iy)| = 0$, g is entire, and $g = f - h$ whence $g(z) \equiv 0$. □

12.22. For f in **12.21**, if $z \notin \mathbf{Z}$, then $f(z) = (-\pi \cot \pi z)'$. On the other hand, $\sum_{n\in\mathbf{Z}\setminus\{0\}} \left(\frac{1}{z-n} + \frac{1}{n}\right) = \sum_{n\in\mathbf{Z}\setminus\{0\}} \frac{z}{n(z-n)} \stackrel{\text{def}}{=} k(z)$ converges uniformly on compact sets not meeting $\mathbf{Z} \setminus \{0\}$. Hence termwise differentiation is permissible and it follows that $-k' = (-p \cot \pi z)'$, i.e., for some constant C, $k(z) = \pi \cot \pi z + C$.

Furthermore, $k(z) = \frac{1}{z} + \sum_{n=1}^{\infty} \frac{2z}{z^2 - n^2}$ whence $k(z) = -k(-z)$. Because $\pi \cot \pi z = -\pi \cot \pi(-z)$, it follows that $C = 0$. □

13
Harmonic Functions

13.1. Basic Properties

13.1. If $g \stackrel{\text{def}}{=} u+iv$, the chain rule for calculating partial derivatives applies to $h(u,v)$. Note that since $g \in H(\Omega)$, the Cauchy-Riemann equations imply that both u and v are harmonic in Ω. □

13.2. Direct calculation shows that if $a \stackrel{\text{def}}{=} b+ic$ and for (x,y) in $N(a)$,

$$v(x,y) \stackrel{\text{def}}{=} \int_c^y h_x(x,t)\,dt - \int_b^x h_y(t,c)\,dt$$

then $h \stackrel{\text{def}}{=} u+iv \in H(N(a))$. □

13.3. Riemann's (conformal) mapping theorem implies there is a biholomorphic bijection $\phi : U \mapsto \Omega$. Thus $h \circ \phi \stackrel{\text{def}}{=} g$ is harmonic in U. From **13.2** it follows that for some F in $H(U)$, $g = \Re(F)$. If $f \stackrel{\text{def}}{=} F \circ \phi^{-1}$ then $f \in H(\Omega)$ and $h = \Re(f)$. □

13.4. If v_1 and v_2 are harmonic conjugates of h then

$$f_j \stackrel{\text{def}}{=} h + iv_j \in H(\Omega), j = 1,2, \text{ whence } (f_1 - f_2)(\Omega) \subset i\mathbb{R}.$$

The open mapping theorem implies $f_1 - f_2$ is a constant, i.e., $v_1 - v_2$ is a constant. □

13.5. Because $L(\Omega)$ contains a v such that $h + iv \stackrel{\text{def}}{=} f \in H(\Omega)$, it follows that $0 \notin f(\Omega)$ and so for some $g \stackrel{\text{def}}{=} p+iq$ in $H(\Omega)$, $f = g^2$. □

13.6. Because $P_{\frac{r}{R}}(\theta - t) = \frac{1}{2\pi}\Re\left\{\dfrac{e^{it} + \dfrac{re^{i\theta}}{R}}{e^{it} - \dfrac{re^{i\theta}}{R}}\right\}$, $P_{\frac{r}{R}}$ is harmonic in $D(0,R)^\circ$. Hence if μ is real and $|z-a| < R$ then h is the real part of

$$f(z) \stackrel{\text{def}}{=} \frac{1}{2\pi i}\int_0^{2\pi} \frac{Re^{it}+z}{Re^{it}-z}\,d\mu\left(a+Re^{it}\right).$$

In the context of **7.33**, X is Ξ, g is id, and if

$$m(z) \stackrel{\text{def}}{=} \frac{1}{2\pi i}\int_{\Xi} \frac{Re^{it}}{Re^{it}-z}\,d\mu\left(a+Re^{it}\right) \text{ and}$$
$$n(z) \stackrel{\text{def}}{=} \frac{1}{2\pi i}\int_{\Xi} \frac{1}{Re^{it}-z}\,d\mu\left(Re^{it}\right)$$

then $m,n \in H\left(D(a,R)^\circ\right)$, and $f(z) = m(z) - zn(z) \in H\left(D(a,R)^\circ\right)$. Thus h is in $L\left(D(a,R)^\circ\right)$. When μ is complex, it is a linear combination of real measures. □

13.7. a) Since h is the real part of a function in $H\left(D(a,r)^\circ\right)$, Cauchy's formula yields the result.

b) *i*. For $A \stackrel{\text{def}}{=} f(a)$, $f^{-1}(A) \stackrel{\text{def}}{=} F \cap \Omega$ is closed in Ω. For b in $F\cap\Omega$, there is a positive r such that $D(b,r) \subset \Omega$. If $D(b,r)^\circ$ contains a c where $f(c) < A$ then for all t in some nonempty open interval J of $[0,2\pi)$, and for some positive ϵ, $f\left(b+|b-c|e^{it}\right) < A-\epsilon$. On the other hand, since $f \in MVP(\Omega)$,

$$\begin{aligned} A = f(b) &= \frac{1}{2\pi}\left(\int_J + \int_{[0,2\pi)\setminus J} f\left(b+|b-c|e^{it}\right)dt\right)\\ &< \frac{1}{2\pi}\left((A-\epsilon)\lambda(J) + A(2\pi - \lambda(J))\right) = A - \frac{\epsilon\lambda(J)}{2\pi}, \end{aligned}$$

a contradiction. Hence $D(b,r)^\circ \subset F\cap\Omega$, i.e., F and $F\cap\Omega$ are open. Because Ω is a region and $F\cap\Omega \neq \emptyset$, $F\cap\Omega = \Omega$.

[**Note s13.1:** If $f \equiv 1$ in U and $f \equiv 2$ in $2+U$ then f is harmonic in the open nonregion $S \stackrel{\text{def}}{=} U \cup (2+U)$. For all z in S, $f(2) \geq f(z)$. Nevertheless, f is not a constant.]

b) *ii*. The result in b)*i* applies to $-f$.

c) If $h \in MVP(\Omega)$, $D(a,R) \subset \Omega$, and $0 \leq r < R$ then from **13.6** it follows that

$$H\left(a+re^{i\theta}\right) \stackrel{\text{def}}{=} \frac{1}{2\pi}\int_0^{2\pi} P_{\frac{r}{R}}(\theta - t)h\left(a+Re^{it}\right)dt \qquad \text{(s13.1)}$$

is harmonic. The properties of P_r (cf. **8.14**) imply that: *i*) H is continuous on $D(a,R)$; *ii*) since h enjoys the MVP, so does H; *iii*)

$$H\left(a+Re^{it}\right) = h\left(a+Re^{it}\right).$$

Thus for z in $D(a,R)$, $H(z) = h(z)$. The right member of (s13.1) is Poisson's formula for h.

For the converse, cf. a). □

13.8. The map

$$\begin{aligned}\mathcal{P} : C(\mathbb{T}, \mathbb{R}) \ni g \\ \mapsto \frac{1}{2\pi}\int_0^{2\pi} P_r(\theta - t) g\left(e^{it}\right) dt \stackrel{\text{def}}{=} \mathcal{P}(g) \in L(U) \cap C(D(0,1), \mathbb{C})\end{aligned}$$

is a bijection. The maximum principle implies $\mathcal{P}(f)(a) \le \max_{|z|=1} |f(z)|$ while the character of $\mathcal{P}$ implies that if $f \ge 0$ then $\mathcal{P}(f) \ge 0$. It follows that $J : C(\mathbb{T}, \mathbb{R}) \ni f \mapsto \mathcal{P}(f)(a)$ is a linear functional to which the theory of the Daniell integral (cf. **Section 1.2**) is applicable. Thus for some nonnegative measure μ_a, $\mathcal{P}(f)(a) = \int_0^{2\pi} f(z)\, d\mu_a(z)$. If $h \in L(U) \cap C(D(0,1), \mathbb{C})$ then the preceding equation applies to both $\Re(h)$ and $\Im(h)$. □

[**Note s13.2:** A Dirichlet region Ω in $\mathbb{C}$ is a region such that for every f in $C(\partial\Omega, \mathbb{R})$ for some h in $L(\Omega) \cap C(\Omega^c, \mathbb{R})$, $h\big|_{\partial\Omega} = f$. If, to boot, Ω^c is compact then for some a in Ω^c, $h(a) = \max_{z \in \Omega^c} h(z)$ and the maximum principle implies $a \in \partial\Omega$. Thus the argument and conclusion above apply to any Dirichlet region Ω such that Ω^c is compact.]

13.9. If $h \stackrel{\text{def}}{=} u + iv$ then $h^2 = u^2 - v^2 + i2uv$ and if $\{h, h^2\} \subset L(\Omega)$ then direct calculation shows that $h_x^2 + h_y^2 = 0$, i.e.,

$$u_x + iv_x = \pm i\left(v_y - iv_y\right), \; u_x = \pm v_y, \text{ and } v_x = \mp u_y.$$

Thus the Cauchy-Riemann equations obtain for h or $\overline{h}$. □

13.10. If $a \in \Omega$ then for some positive r, all z in $D(a,r)^\circ$, and some $g \stackrel{\text{def}}{=} p + iq$ in $H\left(D(a,r)^\circ\right)$,

$$f = e^g, \; |f| = e^p, \; \ln|f| = p \in L\left(D(a,r)^\circ\right).$$

It follows that $\Delta \ln|f(z)| \equiv 0$ in Ω.

Another proof involves the examination of $h \stackrel{\text{def}}{=} |f| = \left(u^2 + v^2\right)^{\frac{1}{2}}$ and a direct calculation of Δh. □

13.11. For $f \stackrel{\text{def}}{=} u + iv$, the observation $|f| = \sqrt{f \cdot \overline{f}}$ and direct calculation of $|f|_{xx}$ and $|f|_{yy}$ shows that if

$$\begin{aligned}S_1 &\stackrel{\text{def}}{=} \{(x,y) \; : \; u_x(x,y) = u_y(x,y) = v_x(x,y) = v_y(x,y) = 0\} \cap \Omega \text{ and} \\ S_2 &\stackrel{\text{def}}{=} \left\{(x,y) \; : \; (u(x,y))^2 + (v(x,y))^2 = 0\right\} \cap \Omega\end{aligned}$$

then $S_1 \cup S_2 = \Omega$. If $f \not\equiv 0$ then $S_2^\circ = \emptyset$, whence $S_1^\circ \neq \emptyset$ in which case f is a constant on some nonempty $D(a,r)^\circ$ contained in Ω. □

13.2. Developments

13.12. If $|a| < R$, Poisson's formula may be written

$$h(a) = \frac{1}{2\pi}\int_{C_0(R)} \frac{R^2 - |a|^2}{|z-a|^2} h(z)\, d\theta = \frac{1}{2\pi}\int_{C_0(R)} \Re\left(\frac{z+a}{z-a}\right) h(z)\, d\theta$$
$$= \Re\left[\frac{1}{2\pi i}\int_{C_0(R)} \frac{\zeta+z}{\zeta-z}\frac{h(\zeta)}{\zeta}\, d\zeta\right] \stackrel{\text{def}}{=} \Re f(z).$$

Thus $f \in H\left(D(0,R)^\circ\right)$ and $\Re(f) = h$. □

13.13. Schwarz's formula (cf. **13.12**) implies that $\mathbb{R}$ contains a sequence $\{k_n\}_{n\in\mathbb{N}}$ such that if $|z| < 1$, $f_n(z) = \dfrac{1}{2\pi i}\displaystyle\int_{C_0(R)} \frac{w+z}{w-z}\frac{u_n(w)}{w}\, dw + ik_n$. The uniform convergence of $\{u_n\}_{n\in\mathbb{N}}$ implies that the integrals converge and since $\lim_{n\to\infty} f_n(0)$ exists, $\lim_{n\to\infty} k_n \stackrel{\text{def}}{=} k$ exists. □

13.14. a) The Poisson formula for v implies

$$v\left(re^{i\theta}\right) = \frac{1}{2\pi}\left(\int_0^\pi + \int_\pi^{2\pi} \frac{2Rr\sin(\theta-t)}{R^2+r^2-2Rr\cos(\theta-t)} u\left(re^{it}\right) dt\right)$$
$$= \frac{1}{2\pi}\left(\int_0^\pi \frac{2Rr\sin(\theta-t)u\left(re^{it}\right)}{R^2+r^2-2Rr\cos(t-\theta)}\, dt\right)$$
$$+ \frac{1}{2\pi}\left(\int_0^\pi \frac{2Rr\sin(t-\theta)u\left(re^{i(t-\pi)}\right)}{R^2+r^2+2Rr\cos(t-\theta)}\, dt\right).$$

Because

$$\frac{d}{dt}\ln\left(R^2+r^2 \pm 2Rr\cos(t-\theta)\right) = \frac{\mp 2Rr\sin(t-\theta)}{R^2+r^2\pm 2Rr\cos(t-\theta)}$$

the condition $|h| \le M$ and the equation

$$\int_0^\pi \left|\frac{2Rr\sin(t-\theta)}{R^2+r^2\pm 2Rr\cos(t-\theta)}\right| dt = \int_0^\pi \frac{2Rr\sin t}{R^2+r^2\pm 2Rr\cos t}\, dt$$

imply

$$\left|v\left(re^{i\theta}\right)\right| \le \frac{M}{\pi}\left(\ln(R+r)^2 - \ln(R-r)^2\right) = \frac{2M}{\pi}\ln\frac{R+r}{R-r}.$$

b) There is a biholomorphic bijection $\phi : U \mapsto \{\, w \;:\; -1 < \Re(w) < 1\,\}$. If $\phi \stackrel{\text{def}}{=} p + iq$ then $|p| \le 1$ whereas $\sup_{z\in U}|q(z)| = \infty$. Furthermore, any two harmonic conjugates of p differ by a constant. □

13.15. If $z \in D(a,R)^\circ$ then $\frac{R-|z|}{R+|z|} \leq \Re\left\{\frac{Re^{it}+z}{Re^{it}-z}\right\} \leq \frac{R+|z|}{R-|z|}$. Since $h \geq 0$, the Schwarz formula (cf. **13.11**) implies the result. □

13.16. The existence of $J \stackrel{\text{def}}{=} \int_0^{2\pi} \ln\left|1-e^{it}\right| dt$ turns on the existence, for ϵ in $(0,1)$, of the (improper) integral, $\int_0^\epsilon \left|\ln\left|1-e^{it}\right|\right| dt$. However, for small positive t,

$$\begin{aligned}\left|\ln\left|1-e^{it}\right|\right| &= \frac{1}{2}\left|\ln|2-2\cos t|\right| \\ &\leq \frac{1}{2}\left(\ln 2 + 2\left|\ln\left|\sin\frac{t}{2}\right|\right|\right) \leq \frac{\ln 2}{2} + \left|\ln\left|\frac{t}{2}\right|\right|\end{aligned}$$

and $\int_0^\epsilon |\ln x|\, dx = -\lim_{\delta\downarrow 0}(x\ln x - x)\big|_\delta^\epsilon = \epsilon - \epsilon\ln\epsilon$. Hence J exists.

Next, if $0 < r < 0.5$ and $z = 1+r$, then $z-w$, as a function of w, is in $H(U)$ and is never zero in U, whence $\ln|z-w| \in L(U) = MVP(U)$ and so $\frac{1}{2\pi}\int_0^{2\pi} \ln\left|z-e^{it}\right| dt = \ln|z|$.

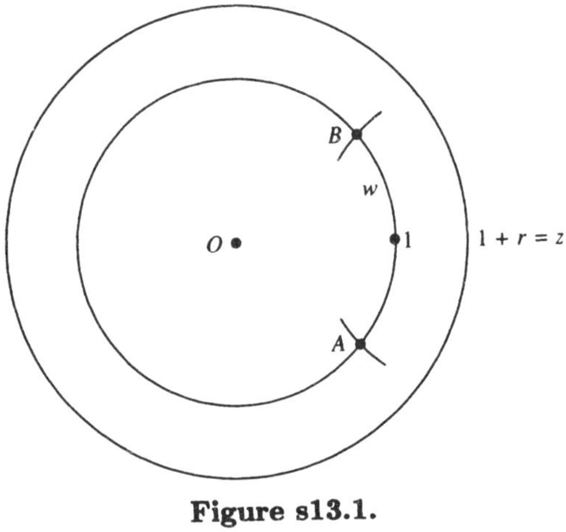

Figure s13.1.

Furthermore, the geometry depicted in **Figure s13.1** indicates that if w lies on the arc $A1B$, then $\ln|w-1|$ and $\ln|w-z|$ are negative but $|w-1| < |w-z|$, whence $\left|\ln|w-1|\right| > \left|\ln|w-z|\right|$. In $\mathbb{T} \setminus \{A1B\}$, $\left|\ln|w-z|\right|$ is bounded. In sum, for w in $\mathbb{T}$ and z in $[1, 1.5]$, $\left|\ln|w-z|\right|$ is dominated by an integrable function. Hence a passage to the limit as $r \downarrow 0$ (and $z \to 1$) is justified. Because $\lim_{z\to 1}\ln|z| = 0$, it follows that

$$\lim_{r\downarrow 0}\frac{1}{2\pi}\int_0^{2\pi} \ln\left|z-e^{it}\right| dt = \frac{1}{2\pi}\int_0^{2\pi} \ln\left|1-e^{it}\right| dt = 0.$$ □

13.17. The basic idea is to replace f by a function F free of zeros in some slightly larger $D(0, R+\epsilon)$ so that $\ln|F| \in L\left(D(0,R+\epsilon)^\circ\right)$.

If some zeros of f are in $D(0,R)^\circ$, the notation may be chosen so that $|z_n| < R, 1 \le n \le m$, and if $m < N$, $|z_n| = R, m+1 \le n \le N$. Then an F as described is given by the formulæ

$$F(z) \stackrel{\text{def}}{=} f(z) \prod_{n=1}^{m} \frac{R^2 - \overline{z_n} z}{R(z - z_n)} \prod_{n=m+1}^{N} \frac{z_n}{z - z_n}$$
$$|F(0)| = |f(0)| \prod_{n=1}^{m} \frac{R}{|z_n|}.$$

If $z_n \stackrel{\text{def}}{=} Re^{i\theta_n}$ and $z \stackrel{\text{def}}{=} Re^{i\theta}$, $\theta_n, \theta \in [0, 2\pi), m+1 \le n \le N$, then

$$\left| \frac{R^2 - \overline{z_n} z}{R(z - z_n)} \right| = \begin{cases} 1 & \text{if } 1 \le n \le m \\ \dfrac{1}{|1 - e^{\theta - \theta_n}|} & \text{if } n+1 \le m \le N. \end{cases} \tag{s13.2}$$

For some positive ϵ, $\ln |F| \in L(D(0, R+\epsilon)^\circ)$ and the mean value property implies $\ln |F(0)| = \dfrac{1}{2\pi} \displaystyle\int_0^{2\pi} \ln \left|F\left(Re^{i\theta}\right)\right| d\theta$. Owing to (s13.2), $\ln \left|F\left(Re^{i\theta}\right)\right| = \ln \left|f\left(Re^{i\theta}\right)\right| - \sum_{n=m+1}^{N} \ln \left|1 - e^{\theta - \theta_n}\right|$, whence by virtue of **13.16**, $\int_0^{2\pi} \ln \left|F\left(Re^{i\theta}\right)\right| d\theta = \int_0^{2\pi} \ln \left|f\left(Re^{i\theta}\right)\right| d\theta$. Note that in the desired formula, $\sum_{n=m+1}^{N} \ln \left|\dfrac{R}{|z_n|}\right| = 0$. □

13.18. The Poisson integral applies to $\ln |F(z)|$ in **Solution 13.17**. □

13.19. a) If $\left|f\left(re^{i\theta}\right)\right| \le M < \infty$ then $\int_0^{2\pi} \ln^+ \left|f\left(re^{i\theta}\right)\right| d\theta \le 2\pi \ln^+ M$.

b) It may be assumed that $f(0) \ne 0$. Jensen's formula and the hypothesis imply that for some M,

$$|f(0)| \prod_{n=1}^{N} \frac{r}{|a_n|} \le \exp\left(\frac{1}{2\pi} \int_0^{2\pi} \ln^+ \left|f\left(re^{i\theta}\right)\right| \right) d\theta$$

and $1 \ge \prod_{n=1}^{N} |a_n| \ge \dfrac{|f(0)| r^N}{M}$. Hence as first, $r \uparrow 1$ and then $N \uparrow \infty$, $1 \ge \prod_{n=1}^{\infty} |a_n| \ge \dfrac{|f(0)|}{M}$. If $\sum_{n=1}^{\infty} (1 - |a_n|) = \infty$, the argument presented in **Solution 8.27**b) may be applied in the present context and yields:

$$\prod_{n=1}^{\infty} |a_n| = \prod_{n=1}^{\infty} (1 - (1 - |a_n|) = 0,$$

a contradiction. □

13.20. a) The dominated convergence theorem implies that for each a in Ω, $h_n(a) = \frac{1}{2\pi}\int_{\partial D(a,r)} h_n\left(a+re^{i\theta}\right)\,d\theta \to \frac{1}{2\pi}\int_{\partial D(a,r)} f_{a,r}\left(a+re^{i\theta}\right))\,d\theta$. It follows that $\widetilde{h}(z) \stackrel{\text{def}}{=} \lim_{n\to\infty} h_n(z)$ exists, $\widetilde{h} = h$ a.e., and $h \in MVP(\Omega)$.

b) See a).

c) Harnack's inequality (cf. **13.15**) implies that if $D(a,r) \subset \Omega$ and $h_n(a) \uparrow \infty$ then $h_n(z)\big|_{D(a,r)} \uparrow \infty$, whence $L_\infty \stackrel{\text{def}}{=} \{\, z \,:\, h_n(z) \uparrow \infty \,\} \cap \Omega$ is open. Harnack's inequality implies as well that if $h_n(a) \le M < \infty$ then for some constant K, $h_n(z)\big|_{D(a,r)} \le K$. Hence $h \in MVP\left(D(a,r)^\circ\right)$, i.e., $h \in L\left(D(a,r)^\circ\right)$. It follows that $(\Omega \setminus L_\infty) \cap \Omega$ is open and since Ω is a region either $L_\infty = \emptyset$ or $L_\infty = \Omega$. $\square$

14
Families of Functions

14.1. Sequences of Functions

14.1. Some subsequence $\{f_{n_k}\}_{k\in\mathbb{N}}$ converges uniformly on compact subsets of Ω. Thus each of the sequences $S_p \stackrel{\text{def}}{=} \{f_{n_k}^{2^p}\}_{k\in\mathbb{N}}$, $p \in \mathbb{Z}^+$, converges on compact subsets of Ω. Furthermore, $S_{p+1} \subset S_p$. The diagonal sequence $S \stackrel{\text{def}}{=} \left\{f_{n_p}^{2^p}\right\}_{p\in\mathbb{N}}$ is a subset of S_0. □

14.2. If $\{\alpha_n\}_{\nu\in\mathbb{N}} \subset U$ and $\lim_{n\to\infty} \alpha_n = 1$, the requirements are met by $\{-\Phi_{\alpha_n}\}_{n\in\mathbb{N}}$. □

14.3. If $\Omega \subset U$, $\mathcal{F}_\Omega$ is normal. If $\Omega \cap (\mathbb{C} \setminus U) \neq \emptyset$, $\{z^n\}_{n\in\mathbb{N}}$ fails to converge somewhere in Ω. Thus $\mathcal{F}_\Omega$ is normal iff $\Omega \subset U$. □

14.4. If $k, n \in \mathbb{N}$, and $z = x + iy \in \Pi_-$ then

$$|\sin nz - \sin(n+k)z|$$
$$= 2\left|e^{inx}e^{-ny}\left(1 - e^{ikx}e^{-ky}\right) - e^{-inx}e^{ny}\left(1 - e^{-iky}e^{ky}\right)\right| \stackrel{\text{def}}{=} 2|I - II|.$$

For large k, n, $|I|$ is large while $|II|$ is small whence if $\Omega \cap \Pi_- \neq \emptyset$, then Ω contains compact sets (points!) on which $\{\sin nz\}_{n\in\mathbb{N}}$ fails to converge. A similar argument shows a similar result if $\Omega \cap \Pi_+ \neq \emptyset$. Because no region is a subset of $\mathbb{R}$, $\mathcal{F}_\Omega$ is normal for no region Ω. □

14.5. If $f_n(z) \stackrel{\text{def}}{=} \sum_{m=0}^{\infty} c_{nm} z^m$ in U, then

$$\int_U |f_n(z)|^2 \, d\lambda_2 = 2\pi \sum_{m=0}^{\infty} \frac{|c_{nm}|^2}{2m+2} \leq 2\pi \int_U |f_n(z)|^q \, d\lambda_2$$

and if $r \in [0,1)$,

$$\left|f_n\left(re^{it}\right)\right| \leq \left(\sum_{m=0}^{\infty} \frac{|c_{nm}|}{\sqrt{2m+2}}\right)\left(\sqrt{2m+2}\right) r^m$$
$$\leq \left(\sum_{m=0}^{\infty} \frac{|c_{nm}|^2}{2m+2}\right)^{\frac{1}{2}} \cdot \left(\sum_{m=0}^{\infty} (2m+2) r^{2m}\right)^{\frac{1}{2}} \stackrel{\text{def}}{=} I_n \cdot II.$$

Because $\lim_{n\to\infty} I_n = 0$ and $II = \left.\frac{d}{dx}\left(\frac{x}{1-x^2}\right)\right|_{x=r} < \infty$, $f_n \stackrel{\text{u}}{\to} 0$ on compact subsets of U. □

14.2. General Families

14.6. If $\{f_n\}_{n\in\mathbb{N}}$ is a $\|\ \|_\infty$-Cauchy sequence then $\lim_{n\to\infty} f_n \stackrel{\text{def}}{=} f$ exists and $f \in C(\Omega, \mathbb{C})$. As the uniform limit of functions holomorphic in Ω, $f \in H(\Omega)$. Cauchy's formula implies $B(\mathbf{O}, 1)$ is equicontinuous. □

14.7. a) Since, for each r, $M_p(r;\cdot)$ is a norm for the vector space

$$\left\{ f\big|_{D(0,r)} : f \in H(U) \right\},$$

it follows that $\|\ \|_p$ is a norm for H^p.

b) Cauchy's formula and Hölder's inequality imply that a $\|\ \|_p$-Cauchy sequence $S \stackrel{\text{def}}{=} \{f_n\}_{n\in\mathbb{N}}$ confined to $D(0,r)$, is a $\|\ \|_\infty$-Cauchy sequence. Thus for a unique f in $H(U)$, the elements of S converge uniformly to f on compact subsets of U. Direct calculation shows that $f \in H^p$ and that $\|f_n - f\|_p \to 0$ as $n \to \infty$.

c) The argument in b) shows that if $T \stackrel{\text{def}}{=} \{f_n\}_{n\in\mathbb{N}}$ is a subset of $B(\mathbf{O},1)$ then T confined to $D(0,r)$ is a normal family. □

14.8. a) Since $M_2(r;f) = \left(\sum_{n=0}^{\infty} |c_n|^2 r^{2n}\right)^{\frac{1}{2}}$ and $f \in H^2$, it follows that $\sum_{n=0}^{\infty} |c_n|^2 < \infty$ and thus $f^*(\theta) \stackrel{\text{def}}{=} \sum_{n=0}^{\infty} c_n e^{in\theta} \in L^2([0,2\pi],\lambda)$. Furthermore, $\|f^* - f_r\|_2 = \sum_{n=0}^{\infty} |c_n|^2 \cdot (1 - r^{2n})$. If $0 \le r_n \uparrow 1$ then from the dominated convergence theorem [applied in $L^1(\mathbb{Z}^+, \zeta)$] it follows that $f_{r_n} \stackrel{\|\ \|_2}{\to} f^*$.

b) The properties listed in **8.14** and **Note 8.1** for the Poisson kernels P_r imply that if $g \in L^1([-\pi,\pi],\lambda)$ and $g_{r_n}(s) \stackrel{\text{def}}{=} \int_{-\pi}^{\pi} g(t-s)P_{r_n}(s)\,ds$ then $g_{r_n} \stackrel{\|\ \|_1}{\to} g$ as $r_n \uparrow 1$. The desired conclusion is somewhat stronger.

For a in $[-\pi,\pi]$, and ϵ in $(0,\pi)$,

$$\begin{aligned} g_r(a) - g(a) &= \int_{-\pi}^{\pi} [g(a-t) - g(a)]\, P_r(s)\,ds \\ &= \int_{-\pi}^{-\epsilon} + \int_{-\epsilon}^{\epsilon} + \int_{\epsilon}^{\pi} [g(a-t) - g(a)]P_r(s)\,ds \\ &\stackrel{\text{def}}{=} I_1 + I_2 + I_3. \end{aligned}$$

Since $\max_{s\in[-\pi,-\epsilon]} P_r(s) = \max_{s\in[\epsilon,\pi]} P_r(s) = P_r(\epsilon)$, reference to the integrals above leads to $|I_1| + |I_3| \le P_r(\epsilon)\left(\|g\|_1 + 2\pi|g(a)|\right)$. It follows that $|I_1| + |I_3| \to 0$ as $r \to 1$.

At this point, recourse to the Lebesgue set L_g of g is helpful. Thus for $G(t;a) \stackrel{\text{def}}{=} \int_0^t |g(a-s) - g(a)|\,ds$ and $R \stackrel{\text{def}}{=} \pi(1-r)$, integration by parts of the right member in (s14.1) below leads to (s14.2):

$$|I_2| \le R\int_{-\epsilon}^{\epsilon} \frac{|g(a-s)-g(a)|}{R^2+4s^2}\,ds \tag{s14.1}$$

$$\le \left(\frac{RG(t;a)}{R^2+4t^2}\right)\Bigg|_{-\epsilon}^{\epsilon} + 8R\int_{-\epsilon}^{\epsilon} \frac{sG(s;a)}{(R^2+4s^2)^2}\,ds. \tag{s14.2}$$

If $a \in L_g$ and ϵ is small, the right member of (s14.2) is small uniformly with respect to r.

Because $H^2 \subset H^1$, it follows that the results in the previous paragraph may be applied to f. □

[**Note s14.1:** The approach is parallel to that used in the proof of Fejér's theorem, where, instead of the existence and properties of the Lebesgue set, the uniform continuity of the function in question is exploited.

The fact that $f_r \overset{\|\ \|_1}{\to} f^*$ permits the conclusion that for some sequence $\{r_n\}_{0\le r_n<1}$, $f_{r_n} \overset{\text{a.e.}}{\to} f^*$. The use of L_f leads to the stronger result: $\lim_{r\uparrow 1} f_r \doteq f^*$.]

14.9. a) Because $f^* \in L^2([0,2\pi],\lambda)$, then in the sense of $\|\ \|_2$-convergence, $f^*(t) = \sum_{n=-\infty}^{\infty} \widehat{f^*}_n e^{int}$, whereas if $0 \le r < 1$,

$$f\left(re^{it}\right)_n^{\widehat{\ }} = \begin{cases} c_n & \text{if } n \in \mathbb{Z}^+ \\ 0 & \text{otherwise} \end{cases}.$$

Since the Fourier coefficients are the values of continuous linear functionals, it follows that $\lim_{n\to\infty} c_n = \widehat{f^*_n}$.

b) Since $\sum_{n=-\infty}^{\infty} |\widehat{g}_n|^2 < \infty$, for z in U, $\sum_{n=0}^{\infty} \widehat{g}_n z^n$ converges and defines a function $\widetilde{g}$ in $H(U)$. The orthogonality of the sequence $\left\{e^{int}\right\}_{n\in\mathbb{Z}}$ implies $\widetilde{g} \in H^2$ and **14.6** implies $\widehat{\widetilde{g}^*_n} = \begin{cases} c_n & \text{if } n \in \mathbb{Z}^+ \\ 0 & \text{otherwise} \end{cases}$. Consequently, $\widetilde{g}^* \doteq g$. □

14.10. If $g(z) \overset{\text{def}}{=} \sum_{n=0}^{\infty} b_n z^n$, $f \in \mathcal{F}$, and $0 \le r < 1$ then

$$\left\| f\big|_{D(0,r)} \right\|_\infty \le \left\| g\big|_{D(0,r)} \right\|_\infty < \infty. \tag{s14.3}$$

If $S \overset{\text{def}}{=} \{f_n\}_{n\in\mathbb{N}} \subset \mathcal{F}$ then (s14.3), Cauchy's formula and the Arzelà-Ascoli theorem imply that S contains a subsequence S_1 converging uniformly on $D\left(0, 1-2^{-1}\right)$. If there are subsequences S_m such that

$$S_m \supset S_{m+1}, m = 1, 2, \ldots, M-2,$$

and the elements of S_m converge uniformly on $D\left(0,1-2^{-m}\right)$, then S_{M-1} contains a subsequence S_M converging on $D\left(0,1-2^{-M}\right)$. Hence for some diagonal sequence $S \stackrel{\text{def}}{=} \left\{f_{n_1}, f_{n_{2_2}}, \ldots\right\}$, $f_{n_1} \in S_1$, $f_{n_{2_2}} \in S_2$, ..., and S converges uniformly on compact sets of U. □

14.11. Let a_0 be zero. If $g \stackrel{\text{def}}{=} \dfrac{f}{\prod_{k=0}^{n} \Phi_{a_k}}$ then the maximum modulus theorem implies $g(U) \subset U$. It follows that

$$f'(0) = \begin{cases} 0 & \text{if the multiplicity of } a_0 \text{ is more than one} \\ g'(0) \prod_{k=1}^{n} (-a_k) & \text{otherwise} \end{cases}$$

Because $g(0) = 0$, Schwarz's lemma implies that if g is not a constant then $|g'(0)| < 1$. Hence $|f'(0)| < \prod_{k=1}^{n} |a_k|$. Furthermore, if $g(z) \equiv e^{i\theta}$ then $|f'(0)| = \prod_{k=1}^{n} |a_k|$. □

14.12. For r in $[0,1)$, let $\mathcal{F}(r)$ be $\sup_{f\in\mathcal{F}} M(r;f)$. If $\mathcal{F}$ is normal and $\mathcal{F}(r) = \infty$ then for some sequence $S \stackrel{\text{def}}{=} \{f_n\}_{n\in\mathbb{N}}$ in $\mathcal{F}$, $M(r;f_n) \uparrow \infty$ in which case S fails to contain a subsequence converging uniformly on compact sets of U, a contradiction.

Hence Cauchy's formula shows that for k in $\mathbb{Z}^+$ and f in $\mathcal{F}$,

$$|c_k(f)| \le \frac{\mathcal{F}(r)}{r^k}$$

and so for all r in $[0,1)$, $\sup_{f\in\mathcal{F}} |c_k(f)|^{\frac{1}{k}} \le \dfrac{(\mathcal{F}(r))^{\frac{1}{k}}}{r}$. It follows that

$$K = \overline{\lim}_{k\to\infty} \sup_{f\in\mathcal{F}} |c_k(f)|^{\frac{1}{k}} \le 1.$$

Conversely, if $K \le 1$ then for each r in $[0,1)$ and some positive $\epsilon(r)$, $(1+\epsilon(r))r \stackrel{\text{def}}{=} \rho < 1$ and for some k_0 and all f in $\mathcal{F}$, if $k > k_0$ then $|c_k(f)|^{\frac{1}{k}} < 1 + \epsilon(r)$. Hence if $f \in \mathcal{F}$ and $|z| \le r$,

$$|f(z)| \le \sum_{n=0}^{\infty} |c_k| r^n \le \frac{1}{1-\rho}.$$

Cauchy's formula implies $\mathcal{F}$ is equicontinuous and the Arzelà-Ascoli theorem implies each sequence in $\mathcal{F}$ contains a subsequence converging uniformly on $D(0,r)$.

The argument at the end of **Solution 14.10** applies. □

14.13. For w in Ω, $\mathsf{N}(w) \stackrel{\text{def}}{=} \dfrac{1}{2\pi i} \displaystyle\int_{\mathbb{T}} \frac{\partial f(w,z)/\partial z}{f(w,z)}\, dz$ is the number of zeros of $f(w,z)$ in U. As a $\mathbb{Z}^+$-valued continuous function, N is a constant and

$\mathsf{N}(0) = n$. Thus for each w in Ω, U contains points, $g_{k,w}, 1 \le k \le n$, not necessarily pairwise different and such that $f(w, g_{k,w}) = 0$. $\square$

14.14. For each f in $\mathcal{F}$ and some g_f in $H(U)$, $f(z) = z^3 g_f(z)$. The maximum modulus theorem implies $g_f(U) \subset U$. Cauchy's formula and the Arzelà-Ascoli theorem imply $\mathcal{F}$ is normal and hence for some f_0 in $\mathcal{F}$, $\left|f_0\left(\frac{1}{2}\right)\right| = \alpha$. Thus if $f_0\left(\frac{1}{2}\right) = \alpha e^{it}$ then $e^{-it} f_0\left(\frac{1}{2}\right) \stackrel{\text{def}}{=} F\left(\frac{1}{2}\right) = \alpha$ and $F \in \mathcal{F}$. Because $g_f(U) \subset U$, $\left|g_f\left(\frac{1}{2}\right)\right| \le 1$. It follows that $\alpha \le \frac{1}{8}$. Since $z^3 \in \mathcal{F}$, $\alpha = \frac{1}{8}$. If g_f is not constant, the maximum modulus theorem implies $\left|g_f\left(\frac{1}{2}\right)\right| < 1$. Thus $F(z) = z^3$. $\square$

14.15. If $S^\bullet \cap \mathbb{C} \ne \emptyset$ then $\mathcal{I} = \{0\} = 0 \cdot \mathcal{E}$. That trivial case aside, for some h in $\mathcal{E}$, $Z(h) = S$ and the order of the zero of h at z_n is q_n. Hence if $f \in \mathcal{I}$ then $\frac{f}{h} \in \mathcal{E}$, i.e., $\mathcal{I} = h \cdot \mathcal{E}$. $\square$

14.16. If $f \in \mathcal{F}$ and $f(0) = a$ then $\Phi_a \circ f \stackrel{\text{def}}{=} g \in \mathcal{F}$, $g(0) = 0$, and Schwarz's lemma implies $|g'(0)| \le 1$. However, $g'(0) = \frac{f'(0) - |a|^2}{(1 - |a|^2)}$. Hence $\left|\frac{\partial \Re(f)(0)}{\partial x}\right| \le |f'(0)| \le 1$. Moreover, if $f_0(z) \stackrel{\text{def}}{=} z$ then $f_0 \in \mathcal{F}$ and

$$\left|\frac{\partial \Re(f_0)(0)}{\partial x}\right| = f_0'(0) = 1. \qquad \square$$

14.17. If $r \in [0,1)$, $f \in \mathcal{F}$, and $z \in D(0,r)$ then Cauchy's formula implies that $|f(z)| \le \frac{1}{2\pi}\frac{1}{1-r}$. Hence Cauchy's formula and the Arzelà-Ascoli technique imply $\mathcal{F}$ is normal. $\square$

14.18. a) For each a in Ω and some positive $r(a)$, $D(a, r(a)) \subset \Omega$. If $f \in \mathcal{F}$ then Cauchy's formula implies

$$|f(a)| \le \int_0^{2\pi} \left|f\left(a + r(a)e^{it}\right)\right| dt \text{ and}$$

$$\int_0^{r(a)} |f(a)|\, s\, ds = \frac{(r(a))^2 |f(a)|}{2} \le \int_{D(a,r(a))} |f(z)|\, d\lambda_2 \le \int_\Omega |f(z)|\, d\lambda_2.$$

Let $M(a)$ be $\frac{2}{(r(a))^2}$.

b) For some sequence $\mathcal{K} \stackrel{\text{def}}{=} \{K_m\}_{m \in \mathbb{N}}$ of compact sets, $K_m \subset K_{m+1}^\circ$ and $\Omega = \bigcup_{m \in \mathbb{N}} K_m$. Hence any compact subset of Ω is a subset of some member of $\mathcal{K}$.

If $S \stackrel{\text{def}}{=} \{f_n\}_{n\in\mathbb{N}} \subset \mathcal{F}$, the result in a) and the Heine-Borel theorem imply that for some constant M_m, $|f_n(z)| \le M_m, n \in \mathbb{N}$, on K_m. Cauchy's formula and the Arzelà-Ascoli theorem imply that S contains a subsequence S_1 uniformly convergent on K_1; S_1 contains a subsequence S_2 uniformly convergent on K_2; ... . Hence there is, by abuse of language, a diagonal subsequence S_∞ uniformly convergent on every compact subset of Ω.

c) For the compact-open topology $\mathcal{T}$ of $\mathcal{F}$, let $\mathcal{O}$ be an open cover of $\mathcal{F}$. If $\mathcal{B}$ is a countable base for the topology of $\mathbb{C}$ and $\mathcal{K}$ is the sequence in b) then the set of all sets $\{f : f(K) \subset O, K \in \mathcal{K}, O \in \mathcal{B}\}$ is a countable base for $\mathcal{T}$. Thus it may be assumed that $\mathcal{O}$ is countable: $\mathcal{O} = \{U_p\}_{p\in\mathbb{N}}$. If for each P in $\mathbb{N}$, $V_P \stackrel{\text{def}}{=} \bigcup_p^P U_p$ fails to cover $\mathcal{F}$, then for some sequence $G \stackrel{\text{def}}{=} \{g_P\}_{P\in\mathbb{N}}, g_P \notin V_P$. But a) ensures that G contains a subsequence $\widetilde{G}$ converging uniformly on each compact subset of Ω to some function g in $H(\Omega)$. For each compact subset K of Ω, $\int_K |g(z)|\,d\lambda_2 \le 1$, whence $g \in \mathcal{F}$. If $g \in U_{p_0}$ then all but finitely many of the elements of $\widetilde{G}$ are in U_{p_0}, hence in V_{p_0}, a contradiction. □

[**Note s14.2:** The method of proof for c) above is essentially that for proving the Heine-Borel theorem.]

14.19. a) Because $I(t) = \#\left(Z(f(\cdot,t)) \cap U\right) \ (\in \mathbb{Z}^+)$ and $I(t)$ is a continuous function of t, it follows that $I(t)$ is a constant.

b) If K is a compact subset of U, the continuity of g implies that the holomorphic functions $F_N(z) \stackrel{\text{def}}{=} \frac{1}{N}\sum_{n=1}^{N} f\left(z, \frac{n}{N}\right) g\left(\frac{n}{N}\right), N \in \mathbb{N}$, are uniformly bounded on K. Cauchy's formula and the Arzelà-Ascoli theorem imply a subsequence $\{F_{N_m}\}_{m\in\mathbb{N}}$ converges uniformly on K to F_g. □

14.3. Defective Functions

14.20. The simple connectedness of Ω justifies each of the following arguments.

a) Since f is $\{0\}$-defective, for some p in $H(\Omega)$, $f = e^{2\pi i p}$. Because f is $\{1\}$-defective p is $\mathbb{Z}$-defective.

b) Since p is $\{0\}$-defective, there is a q such that $p = q^2$.

c) Since $p - 1$ is $\{0\}$-defective, there is an r such that $p - 1 = r^2$.

d) Since $q^2 - r^2 = 1 = (q - r)(q + r)$, $q - r$ is $\{0\}$-defective whence for some s, $q - r = e^s$. □

14.21. In the notations of **14.20**, $f = -e^{\pi i \cosh 2s}$. If $s(z) \in S$ then direct calculation shows $f(z) = 1$. □

[**Note s14.3:** The technique used in **Solutions 14.20, 14.21** may be viewed as producing from a $\{0,1\}$-defective function new functions that are S-defective for various infinite sets S.]

14.22. If $z \stackrel{\text{def}}{=} a + ib \in S$ then $z \pm \frac{\pi i}{2}$ are in S. Because

$$\ln(\sqrt{m+1} + \sqrt{m}) - \ln(\sqrt{m} + \sqrt{m-1}) \begin{cases} = \ln(\sqrt{2}+1) < 1 & \text{if } m = 1 \\ < \ln\sqrt{\dfrac{m+1}{m-1}} \le \ln\sqrt{3} < 1 & \text{if } m > 1 \end{cases},$$

then for any $z \stackrel{\text{def}}{=} x + iy$ and for some $\sigma \stackrel{\text{def}}{=} \rho + i\tau$ in S, $|x - \rho| < \frac{1}{2}$ and $|y - \tau| < \frac{\pi}{4} < \frac{\sqrt{3}}{2}$ and so $|z - s| < 1$. □

14.23. Integration implies $|f(z)| \le RM$ in $D(0, R)$. Because $g \stackrel{\text{def}}{=} 1 - \frac{f}{\gamma}$ is $\{0\}$-defective in $D(0, R)^\circ$, for some h in $H(D(0, R)^\circ)$,

$$(h(z))^2 \stackrel{\text{def}}{=} \left(1 - \frac{f'(0)}{2\gamma}z + \cdots\right)^2 = 1 - \frac{f'(0)}{\gamma}z + \cdots = g(z).$$

In $D(0, R)^\circ$, $\left|(h(z))^2\right| \le 1 + \frac{RM}{|\gamma|}$ whence

$$1 + \frac{a^2}{4|\gamma|^2}R^2 \le 1 + \frac{|f'(0)|^2}{4|\gamma|^2}R^2 \le \|h\|_\infty^2 \le 1 + \frac{RM}{|\gamma|}.$$

Thus $|\gamma| \ge \frac{a^2 R}{4M}$. □

14.24. The sequence $\left\{\dfrac{M\left(1 - \frac{1}{2^k}; f'\right)}{2^k}\right\}_{k \in \mathbb{Z}^+}$ converges to zero and, by hypothesis, $M(0; f') \ge 1$. Let k_0 be the largest k such that

$$\frac{M\left(1 - \frac{1}{2^k}; f'\right)}{2^k} \ge 1.$$

Thus if $r_0 = \frac{1}{2^{k_0}}$ then $0 < r_0 \le 1, r_0 M(1 - r_0; f') \ge 1 > \frac{r_0}{2} M\left(1 - \frac{r_0}{2}; f'\right)$. For some b in $D(0, 1 - r_0)$, $|f'(b)| = M(1 - r_0; f')$. Then

$$g(z) \stackrel{\text{def}}{=} f(z + b) - f(b)$$

behaves as does f in **14.23** when $a \stackrel{\text{def}}{=} \frac{1}{r_0}$, $M = \frac{2}{r_0}$. □

14.25. For s as in **14.20**, if $|b| \leq r$ and $s'(b) \neq 0$ then

$$g(z) \stackrel{\text{def}}{=} \frac{s(b+(1-r)z)}{(1-r)s'(b)} \stackrel{\text{def}}{=} c_0 + z + \cdots$$

is such that $g(U)$ contains no translate of $D\left(0, \frac{1}{(1-r)\,|s'(b)|}\right)$. Thus **14.23** implies $|s'(b)| < \frac{16}{1-r}$ which is valid a fortiori if $s'(b) = 0$.

Since the value of $s(0)$ is determined by the value of $f(0)$, it follows that when $z \in D(0,r)$ then

$$\begin{aligned}
|s(z) - s(0)| &\leq \frac{16}{1-r} r < \frac{16}{1-r}, \\
|s(z)| &< |s(0)| + \frac{16}{1-r} \stackrel{\text{def}}{=} \Phi_1(f(0), r), \text{ and} \\
|f(z)| &< e^{\pi \cosh 2\Phi_1(f(0),r)} \stackrel{\text{def}}{=} \Phi(f(0), r).
\end{aligned}$$

□

14.26. If $f \in \mathcal{F}_{a,b}$ then $g(z) \stackrel{\text{def}}{=} \frac{f(z)-a}{f(z)-b}$ is meromorphic and $\{0,\infty\}$-defective. As a meromorphic function, g is the quotient of two entire functions h, k sharing no zeros: $g = \frac{h}{k}$. Because g is $\{0,\infty\}$-defective, it may be assumed that both h and k are $\{0\}$-defective. Hence g is entire and $\{0,\infty\}$-defective. Picard's theorem implies g is a constant, whence f is a constant. Any family of constant functions is spherically normal. □

14.27. Since S is simply connected, it follows that $\#(\mathbb{C} \setminus S) > 1$. Because the elements of $\mathcal{F}$ are $(\mathbb{C} \setminus S)$-defective, **14.26** implies that $\mathcal{F}$ is spherically normal. □

14.28. A Möbius transformation g maps Π_+ onto $S \stackrel{\text{def}}{=} \{ z : \Re(z) < 0 \}$ and $g(\mathbb{R}) = \{ z : \Re(z) = 0 \}$. By abuse of language, $g \circ \mathcal{G} \stackrel{\text{def}}{=} \mathcal{H}$ consists of functions holomorphic in Ω and mapping Ω into S. If $f \in \mathcal{G}$ then $\left|e^{g \circ f}\right|$ is bounded by one, whence Cauchy's formula and the Arzelà-Ascoli technique imply $e^{g \circ \mathcal{F}}$ is normal, i.e., for some real constants a, b, $\left\{ e^{af+b} : f \in \mathcal{F} \right\}$ is normal. Hence $\left\{ e^{af} : f \in \mathcal{F} \right\}$ is normal. If $a \geq 0$ then $\mathcal{F}$ is normal; if $a < 0$, $\mathcal{F}$ is spherically normal. □

[**Note s14.4:** In the discussion above, Π_+ may be replaced by any half-space (determined by a line).]

14.4. Bergman's Kernel Functions

14.29. a) If $\{f_n\}_{n \in \mathbb{N}} \subset \mathfrak{H}$ is a $\|\ \|_2$-Cauchy sequence then for some f in $L^2(\Omega, \lambda_2)$, $\|f_n - f\|_2 \to 0$ as $n \to \infty$. It follows that $f_n \stackrel{\text{w}}{\to} f$ as $n \to \infty$.

If $a \in \Omega$ and $g \in \mathfrak{H}$ then for some positive r,

$$g(z) = \sum_{m=0}^{\infty} c_m (z-a)^m \stackrel{\text{def}}{=} \sum_{m=0}^{\infty} c_m s^m e^{im\theta}$$

in $D(a,r)$. Thus

$$\begin{aligned} \|g\|_2^2 &\geq \int_0^r \int_0^{2\pi} \left|g\left(se^{i\theta}\right)\right|^2 s\,d\theta\,ds \\ &= 2\pi \sum_{m=0}^{\infty} |c_m|^2 \frac{r^{2m+2}}{2m+2} \geq |g(a)|^2 r^2. \end{aligned} \tag{s14.4}$$

Hence the map $L^2(\Omega, \lambda_2) \ni g \mapsto g(a) \in \mathbb{C}$ is a bounded linear functional. It follows that $f_n(a) \to f(a)$ as $n \to \infty$. The argument shows also that on $D(a,r)$, $f_n \stackrel{\text{u}}{\to} f$. It follows that $f \in H(\Omega)$ whence $f \in \mathfrak{H}$.

b) If $a \in K$ and $0 < \eta < \delta$ then $D(a,\eta)^\circ \subset \Omega$ and if $z \in D(a,\eta)^\circ$,

$$f(z) = \sum_{n=0}^{\infty} \frac{f^{(n)}(a)}{n!}(z-a)^n \text{ and}$$

$$\begin{aligned} \|f\|_2^2 &\geq \int_0^{2\pi} \int_0^{\eta} \left| \sum_{n=0}^{\infty} \frac{f^{(n)}(a)}{n!} r^n e^{in\theta} \right|^2 r\,dr d\theta \\ &\geq \pi \sum_{n=0}^{\infty} \frac{\left|f^{(n)}(a)\right|^2}{(n!)^2} \frac{\eta^{2n+2}}{n+1} \geq \pi\eta^2 |f(a)|^2. \end{aligned}$$

Thus $\dfrac{1}{(\pi\delta^2)}$ may serve for M_K^2.

c) If $\{f_n\}_{n\in\mathbb{N}} \subset \mathcal{F}$, the inequality in (s14.4), the argument in a), Cauchy's formula, and the Arzelà-Ascoli theorem imply that $\mathcal{F}$ is normal. □

14.30. If $z \in \Omega_\delta$ then **14.29**b) implies

$$\left| f(z) - \sum_{k=1}^{n} (f, \phi_k)\, \phi_k(z) \right| \leq \sqrt{\frac{1}{\pi\delta^2}} \left\| f - \sum_{k=1}^{n} (f, \phi_k)\, \phi_k \right\|_2 .$$

□

14.31. If $f \stackrel{\text{def}}{=} \sum_{n=1}^{N} c_n \phi_n \in M_N$ then $1 \leq \|f\|_2^2 \sum_{n=1}^{N} |\phi_n(t)|^2$ and the criterion for equality in Schwarz's inequality implies that for some constant c, $c_n = c\overline{\phi_n(t)}$. Thus $c = \dfrac{1}{\sum_{n=1}^{N} |\phi_n(t)|^2}$ and then $\mu = \sqrt{c}$. □

[**Note s14.5:** In **Solution 14.31**, if r is the distance of t from $\partial\Omega$ then from **14.29** it follows that $\sum_{n=1}^{\infty} |\phi_n(t)|^2 \leq \frac{1}{\pi r^2}$.]

14.32. a) The convergence of the series representing $K(z,w)$ is assured by **14.31** and **Note s14.5**. The Schwarz inequality applies.

b) Because $K(w,w) = \sum_{n=1}^{\infty} |\phi_n(w)|^2$, the argument in **Solution 14.31** implies $K(w,w) > 0$.

c) For the sequence $\left\{f_N \overset{\text{def}}{=} \sum_{n=1}^{N} (f,\phi_n)\,\phi_n\right\}_{N\in\mathbb{N}}$,

$$\left|\int_\Omega (f(w) - f_N(w))\,K(z,w)\,d\lambda_2(w))\right|^2 \leq \|f - f_n\|_2^2\,K(z,z)$$

and the dominated convergence theorem applies. □

14.33. If $g(z,w) \overset{\text{def}}{=} \dfrac{K(z,w)}{K(w,w)}$ then $g(w,w) = 1$ and $\|g\|_2^2 = \dfrac{1}{K(w,w)}$. If $h(z,w) \overset{\text{def}}{=} f(z) - g(z,w)$ then for fixed w, $h \in \mathfrak{H}$ and $h(w,w) = 0$. Since $0 = h(w,w) = \sum_{n=1}^{\infty} \beta_n\phi_n(w)$, direct calculation of the sequence $\left\{\beta_n \overset{\text{def}}{=} (h,\phi_n)\right\}_{n\in\mathbb{N}}$ shows that

$$\|f\|_2^2 = \frac{1}{K(w,w)} + \|h\|_2^2 = \|g\|_2^2 + \|h\|_2^2$$

whence $\|f\|_2 > \|g\|_2$ unless $h = 0$. Hence $\min\{\,\|f\|_2 \;:\; f \in \mathfrak{H}, f(w) = 1\,\}$ is achieved iff $f = g$. In particular, $K(z,w)$ is independent of the choice of the orthonormal system $\{\phi_n\}_{n\in\mathbb{N}}$. □

14.34. Direct calculation shows that

$$\left\{\phi_n(z) \overset{\text{def}}{=} \sqrt{\frac{n}{\pi}}\,\frac{(f(z,w))^{n-1}\,\dfrac{\partial f(z,w)}{\partial z}}{R^n}\right\}_{n\in\mathbb{N}}$$

is a complete orthonormal system for $\mathfrak{H}$, $\phi_1(w) = \dfrac{1}{\sqrt{\pi}R}$, and $\phi_n(w) = 0$ if $n \geq 2$. Because $K(z,w)$ is independent of the choice of a complete orthonormal system, $K(z,w) = \dfrac{\dfrac{\partial f(z,w)}{\partial z}}{\pi R^2}$. □

15
Convexity Theorems

15.1. Thorin's Theorem

15.1. a) If f is monotonely increasing resp. decreasing then

$$\sup_{x\in[a,b]} f(x) = f(b) \text{ resp. } \sup_{x\in[a,b]} f(x) = f(a).$$

If f is convex then the result **3.8** implies the conclusion.

b) For example, $\ln x$ satisfies the hypothesis but $\ln x$ is strictly concave.

□

15.2. If ϕ is convex then

$$\begin{aligned}\phi(t\mathbf{x} + (1-t)\mathbf{y}) - \lambda t &\leq t\phi(\mathbf{x}) + (1-t)\phi(\mathbf{y}) - \lambda t \\ &= t(\phi(\mathbf{x}) - \lambda) + (1-t)\phi(\mathbf{y}) \\ &\leq \max\{\phi(\mathbf{x}) - \lambda, \phi(\mathbf{y})\}.\end{aligned}$$

Conversely, for all λ in $\mathbb{R}$,

$$\sup_{t\in[0,1]} \phi(t\mathbf{x} + (1-t)\mathbf{y}) - \lambda t \leq \max\{\phi(\mathbf{x}) - \lambda, \phi(\mathbf{y})\}$$

iff

$$\phi(t\mathbf{x} + (1-t)\mathbf{y}) \leq \max\{\phi(\mathbf{x}) - \lambda + \lambda t, \phi(\mathbf{y}) + \lambda t\}$$

or

$$\phi(t\mathbf{x} + (1-t)\mathbf{y}) \leq \max\{\phi(\mathbf{x}) - \lambda(1-t), \phi(\mathbf{y}) + \lambda t\}.$$

Hence $\phi(t\mathbf{x} + (1-t)\mathbf{y}) \leq \phi(\mathbf{x}) - \lambda(1-t)$ if

$$\max\{\phi(\mathbf{x}) - \lambda(1-t), \phi(\mathbf{y}) + \lambda t\} = \phi(\mathbf{x}) - \lambda(1-t)$$

and $\phi(t\mathbf{x} + (1-t)\mathbf{y}) \leq \phi(\mathbf{y}) + \lambda t$ if

$$\max\{\phi(\mathbf{x}) - \lambda(1-t), \phi(\mathbf{y}) + \lambda t\} = \phi(\mathbf{y}) + \lambda t.$$

In each case when $\lambda = \phi(\mathbf{x}) - \phi(\mathbf{y})$,

$$\phi(t\mathbf{x} + (1-t)\mathbf{y}) \leq t\phi(\mathbf{x}) + (1-t)\phi(\mathbf{y}).$$

□

15.3. If $t \in [0,1]$ then

$$\begin{aligned}\phi_\lambda(t\mathbf{x} + (1-t)\mathbf{y}) &\le t\phi_\lambda(\mathbf{x}) + (1-t)\phi_\lambda(\mathbf{y})\\ &\le t\phi(\mathbf{x}) + (1-t)\phi(\mathbf{y}).\end{aligned}$$

□

15.4. If $\epsilon > 0$ then for some $\mathbf{b}(\epsilon)$ in B and for all $\mathbf{d}$ in D, $f(\mathbf{b}(\epsilon))+\epsilon > f(\mathbf{d})$ whence, since g is right-continuous, if $\delta > 0$ then

$$\begin{aligned} g(f(\mathbf{b}(\epsilon)) + \delta) &\ge g\circ f(\mathbf{d}),\\ g(f(\mathbf{b}(\epsilon)) + \delta) &\ge \sup_{\mathbf{d}\in D} g\circ f(\mathbf{d}),\\ g\circ f(\mathbf{b}(\epsilon))) &\ge \sup_{\mathbf{d}\in D} g\circ f(\mathbf{d}), \text{ and}\\ \sup_{\mathbf{b}\in B} g\circ f(\mathbf{b}) &\ge \sup_{\mathbf{d}\in D} g\circ f(\mathbf{d}).\end{aligned}$$

□

15.5. As formulated, the result is equivalent to the statement that for all λ, the maximum principle in $[0,1]$ relative to $\{0,1\}$ obtains for

$$f\left(t\mathbf{x}_1 + (1-t)\mathbf{x}_0, y\right) - \lambda t.$$

From **15.2** it follows that for each y, $f(\mathbf{x}, y)$ is a convex function of $\mathbf{x}$. Then **15.3** applies. □

15.6. Since L is linear, $L^{-1}(C)$ is convex. Then **15.5** and **15.3** apply. □

15.2. Applications

15.7. For each λ in $\mathbb{R}$, $e^{\lambda z}f(z) \stackrel{\text{def}}{=} F(z)$ conforms to the hypotheses for f. The maximum modulus principle as discussed in **9.32** implies, in the present context,

$$\widetilde{M}(x;F) \le \max\left(e^{\lambda a}\widetilde{M}(a;f), e^{\lambda b}\widetilde{M}(b;f)\right) \text{ and}$$

$$\ln\widetilde{M}(x;F) = \lambda x + \ln\widetilde{M}(x;f) \le \max(\lambda a + \ln\widetilde{M}(a;f), \lambda b + \ln\widetilde{M}(b;f)).$$

As in the proof of Thorin's theorem (**15.6**), if

$$\lambda = \frac{\ln\widetilde{M}(b;f) - \ln\widetilde{M}(a;f)}{a-b}, \text{ i.e., if } \lambda a + \ln\widetilde{M}(a;f) = \lambda b + \ln\widetilde{M}(b;f),$$

then direct calculation shows that for t in $[0,1]$,

$$\ln\widetilde{M}(ta + (1-t)b; f) \le t\ln\widetilde{M}(a;f) + (1-t)\ln\widetilde{M}(b;f).$$

□

15.8. For Ω as in **15.7** and a suitable real α, the map $\phi : \Omega^c \ni z \mapsto e^{\alpha z}$ carries Ω^c onto $A(a, r : R)$. Hence $f \circ \phi$ conforms to the hypotheses of **15.7**. □

[**Note s15.1:** It is the method of proof of Thorin's theorem rather than the theorem itself that resolves **15.7** and hence **15.8**. In [**Th**] the author attributes his attack on the general theorem to Hadamard's original approach!

For $\ln \widetilde{M}(x; f)$ resp. $\ln M(r; f)$, one may substitute $g \circ \widetilde{M}(x; f)$ resp. $g \circ M(r; f)$ when g is a monotonely increasing function continuous on the right. A notion of the strength of **15.7** and **15.8** and of the last remark can be derived from the following considerations.

The maximum modulus theorem implies merely

$$\widetilde{M}(x; f) \leq \max\left(\widetilde{M}(a; f), \widetilde{M}(b; f)\right)$$

resp.

$$M(t; f) \leq \max\left(M(r; f), M(R; f)\right).$$

These inequalities do not imply the convexity of either $\widetilde{M}(x; f)$ or $M(r; f)$; a fortiori, they do not imply any of the convexity properties of $\ln \widetilde{M}(x; f)$, $\ln M(t; f)$, $g \circ \widetilde{M}(x; f)$, or $g \circ M(r; f)$ (cf. **3.12**).

The appearance of $\ln r$ in **15.8** is occasioned by the map ϕ.]

15.9. a) If $0 \leq r_1 < r_2 < r$ then for some $z \stackrel{\text{def}}{=} r_2 e^{i\theta_2}$,

$$\frac{1}{n} \sum_{k=1}^{n} \left| f\left(r_1 e^{\frac{2k\pi i}{n}}\right)\right|^p \leq \frac{1}{n} \sum_{k=1}^{n} \left| f\left(r_2 e^{\frac{2k\pi i}{n}} e^{i\theta_2}\right)\right|^p$$

(cf. **Note 11.3**, **Note s11.3**, and **11.27**). The inequality persists as $n \to \infty$.

b) Thorin's method of proof as exhibited in **15.7** and as used in **15.8** applies. □

15.10. Write

$$x_j \stackrel{\text{def}}{=} r_j^{\alpha} e^{i\phi_j}, \; y_k \stackrel{\text{def}}{=} s_k^{\beta} e^{i\psi_k}$$
$$0 \leq \phi_j, \; \psi_k < 2\pi, \; 1 \leq j \leq m, \; 1 \leq k \leq n.$$

For λ_1, λ_2 real and fixed and (α_0, β_0) in Q, owing to **15.5** and **15.6**, it suffices to prove that the logarithm of

$$\widetilde{M}\left(\alpha_0 + \lambda_1 t, \beta_0 + \lambda_2 t\right)$$
$$\stackrel{\text{def}}{=} \sup_{(\mathbf{x}, \mathbf{y}, \rho, \sigma) \in S} \left| \sum_{j=1}^{m} \sum_{k=1}^{n} a_{jk} r_j^{\alpha_0 + \lambda_1 t} s_k^{\beta_0 + \lambda_2 t} e^{i(\phi_j + \psi_k)} \right|$$

is a convex function of t on $(0,\infty)$. If the real variable t is replaced by the complex variable $t+iu$ then each ϕ_j resp. ψ_k is translated by $u\lambda_1 \ln r_j$ resp. $u\lambda_2 \ln s_k$ and

$$\begin{aligned}\sup_{(x,y,\rho,\sigma)\in S} &\left|\sum_{j=1}^{m}\sum_{k=1}^{n} a_{jk} r_j^{\alpha_0+\lambda_1(t+iu)} s_k^{\beta_0+\lambda_2(t+iu)} e^{i(\phi_j+\psi_k)}\right| \\ &= \widetilde{M}\left(\alpha_0+\lambda_1 t, \beta_0+\lambda_2 t\right).\end{aligned}$$

If all the variables save t and u are fixed, the three-lines theorem is applicable to $\widetilde{M}\left(\alpha_0+\lambda_1 t, \beta_0+\lambda_2 t\right)$. Thus $\ln \widetilde{M}\left(\alpha_0+\lambda_1 t, \beta_0+\lambda_2 t\right)$ is a convex function of t on $(0,\infty)$. □

Bibliography

The numbers in parentheses following the entries refer to the pages in this text where the citations occur.

Ar — Arens, R., *Note on convergence in topology,* Mathematics Magazine, **23** (1950), 229 - 234 (154).

Be — Berberian, S. K., *Lectures in functional analysis and operator theory,* Springer-Verlag, New York, 1974 (368).

Ber1 — Bergman, S., dissertation, Philosophische Fakultät der Berliner Universität, 1921 (130).

Ber2 — ———, *Über die Entwicklung der harmonischen Funktionen der Ebene und des Raumes nach Orthogonalfunktionen,* Mathematische Annalen, **86** (1922), 238 - 271 (130).

Bes — Besicovitch, A. S., *On the definition and value of the area of surface,* Quarterly Journal of Mathematics, **16** (1945), 86–102 (228).

Bou — Bourbaki, N., *Topologie générale,* Livre III, Chapitre I, Chapitre II, Hermann & Cie., Paris, 1940 (21).

Coh1 — Cohen, P. J., *Factorization in group algebras,* Duke Mathematical Journal, **26** (1959), 199 - 205 (321).

Coh2 — ———, *The independence of the continuum hypothesis,* Proceedings of the National Acadademy of Sciences, **50** (1963), 1143 - 1148; *The independence of the continuum hypothesis,* Proceedings of the National Acadademy of Sciences, **51**, (1964), 105 - 110 (273).

Fin — Fine, N., *On the Walsh functions,* Transactions of the American Mathematical Society, **65** (1949), 372 - 414 (68).

GeK — Gelbaum, B. R. and Kalisch, G. K., *Measure in semigroups,* Canadian Journal of Mathematics, **4** (1952), 396 - 406 (246).

GKO — ———, ———, and Olmsted, J. M. H., *On the embedding of topological semigroups and integral domains,* Proceedings of the American Mathematical Society, **2** (1951), 807 - 821 (247).

GeO — ——— and Olmsted, J. M. H., *Theorems and counterexamples in mathematics,* Springer-Verlag, New York, 1990 (154, 199, 203, 214, 221, 228, 353).

GiJe — Gilman, L. and Jerison, M., *Rings of continuous functions,* D. van Nostrand Company, Inc., New York, 1960 (16).

Halm — Halmos, P. R., *Measure theory,* D. van Nostrand Company, Inc., New York, 1950 (244).

HeR — Hewitt, E. and Ross, K. A., *Abstract harmonic analysis,* Second Edition, 2 vols., Springer-Verlag, New York, 1979 (54, 244, 321).

HeS — Hewitt, E. and Stromberg, K., *Real and abstract analysis,* Springer-Verlag, New York, 1965 (48).

Hi — Hille, E., *Analytic function theory, I,II,* Ginn and Company, Boston, 1962 (130).

KacS — Kaczmarz, S. and Steinhaus, H., *Theorie der Orthogonal Reihen,* Monografje Matematyczne, Warszawa, 1951 (68).

Kak — Kakutani, S., *Ueber die Metrisation der topologischen Gruppen,* Proceedings of the Imperial Academy of Japan, **12** (1936), 82 (191).

Ko — Kolmogoroff, A., *Grundbegriffe der Wahrscheinlichkeitsrechnung,* Chelsea Publishing Company, New York, 1946 (67).

Kö — Köthe, G., *Topological vector spaces I,* Springer-Verlag, New York, 1969 (349).

La — Landau, E., *Darstellung und Begründung einiger neuerer Ergebnisse der Funktionentheorie,* Chelsea Publishing Company, New York, 1946 (129).

Loo — Loomis, L. H., *An introduction to abstract harmonic analysis,* D. van Nostrand Company, Inc., New York, 1953 (54, 346, 368).

Me — Mendelson, E., *Introduction to mathematical logic,* D. van Nostrand Company, Inc., New York, 1964 (243).

NM — Neumann, J. von and Morgenstern, O., *Theory of games and economic behavior,* Princeton University Press, Princeton, 1947 (132).

Rin — Rinow, W., *Lehrbuch der Topologie,* VEB Deutscher Verlag der Wissenschaften, Berlin, 1975 (154).

RoDi — Roelcke, W.and Dierolf, S., *Uniform structures on topological groups and their quotients,* McGraw-Hill, Inc., New York, 1981 (16).

Rud — Rudin, W., *Real and complex analysis,* Third edition, McGraw-Hill, Inc., New York, 1987 (48, 368).

Sch — Schaefer, H. H., *Topological vector spaces,* Springer-Verlag, New York, 1970 (349).

Sier — Sierpinski, W., *Sur une propriété des séries qui ne sont pas absolument convergentes,* Bulletin International de l'Academie Polonaise des Sciences et des Lettres, Classe des Sciences Mathématiques et Naturelles, Cracovie [Cracow] **149** (1911), 149 - 158 (273).

SiW — Singer, I. M. and Wermer, J., *Derivations on commutative Banach algebras,* Mathematische Annalen, **129** (1955), 260 - 264 (368).

Sol — Solovay, R., *A model of set theory in which every set is Lebesgue-measurable,* Annals of Mathematics, **92** (1970), 1 - 56 (243).

Sz — Szpilrajn, E., *O mierzalności i warunku Baire'a* [*On measurability and the condition of Baire*], Comptes Rendus du I Congrès des mathématiciens des Pays Slaves, Varsovie (1929), 299ff. (146).

Th — Thorin, G. O., *Convexity theorems,* Séminaire Mathématique de L'Université de Lund, Tome 9, 1948 (449).

Tu — Tukey, J., *Convergence and uniformity in topology,* Princeton University Press, Princeton, 1940 (21).

Wal — Walsh, J. L., *A closed set of normal orthogonal functions,* American Journal of Mathematics, **55** (1923), 5 – 24 (68).

We1 — Weil, A., *Sur les espaces a structure uniforme et sur la topologie générale,* Hermann & Cie., Paris, 1937 (16, 21, 68).

We2 — ———, *L'intégration dans les groupes topologiques et ses applications,* Hermann & Cie., Paris, 1940 (54, 135, 246).

Wey — Weyl, H., *Über die Gleichverteilung von Zahlen mod. Eins,* Mathematische Annalen, **77** (1916), 313 – 15 (43).

Wi — Widder, D. V., *The Laplace transform,* Princeton University Press, Princeton, 1946 (251).

Zy — Zygmund, A., *Trigonometric series,* 2 vols., Cambridge University Press, Cambridge, 1988 (135).

Symbol List

The notation a.b. d indicates **Chapter** a, **Section** b, page d among the **Problems**; similarly s a.b. c indicates **Chapter** a, **Section** b, page c among the **Solutions** $\aleph_0$, 1.1. 4

$f_n \xrightarrow{\text{u}} f$, 2.3. 25: f_n converges uniformly to f, i.e., if $\epsilon > 0$ then for some $N(\epsilon)$ and all x, $|f_n(x) - f(x)| < \epsilon$ when $n > N(\epsilon)$

$\mathbf{x}_n \xrightarrow{\text{w}} \mathbf{x}$, s 6.1. 308: $\mathbf{x}_n$ converges weakly to $\mathbf{x}$, i.e., for every $\mathbf{x}^*$ in the dual space, $\lim_{n\to\infty} (\mathbf{x}_n, \mathbf{x}^*) = 0$

$f_n \xrightarrow{\| \|_p} f$, 6.1. 73: f_n converges in p-norm to f, i.e., if $\epsilon > 0$ then for some $N(\epsilon)$, $\|f_n - f\|_p < \epsilon$ if $n > N(\epsilon)$

Glossary/Index

The notation a.b. d indicates **Chapter** a, **Section** b, page d among the **Problems**; similarly s a.b. c indicates **Chapter** a, **Section** b, page c among the **Solutions**.

A

B

C

CARATHEODORY, C. measurable, 1.2. 11

cardinality, 1.1. 4: The Axiom of Choice implies every set X can be well-ordered. When X and Y are well-ordered and there is an order- preserving injection $T : X \mapsto Y$, by definition, $X \prec Y$. For a given set Y, $\mathcal{Y} \stackrel{\text{def}}{=} \{X : X \text{ is well-ordered and } X \prec Y\}$ is well-ordered with respect to $\prec$ and $\#(Y) \stackrel{\text{def}}{=} \inf \mathcal{Y}$.

Cartesian product, 1.1. 3, 2.1. 13: for a set $\{X_\gamma\}_{\gamma\in\Gamma}$ of sets, the set of all vectors $\mathbf{x} \stackrel{\text{def}}{=} (\ldots, x_\gamma, \ldots), x_\gamma \in X_\gamma$.

category, 2.3. 24, 29: a) a class $\mathcal{C} \stackrel{\text{def}}{=} \{A\}$ of objects; b) a class of pairwise disjoint sets $[A, B], A, B \in \mathcal{C}$ of morphisms; c) an associative composition $(f, g) \in [A, B] \times [B, C] \mapsto g \circ f \in [A, C]$; d) in each $[A, A]$ an identity $\mathbf{1}_A$ such that for f in $[A, B]$ and g in $[C, A]$, $f \circ \mathbf{1}_A = f, \mathbf{1}_A \circ g = g$.

CAUCHY, A. L. DE, 2.1. 16 et alibi

— complete, 2.2. 16

— completion, 2.1. 16: for a uniformity situation (X, U), the set of equivalence classes of Cauchy nets.

— formula, s 7.3. 377 et alibi: If $f \in H(\Omega)$, $a \in \Omega$, Γ is a cycle, and $\text{Ind}_\Gamma(b) = 0$ for all b in $\mathbb{C} \setminus \Omega$ then

$$f^{(k)}(a) \cdot \text{Ind}_\Gamma(a) = \frac{k!}{2\pi i} \int_\Gamma \frac{f(z)}{(z-a)^{k+1}}\, dz.$$

— -net, 2.1. 16

— -Riemann equations, s 8.1. 381: If $f \stackrel{\text{def}}{=} u + iv \in H(\Omega)$ then

$$u_x = v_y, u_y = -v_x.$$

— theorem, s 9.3. 394, et alibi:If $f \in H(\Omega)$, $a \in \Omega$, Γ is a cycle, and $\text{Ind}_\Gamma(b) = 0$ for all b in $\mathbb{C} \setminus \Omega$ then $\frac{1}{2\pi i} \int_\Gamma f(z)\, dz = 0$.

central index, 10.3. 112

chain (of indices), 3.1. 31

character, 6.1. 72

characteristic function, 1.1. 4

circled, 6.1. 70

closed (with respect to ...), 1.1. 6, 2.1. 13

— ball, 2.1. 16

— curve, 2.1. 12

— graph theorem, s 6.3. 347: If E, F are Banach spaces and $T : E \mapsto F$ is linear (not necessarily continuous) then $T \in [E, F]$ iff the graph of T is closed (in $E \times F$).

— half-line, s 3.2. 224

— interval, 1.1. 3

— set, 2.1. 14

closure, 2.1. 12, 7.1. 95
coarser, 2.1. 14
cofinal, 2.1. 14
compact, 1.1. 4: of a set K in a topological space, that every open cover contains a finite subcover.
— -open,14.2. 127
complete measure situation, 1.2. 11
— metric space, 2.1. 16, 4.1. 42: a metric space (X, d) in which every Cauchy net converges.
— orthonormal set, 6.2. 81: in a Hilbert space $\mathfrak{H}$, an orthonormal set S such that $S^{\perp} = 0$.
—ly regular, 2.2. 21: of a topological space, that for every closed set F and every x in $X \setminus F$, for some f in $C(X, \mathbb{R})$, $f(x) = 0$ and $f(F) = \{1\}$ $(\{x\} \prec 1 - f \prec X \setminus F)$.
completion (of a σ-ring), 1.2. 11
complex measure, 1.1. 4
component, 5.1. 60: a maximal connected subset of a set.
concave (function), 4.1. 41
conditionally convergent, 5.1. 50: of a series $\sum_{n=1}^{\infty} a_n$, that for some permutation π of $\mathbb{N}$, $\sum_{n=1}^{\infty} a_{\pi(n)}$ diverges.
conformal, 10.2. 110: of an f in $H(\Omega)$, that f is biholomorphic.
conjugate bilinear, 6.3. 79
connected, 2.1. 14
continuous, 2.1. 12
— on the right, 15.1. 133: of an f in $X^{\mathbb{R}}$, that $\lim_{\epsilon \downarrow 0} f(x + \epsilon) = f(x)$.
Continuum Hypothesis, s 5.1. 273: $2^{\aleph_\alpha} = \aleph_{\alpha+1}$.
contraction, contractive, 2.1. 19
converge, 2.1. 14
convex function (map), 3.1. 30
— hull, 2.3. 25: for a set S (in a vector space V), the intersection of all convex supersets of S.
— set, 2.3. 25: in a vector space V, a set S such that

$$\{\{\mathbf{x}, \mathbf{y} \in S\} \wedge \{0 \le t \le 1\}\} \Rightarrow \{t\mathbf{x} + (1-t)\mathbf{y} \in S\}.$$

convolution, 6.1. 72
coset representative, 4.2. 45: an element of an equivalence class determined by, e.g., a subgroup of a group, an ideal in a ring, etc.
— space, 2.3. 29: the set of cosets of, e.g., a subgroup of a group, etc.
cover, 2.1. 21: for a set X, in 2^X a subset S such that $\bigcup_{A \in S} A \supset X$.
countably additive, 1.1. 4: of a set function Φ, that

$$\Phi\left(\dot{\bigcup}_{n \in \mathbb{N}} A_n\right) = \sum_{n=1}^{\infty} \Phi(A_n).$$

G

H

I

J

K

L

M

N

O

P

Q

R

V

W

Y

Z